# Finite Fields
and Applications

Springer-Verlag Berlin Heidelberg GmbH

Dieter Jungnickel
Harald Niederreiter
*Editors*

# Finite Fields and Applications

Proceedings of The Fifth International Conference on Finite Fields and Applications $\mathbf{F}_q5$, held at the University of Augsburg, Germany, August 2-6, 1999

Springer

*Editors*

Dieter Jungnickel
Lehrstuhl für Diskrete Mathematik,
Optimierung und Operations Research
Universität Augsburg
86135 Augsburg, Germany
e-mail: jungnickel@math.uni-augsburg.de

Harald Niederreiter
Department of Mathematics
National University of Singapore
2 Science Drive 2
Singapore 117543
Republic of Singapore
e-mail: nied@math.nus.edu.sg

Library of Congress Cataloging-in-Publication Data

International Conference on Finite Fields and Applications (5th : 1999 : University of Augsburg)
Finite fields and applications : proceedings of the Fifth International Conference on Finite Fields and Applications Fq5, held at the University of Augsburg, Germany, August 2-6, 1999 / Dieter Jungnickel, Harald Niederreiter, editors.
p. cm.
Includes bibliographical references.
ISBN 978-3-642-62498-8 ISBN 978-3-642-56755-1 (eBook)
DOI 10.1007/978-3-642-56755-1
1. Finite fields (Algebra)--Congresses. I. Jungnickel, D. (Dieter), 1952- II. Niederreiter, Harald, 1944- III. Title.

QA247.3 .I57 1999
512'.3--dc21
00-054946

Mathematics Subject Classification (2000):
primary: 11Txx; secondary: 05-XX, 51Exx, 94Axx, 94Bxx

ISBN 978-3-642-62498-8



http://www.springer.de


Originally published by Springer-Verlag Berlin Heidelberg New York in 2001
Softcover reprint of the hardcover 1st edition 2001


Typeset by the authors using a Springer TEX macro package
Cover design: *design & production* GmbH, Heidelberg

SPIN: 10699437 46/3142LK - 5 4 3 2 1 0 – Printed on acid-free paper

# Preface

This volume represents the refereed proceedings of the Fifth International Conference on Finite Fields and Applications ($\mathbf{F}_q5$) held at the University of Augsburg (Germany) from August 2–6, 1999, and hosted by the Department of Mathematics. The conference continued a series of biennial international conferences on finite fields, following earlier conferences at the University of Nevada at Las Vegas (USA) in August 1991 and August 1993, the University of Glasgow (Scotland) in July 1995, and the University of Waterloo (Canada) in August 1997. The Organizing Committee of $\mathbf{F}_q5$ comprised Thomas Beth (University of Karlsruhe), Stephen D. Cohen (University of Glasgow), Dieter Jungnickel (University of Augsburg, Chairman), Alfred Menezes (University of Waterloo), Gary L. Mullen (Pennsylvania State University), Ronald C. Mullin (University of Waterloo), Harald Niederreiter (Austrian Academy of Sciences), and Alexander Pott (University of Magdeburg).

The program of the conference consisted of four full days and one half day of sessions, with 11 invited plenary talks and over 80 contributed talks that required three parallel sessions. This documents the steadily increasing interest in finite fields and their applications. Finite fields have an inherently fascinating structure and they are important tools in discrete mathematics. Their applications range from combinatorial design theory, finite geometries, and algebraic geometry to coding theory, cryptology, and scientific computing. A particularly fruitful aspect is the interplay between theory and applications which has led to many new perspectives in research on finite fields. This interplay has always been a dominant theme in $\mathbf{F}_q$ conferences and was very much in evidence at $\mathbf{F}_q5$. Applied or applications-oriented topics accounted for well over half of the program.

These proceedings reflect the wide variety of topics represented at the conference. Most invited talks and a good proportion of the contributed talks are on permanent record here. All contributed talks were screened before the conference and all full papers were carefully refereed. We take this opportunity to thank the members of the Organizing Committee and all referees who helped in these tasks. These colleagues contributed enormously to the quality of the conference presentations and to guaranteeing high standards for these proceedings.

We greatly appreciate the generous financial support received for the conference. A major portion of the funds came from a grant by the German Research Council (DFG). Further sponsors were Andersen Consulting, Certicom Corp., Dresdner Bank AG, Gesellschaft der Freunde der Universität Augsburg, Hewlett Packard Laboratories, and Siemens AG. We are grateful to the University of Augsburg for providing excellent facilities for the conference and to the various officials of the City of Augsburg who helped with organizational questions. Thanks are also due to the Mayor of Augsburg who gave a reception for the participants in the splendid setting of the Great Golden Chamber of the city's renaissance town hall. Last but not least, the highly efficient and friendly manner in which the conference took place would not have been possible without the enthusiasm and hard work by the assistants, secretaries and students who saw to the many details involved in such a major event; we are grateful to all of them.

Regarding the present proceedings, we thank Dr. Martin Peters of Springer Verlag who gave us the opportunity to place this volume with a top publisher and in a form which we find both attractive and practical (no rapidly deteriorating soft covers for once!). The cooperation with him and all the staff at Springer Verlag was always a pleasure.

Finally, we are pleased to confirm that the series of $\mathbf{F}_q$ conferences will continue with $\mathbf{F}_q6$ in Mexico in May 2001. We expect another lively and stimulating conference there, which should, like the previous conferences, serve as an important marketplace of ideas for the finite fields community.

October 2000

Dieter Jungnickel
Harald Niederreiter

# Contents

* invited speaker

# Projective Generalized Reed-Muller Codes over $p$-adic Numbers and Finite Rings

Kanat Abdukhalikov

Institute of Mathematics
Pushkin Str 125
Almaty 480100
Kazakhstan

**Abstract** We consider codes generated over the ring $\mathbb{Z}_p$ of $p$-adic integer numbers and the ring $\mathbb{Z}/p^d\mathbb{Z}$ of integers modulo $p^d$ by incidence vectors of the designs of points and projective subspaces of a projective space $PG_{m-1}(F_p)$, where $F_p$ is a finite field of $p$ elements. Codes invariant under the natural action of the projective group $PGL_m(F_p)$ are described.

## 1 Introduction

We study codes over the ring $\mathbb{Z}_p$ of $p$-adic integer numbers and the ring $\mathbb{Z}/p^d\mathbb{Z}$ of integers modulo $p^d$ arising from a projective space $PG_{m-1}(F_p)$ over a field $F_p$ of $p$ elements, $p$ prime.

Let $P_F(r,m)$ be the code generated over a ring $F$ by incidence vectors of the design of points and projective $(m-1-r)$-dimensional subspaces of $PG_{m-1}(F_p)$. We consider $F = \mathbb{Z}_p$ (resp. $F = \mathbb{Z}/p^d\mathbb{Z}$) and we will call $P_F(r,m)$ a projective generalized Reed-Muller code over $\mathbb{Z}_p$ (resp. over $F = \mathbb{Z}/p^d\mathbb{Z}$). If $F$ is a field then they are classical projective generalized Reed-Muller codes.

Let

$$W = \sum_{z \in PG_{m-1}(F_p)} Fe_z.$$

By a linear code over $F$ we mean a $F$-submodule of $W$. The projective group $G = PGL_m(F_p)$ acts on $W$ by permutations of basic vectors $e_z$. Projective generalized Reed-Muller codes are subcodes in $W$ and they are invariant under $G$. If $F$ is a field then the codes $P_F(r,m)$ and their zero sum subcodes are the only subcodes in $W$ invariant under $G$. We give a classification of $G$-invariant subcodes in $W$ for $F = \mathbb{Z}_p$ and $F = \mathbb{Z}/p^d\mathbb{Z}$. We are going to study these codes from a geometric point of view.

Note that it is sufficient to study codes over $\mathbb{Z}_p$ since $\mathbb{Z}/p^d\mathbb{Z} \cong \mathbb{Z}_p/p^d\mathbb{Z}_p$ and one can get a description of codes over $\mathbb{Z}/p^d\mathbb{Z}$ by considering codes over $\mathbb{Z}_p$ modulo $p^d$.

There is much interest in $\mathbb{Z}/4\mathbb{Z}$ codes as they lead to a systematic way of constructing very good binary codes. For example, the famous Kerdock and Preparata codes are non-linear binary codes that contain more codewords than any comparable linear codes presently known. Recently it was shown in [4] that the Kerdock and Preparata codes can be very simply constructed as binary images under a certain map, called the Gray map, of linear codes over $\mathbb{Z}/4\mathbb{Z}$. This fact stimulated investigations of linear codes over $\mathbb{Z}/p^d\mathbb{Z}$ and $\mathbb{Z}_p$ (see also [2,3]).

## 2 Projective Codes

Let $V$ be an $m$-dimensional vector space with basis $\{v_1, \ldots, v_m\}$ over a field $F_p$, $m \geq 3$. We denote $C_r = P_F(r, m)$ for simplicity, where either $F = \mathbb{Z}_p$ or $F = \mathbb{Z}/p^d\mathbb{Z}$. For a subspace $\pi$ of $P(V) = PG_{m-1}(F_p)$ we consider its incidence vector

$$u_\pi = \sum_{z \in \pi} e_z.$$

Then

$$C_r = \langle u_\pi \ : \ \pi \text{ is an } (m-1-r)\text{-dimensional subspace of } P(V) \rangle$$

is the code of the projective design of points and $(m-1-r)$-dimensional subspaces of $P(V)$, where $0 \leq r \leq m-1$. In particular, $C_0$ is the code generated by all-one vector $w = u_{P(V)} = \sum_{z \in P(V)} e_z$, $C_1$ is the code generated by $p^m - 1$ elements $\sum_{z \in \pi} e_z$, where $\pi$ is a hyperplane in $P(V)$, and $C_{m-1} = W$. Also define zero sum subcodes:

$$W^0 = \left\{ \sum a_z e_z \in W \ : \ \sum a_z = 0 \right\},$$

$$C_r^0 = C_r \cap W^0.$$

It is clear that

$$W = W^0 \oplus \langle w \rangle.$$

The projective group $G = PGL_m(F_p)$ acts in the natural way on $W$:

$$g\left( \sum a_z e_z \right) = \sum a_z e_{g(z)} \quad \text{for } g \in G.$$

Our goal is to study codes $C_r$ and find all subcodes in $W$ invariant under the action of the group $G$. Recall that if $F = F_p$ then the codes $C_r$ are projective generalized codes [1]. In this case $C_r$ and $C_r^0$ are only codes invariant under the group $G$.

The next theorem describes basic properties of the codes $C_r$.

**Theorem 2.1.** *Let $F = \mathbb{Z}_p$. Then the following statements hold:*
*1) $W = C_{m-1} \supset C_{m-2} \supset \cdots \supset C_1 \supset C_0 = \langle w \rangle$;*
*2) $C_r \supset pC_{r+1}$, $C_r \supset p^2 C_{r+2}$, ... , $C_r \supset p^{m-1-r} C_{m-1}$ for $1 \leq r \leq m-1$;*
*3) $C_r \cap p^s W = p^s C_{r+s}$ for $1 \leq r \leq m-1$ and $0 \leq s \leq m-1-r$.*

Now let us define

$$C(r_1, r_2, \dots, r_{m-2}) = C_{r_1} + pC_{r_2} + \dots + p^{m-3}C_{r_{m-2}},$$

where $1 \le r_i \le m-1$, and $r_i < r_{i+1}$ unless $r_{i+1} = m-1$. In other words, $r_i < r_{i+1}$ if $r_{i+1} \neq m-1$ ($i = 1, \dots, m-3$). These restrictions are clear because of Theorem 2.1. We will see later that $C(r_1, r_2, \dots, r_{m-2}) \cap p^{i-1}W + p^iW/p^iW$ is isomorphic to the projective code $P_{F_p}(r_i, m)$ over the field $F_p$.

We set

$$C^0(r_1, r_2, \dots, r_{m-2}) = C(r_1, r_2, \dots, r_{m-2}) \cap W^0.$$

**Theorem 2.2.** *Let $F = \mathbb{Z}_p$. If $C$ is a $G$-invariant code in $W^0$, $C \not\subseteq pW^0$, then*

$$C = C^0(r_1, r_2, \dots, r_{m-2})$$

*for some $(r_1, \dots, r_{m-2})$. If $C$ is a $G$-invariant code in $W$, $C \not\subseteq C_0$ then*

$$C = p^sC^0(r_1, r_2, \dots, r_{m-2}) + p^tC_0$$

*for some $s$, $t \ge 0$ and $(r_1, \dots, r_{m-2})$.*

The standard $G$-invariant inner product is defined by

$$\left(\sum a_z e_z, \sum b_z e_z\right) = \sum a_z b_z.$$

If $F = \mathbb{Z}_p$ then the dual code for $C \subset W$ is

$$C^* = \{a \in \mathbb{Q}_p W \ : \ (a, C) \subseteq \mathbb{Z}_p\},$$

where $\mathbb{Q}_p W = \sum \mathbb{Q}_p e_z$ is a vector space over the field $\mathbb{Q}_p$ of $p$-adic numbers. Moreover, the orthogonal code for $C \subset W$ is the code

$$C^\perp = \{a \in W \ : \ (a, c) = 0 \text{ for all } c \in C\}. \qquad (1)$$

*Remark.* This notion of dual code $C^*$ comes from the theory of integral lattices. In fact, $W$ is a lattice over $\mathbb{Z}_p$. Note that, in general, $C^* \not\subset W$, and $C^*$ is $\mathbb{Z}_p$-module of finite rank if and only if $C$ is a module of full rank in $W$.

**Theorem 2.3.** *Let $F = \mathbb{Z}_p$. Then the following statements hold:*

*1) $(C_0)^\perp = W^0$, $(W^0)^\perp = C_0$;*

*2) $C(r_1, \dots, r_{m-2})^* = (1/p^{m-2})C^0(m-1-r_{m-2}, \dots, m-1-r_1) + C_0$;*

*3) $C^0(r_1, \dots, r_{m-2})^* = (1/p^{m-2})C^0(m-1-r_{m-2}, \dots, m-1-r_1) + \mathbb{Q}_p C_0$,*

*where for the right hand sides of the equations we adopted an agreement that $C^0(0, s_2, \dots, s_{m-2}) = pC^0(s_2, \dots, s_{m-2}, m-1)$ and $C(0, s_2, \dots, s_{m-2}) = pC(s_2, \dots, s_{m-2}, m-1)$.*

The proofs of Theorems 2.1, 2.2 and 2.3 will be given in the next section. Observe that for $C_1 = C(1,2,\dots,m-2)$ Theorem 2.3 implies the equality $(C_1)^* = (1/p^{m-2})C_1^0 + C_0$. Now we consider one more property of the code $C_1$. At first, for any hyperplane $\lambda$ in $P(V)$ we define a linear functional $\delta_\lambda : W \to F$ by

$$\delta_\lambda\left(\sum a_z e_z\right) = \sum_{z\notin\lambda} a_z = \sum_{z\in P(V)} a_z - \sum_{z\in\lambda} a_z.$$

**Proposition 2.4.** *Let $F = \mathbb{Z}_p$. Element $a = \sum a_z e_z$ belongs to the code $C_1$ if and only if $\delta_\lambda(a) \equiv 0 \pmod{p^{m-2}}$ for all hyperplanes in $P(V)$ and $\sum_\lambda \delta_\lambda(a) \equiv 0 \pmod{p^{m-1}}$, where summing is over all hyperplanes.*

*Proof.* If $a \in C_1$ then $a = \sum b_\pi u_\pi$, where $\pi$ is a hyperplane and $b_\pi \in \mathbb{Z}_p$. Then we have

$$\delta_\lambda(u_\pi) = \begin{cases} 0 & \text{if } \lambda = \pi \\ p^{m-2} & \text{if } \lambda \neq \pi \end{cases}$$

and

$$\sum_\lambda \delta_\lambda(u_\pi) = \left(\frac{p^{m-1}-1}{p-1} - 1\right) p^{m-2} \equiv 0 \pmod{p^{m-1}}.$$

Conversely, let $a = \sum b_\pi u_\pi$ (according to Theorem 2.1 we can express $a$ as a linear combination of elements $u_\pi$ with $p$-adic (possibly noninteger) coefficients). We have to prove that $b_\pi \in \mathbb{Z}_p$. Indeed,

$$\delta_\lambda(a) = \sum_{\pi\neq\lambda} b_\pi p^{m-2}.$$

Summing all these equations for all $\lambda$ we have

$$\sum_\lambda \delta_\lambda(a) = \left(\frac{p^{m-1}-1}{p-1} - 1\right) \sum_\pi b_\pi p^{m-2}$$

and

$$\sum_\pi b_\pi = \left(\frac{p^{m-2}-1}{p-1}\right)^{-1} \frac{1}{p^{m-1}} \sum_\lambda \delta_\lambda(a) \in \mathbb{Z}_p.$$

Therefore, $b_\pi \in \mathbb{Z}_p$. □

Now we turn to codes over the finite ring $\mathbb{Z}/p^d\mathbb{Z}$. Theorem 2.1 immediately implies

**Corollary 2.5.** *Let $F = \mathbb{Z}/p^d\mathbb{Z}$. Then the following statements hold:*

*1) $W = C_{m-1} \supset C_{m-2} \supset \cdots \supset C_1 \supset C_0 = \langle w\rangle$;*

*2) $C_r \supset pC_{r+1}, C_r \supset p^2C_{r+2}, \dots, C_r \supset p^{m-1-r}C_{m-1}$ for $1 \le r \le m-1$;*

*3) $C_r \cap p^sW = p^sC_{r+s}$ for $1 \le r \le m-1$ and $0 \le s \le \min(d, m-1-r)$.*

Let $l = \min(d, m-2)$. Then we define

$$C(r_1, r_2, \ldots, r_l) = C_{r_1} + pC_{r_2} + \cdots + p^{l-1}C_{r_l},$$

$$C^0(r_1, r_2, \ldots, r_l) = C(r_1, r_2, \ldots, r_l) \cap W^0,$$

where $1 \leq r_i \leq m-1$, and $r_i < r_{i+1}$ if $r_{i+1} \neq m-1$ ($i = 1, \ldots, l-1$). If $t_i = t_i(r_i)$ denotes the dimension of the projective generalized code $P_{F_p}(r_i, m)$ over the field $F_p$ (exact value of $t_i$ can be found in [1,5]) and $t = |P(V)| = (p^m - 1)/(p-1)$ then the code $C(r_1, r_2, \ldots, r_l)$ has type

$$1^{t_1} p^{t_2 - t_1} (p^2)^{t_3 - t_2} \cdots (p^{l-1})^{t_l - t_{l-1}} (p^l)^{t - t_l}$$

for $d > m-2$, and type

$$1^{t_1} p^{t_2 - t_1} (p^2)^{t_3 - t_2} \cdots (p^{l-1})^{t_l - t_{l-1}}$$

for $d \leq m-2$, in the sense of [2].

**Corollary 2.6.** *Let $F = \mathbb{Z}/p^d\mathbb{Z}$. If $C$ is a $G$-invariant code in $W^0$, $C \not\subseteq pW^0$, then*

$$C = C^0(r_1, r_2, \ldots, r_l)$$

*for some $(r_1, \ldots, r_l)$. If $C$ is a $G$-invariant code in $W$, $C \not\subseteq C_0$ then*

$$C = p^s C^0(r_1, r_2, \ldots, r_l) + p^t C_0$$

*for some $(r_1, \ldots, r_l)$ and $0 \leq s,\ t \leq d-1$.*

Recall that the orthogonal code for $C \subset W$ is given by the formula (1).

**Corollary 2.7.** *Let $F = \mathbb{Z}/p^d\mathbb{Z}$. Then the following statements hold:*
*1) $(C_0)^\perp = W^0$, $(W^0)^\perp = C_0$;*
*2) $C(r_1, \ldots, r_l)^\perp = p^{d-l} C^0(m-1-r_l, \ldots, m-1-r_1)$;*
*3) $C^0(r_1, \ldots, r_l)^\perp = p^{d-l} C^0(m-1-r_l, \ldots, m-1-r_1) + C_0$,*
*where for the right hand sides of the equations we adopted an agreement that $C^0(0, s_2, \ldots, s_l) = pC^0(s_2, \ldots, s_l, m-1)$ and $C(0, s_2, \ldots, s_l) = pC(s_2, \ldots, s_l, m-1)$.*

*Proof.* Assume for the moment that $F = \mathbb{Z}_p$. Then consider

$$C' = \{a \in \mathbb{Q}_p W \ :\ (a, C) \subseteq p^d \mathbb{Z}_p\} = p^d C^*.$$

If $d > m-2$ then $l = m-2$ and our statement follows directly from Theorem 2.3. If $d \leq m-2$ then $l = d$ and we consider a code $C(r_1, \ldots, r_l, m-1, \ldots, m-1)$ for $F = \mathbb{Z}_p$ and apply again Theorem 2.3. □

Corollary 2.7 allows us to calculate orthogonal codes for any $G$-invariant code.

*Example 2.8.* If $d = 2$ and $r < (m-1)/2$ then the code $C = C^0(r, m-1-r) + pC_0$ is a self-dual $G$-invariant code: $C^\perp = C$.

# 3 Proofs

In this section we will prove theorems from the previous section. First of all we consider projective codes over the field $F_p$.

## 3.1 Codes over $F_p$

In this subsection, we will consider the case when $F = F_p$. It is convenient to work with the group $GL_m(p)$ in place of $G = PGL_m(p)$. We assume that the group $GL_m(F_p)$ acts in the natural way on a vector space $V$ with basis $\{v_1, \ldots, v_m\}$. We consider the set of functions

$$\mathcal{F} = \{f : V \setminus \{0\} \to F_p \ : \ f(\alpha v) = f(v) \text{ for all } \alpha \in F_p^*, \ v \in V \setminus \{0\}\}.$$

Let us assume that points in the projective space $P(V) = PG_m(F_p)$ are lines in $V$ without the origin. Then the set of functions

$$f_z(v) = \begin{cases} 1 & \text{if } v \in z, \\ 0 & \text{if } v \notin z \end{cases}$$

defined for all $z \in P(V)$ forms a basis in $\mathcal{F}$ over $F_p$. Define $\widehat{\mathcal{F}} = \mathcal{F}$ as vector spaces. We equip $\mathcal{F}$ and $\widehat{\mathcal{F}}$ with the structure of a $GL_m(F_p)$-module by setting

$$(\rho(g)(f))(v) = f(g^{-1}v) \quad \text{and} \quad (\widehat{\rho}(g)(h))(v) = h({}^t g v),$$

where $g \in GL_m(F_p)$, $f \in \mathcal{F}$, $h \in \widehat{\mathcal{F}}$, $v \in V \setminus \{0\}$, and ${}^t g$ is the transpose of $g$. In fact, we have defined an action of the group $G$. One has

$$\rho(g)(f_z) = f_{gz}, \quad \widehat{\rho}(g)(f_z) = f_{{}^t g^{-1} z}.$$

We will see that $\mathcal{F}$ and $\widehat{\mathcal{F}}$ are modular representations of $GL_m(F_p)$ corresponding to permutations of points and hyperplanes in $P(V)$. We set

$$\mathcal{F}_0 = \widehat{\mathcal{F}}_0 = \left\{ \sum c_z f_z \ : \ c_z \in F_p, \ \sum c_z = 0 \right\}.$$

It is clear that

$$\mathcal{F} = \mathcal{F}_0 \oplus \langle 1 \rangle, \quad \widehat{\mathcal{F}} = \widehat{\mathcal{F}}_0 \oplus \langle 1 \rangle,$$

where 1 denotes a constant function.

Let $\{x_1, \ldots, x_m\}$ be the dual basis for the basis $\{v_1, \ldots, v_m\}$ of $V$. Then each element of $\mathcal{F}$ can be represented as a sum:

$$f = \sum_{i_1, \ldots, i_m} c_{i_1 \ldots i_m} x_1^{i_1} \cdots x_m^{i_m}, \quad \sum_{k=1}^{m} i_k \equiv 0 \pmod{p-1}, \tag{2}$$

where

$$f(a) = \sum_{i_1, \ldots, i_m} c_{i_1 \ldots i_m} x_1(a)^{i_1} \cdots x_m(a)^{i_m}$$

for $a \in V \setminus \{0\}$. Since $x_i(a)^p = x_i(a)$ we can always assume that $i_k < p$ for the exponents in (2).

We set

$$M_0 = \widehat{M}_0 = 0,$$

and

$$M_t = \widehat{M}_t = \left\{ \sum_{i_1,\dots,i_m} c_{i_1\dots i_m} x_1^{i_1} \cdots x_m^{i_m} \in \mathcal{F} \;:\; 0 < \sum_{k=1}^{m} i_k \le t(p-1) \right\}$$

for $t = 1, \dots, m-1$. Then the modules $M_t$ and $\widehat{M}_t$ are $GL_m(F_p)$-submodules of $\mathcal{F}_0$ and $\widehat{\mathcal{F}}_0$. Furthermore,

$$M_0 \subset M_1 \subset M_2 \subset \cdots \subset M_{m-1} = \mathcal{F}_0,$$

$$\widehat{M}_0 \subset \widehat{M}_1 \subset \widehat{M}_2 \subset \cdots \subset \widehat{M}_{m-1} = \widehat{\mathcal{F}}_0.$$

Note that $(1 - x_1^{p-1}) \cdots (1 - x_m^{p-1})$ is the zero function on $V \setminus \{0\}$.

The next fact is well-known (see, for example, [1,8]).

**Proposition 3.1.** *The $G$-modules $M_t/M_{t-1}$ are non-zero, non-isomorphic and irreducible. Furthermore, composition factors of $\mathcal{F}$ are nonisomorphic modules $\langle 1 \rangle$ and $M_t/M_{t-1}$, $t = 1, \dots, m-1$.*

Let us define

$$h_1 = x_1^{p-1}, \qquad h_i = (1 - x_1^{p-1}) \cdots (1 - x_{i-1}^{p-1}) x_i^{p-1}, \quad i = 2, \dots, m-1.$$

Let $U$ be the subgroup of lower unitriangular matrices of $GL_m(F_p)$. By $\mathcal{F}_0^U$ we mean the set of $U$-invariant vectors in $\mathcal{F}_0$.

**Lemma 3.2.** $\mathcal{F}_0^U = \langle h_1, \dots, h_{m-1} \rangle$.

*Proof.* It is easy to see that $h_i \in \mathcal{F}_0^U$. Let $h \in \mathcal{F}_0^U$. It is known that any irreducible $SL_m(F_p)$-module has a unique (up to scalar factor) $U$-invariant vector [7]. If $h \in M_1$ then $h = \alpha h_1$ as $M_1$ is an irreducible module. Assume that $h \in M_i$. Since $M_i/M_{i-1}$ is irreducible, it follows that $h = \alpha_i h_i + f$, where $f \in M_{i-1}$. Consequently, $f \in \mathcal{F}_0^U$ and by induction we obtain that $f = \alpha_{i-1} h_{i-1} + \cdots + \alpha_1 h_1$. □

**Lemma 3.3.** *If $M$ is a non-zero submodule of $\mathcal{F}_0$ then $M \supset M_1$.*

*Proof.* The finite $p$-group $U$ acts on the set $M$ and each $U$-orbit has length equal to a power of $p$. At the same time, the length of the orbit of the element 0 is equal to 1, so the module $M$ contains a non-zero $U$-invariant element. Hence

$$h = h_k + \alpha_{k-1} h_{k-1} + \cdots + \alpha_1 h_1 \in M$$

for some $k$ by Lemma 3.2.

We consider now the matrices $u_{ij}(\gamma) \in GL_m(F_p)$ such that $u_{ij}(\gamma)(x_s) = x_s + \gamma\delta_{is}x_j$, where $\gamma \in F_p$ and $\delta_{is}$ is the Kronecker delta. We have

$$\sum_{\gamma \in F_p} u_{k,k+1}(\gamma)(h) = \sum_{\gamma \in F_p} u_{k,k+1}(\gamma)(h_k) = -(1 - x_1^{p-1}) \cdots (1 - x_{k-1}^{p-1}) x_{k+1}^{p-1}.$$

Using an automorphism mapping $x_{k+1}$ into $x_k$ we obtain that $h_k \in M$. Further,

$$\sum_{\gamma \in F_p} u_{1k}(\gamma)(h_k) = (1 - x_2^{p-1}) \cdots (1 - x_{k-1}^{p-1}) x_k^{p-1}.$$

Hence $h_{k-1} \in M$. In a similar way $h_{k-2}, \ldots, h_1 \in M$. However, the element $h_1$ generates $M_1$. □

**Proposition 3.4.** *If $M$ is a nonzero submodule of $\mathcal{F}_0$ (respectively, of $\widehat{\mathcal{F}}_0$), then $M = M_t$ (respectively, $M = \widehat{M_t}$) for some $t = 1, \ldots, m-1$.*

*Proof.* It is sufficient to prove this statement for $\mathcal{F}_0$. We shall show by induction that if $f \in M_k \setminus M_{k-1}$, then $F_pGf \supset M_k$. Indeed, if $f \in M_1 \setminus \{0\}$, then $F_pGf \neq 0$ and $F_pGf \supset M_1$ by Lemma 3.3.

Further, let $f \in M_k \setminus M_{k-1}$. We set $N = F_pGf$. Since $M_k/M_{k-1}$ is irreducible, there exists an element $h \in N \setminus (N \cap M_{k-1})$ that is $U$-invariant modulo $M_{k-1}$.

If $Uh = h$, then we obtain by Lemma 3.2 that $h = \alpha_k h_k + \alpha_{k-1} h_{k-1} + \cdots + \alpha_1 h_1$, $\alpha_k \neq 0$. Now it can be proved as in Lemma 3.3 that $h_k, h_{k-1} \in M$. Consequently, $M \supset M_{k-1}$ by the inductive hypothesis. Moreover, $M \supset M_k$ since $h_k \in M$ and the module $M_k/M_{k-1}$ is irreducible.

If $Uh \neq h$, then $h = \alpha_k h_k + g$, $\alpha_k \neq 0$, and $g \in M_{k-1}$. For $u \in U$ we have $u(h) - h = u(g) - g \in M_{s(u)} \setminus M_{s(u)-1}$. We choose $s = \max\{s(u) : u \in U\}$; then $M \supset M_s$ by the inductive hypothesis. Moreover, $h = \alpha_k h_k + \cdots + \alpha_{s+1} h_{s+1} + h'$ in this case, where $h' \in M_s$. Since $M \supset M_s$, it follows that $h = \alpha_k h_k + \cdots + \alpha_{s+1} h_{s+1} \in M$. As before, $M \supset M_k$. □

**Proposition 3.5.** *Let $F = \mathbb{Z}_p$. Then $W/pW \cong \mathcal{F}$ and $C_1/pC_1 \cong \widehat{\mathcal{F}}$ as $G$-modules.*

*Proof.* The map $e_z \mapsto f_z$ determines the first isomorphism. Let $\pi$ is a hyperplane in $P(V)$ defined by equation $\alpha_1 x_1 + \cdots + \alpha_m x_m = 0$, $\alpha_i \in F_p$, and $y \in P(V)$ is the point defined by the vector $\alpha_1 v_1 + \cdots + \alpha_m v_m$. Then the map $u_\pi \mapsto e_y$ determines the second isomorphism. □

**Corollary 3.6.** *If $F = F_p$ then $C_r$ and $C_r^0$ are the only codes in $W$ invariant under the group $G$.*

*Proof.* It follows from Propositions 3.4 and 3.5. Note that if $\pi$ is a subspace in $P(V)$ spanned by vectors $v_1, \ldots, v_{m-r}$, then the codeword $u_\pi \in C_r$

corresponds to the element $(1-x_{m-r+1}^{p-1})(1-x_{m-r+2}^{p-1})\cdots(1-x_m^{p-1}) \in M_r+\langle 1\rangle$. Therefore, one has

$$C_r \cong M_r + \langle 1\rangle, \quad C_r^0 \cong M_r,$$

which gives a description of the considered codes. □

It is easy to check the following

**Lemma 3.7.** *For* $i = 1, \ldots, m-2$ *one has* $(M_i)^\perp = M_{m-1-i} + \langle 1\rangle$ *and* $(M_i + \langle 1\rangle)^\perp = M_{m-1-i}$.

### 3.2 Codes over $\mathbb{Z}_p$

*Proof (of Theorem 2.1).* 1) Let a code $C_r$ be generated by the incidence vectors $u_\pi$ of subspaces of $P(V)$ of dimension $m-1-r$. We will prove that if $\lambda$ is a subspace of $P(V)$ of dimension $\dim\lambda = m-1-t > \dim\pi$ then $u_\lambda \in C_r$ (it will follow $C_r \supset C_t$ for $r > t$). Indeed, we have $\dim\lambda - \dim\pi = r-t$ and

$$\sum_{\pi\subset\lambda} u_\pi = \frac{p^{r-t+1}-1}{p-1}\sum_{z\in\lambda} e_z = \frac{p^{r-t+1}-1}{p-1}u_\lambda.$$

But $(p^{r-t+1}-1)(p-1)$ is invertible in $\mathbb{Z}_p$, hence $u_\lambda \in C_r$.

2) Consider again the code $C_r$ generated by the incidence vectors $u_\pi$ of subspaces of dimension $m-1-r$. We prove that if $\dim\lambda = m-2-r = \dim\pi - 1$ for some subspace $\lambda$ then $pu_\lambda \in C_r$ (it will follow $C_r \supset pC_{r+1}$). Indeed,

$$\sum_{\pi\supset\lambda} u_\pi = \frac{p^{r+1}-1}{p-1}\sum_{z\in\lambda} e_z + \sum_{z\notin\lambda} e_z =$$

$$\left(\frac{p^{r+1}-1}{p-1} - 1\right)\sum_{z\in\lambda} e_z + \sum_{z\in P(V)} e_z = p\cdot\frac{p^r-1}{p-1}u_\lambda + w.$$

Therefore, $pu_\lambda \in C_r$, since the all-one vector $w$ belongs to $C_r$ and the number $(p^r-1)(p-1)$ is invertible in $\mathbb{Z}_p$.

3) Consider the following chain of modules:

$$C_1/pC_1 \supset (C_1 \cap pW + pC_1)/pC_1 \supset \cdots$$

$$\supset (C_1 \cap p^{m-3}W + pC_1)/pC_1 \supset (C_1 \cap p^{m-2}W + pC_1)/pC_1.$$

By Propositions 3.4 and 3.5 it corresponds to the chain

$$\widehat{\mathcal{F}} \supset \widehat{M}_{m-2} \supset \cdots \supset \widehat{M}_2 \supset \widehat{M}_1. \tag{3}$$

But the chain

$$C_1/pC_1 \supset (pC_2 + pC_1)/pC_1 \supset \cdots$$

$$\supset (p^{m-3}C_{m-2} + pC_1)/pC_1 \supset (p^{m-2}C_{m-1}W + pC_1)/pC_1$$

also corresponds to the chain (3). Then we have

$$(C_1 \cap p^i W + pC_1)/pC_1 = (p^i C_{i+1} + pC_1)/pC_1$$

for $i = 1, 2, \dots, m-2$, since

$$C_1 \cap p^i W \supset p^i C_{i+1}.$$

Therefore

$$C_1 \cap p^i W = p^i C_{i+1},$$

completing the proof of the theorem. □

*Proof (of Theorem 2.2).* 1) Let $C \subset W^0$ and $C \not\subseteq pW^0$. First of all we prove that $C \supset p^{m-2}W^0$. Indeed, for $x \in P(V)$ let $I_x \subseteq \mathbb{Z}_p$ be the ideal of all $a_x$ occurring in the decompositions $a = \sum a_z e_z$ of elements $a \in C$. Then $I_x$ does not depend on $x$. If $I_x = p^s \mathbb{Z}_p$, $s > 0$, then $C \subset (\oplus_z p^s \mathbb{Z}_p e_z) \cap W^0 = p^s W^0$, which is a contradiction. Hence $I_x = \mathbb{Z}_p$. Therefore, there exists an element $a = e_y + \sum_{z \neq y} a_z e_z \in C$, where $y \in P(V)$ is the point corresponding to the basic vector $v_1 \in V$. Observe that $\sum_{z \neq y} a_z = -1$.

Consider $V' = \langle v_2, \dots, v_m \rangle$ and the projective space $P(V')$. Recall that there exists cyclic subgroup $L'$ of $PGL(V')$ of order $|P(V')|$ acting cyclically on $P(V')$ (Singer cycle [6]). Let $L$ be the corresponding subgroup in $PGL(V)$ which leaves $v_1$ invariant, and let $K \subset G$ be the subgroup generated by the images of the matrices

$$\begin{pmatrix} 1 & \alpha_2 & \alpha_3 & \dots & \alpha_m \\ 0 & 1 & 0 & \dots & 0 \\ 0 & 0 & 1 & \dots & 0 \\ \vdots & \vdots & \vdots & \ddots & \vdots \\ 0 & 0 & 0 & \dots & 1 \end{pmatrix}.$$

Then the subgroup $H = KL$ acts transitively on the set $P(V) \setminus y$. We have

$$\begin{aligned} \sum_{h \in H} h(a) &= |H| \cdot e_y + \sum_{z \neq y} a_z \sum_{h \in H} e_{hz} \\ &= |H| \cdot e_y + \sum_{z \neq y} a_z \frac{|H|}{|P(V)| - 1} \sum_{z \neq y} e_z \\ &= |H| \cdot e_y + \sum_{z \neq y} a_z p^{m-2} \sum_{z \neq y} e_z \\ &= |H| \cdot e_y - p^{m-2} \sum_{z \neq y} e_z \\ &= \frac{p^m - 1}{p - 1} p^{m-2} e_y - p^{m-2} \sum_{z} e_z. \end{aligned}$$

Consequently, the code $C$ contains the codeword $\frac{p^m-1}{p-1}p^{m-2}e_x - p^{m-2}\sum_z e_z$ for each $x \in P(V)$, and also the codewords $\frac{p^m-1}{p-1}p^{m-2}(e_x - e_y)$. Hence $C \supset p^{m-2}W^0$ since $W^0$ is generated by elements $e_x - e_y$.

We consider the series of submodules $W^0 \supset pW^0 \supset \cdots \supset p^{m-2}W^0$. It defines the following sequence of factors $\mathcal{R} = (R_1, \ldots, R_{m-2})$ in $C/p^{m-2}W^0$,

$$\begin{aligned} R_i &= C \cap p^{i-1}W^0 / C \cap p^i W^0 \\ &\cong C \cap p^{i-1}W^0 + p^iW^0/p^iW^0 \subset p^{i-1}W^0/p^iW^0 \cong W^0/pW^0. \end{aligned}$$

By Proposition 3.4 the module $R_i$ is isomorphic to $M_{r_i}$ for some $r_i$. It is clear that $r_1 \le \cdots \le r_{m-2}$. We must prove now that $r_{i-1} < r_i$ for $r_i \ne m-1$. Indeed, assume that $r_{i-1} = r_i < m-1$ for some $i$. It means that the module

$$\begin{aligned} N &= C^0(r_i, r_i+1, m-1, \ldots, m-1)/C^0(r_i-1, r_i, m-1, \ldots, m-1) \\ &= (C^0_{r_i} + p^2W^0)/(C^0_{r_i-1} + p^2W^0) \end{aligned}$$

is decomposable, that is, a sum of two nonzero submodules. Here for $r_i = 1$ we set $C^0(r_i-1, r_i, m-1, \ldots, m-1) = C^0_{r_i-1} + p^2W^0 = pC^0(1, m-1, \ldots, m-1)$. However, $N \cong \widehat{M}_{m-r_i}/\widehat{M}_{m-r_i-2}$ by Propositions 3.4 and 3.5, and the last module is indecomposable by Proposition 3.4.

2) It follows from the part 1), equality $W = W^0 \oplus \langle w \rangle$ and the fact that $W^0$ and $\langle w \rangle$ have no isomorphic composition factors. □

*Proof (of Theorem 2.3).* 1) It follows from definitions.

2) Let us consider

$$C^\# = \{a \in \mathbb{Q}_pW \ : \ (a, C) \subseteq p^{m-2}\mathbb{Z}_p\}$$

to avoid fractions. It is clear that $C^\# = p^{m-2}C^*$. In fact, $C^\#$ means the dual code with respect to the inner product $(*;*) = (1/p^{m-2}) \cdot (*,*)$.

If $C \subset W$ is invariant under the group $G$ then $C^\#$ is also invariant as

$$(ga, c) = (a, g^{-1}c) \in p^{m-2}\mathbb{Z}_p$$

for all $a \in C^\#$ and $c \in C$. We make use of the following simple observation. If

$$D_1 \supset D_2 \supset \cdots \supset D_t$$

is a chain of $G$-invariant codes in $W$ then

$$D_1^\# \subset D_2^\# \subset \cdots \subset D_t^\#.$$

Furthermore, $|D_1/D_2| = |D_2^\#/D_1^\#|, \ldots, |D_{t-1}/D_t| = |D_t^\#/D_{t-1}^\#|$.

One has

$$W^\# = p^{m-2}W, \quad (pW)^\# = p^{m-3}W, \quad \ldots \quad (p^{m-2}W)^\# = W.$$

Theorem 2.2 and Lemma 3.7 imply that for the chain

$$\begin{aligned} W &\supset C(m-2, m-1, \ldots, m-1) \\ &\supset C(m-3, m-1, \ldots, m-1) \supset \cdots \\ &\supset C(1, m-1, \ldots, m-1) \\ &\supset C_0 + pW \supset pW \end{aligned}$$

it corresponds the dual chain

$$\begin{aligned} p^{m-2}W &\subset p^{m-3}C^0(1, m-1, \ldots, m-1) + p^{m-2}W \\ &\subset p^{m-3}C^0(2, m-1, \ldots, m-1) + p^{m-2}W \subset \cdots \\ &\subset p^{m-3}C^0(m-2, m-1, \ldots, m-1) + p^{m-2}W \\ &\subset p^{m-3}C^0(m-1, \ldots, m-1) + p^{m-2}W \subset p^{m-3}W, \end{aligned}$$

which is equal to the chain

$$\begin{aligned} p^{m-2}W &\subset C^0(0, \ldots, 0, 1) + p^{m-2}W \\ &\subset C^0(0, \ldots, 0, 2) + p^{m-2}W \subset \cdots \\ &\subset C^0(0, \ldots, 0, m-2) + p^{m-2}W \\ &\subset C^0(0, \ldots, 0, m-1) + p^{m-2}W \subset p^{m-3}W. \end{aligned}$$

Therefore,

$$\begin{aligned} C(r, m-1, \ldots, m-1)^{\#} &= C^0(0, \ldots, 0, m-1-r) + p^{m-2}W \\ &= C^0(0, \ldots, 0, m-1-r) + p^{m-2}C_0. \end{aligned}$$

Next, for the chain

$$\begin{aligned} C(r, m-1, \ldots, m-1) &\supset C(r, m-2, m-1, \ldots, m-1) \supset \cdots \\ &\supset C(r, r+1, m-1, \ldots, m-1) \supset p^2W \end{aligned}$$

it corresponds the dual chain

$$\begin{aligned} &p^{m-3}C^0(m-1-r, m-1, \ldots, m-1) + p^{m-2}W \\ &\subset p^{m-4}C^0(1, m-1-r, m-1, \ldots, m-1) + p^{m-2}W \subset \cdots \\ &\subset p^{m-4}C^0(m-2-r, m-1-r, m-1, \ldots, m-1) + p^{m-2}W \subset p^{m-4}W, \end{aligned}$$

which is equal to the chain

$$\begin{aligned} &C^0(0, \ldots, 0, m-1-r) + p^{m-2}W \\ &\subset C^0(0, \ldots, 0, 1, m-1-r) + p^{m-2}W \subset \cdots \\ &\subset C^0(0, \ldots, 0, m-2-r, m-1-r) + p^{m-2}W \subset p^{m-4}W, \end{aligned}$$

Therefore,

$$\begin{aligned} &C(r_1, r_2, m-1, \ldots, m-1)^{\#} \\ &= C^0(0, \ldots, 0, m-1-r_2, m-1-r_1) + p^{m-2}W \\ &= C^0(0, \ldots, 0, m-1-r_2, m-1-r_1) + p^{m-2}C_0. \end{aligned}$$

Continuing, we get the required statement.

3) The equality

$$\begin{aligned}(C^0(r_1,\ldots,r_{m-2}))^\# &= (C(r_1,\ldots,r_{m-2})\cap W^0)^\# \\ &= (C(r_1,\ldots,r_{m-2}))^\# + (W^0)^\# \\ &= C^0(m-1-r_{m-2},\ldots,m-1-r_1) + \mathbb{Q}_p C_0\end{aligned}$$

completes the proof. □

## References


1. Assmus, E. F., Key, J. D.: Polynomial codes and finite geometries. In Handbook of Coding Theory, Volume 2, edited by V. S. Pless and W. C. Huffman, Elsevier (1998), Chapter 16, 1269–1343.
2. Calderbank, A. R., Sloane, N. J. A.: Modular and *p*-adic cyclic codes. Designs, Codes and Cryptography **6** (1995) 21–35.
3. Calderbank, A. R., Li, W.-C. W., Poonen, B.: A 2-adic approach to the analysis of cyclic codes. IEEE Trans. Inform. Theory **43**, No. 3 (1997) 977–986.
4. Hammons, R., Kumar, P. V., Calderbank, A. R., Sloane, N. J. A., Solé, P.: The $Z_4$-linearity of Kerdock, Preparata, Goethals, and related codes. IEEE Trans. Inform. Theory **40**, No. 2 (1994) 301–319.
5. Hirschfeld, J. W. P., Shaw, R.: Projective geometry codes over prime fields. In Finite Fields: Theory, Application and Algorithms. 2nd international conference, August 17-21, 1993, Las Vegas, NV, USA. Edited by G. L. Mullen et al., Providence, RI: American Mathematical Society. Contemp. Math. **168** (1994) 151-163.
6. Singer, J.: A theorem in finite projective geometry and some applications to number theory. Trans. Amer. Math. Soc. **43** (1938) 377–385.
7. Steinberg, R.: Lectures on Chevalley groups. Yale University, 1967.
8. Zalesskii, A. E., Suprunenko, I. D.: The modular permutation representations and fragment of the decomposition matrix for the symplectic and special linear groups over a finite field. Preprint No.16 (286), Institute of Mathematics, Minsk 1987. (Russian)

# Towards a Basis for the Space of Regular Functions in a Tower of Function Fields Meeting the Drinfeld-Vladut Bound*

Ilia Aleshnikov[1], Vinay Deolalikar[2], P. Vijay Kumar[3], and Henning Stichtenoth[4]

[1] Universität GH Essen
D-45117 Essen
Germany
e-mail: mat314@uni-essen.de

[2] Hughes Network Systems
San Diego CA
USA 92121
e-mail: vdeolalikar@hns.com

[3] University of Southern California
Los Angeles CA
USA 90089-2565
e-mail: vijayk@usc.edu

[4] Universität GH Essen
D-45117 Essen
Germany
e-mail: mat310@sp2.power.uni-essen.de

**Abstract** In this paper, a convenient set of functions is identified whose span includes all functions in a tower of function fields of Garcia and Stichtenoth that have poles only at the place at infinity. The latter set is of interest in the construction of long and efficient "one-point" AG codes.

## 1 Introduction

The Gilbert-Varshamov bound is commonly used as a yardstick to measure the performance of long codes. By this measure, the commonly used BCH codes are poor at large lengths. The length of a Reed-Solomon code is limited by the size of its alphabet; thus large lengths require very large alphabets, which substantially increases the complexity of encoding and decoding these codes. It is known that there exist long alternant and concatenated codes that meet the Gilbert-Varshamov bound. However, no explicit description of these codes exists.

Around 1980, V. D. Goppa used the theory of algebraic curves to construct a new family of codes, now referred to as algebraic geometric (AG) codes.

* This work was supported by the National Science Foundation under Grant CCR-9714626.

Code performance depends upon the ratio $N/g$ of two parameters of the underlying curve: (i) the genus $g$ of the curve and (ii) the number $N$ of rational points on the curve over $\mathbb{F}_q$ , or equivalently, the number of places of degree one in the function field corresponding to the curve. Good codes result in cases where the ratio $N/g$ is large and the Drinfeld-Vladut bound places an upper bound on the value of this ratio. This bound says that

$$A(q) := \limsup_{g\to\infty} N/g \le \sqrt{q} - 1 \tag{1}$$

In 1982, Tsfasman,Vladut and Zink showed the existence of families of curves whose $N/g$ ratio achieved the Drinfeld-Vladut bound. Moreover, the resulting AG codes had long lengths and performance exceeding that of the Gilbert-Varshamov bound – a feat that until then was widely considered unattainable.

However, the Tsfasman-Vladut-Zink result is existential in nature and does not provide an explicit description of the curves. In 1996, Garcia and Stichtenoth succeeded in showing that two families of curves having an explicit and simple description, also achieve the Drinfeld-Vladut bound. However, an explicit description of the so-called "one-point" AG codes constructed on these curves requires the determination of a basis for the vector spaces $\mathcal{L}(rP)$, which comprise functions having poles only at a specified point $P$, with pole order bounded by $r$.

In this paper, we describe partial results obtained while attempting to construct a basis for these spaces, using techniques borrowed from algebraic number theory.

## 2 Preliminaries and Notation

For an introduction to the theory of function fields, the reader is referred to [5] and [2]. The authors found [1] and [6] very useful for their number-theoretic viewpoint.

Let $K$ be a finite field and let $F = K(x)$ be the rational function field over $K$. The set of places of $F$ is denoted by $\mathbb{P}_F$. In $F$, the place that is the unique zero of $x - \gamma$, where $\gamma \in K$, will be denoted $P_\gamma$. The unique pole of $x$ will be denoted by $P_\infty$. Let $\mathcal{S} = \mathbb{P}_F \setminus \{P_0, P_\infty\}$. Then, $\{F, \mathcal{S}\}$ forms an *ordinary arithmetic field* [6]. Let $E/F$ be a finite separable extension of degree $n$ where $K$ is algebraically closed in $E$. Let $v_P(\cdot)$ denote the discrete valuation associated to a place $P$ in either $E$ or $F$.

**Definition 2.1.** The following rings are defined with respect to the set $\mathcal{S}$:

$$\begin{aligned} \mathcal{O}_F &= \{x \in F : v_P(x) \ge 0, \forall P \in \mathcal{S}\} \\ \text{and } \mathcal{O}_E &= \{x \in E : v_{P'}(x) \ge 0, \forall P' \in \mathcal{S}^E\} \\ \text{where } \mathcal{S}^E &= \{P' \in \mathbb{P}_E \mid P'|P, P \in \mathcal{S}\}. \end{aligned}$$

Then, $\mathcal{O}_E$ is the integral closure of $\mathcal{O}_F$ in $E$ and $\mathcal{O}_F$ (resp. $\mathcal{O}_E$ ) will be called the *ring of regular functions* in $F$ (resp. $E$), w.r.t. $\mathcal{S}$.

**Definition 2.2.** Since $\mathcal{O}_F$ is a principal ideal domain, an integral basis for $\mathcal{O}_E$ over $\mathcal{O}_F$ is guaranteed to exist and will be called a global integral basis.

Let $\overline{F}$ denote the algebraic closure of $F$ and $\sigma_1, \dots, \sigma_n$ denote the $n$ distinct embeddings of $E$ over $F$ into $\overline{F}$. Without loss of generality, we assume that $E \subseteq \overline{F}$ and can thus take $\sigma_1$ to be the identity map.

**Definition 2.3.** The norm $N_{E\to F}$ and trace $Tr_{E\to F}$ are mappings from $E$ to $F$ given by

$$N_{E\to F}(\alpha) = \prod_{i=1}^{n} \sigma_i(\alpha)$$

and

$$Tr_{E\to F}(\alpha) = \sum_{i=1}^{n} \sigma_i(\alpha).$$

**Definition 2.4.** The *discriminant* of an $n$-tuple of elements $(\alpha_1, \dots, \alpha_n)$, with $\alpha_i \in E$, $1 \le i \le n$, denoted $\Delta(\alpha_1, \dots, \alpha_n)$, is defined as

$$\Delta(\alpha_1, \dots, \alpha_n) = det^2 \begin{bmatrix} \alpha_1 & \dots & \alpha_n \\ \sigma_2(\alpha_1) & \dots & \sigma_2(\alpha_n) \\ \vdots & \vdots & \vdots \\ \sigma_n(\alpha_1) & \dots & \sigma_n(\alpha_n) \end{bmatrix}.$$

**Definition 2.5.** Given $P \in \mathbb{P}_F$, let $\mathcal{O}_{E,P}$ denote the integral closure of $\mathcal{O}_P$ in E. The *complementary module* over $\mathcal{O}_P$, denoted $\mathcal{C}_P$, is defined by

$$\mathcal{C}_P := \{z \in E \mid Tr_{E\to F}(z.\mathcal{O}_{E,P}) \subseteq \mathcal{O}_P\}.$$

It is known that there exists a $t_P \in \mathcal{O}_{E,P}$ such that $\mathcal{C}_P = \frac{1}{t_P}\mathcal{O}_{E,P}$. Let $P'$ denote a place in $E$ lying above $P$. If $d(P'|P)$ denotes the *different exponent* for $P'$ over $P$, then $d(P'|P) = v_{P'}(t_P)$. For details, see [5], Section III.4.

## 3 Integral Basis and the Discriminant

A necessary and sufficient condition for a basis $\{\gamma_1, \gamma_2, \dots, \gamma_n\}$ for $E/F$ to be a global integral basis is identified in this section.

**Theorem 3.1.** *$\{\gamma_1, \dots, \gamma_n\}$ is an integral basis for $\mathcal{O}_E$ over $\mathcal{O}_F$ iff it is an integral basis for $\mathcal{O}_{E,P}$ over $\mathcal{O}_P$, $\forall P \in \mathcal{S}$.*

*Proof.* Shown in Prop. 4-8-8 of [6]. □

**Theorem 3.2.** *Let $(\alpha_1, \dots, \alpha_n)$ be any basis for $E/F$ such that all $\alpha_i$ are integral at $P$. Let $(\beta_1, \dots, \beta_n)$ be a known local integral basis for $\mathcal{O}_{E,P}$ over $\mathcal{O}_P$. Then $(\alpha_1, \dots, \alpha_n)$ forms an integral basis at $P$ iff*

$$v_P(\Delta(\alpha_1, \dots, \alpha_n)) = v_P(\Delta(\beta_1, \dots, \beta_n)).$$

*Proof.* Follows from Prop. 4-8-17 (iv) of [6]. □

**Lemma 3.3.** *Let $(\beta_1, \dots, \beta_n)$ be a local integral basis at $P$ and $(\beta_1^*, \dots, \beta_n^*)$ be the trace dual basis. Then*

$$\Delta(\beta_1, \dots, \beta_n)\Delta(\beta_1^*, \dots, \beta_n^*) = 1$$

*Proof.* Follows from Prop. 3-7-4 and 3-7-8 of [6]. □

**Theorem 3.4.** *Let $(\beta_1, \beta_2, \dots, \beta_n)$ be a local integral basis at $P$. Then*

$$v_P(\Delta(\beta_1, \dots, \beta_n)) = v_P\left(N_{E\to F}(t_P)\right)$$

*where $\mathcal{C}_P = \frac{1}{t_P}.\mathcal{O}_{E,P}$, $t_P \in \mathcal{O}_{E,P}$, is the complementary module. If $(\alpha_1, \dots, \alpha_n)$ is a basis of elements, each of whom is integral at $P$, then $(\alpha_1, \dots, \alpha_n)$ is a local integral basis at $P$ iff*

$$v_P(\Delta(\alpha_1, \dots, \alpha_n)) = v_P\left(N_{E\to F}(t_P)\right)$$

*Proof.* Let $(\beta_1, \dots, \beta_n)$ be a local integral basis at $P$. From Proposition III.4.2., [5], we know that $\mathcal{C}_P$ is generated as an $\mathcal{O}_P$ module by elements of the trace dual basis. It follows then from the definition of the complementary module, that the module generated by $\{\frac{1}{t_P}\beta_i\}$ is the same as that generated by $\{\beta_i^*\}$ (both are equal to $\mathcal{C}_P$). Next consider the matrix equation

$$[\sigma_j(\beta_i^*)] = [A][\sigma_j(\frac{1}{t_P}\beta_i)]$$

from which it follows that

$$\Delta(\beta_1^*, \dots, \beta_n^*) = N_{E\to F}^2(\frac{1}{t_P})\Delta(\beta_1, \dots, \beta_n)|A|^2$$

From Lemma 3.3, we get that

$$\Delta(\beta_1, \dots, \beta_n) = u \cdot N_{E\to F}(t_P), \text{ where } u \in \mathcal{O}_P^*$$

and the first part of the theorem follows. The second conclusion now follows from Theorem 3.2. □

The next proposition follows from Proposition 2-5-8 of [6], and will help restate the conclusion of the preceding theorem in terms of different exponents.

**Proposition 3.5.** *Let $\{P_i' \mid i = 1, 2, \ldots, r\}$ be the set of places in $E$ that lie above $P$. Then*

$$v_P(N_{E\to F}(t_P)) = \sum_{i=1}^{r} d(P_i'|P)\, f(P_i'|P)$$

*where $f(P_i'|P)$ denotes the relative degree of $P_i'$ over $P$.*

We now state the main theorem of this section:

**Theorem 3.6.** *Let $\{\gamma_i \mid i = 1, 2, \ldots n\} \subseteq \mathcal{O}_E$. Then $(\gamma_1, \ldots, \gamma_n)$ is a (global) integral basis for $\mathcal{O}_E$ over $\mathcal{O}_F$ iff at every place $P \in \mathcal{S}$, we have*

$$v_P(\Delta(\gamma_1, \ldots, \gamma_n)) = \sum_{i=1}^{r} d(P_i'|P)\, f(P_i'|P)$$

*where $\{P_i' \mid i = 1, 2 \ldots, r\}$ are all the places in $E$ lying above $P$.*

**Note** In the application below, of this proposition to a function field tower of Garcia and Stichtenoth, the relative degree $f(P_i'/P)$ and $r$ both turn out to equal 1, and thus the condition on $v_P(\Delta(\gamma_1, \ldots, \gamma_n))$ that is to be checked simplifies to

$$v_P(\Delta(\gamma_1, \ldots, \gamma_n)) = d(P'|P)$$

where $P'$ is the sole place in $E$ lying over $P$.

# 4 Different Exponents in a Tower of Garcia and Stichtenoth

We apply the observations in the previous section to the second of two towers of function fields proven by Garcia and Stichtenoth [3] to attain the Drinfeld-Vladut bound. The reader is assumed to be familiar with the contents of this paper.

Let $q$ be the power of a prime $p$ and consider the tower of function fields given by:

$$T_1 = \mathbb{F}_{q^2}(x_1),$$

and for $n \geq 2$,

$$T_n = T_{n-1}(x_n),$$

where $x_n$ satisfies the equation

$$x_n^q + x_n = \frac{x_{n-1}^q}{x_{n-1}^{q-1} + 1}.$$

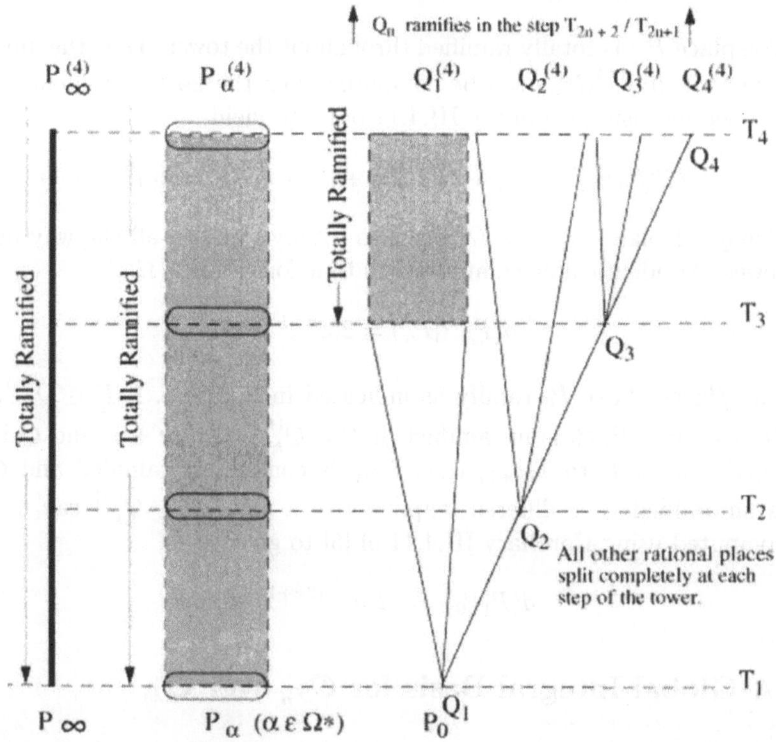


**Figure1.** This diagram shows the ramification in the tower, for the first four stages. The description of the ramification at higher stages is also included. Shaded regions denote sets of totally ramified places. Such places then continue to ramify throughout the tower.

Figure 1 depicts the ramification of places that takes place in this tower.

Let

$$\Omega^* \;=\; \{\alpha \in \mathbb{F}_{q^2} \mid \alpha^q + \alpha = 0, \alpha \neq 0\}.$$

The ramified places in $T_n$ lie above either $P_\infty$, $P_\alpha, \alpha \in \Omega^*$, or else $P_0$. The unique place in $T_n$ lying above $P_\infty$ (resp. $P_\alpha$), will be denoted by $P_\infty^{(n)}$ (resp. $P_\alpha^{(n)}$). Thus the superscript $(n)$ indicates a place in $T_n$. The places lying above $P_0$ in $T_n$ can be divided into the $n$ sets $Q_k^{(n)}, k = 1, 2, \ldots, n$, as shown in the diagram where, if $Q_k^{(k)}$ denotes the unique place in $T_k$ that is the common zero of the functions $x_1, x_2 \ldots, x_k$, we use $Q_k^{(n)}$ to denote the collection of places $Q \in T_n$ that lie above $Q_k^{(k)}$ but not above $Q_{k+1}^{(k+1)}$.

We now determine the different exponent associated with places belonging to the places in $T_n$ lying above $P_\infty$, $\{P_\alpha \mid \alpha \in \Omega^*\}$ and $P_0$.

1. The place $P_\infty$ is totally ramified throughout the tower. Thus the different exponent $d(P_\infty^{(n)}|P_\infty)$ can be computed from the note at the end of the last section using Corollary III.4.11 of [5] to yield
$$d(P_\infty^{(n)}|P_\infty) = 2(q-1)(1+q+\ldots+q^{n-2}) = 2(q^{n-1}-1).$$
2. The $q-1$ places $P_\alpha, \alpha \in \Omega^*$ are also totally ramified all the way up the tower. An identical calculation gives that for any $\alpha \in \Omega^*$,
$$d(P_\alpha^{(n)}|P_\alpha) = 2(q^{n-1}-1).$$
3. The places above $P_0$ ramify as indicated in Figure 1. If $P \in Q_k^{(n)}, n \leq 2k+1$, then $P/P_0$ is unramified. If $P \in Q_k^{(n)}, n > 2k+1$, and $Q$ is the restriction of $P$ to $T_{2k+1}$, then $P/Q$ is completely ramified and $Q/P_0$ is unramified. The different exponent of a place $P \in Q_k^{(n)}$ can now be computed using Corollary III.4.11 of [5] to give
$$d(P|P_0) \;=\; 2(q^{n-(2k+1)}-1).$$

## 5 A Global Integral Basis for $\mathcal{O}_{T_n}$ over $\mathcal{O}_{T_1}$

We first compute the discriminant of the basis of $T_n/T_1$ given by
$$\mathfrak{S} = \{\prod_{i=2}^{n} x_i^{e_i} \mid 0 \leq e_i \leq q-1\}.$$

**Lemma 5.1.**
$$\Delta(\mathfrak{S}) = \pm 1.$$

*Proof.* As is well known (see [6] page 104 for example), the discriminant $\Delta(\mathfrak{S})$ can also be written as the determinant of an $(q^{n-1} \times q^{n-1})$ matrix whose $(\underline{e}, \underline{f})$-th element (we use the lexicographic ordering), where $\underline{e} = (e_2, \ldots, e_n), 0 \leq e_i \leq q-1$ is given by
$$Tr_{T_n \to T_1}(\prod_{i=2}^{n} x_i^{h_i}), \text{ where } h_i = e_i + f_i.$$
From properties of the trace function, (see for example [4] pp.320-321), we can expand
$$Tr_{T_n \to T_1}(\prod_{i=2}^{n} x_i^{h_i}) \;=\; Tr_{T_{n-1} \to T_1}(\prod_{i=2}^{n-1} x_i^{h_i} Tr_{T_n \to T_{n-1}}(x_n^{h_n})).$$

The minimum polynomial of $x_n$ over $T_{n-1}$ is

$$z^q + z = \frac{x_{n-1}^q}{x_{n-1}^{q-1} + 1}.$$

Let $\{u_j, j = 1, 2, \ldots, q\}$, (with $u_1 = x_n$) denote the distinct roots in $\overline{T_1}$ of

$$z^q + z = \frac{x_{n-1}^q}{x_{n-1}^{q-1} + 1}.$$

Then

$$Tr_{T_n \to T_{n-1}}(x_n^{h_n}) \ = \ \sum_{j=1}^{q} u_j^{h_n}.$$

Let $\nu_1, \ldots, \nu_q$ be the elementary symmetric functions of the roots $\{u_j\}$. It follows that

$$\nu_1 = \ldots = \nu_{q-2} = 0 \text{ and } \nu_{q-1} = (-1)^{q-1}.$$

Let $S_r$ denote the $r$th power sum

$$S_r \ = \ \sum_{k=1}^{q} u_k^r.$$

From Newton's identities

$$S_r = \nu_1 S_{r-1} - \nu_2 S_{r-2} + \ldots + (-1)^r \nu_{r-1} S_1 + (-1)^{r+1} r \nu_r \text{ for } 1 \le r \le q,$$
$$S_r = \nu_1 S_{r-1} - \nu_2 S_{r-2} + \ldots + (-1)^{q+1} \nu_q S_{r-q} \text{ for } r > q$$

we obtain

$$\begin{aligned} S_r &= 0 \ \text{ for } \ 0 \le r < q-1 \ \text{ and } \ q \le r < 2(q-1) \\ S_{q-1} &= (-1)^q \nu_{q-1}(q-1) = 1 \\ S_{2(q-1)} &= (-1)^q \nu_{q-1} S_{q-1} = -1. \end{aligned}$$

Since $Tr_{T_n \to T_{n-1}}(x_n^{h_n}) \ = \ S_{h_n}$, it follows that

$$\begin{aligned} Tr_{T_n \to T_{n-1}}(x_n^{h_n}) &= 0 \text{ if } h_n \text{ is not divisible by } q-1 \\ &= 1 \text{ if } h_n = q-1 \\ &= -1 \text{ if } h_n = 2(q-1)\ . \end{aligned}$$

Thus, $Tr_{T_n \to T_{n-1}}(x_n^{h_n})$ is nonzero only if $(q-1)|h_n$ and in this case,

$$\begin{aligned} Tr_{T_n \to T_{n-1}}(x_n^{h_n}) &= \pm 1 \cdot Tr_{T_{n-1} \to T_1}(\prod_{i=2}^{n-1} x_i^{h_i}) \\ &= \pm 1 \cdot Tr_{T_{n-2} \to T_1}(\prod_{i=2}^{n-2} x_i^{h_i} Tr_{T_{n-1} \to T_{n-2}}(x_{n-1}^{h_{n-1}})) \end{aligned}$$

and we can argue similarly with $n$ replaced by $n-1$ and so on to arrive finally at

$$\begin{aligned} Tr_{T_n \to T_1}(\prod_{i=2}^{n} x_i^{h_i}) &= 0 \text{ if some } h_i \text{ is not divisible by } q-1 \\ &= \pm 1 \text{ if each } h_i = q-1 \ . \end{aligned}$$

Thus $\Delta(\mathfrak{S})$ is the determinant of a "reverse triangular" matrix of the form

$$\begin{bmatrix} 0 & 0 & \dots & 0 & \pm 1 \\ 0 & 0 & \dots & \pm 1 & * \\ \dots & \dots & \dots & \dots & \dots \\ 0 & \pm 1 & \dots & * & * \\ \pm 1 & * & * & * & * \end{bmatrix}$$

whose reverse-diagonal terms all equal $\pm 1$, and the entries to the left and above the reverse-diagonal are all zero. Clearly,

$$\Delta(\mathfrak{S}) = \pm 1.$$

□

Next consider the set $\mathfrak{S}_{g_1}$ obtained by multiplying each element in $\mathfrak{S}$ by $g_1$, except for the element '1' where

$$g_1 \ := \ \prod_{\alpha \in \Omega^*} (x_1 - \alpha) \ = \ (x_1^{q-1} + 1),$$

i.e.,

$$\mathfrak{S}_{g_1} = \{1\} \bigcup \{ \ g_1 \prod_{i=2}^{n} x_i^{e_i} \mid 0 \le e_i \le q-1, \text{ at least one } e_i \neq 0\}.$$

Multiplication by $g_1$ has ensured that every element in $\mathfrak{S}_{g_1}$ belongs to $\mathcal{O}_{T_n}$.
Since each embedding of $T_n$ into $\overline{T_1}$, fixes $g_1$ we have:

**Corollary 5.2.**

$$\Delta(\mathfrak{S}_{g_1}) = \pm g_1^{2(q^{n-1}-1)}.$$

**Theorem 5.3.** *The set $\mathfrak{S}_{g_1}$ forms a local integral basis at all places other than $P_\infty$ and $P_0$ and thus is an integral basis for $\mathcal{O}_{T_n}$ over $\mathcal{O}_{T_1}$.*

*Proof.* Follows from Theorem 3.6, Corollary 5.2 and the different exponents derived in Section 4. □

The following lemma describes all functions in $\mathcal{O}_F$.

**Lemma 5.4.**

$$\mathcal{O}_F = \mathbb{F}_{q^2}[x_1]\left[\frac{1}{x_1}\right].$$

This leads to the main theorem of this paper:

**Main Theorem 5.5.** *Every regular function in $T_n$ can be written as a linear combination of functions in the set $\mathfrak{S}_{g_1}$, with coefficients that are of the form $p(x_1)/x_1^i$, where $p(x_1)$ is a polynomial in $x_1$ and $i \geq 0$.*

## 6 Integral Basis for a Second Choice of $\mathcal{S}$

Consider now a different choice of the set $\mathcal{S}$, namely, $\mathcal{S} = \mathbb{P}_F \setminus \{P_0\}$. Of course, by changing $\mathcal{S}$ we have also implicitly changed $\mathcal{O}_{T_n}$ and $\mathcal{O}_{T_1}$. Using similar methods, one can obtain an integral basis for $\mathcal{O}_{T_n}$ over $\mathcal{O}_{T_1}$. Consider the set $\mathfrak{S}_{h_1}$ obtained by multiplying each element in $\mathfrak{S}$ by $h_1 = \frac{g_1}{x_1^q}$, except for the element '1'. i.e.,

$$\mathfrak{S}_{h_1} = \{1\} \bigcup \{\, h_1 \prod_{i=2}^{n} x_i^{e_i} \mid 0 \leq e_i \leq q-1, \text{ at least one } e_i \neq 0\}.$$

Again, multiplication by $h_1$ has ensured that every element in $\mathfrak{S}_{h_1}$ belongs to $\mathcal{O}_E$.

One can similarly prove that in this case:

**Corollary 6.1.**

$$\Delta(\mathfrak{S}_{h_1}) = \pm h_1^{2(q^{n-1}-1)}.$$

and

**Theorem 6.2.** *The set $\mathfrak{S}_{h_1}$ forms a local integral basis at all places other than $P_0$ and thus is an integral basis for $\mathcal{O}_{T_n}$ over $\mathcal{O}_{T_1}$.*

## Acknowledgements

Vinay Deolalikar and P. Vijay Kumar acknowledge the gracious help accorded by the late Prof. Dennis Estes, who very tragically passed away while this work was still being done. Without his help, this work could not have been done. We also thank Prof. Kyeongcheol Yang and Kenneth Shum for many interesting discussions.

## References

1. J.W.Cassels and A Fröhlich (Editors), *Algebraic number theory*, Academic Press, 1967.
2. Claude Chevalley, *Introduction to the theory of algebraic functions of one variable*, American Mathematical Society, 1951.
3. Arnaldo Garcia and Henning Stichtenoth, "On the Asymptotic Behaviour of Some Towers of Function Fields over Finite Fields," Journal of Number Theory, vol. 61, No. 2, December 1996, pp. 248-273.
4. Serge Lang, *Algebra*, Addison-Wesley, 1993.
5. Henning Stichtenoth, *Algebraic function fields and codes*, Springer-Verlag, 1991.
6. Edwin Weiss, *Algebraic number theory*, Dover, 1963.

# Self–Dual Normal Bases and Related Topics

Eva Bayer–Fluckiger

Laboratoire de Mathématiques de Besançon
UMR 6623 du CNRS
16, route de Gray
25030 Besançon
France
e-mail: bayer@math.univ-fcomte.fr

## 1 Introduction

It is known since more than 20 years that an odd degree extension of finite fields has a self–dual normal basis (cf. [14]). This result can be reformulated in terms of $G$–forms, that is forms invariant by the action of a group $G$. It is easy to check that the trace form of a Galois extension with group $G$ is a $G$–form, and the above result is equivalent to saying that this $G$–form is isomorphic to the unit $G$–form. More generally, one can ask for the classification of trace forms of extensions with group $G$ as $G$–forms.

It is more natural, and also more useful, to study this problem in the more general framework of *Galois algebras* rather than just of Galois extensions. This is the point of view of the papers [1]–[5], where some partial results are proved. For instance, if $G$ is a group of odd order, then any $G$–Galois algebra has a self–dual normal basis (cf. [1], [2]). §2–5 of the present paper contains a survey of the basic notions and results in this topic. Further results, obtained for fields of characteristic $\neq 2$, are often formulated in terms of cohomological invariants. This is the subject matter of §6, where the notion of *$G$–discriminant* is introduced (this invariant is implicit in [5]). For the fields of cohomological dimension 1 (for instance, finite fields) the $G$–discriminant is a complete invariant.

In the case of extensions of finite fields, the $G$–discriminant can take at most two values, hence there are at most two possibilities for the isomorphism class of the $G$–trace form. This leads to the notion of *semi–dual normal basis* (cf. §8) and the result that every extension of finite fields of characteristic $\neq 2$ has either a self–dual normal basis or a semi–dual normal basis. Some explicit models for the corresponding $G$–forms are given in §9.

## 2 Which Galois Extensions Have a Self–Dual Normal Basis ?

As pointed out in the introduction, there are several ways of formulating this question. In this section, we start with the most elementary one.

Let $k$ be a field, and let $L$ be a Galois extension of $k$ with Galois group $G$. A *normal basis* of $L$ over $k$ is a basis of the form $(g(e))_{g \in G}$, for some $e \in L^*$. It is well-known that every Galois extension has a normal basis.

The extension $L/k$ being separable, the *trace form*

$$q_L : L \times L \to k$$

$$q_L(x,y) = \mathrm{tr}_{L/k}(xy)$$

is a non-degenerate symmetric bilinear form.

We say that a normal basis $(g(e))_{g \in G}$ is also *self-dual* if

$$q_L(g(e), h(e)) = \delta_{g,h}.$$

As we will see, *not* every Galois extension has a self-dual normal basis. To my knowledge, the question of deciding which Galois extensions have a self-dual normal basis was first raised in the context of *finite fields*, in the book of McWilliams and Sloane [12]. It is proved in [12] that $\mathbf{F}_{2^n}$ has a self-dual normal basis over $\mathbf{F}_2$ provided $n$ is *odd*. In the 80's, this question was taken up by several authors : Imamura [8], Beth-Fumy-Mühlfeld [6], Morii-Imamura [13], Lempel [10], and Lempel-Weinberger [11]. This led to a complete solution to the self-dual normal basis problem for finite fields, that is (cf. [11]) :

**Theorem 2.1.** *Let $k$ be a finite field. Then an extension $L/k$ of degree $n$ has a self-dual normal basis if and only if one of the following holds :*

1. char $(k) \neq 2$, *and $n$ is odd ;*
2. char $(k) = 2$, *and* $n \not\equiv 0 \pmod 4$.

More recently, some other authors – in particular, Conner-Perlis [7], H.W. Lenstra, Jr. – raised the question of the existence of self-dual normal basis for more general fields. The following result of Lenstra extends th. 1.1. to *abelian* extensions of arbitrary fields (cf. [3])

**Theorem 2.2.** *Let $L$ be a Galois extension of finite degree $n$ over $k$, and let* $G = (\mathrm{Gal})(L/k)$. *Suppose that $G$ is abelian.*

1. *Assume that* char $(k) \neq 2$. *Then there exists a self-dual normal basis of $L$ over $k$ if and only if $n$ is odd.*
2. *Assume that* char $(k) = 2$. *Then there exists a self-dual normal basis of $L$ over $k$ if and only if the exponent of $G$ is not divisible by* 4.

In the case of Galois extensions with non-abelian group, no complete criterion for the existence of self-dual normal bases is available. However, the fact that extensions of *odd degree* always have a self-dual normal basis does extend to this case.

**Theorem 2.3.** [1], [3] *Every Galois extension of odd degree has a self-dual normal basis.*

The proof of this result uses a reformulation of the self-dual normal basis problem in terms of *G-forms*. This will be the topic of the next section.

## 3 A Reformulation of the Self–Dual Normal Basis Problem

Let $k$ be a field, and let $L$ be a Galois extension of $k$ with group $G$. Let us denote by $k[G]$ the group algebra. The fact that $L$ has a normal basis over $k$ can be reformulated by saying that the $k[G]$–module $L$ is free of rank one, in other words we have an isomorphism of $k[G]$–modules $L \simeq k[G]$. It is natural to look for a similar "structural" formulation of the existence of a self–dual normal basis. This leads to the notion of *G–form.* Indeed, if $q_L$ is the trace form of $L/k$ then it is immediate that

$$q_L(gx, gy) = q_L(x, y)$$

for all $g \in G$, and for all $x, y \in L$. More generally, one defines the notion of $G$–form as follows :

For any finite group $G$, a *G–form* will be a pair $(V, q)$, where $V$ is a $k[G]$–module and also a finite dimensional $k$–vector space, and $q : V \times V \to k$ a non–degenerate symmetric bilinear form such that

$$q(gx, gy) = q(x, y)$$

for all $g \in G$, and for all $x, y \in V$.

We say that two $G$–forms $(V, q)$ and $(V', q')$ are *isomorphic* if there exists an isomorphism of $k[G]$–modules $f : V \to V'$ such that $q(fx, fy) = q(x, y)$ for all $x, y \in V$. If this is the case, we write $(V, q) \simeq_G (V', q')$, or $q \simeq_G q'$ for short.

The *unit G–form* will be the pair $(k[G], q_0)$, with

$$q_0(g, h) = \delta_{g,h}.$$

The unit $G$–form will often be denoted by $q_0$.

The reformulation of the self–dual basis problem is based on the following fact :

**Proposition 3.1.** *Let $L$ be a Galois extension of $k$ with group $G$. Then $L$ has a self–dual normal basis over $k$ if and only if $q_L \simeq_G q_0$*

*Proof.* This is clear. □

We see that the question of the existence of a self–dual normal basis is equivalent to a question about the structure of the trace form as a $G$–form, namely whether it is isomorphic to the simplest possible $G$–form, the unit $G$–form. More generally, one could ask for the characterisation of trace forms of Galois extensions with group $G$ as $G$–forms. We will come back to this question in a slightly more general context in section 5.

## 4 Odd Degree Extensions

According to th. 3, every Galois extension of odd degree has a self–dual normal basis. The aim of this section is to outline a proof of this result, and to stress the importance of *base change* (extension of the base field) in the process. This again will lead us to look at the self–dual normal basis problem in a different way.

Let $L$ be a Galois extension of $k$ with group $G$. We have seen that the existence of a self–dual normal basis of $L$ over $k$ is equivalent to the existence of an isomorphism $(L, q_L) \simeq_G (k[G], q_0)$.

It is interesting to note that such an isomorphism *always* exists after a finite base change. Indeed, let us look at the field extension $L/k$, and let us base change the $G$–form $(L, q_L)$ to this field extension. We have $L \otimes_k L = L \times \cdots \times L$, the number of copies being equal to $n = [L : k] = |G|$. This is an $L[G]$–module, the action of $G$ being a transitive permutation on the factors. The extension of the $G$–form $q_L$ to $L/k$ is the unit $G$ form $q_0$ over $L$. Hence we obtain an isomorphism $(L, q_L) \otimes_k L \simeq_G (k[G], q_0) \otimes_k L$. The two $G$–forms $(L, q_L)$ and $(k[G], q_0)$ become isomorphic over the field extension $L/k$.

The above isomorphism holds independently of the value of the degree of the field extension $[L : k] = n$. However, if this degree is *odd*, then we have a result that enables us to deduce that the $G$–forms are already isomorphic over $k$ :

**Theorem 4.1.** [3] *Suppose that* char $(k) \neq 2$. *If two $G$–forms become isomorphic over an odd degree extension of $k$, then they are isomorphic over $k$.*

Note that we recover th. 3 as a consequence of th. 5 :

**Corollary 4.2.** [1], [3] *Every Galois extension of odd degree has a self–dual normal basis.*

If char $(k) \neq 2$, then the corollary is an immediate consequence of th. 5. The statement still holds for fields of characteristic 2, but the proof is less direct, cf. [1].

It would be interesting to have an analogue of th. 5 for fields of characteristic 2.

## 5 Galois Algebras

In the last section, we have seen that it is sometimes useful to pass to finite extensions of the ground field. However, this operation does not preserve the property of being a field extension. If $L/k$ is a separable extension of finite degree, and $k'/k$ a finite extension, then $L \otimes_k k'$ is not a field in general. It is a product of separable extensions of $k'$, that is, an *étale algebra*. If moreover $L/k$ is a Galois extension with group $G$, then $L \otimes_k k' = L' \times \cdots \times L'$, where

$L'/k'$ is a Galois extension with group $G' \subset G$. The group $G$ operates on $L \otimes_k k'$, and the $k'[G]$–module $L \otimes_k k'$ is still free of rank one. In other words, we obtain a *G–Galois algebra* over $k'$. This notion can be defined in several ways (cf. for instance [5], [9]). For instance, one can define it as follows :

Let $k_s$ be a separable closure of $k$. For any finite group $G$, we say that a $k$–algebra $L$ of finite rank with an operation of $G$ is a *G–Galois algebra* over $k$ if $L \otimes_k k_s = k_s \times \cdots \times k_s$, and if the group $G$ permutes the factors $k_s$ simply transitively.

The category of $G$–Galois algebras is stable by base change : if $L$ is a $G$–Galois algebra over $k$, then $L \otimes_k k'$ is a $G$–Galois algebra over $k'$.

Let $L$ be a $G$–Galois algebra over $k$. We still denote by $q_L : L \times L \to k$ the trace form, defined as before by $q_L(x, y) = (\mathrm{tr}_{L/k}(xy)$. Then $q_L$ is a $G$–form.

The *split G–Galois algebra* is $L_0 = k \times \cdots \times k$, where $G$ acts by permuting the factors transitively. The trace form $q_{L_0}$ is isomorphic to the unit $G$–form $q_0$.

We see that the self–dual normal basis question is equivalent to asking whether the trace form $q_L$ of a Galois extension – or, more generally, of a $G$–Galois algebra – is isomorphic as a $G$–form to the trace form of a specific $G$–Galois algebra, namely $L_0$. More generally, it is natural to ask the following questions :

*Question 5.1. Let $L_1$ and $L_2$ be two G–Galois algebras over k. When do we have*

$$q_{L_1} \simeq_G q_{L_2} \ ?$$

*Question 5.2. Let q be a G–form. Does there exist a G–Galois algebra L such that*

$$q_L \simeq_G q \ ?$$

In the case of groups of odd order, both questions can be answered :

**Theorem 5.3.** *Let G be a finite group of odd order, and let L be a G–Galois algebra. Then* $q_L \simeq_G q_0$.

The proof is the same as in the case of Galois extensions, cf. section 4. For groups of even order, a few results are known, and the following sections will give a survey of these. However, questions 7 and 8 are far from being solved in general.

# 6 Cohomological Invariants

Classification results of quadratic forms are often formulated in terms of cohomological invariants. It is natural to try this approach also for trace forms of $G$–Galois algebras.

From now on, we always suppose that the field $k$ has *characteristic different from 2.* Let $k_s$ be a separable closure of $k$, and set $\Gamma_k = \mathrm{Gal}(k_s/k)$.

To any $G$–Galois algebra $L$, one associates a continuous homomorphism $\phi_L : \Gamma_k \to G$ (see for instance [5], 1.3.1.). Under this correspondence, the isomorphism classes of $G$–Galois algebras are in bijection with the conjugacy classes of continuous homomorphisms $\Gamma_k \to G$. The algebra $L$ is a field if and only if $\phi_L$ is onto; it is split if and only if $\phi_L = 1$.

The homomorphism $\phi_L$ can be used in several ways to attach cohomological invariants to $L$, and to $q_L$.

## 6.1 Galois Descent

Recall that if $U$ is a smooth linear algebraic group over $k$, then one defines a pointed set $H^1(k, U) = H^1(\Gamma_k, U(k_s))$, cf. [14], I.5. and III.1. This pointed set is very useful in classification questions when $U$ can be identified as an automorphism group. For instance, if $U$ is the group of automorphisms of the unit $G$–form $q_0$, then $H^1(k, U)$ is in bijection with the set of isomorphism classes of $G$–forms that become isomorphic to $q_0$ over $k_s$.

Let $^- : k[G] \to k[G]$ be the canonical involution of the group algebra $k[G]$. Recall that this involution is characterised by the fact that $\overline{g} = g^{-1}$ for all $g \in G$.

For any commutative $k$–algebra $E$, set

$$U_G(E) = \{x \in E[G] \mid x\overline{x} = 1 \}.$$

Then $U_G$ is a smooth linear algebraic group defined over $k$. It is immediate that $U_G$ is the group of automorphisms of $q_0$. Hence the isomorphism class of $q_L$ corresponds to an element $u(L) \in H^1(k, U_G)$.

It is easy to give an explicit description of $u(L)$. Let $f_L : \Gamma_k \to U_G(k_s)$ be the composition of $\phi_L : \Gamma_k \to G$ with the inclusion $G \to U_G(k_s)$. We can regard $f_L$ as a 1–cocycle, and its class is $u(L)$ (cf. [5], 1.5.3.).

Stated in slightly different terms, the map that associates to an isomorphism class of $G$–Galois algebras the isomorphism class of its trace form is given by

$$H^1(k, G) \to H^1(k, U_G).$$

This cohomological description shows that *the trace form of a $G$–Galois algebra determines the trace forms of all subalgebras of fixed points* of normal subgroups of $G$ (see also [5], 1.4.1.). Indeed, let $L$ be a $G$–Galois algebra, and let $H$ be a normal subgroup of $G$. Let $L'$ be the subalgebra $L^H$. Then $L'$ is a $G/H$–Galois algebra. The canonical projection $G \to G/H$ induces a homomorphism $U_G \to U_{G/H}$, which in turn gives rise to a map

$$H^1(k, U_G) \to H^1(k, U_{G/H}).$$

This map sends $u(L)$ to $u(L')$.

### 6.2 Mod 2 Invariants

Set $H^r(G) = H^r(G, \mathbf{Z}/2\mathbf{Z})$, $H^r(k) = H^r(\Gamma_k, \mathbf{Z}/2\mathbf{Z})$. The homomorphism $\phi_L : \Gamma_k \to G$ induces $\phi_L^* : H^r(G) \to H^r(k)$ for all integers $r \geq 0$. Set $x_L = \phi_L^*(x)$.

**Proposition 6.1.** ([5], 2.2.1.) *Let L and L′ be two G–Galois algebras. If $q_L \simeq_G q_{L'}$, then $x_L = x_{L'}$ for all $x \in H^1(G)$.*

**Corollary 6.2.** *Let L be a G–Galois algebra. If L has a self–dual normal basis over k, then $x_L = 0$ for all $x \in H^1(G)$.*

This property does not extend to $r > 1$ in general (see [5], 10.2). However, for certain groups it is possible to associate higher cohomological invariants to trace forms of $G$–Galois algebras in this way. This is done in some cases in [5], §§7 and 9 (see also §7 of the present paper).

Recall that $H^1(k) \simeq k^*/k^{*2}$. Hence for any $x \in H^1(G)$, we obtain an invariant $x_L \in k^*/k^{*2}$. This invariant is easy to describe in concrete terms. Indeed, any non–trivial element of $H^1(G)$ determines a quotient of order 2 of $G$. Let $K$ be the corresponding subalgebra of fixed points. Then $K$ is a quadratic subalgebra of $L$, and $x_L$ is its discriminant. By the preceeding remarks the trace form $q_K$, and hence also its discriminant, are determined by the $G$–form $q_L$.

### 6.3 G–Discriminant

Reformulating the 1-dimensional cohomological invariant defined in the preceding section leads to the notion of *G–discriminant*. Let $G^2$ be the subgroup of $G$ generated by the squares. The quotient $G/G^2$ is an elementary abelian 2–group. The projection $G \to G/G^2$ induces

$$d_G : H^1(k, G) \to H^1(k, G/G^2).$$

Recall that $H^1(k, G)$ is in bijection with the set of isomorphism classes of $G$–Galois algebras. The *G–discriminant* $d_G(L)$ of a $G$–Galois algebra is by definition the image by $d_G$ of the cohomology class corresponding to $L$. With this terminology, we obtain the following reformulations of prop. 10 and cor. 11 :

**Proposition 6.3.** *Let L and L′ be two G–Galois algebras over k. If $q_L \simeq_G q_{L'}$ then $d_G(L) = d_G(L')$.*

**Corollary 6.4.** *Let L be a G–Galois algebra over k. If L has a self–dual normal basis over k then $d_G(L) = 1$.*

Note that $H^1(k, G/G^2) = \mathrm{Hom}(\Gamma_k, G/G^2)$, the set of continous homomorphisms $\Gamma_k \to G/G^2$. As $G/G^2$ is an elementary abelian 2–group of rank $r$, this is also the product of $r$ copies of the square classes $k^*/k^{*2}$.

# 7 Groups of Even Order

As in section 5, we suppose that $k$ is a field of characteristic $\neq 2$, and $G$ is a finite group. We have defined an invariant $x_L \in k^*/k^{*2}$ for all $x \in H^1(G)$. This invariant is an obstruction to the existence of self–dual normal bases. Using this invariant, one can show the following :

**Proposition 7.1.** *Let $L$ be a Galois extension of $k$ with group $G$. Suppose that $G$ has a quotient of order 2. Then $L$ does not have any self–dual normal basis over $k$.*

*Proof.* By hypothesis, $G$ has a quotient of order 2. Let $K$ be the corresponding subalgebra of fixed points, and let $x \in H^1(G)$ be the corresponding cohomology class. As noted in §5, the trace form $q_K$ of the quadratic subalgebra $K$ is determined by $q_L$, and its discriminant is $x_L$. As $L$ is a field, $K$ is also a field. But the discriminant of a quadratic extension of a field of characteristic $\neq 2$ is never trivial. Hence $L$ cannot have a self–dual normal basis.

It is natural to ask whether this property extends to all groups of even order. As the following examples show, this is not the case in general.

If $G = \mathrm{PGL}_2(\mathbf{F}_q)$, $q \equiv \pm 3 \pmod 8$, then $H^1(G) = 0$, and $H^2(G)$ has order 2. Let us denote by $x$ the non–trivial element of $H^2(G)$. □

**Proposition 7.2.** *Let $G = \mathrm{PGL}_2(\mathbf{F}_q)$, with $q \equiv \pm 3 \pmod 8$. Let $L$ and $L'$ be $G$–Galois algebras. Then $q_L \simeq_G q_{L'}$ if and only if $x_L = x_{L'}$.*

*Proof.* see [5], 8.1. and 7.5.4. □

**Corollary 7.3.** *Let $L$ be a Galois extension of $k$ with group $G = \mathrm{PGL}_2(\mathbf{F}_q)$, where $q \equiv \pm 3 \pmod 8$. Then $L$ has a self–dual normal basis over $k$ if and only if $L$ can be embedded in a Galois extension of $k$ with group $\mathrm{SL}_2(\mathbf{F}_q)$.*

*Proof.* [5], th. 8.1.1. and Remarque.

If $G = \mathrm{SL}_2(\mathbf{F}_8)$ or $G = J_1$, the first Janko group, then $H^1(G) = H^2(G) = 0$, and $H^3(G)$ is cyclic of order 2. Let us denote by $x$ the non–trivial element of $H^3(G)$.

**Proposition 7.4.** *Let $G = \mathrm{SL}_2(\mathbf{F}_8)$ or $G = J_1$. Let $L$ and $L'$ be two $G$–Galois algebras over $k$. Then $q_L \simeq_G q_{L'}$ if and only if $x_L = x_{L'}$.*

*Proof.* See [5], 8.2 and 7.5.4.

Note that the groups of prop. 15 and 17 have elementary abelian 2–Sylow subgroups (of rank 2 and 3, respectively). More generally, if $G$ has elementary abelian or quaternionian 2–Sylow subgroups then one obtains complete criteria for the $G$–isomorphism of the trace forms of $G$–Galois algebras : see [5], §§7, 8 and 9. The criteria can often be expressed in terms of higher cohomological invariants. A survey of these results is given in [2], 7.7 and 7.8., as well as in [14], III, Appendice 2.

# 8 Semi-Dual Normal Bases

We have seen in §1 that an extension of finite fields of char $\neq 2$ has a self-dual normal basis if and only if the degree of the extension is odd. The aim of this section is to propose a substitute of self-dual normal bases, called *semi-dual normal bases*, for extensions of even degree.

Several of the results that we need for this are valid in a much more general context. Even though the main emphasis here is on finite fields, we will state the results in a greater generality.

As in the preceeding sections, $k$ will be a field of characteristic $\neq 2$ and $G$ a finite group.

## 8.1 Fields of Dimension $\leq 1$

Let $k_s$ be a separable closure of $k$, and set $\Gamma_k = \mathrm{Gal}(k_s/k)$. We say that the 2-cohomological dimension of $\Gamma_k$ is $\leq 1$, written $\mathrm{cd}_2(\Gamma_k) \leq 1$, if $H^n(\Gamma_k, C) = 0$ for all $n > 1$ and for all finite 2-primary $\Gamma_k$-modules $C$ (cf. [14], II.3.). For instance, finite fields have this property.

It is proved in [5], 2.2., that when $\mathrm{cd}_2(\Gamma_k) \leq 1$, the criteria of prop. 10 and cor. 11 are necessary and sufficient. More precisely, we have :

**Theorem 8.1.** *Suppose that* $\mathrm{cd}_2(\Gamma_k) \leq 1$. *Let* $L$ *and* $L'$ *be two* $G$*-Galois algebras over* $k$. *Then* $q_L \simeq_G q_{L'}$ *if and only if* $x_L = x_{L'}$ *for all* $x \in H^1(G)$.

*Proof.* See [5], th.2.2.3.

This can be reformulated as follows :

**Theorem 8.2.** *Suppose that* $\mathrm{cd}_2(\Gamma_k) \leq 1$. *Let* $L$ *and* $L'$ *be two* $G$*-Galois algebras over* $k$. *Then* $q_L \simeq_G q_{L'}$ *if and only if* $d_G(L) = d_G(L')$.

**Corollary 8.3.** *Suppose that* $\mathrm{cd}_2(\Gamma_k) \leq 1$. *Let* $L$ *be a* $G$*-Galois algebra over* $k$. *Then* $L$ *has a self-dual normal basis over* $k$ *if and only if* $d_G(L) = 1$.

We also obtain a converse of prop. 14 in the case where $\mathrm{cd}_2(\Gamma_k) \leq 1$ :

**Corollary 8.4.** *Suppose that* $\mathrm{cd}_2(\Gamma_k) \leq 1$. *Let* $L$ *be a Galois extension of* $k$ *with group* $G$. *Then* $L$ *has a self-dual normal basis over* $k$ *if and only if* $H^1(G, \mu_2) = 0$, *i.e.* $G$ *has no quotient of order 2.*

The preceding results show that the $G$-discriminant provides a complete invariant when $\mathrm{cd}_2(\Gamma_k) \leq 1$. It remains to see which elements of $H^1(k, G/G^2)$ $= \mathrm{Hom}(\Gamma_k, G/G^2)$ are realised as $G$-discriminants of $G$-Galois algebras. This is done in [4] :

**Theorem 8.5.** *Suppose that* $\mathrm{cd}_2(\Gamma_k) \leq 1$. *Then*

$$d_G : H^1(k, G) \to H^1(k, G/G^2) = \mathrm{Hom}(\Gamma_k, G/G^2)$$

*is onto.*

Combining th. 18 and 22 we obtain (cf. [4]) :

**Corollary 8.6.** *Suppose that* $\mathrm{cd}_2(\Gamma_k) \leq 1$. *Then the set of isomorphism classes of trace forms of G–Galois algebras is in bijection (as pointed sets) with* $\mathrm{Hom}(\Gamma_k, G/G^2)$.

Suppose that $G$ is a cyclic group of even order. Then $G/G^2$ has order 2, we have $H^1(k, G/G^2) \simeq k^*/k^{*2}$, and the $G$–discriminant becomes

$$d_G : H^1(k, G) \to k^*/k^{*2}.$$

Suppose also that $\mathrm{cd}_2(\Gamma_k) \leq 1$. Then by th. 22, the $G$–discriminant is onto. For all $d \in k^*/k^{*2}$, there exist $G$–Galois algebras with $G$–discriminant $d$. Let us choose a $G$–Galois algebra $L_d$ with $d_G(L_d) = d$, and let us denote $Q_d = q_{L_d}$ its trace form. As the $G$–discriminant is a complete invariant (cf. th. 19), the trace form of any $G$–Galois algebra is isomorphic to $Q_d$ for some $d$. We have $Q_1 = q_0$, the trace form of the split $G$–Galois algebra.

Note that in the case of cyclic groups, the $G$–discriminant coincides with the usual discriminant of the étale algebra (cf. [4]).

If moreover $k$ has only two square classes, $k^*/k^{*2} = \{1, D\}$, then there are only two possible discriminants, hence only two possible $G$–trace forms, $Q_1$ and $Q_D$. By definition, $L$ has a self–dual normal basis if and only if $q_L \simeq Q_1$ (=$q_0$). We say that a $G$–Galois algebra has a *semi–dual normal basis* if $q_L \simeq Q_D$. Hence every $G$–Galois algebra has either a self–dual normal basis, or a semi–dual normal basis.

Extensions of even degree of finite fields satisfy all the above hypotheses, hence we obtain

**Proposition 8.7.** *Let $q$ be an odd prime number. Then*

1. *If $n$ is odd,* $\mathbf{F}_{q^n}$ *has a self–dual normal basis;*
2. *If $n$ is even,* $\mathbf{F}_{q^n}$ *has a semi–dual normal basis.*

The above results raise the problem of constructing explicit models for $Q_D$, and hence for semi–dual normal bases. This will be done in the next section.

## 9 Models for $G$–Trace Forms

Let $k$ be a field of char $\neq 2$, and such that $\mathrm{cd}_2(\Gamma_k) \leq 1$. Let $G$ be a cyclic group, and let $L$ be a $G$–Galois algebra over $k$ with discriminant $d$. As usual, we denote by $q_L$ its trace form. If $G$ has odd order, then $L$ has a self–dual normal basis over $k$ and the $G$–form $q_L$ is isomorphic to the split $G$–form $q_0$. If $G$ has even order, then no self–dual normal basis exists in general. In particular, if $L$ is a field, Galois extension of $k$ with group $G$ of even order, then $L$ does not have any self–dual normal basis over $k$. However, we saw in §8 that the $G$–form $q_L$ is uniquely determined by the discriminant of $L$. In other words, $q_L$ is isomorphic to a $G$–form that we denoted by $Q_d$ in §7. The aim of this section is to give explicit models for the $G$–form $Q_d$.

All the above hypothesis are satisfied in the case of extensions of finite fields, which is the main case of interest here. The essential case is the one where $G$ is a cyclic group of prime power order. We consider this case first, and then come back to the case of arbitrary cyclic groups.

### 9.1 Cyclic Groups of 2–Power Order

Suppose that $G$ is cyclic of order $N = 2^n$, and let $\sigma$ be a generator of $G$. Let $V$ be the free $k[G]$–module of rank one with $k$–basis $e_1, \ldots, e_N$, satisfying $\sigma(e_i) = e_{i+1}$ (indices mod $N$). Let $d \in k^*$. Set $a = \frac{1-d}{N}$, and let $Q_d : V \to k$ be defined by $Q_d(e_i, e_i) = a + 1$, $Q_d(e_i, e_j) = (-1)^{|i-j|}a$ if $i \neq j$. It is easy to check that $Q_d$ is a $G$–form and has determinant $d$.

Note that $d = 1$ if and only if $a = 0$, and this happens if and only if $Q_d$ is the unit $G$–form.

Expressed in coordinates, we have

$$Q_d(X_1, \ldots, X_d) = \Sigma_{i=1,\ldots,N} X_i^2 + a(X_1 - X_2 + \cdots - X_N)^2.$$

Another possible model for $Q_d$ gives a diagonal quadratic form, but it is no longer expressed in a normal basis. It is obtained by writing $k[G]$ as $k[X]/(X^N - 1)$, factorising $X^N - 1$ as a product of cyclotomic polynomials, and decomposing $k[G]$ accordingly. We obtain in this way a natural basis $f_1, \ldots, f_N$ together with the action of $\sigma$. We have $Q_d(f_i) = 1$ if $i \neq 2$ and $Q_d(f_2) = d$.

*Example 9.1.* $N = 8$, $\sigma(f_1) = f_1$, $\sigma(f_2) = -f_2$, $\sigma(f_3) = f_4$, $\sigma(f_4) = -f_3$, $\sigma(f_5) = f_6$, $\sigma(f_6) = f_7$, $\sigma(f_7) = f_8$, $\sigma(f_8) = -f_5$.

### 9.2 Arbitrary Cyclic Groups

Let $G$ be a cyclic group of order $mN$, where $m$ is odd and $N$ is a power of 2. Then $G$ has a unique cyclic subgroup $H$ of order $N$. A model of $Q_d$ for the group $G$, denoted by $Q_d^G$, is obtained by taking the orthogonal sum of $m$ copies of a model $Q_d^H$ for $H$.

*Example 9.2.* Let $G$ be of order $2m$, $m$ odd. Let $d = -1$. A model for the cyclic group of order 2 is obtained by $\sigma(e_1) = e_2$, $\sigma(e_2) = e_1$, quadratic form given by the matrix $\begin{pmatrix} 0 & 1 \\ 1 & 0 \end{pmatrix}$. A model for $Q_d$ will be the orthogonal sum of $m$ copies of this model. The quadratic form is then hyperbolic, and one might call this basis a ***hyperbolic normal basis***.

# References


1. Bayer–Fluckiger, E. : Self-dual normal bases. Indag. Math. **51** (1989) 379–383
2. Bayer–Fluckiger, E. : Galois cohomology and the trace form. Jahresber. DMV **96** (1994) 35–55
3. Bayer–Fluckiger, E., Lenstra, H.W., Jr. : Forms in odd degree extensions and self-dual normal bases. Amer. J. Math. **112** (1990) 359–373
4. Bayer–Fluckiger, E., Monsurro, M., Parimala, R., Schoof, R. : Trace forms of $G$–Galois algebras over fields of cohomological dimension $\leq 2$. (in preparation)
5. Bayer–Fluckiger, E., Serre, J.–P. : Torsions quadratiques et bases normales autoduales, Amer. J. Math. **116** (1994) 1-64
6. Beth, T., Fumy, W., Mühlfeld, R. : Zur algebraischen diskreten Fourier–Transformation. Arch. Math. (Basel) **40** (1983) 238–244
7. Conner, P., Perlis, R. : A survey of trace forms of algebraic number fields. World scientific, Singapore (1984)
8. Imamura, K., On self–complementary normal bases of $GF(q^n)$ over $GF(q)$. Trans. IECE Japan **E66** (1983) 717–721
9. Knus, M., Merkurjev, A., Rost, M., Tignol, J.–P. : The book of involutions. AMS Colloquium Publications **44** (1998)
10. Lempel, A. : Characterisation and synthesis of self–complementary normal bases in finite fields. Lin. Alg. Appl. **98** (1988) 331–346
11. Lempel, A., Weinberger, M.J. : Self–complementary normal bases in finite fields. SIAM J. Disc. Math. **1** (1988) 193–198
12. MacWilliams, J., Sloane, N.J.A. : The theory of error–correcting codes. North–Holland, Amsterdam (1977)
13. Morii, A., Imamura, K. : A theorem that $GF(2^{4m})$ has no self–complementary normal basis over $GF(2)$ for odd $m$. Trans. IECE Japan **E67** (1984) 655–656.
14. Serre, J.–P. : Cohomologie galoisienne. 5ème édition, Lecture Notes in Mathematics **5**, Springer–Verlag (1964, 1994) ; english translation, Galois cohomology, Springer–Verlag (1997).

# Permutations amongst the Dembowski-Ostrom Polynomials

Aart Blokhuis[1], Robert S. Coulter*[2], Marie Henderson[2], and Christine M. O'Keefe**[3]

[1] Technische Universiteit
Eindhoven
The Netherlands

[2] Centre for Discrete Mathematics and Computing
The University of Queensland
Brisbane
Australia

[3] Department of Pure Mathematics
The University of Adelaide
5005 Australia

**Abstract** We note that certain Dembowski-Ostrom polynomials can be obtained from the product of two linearised polynomials. We examine this subclass for permutation behaviour over finite fields. In particular, a new infinite class of permutation polynomials is identified.

## 1 Dembowski-Ostrom Polynomials and Linearised Polynomials

Let $p$ be a prime and $q = p^e$. Let $\mathbb{F}_q$ denote the finite field of order $q$ and $\mathbb{F}_q ast$ represent the set of non-zero elements of $\mathbb{F}_q$. The ring of polynomials in the indeterminate $X$ with coefficients from $\mathbb{F}_q$ will be represented by $\mathbb{F}_q X$. A polynomial $f \in \mathbb{F}_q X$ which permutes $\mathbb{F}_q$ under evaluation is called a *permutation polynomial* of $\mathbb{F}_q$. Permutation polynomials have important applications in cryptography. This is because one of the basic requirements of a mapping used to encrypt a message is that it be invertible so that the original message can be recovered. In particular, Dembowski-Ostrom polynomials have been used for a cryptographic application in the public key cryptosystem HFE, see [7]. There the author states that "it seems difficult to choose $f$ (a DO polynomial) such that it is a permutation". It is the purpose of this article to provide some examples of Dembowski-Ostrom permutations. We consider this problem in the purely theoretical spirit of problem P2 of [5]. We do not claim that any of the classes identified in this article could be used to provide a "secure" cryptosystem when implemented in HFE. The class of polynomials of primary interest in this article is now defined.

* Supported by an Australian Research Council grant
** Supported by the Australian Research Council and the University of Ghent

**Definition 1.1.** The polynomial $f \in \mathbb{F}_q X$ is called a *Dembowski-Ostrom polynomial* (DO polynomial) if $f$ has the shape

$$f(X) = \sum_{i,j=0}^{n} a_{ij} X^{p^i+p^j}.$$

This class of polynomials was described by Dembowski and Ostrom in [4]. In that article, the authors considered projective planes of order $n$ which admitted a collineation group of order $n^2$. They introduced the notion of a planar function as an aid for describing these planes. A polynomial $g \in \mathbb{F}_q X$ is called a *planar polynomial* if $g(X + a) - g(X)$ is a permutation polynomial for every $a \in \mathbb{F}_q ast$. It is a simple matter to show that a polynomial can not be planar on $\mathbb{F}_q$ when $q$ is even. In contrast, a DO polynomial $f$ can not be a permutation polynomial of $\mathbb{F}_q$ if $q$ is odd as, in such cases, $f(x) = f(-x)$ for each $x \in \mathbb{F}_q ast$. For this reason, we can restrict to the case $q$ even when searching for DO permutation polynomials.

A set of permutation polynomials is easily obtained from the following class of polynomials.

**Definition 1.2.** A *linearised polynomial* (or *additive polynomial*), $L \in \mathbb{F}_q X$, is a polynomial of the shape

$$L(X) = \sum_{i=0}^{n} a_i X^{p^i}.$$

It is easily seen from this definition that these polynomials are additive in the sense that $L(x + y) = L(x) + L(y)$ for all $x, y \in \mathbb{F}_q$. A permutation condition for this class of polynomials can be directly determined from this property: a linearised polynomial $L(X)$ is a permutation polynomial of $\mathbb{F}_q$ if and only if $x = 0$ is its only root in $\mathbb{F}_q$. Linearised polynomials have many other interesting properties and applications, see Chapter 3 of [6].

It is interesting to note that if $q$ is even then it is possible for a DO polynomial to be a linearised polynomial, for example, take $f = L^2$ (consequently $f$ permutes $\mathbb{F}_q$ if and only if $L$ does). In fact, it is easy to show that a DO polynomial $f$ is a linearised polynomial if and only if $q$ is even and $f = L^2$, where $L$ is a linearised polynomial.

The reduction of a linearised polynomial modulo $X^q - X$ is again a linearised polynomial. We shall call a polynomial *bilinear* if it can be written as the product of two reduced linearised polynomials. (Note that this is not the same definition of bilinear as used in some other articles, for example [1].) The stipulation that the linearised polynomials be reduced makes sense when we consider the following trivial lemma.

**Lemma 1.3.** *Let $B \in \mathbb{F}_q X$ satisfy $B(X) = L_1(X)L_2(X)$ where $L_1$ and $L_2$ are linearised polynomials. Set $L_i'(X) = L_i(X) \bmod (X^q - X)$ for $i = 1, 2$ and define $B' = L_1'L_2'$. Then $B(X) \bmod (X^q - X) = B'(X)$ unless $q = 2^e$*

*and degree*($L_1'$) =*degree*($L_2'$) $= 2^{e-1}$, *in which case* $B(X) \bmod (X^q - X) = B'(X) + \beta(X^q - X)$ *where* $\beta$ *is the coefficient of* $X^q$ *in* $B'(X)$.

Hence, except for one particular case, a bilinear polynomial is still a bilinear polynomial after reduction. This is also the case for DO polynomials with a similar exception: the reduction of a DO polynomial is again a DO polynomial unless there existed a term with degree divisible by $q$ before reduction. This can only occur in characteristic 2 to a term of the shape $X^{2p^i}$ where $i \bmod e = e-1$.

Clearly every bilinear polynomial is a DO polynomial. However, for $e > 1$, a DO polynomial is not necessarily a bilinear polynomial. This can be seen from a simple counting argument. In $\mathbb{F}_q X$, there are $(q^{e(e+1)/2} - 1)/(q-1)$ reduced monic DO polynomials. There are $(q^e - 1)/(q-1)$ monic linearised polynomials and therefore

$$\binom{(q^e-1)/(q-1)}{2} + (q^e-1)/(q-1)$$

monic bilinear polynomials. This is easily seen to be less than the number of DO polynomials for $e > 1$ and any $q$.

## 2 Permutation Behaviour of Dembowski-Ostrom Polynomials

Suppose that $f(X) = L_1(X)L_2(X)$ is a permutation polynomial of $\mathbb{F}_q$. We have $f(x) = 0$ for $x \in \mathbb{F}_q$ if and only if $x = 0$. Therefore, $L_1$ and $L_2$ must also be permutation polynomials of $\mathbb{F}_q$. As $L_1(X)$ permutes $\mathbb{F}_q$, there exists a linearised compositional inverse $L_1^{-1}(X)$ such that $L_1(L_1^{-1}(X)) \bmod (X^q - X) = X$. From this we have $f(L_1^{-1}(X)) \bmod (X^q - X) = XL(X)$ is also a DO permutation polynomial where $L(X) = L_2(L_1^{-1}(X)) \bmod (X^q - X)$ is a linearised permutation polynomial. Hence, for bilinear DO polynomials, we have only to consider the permutation behaviour of DO polynomials with the shape $XL(X)$ where $L(X)$ is a linearised permutation polynomial. All other bilinear DO permutation polynomials can be obtained by composing with linearised permutation polynomials. We now present several classes of permutation polynomials obtained from bilinear DO polynomials. The examples described by the following two theorems were first found using MAGMA [2]. They describe all bilinear DO permutation polynomials over $\mathbb{F}_q$ with $q = 2^e$, $e \leq 6$ known to the authors.

**Theorem 2.1.** *Let* $q = 2^e$ *and* $g$ *be any primitive element of* $\mathbb{F}_q$. *Let* $k$ *be any integer and set* $d = (k, e)$. *Suppose* $f \in \mathbb{F}_q X$ *is a DO polynomial satisfying* $f(X) = XL(X)$ *for some linearised polynomial* $L$. *Then* $f$ *permutes* $\mathbb{F}_q$ *when any of the following conditions are satisfied.*

*(i)* $L(X) = X^{2^k}$ *where* $e/d$ *is odd.*

*(ii)* $L(X) = X^{2^k} + aX^{2^{e-k}}$ *where* $e/d$ *is odd and* $a \neq g^{t(2^d-1)}$ *for any integer* $t$.

*(iii)* $L(X) = X^{2^{2k}} + a^{2^k+1}X^{2^k} + aX$ *where* $e = 3k$ *and* $a \neq g^{t(2^k-1)}$ *for any integer* $t$.

*Proof.* (i) Immediate since $(2^k+1, q-1) = 1$ if and only if $e/d$ is odd, see [3, Lemma 2.1].
(ii) Suppose $a \in \mathbb{F}_q$ satisfies $a \neq g^{t(2^d-1)}$ for any integer $t$. Then $X^{2^k} + aX$ is a permutation polynomial. If $e/d$ is odd then by (i), $X^{2^{e-k}+1}$ also permutes $\mathbb{F}_q$. Composing, and reducing mod $(X^q - X)$, we obtain $X(X^{2^k} + aX^{2^{e-k}})$, which must also permute $\mathbb{F}_q$.
(iii) As $a \neq g^{t(2^k-1)}$ for any integer $t$, the polynomials $X^{2^{2k}} + a^{2^k}X$ and $X^{2^k} + a^{2^k}X$ permute $\mathbb{F}_q$. Now

$$f(X^{2^{2k}} + a^{2^k}X) \equiv (1 + a^{2^{2k}+2^k+1})\big((X^{2^k} + a^{2^k}X) \circ X^{2^k+1}\big) \bmod X^q - X.$$

As $X^{2^k+1}$ also permutes $\mathbb{F}_q$ it follows that $f(X)$ permutes $\mathbb{F}_q$.

We note that all of the DO permutation polynomials given above are constructed by composition from well known classes of permutation polynomials (monomials and linearised binomials). For our next result we set $T = \mathrm{Tr}_{\mathbb{F}_qn/\mathbb{F}_q} : \mathbb{F}_qn \mapsto \mathbb{F}_q$, the trace function from $\mathbb{F}_qn$ to $\mathbb{F}_q$.

**Theorem 2.2.** *Let* $q$ *be even and* $n$ *be odd. The polynomial*

$$f(X) = X\,(T(X) + aX)$$

*is a permutation polynomial over* $\mathbb{F}_qn$ *for all* $a \in \mathbb{F}_q \setminus \{0,1\}$.

*Proof.* Note

$$\begin{aligned} T\,(f(x)) &= T\left(xT(x) + ax^2\right) \\ &= T(x)T(x) + aT(x)^2 \\ &= (1+a)T(x)^2 \end{aligned}$$

for all $x \in \mathbb{F}_qn$. Suppose there exists elements $x, y \in \mathbb{F}_qn$ such that $f(x) = f(y)$. Then $x\,(T(x) + ax) = y\,(T(y) + ay)$. Applying the trace function $T$ gives

$$(1+a)\left(T(x)^2 + T(y)^2\right) = 0,$$

so $T(x) = T(y) = t$, say. Thus $tx + ax^2 = ty + ay^2$ from which $t = a(x+y)$, implying $x + y \in \mathbb{F}_q$. Therefore $T(x+y) = n(x+y) = 0$. So $x = y$.

Without reduction, it is clear that all of the permutation polynomials given in Theorems 2.1 and 2.2 cannot have a linearised decompositional factor since their degrees are never divisible by the characteristic. If we take into consideration the possibility that any of this class could be the reduced form modulo $X^{q^n} - X$ of some other polynomial, then as with all classes of permutation polynomials, it is quite possible that some examples of functionally equivalent polynomials will have linearised decompositional factors. However, it should be stressed that this will not, in general, be the case. To see this, let $f$ be any permutation polynomial over $\mathbb{F}_q$. Then all polynomials of the form $F(X) = f(X) + h(X)(X^q - X)$, where $h$ is any polynomial defined over $\mathbb{F}_q$, are equivalent to $f$ under evaluation and permutations also. However, for every choice of $h$ which provides an $F$ which has a linearised decompositional factor, there are many which will not. In a cryptographic sense, it may be preferable that the permutations (as functions) can not easily be expressed as the composition of monomials and linearised polynomials. In this context, there is a striking difference between the permutations of Theorem 2.1 and 2.2. While those permutations presented in Theorem 2.1 are easily shown to be equivalent under reduction to a simple composition of monomial and linearised binomial permutations (see the proof), the permutations presented in Theorem 2.2 have no such simple representation. Take the simplest case of Theorem 2.2: that is $n = 3$. Let $f_a(X) = X^{4^k+1} + X^{2^k+1} + aX^2$ with $a \in \mathbb{F}_{2^k} \setminus \{0,1\}$. Then $f_a$ is a permutation over $\mathbb{F}_q$ with $q = 8^k$. Let $L_1(X) = aX^{p^i} + bX$, $L_2(X) = cX^{p^j} + dX$ and $s$ be an integer satisfying $(s, q-1) = 1$. Supposing $f_a$ is functionally equivalent to a composition of $L_1$, $L_2$ and $X^s$ (in any combination) always leads to a contradiction. While this does not constitute a rigorous proof, it seems safe to assert that, in general, the permutations given in Theorem 2.2 cannot be expressed as simple compositions of monomials and linearised binomials. Whether any of these permutations are of cryptographic relevance is another matter.

## References


1. H.W. Bao, *On two exponential sums and their applications*, Finite Fields Appl. **3** (1997), 115–130.
2. W. Bosma, J. Cannon, and C. Playoust, *The Magma algebra system I: The user language*, J. Symbolic Comput. **24** (1997), 235–265.
3. R.S. Coulter, *On the evaluation of a class of Weil sums in characteristic 2*, New Zealand J. Math. **28** (1999), 171–184.
4. P. Dembowski and T.G. Ostrom, *Planes of order n with collineation groups of order $n^2$*, Math. Z. **103** (1968), 239–258.
5. R. Lidl and G.L. Mullen, *When does a polynomial over a finite field permute the elements of the field?*, Amer. Math. Monthly **95** (1988), 243–246.
6. R. Lidl and H. Niederreiter, *Finite Fields*, Encyclopedia Math. Appl., vol. 20, Addison-Wesley, Reading, 1983, (now distributed by Cambridge University Press).

7. J. Patarin, *Hidden fields equations (HFE) and isomorphisms of polynomials (IP): two new families of asymmetric algorithms*, Advances in Cryptology – Eurocrypt '96 (U. Maurer, ed.), Lecture Notes in Computer Science, vol. 1070, 1996, pp. 33–48.

# Associative Rational Functions in Two Variables*

Joel V. Brawley[1], Shuhong Gao[1] and Donald Mills[2]

[1] Department of Mathematical Sciences
Clemson University
Clemson, SC 29634-0975 USA
e-mail: brawley@CLEMSON.EDU, sgao@math.clemson.edu

[2] Department of Mathematics
Southeastern Louisiana University
Hammond, LA 70402 USA
e-mail: dmills@selu.edu

**Abstract** A rational function $R(x,y)$ over a field is said to be associative if

$$R(R(x,y),z) = R(x,R(y,z)).$$

Associative rational functions over a field define group laws on subsets of the field (plus the point at infinity). In this paper, all the associative rational functions of two variables over an arbitrary field are determined and consequently all the groups obtainable from such functions are determined as well.

## 1 Introduction and Main Results

Let $\mathbb{F}$ be any field. Trivially the polynomial $x+y$ defines, via substitution, a group operation (or group law) on $\mathbb{F}$, and the polynomial $xy$ defines a group operation on $\mathbb{F}\setminus\{0\}$. Further, it is well-known that the polynomial $x+y-1$ also defines a group on $\mathbb{F}$ and that the polynomial $x+y+xy$ defines a group on $\mathbb{F}\setminus\{-1\}$. It is natural to ask for other such polynomials $f(x,y)\in\mathbb{F}[x,y]$ or even rational functions $R(x,y)\in\mathbb{F}(x,y)$ which define group operations on subsets of $\mathbb{F}$. More generally, one may consider formal power series in two variables and this leads to formal groups of dimension one which are closely related to elliptic curves [6]; see also [4] for formal groups of higher dimension.

We are interested in determining rational functions over $\mathbb{F}$ that can be used to define group laws on some subsets of $\mathbb{F}$. Let $R(x,y)\in\mathbb{F}(x,y)$ be any rational function. We say that $R(x,y)$ is *associative* if the equation

$$R(R(x,y),z) = R(x,R(y,z)) \tag{1}$$

* The second author was supported in part by NSF under Grant #DMS9970637 and NSA under Grant #MDA904-00-1-0048. The current address for the third author: Donald Mills, Department of Mathematical Sciences, United States Military Academy, West Point, NY 10996; E-mail: ddmills@arl.mil.

is valid in $\mathbb{F}(x,y,z)$, the rational function field with distinct variables $x, y, z$. A rational function $R_1(x,y)$ is said to be *equivalent* to $R(x,y)$ if there is a linear fractional map $f(x) = (ax+b)/(cx+d) \in \overline{\mathbb{F}}(x)$, $ad-bc \neq 0$, such that

$$R_1(x,y) = f^{-1}\left(R(f(x), f(y))\right) \tag{2}$$

where $\overline{\mathbb{F}}$ denotes the algebraic closure of $\mathbb{F}$ and $f^{-1}(x) = (-dx+b)/(cx-a)$ is the inverse of $f$. For example, $x+y+xy$ is equivalent to $xy$ under $f(x) = x+1$. It is easy to check that $R_1(x,y)$ is associative iff $R(x,y)$ is. This defines an equivalence relation on associative rational functions over $\mathbb{F}$. The functions $x, y, x+y, x+y+xy$ are associative but not equivalent to each other. One of our main goals is to show that they represent all the associative rational functions over $\mathbb{F}$.

**Theorem 1.1.** *Let $\mathbb{F}$ be any field. Then any nonconstant associative rational function in $\mathbb{F}(x,y)$ is equivalent to exactly one of $x, y, x+y$ and $x+y+xy$.*

**Remark**. In defining the equivalence above, we could have allowed $f$ to be any rational function in $\overline{\mathbb{F}}(x)$ whose compositional inverse is also a rational function in $\overline{\mathbb{F}}(x)$, that is, $f$ is an automorphism of the field $\overline{\mathbb{F}}(x)$ fixing $\overline{\mathbb{F}}$. But the only automorphisms of $\overline{\mathbb{F}}(x)$ that fix $\overline{\mathbb{F}}$ are linear fractional maps.

With this theorem, we can determine all groups whose elements are from $\mathbb{F}$ and whose group laws can be defined by rational functions over $\mathbb{F}$. In the first place, if $G$ is any finite group of order $\leq |\mathbb{F}|$ and $S$ is any subset of $\mathbb{F}$ with $|S| = |\mathbb{F}|$, then it is well-known that any one-to-one correspondence between $S$ and $G$ can be used to define a group law on $S$ for which $S$ and $G$ are isomorphic. Further, this group law for $S$ can always be defined by means of a polynomial $f(x,y)$ obtained by Lagrange interpolation (for two variables); thus, every finite group of order $\leq |\mathbb{F}|$ is possible. However, the polynomial $f(x,y)$ may not be associative in the above sense and classifying such polynomials amounts to classifying finite groups, which is certainly a hard problem with a nature different from that of an infinite $S$. We do not intend to pursue the case of a finite $S$ in this paper, but instead will concentrate on infinite $S$.

Henceforth it is assumed that $S$ is infinite (subset of $\mathbb{F}$) with its group law defined by a rational function $R(x,y)$ (i.e. for $a, b \in S$, $a \cdot b = R(a,b)$). Since a group law is associative, we have

$$R(R(a,b),c) = R(a, R(b,c)) \text{ for all } a,b,c \in S. \tag{3}$$

By using the fact that $S$ is infinite, one can show that (3) implies (1). So $R(x,y)$ must be associative and, by Theorem 1.1, is equivalent to one of $x, y, x+y, x+y+xy$. But none of $x$ or $y$ can define a group so $R(x,y)$ is equivalent to $x+y$ or $x+y+xy$. The linear fractional map that defines the equivalence induces an isomorphism of $S$ to some subgroup of the additive or multiplicative group of $\overline{\mathbb{F}}$. We will show that the linear fractional map can be defined over a quadratic extension of $\mathbb{F}$, so $S$ is in fact isomorphic to a subgroup of the additive or multiplicative group of a quadratic extension of $\mathbb{F}$.

**Theorem 1.2.** *For any infinite subset $S$ of a field $\mathbb{F}$, if $S$ is a group defined by a rational function over $\mathbb{F}$ then $S$ is isomorphic to a subgroup of the additive or multiplicative group of a quadratic extension of $\mathbb{F}$.*

We should remark that when using a rational function to define a group law, it is necessary to use $\infty$, the point at infinity (or use the projective space $\mathbb{P}^1(\mathbb{F})$). Here we adopt the usual convention that, for $a, b, c, d \in \mathbb{F}$,

$$\frac{a\infty + b}{c\infty + d} = \frac{a}{c} \quad \text{for } c \neq 0,$$
$$\frac{a\infty + b}{c\infty + d} = \infty \quad \text{for } ad \neq 0 \text{ and } c = 0,$$
$$\frac{a}{0} = \infty \quad \text{for } a \neq 0.$$

We give below two examples of groups which can be verified directly. We denote by $\mathbb{F}_q$ a finite field of $q$ elements. Example 1.3 comes from the proof of Lemma 3.1, while Example 1.4 is a special case of Lemma 2.7.

*Example 1.3.* Let $\mathbb{F} = \mathbb{F}_2(t^2)$ where $t$ is transcendental over $\mathbb{F}_2$, $S = \mathbb{F}$ and $R = (xy + t^2)/(x + y)$. Then $R$ is equivalent to $x + y$ and $R$ defines a group on $S$ isomorphic to the additive group

$$\left\{ \frac{1}{h+t} : h \in \mathbb{F} \right\}$$

of $\mathbb{F}_2(t)$, a quadratic extension of $\mathbb{F} = \mathbb{F}_2(t^2)$. Note that the group isomorphism is defined by $f = 1/(x + t)$.

*Example 1.4.* Let $R = (xy - 1)/(x + y + 1) \in \mathbb{F}_q(x, y)$ where $q$ is a power of a prime $> 3$. Let $\omega$ be a root of $X^2 + X + 1$. Then $R$ is equivalent to $xy$ under the linear fractional map $f(x) = (1 - \omega x)/(x + \omega + 1)$. Depending on whether $\omega$ lies in $\mathbb{F}_q$, we have the following two cases.

(a) Suppose $q - 1$ is divisible by 3. Let $S = (\mathbb{F}_q \cup \{\infty\}) \setminus \{\omega, \omega^2\}$. Then $R$ defines a group on $S$ isomorphic to the multiplicative group of $\mathbb{F}_q$ with the isomorphism defined by $f(x)$.
(b) Suppose $q - 1$ is not divisible by 3. Let $S = \mathbb{F}_q \cup \{\infty\}$. Then $R$ defines a group on $S$ isomorphic to the unique subgroup of order $q + 1$ of the multiplicative group of $\mathbb{F}_{q^2}$ with the isomorphism defined by $f(x)$.

We should mention the related results in formal groups. If one is allowed to use formal power series for $f$ then $x + y$ and $x + y + xy$ can be obtained from each other. Indeed if the characteristic of $\mathbb{F}$ is zero then

$$x + y + xy = f^{-1}(f(x) + f(y))$$

where

$$f(x) = \log(1+x) = x - \frac{x^2}{2} + \frac{x^3}{3} - \cdots + (-1)^{n-1}\frac{x^n}{n} + \cdots,$$

$$f^{-1}(x) = \exp(x) - 1 = x + \frac{x^2}{2!} + \frac{x^3}{3!} + \cdots + \frac{x^n}{n!} + \cdots.$$

In general, a formal power series $R(x,y) \in \mathbb{F}[[x,y]]$ is called a *formal group* if $R(x,y) = x + y +$ (terms of orders $\geq 2$) and $R(x,y)$ is associative (i.e. satisfying (1). When the characteristic of $\mathbb{F}$ is zero, it can be proved that every formal group $R(x,y) \in \mathbb{F}[[x,y]]$ is equivalent to $x+y$ via a formal power series $f \in \mathbb{F}[[x]]$ defined by some generalized logarithm associated with $R(x,y)$; see [6, Proposition 5.2, p. 122] or [4, Corollary 5.4.4, p. 31]. Our result shows that all the formal groups that are rational functions are equivalent to either $x+y$ or $x+y+xy$ with the equivalence defined by rational functions, instead of formal power series. Note that determining all the formal groups that are polynomials is much easier as indicated by the exercise in [4, p. 2] and Theorem 1.1 in [3, p. 12]; see also Lemma 2.1 in the next section.

Our interest in this paper was motivated by the work in [1,2,5] where it is desirable to construct irreducible polynomials of higher degrees from those of lower degrees. In [1], Brawley and Carlitz introduce a so-called diamond operation on a subset of a field and use this operation to form composition of polynomials. When the diamond operation defines a group law on the subset then composition of irreducible polynomials gives irreducible polynomials under certain conditions. In [2], it is shown that composition of polynomials can be computed efficiently if the diamond operation is defined by a polynomial or a rational function.

For the rest of the paper, we prove Theorem 1.1 in Section 2 and Theorem 1.2 in Section 3 .

## 2 Associative Rational Functions

We start with the easy case when $R(x,y)$ is a polynomial; its proof involves the basic ideas that are used in the general case when $R(x,y)$ is an arbitrary rational function.

**Lemma 2.1.** *Let $R(x,y) \in \mathbb{F}[x,y]$ be a nonconstant polynomial. If $R(x,y)$ is associative then $R(x,y)$ is equivalent to one of $x, y, x+y, x+y+xy$.*

*Proof.* Suppose $R(x,y) = \sum c_{ij}x^iy^j$. The associativity of $R$ implies

$$\sum c_{ij}(R(x,y))^i z^j = \sum c_{ij}x^i(R(y,z))^j.$$

Let $m$ be the degree in $x$ of $R(x,y)$. Then comparing the degrees in $x$ in the above equation gives $m^2 = m$, so $m$ must be 0 or 1. Similarly, the degree in $y$ of $R(x,y)$ is also at most 1. Hence we may assume that $R(x,y) = axy + bx + cy + d$ where $a,b,c,d \in \mathbb{F}$. Then

$$\begin{aligned} &R(R(x,y),z) \\ &\qquad = a^2xyz + abxy + abxz + acyz + b^2x + bcy + (ad+c)z + (b+1)d \\ &R(x,R(y,z)) \\ &\qquad = a^2xyz + abxy + acxz + acyz + (ad+b)x + bcy + c^2z + (c+1)d. \end{aligned}$$

Comparing the coefficients gives

$$ab = ac, \quad b^2 = ad + b, \quad ad + c = c^2, \quad (b+1)d = (c+1)d.$$

First assume $a \neq 0$. Then $b = c$ and $d = (c^2 - c)/a$, so

$$R(x,y) = axy + c(x+y) + (c^2 - c)/a.$$

Let $f(x) = (x - c + 1)/a$. Then $x + y + xy = f^{-1}(R(f(x), f(y)))$. This means that in this case $R(x,y)$ is equivalent to $x+y+xy$ under $f^{-1}(x) = ax + c - 1$.

Assume now $a = 0$. Then $b^2 = b$ and $c^2 = c$. If $b \neq c$ then $d = 0$. Hence the only choices for $R$ are $x, y, x + y + d$. The latter is equivalent to $x + y$ under the map $f = x + d$. □

In general let $R(x,y)$ be any associative rational function over $\mathbb{F}$, say $R(x,y) = F(x,y)/H(x,y)$ where $F(x,y), H(x,y) \in \mathbb{F}[x,y]$ and $\gcd(F(x,y), H(x,y)) = 1$. The associativity of $R$ gives a polynomial equation. Similar to the above argument, we shall compare the degrees of $x$ and $y$ in this equation and prove that both $F(x,y)$ and $H(x,y)$ have degree at most one in $x$ and $y$ separately. Then we reduce $R$ to special forms and eventually to polynomials by linear fractional maps. Since the proof is a little lengthy, we break it into several lemmas.

The following lemma seems interesting by itself, as it says that relatively prime polynomials remain relatively prime when their variables are substituted by other polynomials.

**Lemma 2.2.** *Let $F(x,y), H(x,y) \in \mathbb{F}[x,y]$ with $\gcd(F(x,y), H(x,y)) = 1$ and degrees $f_x, h_x$ in $x$, respectively. Let $x_1, \ldots, x_n, z$ be distinct variables and let $a, b \in \mathbb{F}[x_1, \ldots, x_n]$ with $a/b \notin \mathbb{F}$ and $\gcd(a,b) = 1$. Then*

$$b^{f_x} F(a/b, z), \quad b^{h_x} H(a/b, z)$$

*are polynomials in $\mathbb{F}[x_1, \ldots, x_n, z]$ and are relatively prime.*

*Proof.* We can view $F(x,y), H(x,y)$ as univariate polynomials in $y$ with coefficients in $\mathbb{F}(x)$, the rational function field in $x$. Since $\gcd(F(x,y), H(x,y)) = 1$ in $\mathbb{F}[x,y]$, we still have $\gcd(F(x,y), H(x,y)) = 1$ in $\mathbb{F}(x)[y]$. By the Euclidean algorithm, there are polynomials $U_0, V_0 \in \mathbb{F}(x)[y]$ such that

$$U_0 F + V_0 H = 1.$$

By eliminating the denominators in $U_0, V_0$, we have

$$UF + VH = W \tag{4}$$

for some $U, V \in \mathbb{F}[x, y]$ and $W \in \mathbb{F}[x]$ with $W \neq 0$. Let $\ell = f_x$ and $m = h_x$ be the degrees in $x$ of $F(x, y)$ and $H(x, y)$, respectively. In (4), replacing $x$ by $a/b$ and $y$ by $z$ and eliminating the denominators, which are powers of $b$, we have

$$U_1 A + V_1 B = b^i W(a/b) \tag{5}$$

where $A = b^\ell F(a/b, z)$, $B = b^m H(a/b, z)$, $U_1, V_1 \in \mathbb{F}[x_1, \dots, x_n, z]$ and $b^i W(a/b) \in \mathbb{F}[x_1, \dots, x_n]$.

It follows that $C = \gcd(A, B)$ must divide $b^i W(a/b)$, as polynomials in $\mathbb{F}[x_1, \dots, x_n, z]$. Since $a/b$ is not a constant in $\mathbb{F}$, we have $b^i W(a/b) \neq 0$. So $C \in \mathbb{F}[x_1, \dots, x_n]$ and thus $C$ divides all the coefficients of $A, B$ as polynomials in $\mathbb{F}[x_1, \dots, x_n][z]$.

Let's examine the coefficients. Suppose

$$\begin{aligned} F(x, y) &= F_\ell(x) y^\ell + \cdots + F_1(x) y + F_0(x) \\ H(x, y) &= H_m(x) y^m + \cdots + H_1(x) y + H_0(x). \end{aligned}$$

Then

$$\begin{aligned} A &= b^\ell F_\ell(a/b) z^\ell + \cdots + b^\ell F_1(a/b) z + b^\ell F_0(a/b) \\ B &= b^m H_m(a/b) z^m + \cdots + b^m H_1(a/b) z + b^m H_0(a/b). \end{aligned}$$

Since $\gcd(F(x, y), H(x, y)) = 1$ in $\mathbb{F}[x, y]$, the coefficients

$$F_\ell(x), \dots, F_1(x), F_0(x), H_m(x), \dots, H_1(x), H_0(x)$$

must be relatively prime. Replacing $x$ by $a/b$ in these polynomials and eliminating denominators gives the coefficients of $A$ and $B$. By a similar argument above via the Euclidean algorithm, we see that any common divisor of the coefficients of $A$ and $B$ together must divide a power of $b$. Since $\ell = \deg F_i(x)$ for some $1 \leq i \leq \ell$ and $\gcd(a, b) = 1$, we have $\gcd(b^\ell F_i(a/b), b) = 1$. Hence the coefficients of $A$ and $B$ must be relatively prime and $C$ must be a constant in $\mathbb{F}$. □

**Lemma 2.3.** *Let $R(x, y) = F(x, y)/H(x, y)$ where $F(x, y), H(x, y) \in \mathbb{F}[x, y]$ with $\gcd(F(x, y), H(x, y)) = 1$. Denote by $f_x, f_y$ the degrees of $F(x, y)$ in $x, y$ separately, and similarly $h_x, h_y$ for that of $H(x, y)$. If $R(x, y)$ is associative then $h_x \leq f_x \leq 1$ and $h_y \leq f_y \leq 1$.*

*Proof.* The associativity of $R(x, y)$ means that

$$\frac{F(R(x, y), z)}{H(R(x, y), z)} = \frac{F(x, R(y, z))}{H(x, R(y, z))}$$

i.e.

$$F(R(x,y),z)H(x,R(y,z)) = H(R(x,y),z)F(x,R(y,z)).$$

To eliminate the denominators, let

$$A = H(x,y)^{f_x} F(R(x,y),z), \quad B = H(y,z)^{h_y} H(x,R(y,z)) \tag{6}$$

$$C = H(x,y)^{h_x} H(R(x,y),z), \quad D = H(y,z)^{f_y} F(x,R(y,z)). \tag{7}$$

Then $A, B, C, D \in \mathbb{F}[x,y,z]$ and

$$H(x,y)^{h_x} H(y,z)^{f_y} A\,B = C\,D\,H(x,y)^{f_x} H(y,z)^{h_y}. \tag{8}$$

We shall prove that $h_x \le f_x \le 1$; the proof of $h_y \le f_y \le 1$ is similar.

Since $\gcd(F(x,y), H(x,y)) = 1$, we have

$$\gcd(A, H(x,y)) = 1 \quad \text{and} \quad \gcd(C, H(x,y)) = 1.$$

By Lemma 2.2, we also have $\gcd(A,C) = 1$. It follows from (8) that

$$A \mid D\,H(y,z)^{h_y} \quad \text{and} \quad C \mid B\,H(y,z)^{f_y}. \tag{9}$$

Consider the degrees in $x$ of the polynomials in (9). Certainly,

$$\deg_x D\,H(y,z)^{h_y} = \deg_x D = f_x \quad \text{and} \quad \deg_x B\,H(y,z)^{f_y} = \deg_x B = h_x. \tag{10}$$

To find the degrees of $A$ and $C$ in $x$, suppose

$$F(x,y) = \sum_{i=i_0}^{f_x} a_i(y)x^i, \quad H(x,y) = \sum_{j=j_0}^{h_x} b_j(y)x^j \tag{11}$$

where $a_i(y), b_j(y) \in \mathbb{F}[y]$, $a_i(y) \ne 0$ for $i = i_0$ or $f_x$, and $b_j(y) \ne 0$ for $j = j_0$ or $h_x$. Then

$$A = \sum_{i=i_0}^{f_x} a_i(z)F(x,y)^i H(x,y)^{f_x - i}, \quad C = \sum_{j=j_0}^{h_x} b_j(z)F(x,y)^j H(x,y)^{h_x - j} \tag{12}$$

Now the degree in $x$ of $F(x,y)^i H(x,y)^{f_x-i}$ is

$$if_x + h_x(f_x - i) = h_x f_x + i(f_x - h_x), \quad i_0 \le i \le f_x, \tag{13}$$

and that of $F(x,y)^j H(x,y)^{h_x-j}$ is

$$jf_x + h_x(h_x - j) = h_x^2 + j(f_x - h_x), \quad j_0 \le j \le h_x. \tag{14}$$

First assume $f_x > h_x$. Then $\deg_x A = f_x^2$. By (9) and (10), we have $f_x^2 \le f_x$, so $f_x \le 1$. In this case, we have $h_x = 0$ and $f_x = 1$.

Next assume $f_x = h_x$. Then all the values in (13) and (14) are equal to $h_x^2$. Denote $n = h_x^2$. We need to find the coefficient of $x^n$ in $A$ and $C$. Note that the leading coefficient of $x$ in $F(x,y)$ is $a_{f_x}(y)$, denoted by $a(y)$, and the leading coefficient of $x$ in $H(x,y)$ is $b_{h_x}(y)$, denoted by $b(y)$. By (12), the coefficient of $x^n$ in $A$ is

$$\sum_{i=i_0}^{f_x} a_i(z)a(y)^i b(y)^{f_x-i} = b(y)^{f_x} F\left(\frac{a(y)}{b(y)}, z\right) \tag{15}$$

and the coefficient of $x^n$ in $C$ is

$$\sum_{j=j_0}^{h_x} b_j(z)a(y)^j b(y)^{h_x-j} = b(y)^{h_x} H\left(\frac{a(y)}{b(y)}, z\right). \tag{16}$$

Since $\gcd(F(x,y), H(x,y)) = 1$, the polynomials in (15) and (16) can not both equal to zero. Hence $A$ or $C$ has degree $n$ in $x$. In either case, it follows from (9) and (10) that $n = h_x^2 \leq h_x$. So $f_x = h_x \leq 1$.

Finally assume $f_x < h_x$. We have from (13) that

$$\deg_x A = i_0 f_x + h_x(f_x - i_0). \tag{17}$$

This and (9) imply that

$$i_0 f_x + h_x(f_x - i_0) \leq f_x.$$

Since $h_x > f_x$, it follows that $f_x - i_0 = 0$ and thus $f_x^2 \leq f_x$. Hence $f_x \leq 1$ and $\deg_x A = \deg_x D = f_x$. Considering the degrees of $x$ in (8), we have

$$h_x^2 + \deg_x B = \deg_x C + h_x f_x.$$

By (9), $\deg_x B \geq \deg_x C$, thus $h_x f_x \geq h_x^2$. But $h_x > f_x \geq 0$, so $f_x \geq h_x$, impossible as $f_x < h_x$ by our assumption in this case. □

**Lemma 2.4.** *Suppose*

$$R(x,y) = \frac{u_1xy + u_2x + u_3y + u_4}{v_1xy + v_2x + v_3y + v_4} \in \mathbb{F}(x,y) \tag{18}$$

*is associative. Then $R(x,y)$ is equivalent to a polynomial or one of the following*

$$\frac{u_1xy + u_2x + u_3y + u_4}{v_2x + v_3y}, \qquad \frac{u_1xy + u_2x + u_3y + u_4}{x - y + 1}. \tag{19}$$

*(With possibly different values of u's and v's.)*

*Proof.* We first assume that $v_1 \neq 0$. Let $f(x) = a + 1/x$. Then $f^{-1}(x) = 1/(x-a)$. Note that

$$
\begin{aligned}
&R(f(x), f(y)) \\
&\quad = \frac{u_1(a+1/x)(a+1/y) + u_2(a+1/x) + u_3(a+1/y) + u_4}{v_1(a+1/x)(a+1/y) + v_2(a+1/x) + v_3(a+1/y) + v_4} \\
&\quad = \frac{u_1(a^2xy + a(x+y) + 1) + u_2y(ax+1) + u_3x(ay+1) + u_4xy}{v_1(a^2xy + a(x+y) + 1) + v_2y(ax+1) + v_3x(ay+1) + v_4xy} \\
&\quad = \frac{(u_1a^2 + u_2a + u_3a + u_4)xy + (u_1a + u_3)x + (u_1a + u_2)y + u_1}{(v_1a^2 + v_2a + v_3a + v_4)xy + (v_1a + v_3)x + (v_1a + v_2)y + v_1},
\end{aligned}
$$

and so

$$
\begin{aligned}
&f^{-1}(R(f(x), f(y))) \\
&\quad = \frac{(v_1a^2 + v_2a + v_3a + v_4)xy + (v_1a + v_3)x + (v_1a + v_2)y + v_1}{(-v_1a^3 + (u_1 - v_2 - v_3)a^2 + (u_2 + u_3 - v_4)a + u_4)xy + E}
\end{aligned}
$$

where

$$
E = (-v_1a^2 + (u_1 - v_3)a + u_3)x + (-v_1a^2 + (u_1 - v_2)a + u_2)y + u_1 - av_1. \tag{20}
$$

Taking $a$ to be a root of the polynomial

$$
-v_1X^3 + (u_1 - v_2 - v_3)X^2 + (u_2 + u_3 - v_4)X + u_4 \tag{21}
$$

will yield a new $R$ with $v_1 = 0$.

Hence we may assume that $v_1 = 0$ for our original $R$. We may further assume that $v_4 \neq 0$, otherwise it is of the first type in (19) already. Let $f(x) = ax + b$. Then

$$
\begin{aligned}
R(f(x), f(y)) &= \frac{u_1(ax+b)(ay+b) + u_2(ax+b) + u_3(ay+b) + u_4}{v_2(ax+b) + v_3(ay+b) + v_4} \\
&= \frac{u_1a^2xy + \cdots}{v_2ax + v_3ay + (v_2+v_3)b + v_4}, \\
f^{-1}(R(f(x), f(y))) &= \frac{u_1axy + \cdots}{v_2ax + v_3ay + (v_2+v_3)b + v_4}.
\end{aligned}
$$

If $v_2 + v_3 \neq 0$ then we can take $b = -v_4/(v_2 + v_3)$ and $R$ is equivalent to the first type in (19). Suppose $v_2 + v_3 = 0$, i.e. $v_2 = -v_3$. If $v_2 = 0$ then $R$ is already a polynomial. But if $v_2 \neq 0$ then we can take $a = v_4/v_2$ and $R$ is equivalent to the second type in (19). □

**Lemma 2.5.** *Let $R$ be associative of the first type in (19). If $R$ is not a polynomial then $R$ must be of the form*

$$
\frac{xy + u}{x + y}
$$

*which is indeed associative.*

*Proof.* Assume $v_2 \neq 0$; the proof is similar if $v_3 \neq 0$. Then $R$ can be rewritten as

$$R = \frac{u_1xy + u_2x + u_3y + u_4}{x + vy} = \frac{F(x,y)}{H(x,y)}$$

where $F(x,y) = u_1xy + u_2x + u_3y + u_4$ and $H(x,y) = x + vy$. Since $R$ is not a polynomial, $\gcd(F(x,y), H(x,y)) = 1$. Note that

$$R(R(x,y),z) = \frac{u_1zF(x,y) + u_2F(x,y) + u_3zH(x,y) + u_4H(x,y)}{F(x,y) + vH(x,y)} = \frac{A}{C}$$
$$R(x,R(y,z)) = \frac{u_1xF(y,z) + u_2xH(y,z) + u_3F(y,z) + u_4H(y,z)}{xH(y,z) + vF(y,z)} = \frac{B}{D}$$

where $A, B, C, D$ are the corresponding numerators and denominators. By Lemma 2.2, $\gcd(A,C) = 1$ and $\gcd(B,D) = 1$. So the equation $A/C = B/D$ implies that $C = dD$ for some constant $d \in \mathbb{F}$. If $v = 0$ then $D = xH(y,z) = xy$ and $C = F(x,y) = x$, impossible for $C = dD$. Hence we may assume that $v \neq 0$. As

$$C = u_1xy + vxz + v^2yz + u_2x + u_3y + u_4$$
$$D = xy + vxz + vu_1yz + vu_2y + vu_3z + vu_4.$$

Comparing the coefficients of $x, y$ gives $u_2 = 0$ and $u_3 = 0$. Then by the coefficients of $xy, xz$ and $yz$ we have

$$u_1 = d, \quad v = dv, \quad v^2 = dvu_1.$$

Since $v \neq 0$, it follows that $d = u_1 = 1$ and hence $v = 1$. Therefore $R$ is the required form. □

**Lemma 2.6.** *Let $R$ be associative of the second type in (19). If $R$ is not a polynomial then $R$ must be of the form*

$$\frac{xy + u}{x + y + 1}$$

*which is indeed associative.*

*Proof.* Let $F(x,y) = u_1xy + u_2x + u_3y + u_4$ and $H(x,y) = x - y + 1$. Since $R$ is not a polynomial, $\gcd(F(x,y), H(x,y)) = 1$. Similar to above,

$$R(R(x,y),z)) = \frac{u_1zF(x,y) + u_2F(x,y) + u_3zH(x,y) + u_4H(x,y)}{F(x,y) - zH(x,y) + H(x,y)} = \frac{A}{C}$$
$$R(x,R(y,z)) = \frac{u_1xF(y,z) + u_2xH(y,z) + u_3F(y,z) + u_4H(y,z)}{xH(y,z) - F(y,z) + H(y,z)} = \frac{B}{D}.$$

By Lemma 2.2, $\gcd(A,C) = \gcd(B,D) = 1$. So $C = dD$ for some $d \in \mathbb{F}$. As

$$C = u_1xy - xz + yz + (u_2 + 1)x + (u_3 - 1)y - z + u_4 + 1$$
$$D = xy - xz - u_1yz + x + (1 - u_2)y - (u_3 + 1)z - u_4 + 1,$$

comparing the coefficients of $xy, xz, yz$ gives

$$u_1 = d, \quad -1 = -d, \quad 1 = -du_1.$$

So $d = u_1 = 1$ and $2 = 0$, which means that the characteristic of $\mathbb{F}$ is 2. Now the coefficients of $x, y$ give

$$u_2 + 1 = d, \quad u_3 - 1 = d(-u_2 + 1).$$

So $u_2 = u_3 = 0$ and $R$ is the required type. □

**Lemma 2.7.** *Every rational function of the form*

$$\frac{xy+u}{x+y+v}$$

*is equivalent to $xy$.*

*Proof.* Let $a, b$ be the two roots of the polynomial $X^2 - vX - u$ and $f = (x+a)/(x+b)$. Then one can check directly that the function in the lemma is equal to $f^{-1}(f(x) \cdot f(y))$. □

*Proof of Theorem 1.1.* Let $R \in \mathbb{F}(x, y)$ be any associative rational function. By Lemma 2.3, $R$ must be of the form (18), hence equivalent to a polynomial or one of the two types in (19). In the latter case, by Lemmas 2.5 and 2.6, $R$ must be of the form in Lemma 2.7, which is equivalent to $xy$ thus to $x + y + xy$. Therefore $R$ is equivalent to $x, y, x + y$ or $x + y + xy$ by Lemma 2.1.

## 3 Determining the Groups

This section is devoted to proving Theorem 1.2.

Before we proceed to the proof, we make some general remarks. Suppose that $f : S \longrightarrow G$ is any bijection of sets and $(G, \cdot)$ is a group. Then one can define a group law on $S$ as follows:

$$a \cdot b = f^{-1}(f(a) \cdot f(b)), \quad \forall a, b \in S. \tag{22}$$

Here $f$ is automatically a group isomorphism from $S$ to $G$. On the other hand, if $S$ already has a group law that satisfies (22) for some bijection $f : S \longrightarrow G$ then $f$ is a group isomorphism from $S$ to $G$.

Now let $S$ be an infinite subset of $\mathbb{F} \cup \{\infty\}$ where $\mathbb{F}$ is a field. Suppose that a group law on $S$ is defined by a rational function $R \in \mathbb{F}(x, y)$, i.e. $a \cdot b = R(a, b)$ for $a, b \in S$. By Theorem 1.1, $R$ is equivalent to $x + y$ or $xy$ via some linear fractional map $f \in \overline{\mathbb{F}}(x)$, i.e. $R(x, y) = f^{-1}(f(x) + f(y))$ or $f^{-1}(f(x)f(y))$. Hence

$$a \cdot b = R(a, b) = f^{-1}(f(a) + f(b)), \quad \forall a, b \in S$$

or

$$a \cdot b = R(a, b) = f^{-1}(f(a)f(b)), \quad \forall a, b \in S.$$

Let $G = f(S) \subseteq \overline{\mathbb{F}}$. Then $G$ is an additive subgroup of $\overline{\mathbb{F}}$ in the first case or a multiplicative group of $\overline{\mathbb{F}}$ in the second case. Obviously $G$ is in the same extension of $\mathbb{F}$ as that defining the map $f$. From the polynomial (21), it seems that one may need to go to a cubic extension of $\mathbb{F}$ to find the coefficients of $f$. We prove that this is not the case and indeed a quadratic extension of $\mathbb{F}$ suffices. Consequently $S$ is isomorphic to a subgroup (additive or multiplicative) of a quadratic extension of $\mathbb{F}$, hence Theorem 1.2 is proved.

It remains to prove that the equivalence of $R$ can be realized by a linear fractional map over a quadratic extension of $\mathbb{F}$. This is done by the next two lemmas.

**Lemma 3.1.** *Suppose $R \in \mathbb{F}(x, y)$ is equivalent to $x+y$. Then the equivalence can be defined by a linear fractional map over $\mathbb{F}$, except when the characteristic of $\mathbb{F}$ is 2 and $\mathbb{F}$ is not perfect in which case a required linear fractional map can be found in a quadratic extension of $\mathbb{F}$.*

*Proof.* Suppose $R(x, y) = f^{-1}(f(x) + f(y))$ where $f = (ax + b)/(cx + d)$ with $a, b, c, d \in \overline{\mathbb{F}}$ and $ad - bc \neq 0$. First assume that $c = 0$. Then $f$ can be rewritten as $f = ax + b$ for some $a, b \in \overline{\mathbb{F}}$ with $a \neq 0$. In this case,

$$R = f^{-1}(f(x) + f(y)) = x + y + b/a.$$

As $R \in \mathbb{F}(x, y)$, we have $b/a \in \mathbb{F}$. Let $g = x + b/a$. Then we have $R = g^{-1}(g(x) + g(y))$ so the equivalence is defined by a linear map over $\mathbb{F}$.

Now assume $c \neq 0$. Then $f$ can be rewritten as as $f = (ax + b)/(x + d)$ for some $a, b, d \in \overline{\mathbb{F}}$ with $ad - b \neq 0$. Note that

$$R = f^{-1}(f(x) + f(y)) = \frac{(b - 2ad)xy - ad^2(x + y) - bd^2}{axy + b(x + y) + 2bd - ad^2}.$$

There are two cases depending on $a = 0$ or not. Suppose $a = 0$. Then $b \neq 0$ (as $ad - b \neq 0$), and

$$R = \frac{bxy - bd^2}{b(x + y) + 2bd} = \frac{xy - d^2}{x + y + 2d} \in \mathbb{F}(x, y).$$

So $2d, d^2 \in \mathbb{F}$. If $\operatorname{char}(\mathbb{F}) \neq 2$ then $2d \in \mathbb{F}$ implies that $d \in F$. If $\operatorname{char}(\mathbb{F}) = 2$ and $\mathbb{F}$ is perfect then $d^2 \in \mathbb{F}$ implies that $d \in \mathbb{F}$. In both cases $f$ is defined over $\mathbb{F}$. If $\operatorname{char}(\mathbb{F}) = 2$ but $\mathbb{F}$ is not perfect then $d$ may not be in $\mathbb{F}$ but always in a quadratic extension of $\mathbb{F}$; See Example 1.3 for an instance.

Suppose $a \neq 0$. Then $R$ can be rewritten as

$$R = \frac{(\frac{b}{a} - 2d)xy - d^2(x + y) - \frac{b}{a}d^2}{xy + \frac{b}{a}(x + y) + 2d\frac{b}{a} - d^2} \in \mathbb{F}(x, y).$$

Hence

$$\frac{b}{a},\ \frac{b}{a}-2d,\ d^2 \in \mathbb{F}.$$

This implies that $d \in \mathbb{F}$ except when $\mathrm{char}(\mathbb{F}) = 2$ and $\mathbb{F}$ is not perfect. The proof is finished by noting that $R = g^{-1}(g(x)+g(y))$ where $g = (x+b/a)/(x+d)$. □

**Lemma 3.2.** *Suppose that $R \in \mathbb{F}(x,y)$ is equivalent to $xy$. Then the equivalence is defined by a linear fractional map over a quadratic extension of $\mathbb{F}$.*

*Proof.* Suppose $R(x,y) = f^{-1}(f(x)f(y))$ where $f = (ax+b)/(cx+d)$ with $a,b,c,d \in \overline{\mathbb{F}}$ and $ad-bc \neq 0$. If $c = 0$ then $f$ is of the form $f = ax+b$ with $a \neq 0$ and

$$R = f^{-1}(f(x)f(y)) = axy + b(x+y) + (b^2-b)/a.$$

Since $R \in \mathbb{F}(x,y)$, we have $a, b, \in \mathbb{F}$, so $f \in \mathbb{F}(x)$.

Hence we may assume that $c \neq 0$. Then $f$ can be rewritten as $f = (ax+b)/(x+d)$ for some $a,b,d \in \overline{\mathbb{F}}$ with $ad-b \neq 0$. One can check directly that

$$R = f^{-1}(f(x)f(y)) = \frac{(b-a^2d)xy - bd(a-1)(x+y) - bd(b-d)}{(a^2-a)xy + a(b-d)(x+y) + b^2 - ad^2}. \tag{23}$$

If $a = 0$ then we see easily that $b,d \in \mathbb{F}$ and so $f$ is defined over $\mathbb{F}$. If $a = 1$ then

$$R = \frac{xy - bd}{x+y+b+d} \in \mathbb{F}(x,y).$$

Hence $bd, b+d \in \mathbb{F}$ and thus $b,d$ are in a quadratic extension of $\mathbb{F}$. We may now assume that $a \neq 0$ or 1. Then (23) may be rewritten as

$$R = \frac{\frac{b-a^2d}{a^2-a}xy - \frac{bd}{a}(x+y) - \frac{bd}{a}\frac{b-d}{a-1}}{xy + \frac{b-d}{a-1}(x+y) + \frac{b^2-ad^2}{a^2-a}}.$$

Since $R \in \mathbb{F}(x,y)$, we have

$$\frac{b-d}{a-1},\ \frac{bd}{a},\ \frac{b-a^2d}{a^2-a},\ \frac{b^2-ad^2}{a^2-a} \in \mathbb{F}.$$

If $a = -1$ then $bd, b-d \in \mathbb{F}$ so $b,d$ are in a quadratic extension of $\mathbb{F}$. Hence we may further assume that $a \neq -1$. Let

$$b-d = c_1(a-1),\quad a^2d - b = c_2(a^2-a)$$

where $c_1, c_2 \in \mathbb{F}$. Then

$$b = \frac{a}{a+1}(c_1a+c_2),\quad d = \frac{1}{a+1}(c_1+c_2a).$$

Note

$$\frac{bd}{a} = \frac{1}{(a+1)^2}(c_1 a + c_2)(c_1 + c_2 a) \in \mathbb{F}.$$

Let $c_3 = bd/a \in \mathbb{F}$. Then we see that $a$ is a root of the quadratic polynomial

$$c_3(X+1)^2 - (c_1 X + c_2)(c_1 + c_2 X) \in \mathbb{F}[X].$$

So $a, b, d$ all lie in a quadratic extension of $\mathbb{F}$. This proves that $f$ is always defined over a quadratic extension of $\mathbb{F}$. □

## References


1. J. V. Brawley and L. Carlitz, "Irreducibles and the composed product for polynomials over a finite field", *Discrete Math.* **65** (1987), 115-139.
2. J. V. Brawley, S. Gao and D. Mills, "Computing composed products of polynomials," in *Finite Fields: Theory, Applications, and Algorithms* (Waterloo, ON, 1997, R. C. Mullin and G. L. Mullen, Ed.), 1–15, *Contemporary Mathematics*, **225**, Amer. Math. Soc., Providence, RI, 1999.
3. L. Childs, D. Moss and J. Sauerberg, "Dimension one polynomial formal groups", p. 11–19, in *Hopf Algebras, Polynomial Formal Groups, and Raynaud Orders* (L. N. Childs, C. Greither, D. J. Moss, J. Sauerberg and K. Zimmermann), Memoirs AMS, no. 651, 1998.
4. M. Hazewinkel, *Formal Groups and Applications*, Academic Press, New York, 1978.
5. A. J. Menezes, I. F. Blake, X. Gao, R. C. Mullin, S.A. Vanstone and T. Yaghoobian, *Applications of Finite Fields*, Kluwer Academic Publishers, Boston-Dordrecht-Lancaster, 1993.
6. J. H. Silverman, *The Arithmetic of Elliptic Curves*, Springer-Verlag, New York, 1986.

# A Note on the Minimal Polynomial of the Product of Linear Recurring Sequences

Emrah Çakçak

Middle East Technical University
Department of Mathematics
06531 Ankara
Turkey
e-mail: cakcak@metu.edu.tr

**Abstract** Let $F$ be a field of nonzero characteristic, with its algebraic closure, $\bar{F}$. For positive integers $a, b$, let $J(a, b)$ be the set of integers $k$, such that $(x-1)^k$ is the minimal polynomial of the termwise product of linear recurring sequences $\sigma$ and $\tau$ in $\bar{F}$, with minimal polynomials $(x-1)^a$ and $(x-1)^b$ respectively. This set plays a crucial role in the determination of the product of linear recurring sequences with arbitrary minimal polynomials. Here, we give an explicit formula to determine some of the elements of $J(a, b)$, in the case of characteristic 2. We also give some clues for the extension to arbitrary characteristic. The method given here has produced a family of matrices which are themselves interesting.

## 1 Introduction

This work concerns the minimal polynomial of the product of two linear recurring sequences over a field with nonzero characteristic. There are mainly two results in the literature given by Zierler and Mills [5] and Göttfert and Niederreiter [2]. The author of [1] gave an improvement of the latter. In this note we improve the results of [1], in the sense that we give a formula to determine some of the elements of the set $J(a, b)$, which was defined in [1].

Let us briefly give the basic terminology. Let $F$ be a field with nonzero characteristic. A sequence $(s_n)_{n=0}^{\infty}$ of elements of $F$ is called a *linear recurring sequence* in $F$ with *characteristic polynomial* $f(x) = \sum_{i=0}^{d} c_i x^i \in F[x]$, if $c_d = 1$ and $\sum_{i=0}^{d} c_i s_{n+i} = 0$ for $n = 0, 1, \ldots$. For a linear recurring sequence $\sigma$ in $F$, its *minimal polynomial* $m_\sigma \in F[x]$ is defined to be the (uniquely determined) characteristic polynomial of $\sigma$ of least degree. The *linear complexity* of $\sigma$ is defined to be the degree of $m_\sigma(x)$. Further background on linear recurring sequences can be found in [3, Chapter 6]. Also we refer to [1] for the role of the set $J(a, b)$ in the determination of the minimal polynomial of the product of linear recurring sequences.

We now discuss the main properties of the set $J(a, b)$. The following are dealt by Göttfert and Niederreiter in [2]:

1. $J(a, b) \subseteq \{0, 1, \ldots, a \vee b\}$ ,
2. $J(a, b) = \{a \vee b = a + b - 1\}$ iff $\binom{a+b-2}{a-1} \neq 0$ ,

3. if $\binom{a+b-2}{a-1} = 0$ then $0, a \vee b \in J(a,b)$ .

where $a \vee b$ is defined (in [2]) as:

$$a \vee b = \max\{i+j+1 | \binom{i+j}{i} \neq 0 \quad \text{where } 0 \leq i \leq a-1, 0 \leq j \leq b-1\}.$$

Here we remark that $a \vee b$ can easily be computed by using (the definition above or) the equivalent definition given by Zierler and Mills [5] as follows: for $p \neq 0$, the characteristic of the field $F$, write $a-1 = \sum j_\nu p^\nu$, $0 \leq j_\nu < p$, $b-1 = \sum k_\nu p^\nu$, $0 \leq k_\nu < p$, and let $\lambda$ be the smallest nonnegative integer such that $j_\nu + k_\nu < p$ for all $\nu \geq \lambda$, then

$$a \vee b = p^\lambda + \sum_{\nu \geq \lambda} (j_\nu + k_\nu) p^\lambda. \tag{1}$$

However, there is no efficient algorithm, in general, to find the elements of the set $J(a,b)$, in the case $\binom{a+b-2}{a-1} = 0$. To decide whether an integer $l$ (between 1 and $a \vee b - 1$) is an element of $J(a,b)$, one has to determine the solvability, in $\bar{F}$, of a set of nonlinear equations and inequalities over $F$. In fact, we have:

**Lemma 1.1.** *For $l \in \{0,1,\ldots,a \vee b - 1\}$, $l+1 \in J(a,b)$ if and only if there exist elements $c_0, c_1, \ldots, c_{a-1}, d_0, d_1, \ldots, d_{b-1} \in \bar{F}$ with $c_{a-1}d_{b-1} \neq 0$, satisfying*

$$\sum_{i=\max(0,l-b+1)}^{\min(l,a-1)} \sum_{j=l-i}^{b-1} \binom{l}{i} \binom{i+j}{l} c_i d_j = 1 \tag{2}$$

*and*

$$\sum_{i=\max(0,k-b+1)}^{\min(k,a-1)} \sum_{j=k-i}^{b-1} \binom{k}{i} \binom{i+j}{k} c_i d_j = 0 \tag{3}$$

*$k = l+1, \ldots, a \vee b - 1$. In addition, $0 \in J(a,b)$ iff there are $c_i$'s $d_j$'s in $\bar{F}$ satisfying (3) for all $k = 0, \ldots, a \vee b - 1$.*

The details of the above characterization can be found in [1]. In Sect.3 we will give a more convenient characterization of $J(a,b)$ as Theorem 3.3.

Let's now state our main result that will be proved in Sect.3.

**Theorem 1.2.** *Let $F$ be a field of characteristic 2, and $a,b$ two positive integers with $a \leq b$, satisfying $\binom{a+b-2}{a-1} = 0$. Let $k$ be the smallest integer such that $a \leq 2^k$. We have:*

1. *if $b > 2^k$, then*

$$J(a,b) = K \cup \{l + 2^k | l \in J(a, b-2^k)\}$$

*where $K \subseteq \{0,2,3,\ldots,2^k\}$, in particular,*

$$1 \notin J(a,b);$$

*2. if* $b \leq 2^k$*, then* $a \vee b = 2^k$ *and*

$$\{2^{k-1}+1, 2^{k-1}+2, \ldots, 2^k\} \subset J(a,b).$$

The following result is a direct consequence of Theorem 1.2:

**Corollary 1.3.** *Let $F$ be a field of characteristic 2. If $a, b, k$ and $m$ are positive integers such that $a \leq 2^k$ and $b > m2^k$, then the following integers are not elements of the set $J(a,b)$:*

$$1, 2^k+1, \ldots, t2^k+1, \ldots, (m-1)2^k+1.$$

*In particular, there are infinitely many tuples $(a,b) \in \mathbb{N} \times \mathbb{N}$ with $\binom{a+b-2}{a-1} = 0$, such that*

$$J(a,b) \neq \{0, 1, \ldots, a \vee b\}.$$

We stress that, in case $\binom{a+b-2}{a-1} = 0$, the gaps in the set $J(a,b)$ (the integers between 0 and $a \vee b$ that are not elements of $J(a,b)$), play an important role in the determination of the lower bound on the linear complexity of the product of linear recurring sequences (see [1]). Our result explicitly gives a set of gaps when $F$ has characteristic 2. Note also that Theorem 1.2 does not completely determine the elements of $J(a,b)$ (so, in particular, all the gaps in $J(a,b)$). In Sect.3 we shall give an example that there are still some gaps in the set $J(a,b)$ which are not predicted by Theorem 1.2.

In Sect.2 we construct an interesting family of matrices that will be used in the proof of Theorem 1.2.

## 2 An Interesting Family of Matrices

Let $F$ be a field of characteristic 2. In this section (and Sect.3), $F^{m \times n}$ will denote the set of all $m \times n$ matrices over $F$, also, given $A \in F^{m \times n}$, $A^t$ will be the transpose of $A$.

For each $k = 0, 1, \ldots,$ we define inductively the families of matrices $\mathcal{C}_{2^k}$ and $\mathcal{C}^r_{2^k}$ as follows:

1. $\mathcal{C}_1$ and $\mathcal{C}^r_1$ are equal to $F^{1 \times 1} = F$,
2. if $k \geq 1$ then $\mathcal{C}_{2^k}$ consists of $2^k \times 2^k$ matrices over $F$ of the following form:
$$C = \begin{pmatrix} C_0 & C_0 \\ C_1 & C_0 \end{pmatrix}$$
where $C_0, C_1 \in \mathcal{C}_{2^{k-1}}$,
3. if $k \geq 1$ then $\mathcal{C}^r_{2^k}$ consists of $2^k \times 2^k$ matrices over $F$, of the following form:
$$C = \begin{pmatrix} C_2 & 0 \\ C_1 & C_0 \end{pmatrix}$$
where $C_0, C_1 \in \mathcal{C}^r_{2^{k-1}}$, $C_2 = C_1 + C_0$ and 0 is the $2^{k-1} \times 2^{k-1}$ zero matrix.

From the construction above, it is easy to see that the matrices in these families are uniquely determined by their last rows, as well as by their first columns. The elements of $\mathcal{C}^r_{2^k}$ are lower triangular matrices which are uniquely determined (also) by their diagonals. Although it will not be necessary for our main results, we note that $\mathcal{C}^r_{2^k}$ is a commutative subalgebra of $F^{2^k\times 2^k}$ (endowed with usual matrix operations) with identity, the identity of $F^{2^k\times 2^k}$. Moreover, if $C\in\mathcal{C}^r_{2^k}$ is an invertible matrix (in $F^{2^k\times 2^k}$) then its inverse is also in $\mathcal{C}^r_{2^k}$.

Now, we set the following terminology. Given $a\le 2^k$ a nonnegative integer, and $c_0,c_1,\dots,c_{a-1}\in F$, let

$$C=\mathcal{C}_{2^k}[c_0,\dots,c_{a-1}]\quad (\text{resp. } C=\mathcal{C}^r_{2^k}[c_0,\dots,c_{a-1}])$$

denote the matrix $C\in\mathcal{C}_{2^k}$ (resp. $C\in\mathcal{C}^r_{2^k}$) with last row

$$[0,0,\dots,0,c_{a-1},c_{a-2},\dots,c_0].$$

In that case, we will call the row vector $[c_0,c_1,\dots,c_{a-1}]$, the *base* of $C$. Here we note that, for each vector $c\in F^{1\times 2^k}$ there is an element $C$ of $\mathcal{C}_{2^k}$ (resp. $\mathcal{C}^r_{2^k}$) with base $c$.

For a matrix $C=\mathcal{C}_{2^k}[c_0,\dots,c_{a-1}]$ we define its *reduced form* as the matrix $C^r=\mathcal{C}^r_{2^k}[c_0,\dots,c_{a-1}]$. Indeed $C^r$ is the row reduced form of $C$ by applying the following set of row operations; for each $t=1,2,\dots,k-1$, let $R_t\in F^{2^k\times 2^k}$ be the block diagonal matrix:

$$R_t=\begin{pmatrix} K_t & 0 & \cdots & 0\\ 0 & K_t & \cdots & 0\\ \vdots & \vdots & \ddots & \vdots\\ 0 & 0 & \cdots & K_t\end{pmatrix}$$

with

$$K_t=\begin{pmatrix} I_t & I_t\\ 0 & I_t\end{pmatrix}$$

where $I_t$ is the identity matrix in $F^{2^t\times 2^t}$. Then we have

$$C^r=R_1R_2\cdots R_{k-1}C.$$

We now give the lemmas necessary to establish the main result. We assume that $F$ is algebraically closed.

**Lemma 2.1.** *Let $C$ be an element of the family $\mathcal{C}^r_{2^k}$. Then the matrix equation*

$$Cd=\begin{pmatrix}1\\0\\ \vdots\\ 0\end{pmatrix}\tag{4}$$

*has a solution for $d\in F^{2^k\times 1}$ if and only if $C$ is invertible.*

*Proof.* The sufficiency follows immediately. The necessity is proved by induction on $k$. Let $d \in \bar{F}^{2^k \times 1}$ be a solution of (4). Writing

$$C = \begin{pmatrix} C_1 + C_0 & 0 \\ C_1 & C_0 \end{pmatrix} \text{ and } d = \begin{pmatrix} d_0 \\ d_1 \end{pmatrix},$$

where $C_0, C_1 \in C^r_{2^{k-1}}$ and $d_0, d_1 \in \bar{F}^{2^{k-1} \times 1}$, we have

$$(C_1 + C_0)d_0 = \begin{pmatrix} 1 \\ 0 \\ \vdots \\ 0 \end{pmatrix} \tag{5}$$

and

$$C_1 d_0 + C_0 d_1 = \begin{pmatrix} 0 \\ 0 \\ \vdots \\ 0 \end{pmatrix}. \tag{6}$$

Adding (5) and (6), we get

$$C_0(d_0 + d_1) = \begin{pmatrix} 1 \\ 0 \\ \vdots \\ 0 \end{pmatrix}. \tag{7}$$

As $C_0$ and $C_1 + C_0$ are elements of $C^r_{2^{k-1}}$, the induction hypothesis together with (5) and (7), imply that $C_0$ and $C_1 + C_0$ are invertible. Consequently, $C$ is invertible. □

**Lemma 2.2.** *Let $a, b$ and $k$ be positive integers satisfying*

$$2^{k-1} < a \leq 2^k, 2^{k-1} < b \leq 2^k.$$

*Let $f_1 \in F^{2^{k-1} \times 1}$ be any column vector. Then there are elements*

$$c_0, c_1, \ldots, c_{a-1}, d_0, d_1, \ldots, d_{b-1} \in F$$

*with $c_{a-1}d_{b-1} \neq 0$, and a vector $f_0 \in F^{2^{k-1} \times 1}$ such that $C = C^r_{2^k}[c_0, \ldots, c_{a-1}]$ and $d = [d_0, \ldots, d_{b-1}, 0, \ldots, 0]^t$ satisfy*

$$Cd = \begin{pmatrix} f_0 \\ f_1 \end{pmatrix}.$$

*Proof.* As the elements of $C^r_{2^k}$ are of the form

$$\begin{pmatrix} C_1 + C_0 & 0 \\ C_1 & C_0 \end{pmatrix},$$

we want to find elements

$$c_0, c_1, \ldots, c_{a-1}, d_0, d_1, \ldots, d_{b-1} \in F$$

with $c_{a-1}d_{b-1} \neq 0$ such that if

$$\begin{aligned} C_0 &= C^r_{2^{k-1}}[c_0, \ldots, c_{2^{k-1}-1}], \\ C_1 &= C^r_{2^{k-1}}[c_{2^{k-1}}, \ldots, c_{a-1}], \\ h_0 &= [d_0, \ldots, d_{2^{k-1}-1}]^t, \\ h_1 &= [d_{2^{k-1}}, \ldots, d_{b-1}, 0, \ldots, 0]^t \end{aligned}$$

then

$$C_1 h_0 + C_0 h_1 = f_1 \ .$$

Now, if we set $c_{2^{k-1}}$ and $c_{a-1}$ to be two nonzero distinct elements of $F$ and all other $c_i$'s zero, then $C_0$ will be the zero matrix, and $C_1$ will be invertible. This will finish the proof. To check that, with this choice of $c_i$'s, every entry of the diagonal of $C_1$ is either $c_{2^{k-1}}$ or $c_{2^{k-1}} + c_{a-1}$, and therefore $C_1$ is invertible, is left to the reader. □

We finish this section by giving another interesting property of the families $C_{2^k}$ and $C^r_{2^k}$, which can be derived by using similar arguments as in the proof of Lemma 2.1.

**Proposition 2.3.** *Let* $C = C_{2^k}[c_0, c_1, \ldots, c_{2^k-1}]$, $D = C^r_{2^k}[d_0, d_1, \ldots, d_{2^k-1}]$, *with* $c_0, \ldots, c_{2^k-1}, d_0, \ldots, d_{2^k-1} \in F$. *Then we have*

$$C \begin{pmatrix} d_0 \\ \vdots \\ d_{2^k-1} \end{pmatrix} = D \begin{pmatrix} c_0 \\ \vdots \\ c_{2^k-1} \end{pmatrix}.$$

## 3 Main Results

In this section, we will give another characterization of the set $J(a, b)$ and settle the connection between $J(a, b)$ and the families of matrices that we constructed in Sect.2. Then we shall give a proof of Theorem 1.2.

In the following discussion, we assume $F$ to be an arbitrary field of nonzero characteristic $p$. Let $f_l$ be the left hand side of (2) in Lemma 1.1:

$$f_l = \sum_{i=\max(0,l-b+1)}^{\min(l,a-1)} \sum_{j=l-i}^{b-1} \binom{l}{i} \binom{i+j}{l} c_i d_j. \tag{8}$$

We will find out the nonzero terms of this sum. For simplicity let's write $f_l$ as

$$f_l = \sum_{j=0}^{b-1} \sum_{i=0}^{a-1} \binom{l}{i} \binom{i+j}{l} c_i d_j. \tag{9}$$

Here the equivalence of (8) and (9) can be shown by using the identity

$$\binom{l}{i} \binom{i+j}{l} = \binom{i+j}{i} \binom{j}{l-i}. \tag{10}$$

Now, for $l$ and $j$ fixed, let's find the conditions on $i$ such that the coefficient $\binom{l}{i}\binom{i+j}{l}$ of $c_i d_j$ is nonzero. To this end, we write the $p$-adic digit expansions:

$$\begin{aligned} i &= \sum i_\nu p^\nu, \quad 0 \le i_\nu < p, \\ j &= \sum j_\nu p^\nu, \quad 0 \le j_\nu < p, \\ l &= \sum l_\nu p^\nu, \quad 0 \le l_\nu < p\,. \end{aligned}$$

By the congruence of Lucas

$$\binom{l}{i} \equiv \prod_\nu \binom{l_\nu}{i_\nu} \mod p, \tag{11}$$

the following identity is easily derived (see [4] for a short proof)

$$\binom{i+j}{i} \not\equiv 0 \mod p \iff i_\nu + j_\nu < p \quad \text{for each } \nu. \tag{12}$$

From (10), (11) and (12), we get the following conditions on $i$ (condition on $i$ means conditions on the $p$-adic digits, $i_\nu$, of $i$, in the sequel):

**A1** $i_\nu \le l_\nu$ for each $\nu$,
**A2** $i_\nu + j_\nu < p$ for each $\nu$.

Using the left hand side of (10), $i$ should satisfy

$$l_\nu \le (i+j)_\nu \quad \text{for each } \nu,$$

where $(i+j)_\nu$ are the digits appearing in the $p$-adic expansion of $i + j = \sum (i+j)_\nu p^\nu$, but **A2** shows that $(i+j)_\nu = i_\nu + j_\nu$ (i.e. there is no 'carry' when we add $i$ and $j$ in base $p$). Therefore we have

**A3** $l_\nu \le i_\nu + j_\nu$ for each $\nu$.

These three conditions, **A1**, **A2** and **A3**, are necessary and sufficient for $\binom{l}{i}\binom{i+j}{l}$ to be nonzero. We combine them as

**A4** $\max(0, l_\nu - j_\nu) \le i_\nu \le \min(l_\nu, (p-1) - j_\nu)$ for each $\nu$.

Our aim is to give a suitable characterization of the set $J(a,b)$ in matrix form. Let's define the family of matrices $\mathcal{C}_{p^k}$, $k = 0, 1, \ldots$ as follows:

**Definition 3.1.** A matrix $C \in F^{p^k \times p^k}$

$$C = \begin{pmatrix} \gamma_{0,0} & \cdots & \gamma_{0,p^k-1} \\ \vdots & \ddots & \vdots \\ \gamma_{p^k-1,0} & \cdots & \gamma_{p^k-1,p^k-1} \end{pmatrix} \tag{13}$$

is an element of $\mathcal{C}_{p^k}$, if there are elements $c_0, c_1, \ldots, c_{p^k-1}$ in $F$ such that for each $l = 0, 1, \ldots, p^k - 1$ and $j = 0, 1, \ldots, p^k - 1$,

$$\gamma_{l,j} = \sum_{i=0}^{p^k-1} \binom{l}{i} \binom{i+j}{l} c_i. \tag{14}$$

Clearly, for each $l$ and $j$, $\gamma_{l,j}$ in the definition above, is the sum of those $\binom{l}{i}\binom{i+j}{l} c_i$ such that $i$ satisfies **A4**. This fact gives the following recursive construction of these families.

**Theorem 3.2.** *The family $\mathcal{C}_1$ is equal to $F^{1\times 1} = F$. For $k \ge 1$, let's write $C \in \mathcal{C}_{p^k}$ as*

$$C = \begin{pmatrix} \Gamma_{0,0} & \cdots & \Gamma_{0,p^k-1} \\ \vdots & \ddots & \vdots \\ \Gamma_{p^k-1,0} & \cdots & \Gamma_{p^k-1,p^k-1} \end{pmatrix}$$

*where each block $\Gamma_{l,j}$ is a matrix in $F^{p^{k-1} \times p^{k-1}}$. Then we have*

$$\Gamma_{l,j} = \sum_{i=\max(0,l-j)}^{\min(l,(p-1)-j)} \binom{l}{i} \binom{i+j}{l} \Gamma_{i,0} \quad \text{for } l, j \in \{0, 1, \ldots, p-1\}. \tag{15}$$

*Moreover, each matrix $\Gamma_{l,j}$ is an element of $\mathcal{C}_{p^{k-1}}$.*

*Proof.* The case $k = 0$ is obvious from the definition of $\mathcal{C}_{p^k}$. For $k \ge 1$, take a matrix $C = [\gamma_{l,j}] \in \mathcal{C}_{p^k}$ (where the entries $\gamma_{l,j}$ are labeled as in (13)) and elements $c_0, c_1, \ldots, c_{p^k-1}$ of the field $F$, such that each entry $\gamma_{l,j}$ of $C$ is given by (14). For each $0 \le l, j, i < p^k$, we write

$$\begin{aligned} l &= l_1 p^{k-1} + l_0 \quad 0 \le l_1 < p,\ 0 \le l_0 < p^{k-1}, \\ j &= j_1 p^{k-1} + j_0 \quad 0 \le j_1 < p,\ 0 \le j_0 < p^{k-1}, \\ i &= i_1 p^{k-1} + i_0 \quad 0 \le i_1 < p,\ 0 \le i_0 < p^{k-1}. \end{aligned}$$

With these settings, since $j$ and $i$ (for the nonzero terms of the sum in (14)) should satisfy **A2**, by (11), we have

$$\gamma_{l,j} = \sum_{i_1=0}^{p-1} \binom{l_1}{i_1} \binom{i_1+j_1}{l_1} \sum_{i_0=0}^{p^{k-1}-1} \binom{l_0}{i_0} \binom{i_0+j_0}{l_0} c_{i_1 p^{k-1}+i_0}. \tag{16}$$

Now, $\gamma_{l,j}$ is the entry at $l_0$'th row and $j_0$'th column (where rows and columns are labeled from 0 to $p^{k-1}-1$) of $\Gamma_{l_1,j_1}$. The corresponding entry of $\Gamma_{l_1,0}$ (for some $0 \le l_1 < p$) is obtained by putting $j_1 = 0$ in (16), but since

$$\binom{l_1}{i_1}\binom{i_1}{l_1} \neq 0 \iff l_1 = i_1 \,,$$

this entry becomes the inner sum in (16) (by replacing $i_1$ by $l_1$). This fact, together with **A4**, shows (15). The calculation above also implies that, for each $l_1 = 0, \ldots, p-1$, $\Gamma_{l_1,0}$ is an element of $\mathcal{C}_{p^{k-1}}$, and the result easily follows. □

We now have the following characterization of $J(a,b)$.

**Theorem 3.3.** *Given $F$ an arbitrary field of characteristic $p$, consider the families $\mathcal{C}_{p^k}$ over the algebraic closure, $\bar{F}$, of $F$. Let $a, b$ be two positive integers and $p^k$ be the smallest power of $p$ such that $a, b \le p^k$. Let $l \le p^k$ be a nonnegative integer. Then $l \in J(a,b)$ if and only if there is a matrix $C \in \mathcal{C}_{p^k}$ with its last row*

$$[0, \ldots, 0, (-1)^{a-1}c_{a-1}, \ldots, (-1)^i c_i, \ldots, -c_1, c_0], \quad c_{a-1} \neq 0,$$

*and a column vector*

$$d = [d_0, d_1, \ldots, d_{b-1}, 0, \ldots, 0]^{\mathrm{t}} \in \bar{F}^{p^k \times 1}, \quad d_{b-1} \neq 0,$$

*such that*

$$Cd = f\,,$$

*where $f \in \bar{F}^{p^k \times 1}$ is the zero vector if $l = 0$ and any vector of the following form if $l > 0$:*

$$f = \binom{f_0}{f_1} \qquad \textit{with } f_1 = [1, 0, \ldots, 0]^{\mathrm{t}} \in \bar{F}^{(p^k - l + 1) \times 1}.$$

It can be verified (using **A4**) that, with the settings of Definition 3.1, the first column of the matrix $C \in \mathcal{C}_{p^k}$ is

$$[c_0, c_1, \ldots, c_{p^k-1}]^{\mathrm{t}}$$

and the last row of it is

$$[c_{p^k-1}, \ldots, (-1)^i c_i, \ldots, -c_1, c_0] \quad .$$

We point out also that, if $a, b$ and $k$ are as in Theorem 3.3, then (1) (the equivalent definition of $a \vee b$ given in Sect.1) implies that $a \vee b \le p^k$. Then the proof of Theorem 3.3 follows from Lemma 1.1 and the previous discussions in this section. We note also that, to take $l \le p^k$ in Theorem 3.3 is enough to determine all the elements of $J(a,b)$, as $a \vee b \le p^k$.

The method introduced so far gives an alternative way to calculate $a \vee b$.

**Proposition 3.4.** *For $p$ a prime, let $F = \mathbb{F}_p(c_0, c_1, \ldots, c_{a-1})$ be the field of rational functions in the indeterminates $c_0, \ldots, c_{a-1}$, over $\mathbb{F}_p$. Let $C$ be the element of $\mathcal{C}_{p^k}$, with its first column*

$$[c_0, \ldots, c_{a-1}, 0, \ldots, 0]^{\mathrm{t}}.$$

*Let $C'$ be the matrix formed by taking the first $b$ columns of $C$. Then $a \vee b$ is the number of nonzero rows of $C'$, and $a \vee b = a + b - 1$ (equivalently $\binom{a+b-2}{a-1} \neq 0$) iff the last nonzero row of $C'$ is*

$$[0, \ldots, 0, \alpha c_{a-1}],$$

*where $\alpha$ is a nonzero element in $\mathbb{F}_p$.*

We omit the proof of Proposition 3.4 (that can be carried out by adopting the techniques given in [2]).

Let's now see that the family $\mathcal{C}_{2^k}$ constructed in Sect.2 is indeed $\mathcal{C}_{p^k}$ in Definition 3.1 with $p = 2$. By Theorem 3.2, it is enough to check this, for $k = 1$. When $p = 2$, **A4** reduces to

$$\begin{aligned} l_\nu = 0 &\Rightarrow i_\nu = 0, \\ l_\nu = 1, j_\nu = 1 &\Rightarrow i_\nu = 0, \\ l_\nu = 1, j_\nu = 0 &\Rightarrow i_\nu = 1, \end{aligned}$$

where $l_\nu$, $j_\nu$ and $i_\nu$ are the digits appearing in the $p$-adic expansions $l = \sum l_\nu 2^\nu$, $j = \sum j_\nu 2^\nu$, $i = \sum i_\nu 2^\nu$. So given $l, j$, the integer $i$ satisfying the conditions above is unique. Therefore a typical element $C$ of $\mathcal{C}_2$ (according to Definition 3.1) is of the form

$$C = \begin{pmatrix} c_0 & c_0 \\ c_1 & c_0 \end{pmatrix}, \quad c_0, c_1 \in F,$$

whence the equivalence of definitions follows. In addition, if the characteristic of the field $F$ is 2, we can use the family $\mathcal{C}^r_{2^k}$ (defined in Sect.2) in Theorem 3.3 instead of $\mathcal{C}_{2^k}$. In fact, observe that a matrix $C \in \mathcal{C}_{2^k}$ and a vector $d \in F^{2^k \times 1}$ satisfy

$$Cd = f$$

where $f \in F^{2^k \times 1}$ is as in Theorem 3.3 if and only if the reduced form $C^r = R_1 R_2 \cdots R_{k-1} C$ (see Sect.2 for definitions of reduced form and $R_t$) of $C$, satisfies

$$C^r d = f'$$

where $f' = R_1 R_2 \cdots R_{k-1} f$ has the desired form.

Now, we are ready to prove Theorem 1.2.

*Proof (of Theorem 1.2).* Given a field, $F$, of characteristic 2, let $a, b$ $(a \leq b)$ be two positive integers satisfying $\binom{a+b-2}{a-1} = 0$, and $2^k, 2^t$ be the smallest powers of 2 such that $a \leq 2^k$ and $b \leq 2^t$. The case $b \leq 2^k$ in Theorem 1.2 is a direct consequence of Lemma 2.2 (by suitable choices of $f_1$ in Lemma 2.2) and Theorem 3.3. So let's assume $b > 2^k$ and call $m = 2^t - 2^k$. Recall that a matrix $C \in C^r_{2^k}$ is with base $[c_0, c_1, \ldots, c_{a-1}]$ (with the terminology introduced in Sect.2) if its last row is

$$[0, \ldots, 0, c_{a-1}, \ldots, c_0].$$

By Theorem 3.3 and the remarks above, in order to determine the elements of the set $J(a, b)$, we need to find out existence of matrices $C \in C^r_{2^t}$ (each with base $[c_0, \ldots, c_{a-1}]$ for some $c_i \in \bar{F}$, with $c_{a-1} \neq 0$) and vectors $d \in \bar{F}^{2^t \times 1}$ ($d_{b-1} \neq 0$ with the settings of Theorem 3.3) to satisfy equations of the form $Cd = f$. In fact, the shape of each matrix, $C$, will be in block diagonal form:

$$C = \begin{pmatrix} C_0 & 0 & \cdots & 0 \\ 0 & C_0 & \cdots & 0 \\ \vdots & \vdots & \ddots & \vdots \\ 0 & 0 & \cdots & C_0 \end{pmatrix}$$

with $C_0 \in C^r_{2^k}$ ( with base $[c_0, \ldots, c_{a-1}]$). Write $C$ as

$$C = \begin{pmatrix} C_0 & 0 \\ 0 & C_1 \end{pmatrix},$$

where $C_1 \in \bar{F}^{m \times m}$, and $d$ as

$$d = \begin{pmatrix} d_0 \\ d_1 \end{pmatrix},$$

where $d_0 \in \bar{F}^{2^k \times 1}, d_1 \in \bar{F}^{m \times 1}$. So $1 \in J(a, b)$ implies (by Theorem 3.3) the existence of a matrix $C_0 \in C^r_{2^k}$, and column vectors $d_0 \in \bar{F}^{2^k \times 1}, d_1 \in \bar{F}^{m \times 1}$, with $d_1 \neq 0$, satisfying

$$C_0 d_0 = \begin{pmatrix} 1 \\ 0 \\ \vdots \\ 0 \end{pmatrix}$$

and

$$C_1 d_1 = 0,$$

where $C_1 \in \bar{F}^{m \times m}$ is the block diagonal matrix with $\frac{m}{2^k}$ identical blocks, $C_0$, on its diagonal. Then, by Lemma 2.1, $C_0$ is invertible, whence $C_1$ is invertible and it follows that $d_1$ should be 0. Therefore we have

$$1 \notin J(a, b).$$

Similarly it follows from Theorem 3.3 that

$$\{l + 2^k | l \in J(a, b - 2^k)\} \setminus \{2^k\} \subset J(a, b).$$

So, to finish the proof, we need only to check $2^k \in J(a,b)$. This is done as follows, let $C_0 \in \mathcal{C}^r_{2^k}$ with base $[c_0, \dots, c_{a-1}]$, where $c_0 = \dots = c_{a-2} = 0$ and $c_{a-1} = 1$, $C_1 \in \bar{F}^{m \times m}$ (constructed from $C_0$) as above,

$$d_0 = [0, \dots, 0, 1, 0, \dots, 0]^t \in \bar{F}^{2^k \times 1},$$

where the nonzero entry of $d_0$ is the $a$'th one, and $d_1 \in \bar{F}^{m \times 1}$, (like $d_0$) with 1 at the $(b - 2^k)$'th entry and 0's at remaining entries. Then the properties of the families $\mathcal{C}^r_{2^k}$ and the assumption $\binom{a+b-2}{a-1} = 0$ imply

$$C_0 d_0 = \begin{pmatrix} 0 \\ \vdots \\ 0 \\ 1 \end{pmatrix} \quad \text{and} \quad C_1 d_1 = 0. \quad \square$$

We point out that the results of Theorem 1.2 can be generalized to arbitrary characteristic $p$, by constructing the corresponding family $\mathcal{C}^r_{p^k}$, which consists of lower triangular matrices obtained by applying appropriate row operations (as in Sect.2) to the elements of $\mathcal{C}_{p^k}$. In this case, the calculations will be more involved, and we will no more have the result corresponding to Proposition 2.3.

We conclude by giving an example that Theorem 1.2 does not characterize all the gaps in the set $J(a,b)$.

*Example 3.5.* Let $F$ be a field of characteristic 2. Let $a = 4$ and $b = 6$. Then $\binom{a+b-2}{a-1} = 0$ and the smallest power of 2 greater or equal 6 is 8. We will show that $3 \notin J(a,b)$. Note that this gap in the set $J(a,b)$ is not predicted by Theorem 2. A typical element of $C_0 \in \mathcal{C}^r_4$ is of the form:

$$C_0 = \begin{pmatrix} c_3 + c_2 + c_1 + c_0 & 0 & 0 & 0 \\ c_3 + c_1 & c_2 + c_0 & 0 & 0 \\ c_3 + c_2 & 0 & c_1 + c_0 & 0 \\ c_3 & c_2 & c_1 & c_0 \end{pmatrix}$$

where $c_i \in \bar{F}, i = 0, 1, 2, 3$. So we need to show that there are no elements $c_i \in \bar{F}, i = 0, \dots, 3$ and $d_j \in \bar{F}, j = 0, \dots, 5$, with $c_3 d_5 \neq 0$, such that

$$C_0 \begin{pmatrix} d_0 \\ d_1 \\ d_2 \\ d_3 \end{pmatrix} = \begin{pmatrix} * \\ * \\ 1 \\ 0 \end{pmatrix} \tag{17}$$

and

$$C_0 \begin{pmatrix} d_4 \\ d_5 \\ 0 \\ 0 \\ 0 \end{pmatrix} = 0 \tag{18}$$

where $*$'s in (17) means that the first two rows of $C_0$ are not taken into account in this equation, i.e. if we write $C_0$ as

$$C_0 = \begin{pmatrix} C_{01} \\ C_{02} \end{pmatrix},$$

where $C_{01}, C_{02} \in \bar{F}^{2\times 4}$, then (17) is equivalent to

$$C_{02} \begin{pmatrix} d_0 \\ \vdots \\ d_3 \end{pmatrix} = \begin{pmatrix} 1 \\ 0 \end{pmatrix}. \tag{19}$$

Now, if there is a solution (for $c_i$'s and $d_j$'s) of (17) and (18) with $d_4 = 0$, then (18) implies $c_2 = c_0 = 0$, so that the lower half part of $C_0$ takes the form

$$C_{02} = \begin{pmatrix} c_3 & 0 & c_1 & 0 \\ c_3 & 0 & c_1 & 0 \end{pmatrix},$$

but then (19) yields a contradiction. If $d_4 \neq 0$ then (18) implies $c_3 = c_2$ and $c_1 = c_0$, and in this case $C_{02}$ becomes

$$C_{02} = \begin{pmatrix} 0 & 0 & 0 & 0 \\ c_3 & c_3 & c_1 & c_1 \end{pmatrix},$$

which will not satisfy (19) for any choice of $d_0, \ldots, d_4$. Therefore $3 \notin J(a, b)$.

## References


1. Çakçak, E.: A remark on the minimal polynomial of the product of linear recurring sequences. Finite Fields Appl. **4** (1998) 87–97
2. Göttfert, R., Niederreiter, H.: On the minimal polynomial of the product of linear recurring sequences. Finite Fields Appl. **1** (1995) 204–218
3. Lidl, R., Niederreiter, H.: Introduction to Finite Fields and Their Applications. revised ed., Cambridge Univ. Press, Cambridge 1994
4. McIntosh, R. J.: A generalization of a congruential property of Lucas. Amer. Math. Monthly **99** (1992) 231–238
5. Zierler, N., Mills, W.H.: Products of linear recurring sequences. J. of Algebra **27** (1973) 147–157

# Gelfond-Gramain's Theorem for Function Fields

Mireille Car

Laboratoire de Mathématiques
Faculté des Sciences et Techniques de Saint-Jérôme
Bât H. Poincaré
Avenue Escadrille Normandie-Niemen
13397 Marseille Cédex 20
France
Tel: 04 91 28 89 09 - Fax: 04 91 28 87 41
e-mail: mireille.car@math.u-3mrs.fr

**Abstract** A theorem of G. POLYA, c.f. [8], states that an entire function of order less than 1 or of order 1 and of type less than log 2 which takes integral values on the set N of natural integers is a polynomial. By interpolation on Gauss integers, A. GEL'FOND, c.f. [4], has obtained an analog result for entire functions mapping the ring of Gauss integers in itself. Further, Gel'fond's theorem was improved. The first improvements always used interpolation polynomials and did not give optimal results. For instance, we quote the work of L. GRUMAN, c.f. [6], or the work of D. MASSER, c.f. [7]. Using transcendental methods F. Gramain, c.f. [5], proved the best possible result for the ring of integers of an imaginary quadratic field. In what follows we give analog results for the ring $\mathbb{F}_q[T]$ of polynomials over a finite field, and in the case where $q$ is odd, for the ring $\mathbb{F}_q[T, \sqrt{D}]$, integral closure of $\mathbb{F}_q[T]$ in the quadratic extension $\mathbb{F}_q(T)(\sqrt{D})$. We do not get the best possible results. However, our method has the merit to provide results, even in the case of real extensions.

## 1 Definitions, Statement of the Results

Let $q$ be a power of a prime number. The infinite absolute value is defined on the rational function field $\mathbb{K} = \mathbb{F}_q(T)$ by

$$|A/B| = q^{\deg A - \deg B} \text{ if } A \text{ and } B \text{ are non-zero polynomials.}$$

Let $\mathbb{K}_\infty$ be the completion of $\mathbb{K}$ for the infinite absolute value. We use the same symbol $|\ |$ for the absolute value on the field $\mathbb{K}_\infty$. Let $\overline{\mathbb{K}_\infty}$ be an algebraic closure of the field $\mathbb{K}_\infty$. The absolute value $|\ |$ extends to an absolute value on the field $\overline{\mathbb{K}_\infty}$ still denoted by the same symbol $|\ |$. The field $\overline{\mathbb{K}_\infty}$ is not complete for $|\ |$. Let $\Omega$ be the completion of $\overline{\mathbb{K}_\infty}$ for $|\ |$ and let $|\ |$ denote the extension of the absolute value $|\ |$ of the field $\overline{\mathbb{K}_\infty}$ to the field $\Omega$. The field $\Omega$ is complete for $|\ |$, c.f. [9]. We note here that the absolute value $|\ |$ is ultrametric. The map deg is defined on the field $\Omega$ by:

$$|z| = q^{\deg z} \text{ if } z \neq 0, \deg 0 = -\infty.$$

This map takes real values on the non-zero elements of the field $\Omega$ and extends the degree map defined on the polynomial ring $\mathbb{F}_q[T]$.

**Definition 1.1.** An entire function on $\Omega$ is a power series

$$f(X) = \sum_{n=0}^{+\infty} b_n X^n$$

with coefficients in the field $\Omega$ and such that for any $\xi \in \Omega$, the series

$$\sum_{n=0}^{+\infty} b_n \xi^n$$

converges in $\Omega$.

For such an entire function, let, for any real number $r > 0$,

$$M(f,r) = \sup\{nr + \deg(b_n); n \in \mathbb{N}\}.$$

Polya's theorem for the ring $\mathbb{F}_q[T]$, c.f. [2], may be stated as follows.

**Theorem 1.2.** *Let $f$ be an entire function on $\Omega$ such that*

$$\overline{\lim_{r\to\infty}} \left(\frac{M(f,r)}{q^r}\right) < \frac{1}{\exp(1)\log(q)q^{q/(q-1)}}, \tag{1}$$

$$f(H) \in \mathbb{F}_q[T] \ for \ \ any \ H \in \mathbb{F}_q[T]. \tag{2}$$

*Then $f$ is a polynomial of the ring $\mathbb{F}_q(T)[X]$.*

We quote here a similar result.

**Theorem 1.3.** *Let $f$ be an entire function on $\Omega$ such that*

$$\overline{\lim_{r\to\infty}} \left(\frac{M(f,r)}{q^r}\right) < \frac{q^{q/(q-1)}}{\exp(1)\log(q)}, \tag{3}$$

$$f(H) = b \ for \ any \ H \in \mathbb{F}_q[T]. \tag{4}$$

*Then $f$ is the constant polynomial equal to $b$.*

The statement of the other theorems requires some definitions.

We suppose that $q$ is odd. A quadratic extension of the field $\mathbb{K}$ may be defined as follows. Let $D$ be a square-free polynomial in the ring $\mathbb{F}_q[T]$. Let $\mathbb{L} = \mathbb{K}(\sqrt{D})$ where $\sqrt{D}$ denotes one of the roots of the equation

$$X^2 - D = 0.$$

Then $\mathbb{B} = \mathbb{F}_q[T, \sqrt{D}]$ is the integral closure of the ring $\mathbb{F}_q[T]$ in the field $\mathbb{L}$.

It is easy to verify that the polynomial $X^2 - D$ has its roots in the field $\mathbb{K}_\infty$ if and only if $\deg D$ is even and $sgnD$ is a square in $\mathbb{F}_q$, $sgnD$ denoting the coefficient of the monomial of highest degree of $D$. In this case $\mathbb{L} = \mathbb{K}(\sqrt{D/sgnD})$ and we may suppose that $D$ is monic. Such an extension is a real extension. When $X^2 - D$ has no root in the field $\mathbb{K}_\infty$, the extension $\mathbb{L} = \mathbb{K}(\sqrt{D})$ is an imaginary extension.

In the case of imaginary quadratic extensions, we get the following four results.

**Theorem 1.4.** *Let $D \in \mathbb{F}_q[T]$ be square-free with an odd degree equal to $2g+1$. Let $f$ be an entire function on $\Omega$ such that*

$$\overline{\lim_{r\to\infty}} \left(\frac{M(f,r)}{q^{2r}}\right) < (2\exp(1)\log(q)q^{g+1/(q-1)})^{-1}, \tag{5}$$

$$f(\mathbb{B}) = \{b\}. \tag{6}$$

*Then $f$ is the constant polynomial equal to $b$.*

**Theorem 1.5.** *Let $D \in \mathbb{F}_q[T]$ be square-free with an odd degree equal to $2g+1$. Let $f$ be an entire function on $\Omega$ such that*

$$\overline{\lim_{r\to\infty}} \left(\frac{M(f,r)}{q^{2r}}\right) < (2\exp(1)\log(q)q^{g+(q+1)/(q-1)+3q^{2g+1}/2(q-1)})^{-1}, \tag{7}$$

$$f(\xi) \in \mathbb{B} \text{ for any } \xi \in \mathbb{B}. \tag{8}$$

*Then $f$ is a polynomial in the ring $\mathbb{L}[X]$.*

**Theorem 1.6.** *Let $D \in \mathbb{F}_q[T]$ be square-free with an even degree equal to $2g+2$ and such that $sgn(D)$ is not a square in $\mathbb{F}_q$. Let $f$ be an entire function on $\Omega$ such that*

$$\overline{\lim_{r\to\infty}} \left(\frac{M(f,r)}{q^{2r}}\right) < (2\exp(1)\log(q)q^{g+1+2/(q^2-1)})^{-1}, \tag{9}$$

$$f(\mathbb{B}) = \{b\}. \tag{10}$$

*Then $f$ is the constant polynomial equal to $b$.*

**Theorem 1.7.** *Let $D \in \mathbb{F}_q[T]$ be square-free with an even degree equal to $2g+2$ and such that $sgn(D)$ is not a square in $\mathbb{F}_q$. Let $f$ be an entire function on $\Omega$ such that*

$$\overline{\lim_{r\to\infty}} \left(\frac{M(f,r)}{q^{2r}}\right) < (2\exp(1)\log(q)q^{g+1+2(q^2+1)/(q^2-1)+5q^{2g+3}/2(q-1)})^{-1}, \tag{11}$$

$$f(\xi) \in \mathbb{B} \text{ for any } \xi \in \mathbb{B}. \tag{12}$$

*Then $f$ is a polynomial in the ring $\mathbb{L}[X]$.*

Now, we suppose that $\mathbb{L}$ is real. For every $\alpha \in \mathbb{L}$, let

$$h(\alpha) = \max(\deg(\alpha), \deg(\overline{\alpha})),$$

where $\overline{\alpha}$ denotes the conjugate of $\alpha$. If $f \in \mathbb{L}[[X]]$ is entire on $\Omega$, let

$$M^*(f,r) = \sup\{nr + h(b_n); n \in \mathbb{N}\}.$$

In the case of real quadratic extensions, we have the following two results.

**Theorem 1.8.** *Let $D \in \mathbb{F}_q[T]$ be monic and square-free with an even degree equal to $2g+2$. Let $f$ be an entire function on $\Omega$ such that*

$$\overline{\lim_{r\to\infty}}\left(\frac{M(f,r)}{q^{2r}}\right) < (2\exp(1)\log(q)q^{g+1+2(q^3+1)/(q^2-1)})^{-1}, \qquad (13)$$

$$f(\mathbb{B}) = \{b\}. \qquad (14)$$

*Then $f$ is the constant polynomial equal to $b$.*

**Theorem 1.9.** *Let $D \in \mathbb{F}_q[T]$ be monic and square-free with an even degree equal to $2g+2$. Let $f$ be an entire function on $\Omega$ such that*

$$\overline{\lim_{r\to\infty}}\left(\frac{M(f,r)}{q^{2r}}\right) < (2\exp(1)\log(q)q^{g+1+2(q^3+1)/(q^2-1)})^{-1}, \qquad (15)$$

$$f(\xi) \in \mathbb{B} \textit{ for any } \xi \in \mathbb{B}. \qquad (16)$$

*Then, $f \in \mathbb{L}[[X]]$. Moreover, if*

$$\overline{\lim_{r\to\infty}}\left(\frac{M^*(f,r)}{q^{2r}}\right) < (2\exp(1)\log(q)q^{g+1+2(q^2+1)/(q-1)+5q^{2g+3}/2(q-1)})^{-1}, \qquad (17)$$

*$f$ is a polynomial.*

## 2 Sketches of Proofs

If $f$ is an entire function, we set for $k = 1, 2$,

$$\tau_k(f) = \overline{\lim_{r\to\infty}}\left(\frac{M(f,r)}{q^{kr}}\right).$$

*Proof of theorem 2.* See [2].
*Proof of theorem 3.* It is sufficient to prove this theorem in the case where $b = 0$. This theorem has been proved in [2] for $\mathbb{F}_q$-linear functions. Alain

Thierry proved it in the general case. His proof is unpublished and runs as follows. Let $f$ be an entire function mapping the ring $\mathbb{A} = \mathbb{F}_q[T]$ on $\{0\}$. Then, $f$ admits an expansion as a Weierstrass product

$$f(X) = X^{n_0} \prod_{\substack{H \in \mathbb{A} \\ H \neq 0}} \left(1 - \frac{X}{H}\right)^{n_H},$$

where $n_H$ is the order of the zero $H$ of the function $f$, c.f. [9]. Then

$$f(X) = e(X)g(X),$$

where $e$ is the Carlitz exponential defined by

$$e(X) = X \prod_{\substack{H \in \mathbb{A} \\ H \neq 0}} \left(1 - \frac{X}{H}\right),$$

and where $g$ is an entire function. One may verify that if $g \neq 0$,

$$\tau_1(f) \geq \tau_1(g) + \tau_1(e).$$

According to [2],

$$\tau_1(e) = \frac{q^{q/(q-1)}}{\exp(1)\log(q)}.$$

Then, if $f \neq 0$,

$$\tau_1(f) \geq \frac{q^{q/(q-1)}}{\exp(1)\log(q)}.$$

*Proof of theorems 4, 5, 6, 7, 8 et 9.*. The proof runs in the same way for these six theorems. Computations are simpler in the case where $\deg D$ is odd. We do not deal with this case here. Hence, in the following we will suppose that $\deg D = 2g + 2$.

Let $i \longmapsto a_i$ be a bijective map from the set $\{0, 1, \ldots, q-1\}$ to the field $\mathbb{F}_q$ such that $a_0 = 0$. We associate to this bijection a bijective map $n \longmapsto \alpha_n$ from $\mathbb{N}$ to $\mathbb{B}$ defined as follows. For $s \leq g$, for any integer $n \in \{0, 1, \ldots, q-1\}$, let

$$\mu_{s,n} = a_n T^s. \tag{18}$$

For any integer $s > g$, for any $n \in \{0, 1, \ldots, q^2 - 1\}$, let

$$\mu_{s,n} = a_{n_1} T^s + a_{n_0} T^{s-g-1}\sqrt{D}, \tag{19}$$

where $n = n_1 q + n_0, 0 \leq n_1 < q, 0 \leq n_0 < q$.

Let $n$ be a positive integer. Then, $n$ admits a unique expansion as a sum

$$n = \sum_{i=0}^{g} n_i q^i + \sum_{i=g+1}^{\infty} n_i q^{2i-g-1} \tag{20}$$

with $0 \leq n_i < q$ for $i \in \{0, 1, \ldots, g\}$ and $0 \leq n_i < q^2$ for $i > g$. Let

$$\alpha_n = \sum_{i=0}^{\infty} \mu_{i,n_i}. \tag{21}$$

We note that

$$\alpha_0 = 0, \tag{22}$$

and that for $n > 0$,

$$h(\alpha_n) = \deg(\alpha_n) = s \text{ if } q^s \leq n < q^{s+1} \text{ with } s \leq g, \tag{23}$$

$$h(\alpha_n) = s \text{ if } q^{2s-g-1} \leq n < q^{2s+1-g} \text{ with } s > g. \tag{24}$$

We define the sequence of interpolation polynomials $(H_n(X))$ by

$$H_0(X) = 1, H_n(X) = \prod_{j=0}^{n-1} (X - \alpha_j), \text{ if } n \geq 1. \tag{25}$$

If $f$ is entire on $\Omega$, we define the sequence $(A_n(f))$ of interpolation coefficients of $f$ by

$$A_n(f) = \sum_{m=0}^{n} f(\alpha_m) \prod_{\substack{j=0 \\ j \neq 0}}^{n} (\alpha_m - \alpha_j)^{-1}. \tag{26}$$

Then, let us consider the formal series $\mathbb{I}(f)(X)$ defined by

$$\mathbb{I}(f)(X) = \sum_{n=0}^{+\infty} A_n(f) H_n(X). \tag{27}$$

**Definition 2.1.** Let $f \in \Omega[[X]]$ be entire on $\Omega$. We say that $f(X)$ admits an interpolation series expansion if for any $z \in \Omega$, the series $\mathbb{I}(f)(z)$ is convergent and if its sum is $f(z)$.

Let

$$f(X) = \sum_{n=0}^{+\infty} b_n X^n$$

be an entire function on $\Omega$. We set

$$\eta_{\mathbb{L}} = \begin{cases} 1 & \text{if } \mathbb{L} \text{ is real} \\ 0 & \text{if } \mathbb{L} \text{ is imaginary,} \end{cases} \tag{28}$$

$$\beta_{n,m} = \prod_{\substack{j=0 \\ j \neq m}}^{n} (\alpha_m - \alpha_j). \tag{29}$$

**Lemma 2.2.** *If*

$$\tau_2(f) < (2\exp(1)\log(q)q^{1+g+2(1+\eta_{\mathbb{L}}q^3)/(q^2-1)})^{-1}, \tag{30}$$

*then $f$ admits an interpolation series expansion.*

*Proof.* Let $z \in \Omega$. We get an upper bound for $\deg(H_n(z))$ with (23), (24) and (25). By (30) we get an upper bound for $\deg(b_n)$ for $n$ large enough. Then, again for $n$ large enough, with (24), (25) and (30), we get an upper bound for all

$$\deg\left(\sum_{k=n}^{+\infty} b_k\alpha_m^k\right), 0 \leq m \leq n.$$

More difficult computations give a lower bound for $\deg(\beta_{n,m})$. The results are different according as the extension is imaginary or real. It is the reason for which we have introduced the numbers $\eta_{\mathbb{L}}$. We get

$$\min(\deg(\beta_{n,m}), \deg(\overline{\beta_{n,m}})) \geq \tag{31}$$

$$s(n+2) - \left(\frac{q^2+1}{q^2-1}\right)(1+q\eta_{\mathbb{L}})q^{2s-g+1} + \frac{2q^{g+2}}{q^2-1}(\eta_{\mathbb{L}}-1) + \frac{2q}{q-1} - \kappa_{\mathbb{L},s},$$

where

$$\begin{cases} \kappa_{\mathbb{L},s} = 0 & \text{if } \mathbb{L} \text{ is imaginary,} \\ |\kappa_{\mathbb{L},s}| \leq 2(g+2)(q+1)q^{s-g-1} & \text{if } \mathbb{L} \text{ is real.} \end{cases} \tag{32}$$

Finally, for $n$ large enough, we get

$$2\deg(A_n(f^{(n)}) \leq \tag{33}$$

$$-n\log_q(n) + n\omega + 2(1+q\eta_{\mathbb{L}})\frac{q^2+1}{q^2-1}q^{2s-g-1} + 4(1-\eta_{\mathbb{L}})\frac{q^{g+2}}{q^2-1} + 2\kappa_{\mathbb{L},s},$$

where

$$f^{(n)}(X) = \sum_{k=n}^{+\infty} b_k X^k. \tag{34}$$

But, since $f - f^{(n)}$ is a polynomial of degree $n-1$, $A_n(f - f^{(n)}) = 0$, and

$$A_n(f) = A_n(f^{(n)}). \tag{35}$$

Thus, we may bound $\deg(A_n(f)H_n(z))$ for $n$ large enough and prove the convergence of the series $\mathbb{I}(f)(z)$ for any $z \in \Omega$. Afterwards, using similar arguments, we prove that for any $z \in \Omega$,

$$\mathbb{I}(f)(z) = f(z).$$

□

Now, we can prove theorems 6 and 8. Indeed, suppose that

$$\tau_2(f) < (2\exp(1)\log(q)q^{1+g+2(1+\eta_{\mathbb{L}})/(q^2-1)})^{-1}), \tag{36}$$

$$f(\mathbb{B}) = \{b\}. \tag{37}$$

Let $\phi = f - b$. Then, $\phi(\mathbb{B}) = \{0\}, \tau_2(\phi) = \tau_2(f)$. Condition (36) ensures that $\phi$ admits an interpolation series expansion, so, for any $z \in \Omega$, the series $\mathbb{I}(\phi)(z)$ is convergent and its sum is equal to $\phi(z)$. But, condition (37) together with definition (26) gives $A_n(\phi) = 0$ for any index $n$, whence, $\mathbb{I}(\phi) = 0$. Thus, $\phi$ is equal to 0 and $f$ is the constant polynomial equal to $b$.

Proving theorems 7 and 9 requires much work. We begin proving a lemma.

**Lemma 2.3.** *There exists a sequence* $(\xi_n)$ *of elements of* $\mathbb{B}$ *such that*

$$\xi_n \neq 0 \text{ for any } n, \tag{38}$$

$$\frac{\xi_n}{\beta_{n,m}} \in \mathbb{B} \text{ for any } n \text{ and for any } m \in \{0, 1, \ldots, n\}, \tag{39}$$

$$2h(\xi_n) \leq 2sn + q^{2s+g+1}\left(\frac{5q}{2(q-1)} + \frac{3gq}{2s(q-1)}\right) + \omega_n, \tag{40}$$

*if* $n \in [q^{2s-g-1}, q^{2s+g+1}[$ *with* $s \geq g+1$, *where* $\omega_n \in O(n^{1/2})$ *for* $n$ *tending to* $+\infty$, *the implied constants only depending on the field* $\mathbb{L}$.

*Proof.* Let

$$\mathfrak{B}_n = \bigcap_{m=0}^{n} \mathbb{B}\beta_{n,m}.$$

According to [1, proposition I.2.1], there exists $\xi_n \in \mathfrak{B}_n$ different from 0 and such that

$$q^{2h(\xi_n)} \leq N(\mathfrak{B}_n)q^{g+1}, \tag{41}$$

where $N(\mathfrak{B}_n)$ is the norm of the ideal $\mathfrak{B}_n$.

Such a $\xi_n$ verifies relations (38) and (39). In order to deduce (40) from (41), we need an upper bound for the numbers $N(\mathfrak{B}_n)$. The proof which gives this bound is too long to be summarized. It makes use of the Riemann-Roch theorem, c.f. [3] or [10], and Weil's theorem on the zeta function of the field $\mathbb{L}$, c.f. [10]. □

*Proof of theorem 7.* Let

$$\tau \in ]\tau_2(f), (2\exp(1)\log(q)q^{g+1+2(q^2+1)/(q^2-1)+5q^{2g+3}/2(q-1)})^{-1}[.$$

According to lemma 11, for any $x \in \Omega$,

$$f(x) = \mathbb{I}(f)(x), \tag{42}$$

where $\mathbb{I}(f)(x)$ is the convergent series

$$\mathbb{I}(f)(x) = \sum_{n=0}^{+\infty} A_n(f) H_n(x),$$

with

$$A_n(f) = \sum_{m=0}^{n} f(\alpha_m) \prod_{\substack{j=0 \\ j \neq m}}^{n} (\alpha_m - \alpha_j)^{-1} = \sum_{m=0}^{n} \frac{f(\alpha_m)}{\beta_{n,m}}. \tag{43}$$

Let $(\xi_n)$ be the sequence given by lemma 12. Since $f(\xi) \in \mathbb{B}$ for any $\xi \in \mathbb{B}$,

$$\xi_n A_n(f) \in \mathbb{B} \text{ for any index } n. \tag{44}$$

In view of (33), (35) and (40) we get

$$2\deg(A_n(f)\xi_n) \leq n\left(g+1+\omega+2\frac{q^2+1}{q^2-1}+5\frac{q^{2g+3}}{2(q-1)}+\frac{3gq^{2g+3}}{2s(q-1)}\right)+\Delta_n(f),$$

with $\omega = \log_q(2\exp(1)\tau\log(q))$, $\Delta_n(f) \in O(n^{1/2})$ for $n$ tending to $+\infty$, the implied constants only depending on the field $\mathbb{L}$. The choice of $\tau$ implies that

$$\left(g+1+\omega+2\frac{q^2+1}{q^2-1}+5\frac{q^{2g+3}}{2(q-1)}+\frac{3gq^{2g+3}}{2s(q-1)}\right) < 0.$$

Thus, the sequence $\deg(A_n(f)\xi_n)$ tends to $-\infty$ when $n$ tends to $+\infty$. Since $\mathbb{L}$ is an imaginary extension, 0 is the unique element of $\mathbb{B}$ with a negative degree. Hence, if $n$ is large enough, $A_n(f)\xi_n = 0$ and $A_n(f) = 0$. The series $\mathbb{I}(f)(X)$ is a polynomial. So is the series $f(X)$.

*Proof of theorem 9.* Now, $\mathbb{L}$ is real and $f \in \Omega[[X]]$ verifies the conditions (15) and (16). The proof of theorem 8 gives that

$$f(X) = \mathbb{I}(f)(X).$$

Since (16) is true, the coefficients $A_n(f)$ belong to the field $\mathbb{L}$ and $f \in \mathbb{L}[[X]]$. Now, we suppose that

$$\tau^*(f) < (2\exp(1)\log(q)q^{g+1+2(q^2+1)/(q-1)+5q^{2g+3}/2(q-1)})^{-1}, \tag{45}$$

where $\tau^*(f) = \overline{\lim_{r\to+\infty}}\left(\frac{M^*(f,r)}{q^{2r}}\right)$.

Let

$$\tau \in ]\tau^*(f), (2\exp(1)\log(q)q^{g+1+2(q^2+1)/(q-1)+5q^{2g+3}/2(q-1)})^{-1}[$$

and let $\omega = \log_q(2\exp(1)\tau\log(q))$.

We get an upper bound for the degree of $\overline{b_n}$ as we have bounded the degree of $b_n$. Thus, using relation (31), we get

$$2h(\overline{A_n(f)}) \leq \tag{46}$$

$$-n\log_q(n) + n\omega + 2\left(\frac{q^2+1}{q^2-1}\right)q^{2s-g-1} + 4(q-1)\frac{q^{g+1}}{q^2-1} + 2\kappa_{\mathbb{L},s},$$

with $\kappa_{\mathbb{L},s}$ defined by (32).

Let $(\xi_n)$ be the sequence given by lemma 12. Then,

$$2h(A_n(f)\xi_n) \leq n\left(g+1+\omega+2\frac{q^2+1}{q-1}+\frac{5q^{2g+3}}{2(q-1)}+\frac{3gq^{2g+3}}{2s(q-1)}\right) + \Delta_n(f),$$

with $\Delta_n(f) \in O(n^{1/2})$ for $n$ tending to $+\infty$, the implied constants depending only on the field $\mathbb{L}$. The choice of $\tau$ implies that

$$\left(g+1+\omega+2\frac{q^2+1}{q-1}+\frac{5q^{2g+3}}{2(q-1)}+\frac{3gq^{2g+3}}{2s(q-1)}\right) < 0.$$

So, the sequence $2h(A_n(f)\xi_n)$ tends to $-\infty$ when $n$ tends to $+\infty$, the sequence $\deg(N_{\mathbb{L}/K}(A_n(f)\xi_n)) = (\deg(A_n(f)\xi_n\overline{A_n(f)\xi_n}))$ tends to $-\infty$ when $n$ tends to $+\infty$. For $n$ large enough, $N_{\mathbb{L}/K}(A_n(f)\xi_n) = 0$. Since $\xi_n$ is different from 0, $A_n(f) = 0$. The series $\mathbb{I}(f)(X)$ is a polynomial. So is the series $f(X)$.

*Remarks.* In the case where $q$ is a power of 2, one may consider Artin-Schreier quadratic extensions instead of Kummer quadratic extensions. The method used to prove the Gel'fond-Gramain theorem in the case of Kummer quadratic extensions should work but the proof remains to be written carefully.

The technical nature of the method is the reason for which we have limited ourselves to quadratic extensions. The same ideas adapted to extensions of any degree should lead to too much complexity. Before dealing with the general case it would be worth finding a simpler method, even in the case of quadratic extensions.

## References

1. M. Car, Waring's problem in function fields. Proc. London Math. Soc. (3) 68 (1994), 1-30.
2. M. Car, Polya' s theorem for $\mathbb{F}_q[T]$. J. Number theory, vol 66, n°1 (1997), 148-171.

3. M. Deuring, Lectures on the theory of algebraic functions of one variable, Lecture Notes in Mathematics 314 (Springer, Berlin, 1973).
4. A. Gel'fond, Sur les propriétés arithmétiques des fonctions entières. Tôhoku Math.J., 30 (1929), 280-285.
5. F. Gramain, Sur le théorème de Fukasawa-Gel'fond. Invent. math. 63 (1981), 495-506.
6. L. Gruman, Propriétés arithmétiques des fonctions entières. Bull. Soc. Math. France 108, (1980), 421-440.
7. D. Masser, Sur les fonctions entières à valeurs entières, CRAS Paris, 291, série A, (1980), 1-4.
8. G. Polya, Über ganzwertige ganze Funktionen, Rend. Cir. Math. Palermo, 40 (1915), 1-16.
9. A. Robert. Elliptic curves, Lectures Notes in Mathematics 326 (Springer, Berlin,1973).
10. A. Weil, Basic Number Theory, 3rd edt (Springer, Berlin, 1974).

# On Generalized Bent and $q$-ary Perfect Nonlinear Functions

Claude Carlet and Sylvie Dubuc

GREYC
Université de Caen
France
e-mail: [Claude.Carlet,Sylvie.Dubuc]@info.unicaen.fr

**Abstract** The notion of bent function has been generalized by Kumar et al. to the alphabet $\mathbf{Z}_q = \mathbf{Z}/(q\mathbf{Z})$ and also studied by Nyberg. The classical equivalent definitions of binary bent functions lead, through this generalization, to the notions of generalized bent functions and of $q$-ary perfect nonlinear functions. We show in this paper that a $q$-ary function $f$ is perfect nonlinear if and only if, for every $u$ in $\mathbf{Z}_q^*$, the function $uf$ is generalized bent. We check that, among all known constructions of generalized bent functions, only one (due to Hou) can produce perfect nonlinear functions. This construction works for $n$ even ($n > 2$) and the question of knowing whether there exist perfect nonlinear functions for $n$ odd arises. We introduce a construction of perfect nonlinear functions on $\mathbb{Z}4^n$, for every $n > 1$.

## 1 Introduction

The notion of binary bent function, introduced by Rothaus [16] in the mid-1970's, has received much attention. It plays a role in at least three topics of computer science (cf. [3]): algebraic coding, cryptography and sequences. Bent functions are those Boolean functions on ${F_2}^n$ whose Hamming distance to the Reed-Muller code of order 1 (i.e. the set of all affine functions on ${F_2}^n$) is maximum. When $n$ is even, a Boolean function $f$ is bent if and only if the discrete Fourier transform of the real-valued function $\chi_f(x) = (-1)^{f(x)}$, i.e. the function $\widehat{\chi_f}(s) = \sum_{x\in {F_2}^n} (-1)^{f(x)+x\cdot s}$ where $x \cdot s$ denotes the usual inner product in ${F_2}^n$, has constant magnitude $2^{n/2}$. This is also equivalent to the fact that, for every nonzero word $s$ in ${F_2}^n$, the Boolean function $f(x)+f(x+s)$ is balanced (i.e. takes the values 0 and 1 equally often). This property is often called the *perfect nonlinearity* of $f$ (cf. [13]).

The notions of bentness and perfect nonlinearity have been generalized to residue class rings. Let $q$ be any positive integer. Let $\mathbf{Z}_q = \mathbf{Z}/q\mathbf{Z}$. Denote by $\theta_q$ the $q$-th root of unity $e^{\frac{2i\pi}{q}}$ (where $i = \sqrt{-1}$). For every complex-valued function $\varphi$ defined on $\mathbf{Z_q^n}$, the *discrete Fourier transform* of $\varphi$ is the complex-valued function $\widehat{\varphi}$ whose value at $u \in \mathbf{Z}_q^n$ is:

$$\widehat{\varphi}(u) = \sum_{x\in\mathbf{Z}_q^n} \varphi(x)\theta_q^{-u\cdot x}$$

where "$\cdot$" denotes the usual inner product in $\mathbf{Z}_q^n$.

Let $f$ be any function from $\mathbf{Z}_q^n$ to $\mathbf{Z}_q$ (i.e. *any q-ary function*); we denote by $\chi_f$ the complex-valued function whose value at $x \in \mathbf{Z}_q^n$ is $\chi_f(x) = \theta_q^{f(x)}$. We have then:

$$\widehat{\chi_f}(u) = \sum_{x \in \mathbf{Z}_q^n} \theta_q^{f(x) - u \cdot x}; \; u \in \mathbf{Z}_q^n.$$

**Definition 1.1.** [9] A $q$-ary function $f$ is called generalized bent if, for every $u \in \mathbf{Z}_q^n$, we have $| \widehat{\chi_f}(u) | = q^{\frac{n}{2}}$.

*Remark.* This definition does not depend of the choice of an inner product in $\mathbf{Z}_q^n$.

The original motivation for the introduction of this notion was in the design of sequences (cf. [9,5]). But the notion is also relevant to each of the topics quoted above.

Another generalization is related to the original definition of bent functions as "functions at maximum distance to the Reed-Muller code of order 1" and considers those functions whose Hamming distance to the generalized Reed-Muller code of order 1 is maximum. This leads to a different notion and very little is known about it (cf. [10]).

The third natural generalization is:

**Definition 1.2.** [14] A $q$-ary function $f$ is called perfect nonlinear if, for every nonzero $w \in \mathbf{Z}_q{}^n$, the $q$-ary function $f(x+w) - f(x)$ is balanced, i.e. takes every value in $\mathbf{Z}_q$ the same number of times (this number is then clearly equal to $q^{n-1}$).

This notion is cryptographically more interesting than the notion of generalized bent function because it is related to differental cryptanalysis [2].

A well-known property of Fourier transform (Wiener-Khintchine theorem) being that, given two complex-valued functions $\varphi$ and $\psi$, the Fourier transform of their *cross-correlation function*

$$c(\varphi, \psi)(w) = \sum_{x \in \mathbf{Z}_q^n} \varphi(x+w)\overline{\psi(x)}; \; w \in \mathbf{Z}_q^n$$

is equal to $\widehat{\varphi}(w) \times \overline{\widehat{\psi}(w)}$, we have:

**Proposition 1.3.** *[9]*

*Any q-ary function is generalized bent if and only if* $c(\chi_f, \chi_f)(w) = 0$ *for every* $w \neq 0$.

Thus, as shown in [14], every perfect nonlinear $q$-ary function is generalized bent (see Proposition 2.1 below for the proof of a more precise statement).

Other generalizations of bent functions have been given in [1] over a finite field and in [11] over a finite Abelian group.

## 2 Relationship between the Notions of Generalized Bentness and Perfect Nonlinearity

We first show that the two notions of generalized bent and perfect nonlinear functions are closely related to each other.

**Proposition 2.1.** *Let $f$ be any $q$-ary function.*
*$f$ is perfect nonlinear if and only if, for every nonzero $u \in \mathbf{Z}_q$, the $q$-ary function $uf$ is generalized bent.*

*Proof.* Assume first that $f$ is perfect nonlinear. For every nonzero $u, w \in \mathbf{Z}_q^n$ we have:

$$\begin{aligned} c(\chi_{uf}, \chi_{uf})(w) &= \sum_{x \in \mathbf{Z}_q{}^n} \theta_q^{uf(x+w)-uf(x)} \\ &= \sum_{x \in \mathbf{Z}_q^n} \theta_q^{u(f(x+w)-f(x))} \\ &= q^{n-1} \sum_{k \in \mathbf{Z}_q} \theta_q^{uk}. \end{aligned}$$

Since $\sum_{k \in \mathbf{Z}_q} \theta_q^{uk}$ equals 0 for any $u \neq 0$, we deduce that $uf$ is generalized bent, according to Proposition 1.3.

Conversely, assume that for every nonzero $u \in \mathbf{Z}_q$, the function $uf$ is generalized bent.

For every nonzero $w \in \mathbf{Z}_q^n$ define the integer-valued function:

$$k \in \mathbf{Z}_q \mapsto \phi_w(k) = \mid \{x \in \mathbf{Z}_q^n | f(x+w) - f(x) = k\} \mid .$$

Since $uf$ is generalized bent, $c(\chi_{uf}, \chi_{uf})(w)$ equals 0 for every nonzero $w \in \mathbf{Z}_q^n$, according to Proposition 1.3. We have:

$$\begin{aligned} c(\chi_{uf}, (\chi_{uf}))(w) &= \sum_{x \in \mathbf{Z}_q^n} \theta_q^{u(f(x+w)-f(x))} \\ &= \sum_{k \in \mathbf{Z}_q} \phi_w(k) \theta_q^{uk} \\ &= \widehat{\phi_w}(-u). \end{aligned}$$

Thus, the Fourier transform of $\phi_w$ being null except at 0, $\phi_w$ is constant, i.e. $f$ is perfect nonlinear. □

*Remark 2.2. A function $F$ defined from $\mathbf{Z}_q^n$ to $\mathbf{Z}_q{}^m$ $(n, m \geq 1)$ is perfect nonlinear if for every $w \in \mathbf{Z}_q^n$, $w \neq 0$, the difference $F(x+w) - F(x)$ obtains each value $y \in \mathbf{Z}_q^m$ for $q^{n-m}$ values of $x$ (cf. [14]).*

In [15], a characterization of these functions is given:

*A function $F : \mathbf{Z}_q^n \to \mathbf{Z}_q^m$ is perfect nonlinear if and only if for every $c \in \mathbf{Z}_q^m$, $c \neq 0$, the function $c \cdot F(x)$ is perfect nonlinear.*

But, this characterization is false when $q$ is not prime! Indeed, if we take $c = (k, \cdots, k)$, where $k > 1$ is a divisor of $q$, then the values of $c \cdot F(x)$ are all multiples of $k$. Thus, this function $c \cdot F(x)$ can not be perfect nonlinear. The correct statement is:

*A function $F : \mathbf{Z}_q^n \to \mathbf{Z}_q^m$ is perfect nonlinear if and only if for every $c \in \mathbf{Z}_q^m$, $c \neq 0$, the function $c \cdot F(x)$ is generalized bent.*

This can be easily proved by generalizing Proposition 2.1.

Proposition 2.1 implies the known result (cf. [14]) that, if $q$ is a prime, then the notions of generalized bent and perfect nonlinear functions are equivalent to each other. Indeed, Kumar et al. have shown in [9] that for every positive integer $q$, for every $q$-ary generalized bent function $f$ and for every $u$ co-prime with $q$, the $q$-ary function $uf$ is generalized bent as well.

When $q$ is not a prime, there exist generalized bent functions which are not perfect nonlinear:

**Proposition 2.3.** *If $q$ is not a prime, the constructions of generalized bent functions given in [9], [5] and [7] (cf. Theorems 1 and 3 of [9], Theorem 1 of [5], Theorem 4.2 of [7] and Property 3 of [9]) do not produce perfect nonlinear functions.*

*Proof.*

- Let $g$ be any function from $\mathbf{Z}_q^n$ to $\mathbf{Z}_q$ and $\pi$ any permutation on $\mathbf{Z}_q^n$. The function $f : \mathbf{Z}_q^n \times \mathbf{Z}_q^n \to \mathbf{Z}_q$ defined by:

$$f(x, y) = x \cdot \pi(y) + g(y), \ x, y \in \mathbf{Z}_q^n \tag{1}$$

  is generalized bent (cf. [9]).
  This construction does not lead to perfect nonlinear functions if $q$ is not a prime. Indeed, let $k$ $(1 < k < q)$ be a divisor of $q$. Denote by $w \in \mathbf{Z}_q^n$ the vector $(k, \cdots, k)$ and by $\pi_i(y)$ the $i$-th coordinate of $\pi(y)$. We have:

$$f(x+w, y) - f(x, y) = w \cdot \pi(y) = k \sum_{i=1}^{n} \pi_i(y).$$

  Thus, the values of function $f(x+w, y) - f(x, y)$ are all multiples of $k$. This function can not be balanced.

– Let $r$ be any integer greater than 1. Set $q = r^2$. Let $\pi$ be any permutation on $\mathbf{Z}_r^n$, $(n \geq 1)$ and $g$ any function from $\mathbf{Z}_r^n$ to $\mathbf{Z}_q$.
The function $f$ from $\mathbf{Z}_q^n$ to $\mathbf{Z}_q$ defined for every $(k_1, \cdots, k_n)$ and every $(k'_1, \cdots, k'_n) \in \mathbf{Z}_r{}^n$ by:

$$f(rk_1 + k'_1, \cdots, rk_n + k'_n) = r(k_1, \cdots, k_n) \cdot \pi(k'_1, \cdots, k'_n) + g(k'_1, \cdots, k'_n) \quad (2)$$

is generalized bent. This property is shown by Kumar et al. in [9] for $n = 1$. We check it now in general.
Notation:
For every $w = (w_1, \cdots, w_n) \in \mathbf{Z}_r{}^n$ and $w' = (w'_1, \cdots, w'_n) \in \mathbf{Z}_r{}^n$, we denote by $rw + w'$ the vector $(rw_1 + w'_1, \cdots, rw_n + w'_n) \in \mathbf{Z}_q^n$.
We have:

$$\begin{aligned}
\widehat{\chi_f}(rw + w') &= \sum_{(rk+k') \in \mathbf{Z}_q^n} \theta_q^{f(rk+k') - (rw+w')\cdot(rk+k')} \\
&= \sum_{k \in \mathbf{Z}_r^n} \sum_{k' \in \mathbf{Z}_r^n} \theta_q^{rk\cdot\pi(k') + g(k') - (rw+w')\cdot(rk+k')} \\
&= \sum_{k \in \mathbf{Z}_r^n} \sum_{k' \in \mathbf{Z}_r^n} \theta_q^{rk\cdot\pi(k') + g(k') - r^2 k\cdot w - rk\cdot w' - rk'\cdot w - k'\cdot w'}.
\end{aligned}$$

Since ${\theta_q}^r = \theta_r$ and ${\theta_q}^{r^2} = 1$, we have:

$$\widehat{\chi_f}(rw + w') = \sum_{k' \in \mathbf{Z}_r^n} \theta_q^{g(k') - rk'\cdot w - k'\cdot w'} \left( \sum_{k \in \mathbf{Z}_r^n} \theta_r^{k\cdot(\pi(k') - w')} \right).$$

The inner sum is null except when $\pi(k') = w'$, i.e. when $k' = \pi^{-1}(w')$. Thus:

$$\widehat{\chi_f}(rw + w') = r^n \theta_q^{g(\pi^{-1}(w')) - r\pi^{-1}(w')\cdot w - \pi^{-1}(w')\cdot w'}$$

and $f$ is generalized bent.
This construction does not lead to perfect nonlinear functions either, for similar reasons as above.

– Let $q_1$ and $q_2$ be two co-prime integers. Set $q = q_1 q_2$.
Let $f$ and $g$ be two generalized bent functions on resp. $Z_{q_1}{}^n$ and $Z_{q_2}{}^n$.
The function $h$ :

$$h(q_1 k_1 + q_2 k'_1, \cdots, q_1 k_n + q_2 k'_n) = q_2 f(k_1, \cdots, k_n) + q_1 g(k'_1, \cdots, k'_n)$$

is then generalized bent (cf. [5] for $n = 1$ and [7] for any $n$).
This construction never leads to perfect nonlinear functions. Indeed, let $w'$ be any nonzero element of $\mathbf{Z}_{q_2}{}^n$. We have:

$$\begin{aligned} h(q_2k + q_1k' + q_1w') - h(q_2k + q_1k') \\ &= h(q_2k + q_1(k' + w')) - h(q_2k + q_1k') \\ &= q_2f(k) + q_1g(k' + w') - q_2f(k) - q_1g(k') \\ &= q_1(g(k' + w') - g(k')) \end{aligned}$$

and $h(q_2k + q_1k' + q_1w') - h(q_2k + q_1k')$ cannot be balanced.

- The following secondary construction has been given in [9] as a property of generalized bent functions.
  Let $f$ and $g$ be two $q$-ary functions on resp. $\mathbf{Z}_q^n$ and $\mathbf{Z}_q^m$. Set

  $$h(x_1, \cdots, x_n, x_{n+1}, \cdots, x_{n+m}) = f(x_1, \cdots, x_n) + g(x_{n+1}, \cdots, x_{n+m}).$$

  Then $h$ is generalized bent on $\mathbf{Z}_q^{m+n}$ if and only if $f$ and $g$ are generalized bent (cf. [9]).
  According to Proposition 2.1, this implies that $h$ is also perfect nonlinear if and only if $f$ and $g$ are perfect nonlinear. But this construction needs to know some perfect nonlinear functions to deduce new ones.

□

# 3 Constructions of Perfect Nonlinear Functions.

The results of previous section suggest it is difficult to obtain perfect nonlinear functions. Fortunately, there exists one more construction of generalized bent functions , due to Hou, which can be easily specified to give perfect nonlinear functions.

Before we recount this construction, we recall briefly some background on Galois rings (Hou's construction deals slightly more generally with chain rings but we shall not need this generality).

## 3.1 Galois Rings

Let $p$ be a prime and $k$, $n$ two positive integers. The Galois ring $GR(p^k, n)$ is the Galois extension of degree $n$ of the ring $\mathbf{Z}_{p^k}$ (cf. [12]). There exists in $GR(p^k, n)$ a (non unique) primitive element $\xi$ of order $p^n - 1$. We denote by $\mathcal{T}_n$ the Teichmuller set $\{0, 1, \xi, \cdots, \xi^{p^n-2}\}$.

The elements of $GR(p^k, n)$ can be described in two forms:

- the multiplicative form:

$$x = \sum_{i=1}^{k} p^{i-1} u_i, \text{ where } u_i \in \mathcal{T}_n;$$

– the additive form:

$$x = \sum_{r=0}^{n-1} a_r \xi^r, \text{ where } a_r \in \mathbf{Z}_{p^k}.$$

The Frobenius automorphism $\varphi$ is the ring automorphism defined on $GR(p^k, n)$ by:

$$\varphi\left(\sum_{i=1}^{k} p^{i-1} u_i\right) = \sum_{i=1}^{k} p^{i-1} u_i^p.$$

The trace function $Tr$ is the additive function from $GR(p^k, n)$ to $\mathbf{Z}_{p^k}$:

$$Tr(x) = x + \varphi(x) + \cdots + \varphi^{n-1}(x).$$

The non-units of $GR(p^k, n)$ are the elements of the maximal ideal $pGR(p^k, n)$.

### 3.2 Hou's Construction

In [7] is given a construction of $q$-ary functions, where $q$ is the power of a prime: $q = p^k$ $(k > 1)$. For every $n \geq 1$, the $\mathbf{Z}_q$-module $\mathbf{Z}_q^n$ is identified with the Galois ring $R = GR(q, n)$ and the inner product $x \cdot y$ with $Tr(xy)$ where $Tr$ is the trace function from $GR(q, n)$ to $\mathbf{Z}_q$.

**Proposition 3.1.** *[7]*
*Let $a$ be any element in $\mathbf{Z}_q$ $(q = p^k)$. For every $0 \leq i \leq k-1$, let $\sigma_i$ be any function from the factor-ring $R/p^{k-i}R$ to $\mathbf{Z}_q$ such that*

$$\sum_{\beta \in \mathcal{T}_n} \theta_q^{\sigma_i(z + p^{k-i-1}\beta)} = 0, \text{ for every } z \in R/p^{k-i}R. \tag{3}$$

*Let $f : R \times R \to \mathbf{Z}_q$, be the function:*

$$f(x, y) = \begin{cases} a \text{ if } y = 0 \\ \sigma_i\left(\left[\frac{x}{y'}\right]\right) \text{ if } y = p^i y',\ y' \in R \backslash pR,\ 0 \leq i \leq k-1 \end{cases} \tag{4}$$

*where $\left[\frac{x}{y'}\right]$ is the image of $\frac{x}{y'}$ in the factor-ring $R/p^{k-i}R$.*
*Then $f$ is generalized bent.*

We deduce:

**Proposition 3.2.** *Let $a$ be any element in $\mathbf{Z}_q$. For every $0 \leq i \leq k-1$, let $\sigma_i$ be any function from $R/p^{k-i}R$ to $\mathbf{Z}_q$ such that, for every $z \in R/p^{k-i}R$, the function $\beta \in \mathcal{T}_n \mapsto \sigma_i(z + p^{k-i-1}\beta) \in \mathbf{Z}_q$ is balanced. Then function $f$ defined by (4) is perfect nonlinear.*

*Proof.* According to Proposition 2.1, $f$ is perfect nonlinear if and only if for every nonzero $u \in \mathbf{Z}_q$, function $uf$ is generalized bent.

Since the function $u\sigma_i$ clearly satisfies condition 3, thanks to the hypothesis of Proposition 3.2, $uf$ is generalized bent, according to Proposition 3.1. □

*Remark.* The condition on $\sigma_i$ can be satisfied if the cardinality $q = p^k$ of $\mathbf{Z}_q$ is a divisor of $|\mathcal{T}_n| = p^n$, i.e. if $k \leq n$, only.

Proposition 3.2 permits to construct perfect nonlinear functions on ${\mathbf{Z}_{p^k}}^{2n}$ (for $p$ prime and $n \geq k$). The constructions given in Proposition 2.3 show that there also exist generalized bent functions on $\mathbf{Z}_q^n$ for $n$ odd [1]. The question of knowing whether there exist perfect nonlinear functions for $n$ odd arises now.

### 3.3 Constructions of Perfect Nonlinear Functions on ${\mathbf{Z}_4}^n$ for Every $n$.

We shall need a result on character sums first proved in [20] and later precised in [4] as follows:

**Lemma 3.3.** *Let $n$ be any positive integer and $\mathcal{T}_n$ the Teichmuller set of $GR(4,n)$.*

*The number $\lambda = \sum_{x \in \mathcal{T}_n} i^{\,Tr(x)}$ is equal to $\pm 2^{\frac{n}{2}} e^{i\pi Tr(1)/4}$.*

*For any $a$ in $GR(4,n)$, the sum: $\sum_{x \in \mathcal{T}_n} i^{\,Tr(ax)}$ is equal to $2^n$ if $a = 0$, to $0$ if $a \in 2\mathcal{T}_n^*$ and to $\overline{\lambda} i^{\,Tr((1+u)^2)}$ if $a \in (1+2u)\mathcal{T}_n^*$, $u \in \mathcal{T}_n$.*

We can now prove:

**Proposition 3.4.** *Let $h$ be any function from the Teichmuller set $\mathcal{T}_n$ of $GR(4,n)$ to $\mathbf{Z}_4$ such that*

$$\Big| \sum_{x \in \mathcal{T}_n} i^{h(x) - Tr((2u-1)x)} \Big| = 2^{\frac{n}{2}} \text{ for every } u \in \mathcal{T}_n \tag{5}$$

$$\text{and } \sum_{x \in \mathcal{T}_n} i^{h(x)} = 0. \tag{6}$$

*Define the function $f : GR(4,n) \to \mathbf{Z}_4$ by:*

$$\forall a, b \in \mathcal{T}_n, f(a + 2b) = h(b\, a^{2^n - 2}). \tag{7}$$

*Then, $f$ is generalized bent.*

[1] but there are restrictions on ($q$,$n$), see [9].

*Proof.* :

Denote by $g(x)$ the function $h(x^{2^n-2})$. Let $a+2b$ be any element of $GR(4,n)$. We have:

$$\begin{aligned}
\widehat{\chi_f}(a+2b) &= \sum_{x\in\mathcal{T}_n}\sum_{y\in\mathcal{T}_n} i^{f(x+2y)-Tr((a+2b)(x+2y))} \\
&= \sum_{x\in\mathcal{T}_n}\sum_{y\in\mathcal{T}_n} i^{g(xy^{2^n-2})-Tr(ax+2ay+2bx)} \\
&= \sum_{x\in\mathcal{T}_n} i^{g(0)-Tr(x(a+2b))} + \sum_{x\in\mathcal{T}_n}\sum_{y\in\mathcal{T}_n{}^*} i^{g(xy^{2^n-2})-Tr(ax+2ay+2bx)} \\
&= \sum_{x\in\mathcal{T}_n} i^{g(0)-Tr(x(a+2b))} + \sum_{w\in\mathcal{T}_n}\sum_{y\in\mathcal{T}_n{}^*} i^{g(w)-Tr(awy+2ay+2bwy)} \\
&= \sum_{x\in\mathcal{T}_n} i^{g(0)-Tr(x(a+2b))} + \sum_{w\in\mathcal{T}_n}\sum_{y\in\mathcal{T}_n} i^{g(w)-Tr(awy+2ay+2bwy)} \\
&\quad - \sum_{w\in\mathcal{T}_n} i^{g(w)}.
\end{aligned}$$

Thanks to condition (6), we deduce:

$$\widehat{\chi_f}(a+2b) = \sum_{x\in\mathcal{T}_n} i^{g(0)-Tr(x(a+2b))} + \sum_{w\in\mathcal{T}_n}\sum_{y\in\mathcal{T}_n} i^{g(w)-Tr(awy+2ay+2bwy)}.$$

1. If $a=b=0$, then:

$$\widehat{\chi_f}(0) = 2^n i^{g(0)} + \sum_{w\in\mathcal{T}_n}\sum_{y\in\mathcal{T}_n} i^{g(w)} = 2^n i^{g(0)}.$$

2. If $a=0$ and $b\neq 0$, then:

$$\begin{aligned}
\widehat{\chi_f}(2b) &= \sum_{x\in\mathcal{T}_n} i^{g(0)-Tr(2bx)} + \sum_{w\in\mathcal{T}_n}\sum_{y\in\mathcal{T}_n} i^{g(w)-Tr(2bwy)} \\
&= \sum_{w\in\mathcal{T}_n} i^{g(w)}\Big(\sum_{y\in\mathcal{T}_n} i^{-Tr(2bwy)}\Big)
\end{aligned}$$

since, according to Lemma 3.3, we have $\sum_{x\in\mathcal{T}_n} i^{-Tr(x2b)} = 0$.
The sum $\sum_{y\in\mathcal{T}_n} i^{-Tr(2bwy)}$ is equal to 0, except if $w=0$, in which case it is equal to $2^n$. We deduce:

$$\widehat{\chi_f}(2b) = 2^n i^{g(0)}.$$

3. If $a\neq 0$, we have $a+2b\in(1+2u)\mathcal{T}_n{}^*$ for $u=\frac{b}{a}\in\mathcal{T}_n$. According to Lemma 3.3, we have:

$$\sum_{x\in\mathcal{T}_n} i^{Tr(x(a+2b))} = i^{Tr((1+\frac{b}{a})^2)}\overline{\lambda}.$$

Thus:

$$\widehat{\chi_f}(a+2b) = i^{g(0)-Tr((1+\frac{b}{a})^2)}\lambda + \sum_{w\in\mathcal{T}_n}\sum_{y\in\mathcal{T}_n} i^{g(w)-Tr(y(aw+2a+2bw))}.$$

If $w=0$, then $\sum_{y\in\mathcal{T}_n} i^{-Tr(y(aw+2a+2bw))} = \sum_{y\in\mathcal{T}_n} i^{-Tr(2ya)} = 0.$
If $w \neq 0$, then we have $aw+2a+2bw \in (1+2v)\mathcal{T}_n{}^*$ where $v = \frac{1}{w}+\frac{b}{a}$.
According to Lemma 3.3, we get :

$$\begin{aligned}\widehat{\chi_f}(a+2b) &= i^{g(0)-Tr((1+\frac{b}{a})^2)}\lambda + \sum_{w\in\mathcal{T}_n{}^*} i^{g(w)-Tr((1+\frac{1}{w}+\frac{b}{a})^2)}\lambda\\ &= i^{g(0)-Tr((1+\frac{b}{a})^2)}\lambda + \sum_{w\in\mathcal{T}_n{}^*} i^{h(w)-Tr((1+w+\frac{b}{a})^2)}\lambda\\ &= \sum_{w\in\mathcal{T}_n} i^{h(w)-Tr((1+w+\frac{b}{a})^2)}\lambda.\end{aligned}$$

Expanding $(1+w+\frac{b}{a})^2$ and using $Tr(u^2)=Tr(u)$, $\forall u\in\mathcal{T}_n$ we have:

$$Tr((1+w+\frac{b}{a})^2) = Tr(1-w-\frac{b}{a}+2w\frac{b}{a}).$$

Thus :

$$\begin{aligned}\widehat{\chi_f}(a+2b) &= \sum_{w\in\mathcal{T}_n} i^{h(w)-Tr(1-\frac{b}{a}+w(2\frac{b}{a}-1))}\lambda\\ &= i^{Tr(1-\frac{b}{a})}\sum_{w\in\mathcal{T}_n} i^{h(w)-Tr(w(2\frac{b}{a}-1))}\lambda\end{aligned}$$

and according to Lemma 3.3 and thanks to condition (5):

$$|\widehat{\chi_f}(a+2b)| = 2^{\frac{n}{2}}2^{\frac{n}{2}} = 2^n.$$

Thus $f$ is generalized bent. □

As a direct consequence of Propositions 2.1 and 3.4, we deduce:

**Proposition 3.5.** *Let $f$ be defined by (7).*
*If $h$ is balanced and if $h$ and $2h$ satisfy condition (5), then $f$ is perfect nonlinear.*

We shall give now a construction of functions $h$ satisfying the hypothesis of Proposition 3.5. We need some preliminary results.

$\mathcal{T}_n$ with the operation $x \oplus y = x + y + 2(xy)^{2^{n-1}}$ is an abelian group and an $F_2$-vector-space (cf. [19] and [20]).

The function $(x,y) \in \mathcal{T}_n{}^2 \mapsto x \cdot y = 2Tr(xy) \in 2\mathbf{Z}_2$ is bilinear and nondegenerate on $(\mathcal{T}_n, \oplus)$.

**Lemma 3.6.** *Let $b_1$, $b_2$ be any nonzero elements of $\mathcal{T}_n$ such that $Tr(b_1)$ is even and $Tr(b_1b_2)$ is odd. Denote by $P$ the plane $\{b_0 = 0, b_1, b_2, b_3 = b_1 \oplus b_2\}$ and $E$ its dual: $E = \{x \in \mathcal{T}_n \,|\, 2Tr(xb_1) = 2Tr(xb_2) = 0\}$. Then, $E \oplus P = \mathcal{T}_n$, and $|\, 1 + i^{Tr(b_1)} + i^{Tr(b_2)} + i^{Tr(b_1 \oplus b_2)} \,| = 2$. Moreover:*

$$Tr(b_3) = Tr(b_1) + Tr(b_2) + 2 \tag{8}$$

$$2Tr(b_1b_3) = 2Tr(b_1b_2) = 2 \tag{9}$$

$$2Tr(b_2b_3) = 2Tr(b_2) + 2. \tag{10}$$

*For every $u \in E$, $x \in E$, $b_j \in P$ and $b_k \in P$, we have*

$$Tr((2(u \oplus b_j) - 1)(x \oplus b_k)) = Tr((2u - 1)x) - Tr(b_k) + 2Tr(b_jb_k) \tag{11}$$

*and $\sum_{x \in E} i^{Tr((2u-1)x)}$ has magnitude $2^{\frac{n}{2}-1}$.*

*Proof.* The vectors $b_1$ and $b_2$ do not belong to $E$ because $Tr(b_1b_2)$ is odd. The vector $b_3$ does not belong to $E$ either, since:

$$2Tr((b_1 \oplus b_2)b_1) = 2Tr((b_1 + b_2)b_1) = 2Tr({b_1}^2) + 2tr(b_2b_1) = 2.$$

Thus, $E$ and $P$ have only 0 in common, which implies $E \oplus P = \mathcal{T}_n$.

The cosets $E$, $b_1 \oplus E$, $b_2 \oplus E$ and $b_1 \oplus b_2 \oplus E$ constitute a partition of $\mathcal{T}_n$.

Let us compute the magnitude of $1 + i^{Tr(b_1)} + i^{Tr(b_2)} + i^{Tr(b_3)}$. Relation (8) is straightforward and implies:

- if $Tr(b_1) = 0$ then $Tr(b_3) = Tr(b_2) + 2$ and $1 + i^{Tr(b_1)} + i^{Tr(b_2)} + i^{Tr(b_3)} = 2$.
- if $Tr(b_1) = 2$ then $Tr(b_3) = Tr(b_2)$ and $1 + i^{Tr(b_1)} + i^{Tr(b_2)} + i^{Tr(b_3)} = 2i^{Tr(b_2)}$.

In both cases we have $|\, 1 + i^{Tr(b_1)} + i^{Tr(b_2)} + i^{Tr(b_1 \oplus b_2)} \,| = 2$.

Relation (9) is straightforward. We have $2Tr(b_2b_3) = 2Tr({b_2}^2) + 2Tr(b_1b_2)$ and $Tr((2(u \oplus b_j) - 1)(x \oplus b_k)) = Tr((2(u + b_j) - 1)(x + b_k))$ (since $2Tr(xb_k) = 0$). Thus, relations (10) and (11) are checked.

According to (11) with $j = 0$, and to the fact that $E$, $b_1 \oplus E$, $b_2 \oplus E$ and $b_3 \oplus E$ constitue a partition of $\mathcal{T}_n$, we have for every $u \in E$:

$$\sum_{x \in \mathcal{T}_n} i^{Tr((2u-1)x)} = \sum_{x \in E} i^{Tr((2u-1)x)} (1 + i^{-Tr(b_1)} + i^{-Tr(b_2)} + i^{-Tr(b_3)}).$$

Thus $\left|\sum_{x\in E} i^{Tr((2u-1)x)}\right| = 2^{\frac{n}{2}-1}$, since, according to Lemma 3.3, we have $\left|\sum_{x\in \mathcal{T}_n} i^{Tr((2u-1)x)}\right| = 2^{n/2}$.

According to (11) with $k=0$, we have:

$$\sum_{x\in E} i^{Tr((2(u\oplus b_j)-1)x)} = \sum_{x\in E} i^{Tr((2u-1)x)}.$$

□

**Proposition 3.7.** *Let $a$ be any element of $\mathbf{Z}_4$. Under the hypothesis of Lemma 3.6, let $h : \mathcal{T} \to \mathbf{Z}_4$ be defined by*

- $h(x) = a\ \forall x \in E$.
- $h(x) = a+2\ \forall x \in b_1 \oplus E$.
- $h(x) = a+1,\ \forall x \in b_2 \oplus E$.
- $h(x) = a+3,\ \forall x \in b_3 \oplus E$.

*Then $h$ is balanced and $h$ and $2h$ satisfy condition (5).*

*Proof.* $h$ is clearly balanced.

According to (11) and since $E$, $b_1 \oplus E$, $b_2 \oplus E$ and $b_3 \oplus E$ constitue a partition of $\mathcal{T}_n$, we have for $u \in E$:

$$\sum_{x\in \mathcal{T}_n} i^{h(x)+Tr((2(u\oplus b_j)-1)x)} = \sum_{x\in E} i^{a+Tr((2u-1)x)}\Big(1 + i^{2+2Tr(b_1b_j)-Tr(b_1)} + i^{1+2Tr(b_2b_j)-Tr(b_2)} + i^{3+2Tr(b_3b_j)-Tr(b_3)}\Big)$$

and

$$\sum_{x\in \mathcal{T}_n} i^{2h(x)+Tr((2(u\oplus b_j)-1)x)} = \sum_{x\in E} i^{2a+Tr((2u-1)x)}\Big(1 + i^{2Tr(b_1b_j)-Tr(b_1)} + i^{2+2Tr(b_2b_j)-Tr(b_2)} + i^{2+2Tr(b_3b_j)-Tr(b_3)}\Big)$$

Set

$$S = 1 + i^{2+2Tr(b_1b_j)-Tr(b_1)} + i^{1+2Tr(b_2b_j)-Tr(b_2)} + i^{3+2Tr(b_3b_j)-Tr(b_3)}$$

and

$$S' = 1 + i^{2Tr(b_1b_j)-Tr(b_1)} + i^{2+2Tr(b_2b_j)-Tr(b_2)} + i^{2+2Tr(b_3b_j)-Tr(b_3)}.$$

Let us show that $S$ and $S'$ have magnitude 2. This will complete the proof.

Set $\alpha = i^{2Tr(b_ib_j)-Tr(b_1)}$ and $\beta = i^{2Tr(b_2b_j)-Tr(b_2)}$, we have :

$$i^{2Tr(b_3b_j)-Tr(b_3)} = i^{2Tr((b_1\oplus b_2)b_j)-Tr(b_1)-Tr(b_2)+2}$$
$$= i^{2Tr(b_1b_j)-Tr(b_1)+2Tr(b_2b_j)-Tr(b_2)+2} = -\alpha\beta.$$

Thus:

$$S = 1 + (i^2\alpha) + (i\beta) - (i^3\alpha\beta) = (1-\alpha) + i\beta(1+\alpha).$$

and :

$$S' = 1 + \alpha + (i^2\beta) - (i^2\alpha\beta) = (1+\alpha) + \beta(\alpha - 1).$$

Since $Tr(b_1)$ is even, $\alpha = 1$, $S = 2i\beta$, $S' = 2$ or $\alpha = -1$, $S = 2$, and $S' = -2\beta$.

Since $\beta$ has a magnitude of 1, $S$ and $S'$ have magnitude of 2 in both cases. □

**Conclusion**: We have shown that there exist $q$-ary perfect nonlinear functions on $\mathbf{Z}_q^n$:

- for every even $n > 2$ and every power $q$ of a prime,
- for $q = 4$ and for every $n > 1$.

It would be interesting to extend the construction induced by Propositions 3.5 and 3.7 to every power $q$ of a prime and every $n$. The difficult point is in the computation for every $q$,$n$ of the sum $\sum_{x\in\mathcal{T}_n} {\theta_q}^{Tr(x)}$, which is achievable for small values of $n$ (but not in general) thanks to the result given in [8].

The question "is there a value of $q$ and a value of $n$ for which there exist generalized bent functions but no perfect nonlinear functions" is not completely settled either.

## References


1. A. S. Ambrosimov, "Properties of bent functions of $q$-valued logic over finite fields", *Discrete Math. Appl.* vol. 4, no 4, pp. 341-350 (1994).
2. E. Biham and A. Shamir, "Differential cryptanalysis of DES-like cryptosystems", *CRYPTO'90 Advances in cryptology* Lectures Notes in Computer Science 537, pp. 2-21 (1991).
3. Brualdi, Huffman and Pless, "Handbook on Coding Theory", *Elsevier*, (1998).
4. C. Carlet, "One-weight $\mathbf{Z}_4$-linear codes", *Proceedings of "International conference on Coding Theory, Cryptography and Related Areas*, LNCS (1999), to appear.
5. H. Chung P. V. Kumar, "A new general construction for generalized bent functions", *IEEE Trans. on Inf. Theory*, vol. 35, no 1, pp. 206-209 (1989).

6. A. R. Hammons Jr., P. V. Kumar, A. R. Calderbank, N. J. A. Sloane and P. Solé, "The $Z_4$-linearity of Kerdock, Preparata, Goethals and related codes" *IEEE Trans. on Inf. Theory*, vol. 40, no 2, pp. 301-319 (1994)
7. X.D Hou "q-ary bent functions constructed from chain rings", *Finite Fields and their Applications 4*, pp. 55-61 (1998).
8. P. V. Kumar, T. Helleseth, A. R. Calderbank, "An upper bound for Weil Exponential Sums over Galois Rings and Applications", *IEEE Trans. on Inf. Theory*, vol. 41, no 2, pp. 456-468 (1995).
9. P. V. Kumar, R. A. Scholtz and L. R. Welch, "Generalized bent functions and their properties", *Journal of Combinatorial Theory*, Series A 40, pp. 90-107 (1985).
10. P. Langevin, "On generalized bent functions", *CISM Courses and Lectures 339 (Eurocode)*, pp. 147-157 (1992).
11. O.A. Logachev, A.A Salnikov and V.V Yashchenko, "Bent functions of a finite Abelian group", *Discrete Math. Appl.* vol. 7, no 6, pp. 547-564 (1997).
12. B.R. MacDonald, "Finite rings with identity", *Marcel Dekker*, NY, (1974).
13. W. Meier and O. Staffelbach "Nonlinearity criteria for cryptographic functions" *EUROCRYPT'89 Advances in Cryptology*, Lectures Notes in Computer Science 434, pp. 549-562 (1990).
14. K. Nyberg, "Constructions of bent functions and difference sets", *EUROCRYPT'90 Advances in Cryptology*, Lecture Notes in Computer Science 473, pp. 151-160 (1991)
15. K. Nyberg, "Perfect Nonlinear S-boxes", *EUROCRYPT' 91 Advances in Cryptology*, Lecture Notes in Computer Science 547, pp. 378-386 (1992)
16. O. S. Rothaus, "On bent functions", *Journal of Combinatorial Theory*, Series A 20, pp. 300-305 (1976).
17. F. J. Mac Williams and N. J. Sloane, "The theory of error-correcting codes", *MIT Press* , Cambridge, MA (1961).
18. P. Shankar, "On BCH codes over arbitrary integer rings", *IEEE Trans. on Inf. Theory*, vol. 25, no 4, pp. 480-483 (1979).
19. M. Yamada, "Distance-regular digraphs of Girth 4 over an extension ring of $\mathbf{Z}/4\mathbf{Z}$, *Graphs and combinatorics*, vol. 6, no 4, pp. 381-394 (1990).
20. K. Yang, T. Helleseth, P.V Kumar, A. Shanbhag, "On the Weight Hierarchy of Kerdock Codes Over $\mathbf{Z}_4$", *IEEE Trans. on Inf. Theory*, vol. 42, no 5, pp. 1587-1593 (1996).

# Divisible Designs Admitting, as an Automorphism Group, an Orthogonal Group or a Unitary Group

Cinzia Cerroni[1] and Ralph–Hardo Schulz[2]

[1] Dipartimento di Matematica ed Applicazioni
Università di Palermo
via Archirafi 34
I-90123 Palermo, Italy.
e-mail: cerroni@dipmat.math.unipa.it
[2] Mathematisches Institut
FB Mathematik und Informatik
Freie Universität Berlin
Arnimallee 3
D-14195 Berlin
Germany
e-mail:schulz@math.fu-berlin.de

**Abstract** We construct some divisible designs starting from a projective space. These divisible designs admit an orthogonal group or a unitary group as an automorphism group.

## 1 Introduction

A method to construct divisible designs by $t$-$R$-homogeneous ($t$-$R$-transitive) imprimitive permutation groups $(G, X, R)$ was introduced by A.G. Spera in [Sp1]. The divisible designs which are constructed by such a method admit a large automorphism group. Imprimitive groups which are 2-$R$-homogeneous (2-$R$-transitive) were obtained from translation planes that have a collineation group operating 2-transitively on the line at infinity.(See for example [SchSp1], [SchSp2], [CeSp1], [Ce1].) In this work we introduce a method to construct 2-$R$-transitive imprimitive permutation groups from the desarguesian 3-space, giving rise to the construction of divisible designs. In particular, we consider the affine space $\mathrm{AG}(3,q)$, with plane at infinity $E$ and its translation group $T$, and we choose in $E$ a set of lines $\Omega$ admitting a subgroup $G$ of the collineation group of $E$ which acts 2-transitively. We embed $G$ in $\mathrm{PGL}(4,q)$, and call the embedded group $\widetilde{G}$; then $T \cdot \widetilde{G}$ acts 2-$R$-transitively on $\Pi(\Omega) = \{\pi \text{ a plane of } \mathrm{AG}(3,q) \mid \pi \cap E \in \Omega\}$, if $R$ is the parallelism relation of $\mathrm{AG}(3,q)$. In sections ahead we choose first as $\Omega$ the tangent lines to a conic in $E$, and as $G$ the group $\mathrm{PGO}(3,q)$. Then we take as $\Omega$ the tangent lines to a unital in $E$ and as $G$ the group $\mathrm{PGU}(3,q)$. Thus, we obtain the following imprimitive permutation groups, $(T \cdot \mathrm{GO}(3,q), \Pi(\Omega), R)$ and

$(T \cdot \mathrm{GU}(3,q), \Pi(\Omega), R)$. In this way we construct divisible designs admitting, as an automorphism group, an orthogonal group and a unitary group, respectively. Moreover, in the special case of the conic when $q$ is odd we construct also 3-divisible designs admitting, as an automorphism group, an orthogonal group. We remark that the designs considered here are different from the PBIB designs gained from orthogonal geometries in [YaWei].

Notice that a divisible design is connected with a constant weight code and that the automorphism group of the design induces faithfully an automorphism group of the code, (cf. [Sch1], [SchSp3]). Hence our results can be translated into those on constant weight codes.

## 2 Preliminary Definitions and Results

Let $X$ be a set of size $v$ and $R$ an equivalence relation on $X$. We denote by $[x]$, with $x \in X$, the $R$-equivalence class containing $x$ and by $\mathcal{R}$ the set of all $R$-equivalence classes.

**Definition 2.1.** A subset $Y$ of $X$ is called $R$-transversal if $| Y \cap [x] | \leq 1$ for all $x \in X$.

Consider a family $\mathcal{B}$ of *R-transversal* subsets of $X$, and let $t, s, k, \lambda$ be positive integers with $t \leq k < v$. Then, one has the following definition.

**Definition 2.2.** A triple $D = (X, \mathcal{B}, \mathcal{R})$ is called $t$-$(s, k, \lambda_t)$-divisible design (or simply a $t$-divisible design) if

(i) $| [x] |= s$ for all $x \in X$;
(ii) $| B |= k$ for all $B \in \mathcal{B}$;
(iii) for every $R$-transversal $t$-subset $Y$ of $X$, there exist exactly $\lambda_t$ elements of $\mathcal{B}$ containing $Y$.

The elements of $X$ are called *points*, those of $\mathcal{B}$ *blocks*, and the elements of $\mathcal{R}$ *point classes*.

One can show that, for $t \geq 3$, a $t$-divisible design is as well a $(t-1)$-divisible design with

$$\lambda_{t-1} = \lambda_t(v - st + s)(k - t + 1)^{-1}.$$

Notice that, for $s = 1$, a $t$-divisible design is a $t$-design. If $t = 2$ we call a $t$-divisible design just a divisible design. If $D$ is a $(s, k, \lambda_2)$-divisible design, then the number of blocks through a given point is a constant number $r$ and

$$r(k-1) = \lambda_2(v - s) \quad \text{and} \quad bk = vr,$$

where $b$ denotes the number of blocks of $D$. For examples and background on divisible designs see for instance [BJL1] or [Hal].

**Definition 2.3.** Let $G$ be a group operating on the set $X$ and $R$ an equivalence relation on $X$ which is $G$-admissible, that is

$$x, y \in X \text{ and } xRy \text{ imply } x^g R y^g \text{ for all } g \in G,$$

then the triple $\Lambda = (G, X, R)$ is called an $R$-permutation group (see [Sp1]). The group $G$ induces a permutation group on $X$; in this paper, also nonfaithful representations will occur.

An $R-$permutation group $\Lambda$ is called *t-R-homogeneous* (and *t-R-transitive* respectively) if for every two $R$-transversal subsets of $X$ of size $t$, $S = \{x_1, ..., x_t\}$ and $S' = \{y_1, ..., y_t\}$, there exists an element $g \in G$ such that $S' = S^g$ (and $y_i = x_i{}^g$ for every $i = 1, 2, ..., t$ respectively). Notice that, if $G$ is *t-R-transitive* , then $G$ is *t-R-homogeneous.*

$t$-divisible designs are closely connected to $R$-permutation groups as the following proposition shows.

**Proposition 2.4.** [Sp1] *Let $\Lambda = (G, X, R)$ be a finite t-R-transitive R-permutation group, and let B be an R-transversal subset of X of size k, with $t \leq k < v :=| X |$. Then, the incidence structure $D(\Lambda, B) = (X, B^G, \mathcal{R})$, where $B^G = \{B^g \mid g \in G\}$, is a t-divisible design with the following parameters,*

$$b = \frac{| G |}{| G_B |},\ s = | [x] | \text{ for some } x \in X,\ k = | B | \text{ and}$$

$$\lambda_t = | G | \binom{k}{t} \left( | G_B | \binom{v/s}{t} s^t \right)^{-1};$$

*here $G_B$ denotes the setwise stabilizer of B. Moreover, $D(\Lambda, B)$ admits G as a point- and a block-transitive automorphism group.*

## 3 General Results

The following general theorem helps to find 2-$R$-transitive $R$-permutation groups. Consider the 3-dimensional projective space $\mathrm{PG}(3, q)$ over a finite field $K = \mathrm{GF}(q)$, with $q = p^e$, $p$ a prime, and fix, as plane at infinity, the plane $E := \{X_3 = 0\}$. If $G$ is a subgroup of the collineation group $\mathrm{PGL}(3, q)$ of $E$, then, $G \cong \bar{G}/\bar{G} \cap Z(\mathrm{GL}(3, q))$, where $\bar{G}$ is a suitable subgroup of $\mathrm{GL}(3, q)$ and $Z(\mathrm{GL}(3, q))$ is the centre of $\mathrm{GL}(3, q)$.

We embed $\bar{G}$ in $\mathrm{PGL}(4, q)$ as follows,

$$\widetilde{G} = \left\{ \begin{pmatrix} & & & 0 \\ & A & & 0 \\ & & & 0 \\ 0 & 0 & 0 & 1 \end{pmatrix} \mid A \in \bar{G} \right\};$$

notice that $\widetilde{G} \cap Z(\mathrm{GL}(4,q)) = \begin{pmatrix} 1&0&0&0 \\ 0&1&0&0 \\ 0&0&1&0 \\ 0&0&0&1 \end{pmatrix}$. So, $\widetilde{G}/\widetilde{G} \cap Z(\mathrm{GL}(4,q))$ is isomorphic to $\widetilde{G}$. Define $\mathrm{AG}(3,q)$ as the affine space with plane at infinity $E$ and translation group $T$.

**Proposition 3.1.** *Let $\Omega$ be a set of lines of $E$, $|\ \Omega\ | \geq 2$, and let $\Pi(\Omega)$ be the following set,*

$$\Pi(\Omega) = \{\pi \text{ a plane of } \mathrm{AG}(3,q) \mid \pi \cap E \in \Omega\}.$$

*(i) If $G$ is a subgroup of $\mathrm{PGL}(3,q)$ that acts 2-transitively on the set $\Omega$ then the group $(T \cdot \widetilde{G}, \Pi(\Omega), R)$ is a 2-$R$-transitive $R$-permutation group;*
*(ii) if $G$ is a subgroup of $\mathrm{PGL}(3,q)$ which acts 3-transitively on the set $\Omega$ and if there do not exist three concurrent lines in $\Omega$, then the group $(T \cdot \widetilde{G}, \Pi(\Omega), R)$ is a 3-$R$-transitive $R$-permutation group;*

*here $T$ and $\widetilde{G}$ are defined as above and $R$ denotes the parallelism relation.*

*Proof.* (i) Notice that the group $\widetilde{G}$ is still 2-transitive on $\Omega$ and respects the parallelism relation; let $\{\pi_1, \pi_2\}$ and $\{\pi_1', \pi_2'\}$ be two $R$-transversal 2-subsets of $\Pi(\Omega)$. Then, $\pi_1 \cap \pi_2 = \ell$ and $\pi_1' \cap \pi_2' = \widetilde{\ell}$, where $\widetilde{\ell}$ and $\ell$ are two affine lines. So, if we choose a point $P$ on $\ell$ and a point $P'$ on $\widetilde{\ell}$, there exists a $\tau$ in $T$ such that $P^\tau = P'$. Therefore, we can suppose that $\{\pi_1, \pi_2, \pi_1', \pi_2'\}$ intersect in $O = (0,0,0,1)K$. Moreover, $(T \cdot \widetilde{G})_O = (\widetilde{G})_O = \widetilde{G}$ and $\widetilde{G}$ is 2-transitive on $\Omega$. So, since through a point and a line exists a unique plane, the assertion follows.

(ii) Let $\{\pi_1, \pi_2, \pi_3\}$ and $\{\pi_1', \pi_2', \pi_3'\}$ be two $R$-transversal 3-subsets of $\Pi(\Omega)$; notice that, since $\pi_1 \cap E$, $\pi_2 \cap E$, $\pi_3 \cap E$ and $\pi_1' \cap E$, $\pi_2' \cap E$, $\pi_3' \cap E$ respectively form a triangle, then $\pi_1 \cap \pi_2 \cap \pi_3 = \{P\}$, where $P$ is an affine point. In fact, if $\pi_1 \cap \pi_2 = \{\ell\}$ and $\pi_3$ does not intersect with $\pi_1$ and $\pi_2$, then $\pi_3 \cap E$ will be concurrent with $\pi_1 \cap E$ and $\pi_2 \cap E$. So, we can suppose that $\pi_1$, $\pi_2$, $\pi_3$, $\pi_1'$, $\pi_2'$, $\pi_3'$ intersect in $O = (0,0,0,1)K$. Then, $(T \cdot \widetilde{G})_O = \widetilde{G}_O = \widetilde{G}$ and an element $g \in \widetilde{G}$ maps $\pi_1 \cap E$, $\pi_2 \cap E$, $\pi_3 \cap E$ onto $\pi_1' \cap E$, $\pi_2' \cap E$, $\pi_3' \cap E$. So the assertion follows.

□

*Remark 3.2.* (i) Let $(T \cdot \widetilde{G}, \Pi(\Omega), R)$ be the $R$-permutation group defined above. We show that $T \cdot \widetilde{G}$ acts faithfully on $\Pi(\Omega)$ if not all the lines of $\Omega$ do intersect in a point, and acts not faithfully if all the lines of $\Omega$ are concurrent. Suppose that not all the lines of $\Omega$ do intersect in a point, then every affine point is the intersection of three suitable planes of $\Pi(\Omega)$. Hence $T \cdot \widetilde{G}$ acts faithfully on $\Pi(\Omega)$. Now, suppose that all the lines of $\Omega$ intersect in a point. Then the sets of not parallel affine planes through the lines of $\Omega$ intersect in an affine line, and these affine lines are parallel. So, if $\tau$ is a translation in the direction of these lines and $\tau \neq id$ then $\pi^\tau = \pi$ for all $\pi \in \Pi(\Omega)$. Therefore, at least $\tau$ is in the kernel of the action.

(ii) Let $(H, X, R)$ be a $t$-$R$-transitive $R$-permutation group, and consider the homomorphism of the action on $X$, $\phi : H \longrightarrow \widetilde{H}$ with $\widetilde{H} \subseteq \mathrm{Sym}(X)$. Moreover, suppose that $\mathrm{Ker}\phi \neq id$ and suppose that, using the Proposition 2.4, we have constructed a $t$-$(s, k, \lambda)$ divisible design $\mathcal{D} = (X, B^H, \mathcal{R})$, with $b$ blocks. So, we have

$$\begin{aligned} b &= |\, H \,| \cdot |\, H_B \,|^{-1} = (|\, \widetilde{H} \,| \cdot |\, \mathrm{Ker}\phi \,|)(|\, \widetilde{H}_B \,| \cdot |\, \mathrm{Ker}\phi \,|)^{-1} \\ &= |\, \widetilde{H} \,| \cdot |\, \widetilde{H}_B \,|^{-1}, \end{aligned}$$

where $\widetilde{H} \cong H/\mathrm{Ker}\phi$. So, we can use alternatively $H$ or $H/\mathrm{Ker}\phi$ to determine the parameters of a divisible design. Notice that the automorphism group of $\mathcal{D}$ contains $\widetilde{H}$

*Remark 3.3.* Every $t$-$(s, k, \lambda_t)$-divisible design $\mathcal{D}$ with $n$ point classes corresponds to a code $\mathcal{C}$ having the following properties;

(i) $\mathcal{C} \subseteq (\mathbb{Z}_{s+1})^n$,
(ii) $\mathcal{C}$ has constant weight $k$ and
(iii) For any $t$ different positions $i_1, \ldots, i_t$ and for any elements $j_1, \ldots, j_t \in \mathbb{Z}_{s+1} \setminus \{0\}$ there exist exactly $\lambda_t$ words of $\mathcal{C}$ with coordinate $j_u$ at position $i_u$ for $u = 1, 2, \ldots, t$.

Since Aut$\mathcal{D}$ induces faithfully a subgroup of Aut $\mathcal{C}$ (cf. [Sch1], [SchSp3]) the methods used here also permit the construction of constant weight codes with large automorphism group.

## 4 The Construction

### A) The conic

Consider PG$(2, q)$, $q = p^e$, consider the non singular conic $\widetilde{\mathcal{C}}$ with equation $X_1^2 - X_0X_2 = 0$, and the group $G = \mathrm{PGO}(3, q)$ operating on it. We have the following theorem, see for instance [Hr1] p. 143, [HP1] p. 53 or [Hu1] 10.11 p. 240.

**Theorem 4.1.** PGO$(3, q)$ *on* $\widetilde{\mathcal{C}}$ *is permutation isomorphic to* PGL$(2, q)$ *on* PG$(1, q)$.

The isomorphism from PGL$(2, q)$ onto PGO$(3, q)$ is induced by

$$\sigma : \begin{bmatrix} a\, b \\ c\, d \end{bmatrix} \longrightarrow \begin{bmatrix} a^2 & 2ab & b^2 \\ ac & ad + bc & bd \\ c^2 & 2cd & d^2 \end{bmatrix}, \quad ad - bc \neq 0.$$

An immediate corollary of this theorem is the following,

**Corollary 4.2.** $\mathrm{PGO}(3,q)$ *acts sharply* 3*-transitively on the conic* $\widetilde{\mathcal{C}}$ *with equation* $X_1^2 - X_0X_2 = 0$.

Notice that, since $\mathrm{PGO}(3,q)$ acts 3-transitively on the points of $\widetilde{\mathcal{C}}$, it acts 3-transitively also on the tangent lines to the conic $\widetilde{\mathcal{C}}$. Consider $\mathrm{PG}(3,q)$ and embed the conic $\widetilde{\mathcal{C}}$ as follows,

$$\mathcal{C} : \begin{cases} X_1^2 - X_0X_2 = 0 \\ X_3 \qquad\quad\;\; = 0 \end{cases}.$$

Then the group $\widetilde{G} = \left\{ \begin{pmatrix} & & & 0 \\ & A & & 0 \\ & & & 0 \\ 0 & 0 & 0 & 1 \end{pmatrix} \mid A \in \mathrm{GO}(3,q) \right\}$ is isomorphic to $\mathrm{GO}(3,q)$ and acts 3-transitively on the tangent lines to $\mathcal{C}$. Therefore, by Proposition 3.1, the $R$-permutation group $(T\cdot\widetilde{G}, \Pi(\Omega), R)$, where $\Omega$ is the set of tangent lines to $\mathcal{C}$ in $E$, is at least 2-$R$-transitive.

**Theorem 4.3.** *Let* $q = p^e$, *where* $p$ *is a prime and let* $i$ *be an integer such that* $i \mid e$.

*(i) if* $q$ *is odd, there exists a* $2\text{-}\left(q, p^i+1, \frac{q(q-1)}{p^i-1}\right)$*-divisible design* $\mathcal{D}$ *with* $q^2+q$ *points having as a point- and a block-transitive automorphism group* $T\cdot\widetilde{G}$ *with* $\widetilde{G} \cong \mathrm{GO}(3,q)$;

*(ii) if* $q$ *is even, there exists a* $2\text{-}\left(q, p^i+1, \frac{q-1}{p^i-1}\right)$*-divisible design with* $q^2+q$ *points having, as a point- and a block-transitive automorphism group* $\widetilde{G}\cdot(T/\mathrm{Ker}\phi)$, *with* $\widetilde{G} \cong \mathrm{GO}(3,q)$, *where* $\mathrm{Ker}\phi$ *is the kernel of the action on* $\Pi(\Omega)$, *with* $\Omega$ *the set of tangent lines to* $\mathcal{C}$ *in* $E$.

*Moreover, if* $q$ *is odd the design* $\mathcal{D}$ *constructed here is in fact a* $3\text{-}(q, p^i+1, 1)$*-divisible design with a point- and a block- transitive automorphism group.*

*Proof.* (i) Let $q$ be odd, let $\mathrm{PG}(3,q)$ be the projective space over $K = \mathrm{GF}(q)$, let $\mathcal{C}$ be the conic defined above and let $\Omega$ be the set of tangent lines of $\mathcal{C}$. Consider the $R$-permutation group $(T\cdot\widetilde{G}, \Pi(\Omega), R)$ and notice that, by (*ii*) of Proposition 3.1 and by (*i*) of Remarks 3.2, it is 3-$R$-transitive and its action is faithful. We construct a 3-divisible design and hence a divisible design using Proposition 2.4. Therefore, we choose an $R$-transversal subset of $\Pi(\Omega)$, namely $B$, as follows;

$$B := \Pi_O(\Omega') = \{\pi \text{ a plane of } \mathrm{AG}(3,q) \mid \pi \cap E \in \Omega' \\ \text{and } O = (0,0,0,1)K \in \pi\},$$

where

$$\Omega' = \{\ell \in \Omega \mid \text{ tangent point on } \ell \text{ is } (1,0,0,0)K \\ \text{or } (t^2,t,1,0)K,\ t \in \mathrm{GF}(p^i)\}.$$

Now, we compute the stabilizer $(T\cdot\widetilde{G})_B$ of $B$. Notice that $(T\cdot\widetilde{G})_B = \widetilde{G}_{\Omega'}$, since $T$ is regular on the affine points, $\cap\{\pi \mid \pi \in B\} = O$ and $\widetilde{G}$ fixes $O = (0,0,0,1)K$.
Furthermore, $\widetilde{G} \cong \mathrm{GO}(3,q)$ and

$$\mathrm{PGO}(3,q)_{\Omega'} \cong (\mathrm{GO}(3,q)/Z(\mathrm{GO}(3,q)))_{\Omega'} \cong \mathrm{GO}(3,q)_{\Omega'}/Z(\mathrm{GO}(3,q)).$$

Hence it is sufficient to compute $|\mathrm{PGO}(3,q)_{\Omega'}|$. Moreover, $|\mathrm{PGO}(3,q)_{\Omega'}| = |\mathrm{PGO}(3,q)_{C'}|$, where $C'$ is the set of tangent points on the lines of $\Omega'$. If $g \in \mathrm{PGO}(3,q)$ fixes $C'$, then $g$ maps three points $P_1$, $P_2$, $P_3$ of $C'$ on to three points $P_1'$, $P_2'$, $P_3'$ of $C'$. Furthermore, there exists an element $g' \in \mathrm{PGO}(3,p^i)$ doing this and fixing $C'$. Therefore, $g = g'$ since $\mathrm{PGO}(3,q)$ is 3-regular. Then, $|(T\cdot\widetilde{G})_B| = |\mathrm{PGO}(3,p^i)| \cdot |Z(\mathrm{GO}(3,q))| = p^i(p^{2i}-1)2$. Therefore, by Proposition 2.4, the parameters of the 3-divisible design $\mathcal{D} = (\Pi(\Omega), B^{T\cdot\widetilde{G}}, \mathcal{R})$, with $\widetilde{G} \cong \mathrm{GO}(3,q)$, are the following

$$v = |\Pi(\Omega)| = q(q+1),\ s = q,\ k = |B| = p^i + 1 \text{ and}$$
$$\lambda_3 = \frac{|T||\widetilde{G}|}{|(T\widetilde{G})_B|}\frac{\binom{k}{3}}{\binom{v/s}{3}s^3} = \frac{q^3q(q^2-1)2}{p^i(p^{2i}-1)2}\frac{\binom{p^i+1}{3}}{\binom{q(q+1)/q}{3}q^3} = 1$$

And the parameters of $\mathcal{D}$ as 2-divisible design are

$$v = q(q+1),\ s = q,\ k = p^i + 1 \text{ and } \lambda_2 = \frac{(v-2s)\lambda_3}{k-2} = \frac{q(q-1)}{p^i-1}.$$

Notice that an automorphism group of these divisible designs is $T\cdot\widetilde{G}$, with $\widetilde{G} \cong \mathrm{GO}(3,q)$, and the order of this group is $2(q^2-1)q^4$.

(ii) As in the case (*i*) we consider the conic $C$ and the now 2 -$R$ - transitive $R$-permutation group $(T\cdot\widetilde{G}, \Pi(\Omega), R)$. It is known that the lines of $\Omega$ intersect in the nucleus of the conic, $N = (0,1,0,0)K$, so by (*i*) of Remarks 3.2, the action of this group is not faithful. We want to find the kernel of the action. Notice that $\widetilde{G} \cong \mathrm{GO}(3,q)$ acts faithfully on $\Pi(\Omega)$. In fact, suppose that there exists $\tilde{g} \in \widetilde{G}$ fixing the planes of $\Pi(\Omega)$, then $\tilde{g}$ induces an homology, $\bar{g}$, in $E$ with centre $N$. But $\bar{g}$ has to fix all the points of the conic, since it fixes the tangent lines to the conic. Therefore, it is the identity. Now, let $\tau \in T$ and $g \in \widetilde{G}$, with $\tau g \in \mathrm{Ker}\phi$; then $\tau$ fixes (as $g$) the line $ON$; hence, by Remarks 3.2, $\tau \in \mathrm{Ker}\phi$ and therefore $g = id$. So, only the translations in direction of the line $ON$ form the kernel. Also in this case, we construct a divisible design using Proposition 2.4. We choose the same $R$-transversal subset as in case (*i*), $B = \Pi_O(\Omega')$, and we compute the stabilizer. Then, $|\mathrm{PGO}(3,q)_{\Omega'}|$ is the same as in the case (*i*). Therefore, the 2-divisible design $(\Pi(\Omega), B^{T\cdot\widetilde{G}}, \mathcal{R})$ has got the following parameters;

$v = q(q+1)$, $s = q$, $k = p^i + 1$ and finally, by Remark 3.2 (*ii*),

$$\lambda_2 = \frac{|\, T \cdot \widetilde{G} \,| / |\, \mathrm{Ker}\phi \,|}{|\, (T \cdot \widetilde{G})_B \,| / |\, \mathrm{Ker}\phi \,|} \frac{k(k-1)}{v(v-s)} = \frac{q^3 q(q^2-1)}{qp^i(p^{2i}-1)} \frac{p^i(p^i+1)}{q(q+1)(q^2+q-q)}$$
$$= \frac{q-1}{p^i-1},$$

and has, as a point- and block-transitive automorphism group, the group $\widetilde{G} \cdot (T/\mathrm{Ker}\phi)$, with $\widetilde{G} \cong \mathrm{GO}(3,q)$, of order $q^3(q^2-1)$.

□

**Theorem 4.4.** *Let $q = p^e$, with $p$ an odd prime. Then, there exist a 2-$(q,2,1)$-divisible design and a 3-$(q,3,1)$-divisible design with $q(q+1)$ points each. Both of them admit as a point- and block-transitive automorphism group, $T\widetilde{G}$ with $\widetilde{G} \cong \mathrm{GO}(3,q)$.*

*Proof.* Let $(T \cdot \widetilde{G}, \Pi(\Omega), R)$, where $\widetilde{G} \cong \mathrm{GO}(3,q)$ and $\Omega$ is the set of tangent lines to $\mathcal{C}$, be the 3-$R$-transitive permutation group defined above. Again, we use the Proposition 2.4 to construct a divisible design. In particular, if $\ell_1$ and $\ell_2$ are two fixed tangent lines to $\mathcal{C}$ we choose the $R$-transversal subset $B := \{\pi_1, \pi_2\}$ of $\Pi(\Omega)$ where $\pi_1, \pi_2$ are the two planes of $\mathrm{AG}(3,q)$ through $O = (0,0,0,1)K$ with $\pi_1 \cap E = \ell_1$ and $\pi_2 \cap E = \ell_2$. We compute $(T \cdot \widetilde{G})_B$. Notice that, if $\{P\} = \ell_1 \cap \ell_2$, the translations in the direction of the affine line through $O$ and $P$ fix $B$. We denote this subgroup of $T$ by $\widetilde{T}$. Then, $\widetilde{T} \subseteq (T \cdot \widetilde{G})_B$, and $|\, \widetilde{T} \,| = q$. Moreover, the group $\mathrm{GO}(3,q)_{\{\ell_1,\ell_2\}}$ fixes $B$; note that $\mathrm{PGO}(3,q)_{\{\ell_1,\ell_2\}} = \mathrm{GO}(3,q)_{\{\ell_1,\ell_2\}}/Z(\mathrm{GO}(3,q))$ and $\mathrm{PGO}(3,q)_{\{\ell_1,\ell_2\}} = \mathrm{PGO}(3,q)_P$. In general, $\mathrm{PGO}(3,q)$ acts transitively on the external points of a conic $X_1^2 - X_0X_2 = 0$ in $\mathrm{PG}(2,q)$; in fact the length of the orbit of the point $(0,1,0)K$ is $\frac{q(q+1)}{2}$. Therefore, $|\, \mathrm{PGO}(3,q)_P \,| = 2(q-1)$, and consequently $|\, \mathrm{GO}(3,q)_{\{\ell_1,\ell_2\}} \,| = 2(q-1) \,|\, Z(\mathrm{GO}(3,q)) \,| = 2(q-1)2$. Then, $\widetilde{T} \cdot \mathrm{GO}(3,q)_{\{\ell_1,\ell_2\}} \subseteq (T \cdot \widetilde{G})_B$. Now we show that $\widetilde{T} \cdot \mathrm{GO}(3,q)_{\{\ell_1,\ell_2\}} = (T \cdot \widetilde{G})_B$. Let $\tau g \in (T \cdot \widetilde{G})_B$. Since $\mathrm{GO}(3,q)$ operates 3-transitively on the set of tangent lines of $\mathcal{C}$ the group $\widetilde{T} \cdot \mathrm{GO}(3,q)_{\{\ell_1,\ell_2\}}$ contains elements interchanging $\pi_1$ and $\pi_2$. By norming the element one may assume that $\tau g$ fixes $\pi_1$ and $\pi_2$. Then ${\pi_1}^{\tau g} = {\pi_1}^{g\tau_1} = \pi_1$ for a suitable $\tau_1 \in T$. Since the translations fix $E$ pointwise, ${\pi_1}^g$ is a plane through $O, P$ and $\ell_1$; thus ${\pi_1}^g = \pi_1$, similarly ${\pi_2}^g = \pi_2$ and consequently $g \in \mathrm{GO}(3,q)_{\{\ell_1,\ell_2\}}$ and $\tau_1 \in \widetilde{T}$. So, we obtain a divisible design with the following parameters,

$v = q(q+1)$, $s = q$, $k = 2$ and

$$\lambda_2 = \frac{|\, T \cdot \mathrm{GO}(3,q) \,|}{|\, (T \cdot \mathrm{GO}(3,q))_B \,|} \frac{k(k-1)}{v(v-s)} = \frac{q^3 q(q^2-1)2}{2q(q-1)2} \frac{2}{q(q+1)[q(q+1)-q]} = 1.$$

An automorphism group of this divisible design is $T \cdot \widetilde{G}$ with $\widetilde{G} \cong \mathrm{GO}(3,q)$.

Again, consider the 3-$R$-transitive $R$-permutation group $(T \cdot \widetilde{G}, \Pi(\Omega), R)$. Let $\ell_1$, $\ell_2$, $\ell_3$ be three fixed tangent lines to $\mathcal{C}$. Choose, as an $R$-transversal subset of $\Pi(\Omega)$, $B := \{\pi_1, \pi_2, \pi_3\}$, where $\pi_1$, $\pi_2$, $\pi_3$ are three planes of AG$(3,q)$ through $O = (0,0,0,1)K$ with $\pi_1 \cap E = \ell_1$, $\pi_2 \cap E = \ell_2$ and $\pi_3 \cap E = \ell_3$. We compute $(T \cdot \widetilde{G})_B$. Notice that $(T \cdot \widetilde{G})_B = \widetilde{G}_B = \widetilde{G}_{\{\pi_1,\pi_2,\pi_3\}} = \widetilde{G}_{\{\ell_1,\ell_2,\ell_3\}}$ and $\mathrm{PGO}(3,q)_{\{\ell_1,\ell_2,\ell_3\}} \cong \mathrm{GO}(3,q)_{\{\ell_1,\ell_2,\ell_3\}}/Z(\mathrm{GO}(3,q))$. If $P_1$, $P_2$, $P_3$ are the tangent points on $\ell_1$, $\ell_2$, $\ell_3$, respectively then $\mathrm{PGO}(3,q)_{\{\ell_1,\ell_2,\ell_3\}} = \mathrm{PGO}(3,q)_{\{P_1,P_2,P_3\}}$. Moreover, PGO$(3,q)$ acts sharply 3-transitively on the points of the conic, so the pointwise stabilizer $\mathrm{PGO}(3,q)_{P_1,P_2,P_3} = id$. Therefore, only the permutations on three elements are in $\mathrm{PGO}(3,q)_{\{P_1,P_2,P_3\}}$. So $\mathrm{PGO}(3,q)_{\{P_1,P_2,P_3\}} \cong \mathrm{S}_3$ and $|(T \cdot \widetilde{G})_B| = |\mathrm{S}_3| \cdot |Z(\mathrm{GO}(3,q))| = 12$. Hence, the 3-divisible design $(\Pi(\Omega), B^{T \cdot \widetilde{G}}, \mathcal{R})$ has the following parameters

$$v = q(q+1),\ s = q,\ k = 3,$$

$$\lambda_3 = \frac{|T \cdot \mathrm{GO}(3,q)|}{|(T \cdot \mathrm{GO}(3,q))_B|} \frac{\binom{k}{3}}{\binom{v/s}{3} s^3} = \frac{q^3 q(q^2-1)2}{12} \frac{1}{q^3 q(q+1)(q-1)/3!} = 1,$$

$$\lambda_2 = q(q-1).$$

The constructed 3-divisible design admits $T \cdot \widetilde{G}$, with $\widetilde{G} \cong \mathrm{GO}(3,q)$ as a point- and block- transitive automorphism group. □

Notice that the constructed 2-divisible design is a complete multipartite graph on $(q+1) \times q$ points.

### B) The unital

Let PG$(3,q^2)$, $q = p^e$, be the projective space over $K = \mathrm{GF}(q^2)$ and let $E$ be the plane at infinity defined above. Then, $E$ is a PG$(2,q^2)$. Consider, in PG$(3,q^2)$, the following unital (Hermitian curve), see [Hr1] p.146;

$$\mathcal{U} : \begin{cases} X_0^{q+1} + X_1^{q+1} + X_2^{q+1} & = 0 \\ X_3 & = 0. \end{cases}$$

The group PGU$(3,q^2)$ acts on PG$(2,q^2)$ fixing the set of points with equation $X_0^{q+1} + X_1^{q+1} + X_2^{q+1} = 0$ and is 2-transitive on these points, (see [Lül] p. 156). Now, we embed GU$(3,q^2)$ in the collineation group PGL$(4,q^2)$ of PG$(3,q^2)$, as follows,

$$\widetilde{G} = \left\{ \begin{pmatrix} & & & 0 \\ & A & & 0 \\ & & & 0 \\ 0 & 0 & 0 & 1 \end{pmatrix} \mid A \in \mathrm{GU}(3,q^2) \right\}.$$

The group $\widetilde{G}$ is isomorphic to GU$(3,q^2)$ and fixes the plane $E$. It is known that $\mathcal{U}$ consists of $q^3+1$ points and that, through every point of it, there exist

$q^2$ secant lines and 1 tangent line in $E$, (see [Hr1] p.147). So, we consider the set $\Omega$ of the $q^3+1$ tangent lines to $\mathcal{U}$. By Proposition 3.1, the $R$-permutation group $(T\cdot\widetilde{G}, \Pi(\Omega), R)$ is 2-$R$-transitive. Notice that the tangent lines to $\mathcal{U}$ do not all intersect in a point (see [Hr1] p.147), so, by (*i*) of Remarks 3.2, the action of this group is faithful.

**Theorem 4.5.** *Let $q = p^e$, where $p$ is a prime. Then there exist $(q^2, 2, 1)$-, $(q^2, q^3, q^2(q^3-1))$- and $(q^2, q+1, 1)$-divisible designs with $q^2(q^3+1)$ points each admitting $T\cdot\widetilde{G}$ with $\widetilde{G}\cong \mathrm{GU}(3,q^2)$ as a point- and block-transitive automorphism group; moreover if $q$ is odd there exists a $(q^2, q+1, q^2(q+1))$-divisible design with $q^2(q^3+1)$ points and admitting $T\cdot\widetilde{G}$ with $\widetilde{G}\cong\mathrm{GU}(3,q^2)$ as a point- and block-transitive automorphism group.*

*Proof.* (i) Consider $\mathrm{PG}(3,q^2)$, the projective space over $K = \mathrm{GF}(q^2)$, the unital $\mathcal{U}$ and the 2-$R$-transitive $R$-permutation group $(T\cdot\widetilde{G}, \Pi(\Omega), R)$, where $\widetilde{G}\cong\mathrm{GU}(3,q^2)$ and $\Omega$ is the set of tangent lines of $\mathcal{U}$. Here again, we can construct a divisible design using Proposition 2.4. Let $\Omega'$ be a subset of $\Omega$ as defined by $\Omega' = \{\ell\in\Omega \mid \text{the tangent point on } \ell \text{ is in } \mathcal{U}-\{P\}\}$, where $P$ is a fixed point of $\mathcal{U}$. Let

$$B = \Pi_O(\Omega') = \{\pi \text{ a plane of } \mathrm{AG}(3,q^2) \mid \pi\cap E\in\Omega' \text{ and } O = (0,0,0,1)K\in\pi\};$$

it is an $R$-transversal subset of $\Pi(\Omega)$. We compute the stabilizer of $B$. Then $(T\cdot\widetilde{G})_B = \widetilde{G}_B = \widetilde{G}_{\Omega'}$ since $\widetilde{G}$ fixes $O$ and since $\{\mathrm{O}\}$ is the intersection of the planes in $B$. Moreover, $\widetilde{G}_{\Omega'}\cong\mathrm{GU}(3,q^2)_{\Omega'}$, and $\mathrm{GU}(3,q^2)_{\Omega'} = \mathrm{GU}(3,q^2)_{\mathcal{U}-\{P\}} = \mathrm{GU}(3,q^2)_P$. Therefore,

$$|(T\cdot\widetilde{G})_B| = |\mathrm{GU}(3,q^2)_P| = q^3(q^2-1)(q+1).$$

So, the parameters of the constructed divisible design $(\Pi(\Omega), B^{T\cdot\widetilde{G}}, \mathcal{R})$, with $\widetilde{G}\cong\mathrm{GU}(3,q^2)$, are the following;

$$\begin{aligned} v &= |\Pi(\Omega)| = q^2(q^3+1),\ s = q^2,\ k = q^3 \text{ and}\\ \lambda_2 &= \frac{|T\cdot\mathrm{GU}(3,q^2)|}{|\mathrm{GU}(3,q^2)_P|}\frac{k(k-1)}{v(v-s)}\\ &= \frac{q^6(q^3+1)q^3(q^2-1)(q+1)}{q^3(q^2-1)(q+1)}\frac{q^3(q^3-1)}{q^2(q^3+1)[q^2(q^3+1)-q^2]}\\ &= q^2(q^3-1).\end{aligned}$$

(ii) Now, we take the same $R$-permutation group as above and choose a secant $h$ to $\mathcal{U}$. So $|h\cap\mathcal{U}| = q+1$. Consider the $q+1$ tangent lines having as contact points the points of $h$, they intersect in the pole $S$ of $h$ (see [Hr1] p.147, p.148). Let $(T\cdot\widetilde{G}, \Pi(\Omega), R)$, with $\widetilde{G}\cong\mathrm{GU}(3,q^2)$ be the 2-$R$-transitive

$R$-permutation group used above. We choose $\widetilde{B}$, an $R$-transversal subset of $\Pi(\Omega)$, as follows;

$$\widetilde{B} := \Pi_O(\widetilde{\Omega}) = \Big\{ \pi \text{ a plane of } \mathrm{AG}(3,q^2) \mid \pi \cap E \in \widetilde{\Omega} \text{ and } O = (0,0,0,1)K \in \pi \Big\},$$

where

$$\begin{aligned}\widetilde{\Omega} &= \{\ell \in \Omega \mid \text{the tangent point on } \ell \text{ is a point of } h\} \\ &= \{\ell \in \Omega \mid S \in \ell\}.\end{aligned}$$

We compute the stabilizer of $\widetilde{B}$. Notice that the group of the translations in the direction of the line $OS$, $\widetilde{T}$, is contained in $(T \cdot \widetilde{G})_B$. Also $\widetilde{G}_h$, is contained in $(T \cdot \widetilde{G})_{\widetilde{B}}$, since $\widetilde{G}$ fixes $O$. Thus, $\widetilde{T} \cdot \widetilde{G}_h \subseteq (T \cdot \widetilde{G})_{\widetilde{B}}$. We prove that $\widetilde{T} \cdot \widetilde{G}_h$ is exactly the stabilizer. In fact, suppose that there exists an element $r \in (T \cdot \widetilde{G})_B$, $r = \tau g = g\tau_1$, with $\tau_1 \in T$ and $g \in \widetilde{G}$. Let $\pi_1, \pi_2 \in \widetilde{B}$ with $\pi_1 \neq \pi_2$ and $\pi_i' := \pi_i^{\tau g}$. Then $(OS)^{\tau g} = (\pi_1 \cap \pi_2)^{\tau g} = \pi_1' \cap \pi_2' = OS$ and $S^{\tau g} = S$. Thus $g \in \widetilde{G}_h$ and $\tau \in \widetilde{T}$. Notice that $\widetilde{G} \cong \mathrm{GU}(3,q^2)$ and $\mathrm{PGU}(3,q^2)_h \cong \mathrm{GU}(3,q^2)_h / Z(\mathrm{GU}(3,q^2))$. Therefore, it sufficient to compute $\mid \mathrm{PGU}(3,q^2)_h \mid$. Since $\mathrm{PGU}(3,q^2)$ acts 2-transitively on the points of $X_0^{q+1} + X_1^{q+1} + X_2^{q+1} = 0$, it acts transitively on the secant lines. Moreover, the number of secant lines is $q^2(q^2 - q + 1)$. Thus, $\mid \mathrm{PGU}(3,q^2)_h \mid = q(q^2-1)(q+1)$, and $\mid \widetilde{T} \mid\mid \mathrm{GU}(3,q^2)_h \mid = q^2 q(q^2-1)(q+1)(q+1)$. Therefore, the divisible design has the following parameters

$$\begin{aligned} v &= q^2(q^3+1),\ s = q^2,\ k = q+1 \text{ and} \\ \lambda_2 &= \frac{\mid T \cdot \mathrm{GU}(3,q^2) \mid}{\mid (T \cdot \mathrm{GU}(3,q^2)_B \mid} \frac{k(k-1)}{v(v-s)} \\ &= \frac{q^6(q^3+1)q^3(q^2-1)(q+1)}{q^2 q(q^2-1)(q+1)(q+1)} \frac{q(q+1)}{q^2(q^3+1)[q^2(q^3+1)-q^2]} = 1. \end{aligned}$$

The constructed divisible design admits $T \cdot \widetilde{G}$ with $\widetilde{G} \cong \mathrm{GU}(3,q^2)$ as a point- and block-transitive automorphism group.

(iii) Again, let $(T \cdot \widetilde{G}, \Pi(\Omega), R)$, with $\widetilde{G} \cong \mathrm{GU}(3,q^2)$ be the 2-$R$-transitive $R$-permutation group used above. Let $\ell_1$, $\ell_2$ be two fixed tangent lines to $\mathcal{U}$, and consider the $R$-transversal subset $\hat{B} = \{\pi_1, \pi_2\}$ of $\Pi(\Omega)$, where $\pi_1$, $\pi_2$ are the two planes of $\mathrm{AG}(3,q^2)$ through $O = (0,0,0,1)K$ and such that $\pi_1 \cap E = \ell_1$, $\pi_2 \cap E = \ell_2$. We want to compute $\mid (T \cdot \widetilde{G})_{\hat{B}} \mid$. Notice that, if $\ell_1 \cap \ell_2 = \{S\}$, then $\widetilde{T}$, the group of the translations in the direction of the line $OS$, is contained in $(T \cdot \widetilde{G})_{\hat{B}}$. Moreover, $\widetilde{G}_{\hat{B}} = \widetilde{G}_{\{\ell_1,\ell_2\}} = \widetilde{G}_{\{P,Q\}}$, where $P = \ell_1 \cap \mathcal{U}$ and $Q = \ell_2 \cap \mathcal{U}$, and $\mathrm{PGU}(3,q^2)_{\{P,Q\}} \cong \mathrm{GU}(3,q^2)_{\{P,Q\}} / Z(\mathrm{GU}(3,q^2))$. So, it is sufficient to compute $\mid \mathrm{PGU}(3,q^2)_{\{P,Q\}} \mid$. It is known (see [Lü1] pg. 156) that the pointwise stabilizer $\mathrm{PGU}(3,q^2)_{P,Q}$ of $P$, $Q$ is a cyclic group of order $q^2-1$. Moreover, since $\mathrm{PGU}(3,q^2)$ is 2-transitive on the points of the unital, there is also contained an element in $\mathrm{PGU}(3,q^2)_{\{P,Q\}}$ which permutes the two points. Therefore, $\mid \mathrm{PGU}(3,q^2)_{\{P,Q\}} \mid = 2(q^2-1)$ and $\mid \mathrm{GU}(3,q^2)_{\{P,Q\}} \mid = \mid \mathrm{PGU}(3,q^2)_{\{P,Q\}} \mid\mid Z(\mathrm{GU}(3,q^2)) \mid = 2(q^2-1)(q+1)$. Then, $\widetilde{T}\cdot\widetilde{G}_{\{P,Q\}} \subseteq (T\cdot\widetilde{G})_{\hat{B}}$. We want to prove that $\widetilde{T}\cdot\widetilde{G}_{\{P,Q\}} = (T\cdot\widetilde{G})_{\hat{B}}$. Let $\tau g \in (T\cdot\widetilde{G})_{\hat{B}}$. Since $\widetilde{G}_{\{P,Q\}}$ already interchanges $\pi_1$ and $\pi_2$ we may, without loss of generality, assume that $\tau g$ fixes $\pi_1$ and $\pi_2$ and that $\pi_1^{\tau g} = \pi_1^{g\tau_1} = \pi_1$ for a suitable $\tau_1 \in T$. The translation $\tau_1$ fixes $E$ pointwise; hence $\pi_1^g$ is a plane through $O$, $S$ and $\ell_1$. Therefore $\pi_1^g = \pi_1$ and, similarly, $\pi_2^g = \pi_2$. Hence, $\tau_1 \in \widetilde{T}$ and $g \in \widetilde{G}_{\{P,Q\}}$. So, the obtained divisible design has the following parameters,

$$\begin{aligned} v &= q^2(q^3+1),\ s = q^2,\ k = 2 \text{ and} \\ \lambda_2 &= \frac{\mid T\cdot \mathrm{GU}(3,q^2)\mid}{\mid (T\cdot \mathrm{GU}(3,q^2))_{\hat{B}}\mid}\frac{k(k-1)}{v(v-s)} \\ &= \frac{q^6(q^3+1)q^3(q^2-1)(q+1)}{q^2 2(q^2-1)(q+1)}\frac{2}{q^2(q^3+1)q^5} = 1. \end{aligned}$$

The constructed divisible designs admit, as a point- and block-transitive automorphism group, the group $T\cdot\widetilde{G}$ with $\widetilde{G} \cong \mathrm{GU}(3,q^2)$. (iv) Now, consider the embedded subplane $\mathrm{PG}(2,q)$ of $E$. The unital $\mathcal{U}$ has $q+1$ points in common with the subplane namely (see [Hr1] pg. 148) those satisfying

$$\widetilde{C} = \begin{cases} X_0^2 + X_1^2 + X_2^2 = 0 \\ X_3 \qquad\qquad\quad\; = 0. \end{cases}$$

This is a conic if $p$ is odd and a secant if $p$ is even. Again, we consider the same $R$-permutation group as above, and we use Proposition 2.4 to construct a divisible design. Choose the following $R$-transversal subset of $\Pi(\Omega)$;

$$\begin{aligned} \bar{B} := \Pi_O(\bar{\Omega}_{\widetilde{C}}) = \{\ \pi \text{ a plane of } \mathrm{AG}(3,q^2) \ \mid\ \pi\cap E \in \bar{\Omega}_{\widetilde{C}} \\ \text{and } O = (0,0,0,1)K \in \pi\}, \end{aligned}$$

where

$$\bar{\Omega}_{\widetilde{C}} = \left\{\ell \in \Omega \mid \text{ the tangent point on } \ell \text{ is a point of } \widetilde{C}\right\}.$$

The case for $q$ even is contained in (ii), so let $q$ be odd. Notice that, $(T\cdot\widetilde{G})_{\bar{B}} = \widetilde{G}_{\bar{B}} = \widetilde{G}_{\bar{\Omega}_{\widetilde{C}}}$, since $\widetilde{G}$ fixes $O$, and the tangent lines to $\widetilde{C}$ do not intersect in a point. Moreover, since every point of $\mathcal{U}$ lies on exactly one tangent line, (see [Hr1] p.147), $\widetilde{G}_{\bar{\Omega}_{\widetilde{C}}} = \widetilde{G}_{\widetilde{C}}$. Therefore, $\mathrm{PGU}(3,q^2)_{\widetilde{C}} \cong \mathrm{GU}(3,q^2)_{\widetilde{C}}/Z(\mathrm{GU}(3,q^2))$, and $\widetilde{G}_{\widetilde{C}} \cong \mathrm{GU}(3,q^2)_{\widetilde{C}}$. Thus, we can consider any unital on $\mathrm{PG}(2,q^2)$, with $q$ odd, to find the stabilizer in $\mathrm{PGU}(3,q^2)$

of the points of the unital in PG$(2,q)$. Take on PG$(2,q^2)$ the unital $\widetilde{\mathcal{U}}$ with equation $X_0X_2^q + X_0^qX_2 + X_1^{q+1} = 0$. Then, $\widetilde{\mathcal{U}} \cap \mathrm{PG}(2,q)$ is a conic $\hat{\mathcal{C}}$ with equation $X_1^2 + 2X_0X_2 = 0$. It is known, (see [Hu1] p. 242), that $\mathrm{PSU}(3,q^2)_{\{(0,0,1)K\}}$ is generated by the following elements;

$$Q(a,b) = \left\{ \begin{pmatrix} 1 & a & b \\ 0 & 1 & -a^q \\ 0 & 0 & 1 \end{pmatrix} \mid a,b \in \mathrm{GF}(q^2) \right\} \text{ and}$$

$$H(k) = \left\{ \begin{pmatrix} k^{-q} & 0 & 0 \\ 0 & k^{q-1} & 0 \\ 0 & 0 & k \end{pmatrix} \mid k \in \mathrm{GF}(q^2)^* \right\}.$$

In case of GU$(3,q^2)$, one has to consider as well

$$H(k,c) = \left\{ \begin{pmatrix} k^{-q} & 0 & 0 \\ 0 & c & 0 \\ 0 & 0 & k \end{pmatrix} \mid k \in \mathrm{GF}(q^2)^* \text{ and } c^{q+1} = 1 \right\}.$$

We want to find the elements of $\mathrm{GU}(3,q^2)_{\{(0,0,1)K\}}$ fixing the conic $\hat{\mathcal{C}}$. One can show that they are

$$\left\{ Q\left(a, -\frac{1}{2}a^2\right) \mid a \in \mathrm{GF}(q) \right\},$$

and

$$\left\{ H(h,c) \mid h \in \mathrm{GF}(q^2)^*,\ c = gh \text{ with } g \in \mathrm{GF}(q)^* \text{and } g^2 = (h^{-1})^{q+1} \right\}.$$

Notice that the group generated by these elements acts 2-transitively on the set of points $\hat{\mathcal{C}} \setminus \{(0,0,1)K\}$. Moreover, the following element,

$$A = \begin{pmatrix} 0 & 0 & 1 \\ 0 & -1 & 0 \\ 1 & 0 & 0 \end{pmatrix},$$

is in GU$(3,q^2)$, fixes $\hat{\mathcal{C}}$, and maps the point $(0,0,1)K$ onto the point $(1,0,0)K$. Therefore, $\mathrm{PGU}(3,q^2)_{\hat{\mathcal{C}}}$ is 3-transitive. So, $\mid \mathrm{GU}(3,q^2)_{\widetilde{\mathcal{C}}} \mid = q(q^2-1)(q+1)$. Thus, the constructed divisible design has the following parameters,

$$v = q^2(q^3+1),\ s = q^2,\ k = q+1, \text{ and}$$

$$\lambda_2 = \frac{\mid T \cdot \mathrm{GU}(3,q^2) \mid}{\mid (T \cdot \mathrm{GU}(3,q^2))_{\widetilde{B}} \mid} \frac{k(k-1)}{v(v-s)}$$

$$= \frac{q^6(q^3+1)q^3(q^2-1)(q+1)}{q(q^2-1)(q+1)} \frac{q(q+1)}{q^2(q^3+1)q^5} = q^2(q+1).$$

The constructed divisible design admits $T \cdot \widetilde{G}$ with $\widetilde{G} \cong \mathrm{GU}(3,q^2)$ as a point- and block-transitive automorphism group.

□

## References


[BJL1] T. Beth, D. Jungnickel and H. Lenz, *Design Theory*, BI, Mannheim-Wien-Zürich, 1985[1], and Cambridge Univ. Press 1999[2].

[Ce1] C. Cerroni, *Divisible designs from semifield planes*, to appear.

[CeSp1] C. Cerroni and A.G. Spera, *On divisible designs and twisted field planes*, J. Comb. Designs 7 (1999), 453 - 464.

[Ha1] H. Hanani, *Balanced incomplete block designs and related designs*, Discrete Math. 11, (1975), 255 - 369.

[HP1] D. R. Hughes and F. C. Piper, *Projective planes*, Springer-Verlag, Berlin-Heidelberg-New York, 1973.

[Hr1] J. W. P. Hirschfeld, *Projective Geometries Over Finite Fields*, Clarendon Press, Oxford, 1979.

[Hu1] B. Huppert, *Endliche Gruppen I*, Springer-Verlag, Berlin-Heidelberg-New York, 1967.

[Lü1] H. Lüneburg, *Translation planes*, Springer-Verlag, Berlin-Heidelberg-New York, 1980.

[Sch1] R.-H. Schulz, *Constant weight codes and divisible designs with large automorphism group*, Rendiconti del Cicolo Matematico di Palermo. Serie II, Suppl. 53 (1998) 173-188.

[SchSp1] R.-H. Schulz and A. G. Spera, *Divisible designs admitting a Suzuki group as an automorphism group*, Boll. U.M.I. (8) I - B - 3, (1998), 705-714.

[SchSp2] R.-H. Schulz and A. G. Spera, *Construction of divisible designs from translation planes*, Europ. J. Comb. vol. 19, n.4 (1998), 479-486.

[SchSp3] R.-H. Schulz and A. G. Spera, *Automorphisms of constant weight codes and of divisible designs.* Designs, Codes and Cryptography (DESI) 20, (2000), 89-97.

[Sp1] A. G. Spera, *t - Divisible designs from imprimitive permutation groups*, Europ. J. Comb. 13, (1992), 409 - 417.

[Sp2] A. G. Spera, *On divisible designs and local algebras*, J. Comb. Designs, 3, n.3 (1995), 203 - 212.

[YaWei] B. Yang and W. Wei, *Finite orthogonal geometries with characteristic* $\neq 2$ *and PBIB designs I.* Linear Alg. and its Appl. , Vol. 134 (1990), 1 - 23.

# A Permutation of a Small Infinite Field

Stephen D. Cohen

Department of Mathematics
University of Glasgow, Glasgow G12 8QW
Scotland
e-mail: sdc@maths.gla.ac.uk

**Abstract** Let $\mathbb{F}$ denote the *infinite* field that is the union of all extensions of GF(2) of *odd* degree. Let $q = 2^m$ and let $\mathbb{F}_Q = \mathrm{GF}(2q^2)$. We provide a simple proof of the permutation property of the polynomials $x^{2q+1} + x^3 + x$ on $\mathbb{F}_Q$ for each $m$ by viewing each member of the class as the action of a simply defined permutation of $\mathbb{F}$.

**2000 Mathematics subject classification.** 11T06, 11T30

## 1 Introduction

Let $q = 2^m$ for any positive integer $m$ and set $Q = 2^n = 2q^2$ so that $\mathbb{F}_Q := \mathrm{GF}(2q^2)$ is the typical extension of GF(2) of odd degree $n := 2m+1$. Define $T_n$ to be the trace function from $\mathbb{F}_Q$ to $\mathbb{F}_2$: thus, for $\alpha \in \mathbb{F}_Q$, $T_n(\alpha) = \sum_{i=0}^{n-1} \alpha^{2^i}$. An important conjecture of Welch, quoted in the 1968 paper of Golomb [4], is that the cross-correlation function $a \mapsto \sum_{x \in \mathbb{F}_Q} (-1)^{T_n(x^{q+3}+ax)}$ on $\mathbb{F}_Q$ takes only three values, namely 0 and $\pm 2q$. Equivalently (see [3]), the power function $F(x) := x^{q+3}$ is an *almost perfect nonlinear* function on $\mathbb{F}_Q$, which means that the equation

$$F(x+a) - F(x) = b, \quad a \in \mathbb{F}_Q^*, \ b \in \mathbb{F}_Q,$$

has either 0 or 2 solutions in $\mathbb{F}_Q$. This conjecture remained open until 1999, when Dobbertin [3] proved that $x^{2q+1} + x^3 + x$ is a permutation polynomial on $\mathbb{F}_Q$, whence $x^{q+3}$ is an almost perfect nonlinear function on $\mathbb{F}_Q$. Welch's conjecture follows: see also [2].

Let $\mathbb{F}$ denote the *infinite* field that is the union of all extensions of $\mathbb{F}_2$ of *odd* degree: thus

$$\mathbb{F} := \bigcup_{n \,\mathrm{odd}} \mathbb{F}_{2^n} = \bigcup_{m=1}^{\infty} \mathbb{F}_{2^{2m+1}}.$$

The purpose of this note is to provide a simpler proof of the permutation property of the polynomials $x^{2q+1} + x^3 + x$ for each $m$ by viewing each member of the class as the action of a simply defined permutation of $\mathbb{F}$ when restricted to each finite subfield $\mathbb{F}_Q$. To accomplish this we define the function $x^{\sqrt{2}}$ on $\mathbb{F}$ as follows.

$$x^{\sqrt{2}} = x^{2q} \text{ whenever } x \in \mathbb{F}_{2q^2} \subseteq \mathbb{F}. \tag{1}$$

**Lemma 1.1.** – *$x^{\sqrt{2}}$ is a well-defined function on $\mathbb{F}$;*
– *$x^{\sqrt{2}}$ is an automorphism of $\mathbb{F}$;*
– *$(x^{\sqrt{2}})^{\sqrt{2}} = x^2$.*

*Proof.* (Cf. [1], Theorem 2.15) Suppose that $x$ is a non-zero element of $\mathbb{F}$. Let $n = 2m+1$ be the minimum odd integer such that $x \in \mathbb{F}_{2q^2}$, where $q = 2^m$. Suppose that $N = 2M+1$ is an odd multiple of $n$ and set $Q = 2^M$. Then

$$x^{2Q}/x^{2q} = (x^{2^{(N-n)/2}-1})^{2q} = 1, \tag{2}$$

since $N-n$ is divisible by $2n$. From (1) and (2) it follows that $x^{\sqrt{2}}$ is properly defined on $\mathbb{F}$. Moreover, it is an automorphism since it is one on any finite subfield. Finally, $(x^{\sqrt{2}})^{\sqrt{2}} = x^{4q^2} = x^2$.

Our goal can now be stated very simply.

**Theorem 1.2.** *The function $x^{\sqrt{2}+1} + x^3 + x$ permutes $\mathbb{F}$.*

## 2 The Trace Function

The trace function $T_n$ from $\mathbb{F}_{2^n}$ to $\mathbb{F}_2$ ($n$ odd) is an additive homomorphism and takes (only) the values 0 and 1: in particular, $T(1) = n = 1$, since $n$ is odd. If $x \in \mathbb{F}_{2^n}$ and $N$ is any *odd* multiple of $n$, then, by the transitivity of the trace,

$$T_N(x) = (N/n)T_n(x) = T_n(x)$$

and therefore we can properly define the trace function $T(x)$ as an additive homomorphism on $\mathbb{F}$ by

$$T(x) = T_n(x) \text{ whenever } x \in \mathbb{F}_{2^n} \subseteq \mathbb{F}.$$

Of course, $T(x)$ also takes only the values 0 and 1: in particular, $T(1) = 1$.

**Lemma 2.1.** *Let $x \in \mathbb{F}$. Then $T(x) = 0$ if and only if $T(x+1) = 1$. Moreover, the following are equivalent.*

– *$T(x) = 0$;*
– *$x = y + y^{\sqrt{2}}$ for some $y \in \mathbb{F}$;*
– *$x = y + y^2$ for some $y \in \mathbb{F}$.*

*Proof.* The first part is evident since $T(x+1) = T(x) + T(1) = T(x) + 1$.

Now let $L(x)$ be the additive homomorphism $x + x^{\sqrt{2}}$ on $\mathbb{F}$. Then $L$ is zero on $\mathbb{F}_2$ and $L(y) \neq 1$ on $\mathbb{F}$ (because otherwise $0 = L(1) = L(L(y)) = y^2 + y$, i.e., $y \in \mathbb{F}_2$, a contradiction). Further, if $x \in \mathbb{F}$, then $T(L(x)) = T(x) + T(x^{\sqrt{2}}) = T(x) + (T(x))^{\sqrt{2}} = 0$, since $T(x) \in \mathbb{F}_2$.

Similarly, the additive homomorphism $M(x) := x + x^2$ is zero on $\mathbb{F}_2$ and has trace 0 on $\mathbb{F}$, whence $M(y) \neq 1$ on $\mathbb{F}$ (because the latter comprises elements of *odd* degree over $\mathbb{F}_2$). The equivalences follow from these considerations and the first part.

## 3 Proof of Theorem 1.2

Let $f(x) := x^{\sqrt{2}+1} + x^3 + x$. Since $f$ maps any finite subfield $\mathbb{F}_Q$ of $\mathbb{F}$ into itself it suffices to establish injectivity. For were that to hold, then $f$ would be injective on $\mathbb{F}_Q$ and so a permutation of $\mathbb{F}_Q$. Thus an arbitrary element of $\mathbb{F}$, a member of $\mathbb{F}_Q$, say, would be the image under $f$ of another member of the latter field and the result would follow.

Accordingly, suppose $f(a) = f(b)$, where $a \neq b \in \mathbb{F}$. Put $A = a + b (\neq 0)$, $c = \frac{a}{A}$; thus $c + 1 = \frac{b}{A}$. Observe that by Lemma 2.1 both $1 + c + c^2$ and $1 + c + c^{\sqrt{2}}$ are non-zero. Then

$$((c+1)^{\sqrt{2}+1} + c^{\sqrt{2}+1})A^{\sqrt{2}} + ((c+1)^3 + c^3)A^2 = 1,$$

whence

$$(1 + c + c^{\sqrt{2}})A^{\sqrt{2}} + (1 + c + c^2)A^2 = 1. \tag{3}$$

Next, in (3) set $A^{\sqrt{2}} = \left(\frac{1+c+c^{\sqrt{2}}}{1+c+c^2}\right)^{\sqrt{2}+1} B$. Then

$$\begin{aligned} B + B^{\sqrt{2}} &= \left(\frac{1+c+c^2}{1+c^{\sqrt{2}}+c^2}\right)^{\sqrt{2}+1} \\ &= (1 + d^{\sqrt{2}-1} + d^{\sqrt{2}})^{\sqrt{2}+1}, \end{aligned}$$

where $d = 1/(1 + c + c^{\sqrt{2}})$.

Hence, by Lemma 2.1,

$$T((1 + d^{\sqrt{2}-1} + d^{\sqrt{2}})^{\sqrt{2}+1}) = 0$$

and so, again by Lemma 2.1,

$$T((d^{\sqrt{2}-1} + d^{\sqrt{2}})^{\sqrt{2}+1}) = 1. \tag{4}$$

But by direct expansion

$$(d^{\sqrt{2}-1} + d^{\sqrt{2}})^{\sqrt{2}+1} = d + d^{\sqrt{2}(\sqrt{2}-1)+\sqrt{2}} + d^{\sqrt{2}-1+2} + d^{\sqrt{2}(\sqrt{2}+1)}$$
$$= \{d + d^2\} + \{d^{\sqrt{2}+1} + (d^{\sqrt{2}+1})^{\sqrt{2}}\}.$$

Hence, by Lemma 2.1,

$$T((d^{\sqrt{2}-1} + d^{\sqrt{2}})^{\sqrt{2}+1}) = 0. \tag{5}$$

Evidently, (4) and (5) are contradictory and the theorem is proved.

## References


1. J.V. Brawley and G. E. Schnibben, *Infinite algebraic extensions of finite fields* Contemporary Mathematics, **95**, Amer. Math. Soc., Providence RI, (1989).
2. A. Canteaut, P. Charpin and H. Dobbertin, "Couples de suites binaires de longueur maximale ayant une corrélation croisée à trois valeurs: conjecture de Welch", *C. R. Acad. Sci. Paris Sér. I Math.* **328** (1999), 173–178.
3. H. Dobbertin, "Almost perfect nonlinear power functions on GF($2^n$): the Welch case", *IEEE Trans. Inform. Theory* **45**(1999), 1271–1275.
4. S.W.Golomb, "Theory of transformation groups of polynomials over GF(2) with applications to linear shift register sequences", *Information Sci.* **1** (1968), 87–109.

# Almost Perfect Nonlinear Power Functions on GF($2^n$): A New Case for $n$ Divisible by 5

Hans Dobbertin

University of Klagenfurt
Department of Mathematics
Universitätsstr. 65–67
A-9020 Klagenfurt, Austria
e-mail: hans.dobbertin@uni-klu.ac.at

**Abstract** We prove that for $d = 2^{4s} + 2^{3s} + 2^{2s} + 2^s - 1$ the power function $x^d$ is almost perfect nonlinear (APN) on $L = \mathrm{GF}(2^{5s})$, i.e. for each $a \in L$ the equation $(x+1)^d + x^d = a$ has either no or precisely two solutions in $L$. The proof of this result is based on a new "multi-variate" technique which was recently introduced by the author in order to confirm the conjectured APN property of Welch and Niho power functions.

## 1 Introduction

This note is the third part of a systematic study of APN power functions on GF($2^n$). In the first [5] and second part [6] we showed that Welch and Niho power functions are APN. Here we shall confirm a conjecture on a new class of APN power functions; see Theorem 1 below. This conjecture was found independently by Anne Canteaut [2] and the author [5] based on computer experiments. Table 1 gives an overview about the known APN power functions on finite fields of characteristic 2. (An alternative proof for Case (II) can be found in [5]. For odd $n$, Case (II) is also a consequence of a classical result of Kasami [10].)

**Question.** *Up to equivalence, is Table 1 complete?*

Anne Canteaut confirmed by computer experiments that Table 1 is complete for $n \leq 25$.

A mapping $F$ from GF($p^n$) to GF($p^n$) is called *almost perfect nonlinear (APN)* if each equation

$$F(t+a) - F(t) = b \quad (a \in \mathrm{GF}(p^n)^*,\ b \in \mathrm{GF}(p^n))$$

has at most two solutions $t$ in GF($p^n$). It is obvious that if $x^d$ is APN then $x^{d'}$ is also APN for $d' \equiv p^i d \bmod p^n - 1$ and, in case $\gcd(d, p^n - 1) = 1$, for $d' = d^{-1} \bmod p^n - 1$. If power functions $x^d$ and $x^{d'}$ (or exponents $d$ and $d'$) are related in this way then they are called *equivalent.*

| Case | Exponents | Reference |
|---|---|---|
| (I) | $d = 2^k + 1$, $\gcd(k, n) = 1$ | proof is trivial |
| (II) | $d = 2^{2k} - 2^k + 1$, $\gcd(k, n) = 1$ | Janwa & Wilson [9] |
| (III) | $d = 2^{4s} + 2^{3s} + 2^{2s} + 2^s - 1$ for $n = 5s$ | Dobbertin [this note] |
| (IV) | $d = 2^m + 3$ for $2m + 1 = n$ | Dobbertin [5] (Welch case) |
| (V) | $d = 2^{2r} + 2^r - 1$ for $4r + 1 \equiv 0 \bmod n$ | Dobbertin [6] (Niho case) |
| (VI) | $d = -1$ for odd $n$ | proof is trivial |

**Table1.** APN power functions $x^d$ on $\mathrm{GF}(2^n)$

In this paper we shall exclusively consider the binary case $p = 2$, where minus coincides with plus and therefore the restrictive meaning of the attribute "almost" in APN is not justified, since solutions of $F(t+a)+F(t) = b$ occur always as pairs $t = t_0, t_0+a$. APN power functions on $\mathrm{GF}(2^n)$ and their applications in cryptology were studied e.g. by Beth and Ding [1], Canteaut, Charpin and Dobbertin [3], Carlet, Charpin and Zinoviev [4], Helleseth, Rong and Sandberg [8] and Nyberg [11].

## 2 A New Class of APN Power Functions

**Theorem 2.1.** *Let $L = \mathrm{GF}(2^n)$ with $n = 5s$. Then for*

$$d = 2^{4s} + 2^{3s} + 2^{2s} + 2^s - 1$$

*the power function $x^d$ is almost perfect nonlinear on $L$.*

*Proof.* Let $K = \mathrm{GF}(2^s)$. The Galois group

$$\mathcal{G} = \mathrm{Gal}(L/K)$$

is generated by the Frobenius automorphism

$$\varphi(x) = x^{2^s},\ x \in L.$$

Note that $K = \{x \in L : \varphi(x) = x\}$ and $\mathcal{G} = \{\varphi^i : 0 \leq i < 5\}$. We set

$$\Delta(x) = (x+1)^d + x^d,\ x \in L.$$

In the sequel we suppose

$$\Delta(x_0) = a$$

for some fixed $x_0, a \in L$. We have to show that

$$\{x \in L : \Delta(x) = a\} = \{x_0, x_0 + 1\}.$$

Thus assume $\Delta(x) = a$ for $x \in L$. Using $\varphi$ we can write $\Delta(x)$ as

$$\frac{(\varphi(x)+1)(\varphi^2(x)+1)(\varphi^3(x)+1)(\varphi^4(x)+1)}{x+1} + \frac{\varphi(x)\varphi^2(x)\varphi^3(x)\varphi^4(x)}{x},$$

that is

$$x^2\left(\prod_{i<5}\left(\varphi^i(x)+1\right)\right) + (x^2+1)\left(\prod_{i<5}\varphi^i(x)\right) + a\left(x^4+x^2\right) = 0. \quad (1)$$

As a basic tool in our proof we need the elementary symmetric polynomials with five variables

$$\begin{aligned}
\sigma_0(t_1,t_2,t_3,t_4,t_5) &= t_1t_2t_3t_4t_5,\\
\sigma_1(t_1,t_2,t_3,t_4,t_5) &= t_1t_2t_3t_4 + t_1t_2t_3t_5 + t_1t_2t_4t_5 + t_1t_3t_4t_5 + t_2t_3t_4t_5,\\
\sigma_2(t_1,t_2,t_3,t_4,t_5) &= t_1t_2t_3 + t_1t_2t_4 + t_1t_2t_5 + t_1t_3t_4 + t_1t_3t_5 + t_1t_4t_5\\
&\quad + t_2t_3t_4 + t_2t_3t_5 + t_2t_4t_5 + t_3t_4t_5,\\
\sigma_3(t_1,t_2,t_3,t_4,t_5) &= t_1t_2 + t_1t_3 + t_1t_4 + t_1t_5 + t_2t_3 + t_2t_4 + t_2t_5 + t_3t_4\\
&\quad + t_3t_5 + t_4t_5,\\
\sigma_4(t_1,t_2,t_3,t_4,t_5) &= t_1 + t_2 + t_3 + t_4 + t_5.
\end{aligned}$$

Similar as in the proofs of the Welch and the Niho cases [5,6] we shall consider the conjugates $\varphi^i(x)$ $(i < 5)$ as independent variables ("multi-variate" method, see also [7]). Accordingly our assumption, the equation $\Delta(x) = a$ and its conjugates, forms a five-equation system with five variables $x, y, z, u, v$ and five constants $a, b, c, d, e$; see (2) - (6) below. We shall analyze this system in order to derive the desired conclusion $x = x_0$ or $x = x_0 + 1$.

$$x^2\tilde{P} + (x^2+1)P + a(x^4+x^2) = 0, \quad (2)$$
$$y^2\tilde{P} + (y^2+1)P + b(y^4+y^2) = 0, \quad (3)$$
$$z^2\tilde{P} + (z^2+1)P + c(z^4+z^2) = 0, \quad (4)$$
$$u^2\tilde{P} + (u^2+1)P + d(u^4+u^2) = 0, \quad (5)$$
$$v^2\tilde{P} + (v^2+1)P + e(v^4+v^2) = 0, \quad (6)$$

Here we use the following notations

$$\begin{aligned}
y &= \varphi(x), & b &= \varphi(a),\\
z &= \varphi(y), & c &= \varphi(b),\\
u &= \varphi(z), & d &= \varphi(c),\\
v &= \varphi(u), & e &= \varphi(d),
\end{aligned}$$

$$\begin{aligned}
P &= \sigma_0(x,y,z,u,v), & \tilde{P} &= \sigma_0(x+1,y+1,z+1,u+1,v+1),\\
Q &= \sigma_1(x,y,z,u,v), & \tilde{Q} &= \sigma_1(x+1,y+1,z+1,u+1,v+1),\\
R &= \sigma_2(x,y,z,u,v), & \tilde{R} &= \sigma_2(x+1,y+1,z+1,u+1,v+1),\\
S &= \sigma_3(x,y,z,u,v), & \tilde{S} &= \sigma_3(x+1,y+1,z+1,u+1,v+1),\\
T &= \sigma_4(x,y,z,u,v), & \tilde{T} &= \sigma_4(x+1,y+1,z+1,u+1,v+1).
\end{aligned}$$

The relations between $P, Q, R, S, T$ and $\tilde{P}, \tilde{Q}, \tilde{R}, \tilde{S}, \tilde{T}$ are given by the linear equations:

$$\tilde{P} = P + Q + R + S + T + 1, \tag{7}$$

$$\tilde{Q} = Q + S + 1, \tag{8}$$

$$\tilde{R} = R + S, \tag{9}$$

$$\tilde{S} = S, \tag{10}$$

$$\tilde{T} = T + 1. \tag{11}$$

We have $\varphi(P) = P$, $\varphi(Q) = Q$, ..., $\varphi(\tilde{P}) = \tilde{P}$, $\varphi(\tilde{Q}) = \tilde{Q}$, ...; thus

$$P, Q, R, S, T, \tilde{P}, \tilde{Q}, \tilde{R}, \tilde{S}, \tilde{T} \in K.$$

We note that for $x \notin K$

$$m_x(\xi) = \xi^5 + T\xi^4 + S\xi^3 + R\xi^2 + Q\xi + P$$

is the minimal polynomial of $x$ over $K$, and that

$$m_{x+1}(\xi) = \xi^5 + \tilde{T}\xi^4 + \tilde{S}\xi^3 + \tilde{R}\xi^2 + \tilde{Q}\xi + \tilde{P}$$

is the minimal polynomial of $x + 1$ over $K$. (We have $m_{x+1}(\xi) = m_x(\xi + 1)$, which gives another proof of (7) - (11).)

The case $a \in K$ is trivial. In fact then we conclude from (2) that $x$ is also in $K$, since otherwise the degree of the minimal polynomial of $x$ over $K$ is five, but (2) has degree four. Therefore in this case we have to show that the restriction of $x \longmapsto x^d$ onto $K$ is APN. But this is obvious, because

$$x^d = x^{2^{4s}+2^{3s}+2^{2s}+2^s-1} = x^{1+1+1+1-1} = x^3$$

for $x \in K$, and $x^3$ is clearly always APN (this is Case (I) of Table 1 for $k = 1$).

In what follows we can now assume that

$$x, x_0, a \notin K.$$

Further note that $P, \tilde{P} \neq 0$, since $x \neq 0, 1$. As an additional assumption we require

$$P \neq \tilde{P}.$$

The case that there is a solution of (2) - (6) with $P = \tilde{P}$ is degenerated and can be easily handled.

The next step is to deduce from (2) - (6) a representation of the elementary symmetric polynomials in $a, b, c, d, e$, i.e.

$$A = \sigma_0(a, b, c, d, e),$$

$$B = \sigma_1(a, b, c, d, e),$$

$$C = \sigma_2(a, b, c, d, e),$$

$$D = \sigma_3(a, b, c, d, e),$$

$$E = \sigma_4(a, b, c, d, e),$$

as rational functions of $x, y, z, u, v$. The symmetry of the system (2) - (6) implies that these rational functions are symmetric.

For instance we can use (2) - (6) to represent $A\,P^2\,\tilde{P}^2 = abcde\,P^2\,\tilde{P}^2$ as a polynomial in $x, y, z, u, v$:

$$\begin{aligned} abcde\,P^2\,\tilde{P}^2 &= \left(a(x^4+x^2)\right)\left(b(y^4+y^2)\right)\left(c(z^4+z^2)\right) \\ &\quad\times\left(d(u^4+u^2)\right)\left(e(v^4+v^2)\right) \\ &= \left(x^2\tilde{P}+(x^2+1)P\right)\left(y^2\tilde{P}+(y^2+1)P\right)\left(z^2\tilde{P}+(z^2+1)P\right) \\ &\quad\times\left(u^2\tilde{P}+(u^2+1)P\right)\left(v^2\tilde{P}+(v^2+1)P\right). \end{aligned}$$

The latter polynomial is symmetric in $x, y, z, u, v$. By the famous *fundamental theorem on symmetric polynomials* we known that every symmetric polynomial can be written as a polynomial of elementary symmetric polynomials. Behind this mathematical statement there is a simple algorithm to compute this representation. This means that we can here substitute the variables $x, y, z, u, v$ and replace them by $P, Q, R, S, T$. The same can be done for $B\,P^2\tilde{P}^2$, ..., $E\,P^2\tilde{P}^2$. In this way, after division by the factor $P\tilde{P}$ which occurs on the right hand, we get the following equational system with variables $P, Q, ..., T$, ($\tilde{P}$ stands for $P+Q+R+S+T+1$) and constants $A, B, ..., E$:

$$\begin{aligned} A\,P\,\tilde{P} &= P^3+Q^5+P\,S^4+P\,T^4+P^2\,T^3+Q^3\,R^2+Q^3\,S^2+Q^4\,T \\ &\quad+Q^4\,S+Q^4\,R+Q^3\,T^2+Q^2\,R^3+Q^2\,S^3+Q^2\,T^3+P\,R^2\,S^2 \\ &\quad+P\,Q^2\,S^2+P\,Q^2\,T^2+P^2\,S\,T^2+P^2\,Q\,T^2+P^2\,R\,T^2+Q^2\,R\,S^2 \\ &\quad+Q^2\,R^2\,T+P\,R^2\,T^2+Q^2\,S\,T^2+Q^2\,S^2\,T+Q^2\,R^2\,S+Q^2\,R\,T^2, \\ B\,P\,\tilde{P} &= Q^4+Q^2\,S^2+P\,Q^3+P^2\,T^2+Q^2\,R^2+Q^2\,T^2+Q^4\,T^2+P\,T^5 \\ &\quad+P^2\,T^4+Q^2\,T^4+P\,Q\,T^4+P\,S\,T^4+P\,S^2\,T^3+P\,S^3\,T^2 \\ &\quad+P\,R\,T^4+P\,Q^2\,T+P\,Q^2\,S+P\,Q^2\,R+P\,R\,S^2\,T^2+P\,Q\,S^2\,T^2 \\ &\quad+Q^2\,S^2\,T^2+Q^2\,R^2\,T^2, \\ C\,P\,\tilde{P} &= P^2\,Q+P^2\,R+P^2\,S+P^2\,T+Q^3+Q^2\,R+Q^2\,S+Q^2\,T \\ &\quad+P\,S^4+Q^3\,S^2+Q^3\,T^2+Q^2\,S^3+Q^2\,T^3+P\,R^2+Q^2\,R\,S^2 \\ &\quad+P\,S^2\,T^2+Q^2\,S\,T^2+Q^2\,S^2\,T+Q^2\,R\,T^2, \\ D\,P\,\tilde{P} &= P^2+P\,S^3+P\,Q\,S^2+P\,R\,S^2+P\,S^2\,T+Q^2+Q^2\,R^2+Q^2\,T^2, \\ E\,P\,\tilde{P} &= P+P\,S^2+Q^3+Q^2\,R+Q^2\,S+Q^2\,T. \end{aligned}$$

Since $\tilde{P} = P+Q+R+S+T+1$, we get an equivalent system with variables $P, \tilde{P}, Q, R, S$ if we replace $T$ by $\tilde{P}+P+Q+R+S+1$:

$$\begin{aligned} A P \tilde{P} = {} & P^4\tilde{P} + P\tilde{P}^4 + P^3 + P^3\tilde{P}^2 + P^2\tilde{P}^3 + P^3Q^2 + P^2Q^3 \\ & + P^2\tilde{P}R + P\tilde{P}^2R + P^3S^2 + P^2\tilde{P}R^2, \end{aligned} \tag{12}$$

$$\begin{aligned} B P \tilde{P} = {} & P^4 + P\tilde{P}^5 + \tilde{P}P^5 + P^2\tilde{P}^2 + P\tilde{P}Q^2S^2 + P^2Q^2 + P^2Q^4 \\ & + \tilde{P}^2Q^2 + \tilde{P}^2Q^4 + P^4Q^2 + P\tilde{P}R^2S^2 + P^2R^2 + \tilde{P}^4Q^2 \\ & + P^2S^4 + P^2S^2 + P^4S^2 + P\tilde{P}Q^4 + P\tilde{P}Q^2 + P\tilde{P}R^4 \\ & + P^2Q^2R^2 + P^2\tilde{P}^2S^2 + \tilde{P}^2Q^2R^2 + P^3\tilde{P}S^2 + P\tilde{P}^3S^2 \\ & + \tilde{P}^2Q^2S^2 + P^2R^2S^2 \end{aligned} \tag{13}$$

$$\begin{aligned} C P \tilde{P} = {} & P^3 + P^3Q^2 + \tilde{P}^3Q^2 + PQ^2 + PQ^4 + PR^2 + P^3S^2 \\ & + P^2\tilde{P} + \tilde{P}Q^2 + \tilde{P}Q^4 + P^2\tilde{P}Q^2 + P\tilde{P}^2Q^2 + PQ^2R^2 \\ & + \tilde{P}Q^2R^2 + PQ^2S^2 + PR^2S^2 + P\tilde{P}^2S^2, \end{aligned} \tag{14}$$

$$\begin{aligned} D P \tilde{P} = {} & P^2 + Q^2 + Q^4 + \tilde{P}^2Q^2 + P^2Q^2 + P^2S^2 + Q^2S^2 \\ & + P\tilde{P}S^2, \end{aligned} \tag{15}$$

$$E P \tilde{P} = P + PS^2 + PQ^2 + \tilde{P}Q^2. \tag{16}$$

**Claim** *$x$ is conjugated over $K$ to $x_0$ or $x_0+1$.*

*Proof of the Claim.* We shall confirm this by showing that either $x$ and $x_0$ or $x$ and $x_0+1$ have the same minimal polynomial over $K$, i.e.

$$\begin{array}{lcl} P(x) = P(x_0), & & P(x) = P(x_0+1), \\ Q(x) = Q(x_0), & & Q(x) = Q(x_0+1), \\ R(x) = R(x_0), & \text{or} & R(x) = R(x_0+1), \\ S(x) = S(x_0), & & S(x) = S(x_0+1), \\ T(x) = T(x_0) & & T(x) = T(x_0+1). \end{array}$$

Here the notations $P(x), Q(x), \ldots$, need a comment. For instance $P(x)$ refers to the substitution of $x$ and its conjugates over $K$ in $P$ considered as symmetric polynomial, that is $P(x) = x\varphi(x)\varphi^2(x)\varphi^3(x)\varphi^4(x)$. (Of course $P(x)$ and $T(x)$ are commonly known as the *norm* resp. the *trace* of $x$ over $K$.) The equations $T(x) = T(x_0)$ resp. $T(x) = T(x_0+1)$ can be replaced by $\tilde{P}(x) = \tilde{P}(x_0)$ resp. $\tilde{P}(x) = \tilde{P}(x_0+1)$.

We can state equivalently that we have to show:

*The system of equations (12) - (16) has precisely two solutions*

$$(P, \tilde{P}, Q, R, S) = (P_i, \tilde{P}_i, Q_i, R_i, S_i) \ (i = 1, 2).$$

Accordingly from now on we consider $P, \tilde{P}, Q, R, S$ as unknowns (in $K$) satisfying (12) - (16) in order to prove this statement.

We remark that by (7) - (11) the two solutions of (12) - (16) must be related as follows

$$\begin{aligned}
P_2 &= \tilde{P}_1, \\
\tilde{P}_2 &= P_1, \\
Q_2 &= Q_1 + S_1 + 1, \\
R_2 &= R_1 + S_1, \\
S_2 &= S_1,
\end{aligned}$$

which implies that $P + \tilde{P}, P\tilde{P}$ and $S$ must be uniquely determined by (12) - (16). Of course we have not proved this yet. But it gives us a hint how we should proceed in our proof. The uniqueness of $P+\tilde{P}, P\tilde{P}$ and $S$ is precisely what we shall demonstrate in what follows.

To this end we first eliminate the variable $R$ by computing resultants of (12), (13) and of (12), (14). After division by suitable powers of $P, \tilde{P}$ or $P+\tilde{P}$ we deduce the following two equations with variables $P, \tilde{P}, Q, S$:

$$\begin{aligned}
&P^4 + P^4 Q^2 + \tilde{P}^4 Q^2 + P^4 S^2 + P^2 \tilde{P}^2 S^2 + A P^3 S^2 + A P \tilde{P}^2 S^2 \qquad (17)\\
&+ A P^3 + A P^2 \tilde{P} + A P^2 \tilde{P} Q^2 + A \tilde{P}^3 Q^2 + A P^3 Q^2 + A P \tilde{P}^2 Q^2 \\
&+ A^2 P \tilde{P} + B P \tilde{P}^3 + B P^3 \tilde{P} = 0 \\
&P^4 + P^2 \tilde{P}^4 + P^4 \tilde{P}^2 + P^4 Q^4 + \tilde{P}^4 Q^4 + P^4 S^4 + P \tilde{P}^3 Q^2 S^2 \qquad (18)\\
&+ P^3 \tilde{P} S^4 + P^3 \tilde{P} S^2 + P^3 \tilde{P} Q^2 S^2 + A P^2 \tilde{P} S^2 + A P \tilde{P}^2 Q^2 \\
&+ A P^2 \tilde{P} Q^2 + A P^2 \tilde{P} + C P^2 \tilde{P}^3 + C P^3 \tilde{P}^2 = 0
\end{aligned}$$

By (16) we have

$$Q^2 = \frac{P + P S^2 + P \tilde{P} E}{P + \tilde{P}}$$

Using this equation to eliminate $Q$ in (15), (17) and (18) we get equations which are symmetric in $P, \tilde{P}$:

$$\begin{aligned}
S^4 + S^2 E (P + \tilde{P}) &= E (P + \tilde{P})^3 + (D+1)(P + \tilde{P})^2 + E (P + \tilde{P}) \qquad (19)\\
&\quad + E^2 P \tilde{P} + 1, \\
S^2 (P + \tilde{P})^2 &= (A E + B + 1)(P + \tilde{P})^3 + A (P + \tilde{P}) \qquad (20)\\
&\quad + A^2 + E + P \tilde{P} \\
S^2 (1 + E (P + \tilde{P})) &= (E^2 + 1)(P + \tilde{P})^2 + C (P + \tilde{P}) + A E + 1. \qquad (21)
\end{aligned}$$

We can substitute the variables $P, \tilde{P}$ by

$$\begin{aligned}
\varrho &= P + \tilde{P}, \\
\theta &= P \tilde{P}.
\end{aligned}$$

Eliminating $S$ from all pairs of the preceding three equations (19), (20), (21) we get then three equations in $\varrho$, $\theta$:

$$\theta\,(1+E\,\varrho) = \varrho^4 + (A\,E^2+B\,E+C)\,\varrho^3 + (A\,E+B)\,\varrho^2 + (A^2\,E+A)\,\varrho + A^2, \tag{22}$$

$$\theta\,E^2\,(1+E\,\varrho)^2 = E^3\,\varrho^5 + (1+D\,E^2)\,\varrho^4 + C\,E^2\,\varrho^3 + (A\,E^3+C\,E+C^2+D+1)\,\varrho^2 + A\,E^2\,\varrho + A^2\,E^2. \tag{23}$$

$$\theta\,E\,\varrho^3(1+E\,\varrho)+\theta^2 = E\,\varrho^7 + (D+1)\,\varrho^6 + (A\,E^2+B\,E)\,\varrho^5 + (A\,E+A^2\,E^2+B^2)\,\varrho^4 + A^2\,E\,\varrho^3 + A^2\,\varrho^2 + A^4 \tag{24}$$

From (22) - (24) we can eliminate $\theta$. Thus, covering also the cases $\varrho = 1/E$ and $P = \tilde{P}$, we conclude

$$\varrho^2 = \frac{A^2\,E^4 + A\,E^3 + B\,E^2 + C\,E + C^2 + D + 1}{A\,E^5 + B\,E^4 + C\,E^3 + D\,E^2 + E^2 + 1},$$

i.e. $\varrho$ is unique, since we have represented it as a rational function of the constant $a$. Here $A\,E^5 + B\,E^4 + C\,E^3 + D\,E^2 + E^2 + 1$ is nonzero, because otherwise $1/E \in K$ would be a zero of the minimal polynomial

$$m_a(\xi) = \xi^5 + E\xi^4 + D\xi^3 + C\xi^2 + B\xi + A$$

of $a$ over $K$; a contradition. Now we can substitute $\varrho$ in (22) (resp. (24) if $\varrho = 1/E$) to see that also $\theta$ is unique. Hence in the solutions $(P, \tilde{P}, Q, R, S)$ of (12) - (16) there appear precisely two $(P_i, \tilde{P}_i)$ $(i = 1, 2)$. They satisfy $P_i + \tilde{P}_i = \varrho$ and $P_i\tilde{P}_i = \theta$. We claim that for each of the pairs $(P_i, \tilde{P}_i)$ $(i = 1, 2)$ there are uniquely determined $Q_i, R_i, S_i$ such that we get a solution of the entire system (12) - (16). (The existence of a solution is obvious, since we assume that $a = \Delta(x_0)$ for some fixed $x_0$.) From equation (20) we even conclude that $S$ is unique for all solutions, i.e. $S_1 = S_2$. Given $P_i, \tilde{P}_i, S_i$ we have fixed $Q_i$ by (16). Finally then $R_i$ is uniquely given by (14). This completes the proof of the Claim.

It is now easy to sharpen the Claim and thus to complete the proof of our theorem. In fact we know that $x = \varphi^i(x_0 + \epsilon)$ for some $i < 5$ and $\epsilon \in \{0, 1\}$. Hence

$$a = \Delta(x) = \Delta(\varphi^i(x_0+\epsilon)) = \varphi^i(\Delta(x_0+\epsilon)) = \varphi^i(a),$$

which in view of $a \notin K$ implies that $i = 0$ and therefore $x = x_0 + \epsilon$. □

## References

1. T. Beth and C. Ding, "On almost perfect nonlinear permutations," Advances in Cryptology – EUROCRYPT '93, T. Helleseth (ed.), Lecture Notes in Computer Science, vol. 765, pp. 65-76, Springer-Verlag, 1994.

2. A. Canteaut, private communication.
3. A. Canteaut, P. Charpin and H. Dobbertin, "Divisibility of cyclic codes and highly nonlinear functions on $\mathbf{F}_{2^m}$," SIAM Journal of Discrete Mathematics, to appear.
4. C. Carlet, P. Charpin, V. Zinoviev, "Codes, bent functions and permutations suitable for DES-like cryptosystems," Design, Codes and Cryptography **15** (1998), pp. 125-156.
5. H. Dobbertin, "Almost perfect nonlinear power functions on GF($2^n$): The Welch case," IEEE Trans. Inf. Theory **45** (1999), pp. 1271-1275.
6. H. Dobbertin, "Almost perfect nonlinear power functions on GF($2^n$): The Niho case," Information and Computation **151** (1999), pp. 57-72.
7. H. Dobbertin, "Uniformly representable permutation polynomials," preprint.
8. T. Helleseth, C. Rong and D. Sandberg, "New families of almost perfect nonlinear power mappings," IEEE Trans. Inf. Theory, to appear.
9. H. Janwa and R.M. Wilson, "Hyperplane sections of Fermat varieties in $P^3$ in chararacteristic 2 and some applications to cyclic codes," in Proceedings AAECC-10 (G. Cohen, T. Mora and O. Moreno Eds.), Lecture Notes in Computer Science, vol. 673, Springer-Verlag, 1993, pp. 180-194.
10. T. Kasami, "The weight enumerators for several classes of subcodes of the second order binary Reed-Muller codes," Information and Control **18** (1971), pp. 369-394.
11. K. Nyberg, "Differentially uniform mappings for cryptography," Advances in Cryptology – EUROCRYPT '93, T. Helleseth (ed.), Lecture Notes in Computer Science, vol. 765, pp. 55-64, Springer-Verlag, 1994.

# On Non-Abelian Semi-Regular Relative Difference Sets

Dominic Elvira*[1] and Yutaka Hiramine[2]

[1] Department of Mathematics
Graduate School of Science and Technology
Kumamoto University
Kurokami, Kumamoto
Japan

[2] Department of Mathematics
Faculty of Education
Kumamoto University, Kurokami, Kumamoto
Japan

**Abstract** In their paper [1] J.A. Davis, J. Jedwab and M. Mowbray gave new constructions of abelian semi-regular RDS's by using $(k, m, t)$-building sets on abelian groups. In this article, we generalize this concept by defining a $t$-building set, a certain collection of elements from a group ring. We then study $t$-building sets on cyclic $p$-groups and apply our results to show that there is no non-trivial semi-regular RDS in any dihedral group. We also show that for any dicyclic group, its forbidden subgroup of even order is isomorphic to $\mathbb{Z}_2$.

## 1 Introduction

An $(m, u, k, \lambda)$ *relative difference set* (RDS) in a group $G$ of order $\mu$ relative to a normal subgroup $U$ of order $u$ is a $k$-element subset $R$ of $G$ such that the number of elements in the set $\{(r_1, r_2) \mid r_1, r_2 \in R,\ r_1 r_2^{-1} = g\}$ is $\lambda$ for every $g \in G$, $g \neq 1$ if $g \in G \setminus U$ or 0 if $g \in U$. We often call $U$ the *forbidden subgroup*. If $k = u\lambda$, we call $R$ a *semi-regular* RDS and its parameters are given by $(u\lambda, u, u\lambda, \lambda)$ (see [3] Chapter 4). Moreover, if $u = 1$, $R$ is called a *trivial semi-regular RDS*. Any group $G$ itself is a trivial semi-regular RDS in $G$ relative to $\{1\}$. In this article, we only consider nontrivial semi-regular RDS's.

Recently, J.A. Davis, J. Jedwab and M. Mowbray made two constructions of abelian semi-regular RDS's in [1]. In one of their constructions, they considered $(k, m, t)$-building sets on abelian groups and by using group characters, they were able to construct new sets of semi-regular RDS's in groups of non-prime power order.

In this article, we consider a modification of these building sets and define $t$-building sets, instead. A *$t$-building set* $(t > 1)$ on an abelian group $A$ relative

* The first author is a faculty member of Philippine Normal University (PNU), Manila on study leave at Kumamoto University under a Monbusho grant.

to a subgroup $U$ is a collection $\{s_1, \dots, s_t\}$ of elements of $\mathbb{Z}[A] \setminus \mathbb{Z}\widehat{A}$ such that for each non-principal character $\chi$ of $A$

(i) there exists a unique $i \in \{1, \dots, t\}$ such that $\chi(s_i) \neq 0$ if $\chi$ is non-principal on $U$, and
(ii) $\chi(s_i) = 0$ for every $i \in \{1, \dots, t\}$ if $\chi$ is principal on $U$.

In Section 2, we focus on $t$-building sets in a cyclic $p$-group where $p$ is a prime and show that $|U| \geq p^t$ (Theorem 2.1). We also study semi-regular RDS's in some *non-abelian* groups, especially groups $G$ containing an abelian subgroup $A$ of index 2 inverted by an element of $G \setminus A$. As an application of 2-building sets, we show in Section 3 that there is no non-trivial semi-regular RDS in any dihedral group (Theorem 3.2). We note that in Proposition 7 of [2], N. Ito has shown that any group with a non-abelian dihedral Sylow 2-subgroup has no non-trivial $(n, 2, n, n/2)$ RDS.

For a subset $X$ of $G$, we set $X^{(-1)} = \{x^{-1} \mid x \in X\}$. Throughout this article we identify a subset $X$ of $G$ with a group ring element $\widehat{X} = \sum_{x \in X} x \in \mathbb{Z}[G]$. Set $\mathbb{Z}X = \{mX | m \in \mathbb{Z}\}$.

## 2 A $t$-Building Set in $\mathbb{Z}_{p^n}$

In this section, we study $t$-building sets and semi-regular RDS's in dihedral groups.

**Theorem 2.1.** *If a cyclic $p$-group has a $t$-building set relative to its subgroup $U$, then $|U| \geq p^t$.*

*Proof.* Let $\{s_1, \dots, s_t\}$ be a $t$-building set and $\Omega$ the set of characters of $A$ non-principal on $U$. Set $\Gamma_i = \{\chi \in \Omega \mid \chi(s_i) \neq 0\}$. By definition, $\Omega = \Gamma_1 \cup \cdots \cup \Gamma_t$ and $\Gamma_j \cap \Gamma_k = \phi$ for any distinct $j$ and $k$. As $s_i \notin \mathbb{Z}A$, $\Gamma_i \neq \phi$ for any $i$ by applying the inversion formula (cf. Lemma 1.2.2 of [3]).

Consider an equivalence relation on $\Omega$ defined by $\chi_1 \sim \chi_2$ iff $\chi_1 = \chi_2^\sigma$ for some Galois automorphism $\sigma$ of $\mathbb{Q}(\zeta)$, where $\zeta = \zeta_{p^n}$ and $A = \langle a \rangle \simeq \mathbb{Z}_{p^n}$. Let $\Omega_1, \cdots, \Omega_r$ be the equivalence classes on $\Omega$. For any $s_i$ and $\chi_1, \chi_2 \in \Omega_j$, $\chi_1(s_i) = 0$ iff $\chi_2(s_i) = 0$. Hence each $\Gamma_i$ is a union of some $\Omega_j$'s. Thus $t \leq r$.

It is well known that for any character $\chi \in A^*$ there exists a unique element $g = a^m \in A$ such that $\chi(a^\ell) = \zeta^{m\ell}$. We note that $\chi$ is principal on $U$ iff $\chi(a^{p^{n-u}}) = 1$ iff $p^u$ divides $m$. From this we have $r = u$. Therefore $t \leq u$.

**Lemma 2.2.** *Let $G$ be a group with an abelian normal subgroup $A$ of index 2 inverted by an element of $G \setminus A$. If $G$ has a normal subgroup $U$ not contained in $A$, then $G/(A \cap U)$ is an elementary abelian 2-group.*

*Proof.* Let $s \in G \setminus A$. Then there exist elements $b \in \langle s \rangle$ of odd order and a 2-element $t \in \langle s \rangle$ such that $s = bt$. Clearly $b \in A$ and $t \notin A$. Since $a^{-1} = a^s = a^{bt} = a^t$ for any $a \in A$, $t$ also inverts $A$. Moreover, $t^2$ is inverted by $t$ as $t^2 \in A$. Hence $t^{-2} = (t^2)^t = t^2$. Clearly $G = \langle t \rangle A$ and $t^4 = 1$. Set $B = A \cap U$ $(\lhd G)$ and $\overline{G} = G/B$. Then $\langle \bar{t} \rangle = U/B \simeq \mathbb{Z}_2$ and $\overline{G} = \langle \bar{t} \rangle \times \overline{A}$. Hence $\langle \bar{t} \rangle$ centralizes $\overline{A}$. On the other hand, as $t$ inverts $A$, $\langle \bar{t} \rangle$ also inverts $\overline{A}$. Thus the lemma holds.

**Lemma 2.3.** *Let $G$ be a group with an abelian normal subgroup $A$ of index 2 inverted by an element $t \in G \backslash A$ of order 2 or 4. Suppose $G$ has a $(\lambda u, u, \lambda u, \lambda)$ RDS, say $R$, relative to a normal subgroup $U$ of $G$. Set $S = R \cap A$ and $T = Rt^{-1} \cap A$. Then $R = S + Tt$ and the the following hold.*

(i) *Assume $U \not\subset A$ and set $U_0 = A \cap U$. Then $ST(1+t^2) = \lambda(A - U_0)$ and $SS^{(-1)} + TT^{(-1)} = u\lambda + \lambda(A - U_0)$. Moreover, $A/U_0$ is an elementary abelian 2-group of order $u\lambda = 2^{2e}$ for an integer $e \geq 1$ and $\{|S|, |T|\} = \{2^{2e-1} \pm 2^{e-1}\}$.*
(ii) *Assume $U \subset A$. Then $ST(1+t^2) = \lambda A$ and $SS^{(-1)} + TT^{(-1)} = u\lambda + \lambda(A - U)$. Moreover, $|S| = |T| = u\lambda/2$.*

*Proof.* Since $R$ is a disjoint union of $R \cap A$ and $R \cap At$, $R = S + Tt$. We note that $t^4 = 1$ and $t^2 \in A$. As $R^{(-1)} = S^{(-1)} + t^{-1}T^{(-1)}$, it follows that $RR^{(-1)} = (S + Tt)(S^{(-1)} + t^{-1}T^{(-1)}) = (SS^{(-1)} + TT^{(-1)}) + ST(1+t^2)t$.

Assume $U \not\subset A$. Then $RR^{-1} = \lambda u + \lambda(G-U) = u\lambda + \lambda(A-U_0) + \lambda(A-U_0)t$. Hence we have $ST(1+t^2) = \lambda(A-U_0)$ and $SS^{(-1)} + TT^{(-1)} = u\lambda + \lambda(A-U_0)$. As $|U/U_0| = 2$, $|G/U_0| = u\lambda/2$. By Lemma 2.2, $u\lambda = 2^m$ for an integer $m$. Set $x = |S|$ and $y = |T|$. Then $x + y = u\lambda = 2^m$. Moreover, since $ST(1+t^2) = \lambda(A - U_0)$, $2xy = \lambda(u^2\lambda/2 - u/2) = 2^{2m-1} - 2^{m-1}$. Therefore $\{x, y\} = \{2^{m-1} \pm 2^{\frac{m-2}{2}}\}$. Consequently, $m = 2e$ for some positive integer $e$. Thus (i) holds.

Assume $U \subset A$. As $RR^{-1} = \lambda u + \lambda(G - U) = u\lambda + \lambda(A - U) + \lambda At$, we have $ST(1+t^2) = \lambda A$ and $SS^{(-1)} + TT^{(-1)} = u\lambda + \lambda(A - U)$. As each coset $Ux$ $(x \in G)$ contains exactly one element of $R$, we have $|S| = |A/U| = \lambda u/2$ and $|Tt| = |At|/|U| = \lambda u/2$. Thus (ii) holds.

In the rest of this section, we assume the following.

**Hypothesis.** Let $G$ be a group with an abelian normal subgroup $A$ of index 2 inverted by an element $t \in G \setminus A$ of order 2. Assume that $G$ has a $(\lambda u, u, \lambda u, \lambda)$ RDS, say $R$, relative to a p-subgroup $U$ of $A$ of order $u (= p^b > 1)$. Set $R = S + Tt$, where $S$ and $T$ are subsets of $A$.

**Definition 2.4.** Let $P$ be a Sylow p-subgroup of $A$ and $H$ a Hall $p'$- subgroup of $A$ (i.e. $A = P \times H$, and $p \not| \, |H|$). Set $|P| = p^c$ and $|H| = h$. Let $\sigma$ be a mapping from $\mathbb{Z}[A]$ into $\mathbb{Z}[H]$ induced by the projection $A = P \times H \longrightarrow H$. Similarly, we define a mapping $\tau$ from $\mathbb{Z}[A]$ into $\mathbb{Z}[P]$ by extending the projection $A = P \times H \longrightarrow P$.

**Lemma 2.5.** $\sigma(S) = \sigma(T) = p^{c-b}H$.

*Proof.* Let $\chi$ be any non-principal character of $H$ and $1_P$ a principal character of $P$. As $\widetilde{\chi} = 1_P \times \chi$ is a non-principal character of $A$ such that $(1_P \times \chi)|_P$ is principal, it follows that $\widetilde{\chi}(S) = \widetilde{\chi}(T) = 0$ by Lemma 2.3. On the other hand, $\widetilde{\chi}(S) = \chi(\sigma(S))$ and $\widetilde{\chi}(T) = \chi(\sigma(T))$. Hence $\chi(\sigma(S)) = \chi(\sigma(T)) = 0$ for any non-principal character $\chi \in H^*$. By the inversion formula (see [3]) and Lemma 2.3(ii), $\sigma(S) = \sigma(T) = wH$, where $w = u\lambda/(2h)$. As $u\lambda/(2h) = |A|/(uh) = p^c h/(p^b h) = p^{c-b}$, we have the lemma.

**Lemma 2.6.** $\lambda = 2hp^{c-2b}$.

*Proof.* By Lemma 2.3(ii),

$$SS^{(-1)} + TT^{(-1)} = u\lambda + \lambda(A - U) \tag{1}$$

Hence, for any $s_1, s_2 \in S$, $s_1 \neq s_2$ we have

$$s_1 s_2^{-1} \in P \setminus U \Longleftrightarrow \sigma(s_1) = \sigma(s_2).$$

Similarly, for any $t_1, t_2 \in T$, $t_1 \neq t_2$ we have

$$t_1 t_2^{-1} \in P \setminus U \Longleftrightarrow \sigma(t_1) = \sigma(t_2).$$

We now count the number of elements of $P \setminus U$ in each side of (1). Then $2p^{c-b}(p^{c-b}-1)h = \lambda(p^c - p^b)$ by Lemma 2.5. From this, we have $\lambda = 2hp^{c-2b}$.

**Lemma 2.7.** $\tau(S), \tau(T) \notin \mathbb{Z}P$.

*Proof.* Assume $\tau(S) = kP$ for some positive integer $k$. Then, by Lemma 2.6, $k = u\lambda/(2p^c) = h/p^b$, a contradiction. Therefore $\tau(S) \notin \mathbb{Z}P$. Similarly $\tau(T) \notin \mathbb{Z}P$.

**Theorem 2.8.** *Let $G$ be a group with an abelian normal subgroup $A$ of index 2 inverted by an element $t \in G \setminus A$ of order 2. Assume that $G$ has a $(\lambda u, u, \lambda u, \lambda)$ RDS, say $R$, relative to a $p$-subgroup $U$ of $A$. Set $R = S + Tt$, where $S$ and $T$ are subsets of $A$. Then the following hold:*

(i) *Let $P$ be a Sylow $p$-subgroup of $A$ and let $\tau$ be a natural projection from $A$ into $P$. Then $\{\tau(S), \tau(T)\}$ is a 2-building set on $P$ relative to $U$.*
(ii) *If $P$ is cyclic, then $|U| \geq p^2$*

*Proof.* Let $\chi$ be any non-principal character of $P$ and $1_H$ a principal character of $H$. Set $\widetilde{\chi} = \chi \times 1_H \in A^*$. Then $\widetilde{\chi}(S) = \chi(\tau(S))$ and $\widetilde{\chi}(T) = \chi(\tau(T))$. It follows from Lemma 2.3(ii) that

$$\chi(\tau(S))\chi(\tau(T)) = 0 \tag{2}$$

By Lemma 2.3(ii),

$$|\chi(\tau(S))|^2 + |\chi(\tau(T))|^2 = c \cdot u\lambda \tag{3}$$

where $c = 0$ or $1$ depending on whether $\chi_{|U}$ is principal or non-principal. This fact, together with (2), (3) and Lemma 2.7, gives the proof of (i). By (i) and Theorem 2.1, we have (ii).

# 3 Applications

As an application of Theorem 2.8 we show the following.

**Theorem 3.1.** *Let $G$ be a group with an abelian normal subgroup $A$ of index 2 inverted by an element $t \in G \setminus A$ of order 2. Assume that $G$ has a $(\lambda u, u, \lambda u, \lambda)$ RDS, say $R$, relative to a normal subgroup $U$ of $G$. Then a Sylow $p$-subgroup of $A$ is non-cyclic for any prime $p$ dividing $u$.*

*Proof.* If $U \not\subset A$, then $p = 2$ and $u\lambda = 2$ by Lemma 2.3(i). Hence we have $G \simeq \mathbb{Z}_4$, contrary to the assumption. Thus $U \subset A$.

Let $p$ be a prime divisor of $u$ and $P$ a Sylow $p$-subgroup of $A$. Assume that $P$ is cyclic. Then, there exists a subgroup $N$ of $U$ of index $p$. Consider $\overline{G} = G/N$. Then, by a result of Elliot-Butson (see Lemma 1.1.12 of [3]), $\overline{R}$ is a semi-regular RDS relative to $\overline{U}(\simeq \mathbb{Z}_p)$. Applying Theorem 2.8(ii) we have $|\overline{U}| \geq p^2$, contrary to $[U : N] = p$. Therefore the theorem holds.

As a corollary, we have

**Theorem 3.2.** *There is no semi-regular RDS in any dihedral group.*

We now consider *the dicyclic group*

$$Q_{4m} = \langle x, y \mid x^m = y^2,\ y^4 = 1,\ y^{-1}xy = x^{-1} \rangle \quad (m \geq 2)$$

of order $4m$ and *the semi-dihedral group*

$$SD_{2^{n+1}} = \langle x, y \mid x^{2^n} = y^2 = 1,\ z = x^{2^{n-1}},\ yxy = x^{-1}z \rangle \quad (n \geq 2)$$

of order $2^{n+1}$. We then apply Theorem 3.2 to show existence or non-existence of semi-regular RDS's in these groups.

**Corollary 3.3.** *Let $R$ be a semi-regular RDS in a dicyclic group $Q_{4m}$ relative to a normal subgroup $U$. If $|U|$ is even, then $|U| = 2$.*

*Proof.* Let $G = Q_{4m}$ be as mentioned above and set $A = \langle x \rangle$, $t = y$. By Lemma 2.3(i), $U \subset A$. Let $N = \langle t^2 \rangle (\leq U)$ and consider $\overline{G} = G/N$. Then $\overline{G} \simeq D_{2m}$. Assume $|U| > 2$. Then $\overline{U} \neq \overline{1}$ and so $\overline{R}$ is a semi-regular RDS in $D_{2m}$ relative to $\overline{U}$ by Elliot-Butson's result. This is contrary to Theorem 3.2. Thus $|U| = 2$.

*Remark 3.4.* Let $m = 2^{n-2}$ $(n \geq 3)$. Applying the product construction by J.A. Davis (see Result 2.4 of [4]), an example of a semi-regular RDS, say $R$, in $Q_{2^n}$ is given by the following:

$$R = (1 + x^{2^{n-2}})(1 + y)(1 + x^{2^{n-3}}y)(1 + x^{2^{n-4}}y) \times \cdots \times (1 + x^2y)(1 + xy).$$

**Corollary 3.5.** *Let $R$ be a semi-regular RDS in a semi-dihedral group $SD_{2^{n+1}}$ relative to $U$. Then $|U| = 2$.*

*Proof.* Let $G = SD_{2^{n+1}}$. We note that $Z(G) \simeq \mathbb{Z}_2$ and $G/Z(G) \simeq D_{2^n}$. In particular, $Z(G) \leq U$. Similarly as in the proof of Corollary 3.3, we have the corollary.

**Acknowledgment**

We thank the referees for their valuable suggestions. They showed us a shorter proof of Theorem 2.1.

## References

1. Davis, J.A., Jedwab, J., Mowbray, M. : New Families of Semi-Regular Relative Difference Sets. Designs, Codes and Cryptography **13** (1998) 131–146
2. Ito, N. : On Hadamard groups. J. of Algebra **168** (1994) 981–987
3. Pott, A. : Finite Geometry and Character Theory. Lecture Notes in Mathematics 1601, Springer-Verlag, Berlin (1995)
4. Schmidt, B. : On $(p^a, p^b, p^a, p^{a-b})$ Relative Difference Sets. J. Algebraic Combin. **6** (1997) 279–297

# Applications of Arithmetical Geometry to Cryptographic Constructions

Gerhard Frey*

Institut für Experimentelle Mathematik
Universität GH Essen
Ellernstr. 29
45326 Essen
Germany
e-mail: frey@exp-math.uni-essen.de

**Abstract** Public key cryptosystems are very important tools for data transmission. Their performance and security depend on the underlying crypto primitives. In this paper we describe one such primitive: The Discrete Logarithm (DL) in cyclic groups of prime order (Section 1).
To construct DL-systems we use methods from algebraic and arithmetic geometry and especially the theory of abelian varieties over finite fields. It is explained why Jacobian varieties of hyperelliptic curves of genus $\leq 4$ are candidates for cryptographically "good" abelian varieties (Section 2).
In the third section we describe the (constructive and destructive) role played by Galois theory: Local and global Galois representation theory is used to count points on abelian varieties over finite fields and we give some applications of Weil descent and Tate duality.

## 1 Cryptographic Background

### 1.1 Aim

To ensure a safe environment for data exchange and data storage is a very complex task involving problems from quite different areas like engineering and computer science but also "human factors" and legal problems, and as a small but crucial component, mathematics. It is the task of mathematics to provide the particles called crypto primitives around which cryptographic protocols are developed. To estimate the security of these protocols is a difficult problem which is impossible if one cannot trust the primitives.

In the following paper we shall try to give an overview over one special but most important family of crypto primitives based on the discrete logarithm problem in groups. We shall describe how deep methods from arithmetical geometry and especially the interaction between Galois theory and algebraic

* I would like to thank the organizers for the opportunity to participate in the $\mathbb{F}_{q^5}$-conference, which, because of their great expertise and warm hospitality, became a very interesting and inspiring event.

geometry over special fields can be used to construct them and to give estimates (or hints) for their reliability.

The basic ideas go back to Diffie and Hellman ([DH]). The main aim is the construction of a function $f$ mapping the natural numbers $\mathbb{N}$ (or, in practice, a finite subset of $\mathbb{N}$) to a finite set $A$ of $\mathbb{N}$ satisfying some "functional equations" and with the most important properties:

- The function $f$ can be evaluated rapidly at every element of $\mathbb{N}$ but
- for randomly chosen $y \in A$ the effort needed to compute $x \in \mathbb{N}$ with $f(x) = y$ is very large.

So we have to find functions $f$ together with algorithms to evaluate them with a proven small complexity, and we have to estimate the probability for finding the inverse images by proposing possible "attacks". In contrast to the first task in almost all cases the second one cannot be solved in a satisfactory way. What we can get are estimates for this complexity from above which will lead to reject some functions $f$ and the statement that "to our best knowledge" other functions are safe "at the moment".

Since the security of the whole crypto system depends on this rather vague statement we would feel much better if we could use for $f$ a *one way function* in the sense of information theory, or a *trap door one way function* (for the definition cf. [MOV]).
But since we even do not know whether such functions exist we have to use what we have and to be very sensitive to new developments in technology (e.g. "quantum computers" (cf. [Sho])) and to new results in mathematics. We refer to the rather spectacular progress in recent years in factorizing numbers (by sieve methods in number fields), in counting points on elliptic curves over finite fields (cf. [Sch2] and the references therein) and in applying index-calculus methods to Discrete Logarithm Systems ([Ga] and [EG]).

**Acknowledgement** I would like to thank A. Enge and M. Müller for their help and the many valuable comments they gave.

## 1.2 Exponential and Discrete Logarithm Systems

Let $A$ be a finite subset of $\mathbb{N}$. For normalization reasons we shall assume that $1 \in A$. Let

$$e : \mathbb{N} \times A \to A$$

be a function satisfying a *functional equation*: For all $n_1, n_2 \in \mathbb{N}$ one has

$$e(n_1, e(n_2, 1)) = e(n_1 \cdot n_2, e(1, 1)).$$

This is enough to organize a **key exchange:** Two parties $M_1, M_2$ want to agree on a common *secret key* based on $e$. The communication uses a public channel.

**Key exchange scheme** [1]

i) $M_i$ chooses (as randomly as possible) a number $n_i$ which never leaves the secure environment of $M_i$.
ii) $M_i$ sends $y_i := e(n_i, 1)$.
iii) $M_1$ computes $e(n_1, y_2)$, $M_2$ computes $e(n_2, y_1)$.

Since

$$e(n_1, y_2) = e(n_1, e(n_2, 1)) = e(n_1 n_2, e(1,1)) = e(n_2, e(n_1, 1)) = e(n_2, y_1)$$

$M_1$ and $M_2$ share the common key

$$S = e(n_1 n_2, e(1,1)).$$

It is obvious that the secrecy of $S$ depends on the secrecy of $n_i$ and so on the difficulty to compute $n_i$ from the knowledge of $y_i = e(n_i, 1)$.

So we *assume* that the function $e(n, a)$ as a function of the first variable behaves like a one way function. Then we call

$$(e : \mathbb{N} \times A \to A; e(1,1))$$

an *exponential system with base point* $e(1,1)$.

For many applications one needs more "structure". For instance the *signature* of a message $m$ can use a scheme of *ElGamal type* if we have a rapidly computable function

$$\oplus : A \times A \to A$$

which we require to be *associative*. In other words we require that $A$ is a semigroup.

Using this associativity we define in the usual way for $n \in \mathbb{N}$ and $a \in A$

$$n \circ a$$

as the $n - 1$ fold application of $\oplus$ to $a$.

*Remark 1.1.* Of course the actual computation of $n \circ a$ uses the well known method of (for instance) developing $n$ in the binary system and then "adding and doubling" accordingly. This bounds the number of compositions $\circ$ by 2 log $n$. Several variants can be used to accelerate and sometimes precalculations speed up even more.

Use this to define

$$e : \mathbb{N} \times A \to A$$

[1] We do not go into details of how this protocol has to be refined in order to become as secure as its primitive, the function $e$.

by

$$e(n, a) := n \circ a.$$

Obviously the function $e$ satisfies the functional equation:
For all $n_1, n_2 \in \mathbb{N}$, $a \in A$ we have

$$e(n_1, e(n_2, a)) = e(n_1 n_2, a)$$

and especially

$$e(n_1, e(n_2, 1)) = e(n_1 n_2, e(1, 1)).$$

**Signatures** Let $e$ be as above. $e$ satisfies a second functional equation

$$e(n_1, a) \oplus e(n_2, a) = e(n_1 + n_2, a).$$

This equation is used for signature schemes of El Gamal type. We shall give one variant:

**Aim:** $M$ wants to sign a message $m$ in such a way that everybody can check the authenticity of $m$ but no one can fake the contents of $m$ or the name of $M$.

To *initialize* the system $M$ chooses, again randomly and secretly, his **private key** $x \in \mathbb{N}$ and publishes his **public key** $Y := e(x, 1)$.
This number $Y$ identifies $M$ in public.

**Signing** a message $M$ uses, in addition to the DL-system, a (publicly known) *hash function* $h$ which maps $\mathbb{N}$ to a set of numbers of bounded size. It has to have properties similar to those of one way functions. Especially it has to be impossible in practice to construct a number $z$ such that $h(z)$ is a given value. (For hash functions cf. [MOV].)

**Signature 1.2.** $M$ chooses a second random number $k$ and computes

$$s := h(m)x + h(e(k, 1))k.$$

The signed message consists of

$$(m, e(k, 1), s)$$

To check the authenticity of $(M, m)$ one computes

$$S = e(s, 1),\ P = e(h(m), Y)),\ H = e(h(e(k, 1)), e(k, 1)).$$

Now the functional equations of $e$ imply:

$$S = P \oplus H$$

if the signature is authentic. Otherwise it is rejected.

It is obvious that the security of the signature depends on the quality of $e(., 1)$ as one-way function, but it is easily seen that this is not enough.

### Logarithms

**Definition 1.3.** The notation is as above.
For given $a$ in $A$ and $b \in \{n \circ a, n \in \mathbb{N}\}$ the **logarithm of $b$ with respect to $a$** is a number $\log_a(b) := n_b \in \mathbb{N}$ with $n_b \circ a = b$.

**Definition 1.4.** The notation is as above. Let $C$ be a positive real number.

$$(\oplus : A \times A \to A, 1)$$

is a **Discrete Logarithm System (DL-systems) of exponential security** $C$ if for random elements $a, b \in A$ the computation of $\log_a(b)$ has (probabilistic) complexity[2] $\geq e^{C \cdot \log(|A|)}$.

So DL-systems with exponential security $C$ with $C$ not too small give rise to exponential systems. Our aim is to construct DL-systems which are candidates for exponential security 1/2.

## 1.3 Algebraic DL-Systems

Let $(G, \times)$ be a finite group.
For our purposes the way how to present the group elements and the composition law is essential.

**Definition 1.5.** A **numeration** $(A, f)$ of $G$ is a bijective map

$$f : G \to A$$

where $A$ is a finite subset of $\mathbb{N}$ containing 1.
A **presentation** of an abstract finite group $G$ is an embedding of $G$ into a group with numeration.

Assume that $(A, f)$ is a numeration of the finite group $G$ and that $g_0 \in G$ with $f(g_0) = 1$ is given.

Define

$$\oplus : A \times A \to A$$

by

$$a_1 \oplus a_2 := f(f^{-1}(a_1) \times f^{-1}(a_2)).$$

Then for $n \geq 1$ we have

$$e(n, a) = f(n \circ f^{-1}(a))$$

as one sees easily by induction. Especially we get: $e(n, 1) = f(n \circ g_0)$.

We want to make one thing completely clear: We require that $\oplus$ is rapidly computable *without* the knowledge of $f^{-1}$ and that we can think of $G$ as an abstract group (i.e. we look at it up to isomorphisms) but the security and the efficiency of the DL-System based on $\oplus$ will depend crucially on $f$.

We can refine our signature scheme 1.2:

[2] In the whole paper the measure for complexity is the number of needed bit operations.

**Signature 1.6.** Let $n$ be the order of $G$. Let $(A, f)$ be a numeration of $G$ and $e(.,.)$ as above. The signature procedure of a sender $M$ and a message $m$ is as described in Subsection 1.2 except that we replace in the equation of 1.2 all numbers by their smallest positive residue modulo $n$:
The signature of $m$ given by $M$ is $(m, e(k,1), s)$ with

$$s \equiv h(m)x + h(e(k,1))k \bmod n$$

and $x, k, s, h(.) \in \{0, \cdots, n-1\}$.

*Example 1.7.* We take a prime number $p$ and the set $G := \mathbb{Z}/p$ together with its addition. Hence we deal with "the" group with $p$ elements.
As numeration we take the map

$$f : G \to \{1, \ldots, p\}$$

given by

$$f(r + p\mathbb{Z}) := [r]_p$$

where $[r]_p$ is the smallest positive representative of the class of $r$ modulo $p$. The function $\oplus$ is given by

$$r_1 \oplus r_2 = [r_1 + r_2]_p$$

which is easy to compute from the knowledge of $r_i$.

What about security? We can assume that $a \in \{1, \ldots, p-1\}$ and $b = e(n, a) = [na]_p$. To determine $n$ we have to solve the equation

$$b = na + kp$$

with $k \in \mathbb{Z}$. But by using the *Euclidean algorithm* this task is done in $O(\log(p))$ operations in $\mathbb{Z}$ and so $n$ is computed very rapidly. Hence we do not get a Discrete Logarithm System of exponential security.

*Example 1.8.* Again we take $G = \mathbb{Z}/p$. In addition we choose a prime $q$ such that $p$ divides $q - 1$. As it is well known this implies the existence of an element $\zeta \neq 1$ in $\mathbb{Z}/q$ with $\zeta^p = 1$ (i.e. $\zeta$ is a primitive $p$-th root of unity). Now define for $1 \leq i \leq p$ the number $z_i := [\zeta^i]_q$ and for $\bar{i} = i + p\mathbb{Z} \in G$

$$f(\bar{i}) := [z_i - z_1 + 1]_q.$$

Then $f$ defines an injection of $G$ into $\{1, \ldots, q\}$ with $f(1 + p\mathbb{Z}) = 1$. Let $A$ denote the image of $f$.

Using the addition in $G$ we get for $a_i = f(x_i + p\mathbb{Z}) \in A$

$$a_1 \oplus a_2 = [[\zeta^{x_1 + x_2}]_q - z_1 + 1]_q.$$

As stressed above it is important that one can compute $a_1 \oplus a_2$ without knowing $x_1$ and $x_2$. For this purpose we use the rules for addition and multiplication in $\mathbb{Z}/p$ and get:

$$a_1 \oplus a_2 = [(a_1 + z_1 - 1)(a_2 + z_1 - 1) - z_1 + 1]_q$$

and

$$e(n, 1) = n \circ 1 = [z_1^n - z_1 + 1]_q.$$

What about security in this system? For fixed $a$ and random $b \in A$ we have to find $n$ in $\mathbb{N}$ with

$$b = e(n, a) = n \circ a = [a^n - z_1 + 1]_q.$$

Essentially this means that for one fixed $p$-th root of unity and one random $p$-th root of unity in the multiplicative group of $\mathbb{Z}/q$ one has to determine the exponent needed to transform the fixed root of unity into the random element. This explains the name "discrete logarithm" introduced above.
Indeed to our knowledge at present a system based on this numeration is quite safe though for all constants $C$ there is a number $p_0$ such that for all primes $p \geq p_0$ the corresponding DL-system it not of exponential security $C$. The best known method to compute the discrete logarithm is "subexponential" (cf. [Od]). The suggested size of $p$ is at least 1024 bits.

### 1.4 Generic Systems

In the examples in the last subsection we already had more structure available than necessary for us: The set $A$ was the image of the numeration $f$ of a *finite cyclic group* $G$. This will be so in the following sections, too.

This seems to be reasonable: The part of the numeration we use for key exchange and signatures is given by the numeration of the group generated by $f^{-1}(1)$. But it would be interesting to study whether one can win security by generating this numeration by a more complicated group like a general matrix group.

Next we used in the signature scheme 1.6 the order of $G$. There are crypto systems which are based on the assumption that one cannot compute this order or that one can compute it only by using a secret trapdoor function (RSA-like systems). In this paper we do not enter into these very interesting discussions but assume that the order of $G$ and even its prime factorization are given.
An easy application (by some people called "Pohlig-Hellman attack") of the Chinese remainder theorem and $p$-adic expansion shows that the security of the discrete logarithm attached to $G$ and so a fortiori of the cryptographic schemes is reduced to the difficulty to compute the DL in the subgroups of $G$ with prime order. Hence we have to restrict ourselves to cyclic groups of order $p$ where $p$ is a prime number which is sufficiently large.

We can formulate the mathematical task to be solved more precisely now: We have to find numerations $f$ of groups $\mathbb{Z}/p$ with $p$ large enough which satisfy:

- Time needed (probabilistically) for the computation of the logarithm in $f(\mathbb{Z}/p)$ is **exponential in log($p$).**
- Time needed to write down the elements and to execute the composition $\oplus$ is **polynomial in log($p$).**

But having decided to use the algebraic structure "group" in the construction of the crypto primitive "DL" we have added a lot of structure to our system. So it is important which methods are available to attack crypto systems which use "just" groups and the resulting DL-problems. For a very interesting discussion of different types of information which can weaken the protocols we refer to [MW].
Of course the worst case is that the discrete logarithm can be computed. Here the amazing fact is that we can do much better than brute force attacks: The Baby-Step-Giant-Step method of Shanks as well as the $\rho$- and $\lambda$-methods of Pollard work in every finite group. All of them have time complexity $O(p^{1/2})$ and the methods of Pollard need very little storage space (for more details cf. [MOV]).

These attacks give a first upper bound for the optimal exponential complexity of the Discrete Logarithm in finite groups: The constant $C = 1/2$ is the best one can hope for, and so we are forced to choose $p$ of a size near to $10^{60}$ instead of $10^{30}$.

Our aim is to construct groups of the necessary size for which no attacks of lower complexity than these generic attacks are known.

### 1.5 The Index-Calculus Attack

In reality we shall have to use a concrete presentation of our group. In many examples there are elements in $G$ which are easier to deal with, and this gives rise to the index-calculus attack.

The principle of index-calculus methods in abelian groups $G$ is to find a "factor base" consisting of relatively few elements and to compute $G$ as a $\mathbb{Z}-$module given by the free abelian group generated by the base elements modulo relations.

Then one has to prove that with reasonably high probability every element of $G$ can be written (fast and explicitly) as a sum of elements in the factor base.

The important task in this method is to balance the number of elements in the factor base to make the linear algebra over $\mathbb{Z}$ manageable and to "guarantee" smoothness of enough elements with respect to this base. Typically successes give rise to algorithms for the computation of the DL in $G$ which have **subexponential** complexity and so, for large enough order of $G$, the DL-systems have a poor exponential security.

The effectiveness of this approach is the reason for the large numbers of bits required for safe RSA-systems and for DL-systems based on the multiplicative group of finite fields (Example 2 in Subsection 1.3).

In the second section we shall see more applications of this method, and in [EG] one finds a very nice discussion of the index-calculus attack in a rather general frame. The results found there explain why it is so effective in many cases.

## 2 Numeration by Algebraic Groups

To find the numerations we shall use algebraic groups defined over finite fields $\mathbb{F}_q$ where $q$ is a power of a prime $l_0$.

### 2.1 Algebraic Groups

We begin by a short description of algebraic groups defined over fields $K$.

First of all an algebraic group $\mathcal{G}$ is an algebraic scheme with special properties: It is reduced, non-singular, noetherian, defined over $K$ and hence can be covered by finitely many affine subvarieties (open w.r.t. the Zariski topology, which is used in the following): Locally it is "given" by a system of equations of polynomials. This means that we can cover $\mathcal{G}$ by open subvarieties $U_i$, $i \in I$, on which we can choose coordinate functions $X_1, \ldots, X_{n_i}$ satisfying the relations

$$(F_1(X_1, \ldots, X_{n_i}) = 0, \ldots, F_{m_i}(X_1, \ldots, X_{n_i}) = 0)$$

with $F_j \in K[X_1, \cdots, X_{n_i}]$ ($m_i = 0$ is allowed) which can be glued together along $U_i \cap U_j$. One should keep in mind that the choice and the number of coordinate functions and of relations is not canonically determined by $U_i$.
If $U_i$ is irreducible (w.r.t. the Zariski topology) the polynomials in the coordinate functions generate an integral domain. Its quotient field $F(U_i)$ depends only on $U_i$. Its transcendence degree $d_i$ over $K$ is the dimension of $U_i$.
$\mathcal{G}$ is the disjoint union of irreducible schemes called connected components. It is connected if there is only one connected component. Open non-empty subvarieties inside of one component are dense in this component. All components have the same dimension $d := \dim(\mathcal{G})$.

For extension fields $L$ of $K$ let $\mathcal{G}(L)$ be the set of $L$-rational points of $\mathcal{G}$, i.e. the set of morphisms from Spec($L$) to $\mathcal{G}$. These points can be described explicitly in the following way:

$$\mathcal{G}(L) = \bigcup_{i \in I} U_i(L)$$

(here one has to identifiy points on different open subschemes $U_i$ according to the glueing). For $U_i$ with chosen coordinate functions $X_1, \ldots, X_{n_i}$ and relations $F_1, \ldots, F_{m_i}$ we get:

$$U_i(L) = \{x \in L^{n_i},\ F_1(x) = \cdots = F_{m_i}(x) = 0\}.$$

Next there is an addition law, i.e. there is a morphism (in the category of schemes)

$$m : \mathcal{G} \times \mathcal{G} \to \mathcal{G},$$

an inverse, i.e. a morphism

$$i : \mathcal{G} \to \mathcal{G},$$

and a neutral element, i.e. a morphism

$$e : \mathrm{Spec}(K) \to \mathcal{G},$$

such that these morphisms satisfy the usual group laws. So we get

$$m \circ (id_{\mathcal{G}} \times m) = m \circ (m \times id_{\mathcal{G}}) \text{ (associativity)},$$

$$m \circ (e \times id_{\mathcal{G}}) = pr_2$$

where $pr_2$ is the projection of $\mathrm{Spec}(K) \times \mathcal{G}$ to $\mathcal{G}$, and

$$m \circ (i \times id_{\mathcal{G}}) \circ \delta = j_e$$

where $\delta$ is the diagonal map from $\mathcal{G}$ to $\mathcal{G} \times \mathcal{G}$ and $j_e$ is the map which sends $\mathcal{G}$ to $e(\mathrm{Spec}(K))$.

Down to earth this means that for all extension fields $L$ of $K$ the set $\mathcal{G}(L)$ is a group in which the sum and the inverse of elements are computed by evaluating morphisms which are defined over $K$ and in which the neutral element is the point

$$0 := e(\mathrm{Spec}(K)) \in \mathcal{G}(K).$$

Because of the Noetherian property of $\mathcal{G}$ it follows that it has only finitely many connected components, and it is a fact that the set of these components forms a finite group $Z$ with $\mathcal{G}_0$ as neutral element. We see easily that the unique connected component of $\mathcal{G}$ which contains 0 is a subgroup scheme of $\mathcal{G}$ denoted by $\mathcal{G}_0$.

There are several reasons (both security and complexity are bad) for the fact that $Z$ is of little interest for our cryptographic purposes, and so we shall assume from now on that $\mathcal{G}$ is connected.

Now choose an affine open subvariety $U$ of $\mathcal{G}$ with coordinate functions $X_1, \ldots, X_n$. The morphism $m$ induces a morphism

$$m_U : U \times U \to \mathcal{G}$$

and its image contains a non-empty open subset $V \subset U$. Take $W$ as the inverse image of $V$ in $U \times U$ . Then $m$ induces a map from $W$ to $V$ which can be written as:

$$m_W : W \to V$$

sending pairs of $L$-rational points $(x_1, \dots, x_n) \times (y_1, \dots, y_n)$ in $W$ to

$$(R_1(x_1, \dots, x_n; y_1, \dots, y_n), \dots, R_n(x_1, \dots x_n; y_1, \dots, y_n))$$

with $R_i \in K(X_1, \dots, X_n, Y_1, \dots, Y_n)$.

*Remark 2.1.* This is a birational description of the addition law which is enough for all cryptographic applications. (The set of points where this map is not defined is of small dimension and hence with high probability one will not run into it by chance.) For the performance of the cryptosystem the choice of $(W, m_W)$ is crucial; we require small $n$ and low degree of $R_i$.

If we can take $U = \mathcal{G}$ then $\mathcal{G}$ is an **affine group scheme**. The other important kind of group schemes are projective, i.e. they can be embedded into a projective space $\mathbb{P}^n/K$ and are closed in it. They are called **abelian varieties** and will be in the center of our interest. Abelian varieties have many special properties which are not at all clear from the definition. For instance their addition law has to be **commutative**. (This and much more of the fascinating theory of abelian varieties can be found in [Mu2].)

## 2.2 One-dimensional examples

(i) The **additive group** $G_a$ is the affine line. We can take one variable $X$ as coordinate function, the relation set is empty. $m$ is given by the function $R(X,Y) = X + Y$, $i(X) = -X$ and $e(\text{Spec}(K)) = 0_K$. The set of $L$-rational points is $L$, the group structure is given by the field addition.

(ii) For the **multiplicative group** $G_m$ we choose two variables $X_1, X_2$ and the relation $X_1X_2 - 1 = 0$. So as variety $G_m$ is a curve in the affine plane. The "addition" $m$ is given by

$$(R_1(X_1, X_2; Y_1, Y_2) = X_1Y_1, R_2(X_1, X_2; Y_1, Y_2) = X_2Y_2),$$

$i(X_1, X_2) = (X_2, X_1)$ and $e(\text{Spec}(K)) = (1_K, 1_K)$. The group of $L$-rational points is isomorphic to $L^*$.

(iii) **Elliptic curves** $E$ They are the simplest examples for abelian varieties and at the same time, the most important ones in theory and applications today. Only for the sake of simplicity we shall assume here that char$(K)$ is not equal to $2, 3$ (see [Si] for the general case).
Like in 2.) we use a curve as underlying scheme. But this time it is a curve in the projective plane given by one cubic equation.
If we use homogeneous coordinates $(X, Y, Z)$ the relation is

$$R(X, Y, Z) = Y^2Z - X^3 - AXZ^2 - BZ^3$$

with $A, B \in K$ such that $4A^3 + 27B^2 \neq 0$. As an affine open set we can take $U_f$ given by the inequality $Z \neq 0$. Taking $Z = 1$ in $U_f$ and (ab-)using $X, Y$ as affine coordinates we get the affine equation

$$Y^2 = X^3 + AX + B$$

for $U_f$. The birational version of the addition $m =: \oplus$ is given by the well known addition formulas for elliptic curves: For

$$P_1 = (x_1, y_1)\ , P_2 = (x_2, y_2).$$

we get

$$P_3 = (x_3, y_3) := P_1 \oplus P_2$$

with (in general):

$$x_3 = -(x_1 + x_2) + ((y_1 - y_2)/(x_1 - x_2))^2$$

and $y_3$ such that $(P_1, P_2, (x_3, -y_3))$ are collinear. The inversion is defined by $i(x, y) = (x, -y)$. The neutral element is not inside of $U_f$: We get in homogeneous coordinates: $e(\mathrm{Spec}(K)) = (0, 1, 0) =: \infty$, this is the one point in $E(K) \setminus U_f(K)$.

## 2.3 Birational Numeration

We come back to our task of finding "good" numerations of $\mathbb{Z}/p$. We want to use algebraic groups $\mathcal{G}$ over finite fields $\mathbb{F}_q$. So we first choose (and fix) a numeration of elements of $\mathbb{F}_q$ which provides a numeration of $\mathbb{F}_q^n$ (for instance by the lexicographical order). Next we choose coordinate functions for open affine subvarieties $U$ of $\mathcal{G}$ as described above and get numerations of $U(\mathbb{F}_q)$ depending on these functions. If we want we can use a covering of $\mathcal{G}$ by affine parts $U_i$ to get a numeration of $\mathcal{G}(\mathbb{F}_q)$ but for most applications this will not be necessary (cf. Remark 2.1). Again we stress that there is no canonical way to get such a numeration and a big part of the implementation of a DL-system consists of finding a good numeration (even for $\mathbb{F}_q$).

Now let

$$\alpha : \mathbb{Z}/p \rightarrow \mathcal{G}(K)$$

be a non-trivial homomorphism. Then $\alpha$ induces a numeration $f_i$ of $\mathbb{Z}/p \cap \alpha^{-1}(U_i(\mathbb{F}_q))$ in a natural way and the related map $\oplus_i =: \oplus$ is given by the rational functions defining the group law of $\mathcal{G}$ on $U_i$. The *existence* of $\alpha$ is equivalent with the existence of $\mathbb{F}_q$-rational points $P$ on $\mathcal{G}$ with order $p$, and one such $\alpha$ can be explicitly given if one point $P_0$ of order $p$ is known.

From now on we take $\mathcal{G}$ as connected and commutative.

*Remark 2.2.* It may be be worthwhile to look for other possibilities, e.g. matrix groups, but then structural properties of such groups like the theorem of Maschke about semi-simplicity of representations of groups with order prime to $\mathrm{char}(\mathbb{F}_q)$ have to be taken into account.

A well known structure theorem about commutative and connected algebraic groups yields that without loss of generality we can assume that $\mathcal{G}$ is either an affine group or an abelian variety. Moreover affine connected algebraic commutative groups can be built up by unipotent groups and tori. Unipotent groups have composition factors isomorphic to $G_a$, and tori are up to twists products of copies of $G_m$. So essentially we can try to use copies of $G_m$ and $G_a$ and abelian varieties for our purposes.

It is easily seen that $G_a$ leads to Example 1 and $G_m$ to Example 2 in Subsection 1.3 . So we shall concentrate in the following on discussing numerations induced by abelian varieties.

## 2.4 Admissible Abelian Varieties

It is now clear that we have the

**Aim 2.3.** Construct **explicitly** abelian varieties $A/\mathbb{F}_q$ satisfying simultaneously

(i) A (large) prime $p$ divides $|A(\mathbb{F}_q)|$ and

$$\log p \approx \log(|A(\mathbb{F}_q)|) \approx \dim(A) \cdot \log q \text{ (cf. Theorem 3.3).}$$

(ii) To store a point $P \in A(\mathbb{F}_q)$ one needs only $c_1(\log(p))$ bits of space and to compute $\oplus$ only $c_1(\dim(A)^k)$ operations in $\mathbb{F}_q$ with $c_1, c_2, k$ small (e.g. $c_1 = 1 + \epsilon, k = 2$).

(iii) All **known** algorithms for the computation of the discrete logarithm in $A(\mathbb{F}_q)$ have a large enough probabilistic exponential complexity.

In practice we require that $p \geq 10^{60}$ and the exponential constant is $1/2$. We call abelian varieties satisfying these conditions **"admissible"**.

The embedding of $\mathbb{Z}/p$ into $A(\mathbb{F}_q)$ induces a lot of additional structure from arithmetic geometry in the corresponding DL-system: We can use

- Algebraic Number Theory, especially
  (i) class field theory, CM-fields and cyclotomic fields
  (ii) local and global fields
- Algebraic Geometry
  (i) function fields over finite fields, over global fields and over $\mathbb{C}$
  (ii) the theory of abelian varieties and, more generally, group schemes

and, most importantly, **Galois theory** related to these objects. These theories give tools for constructions, but also for possible (or at least conceivable) attacks.

## 2.5 Manageable Abelian Varieties

The first task to solve is to describe (birationally) abelian varieties and the addition laws in a time and space saving way. This seems to be hopeless as long as we work with **general** abelian varieties $A$: The number of coordinate functions and the degree of the addition formulas both grow exponentially with the dimension of the abelian variety (cf. results of Mumford [Mu1] and Lange-Ruppert [LR]) and so we have to use **special** abelian varieties.

The first specialization is to take $A$ as Jacobian variety $J_C$ of a curve $C$. The big advantage is that both the coordinates and the addition can use objects defined by $C$ since

$$(C \times \ldots \times C)/S_g \quad (g = \text{genus}(C),\ S_g \text{ the symmetric group in } g \text{ letters})$$

is birationally isomorphic to $J_C$ and the points of $J_C$ correspond to divisor classes of degree 0 of the function field of $C$. We can assume that we have an $\mathbb{F}_q$-rational point $P_\infty$ on $C$. The theorem of Riemann-Roch implies that we find "standard " representatives in divisor classes of the form $D - g\,P_\infty$ with $D$ a positive $\mathbb{F}_q$-rational divisor of degree $g$ (i.e. a formal sum of $g$ points of $C$).

The addition in $J_C$ is induced by the addition of divisors and the task is to find in the class of the sum of two divisors standard representatives either by formulas or by an algorithm involving the coordinates of the points occurring in the added divisors. So we need an effective (and very fast) version of the Riemann-Roch theorem.

We consider the following situation: We assume that there is a cover

$$\varphi : C \to \mathbb{P}^1;\ \deg \varphi = d,$$

in which a non singular point $(P_\infty)$ is totally ramified and induces the place $(X = \infty)$ in the function field $\mathbb{F}_q(X)$ of $\mathbb{P}^1$. It is an easy exercise to show:

**Proposition 2.4.** *Let $S$ be the normal closure of $\mathbb{F}_q[X]$ in the function field of $C$, $J(S)$ its ideal class group. Then:*

$$J(S) \cong J_C(\mathbb{F}_q)$$

This allows to present points on $J_C(\mathbb{F}_q)$ by ideals in $S$ and hence as $d$-tuples of polynomials in $K[X]$ whose degree is bounded by $g$. The addition on $J_C$ is reduced to linear algebra and interpolation. The space needed to write down elements of $J_C(\mathbb{F}_q)$ is $O(d \cdot g)$.

There are promising examples of curves with $d = 3$, but here we restrict ourselves to $\mathbf{d = 2}$, i.e. to **hyperelliptic curves**.

**Corollary 2.5.** *Assume that $C$ is a hyperelliptic curve over $\mathbb{F}_q$ having an $\mathbb{F}_q$-rational Weierstraß point $P_\infty$. Let $C$ be given by a separable, monic polynomial $f(X,Y)$ of degree 2 in $Y$ such that $(X)_\infty = 2P_\infty$.*

*Then $J_C(\mathbb{F}_q)$ is isomorphic to the ideal class group of $S = \mathbb{F}_q[X,Y]/f$.*

*The isomorphism is canonical if we fix $P_\infty$ and the birational morphism from $(C \times \cdots \times C)/S_g$ to $J_C$.*

Artin has generalized Gauß's theory of ideal classes of imaginary quadratic number fields to hyperelliptic function fields connecting ideal classes of $S$ with reduced quadratic forms of discriminant $D(f)$ and the addition $\oplus$ with the composition of such forms.

This is the basis for the **Cantor algorithm** which can be written down "formally" and then leads to addition **formulas** or can be implemented as **algorithm**. In both cases we need an efficient arithmetic in polynomial rings over $\mathbb{F}_q$ (and nothing more).

Implementations of both variants and data concerning the performance can be found in [Sp] (for $g = 2$) and in [Kr] in general. It is not surprising that for elliptic curves the formulas are faster than the algorithm, for genus 2 both methods seem to be equivalent, and for higher genus the algorithm is the only feasible variant.

Using these algorithms it turns out that the computing time for an addition grows like twice the genus of $C$ and so elliptic curves are optimal. But this assertion is a little bit unfair: For $g = 1$ a lot of work has been invested to get a "best implementation", for $g \geq 2$ we did not optimize the algorithm. Moreover we shall see below that the bit size of $J_C(\mathbb{F}_q)$ grows linearly with the genus and so we can realize a large group over a relatively small field and so we may take advantage of special features of the hardware configuration.

### 2.6 The Index-Calculus Attack on Hyperelliptic Curves

But we have to pay a price: We have left the abstract realm of abelian varieties and we use the ideal class groups of orders $S$ over the polynomial ring in one variable for the DL-system. This approach is well known in the analogous situation in number theory, and there exists a subexponential attack based on the index-calculus principle. So it is not surprising that we are confronted with the same kind of attack in the geometric situation, too.

The general principle of the index-calculus attack was described in Subsection 1.5. One of the essential parts of it is the choice of a factor base. The ideal classes of $S$ can be represented by polynomials of degrees bounded by $g = g(C)$. This suggests a rather natural way of choosing a factor base for the index-calculus attack: Take the ideal classes which can be represented by polynomials of small degrees as factor base!

It is to be expected that (for fixed field $\mathbb{F}_q$) this method becomes more and more effective with increasing $g$. And indeed Adleman, DeMarrais and Huang ([ADH]) proposed an index-calculus algorithm for hyperelliptic curves with subexponential complexity in $g$. This was made precise in [MST] and [E] and finally settled by a smoothnessness result in [ES]:

**Proposition 2.6.** *([E]) For $g/\log(q) > t$ the discrete logarithm in the divisor class group of a hyperelliptic curve of genus $g$ defined over $\mathbb{F}_q$ can be computed with complexity bounded by $L_{q^g}[\frac{5}{\sqrt{6}}((1+\frac{3}{2t})^{1/2}+(\frac{3}{2t})^{1/2})]$.*

For the definition of the subexponential function $L$ we refer to [E] and we do not go into more details since this result predicts the behaviour of the

DL-systems for fairly large $g$ which is not in the centre of our (practical) interest.

But we notice that because of this result (and, by the way, because of the difficulty of finding hyperelliptic curves for which we can count points on their Jacobian) we have to restrict ourselves to small $g$ ($\leq 7$ in the range of security parameters used today) if we want to have secure DL-systems.

In a new paper of Gaudry (cf. [Ga]) things become more serious for practical use: For hyperelliptic curves of relatively small genus (in practice: $g \leq 8$) with known structure and order of the group of $\mathbb{F}_q$−rational points of their Jacobian variety one can proceed in a much simpler way: It is "easy" (with complexity $O(q\log(q))$ to decide whether a positive divisor of degree $\leq g$ is a sum of divisors of degree 1 (i.e. of $\mathbb{F}_q$−rational points of $C$) and if this is the case one gets the explicit expression of this sum with the same time complexity (probabilistically). (One has to factorize polynomials of degree $g$ over $\mathbb{F}_q$.) On the other side such an event occurs with probability $\approx 1/g!$. Using this Gaudry (cf. [Ga]) succeeds to prove:[3]

**Theorem 2.7.** *Assume that $C$ is a hyperelliptic curve of genus $g \leq 8$ defined over $\mathbb{F}_q$ such that $|J_C(\mathbb{F}_q)|$ is known.*

*Then the DL in $J_C(\mathbb{F}_q)$ can be computed with complexity $O(q^2(\log(q))^\gamma)$, with "reasonably small" constants.*

Of course by this result only a specific presentation of $\mathbb{Z}/p$ as subgroup of the group of rational points of $J_C$ is attacked, but since we do not know a different manageable presentation till now we get the

**Result: The Jacobian varieties attached to hyperelliptic curves of genus $\leq 4$ or closely related abelian varieties[4] are the only known candidates for admissible abelian varieties.**

*Remark 2.8.* It is obvious that Gaudry's method works in all situations in which we embed $\mathbb{Z}/p$ into the ideal class group of an order over a polynomial ring in which addition is made effective by a "reduction theory".

## 3 Galois Theory

### 3.1 Counting Points

In order to use manageable abelian varieties $A$ of dimension $d$ for DL-systems we have to determine the order of the Mordell-Weil group $A(\mathbb{F}_q)$. Since the desired size of $|A(\mathbb{F}_q)|$ is at least $10^{60}$ we cannot count directly.

Instead we shall use the **Galois group** of $\mathbb{F}_q$ and the known properties of

[3] A more precise formulation of the result and examples can be found in the paper of Gaudry, for a general point of view cf. [EG].

[4] for instance factors or abelian varieties constructed by Weil descent (cf. Section 3.2)

its action on torsion points of abelian varieties. We can exploit a rich theory due to Hasse, Weil, Tate and others.

The absolute Galois group $G_q$ of $\mathbb{F}_q$ is topologically generated by the **Frobenius automorphism** $\pi_q$ which maps elements of $\overline{\mathbb{F}}_q$ to their $q$–th power.

Let $n$ be a natural number. For all numbers $n$ $\pi_q$ acts on the points $A(\overline{\mathbb{F}}_q)[n]$, the group of elements of $A(\overline{\mathbb{F}}_q)$ whose order divides $n$. Hence it induces a linear map on $A(\overline{\mathbb{F}}_q)[n]$ which is, if $n$ is prime to char$(K)$, as $\mathbb{Z}/n$–module isomorphic to $(\mathbb{Z}/n)^{2d}$.

**Fact (Weil):**
The characteristic polynomial of $\pi_q$ w.r.t. this action is the reduction modulo $n$ of a monic polynomial with integer coefficients of degree $2d$. This polynomial is independent of $n$ and is called the **L-series** $L_A(T)$ of $A$ over $\mathbb{F}_q$.

Since $A(\mathbb{F}_q)$ is the kernel of the map $\pi_q - \mathrm{id}$ we get by elementary linear algebra:

**Proposition 3.1.** $|A(\mathbb{F}_q)| = | L_A(1) |$.

A trivial but crucial consequence is:

**Corollary 3.2.** *$\mathbb{Z}/p$ is embeddable into $A(\mathbb{F}_q)$ iff $L_A(1) \equiv 0 \bmod p$.*

A fundamental theorem of Weil states that the absolute value of the zeroes of $L_A(T)$ is equal to $\sqrt{q}$ and so we get the

**Theorem 3.3.**

$$\dim A \cdot \log q \approx \log |A(\mathbb{F}_q)|.$$

**Conclusion:** Task 2. of Aim 2.3 can be solved if we can find enough manageable abelian varieties A of dimension $d$ for which we can

(i) compute $L_A(1)$ rapidly, and
(ii) prove that with a not too small probability a prime of size $\approx q^d$ divides $L_A(1)$.

The second item can be discussed by **global Galois theory** using analytic and algebraic number theory. These theories provide tools like effective versions of Chebotarev's density theorem and conjectures about the distribution of traces of Frobenius elements (Lang-Trotter) and about the distribution of class groups generalizing heuristics of Cohen-Lenstra. Recent interesting work on these topics can be found in [Mur].

We shall present some methods to solve the first item. They all use more or less sophisticated properties of $L$-series and Frobenius automorphisms.

**Constant Field Extensions** This approach is the simplest method: We begin with a small field $\mathbb{F}_{q_0}$ and $A$ defined over this field such that we *can* count the $\mathbb{F}_{q_0}$−rational points of $A$ [5]. We use this to compute the zeroes $\omega_i$ of the L-series of $A$ over $\mathbb{F}_{q_0}$.
For $m \in \mathbb{N}$ and $q = q_0^m$ the Frobenius automorphism $\pi_q$ is the $m$-th power of $\pi_{q_0}$.
Hence the zeroes of the $L$-series of $A \times \mathbb{F}_q$ are the $m$−th powers of $\omega_i$ and so the order of $A(\mathbb{F}_q)$ can be computed easily.

There is one obvious disadvantage: The group $A(\mathbb{F}_{q_0})$ is embedded into $A(\mathbb{F}_q)$ and so the maximal prime $p$ dividing $|A(\mathbb{F}_q)|$ and useful for DL-systems divides

$$|A(\mathbb{F}_q)|/|A(\mathbb{F}_{q_0})|.$$

We summarize:

**Proposition 3.4.** *Let the notation be as above.*

*Let p be a prime larger than $|A(\mathbb{F}_{q_0})|$. Then $\mathbb{Z}/p$ is embeddable into $A(\mathbb{F}_q)$ iff p divides*

$$\Pi_{i=1,\cdots,2d}\,(1-\omega_i^m)(1-\omega_i)^{-1}$$

*and so the size of p is at most $\approx q_0^{d(m-1)}$.*

This means that the size of the DL-system compared with the key length and the time needed for an addition (roughly proportional to log($q$)) is not optimal.

So mostly the method of constant field extensions is used only for very small $q_0$ and large prime $m$. The typical examples are Koblitz curves defined over fields with 2-power order (cf. [Ko]).

In Subsection 3.2 we shall present another method to construct admissible abelian varieties using constant field extensions in the situation that $q_0$ is as big as possible and $m$ small. It uses the observation stated in

**Proposition 3.5.** *Let $A_0(\mathbb{F}_q)$ denote the subgroup of $A(\mathbb{F}_q)$ annihilated by the trace map*

$$Tr_{\mathbb{F}_q/\mathbb{F}_{q_0}} = \sum_{j=0,\cdots,m-1} \pi_{q_0}^j.$$

*Then*

$$|A_0(\mathbb{F}_q)| = \prod_{i=1,\cdots,2d} (1-\omega_i^m)(1-\omega_i)^{-1}.$$

The proof of the proposition uses the fact that the trace map is a surjective homomorphism from $A(\mathbb{F}_q)$ to $A(\mathbb{F}_{q_0})$.

[5] to give a rough idea: small means that $q^d \leq 10^{10}$

**Schoof's Algorithm** Remember: $L_A(T)$ is a polynomial with integral coefficients which simultaneously for all natural numbers $n$ is the characteristic polynomial of $\pi_q$ acting on torsion points of order $n$ of $A$. Since it has integral coefficients (of size depending on $q$ and $\dim(A)$ only) it is determined by this action for small $n$.

This is the starting point of Schoof's algorithm (cf. [Sch1]) for computing the number of points of elliptic curves over $\mathbb{F}_q$. It is made effective by another ingredient: It is well known that there is a linear recurrency between the $n$-division polynomials of elliptic curves (cf. [Si], p.105).

**Theorem 3.6. (Schoof)**
*For elliptic curves $E$ the complexity to compute $L_E(T)$ is bounded by a polynomial function in* $\log(q)$.

One can try to generalize this idea to arbitrary abelian varieties. One main task is to find a way of computing division polynomials. Kampkötter has done this in the case of Jacobians of hyperelliptic curves (cf. [Ka]) proving that for fixed $g$ the computation of the $L$-series has polynomial complexity (similar results are due to Pila).
But Schoof's algorithm is too slow for practical applications, and the same is true in a much worse way for Kampkötter's algorithm.

For **elliptic curves** things have become much better by observations and refinements due to Atkin and Elkies: Instead of using the kernel of the multiplication by small $n$ on elliptic curves $E$ one can use the kernel of endomorphisms of small norm and determine $L_E(T)$ modulo ideals in the endomorphism ring of $E$ (which is an *order $R_E$ in an imaginary quadratic field or in a quaternion algebra* (cf. [Si] p.100 ff)). This explains why the determination of $L_E(T)$ modulo $l$ is easier if the prime $l$ splits in $R_E$.

For the actual computation one has to find convenient methods to describe isogenies of elliptic curves. Here enter the *modular curves* $X_0(l)$. These curves parametrize pairs of elliptic curves with cyclic isogenies of degree $l$. Their rich theory is the key to many of the important new results in arithmetic geometry. We recall as the most spectacular result the proof of Fermat's Last Theorem by Wiles which was obtained by proving the Taniyama conjecture for elliptic curves.

Our interest are elliptic curves over finite fields. To handle them we are led in a natural way to **global** objects in number theory: Orders in number fields, modular curves and corresponding Galois representations!

Based on these considerations Schoof's algorithm for elliptic curves $E$ has been refined ([Mo]):

**Proposition 3.7.** *(cf. [Mo]) Let $\epsilon$ be a positive real number.*

*Let $E$ be an elliptic curve over $\mathbb{F}_q$ and $l$ a prime which is split in $R_E$. Then $L_E(T)$ modulo $l$ can be computed with probabilistic complexity $O((l^2 \cdot \log(q))^{1+\epsilon})$ operations in $\mathbb{F}_q$.*

To determine $L_E(T)$ one uses the information modulo different primes $l$ and applies the Chinese remainder theorem. The estimates due to Hasse-Weil imply that $O(\log(q))$ different primes $l$ are sufficient. To use Proposition 3.7 we want to use split primes only and so we need bounds which ensure that we have found enough of them.

An easy check decides whether $E$ is supersingular (and then $L_E(T)$ is known). So assume that $E$ is not supersingular. We observe that $q$ is a non-trivial norm with respect to $R_E/\mathbb{Z}$ and so the size of the discriminant of $R_E$ is bounded by $O(|q|)$.

This implies conjecturally (and in practice) that it is enough to use primes $l$ up to a bound of size $O(\log(q))$. Under the assumption of the Generalized Riemann Hypothesis (GRH) it can be proved[6] that the bound $O((\log(q))^2)$ (with explicitly computable constants) is big enough. So we get

**Theorem 3.8.** *Assume that GRH is true. Let $\epsilon$ be a positive real number. Let $E$ be an elliptic curve defined over $\mathbb{F}_q$. The order of $E(\mathbb{F}_q)$ can be computed with (probabilistic) complexity $O((\log(q))^\delta)$ with $\delta \leq 6+\epsilon$ and conjecturally $\delta \leq 4+\epsilon$ operations in $\mathbb{F}_q$.*

Using this theorem and taking into account the predictions on the structure of $E(\mathbb{F}_q)$ we can solve Task 1. and Task 2. of Aim 2.3 for **random** elliptic curves in a range which is sufficient for cryptography.

**Real and Complex Multiplication** The following is a global construction, i.e. one constructs an abelian variety $A$ over a number field such that one can compute the number of points of the reduction of $A$ modulo primes of this field.

(i) We use the arithmetic theory of the Galois groups of special number fields $K$ (called **CM-fields**) which are totally imaginary quadratic extensions of totally real number fields.
The simplest and well known case deals with **elliptic curves with complex multiplication** whose endomorphism ring is an order in a field $K = \mathbb{Q}(\sqrt{d})$ with $d < 0$.
More generally: **Class field theory** relates endomorphisms of special abelian varieties $A_K$ to elements in orders $\mathcal{O}_K$ in CM-fields $K$ **(Shimura - Taniyama)** ([ST]):
We can determine (in principle explicitly, see examples) an abelian variety $A_K$ defined over a known finite field extension $L$ of $\mathbb{Q}$ such that the reduction of $A_K$ modulo (suitably chosen) prime ideals $\mathfrak{q}$ in the ring of integers of $L$ leads to abelian varieties defined over the residue field of $\mathfrak{q}$ with known $L$-series. The necessary computations can be done (after a precalculation) in $K$!

[6] We thank K. Murty and R. Murty for the communication of this result (e-mails February 2000).

So the calculation of $A_K$ mod $\mathfrak{q}$ and of $L_{A_K \bmod \mathfrak{q}}(T)$ is very fast and one variety $A_K$ gives rise to many abelian varieties over finite fields for which Task 1. and Task 2. of Aim 2.3 are solved. Even after Morain's results this method remains interesting for elliptic curves (for instance for statistical studies about the behaviour of class groups), and for higher dimensions it is the most effective method known till now.
One can object that this method produces very special abelian varieties with an explicitly known lifting to a $CM$−variety over a number field. So it may be wise to make $A_K$ complicated enough, e.g. by using only fields $K$ whose class number is between 200 and 1000.

*Example 3.9.*

(a) $g = 1$ : *Class field theory* of imaginary fields applied to elliptic curves and especially to the $j$-invariant (cf. [Si], pp.338 ff) was proposed by Atkin and implemented by *A. M. Spallek 1992* (diploma thesis) and is used in practice till today. It works very efficiently, the hardest computational problem is the factorization of polynomials of degree $\leq 1000$ over $\mathbb{F}_q$.

(b) $g = 2$ : This is implemented by A.M. Spallek in her thesis, Essen 1994, and uses
  i. *class field theory* of fields of degree 2 over real quadratic fields (non-Galois over $\mathbb{Q}$),[7]
  ii. *Invariant theory* which is explicit and "easy" (cf [Ig]) and
  iii. either elimination theory to solve a system of three polynomial equations in 6 variables of degree $> 1$ over the ring of integers of a number field or
  iv. *much better:* Mestre's method intersecting invariant forms (with one of them a conic, cf. [Me]).

The implementation of the algorithm is relatively easy, it works efficiently. We give one example computed by Spallek in 1993:
As CM field take $\mathbb{Q}(\sqrt{-2-\sqrt{2}})$. The resulting curve is

$$C : Y^2 = X^5 - 140X^3 + 240X^2 + 3810X - 6928.$$

Reduce modulo

$$p = 153946287550700989943 \approx 1.5 \cdot 10^{20}.$$

The point

$$(7550700989929, 49, 31694823907497262594, 86028807748921141745)$$

on $(J_C)(\mathbb{F}_p)$ has order

$$l = 64570868647934186550539174412679.$$

[7] to avoid non-necessary automorphisms

For $g \geq 3$ things become more complicated.
First the invariant theory is not so well understood and the determination of the equation of the curve from the knowledge of the invariants is too complicated. The way out is to use the period matrix of the Jacobian from the beginning and Mestre's method.
The given CM-field can be used to compute the period matrix of an abelian variety. It can be tested whether it is principally polarized (cf. [Mu3]). But a principally polarized abelian variety of dimension $\geq 4$ need not to be the Jacobian of a curve (Schottky problem) and even if so, the chance of finding a *hyperelliptic* curve becomes dramatically smaller if the genus increases.
In his thesis (Essen 1995)(cf. [We]) Weber used results of Mumford and Poor and succeeded in overcoming the computational difficulties for $g \leq 5$. He implemented an algorithm which decides whether a given period matrix belongs to a hyperelliptic curve and if so, computes the equation of the curve. He applied his algorithm to abelian varieties with real multiplication (see below) and found, as expected, many examples with $g = 3$, some with $g = 4$ and only one with $g = 5$ (which is a modular curve)[8].

(ii) As explained above the main task in the construction of curves is to find the period matrix of an abelian variety which is a candidate to be the Jacobian of a hyperelliptic curve. As first source we used CM-Theory.
A second source uses **Real multiplication** (i.e. the ring of endomorphisms of $A$ contains an order of a totally real field $K$ with $[K : \mathbb{Q}] = \dim(A)$) and its relation with modular forms: Over $\mathbb{Q}$ the conjecture of Taniyama (proved for elliptic curves) says that we should work with abelian varieties $A$ which are factors of the Jacobian of modular curves $X_0(N)$. For these factors we know by work of Shimura how their period matrices are related to cusp forms.
To use the Jacobians of the constructed curves for cryptography one needs their $L$-series over residue fields. In other words one needs the **local factors** of the global $L$-series. Again the theory of cusp forms and their relation to representations of the Galois group of $\mathbb{Q}$ (cf. [Sh]) becomes crucial:
For all prime numbers $l$ not dividing $N$ there are special endomorphisms, Hecke operators $T_l$, of $A$, and knowing $T_l$ is as good as knowing the Frobenius $\pi_l$ because of the **Eichler-Shimura relation**:

$$\pi_l^2 + T_l \pi_l + l = 0.$$

So the fast computation of $T_l$ is crucial both for the computation of the period matrix and for the local $L$-series of $A$.
By methods of Basmaji (thesis Essen 1996) [Ba] and M. Müller this can be done for $N \leq 10000$ and $l \leq 10^{11}$.

[8] In view of Gaudry's result 2.7 this is sufficient for cryptographic applications.

*Example 3.10.* (*Weber*) Take $N = 284$. The *new part* of $J_0(284)$ has dimension 6 and splits into two simple factors of dimension 3, one of them, $A_f$, corresponds to the cusp form $f$ determined by

$$f = q + \beta q^3 + (-\beta^2 - 3\beta - 1)q^5 + \dots$$

with $\beta^3 + 3\beta^2 - 3 = 0$.
$A_f$ is the Jacobian of the curve

$$Y^2 = X^7 + 3X^6 + 2X^5 - X^4 - 2X^3 - 2X^2 - X - 1.$$

This curve has good reduction modulo primes $l$ which do not divide $N$ and so the same is true for $A_f$. We denote this reduction by $A_f^{(l)}$. Using Basmaji's algorithm one can compute the Hecke operator $T_l$ and so the order of $A_f^{(l)}(\mathbb{Z}/l)$ for primes up to $10^{11}$. The size of this group is $\approx 10^{33}$ which is too small for DL-systems. But we hope that we can combine this construction with Weil descent methods (Subsection 3.2) to find interesting candidates for DL-systems.

### 3.2 Weil Restriction

We recall that one of the methods to determine the $L$-series of abelian varieties $A$ of dimension $d$ was related to constant field extensions. More generally one can look at the situation that $q = l_0^f$ with $f \leq 2$. Hence we have the Frobenius automorphism $\pi_{l_0}$ acting non-trivially on $\mathbb{F}_q$. One may ask what *geometric* consequences this has for abelian varieties $A$ defined over $\mathbb{F}_q$ and whether there are hidden structures which weaken the DL.

In [Fr] I have suggested that one should study the geometric method of restricting scalars (Weil-descent) at least in special cases.

**Scalar Restriction** Here is a short description of this procedure (for a detailed exposition and the strict geometric construction cf. [Na]):

Take a field $K$ and a finite separable field extension $L/K$. Let $V$ be a quasi-projective variety (i.e. $V$ can be embedded into a projective space) defined over $L$. Then there is a quasi-projective variety $W_V$ defined over $K$ with

$$W_V(K) = V(L) \text{ and } W_V \times L \simeq V^{[L:K]}.$$

The recipe for the construction of $W_V$ for affine $V$ is:
Choose coordinate functions $X_1, \dots, X_n$ of $V$ over $L$ and a basis $(u_1, \dots, u_m)$ of $L/K$. Define the $n \cdot m$ variables $Y_{i,j}$ by

$$X_i = u_1 Y_{1,i} + \dots + u_m Y_{m,i},$$

and plug these expressions into the relations defining $V$. Next express the coefficients of the resulting relations as linear combinations of the basis $(u_1, \dots,$

$u_m$) and order these relations according to this basis to get the relations of the coordinate functions $Y_{1,1}, \ldots, Y_{m,n}$ of $W_V$ over $K$.

For quasi-projective varieties one has to choose an appropriate cover of $V$ by affine varieties, apply the descent recipe to them and then glue together the resulting affine varieties over $K$.

**Facts:** If $V$ is a projective variety then $W_V$ is projective, and if $V$ is an abelian variety $A$ then $W_A$ gets the structure of an abelian variety over $K$ in a natural way: The addition law is the descent of the addition law of $A$.

**The Case of Finite Fields** Take $K = \mathbb{F}_q$ and $L = \mathbb{F}_{q^m}$ and let $A$ be an abelian variety of dimension $d$ defined over $\mathbb{F}_{q^m}$.

The Weil restriction $W_A$ is an abelian variety defined over $\mathbb{F}_q$ and its group of $\mathbb{F}_q$-rational points is in a natural way isomorphic to $A(\mathbb{F}_{q^m})$.

So the DL-problem on $A/\mathbb{F}_{q^m}$ is equivalent with the DL-problem on the $m \times d-$dimensional abelian variety $W_A$ over $\mathbb{F}_q$. But $W_A$ has more geometric structure: We find $\mathbb{F}_q$-rational subvarieties like curves and hypersurfaces on $W_A$ which are not on $A$.

It seems that "in general" these subvarieties are rather complicated. For instance take an elliptic curve $E$. Then "in general" one expects that the minimal genus of curves on $W_E$ is $\approx 2^m$ and so at least till now we cannot apply this additional information (positively or negatively) to the DL-problem if $m$ is large. But for small $m$ or special fields $\mathbb{F}_q$ one has to expect a different picture.

Here is one such example. It is due to Gaudry, Hess and Smart (cf. [GHS]).

**Theorem 3.11.** *Take $q = 2^n$ and $m$ prime to $n$. Let $E$ be an elliptic curve defined over $\mathbb{F}_{q^m}$ and assume that $E$ is not defined over a proper subfield of $\mathbb{F}_{q^m}$. Assume that $\mathbb{Z}/p$ can be embedded into $E(\mathbb{F}_{q^m})$.*
*Then there is an (explicitly given) embedding of $\mathbb{Z}/p$ into the Jacobian variety of a hyperelliptic curve $C$ of genus $\leq 2^{m-1}$ defined over $\mathbb{F}_q$ such that the discrete logarithm problems (corresponding to these embeddings) of $\mathbb{Z}/p$ in $E(\mathbb{F}_{q^m})$ and in $J_C(\mathbb{F}_q)$ are equivalent.*

The proof of this theorem uses Weil descent applied to $E$ from $\mathbb{F}_{q^m}$ to $\mathbb{F}_q$ and constructs on (an affine part of) $W_E$ a hyperelliptic curve by using a convenient basis of $\mathbb{F}_{q^m}/\mathbb{F}_q$ and hyperplane sections.

Typically for our present knowledge about Weil descent are the applications of this result:

(i) If $m$ is too large (e.g. $\geq 7$) the resulting objects (here the hyperelliptic curves) are too complex to be used.
(ii) If $m$ is small (here $m \leq 3$) we can use the Weil descent to produce new admissible abelian varieties, here Jacobians of hyperelliptic curves $C$: By construction we know that the group of $\mathbb{F}_q-$rational points on $J_C$ contains a subgroup of order $p$ if the elliptic curve $E$ we begin with has such a subgroup in $E(\mathbb{F}_{q^m})$.

(iii) In a medium range ($4 \leq m \leq 6$) the discrete logarithm system attached to $E$ is weakened.
This follows from Theorem 2.7 at least for $m \geq 5$. Numerical experiments made in [GHS] rise the suspicion that the same is true for $m = 4$.

The work of Gaudry, Hess and Smart is a clear signal to designers of DL-systems based on elliptic curves:

**The degree $d$ of the finite field over its prime field over which the numeration of $\mathbb{Z}/p$ is realized should be equal to $k \cdot l$ where $l$ is a large prime and $k \leq 3$.**

In Theorem 3.11 it is assumed that $E$ is not defined over a proper subfield. One reason for this is the following fact:

**Proposition 3.12.** *Assume that $A$ is a simple abelian variety defined over $\mathbb{F}_{q^m}$. In general $A$ is defined over a proper subfield of $\mathbb{F}_{q^m}$ iff $W_A$ is not a simple abelian variety over $\mathbb{F}_q$.*
*Here "in general" means that we can give (for any dimension of $A$ and given $m$) a complete list of the exceptions.*

For details and the explicit discussion of the case that $A$ is an elliptic curve we refer to [Na].

**Trace-0-Variety** Assume now that $A$ **is** defined over $\mathbb{F}_q$. We continue to assume that $A$ is simple. Again in general $W_A$ has two (over $\mathbb{F}_q$) irreducible abelian subvarieties. One of them is isomorphic to $A$, we call the other one $B$.

$B$ has dimension $d(m-1)$ and is the kernel of the **trace projection** morphism which is the sum of the first $m$ powers of the Frobenius automorphism. Its $L$-series over $\mathbb{F}_q$ is equal to the $L$-series of $A \times \mathbb{F}_{q^m}$ divided by the $L$-series of $A$ over $\mathbb{F}_q$ and so (up to points of order dividing $m$) we get rid of points of $A$ with coordinates in $\mathbb{F}_q$.

Of course this procedure is closely related to the method of constant field extensions we described in Subsection 3.1. But we have set the right geometric context for this method by interpreting the points of $A(\mathbb{F}_{q^m})$ lying in the kernel of the trace map as $\mathbb{F}_q$-rational points of an abelian variety $B$ defined over $\mathbb{F}_q$.

As one positive consequence we can reduce the key length. There is a second advantage: The addition on $B$ is induced by the group law of $A/\mathbb{F}_{q^m}$ but multiplication by numbers $k$ have shorter expressions if developed w.r.t. powers of the Frobenius automorphism (cf. analogous advantages of Koblitz curves).

This speeds up the addition considerably and so one can expect that for small $m$ we find new admissible abelian varieties of higher dimension by using "small" abelian varieties over $\mathbb{F}_q$ which are manageable but not secure. Below we describe in detail an example.

*Example 3.13.* The Weil descent of elliptic curves with $m = 4$ was worked out by Galbraith and Smart in [GS].

We take $m = 3$ and $A = E$ an elliptic curve given by the equation

$$Y^2 = X^3 + AX + B.$$

We are interested in the case that $q$ is odd.
We assume that $q \equiv 1 \bmod 3$. So we can use Kummer theory to describe $\mathbb{F}_{q^3}$: Let $\mu$ be an element of $\mathbb{F}_q$ which is no third power and $\zeta$ be an element of $\mathbb{F}_{q^3}$ with $\zeta^3 = \mu$. So $(1, \zeta, \zeta^2)$ is a basis of $\mathbb{F}_{q^3}/\mathbb{F}_q$.

We choose coordinates for the Weil descent

$$Y = y_0 + y_1\zeta + y_2\zeta^2$$
$$X = x_0 + x_1\zeta + x_2\zeta^2$$

and plug this into the equation of the elliptic curve $E$:

$$(y_0 + y_1\zeta + y_2\zeta^2)^2 = (x_0 + x_1\zeta + x_2\zeta^2)^3 + A(x_0 + x_1\zeta + x_2\zeta^2) + B.$$

This produces three equations over $\mathbb{F}_q$:

$$y_0^2 + 2\mu y_1 y_2 = x_0^3 + \mu x_1^3 + \mu^2 x_2^3 + 6\mu x_0 x_1 x_2 + Ax_0 + B,$$

$$2y_0y_1 + y_2^2\mu = 3x_1^2x_2\mu + 3x_0^2x_1 + 3x_0x_2^2\mu + Ax_1$$

and

$$2y_0y_2 + y_1^2 = 3x_1x_2^2\mu + 3x_0^2x_2 + 3x_0x_1^2 + Ax_2$$

which describe an open affine part of $W_E$.

Now we want to determine the subvariety $B$ given by the condition

$$\mathrm{Trace}(\pi_q) = 0.$$

We find on $B$ the relations:

$$3x_0 = \left(\frac{y_1}{x_1}\right)^2 = \left(\frac{y_2}{x_2}\right)^2 = \frac{y_1y_2}{x_1x_2}.$$

Hence we get one equation:

$$(3x_1x_2\mu + 3x_0^2 + A)^2 = 12x_0(x_1^3\mu + x_2^3\mu^2 + x_0^3 + Ax_0 + B).$$

which together with the relation

$$y_0^2 = x_0^3 + \mu x_1^3 + \mu^2 x_2^3 + Ax_0 + B$$

describes an open affine part of $B$. We find curves on $B$ by intersecting by hyperplanes, e.g. we choose $x_0 = \text{const}$. In general these curves will have genus $\geq 4$ but for the special choice

$$x_0 = 0$$

we find a hyperelliptic curve with equation:

$$C_0 : y^2 = \mu x_1^7 + Bx_1^4 + A^3/(27\mu)x_1$$

with $y = y_0 x_1^2$. This curve[9] has genus 3. Its Jacobian has a factor which is isogenous to $B$ and hence is manageable if $E/\mathbb{F}_q$ is manageable. In [Na] one finds choices of $E, q$ which lead to rather good DL-systems.

Next assume that $q$ is a power of 3. Take $m = 3$ and an elliptic curve given by

$$E : Y^2 = X^3 + AX^2 + B.$$

We generate $\mathbb{F}_{q^3}$ by an element $\alpha$ satisfying the equation

$$\alpha^3 - \alpha = a.$$

By analogous computations as above Diem finds the curve

$$y_0^2 = A^{-3}x_1^6 + ax_1^3 + Ax_1^2 - A^2x_1 + aA^3 + A^{-1} + B$$

on $B/\mathbb{F}_q$ which has genus 2.

One could speculate that as in the case exposed in [GHS] Weil descents of degree $m \geq 5$ can weaken DL-systems over $\mathbb{F}_{q^m}$.

### 3.3 Tate Duality and Brauer Groups

**The Tate Pairing** In the last subsections we have already seen how the operation of Galois groups plays a key role (constructively and partly destructively) for the realization of discrete logarithm systems. In the centre (even in the global constructions via CM-theory or modular varieties) were the Frobenius automorphisms.

In this subsection we shall describe a duality structure intrinsically related to abelian varieties over local and global fields for which ramified extensions play an essential role, too. Applications of this duality can be seen as prototypes of "lifting attacks". Typically we shall not only lift abelian varieties and points on them from finite fields to $\mathfrak{l}$-adic fields and global fields but also use Galois cohomology groups related to these liftings which become trivial over finite fields.

We begin with the general setting. Let $K$ be a field and $A$ a (principally polarized) abelian variety over $K$, $n$ prime to char($K$). Let $M$ be a $\mathbb{Z}$-module.

[9] This curve as well as the next one have been found by C. Diem.

By $M[n]$ we denote the subgroup of $M$ consisting of the elements of order dividing $n$. Let $K_s$ be the separable closure of $K$ and $G_K = \mathrm{Aut}(K_s/K)$ the absolute Galois group of $K$.

$G_K$ acts continously on $A(K_s)$ and we have the exact sequence of $G_K-$modules (Kummer sequence)

$$0 \to A(K_s)[n] \to A(K_s) \overset{\cdot n}{\to} A(K_s) \to 0$$

Application of Galois cohomology gives the exact sequence

$$0 \to A(K)/nA(K) \overset{\delta}{\to} H^1(G_K, A(K_s)[n]) \overset{\alpha}{\to} H^1(G_K, A(K_s))[n] \to 0.$$

Next we use that $A(K_s)[n]$ is as a $G_K-$module isomorphic to its dual group (since $A$ is assumed to be principally polarized) and so we can use the cup product to get the **Tate-pairing**

$$<,>_K: A(K)/nA(K) \times H^1(G_K, A(K_s))[n] \to H^2(G_K, \mu_n)$$

(where $\mu_n$ is the group generated by the $n$-th roots of unity in $K_s$) given by

$$< P + nA(K), \gamma >_K = \delta(P + nA(K)) \cup \alpha^{-1}(\gamma).$$

For computational reasons it is important that **Lichtenbaum** [Li] gave a definition of $<,>_K$ in the case that $A$ is the Jacobian of a curve $C$ which uses functions and points on $C$ for the computation of $<,>_K$ (see 3.3).

The group $H^2(G_K, \mu_n)$ is very important for the arithmetic of $K$: It consists of the elements of order dividing $n$ of the **Brauer group** $Br(K)$ of $K$. The information we can get out of the Tate-pairing depends on the information given by the Brauer group and on its degree of non-degeneracy.

**Brauer Groups of Local Fields** The Brauer group of a finite field is $\{0\}$.

If $K$ is an $\mathfrak{l}-$adic field (i.e. $K$ is a finite algebraic extension of the field of $l-$adic numbers $\mathbb{Q}_l$) the Brauer group of $K$ is canonically isomorphic to $\mathbb{Q}/\mathbb{Z}$, the isomorphism is given by the so called **invariant map** $\mathrm{inv}_K$. This and the explicit determination (called Artin's reciprocity law) of the invariant are the key to **local class field theory** (cf. [Ne]): One can describe the Brauer group of $K$ by using cocycles defined by cyclic extensions $L$ of $K$ of degree $n$ in the following way:

**Theorem 3.14.** *Let $K$ be an $\mathfrak{l}-$adic field with normed valuation $v_\mathfrak{l}$. Let $L/K$ be a cyclic extension of degree $n$ and let $\tau$ be a generator of its Galois group.*

*(i)* $Br(K)[n] = H^2(G(L/K), L^*)$.
*(ii)* $H^2(G(L/K), L^*)$ *is generated by the cohomology classes of the 2-cocycles $\gamma$ defined by*

$$\gamma(\tau^i, \tau^j) = \begin{cases} a \in K^* & \text{if } i + j \geq n \\ 1 & \text{otherwise} \end{cases}$$

*(iii) Two cocycles $\gamma$ and $\gamma'$ are in the same class iff the corresponding elements $a$ and $a'$ satisfy:*

$$a \cdot {a'}^{-1} \in N_{L/K}(L^*).$$

*(iv) Let $L/K$ be unramified of degree $n$ and let $\tau$ be the automorphism of $L/K$ which induces the Frobenius automorphism of the residue field extension. Then*

$$\operatorname{inv}(c) = v_{\mathfrak{l}}(a)/n \bmod \mathbb{Z}.$$

There are two problems if one wants to use Theorem 3.14 to compute the invariant function in the Brauer group of $K$:

(i) In general a cohomology class in $Br(K)[n]$ is not given as a cocycle of an unramified extension. We shall see below that in the cases of interest for us we have to deal with ramified cyclic extensions $L/K$.
(ii) Even if $c$ is given by an unramified field extension $L/K$ the given generator of $G(L/K)$ need not be the Frobenius automorphism.

In both cases one checks that in order to compute $\operatorname{inv}(c)$ we have to compute discrete logarithms in the residue field of $L$. (Details of these considerations will be found in [Ng]). Hence there is a close connection between the discrete logarithm in this residue field and $Br(K)[n]$ which is especially easy in the following case:

**Proposition 3.15.** *Assume that $n$ is a number prime to the characteristic of the residue field $k$ of $K$ and that $L/K$ is a totally ramified Galois extension of degree $n$. Then $Br(K)[n]$ is isomorphic to $H^2(G(L/K), L^*)$ and the DL-problem in the group $k^*[n] = \mu_n$ is equivalent with the DL-problem in $H^2(G(L/K), L^*)$ and hence with the computation of the invariant.*

*Proof.* $L/K$ is tamely ramified by assumption and hence by Hilbert theory (cf. [Ne]) it is a cyclic extension and the $n-$th roots of unity are contained in $k$. Hence there is an element $\pi$ in $K$ with $v_{\mathfrak{l}}(\pi) = 1$ such that $L = K(\pi^{1/n})$ and so $\pi$ is in the image of the norm map from $L$ to $K$. By this and Hensel's lemma we see that $N_{L/K}(L^*)$ is isomorphic to $k^*/k^{*n}$. Now use Theorem 3.14 to get the result.

*Remark 3.16.* Let $l_0$ be the characteristic of $k$ and take $n = l_0$. By using Artin-Schreier theory in power series fields over $k$ it is not difficult to get results for extensions of degree $l_0$ analoguous to the results in Theorem 3.14.

**Tate Duality over Local Fields** Let $K$ be an $\mathfrak{l}-$adic field with residue field characteristic $l_0$ and let $A$ be a principally polarized abelian variety defined over $K$.

**Theorem 3.17. (Tate,** *cf.[Li]***)**

$$<,>_K \text{ is non-degenerate.}$$

Now assume that $n$ is prime[10] to $l_0$ and that $A$ has good reduction modulo $\mathfrak{l}$.

[10] For $n = l_0$ cf. [Rü].

What about $H^1(G_K, A(K_s))[n]$?

Let $\pi$ be an element in $K$ with $v_{\mathfrak{l}}(\pi) = 1$ and let $\zeta_n$ be a primitive $n$−th root of unity. Let $L_n$ be equal to $K(\zeta_n, \pi^{1/n})$. The Galois group $G_n$ of $K(\zeta_n)/K$ acts on the Galois group of $L_n/K(\zeta_n)$ by conjugation.

The field $K_n := K(A(K_s)[n])$, the smallest field containing the coordinates of all points of order $n$ of $A$, is an unramified cyclic extension of $K$ containing $\zeta_n$. Hence the Galois group of $K(\zeta_n)/K$ has a canonical lifting to the Galois group $U_n$ of $K_n/K$ and so $U_n$ acts on $A(K_s)[n]$ and on $G(L_n/K(\zeta_n))$. Using standard arguments from Galois cohomology and the fact that abelian varieties are cohomologically trivial over finite fields we get:

**Lemma 3.18.** *In the situation above*

$$H^1(G_K, A(K_s))[n] = \mathrm{Hom}_{U_n}(G(L_n/K(\zeta_n)), A(K_s)[n]).$$

Hensel's lemma implies that $A(K)/nA(K) \simeq A(K)[n]$. Hence by Theorem 3.17 we get a non-degenerate pairing

$$\overline{<,>_K} : A(K)[n] \times \mathrm{Hom}_{U_n}(G(L_n/K(\zeta_n), A(K_s)[n]) \to Br(K)[n] \overset{\mathrm{inv}_K}{\to} \mathbb{Z}/n\mathbb{Z}.$$

If we could compute $\overline{<,>_K}$ we would have solved the computation of the discrete logarithm in $A(K)[n]$.

In the special case that $\zeta_n \in K$ we get from Theorem 3.17 the

**Corollary 3.19.** *Assume that $\zeta_n \in K$ (or equivalently that $|k^*| \equiv 0 \bmod n$). There is a non-degenerate pairing*

$$<,>_K : A(K)[n] \times \mathrm{Hom}(G(L_n/K), A(K)[n]) \to Br(K)[n]$$

*induced by the Tate pairing.*

We describe the pairing in the simplest case: Assume that $n$ is a prime and that there is no point of order $n^2$ in $A(K)$. Let $P_1$ be a point of order $n \in A(K)$ and $Q_1 \in A(K_s)$ with $nQ_1 = P_1$. Let $\pi$ be the generator of the Galois group of the maximal unramified extension of $K$ which induces the Frobenius automorphism modulo $\mathfrak{l}$. Let $\tau$ be a generator of the Galois group of $L_n/K$ and $\phi_{P_2}$ the homomorphism of $G(L_n/K)$ to $A(K)[n]$ given by $\phi_{P_2}(\tau) = P_2$.

Then $< P_1, \phi_{P_2} >_K$ is the cohomology class of the 2-cocycle $\gamma$ of the group generated by $\tau$ and $\pi$ given by

$$\gamma(\tau^i \cdot \pi^r, \pi^j \cdot \tau^s) = i \cdot P_2 \cup (\pi^j(Q_1) - Q_1).$$

**Application to Jacobian Varieties over Finite Fields** We shall assume that $p$ is a prime dividing $q - 1$. Let $A$ be defined over $\mathbb{F}_q$. We lift $A$ to an abelian variety $\tilde{A}$ over an $\mathfrak{l}$-adic field $K$ with residue field $\mathbb{F}_q$, assume the same notations as above and apply Corollary 3.19 to $\tilde{A}$.

Since Hensel's lemma implies that $\tilde{A}(K)[p]$ is canonically isomorphic to $A(\mathbb{F}_q)[p]$ we get a non-degenerate pairing

$$<,>_{\mathbb{F}_q}: A(\mathbb{F}_q)[n] \times \mathrm{Hom}(G(L_n/K), A(\mathbb{F}_q)[n]) \to Br(K)[n].$$

It remains to give a definition of this pairing which makes its computation fast. For this we assume in addition that $A$ is the Jacobian $J_C$ of a curve $C$ and lift the pair $(C, J_C)$ to $K$.

**Proposition 3.20. (Lichtenbaum)** *Let $P_1, P_2$ be points of order $n$ of $J_C(\mathbb{F}_q)$. Represent $P_i$ by coprime divisors $D_i$ in the divisor class group, and let $f_1$ be a function on $C$ with divisor $n \cdot D_1$.*

*Then $< P_1, P_2 >_{\mathbb{F}_q}$ is the cohomology class $c$ represented by the cocycle*

$$\gamma(\tau^i, \tau^j) := \begin{cases} f_1(D_2) & \text{if } i + j \geq n \\ 1 & \text{otherwise} \end{cases}$$

We call this version of the pairing the Tate-Lichtenbaum pairing.

Since $\gamma$ represents $c$ modulo the image of the norm map from $L_n$ to $K$ and because of the divisibility of the principal units of $K$ by $n$ we get (with the notation of Proposition 3.20)

**Corollary 3.21.** *The Tate-Lichtenbaum pairing induces a non-degenerate pairing*

$$<,>: J_C(\mathbb{F}_q)[n] \times J_C(\mathbb{F}_q)[n] \to \mu_n(\mathbb{F}_q)$$

*given by*

$$< P_1, P_2 > := f_1(D_2)^{\frac{q-1}{n}}.$$

One finds in [FR] how this pairing can be computed in $O(\log(q))$ addition steps and so it reduces the discrete logarithm in $J_C(\mathbb{F}_q)[n]$ to the discrete logarithm in $\mu_n(\mathbb{F}_q)$.

In general the conditions that $\mathbb{F}_q$ contains $p$-th roots of unity **and** a manageable abelian variety $A$ has points of order $p$ rational over $\mathbb{F}_q$ which are cryptographically interesting will not be satisfied at the same time. For elliptic curves we can formulate this more precisely:

**Corollary 3.22.** *Let $E$ be an elliptic curve defined over $\mathbb{F}_q$ and $p$ a prime. Let $\pi_q$ be the Frobenius automorphism of $\mathbb{F}_q$.*
*Then $\mathbb{Z}/p$ can be embedded into $E(\mathbb{F}_{q^f})$ iff the trace of $\pi_q^f$ is congruent to $q^f + 1$ modulo $p$. In this case the corresponding discrete logarithm in $E(\mathbb{F}_{q^f})$ can be reduced to the discrete logarithm in $\mu_p$ in the field $\mathbb{F}_{q^{fm}}$ where $m$ is the smallest integer such that the trace of $\pi_q^{fm}$ becomes congruent to 2 modulo $p$.*

Hence to satisfy Task 3. of Aim 2.3 we have to make sure that $f$ is rather large. This excludes supersingular elliptic curves and very few others. We end this subsection by remarking that analogous arguments make it even easier to compute the DL if $p = \mathrm{char}(\mathbb{F}_q)$.

**Outlook** There are two natural **Questions**:

(i) Can one compute $\overline{<,>}$ rapidly?
(ii) Can one compute $<,>$ without using the condition that $K$ contains all $n$-th roots of unity?

To answer these questions (positively or negatively) we need much more information about the arithmetic inside of Brauer groups.

We recall that to use the Tate pairing we had to lift the abelian variety from finite fields to $\mathfrak{l}$−adic fields. **Hasse's** results on Brauer groups make it possible, at least in theory, to lift the problem to global fields (i.e. number fields and function fields in one variable over finite fields) and it may well be that his celebrated sequence for global fields $K$ and their completions $K_{\mathfrak{l}}$:

$$0 \to Br(K) \to \bigoplus_{\mathfrak{l}} Br(K_{\mathfrak{l}}) \stackrel{\oplus \mathrm{inv}_{\mathfrak{l}}}{\to} \mathbb{Q}/\mathbb{Z} \to 0$$

can play an important role.

## References


[ADH] L. Adleman, J. DeMarrais and M. Huang, A Subexponential Algorithm for Discrete Logarithm over the Rational Subgroup of the Jacobians of Large Genus Hyperelliptic Curves over Finite Fields, *Algorithmic Number Theory - Proceedings of the First International Symposium, ANTS–I*, Springer Verlag (1994), 28–40.

[Ba] J. Basmaji, Ein Algorithmus zur Berechnung von Hecke-Operatoren und Anwendung auf modulare Kurven, *Dissertation Essen* (1996).

[DH] W. Diffie and M. Hellman, New directions in cryptography, *IEEE Trans. Inform. Theory* **22** (1976), 644–654.

[E] A. Enge, Computing discrete logarithms in high-genus hyperelliptic Jacobians in provably subexponential time, *CORR* **99-04**, *Univ. of Waterloo*, (Febr. 1999), to appear in *Math. Comp.*

[ES] A. Enge and A. Stein, Smooth Ideals in Hyperelliptic Function Fields, *CORR* **2000-08**, *Univ. of Waterloo*, (Febr. 2000).

[EG] A. Enge and P. Gaudry, A General Framework for Subexponential Discrete Logarithm Algorithms, Manuscript 19 pp (Febr. 2000).

[Fr] G. Frey, Weil Restriction , *ECC Waterloo* (1998).

[FMR] G. Frey, M. Müller and H.-G. Rück, The Tate Pairing and the Discrete Logarithm Applied to Elliptic Curve Cryptosystems, *Trans. on Inf. Th., IEEE*, **45** (1999), 1717-1719.

[FR] G. Frey and H.-G. Rück, A Remark Concerning m-Divisibility and the Discrete Logarithm in the Divisor Class Group of Curves, *Math. of Comp.*, **62** (1994), 865-874.

[Ga] P. Gaudry, A variant of the Adleman–DeMarrais–Huang algorithm and its application to small genera, *Laboratoire d' Informatique* Preprint LIX/-RR/99/04 (1999).

[GHS] P. Gaudry, F. Hess and N. Smart, Constructive and Destructive Facets of Weil Descent on Elliptic Curves, Preprint (2000).

[GS] S. Galbraith and N. Smart, A cryptographic application of Weil descent, in: *Codes and Cryptography, M. Walker ed., Springer LNCS* **1746** (1999), 191–200.

[Ig] J.-I. Igusa, The arithmetic variety of genus two, *Ann. Math.* **72** (1960), 612–649.

[Ka] W. Kampkötter, Explizite Gleichungen für Jacobische Varietäten hyperelliptischer Kurven, *Dissertation Essen* (1991).

[Ko] N. Koblitz, Hyperelliptic Cryptosystems, *J. Cryptology* **1** (1989),139–150.

[Kr] U. Krieger, signature.c–Anwendungen hyperelliptischer Kurven in der Kryptographie, *Diplomarbeit Essen* (1997).

[LR] H. Lange and W. Ruppert, Complete systems of addition laws on abelian varieties, *Invent. Math.*, **79** (1985), 603-610.

[LV] A.K. Lenstra and E.R. Verheul, Selecting Cryptographic Key Sizes, manuscript, (Oct.1999).

[Li] S. Lichtenbaum, Duality Theorems for Curves over p-adic fields, *Inv. math.*, **7** (1969), 120-136.

[Me] J.-F. Mestre, Construction de courbes de genre 2 à partir de leurs modules, in: *Progress in Math.*, **94**, Birkhäuser Verlag (1991), 313–334.

[Mo] F. Morain, Calcul du nombre de points sur une courbe elliptique dans un corps fini: aspects algorithmiques,*J. Th. des Nombres de Bordeaux*, **7** (1995), 111-137.

[MOV] A. Menezes, P. van Oorschot and S. Vanstone, *Handbook of Applied Cryptography*, CRC Press (1995).

[MST] V. Müller, A. Stein and A. Thiel, Computing discrete logarithms in real quadratic congruence function fields, *Math. Comp.*, **68** (226) (1999), 807–822.

[Mu1] D. Mumford, On the equations defining abelian varieties I, *Invent. Math.*, **1** (1966), 287–354.

[Mu2] D. Mumford, *Abelian Varieties*, Oxford University Press (1974).

[Mu3] D. Mumford, *Tata lectures on Theta I,II + III*, Birkhäuser Verlag (1982).

[Mur] R. Murty, Exponents of Class Groups of Quadratic Fields, in: *Proc. Conf. Number Th. and Arith. Geom. 19977, ed. G.Frey*, Preprint **14** (1998), IEM, Essen.

[MW] U. Maurer and S. Wolf, Lower Bounds on Generic Algorithms in Groups, in: *Advances in Cryptology-EUROCRYPT 98, K.Nyberg ed., Lecture Notes in Computer Science* **1403**, Springer Verlag (1998), 72-84.

[Na] N. Naumann, Weil-Restriktion Abelscher Varietäten, *Diplomarbeit Essen* (1999).

[Ne] J. Neukirch, *Algebraische Zahlentheorie*, Springer (1992).

[Ng] K. Nguyen, Brauer Groups of Local and Global Fields and Discrete Logarithms, in preparation.

[Od] A.M.Odlyzko, Discrete Logarithms and Smooth Polynomials, in *Finite Fields: Theory, Applications and Algorithms, G.L.Mullen and P. Shiue,eds. Cont. Math.* **168** *AMS* (1994), 269-278.

[Rü] H.-G. Rück, On the Discrete Logarithm in the Divisor Class Group of Curves,*Math. Comp.*, **68** (1999), 1233-1241.

[Sch1] R. Schoof, Elliptic curves over finite fields and the computation of square roots mod *p*, *Math. Comp.*, **43** (1985), 483–494.

[Sch2] R. Schoof, Counting Points on elliptic curves over finite fields, *J. de Th. des Nombres de Bordeaux*, **7** (1995), 219–254.

[Sh] G. Shimura , *Introduction to the Arithmetic Theory of Automorphic Functions*, Princeton University Press (1971)

[ST] G. Shimura and Y. Taniyama, *Complex Multiplication of Abelian Varieties and its applications to Number Theory*, Publ. Math. Soc. Japan (1961).

[Sho] P. W. Shor, Quantum Computing, *Doc.Math.J.DMV Extra Volume ICM I* (1998), 467–486.

[Si] J. Silverman, *The Arithmetic of Elliptic Curves*, Springer Verlag (1992).

[Sp] A. M. Spallek, Kurven vom Geschlecht 2 und ihre Anwendungen in Public–Key–Kryptosystemen, *Dissertation Essen* (1994).

[We] H. J. Weber, Hyperelliptic Simple Factors of $J_0(N)$ with dimension at least 3, *Experimental Math.*, **6** (1997), 273-287.

# Gauß Periods in Finite Fields

Joachim von zur Gathen[1] and Igor Shparlinski[2]

[1] Fachbereich 17 Mathematik-Informatik
Universität Paderborn
D-33095 Paderborn
Germany
e-mailgathen@upb.de

[2] Department of Computing
Macquarie University
Sydney, NSW 2109
Australia
e-mail: igor@comp.mq.edu.au

## 1 Introduction

In this survey, we review two recent applications of a venerable tool: Gauß periods.

In Section 2, we describe Gauß' original construction, and how it can be used to generate normal bases in extensions of finite fields.

Section 3 contains the first application: finding elements of exponentially large order in certain finite fields. This can be viewed as a step towards solving the famous open problem of finding efficiently a primitive element in a given finite field. A pleasant feature is that the prime factorization of the order of the multiplicative group is not required. In Section 4 we give another example of the method, yielding a different kind of bound: among the $q$ shifts $\beta + a$ of an element $\beta$ of an extension of $\mathbb{F}_q$, where $a$ runs through $\mathbb{F}_q$, at most one has "small" order.

The second application, in Section 5, deals with efficient exponentiation in finite fields, an important subroutine in some cryptographic systems. Gauß periods lead to the fastest algorithms for this problem, both in theory and in practice. In Section 6 we describe a recent generalization of Gauß construction.

Our two applications are, in turn, useful in several areas: cryptography, coding theory, pseudo-random element generation, and combinatorial designs. Actually, the two work in tandem: where one of them is required, usually the other one is as well.

The results of Section 4 are new, while the others report on the recent literature.

## 2 Normal Bases via Gauß Periods

[16] introduced his periods as follows. He lets $n, k$, and $r$ be integers, with $r$ prime, and

$$nk = \varphi(r) = r - 1.$$

Furthermore, $\zeta \in \mathbb{C}$ is a primitive $r$th root of unity, and $\mathcal{K} \subseteq \mathbb{Z}_r^\times = \mathrm{Gal}\,(\mathbb{Q}(\zeta)\colon \mathbb{Q})$ the unique subgroup of order $k$ of the cyclic group $\mathbb{Z}_r^\times$. Then

$$\eta = \sum_{i \in \mathcal{K}} \zeta^i \in \mathbb{Q}(\zeta)$$

is the **Gauß period** of type $(n, \mathcal{K})$ over $\mathbb{Q}$.

The fields involved in this construction are as follows:

$$\begin{array}{rl} \text{primitive } r\text{th root of unity } \zeta \in & \mathbb{Q}(\zeta) \\ & |_k \\ \eta = \sum_{i\in\mathcal{K}} \zeta^i \in & \mathbb{Q}(\eta) \\ & |_n \\ & \mathbb{Q} \end{array}$$

As illustrated in Figure 1, Gauß used the three subgroups of orders 2 (solid line), 4 (dashed line), and 8 (dotted line) of $\mathbb{Z}_{17}^\times = \mathrm{Gal}(\mathbb{Q}(\zeta)\colon \mathbb{Q})$ whose periods describe a tower of field extensions from $\mathbb{Q}$ to the field generated by the primitive 17th root of unity $\zeta = e^{2i\pi/17}$, where each extension degree equals 2. This shows that the regular 17-gon can be constructed by ruler and compass.

Over a finite field $\mathbb{F}_q$ with $q$ elements, we can consider the analogous situation: we have $n, k, r$, and $\mathcal{K} \subseteq \mathbb{Z}_r^\times$ as above, and also require $\gcd(q, r) = 1$:

$$\begin{array}{rl} \text{primitive } r\text{th root of unity } \beta \in & \mathbb{F}_{q^{r-1}} \\ & |_k \\ \alpha = \sum_{i\in\mathcal{K}} \beta^i \in & \mathbb{F}_{q^n} \\ & |_n \\ & \mathbb{F}_q \end{array}$$

Then $\alpha$ is a Gauß period of type $(n, \mathcal{K})$ over $\mathbb{F}_q$. (The required $\beta$ exists in $\mathbb{F}_{q^{r-1}}$, but $\mathbb{F}_q(\beta)$ may be a proper subfield of $\mathbb{F}_{q^{r-1}}$.)

A **normal basis** for $\mathbb{F}_{q^n}$ over $\mathbb{F}_q$ is a basis of the form $(\gamma, \gamma^q, \ldots, \gamma^{q^{n-1}})$ of $\mathbb{F}_{q^n}$ as a vector space over $\mathbb{F}_q$. Then $\gamma \in \mathbb{F}_{q^n}$ is called *normal* over $\mathbb{F}_q$. We will see in Section 5 that this data structure for representing the elements of $\mathbb{F}_{q^n}$ is particularly useful for exponentiation. [27] shows that Gauß periods often generate normal bases, namely that $\alpha$ is normal in $\mathbb{F}_{q^n}$ over $\mathbb{F}_q$ if and only if $q$ and $\mathcal{K}$ together generate the multiplicative group $\mathbb{Z}_r^\times$: $\langle q \bmod r, \mathcal{K}\rangle = \mathbb{Z}_r^\times$.

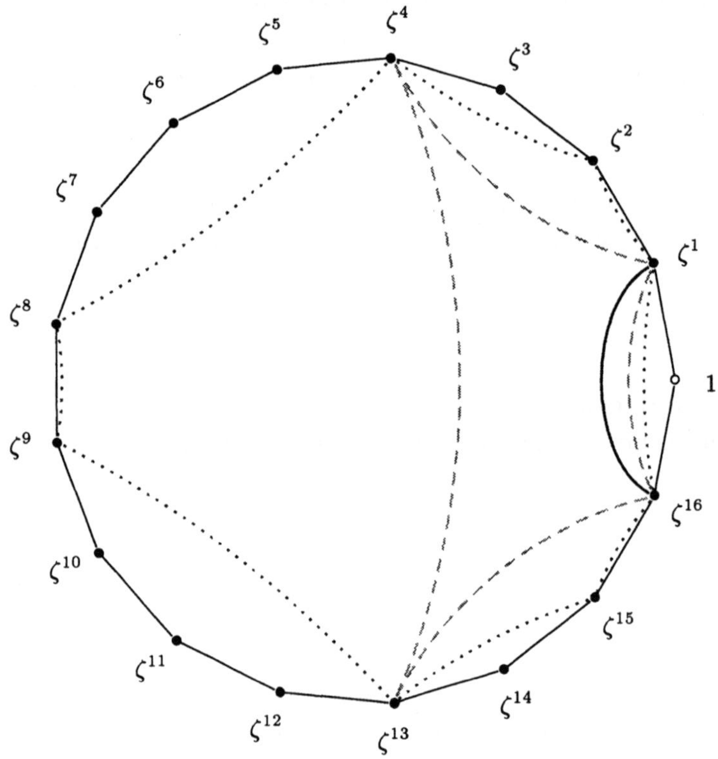


**Figure1.** 17-gon with Gauß periods of order 2, 4, and 8

## 3 Elements of Large Order

The group $\mathbb{F}_q^\times$ of units of $\mathbb{F}_q$ is cyclic of order $q-1$. An element $\gamma \in \mathbb{F}_q^\times$ is **primitive** if and only if its order is $q-1$.

Such elements are useful for a variety of tasks. How can we find one? Gauß writes in article 73 of his *Disquisitiones Arithmeticae* (1801): *Methodi radices primitivas inveniendi maximam partem tentando innituntur. [ ... ] Ill. Euler confitetur, maxime difficile videri, hos numeros assignare, eorumque indolem ad profundissima numerorum mysteria esse referendam. At tentando satis expedite sequenti modo determinari possunt. Exercitatus operationis prolixitati per multifaria artificia particularia succurrere sciet: haec vero per usum multo citius quam per praecepta ediscuntur.*[1]

[1] Methods for finding primitive roots are mainly reduced to trial and error. [ ... ] The illustrious Euler admits that it is extremely difficult to find these numbers and that their nature is one of the deepest mysteries of numbers. But they can be determined easily enough by the following method. The experienced computer

Gauß' method is a clever variant of exhaustive search and, in modern parlance, uses exponential time and space.

The importance of primitive elements for the human computers of the 19th century is illustrated by several publications. The *Canon arithmeticus* of [19] contains 240 pages of tables of primitive elements and discrete logarithms, one table for each prime number up to 1000. Euler had previously given such a table for $3 \le p \le 37$, and [4] for $3 \le p \le 101$. The tables of [1,2] give primitive elements in nonprime fields $\mathbb{F}_q$ for all relevant $q \le 1000$.

To test whether a given $\gamma \in \mathbb{F}_q^\times$ is primitive, we only have to check if $\gamma^{(q-1)/t} \ne 1$ for all prime divisors $t$ of $q-1$. There are $\varphi(q-1)$ many primitive elements, so taking a random element of $\mathbb{F}_q^\times$ will give a primitive one with fairly high probability, namely $\varphi(q-1)/(q-1) \ge c/\operatorname{loglog} q$ for some absolute constant $c > 0$, see Theorem 5.1 in Chapter 1 of [23].

Computationally, the bottleneck in this procedure is the prime factorization of $q-1$. No polynomial-time algorithm is known for this, where polynomial time means $(\log q)^{O(1)}$ operations, since the size of a reasonable representation of an element of $\mathbb{F}_q$ is proportional to $\log_2 q$. We have the following tasks:

(i) test whether a given $\gamma \in \mathbb{F}_q^\times$ is primitive,
(ii) find a primitive element,
(iii) find an element of "large order".

To solve any of them in (random) polynomial time is an important open problem.

Now a further relaxation of our requirements allows a successful solution of the last problem. Namely, for a given $N$, we do not insist on finding a good element in $\mathbb{F}_{q^N}$, but are content if we find one in $\mathbb{F}_{q^n}$ for some $n$ close to $N$, as in Theorem 3.1 below. A variation, mentioned at the end of this section, works when we want to work in a small extension of $\mathbb{F}_{q^N}$.

The progress here can be viewed as follows: finding a primitive element in polynomial time is a black and white question: either you have such an algorithm or you don't. Now we have a question with many shades of gray where incremental progress is possible: increase the order of the elements and decrease the degree of the field extension.

We will make use of the following famous conjecture.

**Artin's conjecture:** *For $a \in \mathbb{Z}$, not $-1$ or a square, there exists $c(a) > 0$ such that*

$$\#\{p \le x\colon (a \bmod p) \in \mathbb{Z}_p^\times \textit{ is primitive }\} \sim c(a)\frac{x}{\log x}.$$

[18] proved this under the Extended Riemann Hypothesis, and also determined $c(a)$ explicitly. Then [17] showed, without any assumption, that a

will know how to reduce the tediousness of the calculation by a variety of devices, but these are better taught by practice than by rules.

slightly weaker lower bound of the form $c(a)x/\log^2 x$ for the cardinality of the above set holds for all prime powers $a$, except maybe powers of at most three primes. (It is conjectured that these three potential exceptions are an artifact of the proof and actually do not exist.) In fact, even more general results have been established in [17], but for our purposes prime powers are of primal interest. Furthermore, even a weaker version, say cardinality at least $c(a)x/\log^3 x$ for the set of prime powers $a$ would be sufficient for our applications.

This question is already implicit in Jacobi's *Canon*, where he counts the primes modulo which 10 is primitive in the first 25 blocks of 100 integers each.

**Theorem 3.1.** *Suppose that Artin's conjecture holds for $q$. Then for any $N$ there exists $n \geq N$ with $n \in O(N \log N)$ such that the Gauß period $\alpha$ in $\mathbb{F}_{q^n}$ of type $(n, \{\pm 1\})$ over $\mathbb{F}_q$ is normal and has order at least*

$$2^{(2n)^{1/2}-2}.$$

*These $n$ and $\alpha$ can be computed (probabilistically) in time polynomial in $N$ and* $\log q$.

We sketch the idea of the *proof*. Under the assumptions, we can find a prime $r$ so that $n = (r-1)/2$ is sufficient and $q$ is primitive in $\mathbb{Z}_r^\times$. We take a primitive $r$th root of unity $\beta$ in $\mathbb{F}_{q^{2n}}$; then $\beta$ has degree $r-1$ over $\mathbb{F}_q$. Furthermore, we consider the quantities

$$\begin{aligned}
&\alpha = \beta + \beta^{-1}, \\
&h = \lfloor r^{1/2} \rfloor - 1, \\
&S = \{i\colon 0 \leq i \leq r-1 \text{ and } 1 \leq (q^i \text{ rem } r) \leq h\} \subseteq \mathbb{F}_r^\times, \\
&U \neq U' \subseteq S, \\
&u = \sum_{s \in U} q^s, \quad u' = \sum_{s \in U'} q^s.
\end{aligned}$$

We now claim that $\alpha^u \neq \alpha^{u'}$. If the claim is true, then we have at least $2^h$ different powers of $\alpha$, and the lower bound on the order of $\alpha$ follows with an easy calculation.

If the claim is false, the following calculation yields a nonzero $f \in \mathbb{F}_q[x]$ of degree less than $r-1$ with $f(\beta) = 0$. This contradiction to the fact that $\beta$ has degree $r-1$ over $\mathbb{F}_q$ proves the claim.

We may suppose that $U \cap U' = \emptyset$. We assume that $\alpha^u = \alpha^{u'}$, and thus

$$\begin{aligned}
0 = \alpha^u - \alpha^{u'} &= \prod_{s \in U} (\beta + \beta^{-1})^{q^s} - \prod_{s \in U'} (\beta + \beta^{-1})^{q^s} \\
&= \beta^{-u} \prod_{s \in U} (\beta^{2q^s} + 1) - \beta^{-u'} \prod_{s \in U'} (\beta^{2q^s} + 1).
\end{aligned}$$

Since $\beta$ is an $r$th root of unity, we may reduce the exponents modulo $r$. We define

$$E = \{q^s \text{ rem } r \colon s \in U\}, E' = \{q^s \text{ rem } r \colon s \in U'\} \subseteq \{1, \ldots, r-1\},$$

$$e = \sum_{t \in E} t, \quad e' = \sum_{t \in E'} t.$$

Then $E \cap E' = \emptyset$, and

$$0 = \beta^{-e} \prod_{t \in E} (\beta^{2t} + 1) - \beta^{-e'} \prod_{t \in E'} (\beta^{2t} + 1).$$

We may assume that $e' \geq e$, and let

$$f(x) = x^{e'-e} \prod_{t \in E} (x^{2t} + 1) - \prod_{t \in E'} (x^{2t} + 1) \in \mathbb{F}_q[x].$$

Then $f(\beta) = 0$, and

$$\deg f \leq 2e' \leq 2 \sum_{1 \leq j \leq h} j = h(h+1) \leq r - \sqrt{r} < r - 1.$$

If $e' > e$, then $f(0) = -1$. Thus $e' = e$. But then the monomial $x^{2\tau}$ occurs in $f$ with nonzero coefficient, where $\tau = \min(E \cup E')$, and we have the desired contradiction. □

Similar arguments yield:

- a denser sequence of $n$ in Theorem 3.1,
- for each $\mathbb{F}_q$, a small extension $\mathbb{F}_{q^n}$ and an element of exponential order in it,
- unconditional results,
- deterministic algorithms.

These results have appeared, with complete proofs, in [14,15].

## 4 A New Lower Bound on Multiplicative Orders

Applying the same method as in the previous section, we obtain a new result which is of independent interest, while its proof exhibits the main tool of our method—information about the distribution of exponential functions in residue classes.

Let $\beta \in \mathbb{F}_{q^n}^\times$ be a root of an irreducible polynomial in $\mathbb{F}_q[x]$ of degree $n \geq 2$. For the order $t$ of $\beta$, we obviously have $t \geq n$. In fact, this can be strengthened a little as $\varphi(t) \geq n$. This bound is tight, since it is attained if $t = n + 1$ is prime, $q$ is a primitive root modulo $t$, and $\beta$ is a primitive $t$th root of unity. Below we show that for any $\varepsilon > 0$ there is a constant $c > 0$ such that among the $q$ shifts $\beta + a$, with $a \in \mathbb{F}_q$, at most one is of order less than $c \cdot n^{4/3-\varepsilon}$.

We need some results about exponential sums and their distribution in residue classes of exponential functions.

**Lemma 4.1.** *Let $q$ and $r$ be positive integers with $\gcd(q,r) = 1$, and let $\tau$ be the order of $q$ modulo $r$. Then for any integer $c$ we have*

$$\left| \sum_{1 \le k \le \tau} \exp(2\pi i c q^k / r) \right| \le \delta^{1/2} r^{1/2},$$

*where $\delta = \gcd(c,r)$.*

*Proof.* If $\delta = 1$, then this bound is essentially Theorem 10 in Chapter 1 of [21]; see also the proof of Lemma 2 in [20]. For $\delta > 1$, we denote by $\tau_\rho$ the order of $q$ modulo $\rho = r/\delta$, and we also put $c/\delta = \gamma$. Thus $\gcd(\gamma, \rho) = 1$ and we obtain

$$\left| \sum_{1 \le k \le \tau} \exp(2\pi i c q^k / r) \right| \le \frac{\tau}{\tau_\rho} \left| \sum_{1 \le k \le \tau_\rho} \exp\left(2\pi i \gamma q^k / \rho\right) \right| \le \frac{\tau}{\tau_\rho} \rho^{1/2}.$$

Finally, we have $\tau \le \delta \tau_\rho$ by Lemma 3 of [25]. □

We also need the following well-known identity (see Problem 11.a of Chapter 3 of [26])

$$\sum_{0 \le c < r} \exp(2\pi i c u / r) = \begin{cases} 0 \text{ if } u \not\equiv 0 \bmod r, \\ r \text{ if } u \equiv 0 \bmod r, \end{cases} \tag{1}$$

and the bound

$$\sum_{1 \le c < r} \left| \sum_{0 \le u \le h} \exp\left(2\pi i c u / r\right) \right| = O(r \log r), \tag{2}$$

which hold for any integers $r \ge 1$ and $h \ge 0$; see Problem 11.c of Chapter 3 of [26].

**Lemma 4.2.** *Let $q, h$, and $r$ be positive integers, $\tau$ the order of $q$ modulo $r$, and $T$ be the number of elements in $\{0, 1, \ldots, h-1\}$ that are powers of $q$ modulo $r$. Then for any $\varepsilon > 0$, we have*

$$T = \frac{\tau h}{r} + O\left(r^{1/2+\varepsilon}\right).$$

*Proof.* From (1) we derive

$$\begin{aligned} T &= \frac{1}{r} \sum_{1 \le k \le \tau} \sum_{0 \le x < h} \sum_{0 \le c < r} \exp\left(2\pi i c (q^k - x)/r\right) \\ &= \frac{1}{r} \sum_{0 \le c < r} \sum_{1 \le k \le \tau} \exp\left(2\pi i c q^k / r\right) \sum_{0 \le x < h} \exp\left(-2\pi i c x / r\right). \end{aligned}$$

The contribution of the term corresponding to $c = 0$ is $\tau h/m$. Therefore, from Lemma 4.1 and the bound (2), we obtain

$$\begin{aligned}
\left|T - \frac{\tau h}{r}\right| &\leq \frac{1}{r} \sum_{1\leq c<r} \left| \sum_{1\leq k\leq \tau} \exp\left(2\pi i c q^k/r\right) \right| \cdot \left| \sum_{0\leq x<h} \exp\left(2\pi i c x/r\right) \right| \\
&\leq \frac{1}{r} \sum_{\substack{\delta|r\\ \delta<r}} \sum_{\substack{1\leq c<r\\ \gcd(c,r)=\delta}} \left| \sum_{1\leq k\leq \tau} \exp\left(2\pi i c q^k/r\right) \right| \left| \sum_{0\leq x<h} \exp\left(-2\pi i c x/r\right) \right| \\
&\leq \frac{1}{r^{1/2}} \sum_{\substack{\delta|r\\ \delta<r}} \delta^{1/2} \sum_{\substack{1\leq c<r\\ \gcd(c,r)=\delta}} \left| \sum_{0\leq x<h} \exp\left(2\pi i c x/r\right) \right| \\
&\leq \frac{1}{r^{1/2}} \sum_{\substack{\delta|r\\ \delta<r}} \delta^{1/2} \sum_{1\leq c<r/\delta} \left| \sum_{0\leq x<h} \exp\left(2\pi i c \delta x/r\right) \right| \\
&\in O\left( r^{1/2} \log r \sum_{\substack{\delta|r\\ \delta<r}} \delta^{-1/2} \right).
\end{aligned}$$

Taking into account that

$$\sum_{\substack{\delta|r\\ \delta<r}} \delta^{-1/2} \leq \sum_{\delta|r} 1 \in O(r^{\varepsilon/2}),$$

by Theorem 5.2 in Chapter 1 in [23], we obtain the desired result. □

Now we are prepared to establish the main result of this section.

**Theorem 4.3.** *For any positive integers $d$ and $n$ and real $\varepsilon > 0$, there exists $c > 0$ such that for any nonconstant rational function $R$ over $\mathbb{F}_q$ of the form*

$$R = \sum_{-d\leq\nu\leq d} R_\nu x^\nu \in \mathbb{F}_q(x),$$

*with all $R_\nu$ in $\mathbb{F}_q$, and for any root $\beta$ of an irreducible polynomial of degree $n$ with $R(\beta) \neq 0$, at least one of the elements $\beta$ and $R(\beta)$ has order at least $cn^{4/3-\varepsilon}$.*

*Proof.* Let $r$ be the order of $\beta$. Because $\beta$ is a root of an irreducible polynomial of degree $n$, the order of $q$ modulo $r$ is $n$. Indeed, if $q^k \equiv 1 \bmod r$ for some positive integer $k$, then $\beta^{q^k} = \beta$ and thus $\beta \in \mathbb{F}_{q^k}$, which is true for $k = n$ and false for $k < n$.

We consider the set $K$ of integers $k \in \{1, \dots, n\}$ such that the remainders $e = q^k \text{ rem } r$ with $0 \leq e < r$ satisfy $e < n/8d$. We denote by $M = \#K$ its size, and claim that the $M(M+1)/2$ powers

$$R(\beta)^{q^k+q^{k'}} Quad \text{with } k, k' \in K \text{ and } k \le k'$$

are pairwise distinct. Indeed

$$R(\beta)^{q^k+q^{k'}} = R(\beta)^{q^k} R(\beta)^{q^{k'}} = R\left(\beta^{q^k}\right) R\left(\beta^{q^{k'}}\right) = R\left(\beta^{e}\right) R\left(\beta^{e'}\right),$$

where $e = q^k$ rem $r$ and $e' = q^{k'}$ rem $r$. We take two pairs $(s,t)$ and $(u,v)$ of integers with $0 \le s \le t < n/8d$, $0 \le u \le v < n/8d$, and $R(\beta^s) R(\beta^t) = R(\beta^u) R(\beta^v)$, and claim that $(s,t) = (u,v)$. We write $R = f \cdot x^l$ with some integer $l$, where $|l| \le d$ and the polynomial $f \in \mathbb{F}_q[x]$ satisfies $\deg f \le 2d$ and $f(0) \ne 0$. Then the above equation implies that

$$f(\beta^s) f(\beta^t) \beta^{-l(s+t)} = f(\beta^u) f(\beta^v) \beta^{-l(u+v)}.$$

If $l \ge 0$, we consider the polynomial

$$G = f(x^s) f(x^t) x^{l(u+v)} - f(x^u) f(x^v) x^{l(s+t)} \in \mathbb{F}_q[x]$$

of degree

$$\deg G \le 2d \cdot (s+t+u+v) < n.$$

Furthermore, $G(\beta) = 0$, and hence $G = 0$. This implies that $s+t = u+v$, and thus $F = f(x^s) f(x^t) - f(x^u) f(x^v)$ is the zero polynomial.

If $l < 0$, we consider the polynomial

$$G = f(x^s) f(x^t) x^{-l(s+t)} - f(x^u) f(x^v) x^{-l(u+v)}$$

to derive that $s+t = u+v$ and $F = 0$.

Since $s+t = u+v$, we may assume that $s \ne u$, because otherwise $(s,t) = (u,v)$. Without loss of generality we may assume that $s < u$. We let $a_m x^m$ be the term of smallest degree in $f - f(0)$ with $a_m \ne 0$. Then the first product contains the term $a_m f(0) x^{ms} \ne 0$. Moreover, this term is unique and does not cancel with any other term unless $s = u$. So we conclude that $s = u$ and $t = v$.

This proves our claim about the existence of $M(M+1)/2$ pairwise distinct powers of $R(\beta)$, and thus the order of $R(\beta)$ is at least $M(M+1)/2$.

From Lemma 4.2, we know that for any $\varepsilon > 0$

$$M = \frac{n^2}{8dr} + O(r^{1/2+\varepsilon/2}).$$

We let $c_1$ be the constant implicit in the "$O$", and set $c_2 = (8c_1 d)^{-2/(3+\varepsilon)}$. Then for $r \le c_2 \cdot n^{4/3-\varepsilon/2}$ we have

$$c_1 r^{(1+\varepsilon)/2} \le \frac{n^2}{8dr},$$

$$M \geq \frac{n^2}{8dr} - c_1 r^{1/2+\varepsilon/2} \geq \frac{n^2}{16dr} \geq (16c_2 d)^{-1} n^{2/3-\varepsilon/2}.$$

Thus in this case the order of $R(\beta)$ is at least $(16c_2d)^{-2}n^{4/3-\varepsilon}$. Otherwise the order of $\beta$ is $r$ larger than $c_2 \cdot n^{4/3-\varepsilon/2}$. □

**Corollary 4.4.** *For any $\varepsilon > 0$ there exists $c > 0$ such that for any root $\beta$ of an irreducible polynomial of degree $n$, at most one of the shifts $\beta + a$ with $a \in \mathbb{F}_q$ is of order less than $cn^{4/3-\varepsilon}$.*

We see from the proof of Theorem 4.3 that the more information about the order $r$ of $\beta$, the order $n$ of $q$ modulo $r$ and the distribution of powers of $q$ modulo $r$ is available, the stronger are the results produced by this method. In particular, this explains how Artin's conjecture comes into play: if $q$ is a primitive root modulo $r$, then we have full control over the distribution of powers of $q$ modulo $r$ and, as we have seen in Section 3, this implies much stronger results about the order of $R(\beta)$. Another case when we have good information about this distribution is the case when $r$ is a product of large powers of small primes, see [20,21]. Accordingly, these two cases are the main sources of the results in [14,15].

# 5 Exponentiation in Finite Fields

We consider the following task: Given are finite fields $\mathbb{F}_q \subseteq \mathbb{F}_{q^n}$, an integer $e$ with $1 \leq e < q^n$, and $u \in \mathbb{F}_{q^n}$. The objective is to compute $u^e \in \mathbb{F}_{q^n}$. An important special case is $q = 2$. This is a basic operation in several cryptosystems: Diffie-Hellman and ElGamal.

The "classical" method is to use repeated squaring, with at most $2\log_2 e \leq 2n\log_2 q$ multiplications in $\mathbb{F}_{q^n}$, and classical multiplication, with $O(n^2)$ operations in $\mathbb{F}_q$. This gives a total of $O(n^3 \log q)$ operations in $\mathbb{F}_q$ for one exponentiation in $\mathbb{F}_{q^n}$.

In the *polynomial representation*, we write $\mathbb{F}_{q^n} = \mathbb{F}_q[x]/(\varphi)$, where $\varphi \in \mathbb{F}_q[x]$ is irreducible of degree $n$, and use the basis $(1, x \bmod \varphi, \ldots, x^{n-1} \bmod \varphi)$ of $\mathbb{F}_{q^n}$ over $\mathbb{F}_q$.

**Table1.** Multiplication in $\mathbb{F}_{q^n}$

| | time |
|---|---|
| classical | $O(n^2)$ |
| FFT ([24]) | $O(n \log n \operatorname{loglog} n)$ |
| additive subspaces ([3]) | $O(n \log^2 n)$ |

Three multiplication algorithms are given in Table 1. The running time is the number of operations in $\mathbb{F}_q$. Using the fastest method, the cost for one exponentiation in $\mathbb{F}_{q^n}$ becomes $O(n^2 \log n \operatorname{loglog} n \log q)$ operations in $\mathbb{F}_q$.

A faster method is obtained from the *polynomial representation* of the Frobenius map, using *modular composition*, as introduced in [13]. Then exponentiation can be performed with $O(n^2 \operatorname{loglog} n \log q)$ operations in $\mathbb{F}_q$, as proved in [8].

Now we consider a normal basis $(\alpha, \alpha^q, \ldots, \alpha^{q^{n-1}})$ of $\mathbb{F}_{q^n}$ over $\mathbb{F}_q$, and the coordinates $u_0, \ldots, u_{n-1} \in \mathbb{F}_q$ of a general element $\sum_{0 \le i < n} u_i \alpha^{q^i}$ of $\mathbb{F}_{q^n}$. Then

$$\Big(\sum_{0 \le i < n} u_i \alpha^{q^i}\Big)^q = \sum_{0 \le i < n} u_i \alpha^{q^{i+1}} = \sum_{0 \le i < n} u_{i-1} \alpha^{q^i},$$

where $u_{-1} = u_{n-1}$. Thus a $q$th power corresponds to a cyclic shift of coordinates, and has no arithmetic cost. It is plausible that when a specific power, as the $q$th one here, becomes cheaper, then this may also reduce the cost of a general exponentiation. Indeed, the number of multiplications in $\mathbb{F}_{q^n}$ becomes

$$\text{at most } \Big(1 + o(1)\Big)\frac{n}{\log_q n} \quad \text{for any } e,$$

$$\text{at least } \Big(\frac{1}{3} + o(1)\Big)\frac{n}{\log_q n} \quad \text{for most } e$$

([9]). The crucial question now is: How expensive is one multiplication in $\mathbb{F}_{q^n}$?

We begin by considering multiplication via linear algebra in general. Given any basis $\alpha_0, \ldots, \alpha_{n-1}$ of $\mathbb{F}_{q^n}$ as vector space over $\mathbb{F}_q$, we can represent "multiplication by $\alpha_i$" by an $n \times n$ matrix $A_i$. This yields the *multiplication tensor* $(A_0, \ldots, A_{n-1})$. We then have the following estimates:

- cost for multiplication by $A_i$ : $O(n^2)$,
- cost for a general multiplication: $O(n^3)$,
- cost for exponentiation: $O(n^4 \log q)$.

In a normal basis $\alpha_i = \alpha^{q^i}$, each matrix $A_i$ is a shift of $A_0$. This corresponds to the *Massey-Omura multiplier*. The time for a multiplication is still $O(n^3)$, but the storage requirement is only $O(n^2)$, rather than $O(n^3)$. An exponentiation costs $O(n^4 \log q)$ operations in $\mathbb{F}_q$.

[22] proved that $A_0$ has at least $2n-1$ nonzero entries. A fruitful suggestion of theirs was to consider *"optimal" normal basis*, which have exactly this minimal number $2n - 1$ of nonzero entries. Then the costs drop by a factor of $n$:

- cost for one multiplication: $O(n^2)$,
- cost for exponentiation: $O(n^3/\log n)$.

They had actually rediscovered a special case of Gauß periods. Namely, [6] showed that optimal normal bases correspond to Gauß periods of type $(n, \mathcal{K})$ with $\#\mathcal{K} = 1$ or $\#\mathcal{K} = 2$, for $q = 2$.

But even in this improved situation, the computation is slower by a factor of almost $n$ than those discussed in the above. The question is: can we reconcile the advantages of normal bases with those of fast multiplication? [7] have shown that this is indeed possible. In the notation from Section 2, we have the tower of fields:

$$\mathbb{F}_q \subseteq \mathbb{F}_{q^n} = \mathbb{F}_q(\alpha) \subseteq \mathbb{F}_q(\beta).$$

We can represent the right hand field in a polynomial basis, using an appropriate polynomial which vanishes at $\beta$, and for the left hand extension we have the advantages of the normal basis. Now the algorithms become quite efficient:

- cost for one multiplication: $O(kn \log n \operatorname{loglog} n)$,
- cost for exponentiation: $O(kn^2 \operatorname{loglog} n \log q)$

operations in $\mathbb{F}_q$, with $k$ as in Section 2.

The algorithms mentioned here work very well also in practice, as reported in the experimental results of [10,11].

**Table2.** Cost of exponentiation algorithms, for $q = 2$

| | |
|---|---|
| classical | $n^3$ |
| fast multiplication | $n^2 \log n \operatorname{loglog} n$ |
| optimal normal basis | $n^3 / \log n$ |
| modular composition | $n^2 \operatorname{loglog} n$ |
| optimal normal basis plus fast multiplication | $n^2 \operatorname{loglog} n$ |

## 6 Construction of Gauß Periods

When do optimal normal bases exist? Or, more generously, when do we have Gauß periods with $k$ "small"? Or any Gauß period at all? In the terminology of Section 2, we have $r$ prime and $\varphi(r) = nk$, and want $k$ as small as possible. Thus we study the following function.

$$\kappa_p(q,n) = \begin{cases} \min \#\mathcal{K} & \text{prime Gauß period of type } (n,\mathcal{K}) \text{ over } \mathbb{F}_q \text{ exists,} \\ \infty & \text{if none exists.} \end{cases}$$

[28], Theorem 3.3.4, showed that if $p = \operatorname{char}(\mathbb{F}_q)$, $q = p^m$ and $n \in \mathbb{N}$ is positive, then $\kappa_p(q,n) < \infty$ if and only if the following conditions hold:

(i) $\gcd(m,n) = 1$,
(ii) $2p \nmid n$, if $p \equiv 1 \bmod 4$, and $4p \nmid n$, if $p = 2$ or $p \equiv 3 \bmod 4$.

The "prime" (and the index $p$ of $\kappa$) refers to the fact that $r$ is prime in the above construction. [16] suggested in article 356 of his *Disquisitiones Arithmeticae* to remove this condition: *Haecce theoremata salva vel potius aucta elegantia sua etiam ad valores quosvis compositos ipsius n extendi posse: sed de his rebus, quae altioris sunt indaginis, hoc loco tacere earumque considerationem ad aliam occasionem nobis reservare oportet.*[2] It seems that the other occasion never arose.

Following Gauß' suggestion, we now drop this requirement, and take positive integers $n, k$, and $r$, with $nk = \varphi(r)$, a subgroup $\mathcal{K} \subseteq \mathbb{Z}_r^\times$ of order $k$, a primitive $r$th root of unity $\beta \in \mathbb{F}_{q^{\varphi(r)}}$, and $\alpha = \sum_{i \in \mathcal{K}} \beta^i$.

**Theorem 6.1.** *The* Gauß period $\alpha$ *is normal in* $\mathbb{F}_{q^n}$ *over* $\mathbb{F}_q$ *if and only if* $r$ *is squarefree,* $\gcd(r, q) = 1$, *and* $\langle q, \mathcal{K} \rangle = \mathbb{Z}_r^\times$.

For arbitrary $r$, the expression for $\alpha$ reads as follows. For a prime divisor $\ell$ of $r$, we let $\nu_\ell(r)$ be the multiplicity of $\ell$ in $r$. We write $r = r_1 r_2$ where $r_2$ is the product of all primes $\ell$ such that $\nu_\ell(r) = 1$. For any prime $\ell$ dividing $r$, let $\ell' = r/\ell^{\nu_\ell(r)}$, and set

$$g = x^{r_1} \prod_{\ell | r_1} \sum_{0 \le i < \nu_\ell(r_1)} x^{\ell' \ell^i} \in \mathbb{Z}[x].$$

The *general Gauß period of type* $(n, \mathcal{K})$ is defined as

$$\alpha = \sum_{a \in \mathcal{K}} g(\beta^a).$$

If $r$ is squarefree, then a general Gauß period is given by the same formula as a Gauß period. Theorem 6.1, without the squarefreeness condition, also holds for these Gauß periods ([5]). The proof argues mainly in algebraic number fields, which is of course Gauß' original setting. It would be nice to have an argument working just in finite fields.

The corresponding variant of $\kappa_p(q, n)$ is the following:

$$\kappa_g(q,n) = \begin{cases} \min \#\mathcal{K} & \text{general Gauß period of type } (n,\mathcal{K}) \\ & \text{\textit{Quad} over } \mathbb{F}_q \text{ exists,} \\ \infty & \text{if none exists.} \end{cases}$$

Then $\kappa_g(q, n) \le \kappa_p(q, n)$.

There are considerably more of these general Gauß periods than of the ones with $r$ prime, and the smallest value of $k$ often gets reduced. For $q = 2$, this is illustrated in Table 3, where the "□" means that $r$ has a square factor.

[2] These theorems can be extended to arbitrary composite values of $n$ [in our notation: $r$], retaining or even enhancing their elegance; but these matters, which are at a higher level of research, are best left unsaid in this place, and we reserve their consideration for another occasion.

**Table3.** Improvements for $q = 2$ and $2 \leq n \leq 156$:

| $n$ | $\kappa_p(2,n)$ | $\kappa_g(2,n)$ | ratio | $r$ | | $\mathcal{K}$ |
|---|---|---|---|---|---|---|
| 6 | 2 | 1 | 2.0 | 9 | □ | {1} |
| 20 | 3 | 1 | 3.0 | 25 | □ | {1} |
| 21 | 10 | 2 | 5.0 | 49 | □ | {1, 48} |
| 22 | 3 | 2 | 1.5 | 69 | | {1, 68} |
| 27 | 6 | 2 | 3.0 | 81 | □ | {1, 80} |
| 34 | 9 | 6 | 1.5 | 309 | | {1, 46, 47, 262, 263, 308} |
| 42 | 5 | 2 | 2.5 | 147 | □ | {1, 146} |
| 44 | 9 | 2 | 4.5 | 115 | | {1,91} |
| 46 | 3 | 2 | 1.5 | 141 | | {1, 140} |
| 54 | 3 | 1 | 3.0 | 81 | □ | {1} |
| 55 | 12 | 2 | 6.0 | 121 | □ | {1, 120} |
| 57 | 10 | 6 | 1.67 | 361 | □ | {1, 68, 69, 292, 293, 360} |
| 68 | 9 | 6 | 1.5 | 515 | □ | {1, 46, 56, 356, 366, 411} |
| 70 | 3 | 2 | 1.5 | 213 | | {1, 212} |
| 75 | 10 | 8 | 1.25 | 707 | | {1, 111, 293, 302, 405, 414, 596, 706} |
| 78 | 7 | 2 | 3.5 | 169 | □ | {1, 168} |
| 84 | 5 | 2 | 2.5 | 203 | | {1, 202} |
| 92 | 3 | 2 | 1.5 | 235 | | {1, 46} |
| 102 | 6 | 2 | 3.0 | 309 | | {1, 308} |
| 108 | 5 | 2 | 2.5 | 405 | □ | {1, 404} |
| 110 | 6 | 1 | 6.0 | 121 | □ | {1} |
| 111 | 20 | 8 | 2.5 | 1043 | | {1, 148, 342, 491, 552, 701, 895, 1042} |
| 114 | 5 | 3 | 1.67 | 361 | □ | {1, 68, 292} |
| 116 | 3 | 2 | 1.5 | 295 | | {1, 176} |
| 123 | 10 | 4 | 2.5 | 581 | | {1, 167, 414, 580} |
| 125 | 6 | 4 | 1.5 | 625 | □ | {1, 182, 443, 624} |
| 132 | 5 | 2 | 2.5 | 299 | | {1, 298} |
| 140 | 3 | 2 | 1.5 | 319 | | {1, 318} |
| 145 | 10 | 4 | 2.5 | 649 | | {1, 296, 353, 648} |
| 147 | 6 | 2 | 3.0 | 343 | □ | {1, 342} |
| 150 | 19 | 4 | 4.75 | 707 | | {1, 302, 405, 706} |
| 154 | 25 | 4 | 6.25 | 667 | | {1, 231, 505, 597} |
| 156 | 13 | 1 | 13.0 | 169 | □ | {1} |

Fast arithmetic can also be used with these more general Gauß periods; see [11].

From a different point of view, [12] have determined the density of primes $r = nk + 1$ that yield a Gauß period over $\mathbb{F}_q$, in terms of $n$ and $q$.

## Acknowledgements

We thank Michael Nüsken for help with the typesetting.

## References

1. W. H. Bussey (1906). Galois field tables for $p^n \leq 169$. *Bulletin of the American Mathematical Society* **12**, 22–38.
2. W. H. Bussey, Tables of Galois fields of order less than 1,000. *Bulletin of the American Mathematical Society* **16** (1910), 188–206.
3. David G. Cantor, On arithmetical algorithms over finite fields. *Journal of Combinatorial Theory, Series A* **50** (1989), 285–300.
4. August L. Crelle, Table des racines primitives etc. pour les nombres premiers depuis 3 jusqu'à 101, precédée d'une note sur le calcul de cette table. *Journal für die Reine und Angewandte Mathematik* **9** (1832), 27–53.
5. Sandra Feisel, Joachim von zur Gathen, and M. Amin Shokrollahi, Normal bases via general Gauß periods. *Mathematics of Computation* **68**(225) (1999), 271–290.
6. S. Gao and H. W. Lenstra, Jr., Optimal normal bases. *Designs, Codes, and Cryptography* **2** (1992), 315–323.
7. Shuhong Gao, Joachim von zur Gathen, and Daniel Panario, Gauss periods and fast exponentiation in finite fields. In *Proceedings of LATIN '95*, Valparaíso, Chile, Lecture Notes in Computer Science **911**. Springer-Verlag, 1995, 311–322.
8. Shuhong Gao, Joachim von zur Gathen, Daniel Panario, and Victor Shoup, Algorithms for exponentiation in finite fields. *Journal of Symbolic Computation* **26**(6) (2000), 879–889.
9. Joachim von zur Gathen, Efficient and optimal exponentiation in finite fields. *computational complexity* **1** (1991), 360–394.
10. Joachim von zur Gathen and Michael Nöcker, Exponentiation in finite fields: Theory and practice. In *Applied Algebra, Algebraic Algorithms and Error-Correcting Codes: AAECC-12*, Toulouse, France, ed. Teo Mora and Harold Mattson, Lecture Notes in Computer Science **1255**. Springer-Verlag, 1997, 88–113.
11. Joachim von zur Gathen and Michael Nöcker, Normal bases, Gauss periods, and fast arithmetic. In *Abstracts of the Fifth International Conference on Finite Fields and Applications, August 2-6, 1999*, University of Augsburg, 1999, p. 70.
12. Joachim von zur Gathen & Francesco Pappalardi (2000). Density estimates for Gauß periods. In *Proc. Workshop on Cryptography and Computational Number Theory*. Birkhäuser Verlag. To appear.

13. Joachim von zur Gathen and Victor Shoup, Computing Frobenius maps and factoring polynomials. *computational complexity* **2** (1992), 187–224.
14. Joachim von zur Gathen and Igor Shparlinski, Orders of Gauss periods in finite fields. *Applicable Algebra in Engineering, Communication and Computing* **9** (1998), 15–24.
15. Joachim von zur Gathen and Igor Shparlinski, Constructing elements of large order in finite fields. In *Applied Algebra, Algebraic Algorithms and Error-Correcting Codes: AAECC-13*, Hawaii, Lecture Notes in Computer Science **1719**, Springer-Verlag, 1999, 404–409.
16. Carl Friedrich Gauss, *Disquisitiones Arithmeticae.* Gerh. Fleischer Iun., Leipzig, 1801. English translation by Arthur A. Clarke, Springer-Verlag, New York, 1986.
17. D. R. Heath-Brown, Artin's conjecture for primitive roots. *Quarterly Journal of Mathematics* **37** (1986), 27–38.
18. C. Hooley, On Artin's conjecture. *Journal für die Reine und Angewandte Mathematik* **225** (1967), 209–220.
19. C. G. J. Jacobi, *Canon Arithmeticus sive tabulae quibus exhibentur pro singulis numeris primis vel primorum potestatibus infra 1000 numeri ad datos indices et indices ad datos numeros pertinentes.* Typus Academicus, Berlin, 1839.
20. Nikolai M. Korobov, On the distribution of digits in periodic fractions. *Math. USSR - Sbornik* **18** (1972), 659–676. (In Russian).
21. Nikolai M. Korobov, *Exponential sums and their applications.* Kluwer Acad. Publ., Dordrecht, 1992.
22. R. C. Mullin, I. M. Onyszchuk, S. A. Vanstone, and R. M. Wilson, Optimal normal bases in GF($p^n$). *Discrete Applied Mathematics* **22** (1989), 149–161.
23. K. Prachar, *Primzahlverteilung.* Springer-Verlag, Berlin, 1957.
24. A. Schönhage and V. Strassen, Schnelle Multiplikation großer Zahlen. *Computing* **7** (1971), 281–292.
25. Igor E. Shparlinski, On the dimension of BCH codes. *Problemy Peredachi Inform.* **25**(1) (1988), 100–103. (In Russian).
26. I. M. Vinogradov, *Elements of number theory.* Dover Publications, Inc., New York, 1954.
27. Alfred Wassermann, Konstruktion von Normalbasen. *Bayreuther Math. Schriften* **31** (1990), 155–164.
28. Alfred Wassermann, Zur Arithmetik in endlichen Körpern. *Bayreuther Math. Schriften* **44** (1993), 147–251.

# Constructions of Orthomorphisms of $\mathbb{Z}_2^n$

Solomon W. Golomb[1], Guang Gong[2], and Lothrop Mittenthal[3]

[1] Communication Sciences Institute
University of Southern California
Electrical Engineering-Systems, EEB#500
Los Angeles, CA 90089-2565
USA
e-mail: milly@usc.edu

[2] Department of Combinatorics and Optimization
University of Waterloo
Waterloo, Ontario N2L 3G1
Canada
e-mail: ggong@cacr.math.uwaterloo.ca

[3] Teledyne Electronic Technologies
8920 Quartz Ave.
Northridge, CA 91324-3339
USA

**Abstract** A permutation $\sigma$ on $\mathbb{Z}_2^n$, the linear space over $\mathbb{Z}_2$ of dimension $n$, is an orthormorphism iff the mapping $x \mapsto \sigma(x)+x$ is also a permutation on $\mathbb{Z}_2^n$, as $x$ takes all values in $\mathbb{Z}_2^n$. It is a linear orthomorphism iff $\sigma$ is a linear transformation on $\mathbb{Z}_2^n$. This paper contains two parts. In the first part, in terms of the isomorphism between the linear space $\mathbb{Z}_2^n$ and the finite field $GF(2^n)$ , an algebraic method of constructing linear orthomorphisms with maximal length cycles is provided. Then two algorithms to implement these linear orthomorphisms are presented. In the second part, by using this type of linear orthomorphisms, special types of Latin squares, called *shift Latin squares* are constructed and nonlinear orthomorphisms, which can be represented as transversals of such Latin squares, are obtained. Some discussion on nonlinearity of the resulting nonlinear orthomorphisms and a construction of arbitrary nonlinear orthormorphisms are also included in this part. A motivation is to use such mappings for encryption of digital data.

## 1 Introduction

Encryption and decryption algorithms play an important role in the information age. Mathematically, to design these kinds of algorithms is to construct permutations of a finite set with specific properties. In [5] [6], Mittenthal discussed this concept by using the orthomorphisms of $\mathbb{Z}_2^n$ to design cryptographic algorithms. This paper will discuss how to construct and implement the linear orthomorphisms with the maximal length cycles when they are regarded as permutations of $\mathbb{Z}_2^n$ and how to construct nonlinear orthomorphisms from these linear orthomorphisms. The paper contains two parts. In the first part, in terms of the isomorphism between the linear space $\mathbb{Z}_2^n$ and the finite field $GF(2^n)$ [4] [9], we provide an algebraic method of constructing linear orthomorphisms with maximal length cycles. Then we present two algorithms to implement this type of linear orthomorphisms. In the second part, by using this type of linear orthomorphisms, we construct a special Latin square, called *shift Latin square*, as opposed to a Latin square which is a Cayley table. Finally, nonlinear orthomorphisms, which can be represented as transversal, are constructed in terms of the shift Latin squares.

This paper is organized as follows. In Section 2, we will give an algebraic method for constructing linear orthomorphisms. In Sections 3, we will give two algorithms to implement this type of linear orthomorphisms of $\mathbb{Z}_2^n$. One is to use matrix transformation where a basis of $GF(2^n)/GF(2)$ is fixed. The other is to use the characteristic m-sequences of period $2^n - 1$ where altering bases of $GF(2^n)/GF(2)$ is possible. In Sections 4 and 5, in terms of these linear orthomorphisms, we construct *shift Latin squares* and show a strategy to construct highly nonlinear orthomorphisms from them. Some discussion on constructing arbitrary nonlinear orthomorphisms is provided in Section 6.

## 2 A Class of Linear Orthomorphisms of $\mathbb{Z}_2^n$

We will specify the following notation used through out this paper.

- $\mathbb{Z}_v$, the integer ring modulo $v$,
- $\mathbb{F}_q = GF(q)$,
- $\mathbb{Z}_2^n = \{(x_0, x_1, \cdots, x_{n-1}) | x_i \in F_2, 0 \leq i < n\}$, and
- $p = 2^n - 1$ and $q = 2^n$.

**Definition 2.1.** A permutation $\sigma$ on $\mathbb{Z}_2^n$ is an orthormorphism iff the mapping $x \mapsto x + \sigma(x)$ is also a permutation on $\mathbb{Z}_2^n$ , as $x$ takes all values in $\mathbb{Z}_2^n$. It is a linear orthomorphism iff $\sigma$ is a linear transformation on $\mathbb{Z}_2^n$.

Define $S(x) = x + \sigma(x)$. Then $S(x)$ is a permutation of $\mathbb{Z}_2^n$ if $\sigma$ is an orthomorphism of $\mathbb{Z}_2^n$. Therefore $S$ is a linear permutation of $\mathbb{Z}_2^n$ if and only if $\sigma$ is a linear permutation of $\mathbb{Z}_2^n$.

### 2.1 An Isomorphism between $\mathbb{Z}_2^n$ and $\mathbb{F}_{2^n}$

Let $\{\alpha_0, \alpha_1, \cdots, \alpha_{n-1}\}$ be a basis of $\mathbb{F}_{2^n}$ over $\mathbb{F}_2 (= \mathbb{Z}_2)$. Let $x = (x_0, x_1, \cdots, x_{n-1}) \in \mathbb{Z}_2^n$, then the following map

$$x = (x_0, x_1, \cdots, x_{n-1}) \mapsto x_0\alpha_0 + x_1\alpha_1 + \cdots + x_{n-1}\alpha_{n-1}$$

is an isomorphism between $\mathbb{Z}_2^n$ and $\mathbb{F}_{2^n}$ when both of them are regarded as vector spaces over $\mathbb{F}_2 = \mathbb{Z}_2$. Note that there is a one-to-one correspondence between $\mathbb{Z}_2^n$ and $\mathbb{Z}_{2^n}$ via

$$x = (x_0, x_1, \cdots, x_{n-1}) \mapsto x_0 + x_1 2 + x_2 2^2 + \cdots + x_{n-1} 2^{n-1} \in \mathbb{Z}_{2^n}.$$

Thus in this paper, we use the same symbol $x$ for $x \in \mathbb{F}_{2^n}$, $\mathbb{Z}_2^n$ or $x \in \mathbb{Z}_{2^n}$. So, the exact meaning of $x$ depends on the context.

Let $\{\eta_0, \eta_1, \cdots, \eta_{n-1}\}$ be the dual basis of $\{\alpha_0, \alpha_1, \cdots, \alpha_{n-1}\}$. Then

$$(x_0, x_1, \cdots, x_{n-1}) = (Tr(\eta_0 x), Tr(\eta_1 x), \cdots, Tr(\eta_{n-1} x)), \forall x \in \mathbb{F}_{2^n}$$

where $Tr(x) = x + x^2 + \cdots + x^{2^{n-1}}$ is the trace function from $\mathbb{F}_{2^n}$ to $\mathbb{F}_2$.

A permutation $\sigma(x)$ of $\mathbb{F}_{2^n}$ (or $\mathbb{Z}_2^n$) is said to be *maximal* if it contains a cycle of length $p$.

### 2.2 An Algebraic Construction

**Proposition 2.2.** *Let $\alpha$ be a primitive element in $\mathbb{F}_{2^n}$. Let $s$ be a fixed positive integer in the range of $1 \leq s < p$ with $gcd(s,p) = 1$. Let*

$$\sigma(x) = \alpha^s x, x \in \mathbb{F}_{2^n}.$$

*Then $\sigma(x)$ is a linear orthomorphism of $\mathbb{F}_{2^n}$. Moreover, $\sigma(x)$ has a maximal length cycle when it is regarded as a permutation.*

*Proof.* Let $\alpha$ be a primitive element of $\mathbb{F}_{2^n}$. Then we can write $\mathbb{F}_{2^n}^*$, the multiplicative group of $\mathbb{F}_{2^n}^*$, as follows

$$\mathbb{F}_{2^n}^* = \{\alpha^i | 0 \leq i < p\}.$$

For any $0 \leq i \neq j < p$, there is an integer $k$: $0 \leq k < p$ such that

$$\alpha^i + \alpha^j = \alpha^k.$$

Therefore, there exists $\tau$: $0 \leq \tau < p$ such that $1 + \alpha^s = \alpha^\tau$. Consequently,

$$S(x) = x + \sigma(x) = (1 + \alpha^s)x = \alpha^\tau x$$

which is a linear permutation of $\mathbb{F}_{2^n}$. Note that $\sigma(\alpha^i) = \alpha^{s+i}, 0 \leq i < p$ with $gcd(s,p) = 1$. This shows that $\sigma(x)$ has a maximal length cycle when it is regarded as a permutation.

□

Let $S(x) = x + \sigma(x)$. Then $S(x)$ can be represented as a number in $Z_2^n$ for any $x \in \mathbb{F}_{2^n}$ as follows

$$S(x) = (Tr(\eta_0 \alpha^\tau x), Tr(\eta_1 \alpha^\tau x), \cdots, Tr(\eta_{n-1} \alpha^\tau x)) \in Z_2^n \quad (1)$$

where $1 + \alpha^s = \alpha^\tau$. So the *implementation problem* is how to compute $S(x)$ from a given $x$ in $Z_2^n$ using the formula (1).

# 3 Implementation of Linear Orthomorphisms

In this section, we will give two algorithms to implement $S(x)$ represented by (1).

## 3.1 Implementation in Terms of Matrix Transformations

Let $(a_0, a_1, \cdots, a_s)^T$ be the transpose of the vector $(a_0, a_1, \cdots, a_s)$.

**Algorithm 3.1.** Procedure for Implementation of $S(x) = x + \sigma(x)$ on $Z_2^n$.
*Input:* $m = (m_0, m_1, \cdots, m_{n-1}) \in Z_2^n$.
*Output:* $S(m) \in Z_2^n$.

(i) Generate a finite field $\mathbb{F}_{2^n}$ and choose $\alpha$ a primitive element of $\mathbb{F}_{2^n}$.
(ii) Compute $\tau : 0 < \tau < p$ such that $1 + \alpha^s = \alpha^\tau$. (I.e., compute the discrete logarithm of $\tau = log_\alpha \beta$ where $\beta = 1 + \alpha^s$.)
(iii) Choose a basis of $\mathbb{F}_{2^n}/\mathbb{F}_2 : \{1, \alpha, \cdots, \alpha^{n-1}\}$.
(iv) Compute the matrix form $B = (b_{ij})_{n\times n}$ of $S(x)$ by

$$(\alpha^\tau, \alpha^{\tau+1}, \cdots, \alpha^{\tau+n-1})^T = B(1, \alpha, \cdots, \alpha^{n-1})^T.$$

Note that $B$ is a $(0,1)$ matrix.
(v) Evaluate $mB$ which is $S(m)$.

*Note.* This algorithm can only easily change the value of $s$, but the basis involved here can not be easily changed.

*Example 3.2.* Let $n = 4$, $s = 1$, and $\alpha$ be a primitive element of $\mathbb{F}_{2^4}$ with $f(\alpha) = 0$ where $f(x) = x^4 + x + 1$.
*Input:* $m = 12 = (0, 0, 1, 1) \in \mathbb{Z}_2^4$.
*Output:* $S(12) \in Z_2^4$.

(i) Generate $\mathbb{F}_{2^4}$ in Table 3.2.
(ii) From Table 1, $1 + \alpha = \alpha^4 \Longrightarrow \tau = 4$.
(iii) Choose the basis of $\mathbb{F}_{2^4}/\mathbb{F}_2$: $\{1, \alpha, \alpha^2, \alpha^3\}$.
(iv) Compute the matrix form of $S(x)$ by $(\alpha^4, \alpha^5, \alpha^6, \alpha^7)^T = B(1, \alpha, \alpha^2, \alpha^3)^T$. So

$$B = \begin{pmatrix} 1\,1\,0\,0 \\ 0\,1\,1\,0 \\ 0\,0\,1\,1 \\ 1\,1\,0\,1 \end{pmatrix}.$$

(v) Compute

$$mB = (0, 0, 1, 1)B = (1, 1, 1, 0)$$

which is a binary representation of the integer 7. I.e., $S(m) = 7$ when it is regarded as an integer.

**Table 1.** $\mathbb{F}_{2^4}$ Generated by $\alpha^4 + \alpha + 1 = 0$

| Index $i = log_\alpha^\beta$ | $1$ | $\alpha$ | $\alpha^2$ | $\alpha^3$ |
|---|---|---|---|---|
| 0 | 1 | 0 | 0 | 0 |
| 1 | 0 | 1 | 0 | 0 |
| 2 | 0 | 0 | 1 | 0 |
| 3 | 0 | 0 | 0 | 1 |
| 4 | 1 | 1 | 0 | 0 |
| 5 | 0 | 1 | 1 | 0 |
| 6 | 0 | 0 | 1 | 1 |
| 7 | 1 | 1 | 0 | 1 |
| 8 | 1 | 0 | 1 | 0 |
| 9 | 0 | 1 | 0 | 1 |
| 10 | 1 | 1 | 1 | 0 |
| 11 | 0 | 1 | 1 | 1 |
| 12 | 1 | 1 | 1 | 1 |
| 13 | 1 | 0 | 1 | 1 |
| 14 | 1 | 0 | 0 | 1 |

### 3.2 Implementation in Terms of M-sequences

Let $f(x) = x^n + c_{n-1}x^{n-1} + \cdots + c_1 x + 1$ be a primitive polynomial of degree $n$ over $\mathbb{F}_2$ and $\alpha$ be a root of $f(x)$ in the extension $\mathbb{F}_{2^n}$. A sequence $A = \{a_i\}$ whose elements are given by

$$a_i = Tr(\beta\alpha^i), \beta \in \mathbb{F}_{2^n}, i = 0, 1, \cdots$$

is an m-sequence of degree $n$. I.e., $A$ satisfies that following recursive relation

$$a_{i+n} = \sum_{k=0}^{n-1} c_k a_{i+k}, i \geq 0.$$

If $\beta = 1$, then $A$ is called *the characteristic m-sequence.*

Recall that $p = 2^n - 1$ and $q = 2^n$.

**Algorithm 3.3.** Procedure for Implementation of $S(x) = x + \sigma(x)$ on $Z_q$.
*Input:* $m = m_0 + m_1 2 + \cdots + m_{n-1}2^{n-1}$, an integer in $\mathbb{Z}_q$, i.e., $0 \leq m < q$.
*Output:* $S(m)$, an integer in $\mathbb{Z}_q$.

(i) Generate a finite field $\mathbb{F}_q$ by a primitive polynomial $f(x)$ and set $\alpha$ be a root of $f(x)$.
(ii) Compute $\tau : 0 < \tau < p$ such that $1 + \alpha^s = \alpha^\tau$.
(iii) Choose $t = (t_0, t_1, \cdots, t_{n-1}) \in \mathbb{Z}_p^n$ so that $\{\alpha^{t_0}, \alpha^{t_1}, \cdots, \alpha^{t_{n-1}}\}$ is a basis of $\mathbb{F}_{2^n}/\mathbb{F}_2$.
(iv) Generate the characteristic $m$-sequence $A = \{a_i\}$ by $f(x)$.

(v) Compute the one-to-one correspondence table from $\mathbb{Z}_p$ to $\mathbb{Z}_q$:

$$\begin{pmatrix} 0 & 1 & \cdots & p-1 \\ r_0 & r_1 & \cdots & r_{p-1} \end{pmatrix}$$

by evaluating $r_i = a_{i+t_0} + a_{i+t_1}2 + \cdots + a_{i+t_{n-1}}2^{n-1} \in Z_q, i = 0, 1, \cdots, p-1$.

(vi) Look up the table to find an $i$ such that $m = r_i$.

(vii) Output $S(m) = r_{i+\tau}$.

*Example 3.4.* With the same notation as Example 3.2.

(i) See Table 1.

(ii) $\tau = 4$ from Example 1.

(iii) Choose $t = (t_0, t_1, t_2, t_3) = (0, 1, 2, 3)$. It is clear that $\{1, \alpha, \alpha^2, \alpha^3\}$ is a basis of $\mathbb{F}_{2^4}/\mathbb{F}_2$.

(iv) From Table 1, we have $Tr(1) = 0, Tr(\alpha) = 0, Tr(\alpha^2) = 0$, and $Tr(\alpha^3) = 1$. Therefore the characteristic m-sequence generated by $f(x) = x^4 + x + 1$ is as follows.

$$A = 000100110101111.$$

(v) Compute $r_i = a_i + a_{i+1}2 + a_{i+2}2^2 + a_{i+3}2^3$, $0 \le i < 15$. Then we have the following 1-1 correspondence from $\mathbb{Z}_{15}$ to $\mathbb{Z}_{16}$.

**Table 2**

| | | | | | | | | | | | | | | |
|---|---|---|---|---|---|---|---|---|---|---|---|---|---|---|
| 0 | 1 | 2 | 3 | 4 | 5 | 6 | 7 | 8 | 9 | 10 | 11 | 12 | 13 | 14 |
| 8 | 4 | 2 | 9 | 12 | 6 | 11 | 5 | 10 | 13 | 14 | 15 | 7 | 3 | 1 |

(vi) Look up Table 2, we have $m = 12 = r_4 \Longrightarrow i = 4$.

(vii) Output $S(m) = S(12) = r_{4+4} = r_8 = 10$.

*Remark 3.5.* Algorithm 3.3 can generate all linear orthomorphisms when different bases are plugged in at step 3. Since different bases may result in the same linear orthomorphisms. This procedure can not distinguish them. A solution for this problem and the number of different linear orthomorphisms on $\mathbb{Z}_2^n$ are given by Dai, Golomb and Gong in a separate paper [1].

*Remark 3.6.* In applications of encryption, the steps 1-4 of Algorithm 3.1 and the steps 1-5 of Algorithm 3.3 can be put into pre-computing units. The parameter $s$ should be kept as secret key and used for a short time when the orthomorphism is used as an enciphering function. If $s$ is exposed, the system will be broken. In fact, as an attacker has a cipher $c = S(m)$ where $m$ is a message, and if the attacker knows $s$, then he can obtain the message $m$ by using the same primitive polynomial used for generating the finite field. In other words, from $c = \alpha^k$ the attacker can obtain the message $m$ by computing $m = \alpha^{k-\tau}$ . Notice that the process for computing the cipher involves solving a discrete logarithm. So, the parameter $n$ can not be large. Therefore it is difficult to keep the generating polynomial $f(x)$ as a secret key.

# 4 Construction of Nonlinear Orthomorphisms

For many applications, linear mappings including orthomorphisms, are unsatisfactory. To construct nonlinear orthomorphisms directly requires discarding some of the tools that were so useful in the linear case.

We list the elements of $\mathbb{Z}_2^n$ as

$$Z_2^n = \{0, x_i, i = 1, 2, \cdots, p\}(p = 2^n - 1).$$

Any orthomorphism on $\mathbb{Z}_2^n$ with the maximal cycle can be expressed as a set of equations and without loss of generality, the necessary single fixed element can be 0. The equations are of the form:

$$x_{k-1} + x_k = z_k$$

except for $0 + 0 = 0$. (Note that 0 is zero in $\mathbb{F}_{2^n}$ or in $\mathbb{Z}_{2^n}$ or a vector in $\mathbb{Z}_2^n$ depends on the context.) $x_{k-1}$ and $x_k$ are successive elements in the permutation of $\mathbb{Z}_2^n$ defined by the orthomorphism. In other words, $x_k = \sigma(x_{k-1})$ if $\sigma$ is the referred orthomorphism.

A non trivial pair of equations when added component-wise will not be another equation in the same orthomorphism unless the orthormorphism is linear or linear when restricted to some subgroup of the domain. However, a linear or nonlinear orthomorphism can be expressed as an ordered set of equations:

| Row Number | L | M | R |
|---|---|---|---|
| 0 | $0$ | $+\,0$ | $=0$ |
| 1 | $x_p$ | $+\,x_1$ | $=z_1$ |
| 2 | $x_1$ | $+\,x_2$ | $=z_2$ |
| $\vdots$ | | | |
| $k$ | $x_{k-1}$ | $+\,x_k$ | $=z_k$ |
| $\vdots$ | | | |
| $p$ | $x_{p-1}$ | $+\,x_p$ | $=z_p$ |

**Figure 1**

As mentioned above, an orthomorphic permutation is one in which the pairs of successive numbers in the permutation have distinct sums.

The left (L) and middle (M) columns represent an orthomorphic permutation but the right (R) column in general is not an orthomorphic permutation. In most cases there are also some cycles when they are regarded as permutations of $\mathbb{Z}_2^n$. In what follows, we show a method of deriving nonlinear orthomorphisms as illustrated in Figure 1, from a linear orthomorphism $S(x) = x + \sigma(x)$ which is defined in (1) in Section 2, as illustrated in Figure

2. As before, a linear orthomorphism on $\mathbb{Z}_2^n$ can be written as:

| Row Number | L | | M | | R |
|---|---|---|---|---|---|
| 0 | $0$ | $+$ | $0$ | $=$ | $0$ |
| 1 | $x_p$ | $+$ | $x_1$ | $=$ | $x_{1-\tau}$ |
| 2 | $x_1$ | $+$ | $x_2$ | $=$ | $x_{2-\tau}$ |
| $\vdots$ | | | | | |
| $k$ | $x_{k-1}$ | $+$ | $x_k$ | $=$ | $x_{k-\tau}$ |
| $\vdots$ | | | | | |
| $p$ | $x_{p-1}$ | $+$ | $x_p$ | $=$ | $x_{p-\tau}$ |

**Figure 2**

($p = 2^n - 1$.) For our purposes, namely the generation of nonlinear orthomorphisms, it is sufficient that the linear orthomorphisms are maximal when they are regarded as permutations of $\mathbb{Z}_2^n$. It will have the single fixed element 0, and a cycle of length $p$ as illustrated in Figure 2. These linear orthomorphisms can be generated by the methods presented in Section 2. Figure 2 represents a family of linear orthomorphisms corresponding to a particular primitive polynomial over $\mathbb{F}_2$ of degree $n$. Any complete linearly independent set in $\mathbb{Z}_2^n$ can be used to generate a particular orthomorphism in that family. The shift $\tau$ is uniquely associated with the primitive polynomial (we will give a construction later).

For simplicity of notation, we can use $x_{k-1} = L_k$ in the left column; $x_k = M_k$ in the middle column; and $x_{k-\tau} = R_k$ in the right right column. In that case, Fig. 2 can be rewritten as:

| Row Number | L | | M | | R |
|---|---|---|---|---|---|
| 0 | $\underline{0}$ | $+$ | $\underline{0}$ | $=$ | $\underline{0}$ |
| 1 | $L_1$ | $+$ | $M_1$ | $=$ | $R_1$ |
| 2 | $L_2$ | $+$ | $M_2$ | $=$ | $R_2$ |
| $\vdots$ | | | | | |
| $k$ | $L_k$ | $+$ | $M_k$ | $=$ | $R_k$ |
| $\vdots$ | | | | | |
| $p$ | $L_p$ | $+$ | $M_p$ | $=$ | $R_p$ |

**Figure 3**

If columns $L$ and $M$ can be permuted, leaving column $R$ fixed, and maintaining equality, a new and generally nonlinear orthomorphism can be generated. This means finding a consistent set of equations:

$$L_i + M_j = R_k \tag{2}$$

where the indices refer to the row numbers in Fig. 3. The index $k$ is a function of indices $i$ and $j$, so that one can write:

$$k_{ij} = k(i,j) \in Z_{2^n}, 0 \leq i,j \leq p. \tag{3}$$

A construction of nonlinear orthomorphisms from the set of equations shown in Figure 3 depends upon the structure of the equations. I.e., it depends upon the evaluation of $k_{ij}$ in (3). In the following, we will give an algebraic construction for $k_{ij}$ by using a linear orthomorphism discussed in Section 2.

*Construction A.*

Recall that from Section 2 $\alpha$ is a primitive element of $\mathbb{F}_{2^n}$ and $\sigma(x) = \alpha^s x$ for a fixed $s : 0 < s < p$ is a linear orthomorphism. For $0 \neq \beta \in \mathbb{F}_{2^n}$, we have $\beta = \alpha^i$. We will write the discrete logarithm of $\beta$ with respect to base $\alpha$ as $i = log_\alpha \beta$ or simply, $i = log\beta$. Let $\tau(i)$ be defined by

$$\tau(i) = log(1 + \alpha^i), 0 < i < p.$$

I.e.,

$$\alpha^{\tau(i)} = 1 + \alpha^i, 0 < i < p.$$

Let

$$f(x,y) = \tau(s)^{-1}(\alpha^s x + y), x, y \in \mathbb{F}_{2^n}. \tag{4}$$

Then we have

$$f(\alpha^i, \alpha^j) = \tau(s)^{-1}(\alpha^s \alpha^i + \alpha^j) = \alpha^{j+\tau(s+i-j)-\tau(s)} \tag{5}$$

for $\alpha^s \alpha^i + \alpha^j \neq 0$. Note that

$$f(0,\alpha^j) = \alpha^{j-\tau(s)} \neq 0, \text{ for } 1 \leq j \leq p,$$
$$f(\alpha^i,0) = \alpha^{s+i-\tau(s)} \neq 0, \text{ for } 1 \leq i \leq p.$$

We define

$$k_{00} = 0$$
$$\text{for } 1 \leq j \leq p, k_{0j} = \begin{cases} logf(0,\alpha^j) & \text{if } f(0,\alpha^j) \neq 1 \\ p & \text{if } f(0,\alpha^j) = 1 \end{cases}$$
$$\text{for } 1 \leq i \leq p, k_{i0} = \begin{cases} logf(\alpha^i,0) & \text{if } f(\alpha^i,0) \neq 1 \\ p & \text{if } f(\alpha^i,0) = 1. \end{cases}$$

For $1 \leq i,j \leq p$,

$$k_{ij} = \begin{cases} logf(\alpha^i,\alpha^j) & \text{if } f(\alpha^i,\alpha^j) \neq 0,1 \\ 0 & \text{if } f(\alpha^i,\alpha^j) = 0 \\ p & \text{if } f(\alpha^i,\alpha^j) = 1. \end{cases}$$

**Table 3: A Shift Latin Square Defined by A Family of Maximal Linear Orthomorphisms on $\mathbb{Z}_2^n$**

| | | Index $j$ of $M_j$, the Middle Column Entry | | | | | | |
|---|---|---|---|---|---|---|---|---|
| | | 0 | 1 | 2 | $\cdots$ | $j$ | $\cdots$ | $p$ |
| | 0 | $k_{00}$ | $k_{01}$ | $k_{02}$ | $\cdots$ | $k_{0j}$ | $\cdots$ | $k_{0p}$ |
| | 1 | $k_{10}$ | $k_{11}$ | $k_{12}$ | $\cdots$ | $k_{1j}$ | $\cdots$ | $k_{1p}$ |
| | $\vdots$ | | | | | | | |
| Index $i$ of $L_i$, the Left Column Entry | $i$ | $k_{i0}$ | $k_{i1}$ | $k_{i2}$ | $\cdots$ | $k_{ij}$ | $\cdots$ | $k_{ip}$ |
| | $\vdots$ | | | | | | | |
| | $p$ | $k_{p0}$ | $k_{p1}$ | $k_{p2}$ | $\cdots$ | $k_{pj}$ | $\cdots$ | $k_{pp}$ |

The entries $k_{ij} = k(i, j)$ are the indices of the entries $R_k$ in the right column in the linear orthomorphism in Figures 2 and 3 defined by Construction A.

The possible combinations in equation (2) can be expressed in Table 3 whose entries are $k_{ij}$ defined by Construction A.

Each row and each column in Table 3 contains each index in the set $\mathbb{Z}_{2^n}$ exactly once. An orthormorphism will be generated by selecting one entry from each row and column without duplicates. That is, by selecting a transversal. It is known that for a group which has an orthomorphism, its multiplication table (Cayley table) has a transversal. (See reference [2]). However, Table 3, while related to $\mathbb{Z}_2^n$, is not its Cayley Table but a table of indices for a specific maximal linear orthomorphism. The principle diagonal which is a transversal, represents that linear orthomorphism. Every other othormorphism can be represented by a transversal in this table. From the Construction A, it is easy to see that the table represents the multiplicative table of a quasi-group as the operation represented is non-associative.

The other main diagonal running from lower left to upper right, is not a transversal. But it turns out, as shown below, to be of particular interest. We will refer Table 3 as a *shift Latin square*.

We have the following procedure to find nonlinear orthomorphisms from a shift Latin square.

*Procedure for Finding Nonlinear Orthomorphisms from the Shift Latin square*:

(i) Choose a diagonal from upper right to lower left and start with some pair

$$L_a + M_b = R_k \text{ with } k = k(a, b).$$

For $i = a$ and $j = b$ the entry $k(a,b)$ is in the intersection of row $a$ and column $b$, which corresponds to the equation

$$x_{a-1} + x_b = x_{k-\tau}.$$

So, $k = k(a,b) = b + \tau(s + a - b) - \tau(s)$.

(ii) Proceeding downward from right to left with successive values:

$$k(a+i, b-i) = b - i + \tau(s + a - b + 2i) - \tau(s)$$

and equations of the form

$$x_{a+i-1} + x_{b-i} = x_{b-i-\tau(s)}.$$

(iii) Proceeding along such diagonals from upper right to lower left will eventually lead to a repeated value of $k(i,j)$. This collision will occur when

$$k(a+i, b-i) = k(a,b) \text{ or equivalently}$$
$$b - i + \tau(s + a - b + 2i) - \tau(s) = b + \tau(s + a - b) - \tau(s).$$

so that $i = \tau(s + a - b + 2i) - \tau(s + a - b)$. One skips over the repeated value or goes to a parallel diagonal.

This process is completed when a full transversal is found.

*Note.* This procedure is not optimized. But it is much more efficient than exhausted search for nonlinear orthomorphisms. The latter stages of this process can be expedited to minimize backtracking by estimating which one of the remaining choices for the next step in the transversal will maximize the probability of success.

## 5 Examples and Some Discussions

In the following examples, we always suppose that a finite field $\mathbb{F}_{2^4}$ is defined by the primitive polynomial $f(x) = x^4 + x^3 + 1$ and $\alpha$ is a primitive element of $\mathbb{F}_{2^4}$ with $f(\alpha) = 0$. The elements of $\mathbb{F}_{2^4}$ are listed in Table 4.

*Example 3.* We will use Construction A to construct a shift Latin square. Let $\sigma(x) = \alpha^{14}x$. Then $s = 14$ and $\tau(s) = 11$. Hence

$$f(x,y) = \alpha^{-11}(\alpha^{14}x + y), x, y \in \mathbb{F}_{2^4}.$$

Using Construction A, we have Table 5 whose entries are $k_{ij}$.

The table shows compatible rearrangements of the three columns. For example, $L_{10}$ and $M_6$ have the corresponding entry $k(10,6) = 14$ which means that $L_{10} + M_6 = R_{14}$ or that

$$x_{10-1} + x_6 = x_{14-4} \text{ i.e., } x_9 + x_6 = x_{10}.$$

**Table 4.** $\mathbb{F}_{2^4}$ defined by $\alpha^4 + \alpha^3 + 1 = 0$

| Index $i = log_\alpha^\beta$ | 1 | $\alpha$ | $\alpha^2$ | $\alpha^3$ |
|---|---|---|---|---|
| 0 | 1 | 0 | 0 | 0 |
| 1 | 0 | 1 | 0 | 0 |
| 2 | 0 | 0 | 1 | 0 |
| 3 | 0 | 0 | 0 | 1 |
| 4 | 1 | 0 | 0 | 1 |
| 5 | 1 | 1 | 0 | 1 |
| 6 | 1 | 1 | 1 | 1 |
| 7 | 1 | 1 | 1 | 0 |
| 8 | 0 | 1 | 1 | 1 |
| 9 | 1 | 0 | 1 | 0 |
| 10 | 0 | 1 | 0 | 1 |
| 11 | 1 | 0 | 1 | 1 |
| 12 | 1 | 1 | 0 | 0 |
| 13 | 0 | 1 | 1 | 0 |
| 14 | 0 | 0 | 1 | 1 |

The small squares in Table 5 are a selection of values of $k(i,j)$ which are all different and taken from each row and column exactly once. This is a transversal of the Latin square. Equivalently, this represents a 16 x 16 x 16 cube in which each row and column in the $X, Y$ plane and each parallel plane along the $Z$ axis are sampled once. As far as possible, a lower left to upper right diagonal was used.

**Table 5.** Shift Latin Square for $n = 4$

| | | $j$ index of $\underline{x}_j$ | | | | | | | | | | | | | | | |
|---|---|---|---|---|---|---|---|---|---|---|---|---|---|---|---|---|---|
| | | 0 | 1 | 2 | 3 | 4 | 5 | 6 | 7 | 8 | 9 | 10 | 11 | 12 | 13 | 14 | 15 |
| | 0 | 0 | 5 | 6 | 7 | 8 | 9 | 10 | 11 | 12 | 13 | 14 | 15 | 1 | [2] | 3 | 4 |
| | 1 | 4 | 1 | 13 | 8 | 7 | 14 | 12 | 2 | 10 | 6 | 9 | 3 | 5 | 11 | 15 | [0] |
| | 2 | 5 | 0 | 2 | 14 | 9 | 8 | 15 | 13 | 3 | 11 | [7] | 10 | 4 | 6 | 12 | 1 |
| | 3 | 6 | 2 | 0 | 3 | 15 | 10 | 9 | 1 | 14 | [4] | 12 | 8 | 11 | 5 | 7 | 13 |
| $i$ | 4 | 7 | 14 | 3 | 0 | 4 | 1 | 11 | 10 | 2 | 15 | 5 | 13 | 9 | 12 | [6] | 8 |
| Index | 5 | 8 | 9 | 15 | 4 | 0 | 5 | 2 | 12 | 11 | 3 | 1 | 6 | [14] | 10 | 13 | 7 |
| of | 6 | 9 | 8 | 10 | 1 | 5 | 0 | 6 | 3 | [13] | 12 | 4 | 2 | 7 | 15 | 11 | 14 |
| $\underline{x}_{i-1}$ | 7 | 10 | 15 | 9 | 11 | 2 | 6 | 0 | 7 | 4 | 14 | 13 | [5] | 3 | 8 | 1 | 12 |
| | 8 | 11 | 13 | 1 | 10 | 12 | [3] | 7 | 0 | 8 | 5 | 15 | 14 | 6 | 4 | 9 | 2 |
| | 9 | 12 | 3 | 14 | 2 | [11] | 13 | 4 | 8 | 0 | 9 | 6 | 1 | 15 | 7 | 5 | 10 |
| | 10 | 13 | 11 | 4 | [15] | 3 | 12 | 14 | 5 | 9 | 0 | 10 | 7 | 2 | 1 | 8 | 6 |
| | 11 | 14 | 7 | [12] | 5 | 1 | 4 | 13 | 15 | 6 | 10 | 0 | 11 | 8 | 3 | 2 | 9 |
| | 12 | 15 | [10] | 8 | 13 | 6 | 2 | 5 | 14 | 1 | 7 | 11 | 0 | 12 | 9 | 4 | 3 |
| | 13 | [1] | 4 | 11 | 9 | 14 | 7 | 3 | 6 | 15 | 2 | 8 | 12 | 0 | 13 | 10 | 5 |
| | 14 | 2 | 6 | 5 | 12 | 10 | 15 | [8] | 4 | 7 | 1 | 3 | 9 | 13 | 0 | 14 | 11 |
| | 15 | 3 | 12 | 7 | 6 | 13 | 11 | 1 | [9] | 5 | 8 | 2 | 4 | 10 | 14 | 0 | 15 |

The numbers framed by suares represent a transversal. The linear orthmorphism defining Table 5 is $\sigma(x) = \alpha^{14}x, x \in \mathbb{F}_{2^4}$ and the nonlinear orthomorphism $g(x)$ derived from Table 5 is as follows.

$$
\begin{aligned}
g(x) + x &= F(x)\\
\alpha^{11} + 0 &= \alpha^{11}\\
\alpha^{10} + 1 &= \alpha^{5}\\
\alpha^{9} + \alpha &= \alpha^{7}\\
\alpha^{8} + \alpha^{2} &= \alpha^{10}\\
\alpha^{7} + \alpha^{3} &= \alpha^{6}\\
\alpha^{6} + \alpha^{4} &= \alpha^{13}\\
\alpha^{12} + \alpha^{5} &= \alpha^{3}\\
\alpha^{13} + \alpha^{6} &= \alpha^{4}\\
\alpha^{4} + \alpha^{7} &= \alpha^{8}\\
\alpha + \alpha^{8} &= \alpha^{14}\\
1 + \alpha^{9} &= \alpha^{2}\\
\alpha^{5} + \alpha^{10} &= 1\\
\alpha^{3} + \alpha^{11} &= \alpha^{9}\\
0 + \alpha^{12} &= \alpha^{12}\\
\alpha^{2} + \alpha^{13} &= \alpha\\
\alpha^{14} + \alpha^{14} &= 0
\end{aligned}
$$

**Figure 4**

Let the elements in $\mathbb{F}_{2^4}$ be represented by the exponents of $\alpha$ where we write $0 = \alpha^{\infty}$ as convention. The mapping from the left column to the right column of Figure 4 can be represented by the following permutation:

$$(11)(10, 5, 0, 2, 1, 14, \infty, 12, 3, 9, 7, 6, 13, 4, 8).$$

Thus it is maximal except for the single fixed element 11. The orthomorphism in Figure 4 can also be represented by a permutation in which each number is added to its successor to obtain a unique sum in that same permutation. This permutation is as follows:

$$(14)(10, 0, 9, 1, 8, 2, 13, 6, 4, 7, 3, 11, \infty, 12, 5)$$

Once again, it is maximal.

This orthomorphism can also be examined for nonlinearity [7] [8]. A linear function $f(x)$ over a $R$-module $M$ is defined to be one for which:

$$f(ax + by) = af(x) + bf(y), \forall x, y \in M, a, b \in R.$$

$\mathbb{Z}_2^n$ can be considered as a $\mathbb{Z}_2$ module. We define a *discrepancy function* of a function $f(x)$ of $\mathbb{Z}_2^n$ as

$$D_f(x,y) = f(x) + f(y) + f(x+y), \forall x, y \in \mathbb{Z}_2^n.$$

Then a linear function over $\mathbb{Z}_2^n$ is one for which:

$$D_f(x,y) = 0, \ \forall x, y \in \mathbb{Z}_2^n.$$

$D_f(x,y)$ as defined is a mapping of subgroups of order $\leq 4$ into $\mathbb{Z}_2^n$. That is, any additive triple supplemented by 0, $\{x, y, x+y, 0\}$ is a subgroup of $\mathbb{Z}_2^n$. The range of $D_f(x,y)$, denoted by $Im(D_f)$, satisfies that $Im(D_f) \in \mathbb{Z}_2^n$. A reasonable goal for a nonlinear mapping is that

$$Im(D_f) = \mathbb{Z}_2^n.$$

That is, all possible values in $\mathbb{Z}_2^n$ are taken on. In addition, the nonzero values should be approximately equally distributed. This correlates well with the $L_1$ and $L_4$ norms using the Walsh-Fourier transform. In the example above, calculating all 16 x 16 $= 256$ values of $D_f(x,y)$ shows that $D_f(x,y)$ takes on all values in $\mathbb{Z}_2^4$. In other words, we have

$$Im(D_f) = \mathbb{Z}_2^4$$

where $f(x)$ is the nonlinear orthomorphism given above. Note that if $y$ is fixed in $D_f(x,y)$, then this is the same function discussed in [7].

## 6 Finding a Transversal for an Arbitrary Orthomorphism

An arbitrary orthomorphism with the maximal cycle, in particular an nonlinear one, can be written as in Figure 1. Without changing the mapping defined by the orthomorphism, the set of equations defining the orthomorphism can be permuted so that the right column is identical to the right column of the linear orthormorphism in Figure 2, as follows:

| Row Number | $L$ | $M$ | $R$ |
|---|---|---|---|
| 0 | $0$ | $+\,0$ | $=0$ |
| 1 | $u_1$ | $+\,v_1$ | $=x_{1-\tau}$ |
| 2 | $u_2$ | $+\,v_2$ | $=x_{2-\tau}$ |
| $\vdots$ | | | |
| $k$ | $u_k$ | $+\,v_k$ | $=x_{k-\tau}$ |
| $\vdots$ | | | |
| $p$ | $u_p$ | $+\,v_p$ | $=x_{p-\tau}$ |

**Figure 5**

The resultant $L$ and $M$ columns in Figure 5 are, respectively, permutations of the $L$ and $M$ columns of Figure 2. $(x_{1-\tau}, x_{2-\tau}, \cdots, x_{p-\tau})$ in column $R$ is a permutation of the original $(z_1, z_2, \cdots, z_p)$ in column 1 and is identical to column $R$ of Figure 2. Consider an arbitrary equation in Figure 5:

$$u_k + v_k = x_{k-\tau}$$

where $u_k$ occurs at some index $i$ and $v_k$ occurs at some index $j$, respectively, in columns $L$ and $M$ of Figure 3. Thus

$$L_i + M_j = R_k \text{ or } x_{i-1} + x_j = x_{k-\tau}.$$

Note that $k_{ij} = k(i,j) = k$ defines an integer in Table 3. The set of $2^n$ such integers defined by comparing Figure 5 with Figure 2 specifies a transversal in Table 3. The following are two examples.

*Example 4.* Let $g(x)$ be a nonlinear orthomporphism derived by nonlinear shift register logic I [3] which is given as below. The transversal in Table 6, framed numbers, representing this nonlinear orthomorphism.

$$\begin{aligned}
g(x) + x &= F(x)\\
0 + 0 &= 0\\
\alpha^3 + 1 &= \alpha^4\\
\alpha^{12} + \alpha &= 1\\
\alpha^5 + \alpha^2 &= \alpha^6\\
\alpha^9 + \alpha^3 &= \alpha^{11}\\
\alpha^7 + \alpha^4 &= \alpha^8\\
\alpha^8 + \alpha^5 &= \alpha^9\\
\alpha^4 + \alpha^6 &= \alpha^{13}\\
\alpha^{14} + \alpha^7 &= \alpha^5\\
\alpha^{11} + \alpha^8 &= \alpha^{12}\\
\alpha^6 + \alpha^9 &= \alpha^{10}\\
\alpha^{13} + \alpha^{10} &= \alpha^{14}\\
\alpha^{10} + \alpha^{11} &= \alpha^7\\
1 + \alpha^{12} &= \alpha\\
\alpha + \alpha^{13} &= \alpha^2\\
\alpha^2 + \alpha^{14} &= \alpha^3
\end{aligned}$$

*Example 5.* The transversal marked in Table 7 is the transversal representing the nonlinear orthomorphism generated by computer search.

Note that Example 4 is clearly "not very nonlinear" as 10 of 16 elements in the transversal are on the main diagonal corresponding to a linear ortho-

**Table 6.** The Same Shift Latin Square in Example 3

| | | $j$ index of $x_j$ | | | | | | | | | | | | | | | |
|---|---|---|---|---|---|---|---|---|---|---|---|---|---|---|---|---|---|
| | | 0 | 1 | 2 | 3 | 4 | 5 | 6 | 7 | 8 | 9 | 10 | 11 | 12 | 13 | 14 | 15 |
| | **0** | 0 | 5 | 6 | 7 | 8 | 9 | 10 | 11 | 12 | 13 | 14 | 15 | 1 | 2 | 3 | 4 |
| | 1 | 4 | 1 | 13 | 8 | 7 | 14 | **12** | 2 | 10 | 6 | 9 | 3 | 5 | 11 | 15 | 0 |
| | 2 | 5 | 0 | **2** | 14 | 9 | 8 | 15 | 13 | 3 | 11 | 7 | 10 | 4 | 6 | 12 | 1 |
| | 3 | 6 | 2 | 0 | **3** | 15 | 10 | 9 | 1 | 14 | 4 | 12 | 8 | 11 | 5 | 7 | 13 |
| $i$ | 4 | 7 | 14 | 3 | 0 | **4** | 1 | 11 | 10 | 2 | 15 | 5 | 13 | 9 | 12 | 6 | 8 |
| Index | 5 | 8 | 9 | 15 | 4 | 0 | **5** | 2 | 12 | 11 | 3 | 1 | 6 | 14 | 10 | 13 | 7 |
| of | 6 | 9 | 8 | 10 | 1 | 5 | 0 | 6 | 3 | 13 | 12 | 4 | 2 | 7 | 15 | **11** | 14 |
| $x_{i-1}$ | 7 | 10 | 15 | 9 | 11 | 2 | 6 | 0 | **7** | 4 | 14 | 13 | 5 | 3 | 8 | 1 | 12 |
| | 8 | 11 | 13 | 1 | 10 | 12 | 3 | 7 | 0 | 8 | 5 | 15 | **14** | 6 | 4 | 9 | 2 |
| | 9 | 12 | 3 | 14 | 2 | 11 | 13 | 4 | 8 | 0 | **9** | 6 | 1 | 15 | 7 | 5 | 10 |
| | 10 | 13 | 11 | 4 | 15 | 3 | 12 | 14 | 5 | 9 | 0 | **10** | 7 | 2 | 1 | 8 | 6 |
| | 11 | 14 | 7 | 12 | 5 | 1 | 4 | 13 | 15 | 6 | 10 | 0 | 11 | **8** | 3 | 2 | 9 |
| | 12 | 15 | 10 | 8 | 13 | 6 | 2 | 5 | 14 | **1** | 7 | 11 | 0 | 12 | 9 | 4 | 3 |
| | 13 | 1 | 4 | 11 | 9 | 14 | 7 | 3 | 6 | 15 | 2 | 8 | 12 | 0 | **13** | 10 | 5 |
| | 14 | 2 | **6** | 5 | 12 | 10 | 15 | 8 | 4 | 7 | 1 | 3 | 9 | 13 | 0 | 14 | 11 |
| | 15 | 3 | 12 | 7 | 6 | 13 | 11 | 1 | 9 | 5 | 8 | 2 | 4 | 10 | 14 | 0 | **15** |

**Table 7.** The Same Shift Latin Square in Example 3

| | | $j$ index of $\underline{x}_j$ | | | | | | | | | | | | | | | |
|---|---|---|---|---|---|---|---|---|---|---|---|---|---|---|---|---|---|
| | | 0 | 1 | 2 | 3 | 4 | 5 | 6 | 7 | 8 | 9 | 10 | 11 | 12 | 13 | 14 | 15 |
| | 0 | **0** | 5 | 6 | 7 | 8 | 9 | 10 | 11 | 12 | 13 | 14 | 15 | 1 | 2 | 3 | 4 |
| | 1 | 4 | 1 | 13 | 8 | 7 | 14 | 12 | 2 | 10 | **6** | 9 | 3 | 5 | 11 | 15 | 0 |
| | 2 | 5 | 0 | 2 | 14 | 9 | 8 | 15 | 13 | 3 | 11 | 7 | 10 | 4 | 6 | **12** | 1 |
| | 3 | 6 | 2 | 0 | 3 | 15 | 10 | 9 | 1 | 14 | 4 | 12 | 8 | 11 | 5 | 7 | **13** |
| $i$ | 4 | 7 | 14 | 3 | 0 | 4 | 1 | 11 | 10 | 2 | 15 | 5 | 13 | **9** | 12 | 6 | 8 |
| Index | 5 | 8 | 9 | 15 | 4 | 0 | 5 | 2 | 12 | **11** | 3 | 1 | 6 | 14 | 10 | 13 | 7 |
| of | 6 | 9 | 8 | 10 | 1 | 5 | 0 | 6 | **3** | 13 | 12 | 4 | 2 | 7 | 15 | 11 | 14 |
| $x_{i-1}$ | 7 | 10 | **15** | 9 | 11 | 2 | 6 | 0 | 7 | 4 | 14 | 13 | 5 | 3 | 8 | 1 | 12 |
| | 8 | 11 | 13 | 1 | 10 | 12 | 3 | 7 | 0 | 8 | 5 | 15 | 14 | 6 | **4** | 9 | 2 |
| | 9 | 12 | 3 | 14 | 2 | 11 | 13 | 4 | 8 | 0 | 9 | 6 | **1** | 15 | 7 | 5 | 10 |
| | 10 | 13 | 11 | 4 | 15 | 3 | 12 | 14 | 5 | 9 | 0 | **10** | 7 | 2 | 1 | 8 | 6 |
| | 11 | 14 | 7 | 12 | **5** | 1 | 4 | 13 | 15 | 6 | 10 | 0 | 11 | 8 | 3 | 2 | 9 |
| | 12 | 15 | 10 | 8 | 13 | 6 | **2** | 5 | 14 | 1 | 7 | 11 | 0 | 12 | 9 | 4 | 3 |
| | 13 | 1 | 4 | 11 | 9 | **14** | 7 | 3 | 6 | 15 | 2 | 8 | 12 | 0 | 13 | 10 | 5 |
| | 14 | 2 | 6 | 5 | 12 | 10 | 15 | **8** | 4 | 7 | 1 | 3 | 9 | 13 | 0 | 14 | 11 |
| | 15 | 3 | 12 | **7** | 6 | 13 | 11 | 1 | 9 | 5 | 8 | 2 | 4 | 10 | 14 | 0 | 15 |

morphism. Example 5 has only 2 elements on the main diagonal and thus is "more nonlinear".

## 7 Conclusion

In terms of the isomorphism between the linear space $\mathbb{Z}_2^n$ and the finite field $\mathbb{F}_{2^n}$, an algebraic method of constructing linear orthomorphisms on $\mathbb{Z}_2^n$ with maximal length cycles and two algorithms for implementation are provided. By using the resulting linear orthomorphisms, a special type of Latin squares are constructed, which are called *shift Latin squares*, from which all orthomorphisms, especially, nonlinear orthomorphisms on $\mathbb{Z}_2^n$, can be represented as transversals. Those transversals which are parallel to the principal diagonal, from upper left to lower right, represent linear orthomorphisms while those transversals with segments generally parallel to the diagonal from lower left to upper right represent nonlinear orthomorphisms. If one can count the number of transversals in the shift Latin squares, one can determine the number of orthomorphisms in $\mathbb{Z}_2^n$.

## References


1. Z.T. Dai, S.W. Golomb, and G. Gong, Generating all linear orthomorphisms without repetitions, *Discrete Mathematics*, 205 (1999) pp. 47-55.
2. J. Denes and A.D. Keedwell, *Latin Squares: New Developments in the Theory and Applications*, North Holland, 1991.
3. S.W. Golomb, *Shift Register Sequences*, Revised Edition, Aegean Park Press, May 1982, pp. 142.
4. R. Lidl and H. Niederreiter, *Finite Fields*, Encyclopaedia of Mathematics and its Applications, Volume 20, Addison-Wesley, 1983.
5. Lothrop Mittenthal, Block substitutions using orthomorphic mappings, *Advances in Applied Mathematics*, Vol. 16, No. 1, March 1995.
6. Lothrop Mittenthal, The algebraic structure of nonlinear orthomorphisms of binary numbers, *Technical report, Teledyne Electronic Technologies*, Feb. 1994.
7. K. Nyberg, Differentially uniform mappings for cryptography, *Advances in Cryptology, Proceedings of EuroCrypt'93*, Lecture Notes in Computer Science, 1993.
8. J.P. Pieprzyk, Nonlinearity of exponent permutations, *Advances in Cryptology, Proceedings of EuroCrypt'89*, pp. 80-92, Lecture Notes in Computer Science, 1989.
9. F.J. MacWilliams and N.J.A. Sloane, *The Theory of Error-Correcting Codes*, North-Holland, New York, 1977.
10. J.H. van Lint and R.M. Wilson, *A Course in Combinatorics*, Cambridge University Press 1992, pp. 264.

# Period Polynomials for $F_{p^2}$ of Fixed Small Degree

Stanley Gurak

Department of Mathematics and Computer Science
University of San Diego
Alcala Park
San Diego, CA 92110-2492
USA
e-mail: gurak@acusd.edu

**Abstract** Let $q = p^a$ with $p$ a prime, and $e$ and $f$ be positive integers satisfying $ef + 1 = q$. Let $\mathbf{F}_q$ be the finite field of $q$ elements. The Gauss periods of order $e$ satisfy a period polynimial $F(x)$ of degree $e$ over the rational field $\mathbf{Q}$. In the classical case $q = p$, Gauss showed that $F(x)$ is irreducible over $\mathbf{Q}$ and determined the polynomial for $e = 2, 3$ and 4. Using the methods and results from the theory of cyclotomy, Dickson, Lehmer, Muskat and Whiteman, among others, determined $F(x)$ for $e = 5, 6, 8, 10, 12, 16, 24$ in the case $q = p$. For the general case, Myerson determined $F(x)$ for $e = 2, 3$ and 4, but beyond this very little is explicitly known. Here I determine $F(x)$ explicitly for the case $q = p^2$ when $e$ divides 8 or 12, by applying work of Berndt and Evans on Gauss sums for $F_{p^2}$ with characters of order 6, 8 and 12. The results settle an unanswered question concerning the splitting of the middle factor of $F(x)$ when $p \equiv 7 (\mathrm{mod}\ 24)$.

## 1 Introduction

Let $q = p^a$ with $p$ a prime, and $e$ and $f$ be positive integers such that $ef + 1 = q$. Let $\mathbf{F}_q$ denote the field of $q$ elements, $\mathbf{F}_q^*$ its multiplicative group and $g$ a fixed generator of $\mathbf{F}_q^*$. Then $G = g^{\frac{q-1}{p-1}}$ generates $\mathbf{F}_p^*$. Let $\mathrm{Tr} : \mathbf{F}_q \to \mathbf{F}_p$ be the usual trace map and set $\zeta_m = exp(2\pi i/m)$ for any positive integer $m$. Put $\delta = GCD(\frac{q-1}{p-1}, e)$ and $R = \frac{q-1}{\delta(p-1)}$. The (modified) Gauss periods are

$$H_j = \sum_{x \in \mathbf{F}_q} \zeta_p^{Tr\ g^j x} \quad (1 \leq j \leq e) \tag{1}$$

(if $j > e$ then the subscripts in $H_j$ are taken modulo $e$), and satisfy the period polynomial

$$F(x) = \prod_{j=1}^{e} (x - H_j). \tag{2}$$

G. Myerson[8] showed that $F(x)$ splits over $\mathbf{Q}$ into $\delta$ factors

$$F(x) = \prod_{w=1}^{\delta} F^{(w)}(x), \tag{3}$$

each of degree $e/\delta$, where

$$F^{(w)}(x) = \prod_{k=0}^{e/\delta-1} (x - H_{w+k\delta}) \quad (1 \leq w \leq \delta)$$

(chiefly, a consequence of (10) later).

In the classical case $q = p$ (so $\delta = 1$), Gauss showed that $F(x)$ is irreducible over $\mathbf{Q}$ and determined the polynomial for $e = 2, 3$ and 4. Using the methods and results from the theory of cyclotomy, Dickson, E. Lehmer, Muskat and Whiteman, among others, determined $F(x)$ for $e = 5, 6, 8, 10, 12, 16$ and 24 in the case $q = p$. For the general case $q = p^a$, Myerson determined the factors $F^{(w)}(x)$ of $F(x)$ for $e = 2, 3$ and 4, but beyond this very little is known. When $q = p^2$, Berndt and Evans [3] essentially determined the last Gauss period $H_e$ for $e = 6, 8$ and 12. Extending their methods, it is possible to determine each of the factors $F^{(w)}(x)$ of $F(x)$ in (3) for $q = p^a$ when $e|24$. Here we are content to explicitly determine the factors for the case $q = p^2$, and then apply the results to decide an unanswered question posed in [6] concerning the splitting of the middle factor when $p \equiv 7 \pmod{24}$. A full treatment of the general case $q = p^a$ when $e|24$ would be much too cumbersome to describe here, but will be reported later in detail elsewhere.

## 2 Preliminaries

The Gauss periods (1) are intimately connected with Gauss sums. For $\chi$ a character of $\mathbf{F}_q^*$ of order dividing $e$, define the Gauss sum $G_a(\chi)$ by

$$G_a(\chi) = \sum_{x \in \mathbf{F}_q^*} \chi(x)\zeta_p^{\mathrm{Tr}\, x}, \tag{4}$$

which satisfies

$$G_a(\chi)G_a(\chi^{-1}) = \chi(-1)p^a \quad for \ \chi \neq 1. \tag{5}$$

(We will write $G(\chi)$ for $G_1(\chi)$. ) For $a > 1$, the value of $G_a$ can be expressed in terms of the Eisenstein sum

$$E_a(\chi) = \sum_{x \in \mathbf{F}_q^*, \mathrm{Tr}\, x = 1} \chi(x); \tag{6}$$

namely (chiefly, Theorem 12.1.1 of [4]),

$$G_a(\chi) = \begin{cases} E_a(\chi)G(\chi^*) & \text{if } \chi^* \text{ is nontrivial} \\ -pE_a(\chi) & \text{if } \chi^* \text{ is trivial,} \end{cases} \tag{7}$$

where $\chi^*$ denotes the restriction of $\chi$ to $\mathbf{F}_p^*$. Normalizing such characters $\chi$ so $\chi_j(g) = \zeta_e^j$ for $1 \le j \le e$, one has

$$G_a(\chi_j) = \frac{1}{e}\sum_{i=1}^{e} \zeta_e^{ij} H_i \;\; (j \neq e); G_a(\chi_e) = -1, \tag{8}$$

and

$$H_i = \sum_{j=1}^{e-1} \zeta_e^{-ij} G_a(\chi_j) \;\; (1 \le i \le e) \tag{9}$$

For later use, we list the values of $E_2(\chi)$ compiled by Berndt and Evans [3,4] for $\chi$ of orders $e|8$ or 12 with $\chi(g) = \zeta_e$.

**Proposition 2.1.** *The Eisenstein sums $E_2(\chi)$, where $\chi(g) = \zeta_e$ determines a character of order e with $e|8$ or 12, are evaluated as follows:*
*For e=3 with p=3k+1,* $E_2(\chi) = -\frac{1}{2}(r_3 + is_3\sqrt{3})$
*For e=4 with p=4k+1,* $E_2(\chi) = (-1)^{k+1}(a_4 + ib_4)$
*For e=6 with p=6k+1,* $E_2(\chi) = (-1)^{k+1}\chi^*(2)\frac{1}{2}(r_3 + is_3\sqrt{3})$
*For e=8 with p=8k+1,* $E_2(\chi) = -\chi^*(2)(a_8 + ib_8\sqrt{2})$*; with p=8k+3,* $E_2(\chi) = -a_8 - ib_8\sqrt{2}$ *and with p=8k+5,* $E_2(\chi) = -\chi^*(2)(a_4 + ib_4)$
*For e=12 with p=12k+1,* $E_2(\chi) = -\bar{\chi}^*(2)(a_{12} + ib_{12})$*; with p=12k+5,* $E_2(\chi) = \chi^*(2)\epsilon(-b_4 + ia_4)$ *and with p=12k+7,* $E_2(\chi) = \bar{\chi}^*(2)(-1)^k(a_3 - ib_3\sqrt{3})$*;*
*where* $p = a_2 + 3b_3^2$, $a_3 \equiv -1 (\mathrm{mod}\ 3)$, $3b_3 \equiv (2G^{(p-1)/3} + 1)a_3 (\mathrm{mod}\ p)$ *and* $r_3 + is_3\sqrt{3} = 2\zeta_3^Z(a_3 + ib_3\sqrt{3})$ *where* $Z = ind_G\, 2$*;*
$p = a_4^2 + b_4^2$, $a_4 \equiv -(\frac{2}{p}) (\mathrm{mod}\ 4)$, *and* $b_4 \equiv a_4 G^{(p-1)/4} (\mathrm{mod}\ p)$
$p = a_8^2 + 2b_8^2$ *with* $a_8 \equiv -1 (\mathrm{mod}\ 4)$ *and* $2b_8 \equiv (G^{(p-1)/8} + G^{3(p-1)/8})a_8$ $(\mathrm{mod}\ p)$ *if* $p \equiv 1 (\mathrm{mod}\ 8)$ *or with* $a_8 \equiv (-1)^{(p-3)/8} (\mathrm{mod}\ 4)$ *and* $2b_8 \equiv (g^{(p^2-1)/8} - g^{(1-p^2)/8})a_8\ (\mathrm{mod}\ p)$ *if* $p \equiv 3 (\mathrm{mod}\ 8)$*;*
$a_{12} = a_4, b_{12} = b_4$ *if* $3 \nmid a_4$ *or* $a_{12} = -a_4, b_{12} = -b_4$ *if* $3|a_4$.
*Also, if* $p \equiv 5 (\mathrm{mod}\ 12)$, $\epsilon = \pm 1$ *is chosen so* $\epsilon \equiv a_4 b_4 (\mathrm{mod}\ 3)$.

To describe the Galois action on the Gauss periods $H_j$, for any integer $\mu$, let $\phi_\mu$ denote the map of $\mathbf{Q}(\zeta_{pe})$ which fixes $\zeta_e$ and sends $\zeta_p$ to $\zeta_p^{G^\mu}$. From (4) one finds that

$$\phi_\mu(G_a(\chi_j)) = \zeta_e^{-\mu j \delta R} G_a(\chi_j),$$

in particular, $\phi_\mu(H_i) = H_{i+\delta R\mu}$ from (9). Choosing $\bar{\mu}$ so that $R\bar{\mu} \equiv 1$ $(\mathrm{mod}\ e/\delta)$ one obtains

$$\phi_{\bar{\mu}k}(H_w) = H_{w+\delta k} \;\; (1 \le w \le \delta, 0 \le k < e/\delta). \tag{10}$$

One also finds repetitions among the Gauss periods due to the action of the Frobenius map $x \to x^p$ in $\mathbf{F}_q$. In particular, one has

$$G_a(\chi_{jp}) = G_a(\chi_j), \;\; \textit{or equivalently}, \; H_{jp} = H_j \tag{11}$$

for any integer $j$. See [5,6] for a detailed account of this and a discussion of other patterns of duplications.

As in [4], we set $g_a(e) = \sum_{x\in \mathbf{F}_p^a} \zeta_p^{Tr\, x^e}$ so $H_e = g_a(e)$ from (9). (We write $g(e)$ for $g_1(e)$.) Set $i^* = 1$ or $i$ as $p \equiv 1$ or $3 (\mathrm{mod}\ 4)$. For small values of $e$, it is known that $g(2) = i^*\sqrt{p}$ whereas $g_2(2) = -g(2)^2 = (-1)^{\frac{p+1}{2}}p$.

The value $g(3)$ satisfies $x^3 - 3px - pr_3 = 0$ with $r_3$ and $s_3$ as in Proposition 1. The correct choice of root for $g(3)$ was determined by Matthews [7] and is decribed in [4]. Now, for an appropriate choice of generator $G$ of $\mathbf{F}_p^*$ one has

$$G(\chi_1) = \frac{1}{2}(g(3) + i\nu\sqrt{4p - g(3)^2}), \text{with } \nu = sgn\left(s_3(g(3)^2 - p)\right). \quad (12)$$

Then the real numbers

$$g(3)' = \frac{-g(3) + \nu\sqrt{12p - 3g(3)^2}}{2} = \frac{1}{s_3}(g(3)^2 - 2p - \frac{1}{2}(s_3 + r_3)g(3)) \quad (13)$$

and

$$g(3)'' = \frac{-g(3) + \nu\sqrt{12p - 3g(3)^2}}{2} = \frac{1}{s_3}(2p - g(3)^2 - \frac{1}{2}(s_3 - r_3)g(3))$$

are the conjugates of $g(3)$. In particular, one has

$$\bar{\chi}_1(2)G(\chi_1) + \chi_1(2)G(\bar{\chi}_1) = g(3), g(3)' \text{ or } g(3)'' \quad (14)$$

according as $\chi_1(2) = \zeta_3^\alpha$ with $\alpha = 0, 1$ or $-1$. >From Proposition 1, one finds $r_3 = 2a_3$, $s_3 = 2b_3$ if $\chi_1(2) = 1$ or $r_3 = -a_3 - 3\alpha b_3$, $s_3 = \alpha a_3 - b_3$ if $\chi_1(2) \neq 1$. In particular, the quantity

$$s_3 b_3 = 2b_3^2 \text{ or } -\frac{1}{2}(p + a_3 r_3) \text{ according as } \chi_1(2) = 1 \text{ or not.} \quad (15)$$

In the case $q = p^2$, one has $g_2(3) = 2p$ if $p \equiv 5 (\mathrm{mod}\ 6)$; otherwise if $p \equiv 1 (\mathrm{mod}\ 6)$, one finds that

$$g_2(3) = \begin{cases} -a_3 g(3) - b_3(g(3)' - g(3)'') & \text{if } \chi_1^*(2) = 1 \\ -a_3 g(3)' - b_3(g(3)'' - g(3)) & \text{if } \chi_1^*(2) = \zeta_3 \\ -a_3 g(3)'' - b_3(g(3) - g(3)') & \text{if } \chi_1^*(2) = \zeta_3^{-1} \end{cases} \quad (16)$$

When $e = 6$ and $p \equiv 1 (\mathrm{mod}\ 6)$, the value $g(6)$ is seen to satisfy (chiefly, Theorem 4.1.4 in [4])

$$g(6) = \begin{cases} g(3) + \frac{i^*}{\sqrt{p}}(g(3)^2 - p) & \text{if } \chi_1(4) = 1 \\ g(3) + \frac{i^*}{\sqrt{p}}(g(3)''^2 - p) & \text{if } \chi_1(4) = \zeta_3 \\ g(3) + \frac{i^*}{\sqrt{p}}(g(3)'^2 - p) & \text{if } \chi_1(4) = \zeta_3^{-1} \end{cases} \quad (17)$$

(Here and later for $3|e$, $g$ is chosen so that $G = g^{p+1}$ consistent with that in (12). )

In the case $q = p^2$ with $p = 6k+5$, $g_2(6) = 2p - 3(-1)^k p$. If $p = 6k+1$ with $k$ even,

$$g_2(6) = \begin{cases} -p - 2a_3 g(3) & \text{if } \chi_1^*(4) = 1 \\ -p - 2a_3 g(3)' & if \chi_1^*(4) = \zeta_3 \\ -p - 2a_3 g(3)'' & if \chi_1^*(4) = \zeta_3^{-1} \end{cases} \tag{18}$$

With $k$ odd,

$$g_2(6) = \begin{cases} p - 2b_3(g(3)' - g(3)'') & \text{if } \chi_1^*(4) = 1 \\ p - 2b_3(g(3)'' - g(3)) & \text{if } \chi_1^*(4) = \zeta_3 \\ p - 2b_3(g(3) - g(3)') & \text{if } \chi_1^*(4) = \zeta_3^{-1} \end{cases} \tag{19}$$

When $e = 4$ and $p \equiv 1 (\mathrm{mod}\ 4)$, the value $g(4)$ is given by (chiefly, Theorem 4.2.1 in [4])

$$g(4) = \sqrt{p} \pm \left(2(\frac{2}{p})(p + a_4\sqrt{p})\right)^{1/2} \tag{20}$$

with $a_4$ as in Proposition 1.

The choice of sign was determined by Matthews [7] and is described in [4]. When $q = p^2$ (chiefly, Theorem 12.10.4 in [4]),

$$g_2(4) = \begin{cases} -p - 2(\frac{2}{p})a_4\sqrt{p} & \text{if } p \equiv 1 (\mathrm{mod}\ 4) \\ -p + 2(-1)^{(p+1)/4} p & \text{if } p \equiv 3 (\mathrm{mod}\ 4) \end{cases} \tag{21}$$

with $a_4$ as in Proposition 1. When $e = 8$ or 12, we will require only the values of $g_2(e)$ for our purposes, but not those of $g(e)$. For $e = 8$, one has (chiefly, Theorem 12.7.1 in [4])

$$g_2(8) = \begin{cases} -p - 2a_4\sqrt{p} \pm 2a_8(2p + 2a_4\sqrt{p})^{1/2} & \text{if } p = 8k+1 \\ -p - 4ia_8\sqrt{p} & \text{if } p = 8k+3 \\ -p + 2a_4\sqrt{p} \pm 4ib_4 p(2p + 2a_4\sqrt{p})^{-1/2} & \text{if } p = 8k+5 \\ 3p - 4(-1)^k p & \text{if } p = 8k+7 \end{cases} \tag{22}$$

with $a_4$, $b_4$ and $a_8$ as in Proposition 1.

For $e = 12$ one has (chiefly, Theorems 4.8-10 in [3])

$$g_2(12) = \begin{cases} g_2(6) - 2(-1)^k a_4\sqrt{p} - 2(-1)^k a_{12}\frac{g(3)^2 - 2p}{\sqrt{p}} & \text{if } p = 12k+1 \\ -p + 2(-1)^k\sqrt{p}(a_4 + 2\epsilon b_4) & \text{if } p = 12k+5 \\ g_2(6) + 2(-1)^{k+1} g_2(3) + 2(-1)^k p & \text{if } p = 12k+7 \\ 5p - 6(-1)^k p & \text{if } p = 12k+11, \end{cases} \tag{23}$$

with $a_4$, $b_4$, $a_{12}$ and $\epsilon$ as in Proposition 1.

## 3 Factors $F^{(w)}(x)$ of $F(x)$ for $e = 6, 8, 12$

Our interest here is to explicitly evaluate the factors $F^{(w)}(x)$ of the period polynomial $F(x)$ for the finite field $\mathbf{F}_{p^2}$ for $e|24$ (or at least give a zero of each irreducible factor.) To simplify the discussion we shall give complete results for the values $e = 6, 8$ and 12. We assume $q = p^2$ here and throughout the remainder of the paper.

For $e = 6$ the factors of $F(x)$ are given by

**Proposition 3.1.** *(i) If $p = 6k+5$, then $F(x) = (x+p)^5(x-5p)$ (ii) If $p = 6k+1$, then $F(x) = (x^3+3px^2+p(3p-12a_3^2)x+(p^3-12a_3^2p^2+8a_3^2pr_3))\cdot(x^3-3px^2+p(3p-36b_3^2)x-(p^3-36b_3^2p^2+72b_3^3s_3p))$, with $b_3s_3$ as in (15).*

*Proof.* When $p = 6k+5$, $\delta = 6$ and Prop. 5 of [5, p. 322] applies. If $k$ is even, $H_3 = 5p$ and $H_i = -p$ for $i \neq 3$. If $k$ is odd, $H_6 = 5p$ and $H_i = -p$ for $i \neq 6$. Thus statement (i) follows.

When $p = 6k+1$, $\delta = 2$ so $F = F^{(1)}\cdot F^{(2)}$. To verify (ii) when $k$ is odd, first note from (16) and (19) that $H_6 = g_2(6)$ is a conjugate of $p-2b_3(g(3)'-g(3)'')$ and $H_3 = 2g_2(3)-g_2(6)$ is a conjugate of $-p-2a_3g(3)$ so $F^{(1)}(x) = x^3+3px^2+p(3p-12a_3^2)x+(p^3-12a_3^2p^2+8a_3^3pr_3)$ and $F^{(2)}(x) = x^3-3px^2+p(3p-36b_3^2)x-(p^3-36b_3^2p^2+72b_3^3ps_3)$ where $s_3b_3$ satisfies (15). Now, if $k$ is even, note from (16) and (18) that $H_6 = g_2(6)$ is a conjugate of $-p-2a_3g(3)$ and $H_3 = 2g_2(3)-g_2(6)$ is a conjugate of $p-2b_3(g(3)'-g(3)'')$ so the expressions just given for $F^{(1)}(x)$ and $F^{(2)}(x)$ are interchanged.

For $e = 8$, the factors of $F(x)$ are given by

**Proposition 3.2.** *(i) If $p = 8k+7$ then $F(x) = (x+p)^7(x-7p)$*

*(ii) If $p = 8k+5$ then $F = F^{(1)}\cdot F^{(2)}$, where $F^{(1)}(x) = (x^2-2px+p(p-4b_4^2))^2$ and $F^{(2)}(x) = x^4+4px^3+p(22p-8a_4^2)x^2+4p^2(9p-20a_4^2)x+p^2(9p-4a_4^2)^2$*

*(iii) If $p = 8k+3$ then $F = F^{(1)}F^{(2)}F^{(3)}F^{(4)}$, where $F^{(1)}(x) = F^{(3)}(x) = x^2+2px+p(p+16b_8^2)$, $F^{(2)}(x) = (x-3p)^2$ and $F^{(4)}(x) = x^2+2px+p(p+16a_8^2)$.*

*(iv) If $p = 8k+1$ then $F = F^{(1)}\cdot F^{(2)}$, where $F^{(1)}(x) = x^4-4px^3+(6p^2-8b_4^2p-32b_8^2p)x^2-(4p^3-16p^2b_4^2-64p^2b_8^2+128b_4^2b_8^2p)x+(p^4-8b_4^2p^3+16b_4^4p^2+256b_8^4p^2-32b_8^2p^3-256b_4^2b_8^4p)$*

*$F^{(2)}(x) = x^4+4px^3+(6p^2-8a_4^2p-16a_8^2p)x^2+(4p^3-16p^2a_4^2-32p^2a_8^2+64a_4^2a_8^2p)x+(p^4-8a_4^2p^3+16a_4^4p^2+64a_8^4p^2-16a_8^2p^3-64a_4^2a_8^4p)$.*

*(Here $p = a_4^2+b_4^2 = a_8^2+2b_8^2$ as applicable, with $a_4 \equiv (-1)^k \pmod 4$ and $a_8 \equiv (-1)^k \pmod 4$.)*

*Proof.* When $p = 8k+7$, $\delta = 8$ and Prop. 5 of [5, p. 322] again applies. If $k$ is even $H_4 = 7p$ and $H_i = -p$ for $i \neq 4$. If $k$ is odd, $H_8 = 7p$ and $H_i = -p$ for $i \neq 8$. Statement (i) follows.

When $p = 8k+5$, $\delta = 2$ so $F = F^{(1)}\cdot F^{(2)}$. To verify (ii) first note that $H_8 = g_2(8) = -p+2a_4\sqrt{p} \pm 4ib_4p(2p+2a_4\sqrt{p})^{-1/2}$ from (22). Also, from

(9) and the fact that $G_2(\chi) = G_2(\chi^5)$ here for any $\chi$ of order dividing 8, $H_1 = -iG_2(\chi_2) - G_2(\chi_4) + iG_2(\chi_6)$ which equals $-i(a_4+ib_4)\sqrt{p}+p+i(a_4-ib_4)\sqrt{p} = p + 2b_4\sqrt{p}$, since $\chi_2^*(2) = -1$, $E_2(\chi_2) = a_4 + ib_4$ and $G(\chi_2) = \sqrt{p}$ from (7) and Proposition 1. A routine calculation now yields the expression for the factors $F^{(1)}$ and $F^{(2)}$ in (ii).

When $p = 8k+3$, $\delta = 4$ so $F = F^{(1)}F^{(2)}F^{(3)}F^{(4)}$. To verify (iii) first note that $H_8 = g_2(8) = -p - 4ia_8\sqrt{p}$ from (22). Also, from (9) and the fact that $G_2(\chi) = G_2(\chi^3)$ for any $\chi$ of order dividing 8 here, $H_3 = H_1 = -i\sqrt{2}G_2(\chi_1) - G_2(\chi_4) + i\sqrt{2}G_2(\chi_7)$ equals $-p - 4ib_8\sqrt{p}$ by Proposition 1 and (7) as $G(\chi_1) = i\sqrt{p}$ and $E_2(\chi_1) = -a_8 - ib_8\sqrt{2}$ here. In addition, from (9) $H_2 = -G_2(\chi_2) + G_2(\chi_4) - G_2(\chi_6) = 3p$ by (7). A routine calculation now yields the expressions for $F^{(1)} = F^{(3)}$, $F^{(2)}$ and $F^{(4)}$ given in (iii).

When $p = 8k+1$, $\delta = 2$ so $F = F^{(1)} \cdot F^{(2)}$. To verify (iv), first note that $H_8 = g_2(8) = -p - 2a_4\sqrt{p} \pm 2a_8(2p + 2a_4\sqrt{p})^{1/2}$ from (22). Also, from (9) one has $H_1 = \zeta_8^{-1}G_2(\chi_1) + \zeta_4^{-1}G_2(\chi_2) + \zeta_8^{-3}G_2(\chi_3) - G_2(\chi_4) + \zeta_8^{3}G_2(\chi_5) + \zeta_4 G_2(\chi_6) + \zeta_8 G_2(\chi_7)$, which equals

$$\zeta_8^{-1}\chi_1(2)(-a_8 - ib_8\sqrt{2})G(\chi_1) + \zeta_8^{-3}\chi_1(2)(-a_8 - ib_8\sqrt{2})G(\bar{\chi}_1)$$
$$-\zeta_8^{-1}\chi_1(2)(-a_8 + ib_8\sqrt{2})G(\chi_1) - \zeta_8^{-3}\chi_1(2)(-a_8 + ib_8\sqrt{2})G(\bar{\chi}_1) - 2b_4\sqrt{p} +$$

$p$, by (7) and Proposition 1 as $\chi_1(2) = \pm 1$ here with $E_2(\chi_1) = -\chi_1(2)(a_8 + ib_8\sqrt{2})$ and $E_2(\chi_2) = -a_4 - ib_4$ . Now

$$\zeta_8^{-1}\chi_1(2)(-a_8 - ib_8\sqrt{2})G(\chi_1) + \zeta_8^{-3}\chi_1(2)(-a_8 - ib_8\sqrt{2})G(\bar{\chi}_1)$$
$$-\zeta_8^{-1}\chi_1(2)(-a_8 + ib_8\sqrt{2})G(\chi_1) - \zeta_8^{-3}\chi_1(2)(-a_8 + ib_8\sqrt{2})G(\bar{\chi}_1)$$

equals $-2\sqrt{2}\zeta_8^{-1}b_8(iG(\chi_1) + G(\bar{\chi}_1))$ whose square is $-8ib_8^2(-G(\chi_1)^2 + G(\bar{\chi}_1)^2 + 2iG(\chi_1)G(\bar{\chi}_1)) = -8ib_8^2(-2ib_4\sqrt{p} + 2ip) = 16b_8^2(p - b_4\sqrt{p})$ since $G(\chi_1)^2 - G(\bar{\chi}_1)^2 = 2ib_4\sqrt{p}$ as $p \equiv 1 \pmod 8$ (chiefly, Berndt and Evans [3 ,p. 431].) Thus up to conjugate, $H_1$ equals $p - 2b_4\sqrt{p} + \sqrt{p - b_4\sqrt{p}}$. A routine calculation now gives the expressions for $F^{(1)}$ and $F^{(2)}$ in (iv).

For $e = 12$ the factors of $F(x)$ are given by

**Proposition 3.3.** *(i) If $p = 12k + 11$ then $F(x) = (x+p)^{11}(x - 11p)$.*

*(ii) If $p = 12k+5$ then $F = F^{(1)}F^{(2)}F^{(3)}F^{(4)}F^{(5)}F^{(6)}$ where*
*$F^{(1)}(x) = F^{(2)}(x) = F^{(4)}(x) = F^{(5)}(x) = (x+p)^2 - 4p(a_4 - \epsilon b_4)^2$,*
*$F^{(3)}(x) = (x-5p)^2 - 4p(2a_4 + b_4\epsilon)^2$, and $F^{(6)}(x) = (x+p)^2 - 4p(a_4 + 2\epsilon b_4)^2$.*

*(iii) If $p = 12k+7$ then $F = F^{(1)}F^{(2)}F^{(3)}F^{(4)}$, where if $k$ is odd,*
*$F^{(1)}(x) = F^{(3)}(x) = x^3 + 3px^2 + p(3p - 12a_3^2)x + (p^3 - 12a_3^2p^2 + 8a_3^3pr_3)$*
*$F^{(2)}(x) = x^3 - 9px^2 + (27p^2 - 12pa_3^2)x - (27p^3 - 36p^2a_3^2 + 8pr_3a_3^3)$*
*$F^{(4)}(x) = x^3 + 3px^2 - p(9a_3^2 + 135b_3^2)x + (8a_3^3pr_3 + 144a_3^2b_3s_3p - 12a_3^2p^2 - 144b_3^2p^2 - 288pr_3a_3b_3^2 - 576ps_3b_3^3 + p^3)$ where $b_3s_3$ satisfies (15).*

*(If $k$ is even, then the expressions for the factors $F^{(2)}(x)$ and $F^{(4)}(x)$ are interchanged.)*

*(iv) If $p = 12k+1$ then $F = F^{(1)} \cdot F^{(2)}$, where $F^{(2)}(x)$ is satisfied by $-p - 2a_3g(3) - 2(-1)^k\sqrt{p}(a_4 - a_{12}) - 2(-1)^k a_{12}(g(6) - g(3))$ and $F^{(1)}(x)$ is satisfied by $-p - 2a_3g(3) - 2(-1)^k b_{12}(g(6) - g(3)) + 2(-1)^k(b_{12} + b_4)\sqrt{p}$.*

*Here $a_3$, $b_3$, $r_3$, $s_3$, $a_4$, $b_4$, $a_{12}$, $b_{12}$ and $\epsilon$ are determined by the conditions in Proposition 1.*

*Proof.* When $p = 12k+11$, Prop. 5 of [5,p. 322] again applies to give (i) with $H_6 = 11p$ and $H_i = -p$ for $i \neq 6$ when $k$ is even, or $H_{12} = 11p$ and $H_i = -p$ for $i \neq 12$ when $k$ is odd.

When $p = 12k + 5$, $\delta = 6$ so $F = F^{(1)}F^{(2)}F^{(3)}F^{(4)}F^{(5)}F^{(6)}$. From (23), $H_{12}$ is a conjugate of $-p + 2\sqrt{p}(a_4 + 2\epsilon b_4)$. From (9) and the fact that $G_2(\chi) = G_2(\chi^5)$ for any character $\chi$ of order dividing 12 here, $H_9 = -i(G_2(\chi^3) - G_2(\chi^9)) - G_2(\chi^6) + 2i(G_2(\chi_1) - G_2(\chi_{11})) + 2(G_2(\chi_4) - G_2(\chi_2))$ which equals $-i(-1)^k(2ib_4\sqrt{p}) - (-p) + 2i(-1)^{k+1}(2ia_4\epsilon\sqrt{p}) + 2(2p) = 5p$ $+2(-1)^k\sqrt{p}(2a_4\epsilon + b_4)$ from (7) and Proposition 1. In addition, from (9)

$$H_8 = H_4 = (\zeta_3^{-1} + \zeta_3)G_2(\chi_1) + (\zeta_3 + \zeta_3^{-1})G_2(\chi_2) + (\zeta_3^{-1} + \zeta_3)G_2(\chi_4)$$

$$+(\zeta_3^{-1} + \zeta_3)G_2(\chi_{11}) + (G_2(\chi_3) + G_2(\chi_9)) + G_2(\chi_6)$$

which equals

$$-(G_2(\chi_1) + G_2(\chi_{11})) - G_2(\chi_2) - G_2(\chi_4) + (G_2(\chi_3) + G_2(\chi_9)) + G_2(\chi_6)$$

or $-(-1)^k2\epsilon b_4\sqrt{p} - (-p) - p + (-1)^k(2a_4\sqrt{p}) - p = -p + 2(-1)^k\sqrt{p}(a_4 - \epsilon b_4)$ from (7) and Proposition 1. Finally from (9),

$$H_5 = H_1 = (\zeta_{12}^{-1} + \zeta_{12}^{-5})G_2(\chi_1) + (\zeta_{12}^{5} + \zeta_{12})G_2(\chi_{11}) + (\zeta_{12}^{-2} + \zeta_{12}^{2})G_2(\chi_2)$$

$$+(\zeta_{12}^{-4} + \zeta_{12}^{4})G_2(\chi_4) - i(G_2(\chi_3) - G_2(\chi_9)) - G_2(\chi_6)$$

which equals

$$-i(G_2(\chi_1) - G_2(\chi_{11})) + G_2(\chi_2) - G_2(\chi_4) - i(G_2(\chi_3) - G_2(\chi_9)) - G_2(\chi_6)$$

or $i(-1)^k(2ia_4\epsilon\sqrt{p}) - p - p - i((-1)^k2ib_4\sqrt{p}) - (-p) = -p + 2(-1)^k\sqrt{p}(b_4 - \epsilon a_4)$ from (7) and Proposition 1. Now a routine calculation yields the expressions for the factors of $F(x)$ in (ii).

When $p = 12k + 7$, $\delta = 4$, so $F = F^{(1)}F^{(2)}F^{(3)}F^{(4)}$. From (22), $H_{12} = g_2(6) + 2(-1)^{k+1}g_2(3) + 2(-1)^kp$ and from (9),

$$H_6 = -(G_2(\chi_1) + G_2(\chi_3) + \cdots + G_2(\chi_{11})) + (G_2(\chi_2) + G_2(\chi_4) + \cdots + G_2(\chi_{10}))$$

$= -g_2(12) + 2g_2(6) = g_2(6) + 2(-1)^kg_2(3) + 2(-1)^{k+1}p$. For $k$ odd, one finds from (16) and (19) that $H_{12}$ is a conjugate of $p - 2b_3(g(3)' - g(3)'') +$ $2(-a_3g(3) - b_3(g(3)' - g(3)'') - 2p = -p - 2a_3g(3) - 4b_3(g(3)' - g(3)'')$, and $H_6$ is a conjugate of $p - 2b_3(g(3)' - g(3)'') - 2(-a_3g(3) - b_3(g(3)' - g(3)'')) + 2p$ $= 3p + 2a_3g(3)$. A routine calculation yields the expressions for the factors $F^{(2)}(x)$ and $F^{(4)}(x)$ appearing in (iii), where $b_3s_3$ satisfies (15). If $k$ is even, the expressions for the factors $F^{(2)}(x)$ and $F^{(4)}(x)$ are interchanged since (19) again applies (noting the relation between the values of $k$ in $p = 12k+7$ and $p = 6k + 1$.) Finally, whether $k$ is odd or even, one finds from (9) and the fact that $G_2(\chi) = G_2(\chi^7)$ for any character $\chi$ of order dividing 12 here, that

$$H_3 = -G_2(\chi_2) + G_2(\chi_4) - G_2(\chi_6) + G_2(\chi_8) - G_2(\chi_{10})$$

equals $2g_2(3) - g_2(6)$ which is conjugate to $-p - 2a_3g(3)$ as in the proof of Proposition 2. In particular,

$$F^{(1)}(x) = F^{(3)}(x) = x^3 + 3px^2 + p(3p - 12a_3^2)x + (p^3 - 12a_3^2p^2 + 8a_3^3pr_3).$$

When $p = 12k + 1$, $\delta = 2$ so $F = F^{(1)} \cdot F^{(2)}$ in (3). In view of (17),

$$\frac{1}{\sqrt{p}}(g(3)^2 - p) = g(6) - g(3), g(6)' - g(3)' \text{ or } g(6)'' - g(3)'' \tag{24}$$

according as $\alpha = 0, 1$ or $-1$, where $\chi_1^*(4) = \zeta_3^\alpha$. In particular, $H_{12} = g_2(12)$ in (23) is a conjugate of $-p - 2a_3g(3) - 2(-1)^k(a_4 - a_{12})\sqrt{p} - 2(-1)^k a_{12}(g(6) - g(3))$ from (18) and (24). From (9), one has

$$H_3 = -i(G_2(\chi_1) + G_2(\chi_5)) + i(G_2(\chi_7) + G_2(\chi_{11})) + i(G_2(\chi_3) - G_2(\chi_9))$$

$$+(-G_2(\chi_2) + G_2(\chi_4) - G_2(\chi_6) + G_2(\chi_8) - G_2(\chi_{10}))$$

which equals

$$2(-1)^{k+1}b_{12}\frac{g(3)^2 - 2p}{\sqrt{p}} + 2(-1)^k b_4\sqrt{p} + 2g_2(3) - g_2(6)$$

from Proposition 1 since

$$G_2(\chi_1) = \bar{\chi}_1(2)(-a_{12} - ib_{12})(\frac{\bar{\chi}_1(4)}{2\sqrt{p}}(g(3)^2 - 2p + i\nu g(3)\sqrt{4p - g(3)^2})),$$

$$G_2(\chi_5) = \chi_1(2)(-a_{12} - ib_{12})(\frac{\chi_1(4)}{2\sqrt{p}}(g(3)^2 - 2p - i\nu g(3)\sqrt{4p - g(3)^2})),$$

$$G_2(\chi_7) = \bar{\chi}_1(2)(-a_{12} + ib_{12})(\frac{\bar{\chi}_1(4)}{2\sqrt{p}}(g(3)^2 - 2p + i\nu g(3)\sqrt{4p - g(3)^2})),$$

and

$$G_2(\chi_{11}) = \chi_1(2)(-a_{12} + ib_{12})(\frac{\chi_1(4)}{2\sqrt{p}}(g(3)^2 - 2p - i\nu g(3)\sqrt{4p - g(3)^2}))$$

from Lemma 4.1.3 in [4]. In particular, $H_3$ is a conjugate of $p - 2b_3(g(3)' - g(3)'') + 2(-1)^k(b_4 + b_{12})\sqrt{p} - 2(-1)^k b_{12}(g(6) - g(3))$ from (24), (16) and (18).

The result in (ii) of Proposition 4 that $F^{(1)} = F^{(2)} = F^{(4)} = F^{(5)}$ when $p = 12k + 5$ was a surprise, since the only repetition predicted by the Frobenius action (11) should be $F^{(1)} = F^{(5)}$ and $F^{(2)} = F^{(4)}$. In the next section, we shall apply the results found here to explain instances of additional reducibility for the factor $F^{(4)}(x)$ when $e = 24$ and $p = 24k + 7$ with $k$ even; thereby settling a question posed by the author in [Example 2, 6]. But for these two exceptions, the patterns of duplications and reducibility among the factors of $F^{(w)}(x)$ in (3) when $e|24$ are precisely those predicted as a consequence of the Frobenius action (11).

# 4 Explanation of Some Unexpected Reducibility

In previous work [5,6], I have found certain patterns of reducibility and duplication among the factors $F^{(w)}(x)$ of the period polynomial $F(x)$ in (3) that were not simply the result of the Frobenius action $x \to x^p$ in $\mathbf{F}_q$. For the case $q = p^2$ studied here, $F^{(w)}(x)$ was found to be irreducible except possibly when $w = \delta/2$, $e/\delta > 1$ and $R \equiv 1(\text{mod } 2e/\delta)$ where $R = (p+1)/\delta$ (see [6]). There I found a rather curioius example of the exceptional case, where the middle factor $F^{(\delta/2)}(x)$ sometimes split, but for which I had no explanation. Namely, set $e = 24$ and consider primes of the form $p = 24k + 7$ with $k$ even. Then $\delta = (24, p+1) = 8$ so $e/\delta = 3$, and $R = (p+1)/8 = 3k+1 \equiv 1(\text{mod } 6)$ since $k$ is even. The middle factor $F^{(4)}(x)$ is computed below for the first few primes $p \equiv 7(\text{mod } 48)$

$p = 7,\quad F^{(4)}(x) = (x-49)^3$

$p = 103,\ F^{(4)}(x) = x^3 - 2163x^2 + 1554579x - 371155041$

$p = 151,\ F^{(4)}(x) = (x-1057)^3$

To explicitly compute $H_{12}$ using the results of the previous sections, I obtain the following criteria for the reducibility of $F^{(4)}(x)$. Write $p = 24k+7$ above in the form $p = a_3^2 + 3b_3^2 = a_{24}^2 + 6b_{24}^2$ where $a_3 \equiv -1$ and $a_{24} \equiv 1(\text{mod } 3)$, and express $4p = r_3^2 + 3s_3^2$ with $r_3 \equiv 1$, $s_3 \equiv 0(\text{mod } 3)$. Then

**Proposition 4.1.** *For the situation at hand, the middle factor $F^{(4)}(x)$ equals*

$$(x-7p)^3 \text{ or } x^3 - 21px^2 + p(147p - 12A^2)x - p(343p^2 - 84A^2p + 8A^3r_3)$$

*according as $a_3 = 2a_{24}$ or not, where $A = \frac{1}{12}(a_3 - 2a_{24})$ is an integer.*

*Proof.* First note that since $p \equiv 7(\text{mod } 16)$, congruence considerations imply that $2||a_3$ and $a_{24}$ is odd, so $12|(a_3 - 2a_{24})$ confirming that $A$ is an integer. Proceeding to compute $H_{12}$, one has from (9) that

$$H_{12} = \sum 1_{j=1}^{23}(-1)^j G_2(\chi_j) = -g_2(24) + 2g_2(12) = g_2(12) - g_2(8) + g_2(4) - S$$

since $g_2(24) = g_2(12) + g_2(8) - g_2(4) + S$, where $S = \sum_{j=1,j\ odd}^{24} G_2(\chi_j) = 4Re\{G_2(\chi_1) + G_2(\chi_5)\}$ as $G_2(\chi_1) = G_2(\chi_7)$ and $G_2(\chi_5) = G_2(\chi_{11})$ from (11).

Now from (7) and Theorem 4.11 in [3], one finds $G_2(\chi_1) = \bar{\chi}_1(2)G(\chi_1)(a_{24} + ib_{24}\sqrt{6})$ and $G_2(\chi_5) = \chi_1(2)G(\bar{\chi}_1)(a_{24} + ib_{24}\sqrt{6})$. Thus $S = 4a_{24}(\bar{\chi}_1(2)G_1(\chi_1) + \chi_1(2)G_1(\bar{\chi}_1)) = 4a_{24}g(3), 4a_{24}g(3)'$ or $4a_{24}g(3)''$ as $\chi_1(2) = \zeta_3^\alpha$ with $\alpha = 0$, 1 or $-1$ by (14). In particular, $H_{12}$ is a conjugate of $7p + 2g(3)(a_3 - 2a_{24})$ and satisfies

$$x^3 - 21px^2 + (147p^2 - 12pA^2)x - (343p^3 - 84A^2p^2 + 8A^3pr_3) = 0$$

where $A = \frac{1}{12}(a_3 - 2a_{24})$. This polynomial is irreducible if $a_3 \neq 2a_{24}$ and otherwise equals $(x - 7p)^3$.

Those primes for which $F^{(4)}(x)$ splits in Proposition 5 have a rather interesting characterization in terms of representation by a certain norm form. We close this discussion with a description of that characterization, but first require the following elementary fact concerning the ternary form $X^2 + Y^2 - 2Z^2$.

**Lemma 4.2.** *The positive integral solutions of $X^2+Y^2 = 2Z^2$ with $(X,Y,Z) = 1$ and $X \neq Y$ are parametrized by $X = |s^2 - 2st - t^2|, Y = |s^2 + 2st - t^2|$ and $Z = s^2 + t^2$ for positive integers $(s,t) = 1$ with $s \not\equiv t \pmod 2$.*

*Proof.* First note that for any given parametrization of the form above, the equation $X^2 + Y^2 = 2Z^2$ is clearly satisfied. Now it is well-known that the positive integral solutions of $A^2+B^2 = C^2$ with $(A, B, C) = 1$ and $B$ even are parametrized by $A = s^2 - t^2$, $B = 2st$ and $C = s^2 + t^2$ for positive integers $(s,t) = 1$ with $s \not\equiv t \pmod 2$ and $s > t$. Now if $X$, $Y$ and $Z$ are positive integers satisfying $X^2 + Y^2 = 2Z^2$ with $(X,Y,Z) = 1$ and $X < Y$, then $X, Y$ and $Z$ are odd with $(X + Y)/2)^2 + ((-X + Y)/2)^2 = Z^2$. From the above characterization of the solutions of $A^2 + B^2 = C^2$, if $X \equiv Y \pmod 4$, then $(X + Y)/2 = s^2 - t^2$, $(-X + Y)/2 = 2st$ and $Z = s^2 + t^2$ for positive integers $(s,t) = 1$ with $s > t$ and $s \not\equiv t \pmod 2$, yielding $X = s^2 - 2st - t^2$ and $Y = s^2 + 2st - t^2$. If $X \equiv -Y \pmod 4$ then $(X+Y)/2 = 2st, ((-X+Y)/2 = s^2 - t^2$ and $Z = s^2 + t^2$ for positive integers $(s,t) = 1$ with $s > t$ and $s \not\equiv t \pmod 2$, yielding $X = -s^2 + 2st + t^2$ and $Y = s^2 + 2st - t^2$. In either case, $X, Y$ and $Z$ are of the desired form.

On the other hand, if $X, Y$ and $Z$ are positive integers satisfying $X^2 + Y^2 = 2Z^2$ with $(X,Y,Z) = 1$ and $X > Y$, then reasoning as above one finds that $X = s^2 + 2st - t^2$ and $Y = s^2 - 2st - t^2$ if $X \equiv Y \pmod 4$ or $X = s^2 + 2st - t^2$ and $Y = -s^2 + 2st + t^2$ if $X \equiv -Y \pmod 4$ for positive integers $(s,t) = 1$ with $s > t$ and $s \not\equiv t \pmod 2$. Upon interchanging the role of $s$ and $t$, one finds that the positive integer solutions of $X^2 + Y^2 = 2Z^2$ with $(X,Y,Z) = 1$ and $X > Y$ are parametrized by $X = t^2 + 2st - s^2$, $Y = t^2 - 2st - s^2$ and $Z = s^2 + t^2$ if $X \equiv Y \pmod 4$ or $X = t^2 + 2st - s^2$, $Y = -t^2 + 2st + s^2$ and $Z = s^2 + t^2$ if $X \equiv -Y \pmod 4$ for positive integers $(s,t) = 1$ with $s < t$ and $s \not\equiv t \pmod 2$. Again these are of the desired form.

**Proposition 4.3.** *For the situation at hand, $F^{(4)}(x)$ splits if and only if $p$ is integrally represented by the decomposable norm form $N_{\mathbf{Q}(\sqrt{2},\sqrt{-3})/\mathbf{Q}}((1-\zeta_3\sqrt{2})s + (\sqrt{-3}-\sqrt{2}\zeta_3^2)t) = 7s^4 - 4s^3t + 14s^2t^2 + 4st^3 + 7t^4$.*

*Proof.* The result is clear if $p = 7$, so we may assume $p > 7$. Suppose $F^{(4)}(x)$ splits, so $a_3 = 2a_{24}$ by Proposition 5. Then $p = 4a_{24}^2 + 3b_3^2 = a_{24}^2 + 6b_{24}^2$ with $a_{24} \equiv 1 (\mathrm{mod}\ 3)$, so $a_{24}^2 + b_3^2 = 2b_{24}^2$ and $|a_{24}| \neq |b_3|$ since $p > 7$. From the lemma, one finds that $a_{24} = \pm(s^2 - 2st - t^2)$, $b_{24} = \pm(s^2 + t^2)$ and $b_3 = \pm(s^2 + 2st - t^2)$ for some positive integers $(s, t) = 1$ with $s \not\equiv t (\mathrm{mod}\ 2)$. In particular, $p = a_{24}^2 + 6b_{24}^2 = 7s^4 - 4s^3t + 14s^2t^2 + 4st^3 + 7t^4$ as desired. For the reverse implication suppose $p > 7$ equals $7s^4 - 4s^3t + 14s^2t^2 + 4st^3 + 7t^4$ for some integers $s$ and $t$. Set $a_{24} = \epsilon|s^2 - 2st - t^2|$, $b_{24} = \pm(s^2 + t^2)$, $b_3 = \pm(s^2 + 2st - t^2)$ and $a_3 = 2\epsilon|s^2 - 2st - t^2|$, where $\epsilon = \pm 1$ chosen so $a_{24} \equiv 1 (\mathrm{mod}\ 3)$. Then $p = a_{24}^2 + 6b_{24}^2 = a_3^2 + 3b_3^2$ with $a_3 \equiv -1$ and $a_{24} \equiv 1 (\mathrm{mod}\ 3)$ as in Proposition 5. Since $a_3 = 2a_{24}$, $F^{(4)}(x)$ splits.

## References


1. Z. Borevich and I. Shafarevich. *Number Theory*, Academic Press, New York, (1966).
2. B.C Berndt and R.J. Evans, "Sums of Gauss, Jacobi and Jacobsthal", *Journal No. Theory* **11** (1979), 349-398.
3. B.C. Berndt and R.J. Evans, "Sums of Gauss, Eisenstein, Jacobi, Jacobsthal and Brewer", *Ill. Journal of Math.* **23**, no. 3 (1979), 374-437.
4. B.C. Berndt, R.J. Evans, and K.S. Williams, *Gauss and Jacobi Sums*, Wiley, New York, (1997).
5. S. Gurak, "Factors of period polynomials for finite fields, I," *Contemporary Math.*, Amer. Math. Soc., **166** (1994), 309-333.
6. S. Gurak,"Factors of period polynomials for finite fields, II," *Contemporary Math.*, Amer. Math. Soc., **168** (1994), 127-138.
7. C.R. Matthews, "Gauss sums and elliptic functions I, II," *Invent. Math.* **52** (1979), 163-185; **54** (1979), 23-52.
8. G. Myerson, "Period polynomials and Gauss sums for finite fields," *Acta Arith.* **39** (1981), 251-264.

# Universal Generators for Primary Closures of Galois Fields

Dirk Hachenberger

Institut für Mathematik der Universität Augsburg
D-86135 Augsburg
Germany
e-mail: hachenberger@math.uni-augsburg.de

Dedicated to Heinz Lüneburg on the occasion of his 65th birthday.

**Abstract** If $\bar{F}$ is an algebraic closure of a Galois field $F$, then for each integer $n \geq 1$ there is exactly one subfield $E_n$ of $\bar{F}$ containing $F$ and having degree $n$ over $F$. For a prime number $r$, we consider the $r$-primary closure $\bar{F}_r := \bigcup_{m \geq 0} E_{r^m}$ over $F$ and prove, under the assumption that $r \geq 7$, but without any restriction on the cardinality $q$ of $F$, the existence of a *universal generator* for $\bar{F}_r$ over $F$: this is a sequence $w = (w_{r^m})_{m \geq 0}$ in $\bar{F}_r$ which satisfies all the following properties:

(1) $w_{r^m}$ is a primitive element of $E_{r^m}$ (for all $m \geq 0$),
(2) $w_{r^m}$ generates a normal basis for $E_{r^m}$ over $F$ (for all $m \geq 0$),
(3) $w$ is norm-compatible,
(4) $w$ is trace compatible.

We prove furthermore that (2) can be strengthened to

($2^c$) $w_{r^m}$ is completely free in $E_{r^m}$ over $F$ (for all $m \geq 0$),

which means that $w_{r^m}$ simultaneously generates a normal basis for $E_{r^m}$ over $E_{r^i}$ for all $i = 0, 1, ..., m$, whence $w$ is called a *complete universal generator* for $\bar{F}_r$ over $F$. The results establish a (complete) primitive normal basis theorem for $\bar{F}_r$ over $F$.

## 1 (Complete) Universal Generators

It is well-known that the multiplicative group $E^*$ of a Galois field $E$ is cyclic, i.e., free on one generator as a module over the ring of integers. If $u \in E$ is a generator of $E^*$, then $u$ is called *primitive in* $E$.

A further classical result is the *normal basis theorem*: for every extension $E/F$ of Galois fields, there exists $v \in E$ such that $v$ generates the additive group of $E$ as a module over the group algebra $FG$, i.e., $v^G = \{v^g | g \in G\}$ is a basis of $E$ as $F$-vector space (here, $G$ denotes the (cyclic) Galois group of $E/F$); $v$ is called *free* or *normal in* $E$ *over* $F$. (Unfortunately, the terminology is not consistent. Since we also work with the norm-mapping, we shall here use the term *free*.) In this generality the normal basis theorem for finite fields was first proved by Hensel [He] in 1888.

Besides their theoretical importance, primitive and free elements are interesting for some practical purposes as well, because they allow presentations of Galois fields, which are useful for applications where the arithmetic in the underlying fields has to be performed efficiently (such as the decoding for error-correcting codes, the encryption and key-exchange in public-key-cryptosystems, or the generation of pseudorandom numbers). For the basic theory, applications and the arithmetic of finite fields we refer to Jungnickel [Ju] and Lidl and Niederreiter [LiNi]. For the early history of Galois fields we refer to Lüneburg [Lü]. For the theory of normal bases we refer to Hachenberger [Ha1].

It is natural to ask, whether, for a Galois field extension $E/F$, there exists a primitive element of $E$ which is additionally free over $F$. Completing previous work of Carlitz [Ca] and Davenport [Da], the final answer was only given in 1987, when Lenstra and Schoof [LeSc] proved the *primitive normal basis theorem*, which states that this is indeed the case for every extension $E/F$.

In the present paper we are concerned with an (infinite) version of the primitive normal basis theorem for the $r$-primary closure $\bar{F}_r$ of a Galois field $F$. Before stating our main results, Theorem 1.2 and its *complete* version Theorem 1.2$^c$, we have to introduce some terminology.

Working in an algebraic closure $\bar{F}$ of a Galois field $F$, for a nonempty set $N$ of positive integers, we denote by $E_N$ the set $\{E_n | n \in N\}$ of extensions over $F$ (where $E_n$ is as in the abstract). Let $w = (w_n)_{n \in N}$ be a sequence in $\bar{F}$ such that $w_n \in E_n$ for each $n$. Then $w$ is called *norm-compatible*, if for all $l, n \in N$ with $l$ dividing $n$, the $(E_n, E_l)$-norm $\prod_{\gamma \in G_{n,l}} w_n^\gamma$ of $w_n$ is equal to $w_l$ (here $G_{n,l}$ denotes the Galois group of $E_n$ over $E_l$). Thus, norm-compatibility just means that $w$ belongs to the (multiplicative) projective limit of $E_N$ with respect to the norm-mappings of the extensions associated with $N$. Similarly, $w$ is called *trace-compatible*, if for all $l, n \in N$ with $l$ dividing $n$, the $(E_n, E_l)$-trace $\sum_{\gamma \in G_{n,l}} w_n^\gamma$ of $w_n$ is equal to $w_l$, i.e., $w$ is a member of the (additive) projective limit of $E_N$ with respect to the trace-mappings of the extensions associated with $N$.

A sequence $w = (w_n)_{n \in N}$ is called a *generator for* $E_N$, if for all $n \in N$, $F(w_n)$, the field obtained by adjoining $w_n$ to $F$, is equal to $E_n$. A generator $w$ is called *primitive for* $E_N$, if for each $n \in N$, $w_n$ is primitive in $E_n$. Finally, a generator $w$ is called *free for* $E_N$ *over* $F$, if for each $n \in N$, $w_n$ is free in $E_n$ over $F$.

**Definition 1.1.** Let $N$ be a non-empty set of positive integers, $F$ a Galois field and $E_N$ the set of finite extensions of $F$ (in $\bar{F}$) corresponding to $N$. A generator $w$ of $E_N$ which is primitive and free over $F$ as well as norm- and trace-compatible is called a *universal generator for* $E_N$ *over* $F$. □

From an algebraic point of view the most interesting sets $N$ are those for which $F_N := \cup_{n \in N} E_n$ is a subfield of $\bar{F}$ (infinite, if $N$ is infinite): a norm-compatible primitive sequence can then be seen as a primitive element of $F_N$, while a trace-compatible free sequence can be interpreted as a free element for $F_N$ over $F$ (we refer to Lenstra [Le] for a rigorous justification of the latter statement, see also Scheerhorn [Sche]). It holds that $F_N$ is a field if and only if for all $n, m \in N$ there exists an $l \in N$ such that $n$ and $m$ divide $l$. The latter is in particular the case, when $N$ is *closed*, i.e., if for any two members $n$ and $m$ of $N$, *every* divisor of the least common multiple of $n$ and $m$ is again a member of $N$. Hence, for a closed set $N$, a generator for $E_N$ contains information for *every* finite subfield of $F_N$.

It is not known at all for which closed sets there exist universal generators. In the present contribution, however, we prove the existence of such an interesting object for certain *infinite* closed sets. For a prime $r$, let $N_r$ be the closed set $\{r^m | m \geq 0\}$, whence $E_{N_r}$ is a tower of extensions over $F$ and $F_{N_r} = \bar{F}_r$ is the $r$-primary closure of $F$ (in $\bar{F}$).

**Theorem 1.2.** *Let $F$ be any Galois field and let $r \geq 7$ be any prime. Then there exists a universal generator for $E_{N_r}$ over $F$.*

We shall be able to strengthen the assertion of Theorem 1.2 even further by requiring that all members $w_{r^m}$ of the universal generator $w$ are completely free over $F$ (an element $v$ of $E_n$ is called *completely free over $F$* if $v$ is simultaneously free over $E_d$ for every divisor $d$ of $n$).

**Definition 1.3.** Let $N$ be a non-empty set of positive integers and $F$ a Galois field. A generator $w$ for $E_N$ is called *completely free for $E_N$ over $F$*, if for each $n \in N$, $w_n$ is completely free in $E_n$ over $F$. A universal generator for $E_N$ over $F$ which is completely free is called a *complete universal generator for $E_N$ over $F$.* □

**Theorem 1.2$^c$.** *Let $F$ be any Galois field and let $r \geq 7$ be any prime. Then there exists a complete universal generator for $E_{N_r}$ over $F$.*

We give an outline of the content of the present paper.

The basic technique for proving Theorem 1.2 is by induction (on the exponent $m$ of $r$). In order to accomplish the induction step (see Section 2) we consider a problem which is interesting in itself, namely, whether for a quadruple $(F, K, L, E)$ of Galois fields ($E$ an extension of $F$ and $K, L$ intermediate fields of $E/F$) and given elements $a \in K$ (free over $F$) and $b \in L$ (primitive) there exists an element $w_{a,b} \in E$ which is primitive and free over $F$, whose $(E, K)$-trace is equal to $a$ and whose $(E, L)$-norm is equal to $b$. In this context, our main result is Theorem 2.3 (from which Theorem 1.2 immediately follows). Observe that Theorem 2.3 asserts more than is necessary for proving Theorem 1.2, because we show that (for all $m$) *every* universal generator for $E_{\{1,r,r^2,\ldots,r^m\}}$ can be *extended* to one of $E_{N_r}$.

In order to prove Theorem 2.3, a sufficient criterion for the existence of an element $w_{a,b}$ as above is derived in Section 3 by means of characters and Gauss sums for $E/F$ (see Proposition 3.1). The proof of Theorem 2.3 is completed in Section 4 by applying concrete versions of that criterion to the case where $E/F$ has prime power degree and $K = L$ is the unique maximal intermediate field of that extension (see Section 4).

We have assumed that $r \geq 7$, because we seeked to obtain a result which holds independently from the cardinality $q$ of the ground field $F$. Our criterion is less effective when $r = 5$: in that case it yields only the *asymptotic* existence of universal generators, i.e., as long as $q$ is large enough. If $r = 2$ or $r = 3$ the criterion does not work at all with $K = L$ being the maximal intermediate field. However, when seeking for results with $r = 2, 3, 5$, unrestricted in $q$, one can study towers with relative degrees $8, 9, 25$, i.e., with larger gaps in the relative degrees. We shall not work out the latter here.[1]

In Section 6, we shall prove Theorem $1.2^c$. We postpone a more detailed discussion of completeness to Section 5.

Finally, in Section 7 we conclude with some remarks.

## 2 Universal Quadruples

Let $(q, k, l, n)$ be a quadruple of nonnegative integers, where $q > 1$ is a prime power and $k, l$ are proper divisors of $n$, and let $(F, K, L, E)$ be the corresponding quadruple of Galois fields, i.e., $F = \mathrm{GF}(q)$, and, in a fixed algebraic closure $\bar{F}$ of $F$, $K = E_k$, $L = E_l$ and $E = E_n$.

**Definition 2.1.** $(q, k, l, n)$ is called *universal*, if for the corresponding quadruple $(F, K, L, E)$ it holds that for **every** $b \in L$ which is primitive in $L$ and for **every** $a \in K$ which is free in $K$ over $F$, there exists an element $w = w_{a,b}$ in $E$ which satisfies all the following properties:

(1) $w$ is primitive in $E$,
(2) $w$ is free in $E$ over $F$,
(3) $\mathrm{N}_{E,L}(w)$, the $(E, L)$-norm of $w$, is equal to $b$,
(4) $\mathrm{Tr}_{E,K}(w)$, the $(E, K)$-trace of $w$, is equal to $a$. □

As norms of primitive elements are primitive and traces of free elements are free, the assumptions on $a \in K$ and $b \in L$ are necessary.

So far, the universality of quadruples has only been considered for the case where $k = l = 1$, which in view of Proposition 2.4 is the easiest instance of the problem. (In this case $K = L = F$, $a$ is any nonzero element of $F$ and $b$ is any primitive element of $F$.) First, in [CoHa], Cohen and Hachenberger have proved that, for $n \geq 9$, $(q, 1, 1, n)$ is universal for every prime power $q$; moreover, for $n = 7, 8$ the universality of $(q, 1, 1, n)$ is proved except for 8 values of $q$.[2] The main tool in [CoHa] is Proposition 3.1 of the present

[1] We refer to Hachenberger [Ha2]; in that paper, a similar problem is studied, namely the existence of trace-compatible primitive free generators for $E_{N_r}$ over $F$.

[2] In [CoHa], $(q, n)$ is called a *PFNT-pair*, if, in the present notation, $(q, 1, 1, n)$ is universal.

paper, applied to the case $k = l = 1$ (the corresponding result is not proved in [CoHa] and neither is proved a character sum formulation, which, for the general case is carried out in Section 3 of the present paper). By improving the estimates for Gauss sums and by developing a sieving technique (both are available for the case $K = L = F$, only), Cohen [Co] was able to prove the universality of $(q, 1, 1, n)$ for all $n \in \{5, 6, 7, 8\}$ and all $q$, whence altogether the following holds.

**Theorem 2.2.** *If $n \geq 5$ then $(q, 1, 1, n)$ is universal for all prime powers $q \geq 2$.* □

Since we want to prove the existence of universal generators for primary closures of Galois fields, we consider here the case where the degree of $E/F$ is a power of a prime. The main result in this direction is the following theorem (which is proved in Section 3 and Section 4).

**Theorem 2.3.** *Assume that $r \geq 7$ is a prime. Let $q > 1$ be any prime power and $m \geq 0$ be any integer. Then $(q, r^m, r^m, r^{m+1})$ is universal.*

Observe that the case $m = 0$ is covered by Theorem 2.2, and we therefore restrict our attention to the case $m \geq 1$, later. Now, Theorem 1.2 follows at once by induction from Theorem 2.3.

*Proof of Theorem 1.2.* If $w_1 \in F$ is primitive, then, trivially, $(w_n)_{n \in \{1\}}$ is a universal generator for $E_1$ over $F$. For $m \geq 0$ assume that $(w_n)_{n \in \{1, r, \ldots, r^m\}}$ is a universal generator for $E_{\{1, r, \ldots, r^m\}}$ over $F$. As $(q, r^m, r^m, r^{m+1})$ is universal, there exists $v = w_{m+1} \in E = E_{r^{m+1}}$ which is primitive in $E$, free over $F$ and satisfies $\mathrm{Tr}_{E, E_{r^m}}(v) = w_m = \mathrm{N}_{E, E_{r^m}}(v)$. The transitivity of the trace- and norm-mappings now implies that $(w_n)_{n \in \{1, r, \ldots, r^{m+1}\}}$ is a universal generator for $E_{\{1, r, \ldots, r^{m+1}\}}$ over $F$, and everything is proved. □

We finally mention that Theorem 2.3 can be strengthened as follows.

**Theorem 2.3'.** *Assume that $r \geq 7$ is a prime. Let $q > 1$ be any prime power. Then $(q, r^\alpha, r^\beta, r^\gamma)$ is universal for all $\alpha, \beta \geq 0$ and all $\gamma > max\{\alpha, \beta\}$.* □

The latter is an immediate consequence of the following proposition.

**Proposition 2.4.** *Assume that $(q, k, l, n)$ is universal. Let $k'$ be a divisor of $k$ and $l'$ a divisor of $l$. Then $(q, k', l', n)$ likewise is universal.*

*Proof.* If $a' \in E_{k'}$ is free over $F = \mathrm{GF}(q)$, then there exists an element $a$ in $E_k$, free over $F$, whose $(E_k, E_{k'})$-trace is equal to $a'$. Analogously, if $b' \in E_{l'}$ is primitive, then there exists a primitive $b \in E_l$ such that the $(E_l, E_{l'})$-norm of $b$ is equal to $b'$. Since $(q, k, l, n)$ is universal by assumption, the result follows using the transitivity of the norm- and trace-mappings. □

# 3 Character Sum Formulation and a Sufficient Criterion

In the present section we shall prove a sufficient criterion for a quadruple $(q, k, l, n)$ to be universal (see Proposition 3.1). We have to introduce some notation. Let $P = P(q, n, l)$ be the largest divisor of $q^n - 1$ which is relatively prime to $q^l - 1$ and let $\omega = \omega(P)$ be the number of distinct prime divisors of $P$. Let $t = t(q, n, k)$ be the largest monic divisor of $x^n - 1$ which is relatively prime to $x^k - 1$ and let $\Omega = \Omega(q, n, k)$ be the number of distinct monic irreducible $F$-divisors of $t$.

**Proposition 3.1.** *Assume that $(q, k, l, n)$ is as at the beginning of Section 2. If*

$$\frac{q^{\frac{n}{2}}}{q^k(q^l - 1)} > \left(2^\Omega - \frac{1}{q^k}\right) \cdot \left(2^\omega - \frac{1}{q^l - 1}\right), \tag{3.1}$$

*then $(q, k, l, n)$ is universal.*

The proof of Proposition 3.1 is based on the character-sum formulation of the four conditions in Definition 2.1 and on an estimate for Gauss sums. We let $(F, K, L, E)$ be the quadruple of fields corresponding to $(q, k, l, n)$, throughout.

We first investigate *the multiplicative part* which comprises the primitivity and the condition of prescribed $(E, L)$-norm, i.e., (1) and (3) in Definition 2.1. For simplicity let $N$ denote the $(E, L)$-norm mapping.

For a divisor $d$ of $q^n - 1$ let $C_d$ be the unique (cyclic) subgroup of the multiplicative group of $E^*$ having cardinality $d$. Let furthermore $D$ denote the cofactor of $P$ in $q^n - 1$ and let $\delta$ be the cofactor of $q^l - 1$ in $D$. Thus, $L^*$ is a subgroup of $C_D$, and, as $D$ and $P$ are relatively prime, $E^*$ decomposes into the direct product of $C_D$ with $C_P$. This decomposition is crucial for the subsequent characterization of primitive elements in $E$ having $(E, L)$-norm equal to $b$.

Since, for $x \in E$, $N(x) = x^{\delta P}$, $C_{\delta P}$ is the kernel of $N$. Restricting $N$ onto $C_D$ induces an epimorphism onto $L^*$ with kernel equal to $C_\delta$. Now, with $\varphi$ denoting the Euler totient function, it holds that $C_D$ has exactly $\varphi(D) = \varphi(q^l - 1) \cdot \delta$ generators (i.e., elements of $E$ whose multiplicative order is equal to $D$), whence for a primitive $b$ of $L^*$, the entire preimage of $b$ under $N$ in $C_D$ consists of generators of the group $C_D$. Consequently, if $w \in E^*$ with $N(w) = b$, letting $w = w_D w_P$ (with $w_D \in C_D$ and $w_P \in C_P$), we have $b = N(w_D)$, and the above argument yields that $w_D$ is a generator of $C_D$. We have proved the following.

**Lemma 3.2.** *An element $w \in E$ is primitive in $E$ if and only if the $(E, L)$-norm of $w$ is primitive in $L$ and the multiplicative order of $w$ is divisible by $P$.* □

From Cohen and Hachenberger [CoHa1] (or Carlitz [Ca], Davenport [Da], Lenstra and Schoof [LeSc]) we know that the function $M_P$ in (3.2) is the characteristic function of the set of elements of $E^*$ whose multiplicative order is divisible by $P$.

$$M_P(w) = \frac{\varphi(P)}{P} \sum_{d|P} \frac{\mu(d)}{\varphi(d)} \sum_{(\eta,d)} \eta(w), \qquad w \in E^* \tag{3.2}$$

($\mu$ denotes the Möbius function and the first sum runs over all positive divisors of $P$, while the second one runs over all $\varphi(d)$ multiplicative characters $\eta \in E^*$ whose multiplicative order is equal to $d$).

The set of elements $w \in E^*$ having $(E, L)$-norm equal to $b$ can be described by its characteristic function $N_b$ which in (3.3) is given in terms of the group $\hat{L^*}$ of multiplicative characters of $L$.

$$N_b(w) := \frac{1}{q^l - 1} \sum_{\nu \in \hat{L^*}} \nu(N(w)b^{-1}), \qquad w \in E^*. \tag{3.3}$$

Thus, combining (3.2) with (3.3), we conclude that $w \in E^*$ satisfies the conditions (1) and (3) of Definition 2.1 if and only if $M_P(w)N_b(w) = 1$.

We next investigate *the additive part*, comprising the freeness and the prescribed $(E, K)$-trace, i.e., the conditions (2) and (4) of Definition 2.1. For simplicity, let $T$ denote the $(E, K)$-trace mapping. Then, with $\hat{K}$ being the group of additive characters of $K$, the mapping $T_a$ in (3.4) is the characteristic function of the set of all $w \in E$ such that $T(w) = a$.

$$T_a(w) := \frac{1}{q^k} \sum_{\lambda \in \hat{K}} \lambda(T(w) - a), \qquad w \in E. \tag{3.4}$$

In order to cope with the freeness, we have to recall that the additive group of $E$ is equipped with a module structure over the polynomial ring $F[x]$ by defining $f \circ w := f(\sigma)(w)$ (for $w \in E$ and $f \in F[x]$), where $\sigma$ is a generator of the Galois group of $E/F$, the Frobenius automorphism for instance. The *$F$-order* of $w \in E$ is the monic polynomial $f \in F[x]$ of least degree such that $f \circ w = 0$. The group $\hat{E}$ of additive characters of $E$ likewise carries the structure of an $F[x]$-module by defining $(f \circ \chi)(w) := \chi(f \circ w)$ (where $\chi \in \hat{E}$, and $f, w$ are as above). Similarly, the $F$-order of $\chi \in \hat{E}$ is defined to be the monic polynomial $f \in F[x]$ of least degree such that $f \circ \chi = \chi_0$, the trivial additive character. As $\hat{E}$ and the additive group of $E$ are isomorphic as $F[x]$-modules (namely free on one generator with minimal polynomial $x^n - 1$), for each monic divisor $g \in F[x]$ of $x^n - 1$ there are exactly $\phi_q(g)$ characters $\chi \in \hat{E}$ whose $F$-order is equal to $g$, where $\phi_q(g)$ is the number of units in the ring $F[x]/gF[x]$ (i.e., $\phi_q$ is the $q$-analogue of the Euler totient function).

Now, compare again with [CoHa1] (or [Ca], [Da], [LeSc]), with $t$ as in Proposition 3.1, it holds that $A_t$ in (3.5) is the characteristic function of the set of elements in $E$ whose $F$-order is divisible by $t$.

$$A_t(w) := \frac{\phi_q(t)}{q^{\deg t}} \sum_{g|t} \frac{\mu_q(g)}{\phi_q(g)} \sum_{(\chi,g)} \chi(w), \tag{3.5}$$

where, the first sum runs over all monic $F$-divisors of $t$ and the symbol $(\chi, g)$ indicates that the second sum runs over all additive characters of $E$ having $F$-order equal to $g$.

The following is the additive analogue of Lemma 3.2. It implies that $w \in E$ satisfies (2) and (4) of Definition 2.1 if and only if $A_t(w)T_a(w) = 1$.[3]

**Lemma 3.3.** *An element $w \in E$ is free in $E$ over $F$ if and only if its $(E, K)$-trace is free in $K$ over $F$ and its $F$-order is divisible by $t$.*

*Proof.* Let $p$ be the characteristic of $F$ and let $\pi$ be the largest power of $p$ dividing $n/k$. Then $t$ is equal to $(x^n - 1)/(x^{k\pi} - 1)$ (which is relatively prime to $x^k - 1$). This leads to a decomposition of the additive group of $E$ into the direct sum $\tilde{K} \oplus \mathcal{T}$, where $\tilde{K}$ is the extension of degree $k\pi$ over $F$, and where $\mathcal{T}$ is the kernel of the $(E, \tilde{K})$-trace mapping. For $w \in E$, let $w_{\tilde{K}} + w_{\mathcal{T}}$ be the decomposition of $w$ corresponding to that of $E$. By Theorem 8.6 in Hachenberger [Ha1], it holds that $w$ is free over $F$ if and only if $w_{\tilde{K}}$ is free in $\tilde{K}$ over $F$ and the $F$-order of $w_{\mathcal{T}}$ is equal to $t$ (the latter means that $w_{\mathcal{T}}$ is a generator of $\mathcal{T}$ as $F[x]$-module). An application of Lemma 7.4 of [Ha1] shows that the $F$-order of the $(E, \tilde{K})$-trace of $w$ and the $F$-order of $w_{\tilde{K}}$ are equal. Finally, since $\tilde{K}/K$ has degree a power of $p$, an application of Theorem 10.5 in [Ha1] gives that $u \in \tilde{K}$ is free over $F$ if and only if the $(\tilde{K}, K)$-trace of $u$ is free in $K$ over $F$. From all that the assertion follows. □

Combining *the multiplicative and the additive part*, we obtain that the total number $Y$ of elements in $E$ which are primitive and free over $F$ with $N(w) = b$ and $T(w) = a$, i.e., satisfying all conditions in Definition 2.1, is equal to

$$Y = \sum_{w \in E} M_P(w) N_b(w) A_t(w) T_a(w). \tag{3.6}$$

(As usual, we let $\eta(0) := 0$ if $\eta$ is a nontrivial multiplicative character, and $\eta_0(0) := 1$ for the trivial multiplicative character $\eta_0$.) Letting

$$\theta := \frac{\varphi(P)}{P}, \quad \Theta_q := \frac{\phi_q(t)}{q^{\deg t}},$$

and using (3.2)-(3.5), we have

[3] Lemma 3.2 is therefore also crucial for proving Proposition 3.1 in [Ha2].

$$\frac{q^k \cdot (q^l - 1)}{\theta \Theta_q} \cdot Y = \sum_{d|P} \sum_{g|t} \frac{\mu(d)}{\varphi(d)} \frac{\mu_q(g)}{\phi_q(g)} \sum_{(\eta,d)} \sum_{(\chi,g)} \sum_{\nu \in \hat{L^*}} \sum_{\lambda \in \hat{K}} \frac{G_{a,b}(\eta\hat{\nu}, \chi\hat{\lambda})}{\nu(b)\lambda(a)}, \quad (3.7)$$

where $G_{a,b}(\eta\hat{\nu}, \chi\hat{\lambda})$ denotes the *Gauss sum*

$$G_{a,b}(\eta\hat{\nu}, \chi\hat{\lambda}) = \sum_{w \in E} (\eta\hat{\nu})(w)(\chi\hat{\lambda})(w), \quad (3.8)$$

and where $\hat{\nu} \in \hat{E}^*$ and $\hat{\lambda} \in \hat{E}$ denote the lifted characters of $\nu \in \hat{L^*}$ and $\lambda \in \hat{K}$, respectively, i.e,

$$\hat{\nu}(w) := \nu(N(w)), \quad \hat{\lambda}(w) := \lambda(T(w)), \quad w \in E.$$

The decompositions of $E^*$ and $E$ in Lemma 3.2 and Lemma 3.3, respectively, yield corresponding decompositions of the character groups $\hat{E}^*$ and $\hat{E}$. These are crucial for the analysis of (3.7), because $\eta\hat{\nu}$ is trivial if and only if $\eta$ and $\hat{\nu}$ are both trivial, and similarly, $\chi\lambda$ is trivial if and only if $\chi$ and $\hat{\lambda}$ are both trivial. We are now able to complete the proof of Proposition 3.1.[4]

*Proof of Proposition 3.1.* If $\eta\hat{\nu}$ and $\chi\hat{\lambda}$ are both trivial, then

$$G_{a,b}(\eta\hat{\nu}, \chi\hat{\lambda}) = q^n.$$

If either $\eta\hat{\nu}$ or $\chi\hat{\lambda}$ is trivial, then $G_{a,b}(\eta\hat{\nu}, \chi\hat{\lambda}) = 0$. Finally, if both $\eta\hat{\nu}$ and $\chi\hat{\lambda}$ are nontrivial, then the absolute value of $G_{a,b}(\eta\hat{\nu}, \chi\hat{\lambda})$ is equal to $q^{n/2}$.

Now, using properties of the Möbius functions as well as the triangle inequality in combination with the absolute values of the Gauss sums, we obtain

$$\left| \frac{q^k \cdot (q^l - 1) \cdot Y}{\theta \Theta_q} - q^n \right| \leq q^{\frac{n}{2}} \cdot (U + V + W + X),$$

where $U = (2^\omega - 1)(q^l - 1)(2^\Omega - 1)q^k$, $V = (2^\omega - 1)(q^l - 1)(q^k - 1)$, $W = (q^l - 2)(2^\Omega - 1)q^k$ and $X = (q^l - 2)(q^k - 1)$. Thus, $Y = 0$ implies

$$\frac{q^{n/2}}{q^k \cdot (q^l - 1)} \leq (2^\Omega - \frac{1}{q^k}) \cdot (2^\omega - \frac{1}{q^l - 1}),$$

and everything is proved. □

[4] For the basic properties of Gauss sums we refer to Section 5.2 of Lidl and Niederreiter [LiNi] or Section 7.2 of Jungnickel [Ju].

# 4 Proof of Theorem 2.3

In the present section, in order to complete the proof of Theorem 2.3, we analyse the sufficient condition in Proposition 3.1 for quadruples $(q, r^m, r^m, r^{m+1})$, where $r \geq 7$ is a prime number and where $m \geq 1$ (recall that $m = 0$ is covered by Theorem 2.2). Using the notation of the previous section, we have $t = 1$ if $r = p$ is the characteristic of $F$ =GF$(q)$, whereas $t = \Phi_{r^{m+1}}$ is the $r^{m+1}$st cyclotomic polynomial if $r \neq p$. In the following, $|\Lambda|$ denotes the cardinality of the set $\Lambda$.

**Proposition 4.1.** *Let $q > 1$ be a prime power and $r$ a prime. Let $\lambda \geq 1$ be an integer, $\Lambda$ a set of primes $s < \lambda$ such that each prime divisor of $P$ which is less than $\lambda$ is contained in $\Lambda$, and let $L(\Lambda) = \prod_{s \in \Lambda} s$ (being equal to 1 if $\Lambda$ is empty). Finally, if $r \neq p$, let $\delta = \operatorname{ord}_{r^{m+1}}(q)$; if $r = p$, let $\delta = \varphi(r^{m+1})$. If*

$$\left(\frac{r-4}{\log 4} - \frac{r-1}{\log \lambda}\right) \cdot \log q \geq \frac{r-1}{\delta} + \frac{1}{r^m} \cdot \left(|\Lambda| - \frac{\log L(\Lambda)}{\log \lambda}\right), \tag{4.1}$$

*then $(q, r^m, r^m, r^{m+1})$ is universal.*

*Proof.* With $\omega$ as in Section 3, it follows from Lemma 2.6 of Lenstra and Schoof [LeSc] that

$$\omega \leq \frac{\log P - \log L(\Lambda)}{\log \lambda} + |\Lambda|.$$

Furthermore, $\Omega = 1$ if $r = p$ and

$$\Omega = \varphi(r^{m+1})/\operatorname{ord}_{r^{m+1}}(q) \leq (r-1)r^m$$

if $r \neq p$, where $\operatorname{ord}_{r^{m+1}}(q)$ denotes the order of $q$ modulo $r^{m+1}$. Using these upper bounds, an easy calculation shows that the validity of (4.1) implies that of (3.1) (with $k = l = r^m$), whence everything is proved. □

We next return to the case $r \geq 7$. In that case, a choice of $\lambda \geq 17$ asserts that the factor

$$N(r, \lambda) := \frac{r-4}{\log 4} - \frac{r-1}{\log \lambda} \tag{4.2}$$

of $\log q$ in the left hand side of (4.1) is greater than zero. In what follows, we have always chosen $\lambda = 300$. For $r \neq p$, we let $\alpha(q, r, m, \lambda, \Lambda)$ denote the right hand side of (4.1), and for $r = p$, let $\beta(r, m, \lambda, \Lambda)$ be the right hand side of (4.1). For an integer $d \geq 1$, let

$$\gamma(r,m,d,\lambda,\Lambda) := \frac{r-1}{d} + \frac{1}{r^m} \cdot \left( |\Lambda| - \frac{\log L(\Lambda)}{\log \lambda} \right). \tag{4.3}$$

*Step 1.* Assume first that $q-1$ is divisible by $r$.

Then, independently from $m$, each prime divisor $s$ of $P$ is congruent to 1 modulo $2r$, whence Proposition 4.1 is applicable with $\Lambda$ being the set of all primes $s < \lambda$ such that $s-1$ is divisible by $2r$. Now, $\log q < \gamma(r,1,1,\lambda,\Lambda)/N(r,\lambda)$ implies $q \leq 267$ if $r=7$; $q \leq 21$ if $r = 11$ and $q \leq 13$ if $r \geq 13$. Since $\gamma(r,1,1,\lambda,\Lambda)$ is an upper bound for $\alpha(q,r,m,\lambda,\Lambda)$ for all $m \geq 1$, we are therefore left with the cases where $r = 7$ and $q \in \{8,29,43,64,71,113,127,169,197,211,239\}$. For each $q$ of the latter set we apply Proposition 4.1 with $\Lambda = \Lambda_\lambda(q,7^\infty)$ being the set of all primes $s < \lambda$ such that $\mathrm{ord}_s(q)$ is a power of 7: we obtain that (4.1) is satisfied for all these $q$ with $m = 1$, except when $q = 197$. But (4.1) is satisfied for $q = 197$ and $m = 2$. Now, it is important to observe that

$$\alpha(q,r,m,\lambda,\Lambda_\lambda(q,7^\infty)) \geq \alpha(q,r,M,\lambda,\Lambda_\lambda(q,7^\infty)), \quad \text{if } M \leq m. \tag{4.4}$$

Since the left side of (4.1) is independent of $m$, it therefore remains to show that $(197,7,7,49)$ is universal. We apply Proposition 3.1: $P$ has only one prime divisor which is smaller than 10000 (namely 1373), whence $\omega \leq 29$; since $\Omega = 42$, (3.1) is valid for $(197,7,7,49)$. Altogether, we have shown that $(q,r^m,r^m,r^{m+1})$ is universal for all $m \geq 1$, all $r \geq 7$ and all $q$ such that $q-1$ is divisible by $r$.

*Step 2.* Assume that $q-1$ is not divisible by $r$, that $r \neq p$ and that $q \geq 4$ if $r \geq 17$.

Independently from $m$, we first choose $\Lambda$ to be the set of all odd primes $s < \lambda$ (recall that $P$ is odd). Here, $\gamma(r,1,2,\lambda,\Lambda)$ is an upper bound for $\alpha(q,r,m,\lambda,\Lambda)$ for all $m \geq 1$, and $\log q < \gamma(r,1,2,\lambda,\Lambda)/N(r,\lambda)$ implies $q \leq 74$ if $r = 7$, $q \leq 6$ if $r = 11$, $q \leq 4$ if $r = 13$ and $q \leq 3$ if $r \geq 17$. For the remaining values for $q$ and $r$, we apply Proposition 4.1 with $\Lambda = \Lambda_\lambda(q,r^\infty)$ (see Step 1). Here, (4.1) is satisfied for all the remaining cases with $m = 1$, and thus, using again (4.4), we conclude that $(q,r^m,r^m,r^{m+1})$ is universal for all $m \geq 1$, all $r \geq 7$ and all $q$ as in the assumption.

*Step 3.* Assume that $q = 2$ or $q = 3$ and that $r \geq 17$.

In either case for $q$, and independently from $m$, we first take $\Lambda$ to be the set of odd primes $s < \lambda$ different from $p$. As $r \geq 17$, we have $\mathrm{ord}_r(q) \geq 5$, whence $\gamma(r,1,5,\lambda,\Lambda)$ is an upper bound for $\alpha(q,r,m,\lambda,\Lambda)$ for all $m \geq 1$. Since $\log q \geq \gamma(r,5,1,\lambda,\Lambda)/N(r,\lambda)$ for all $r \geq 17$ and $q \in \{2,3\}$, Proposition 4.1 yields the universality of $(q,r^m,r^m,r^{m+1})$ for all quadruples under consideration.

*Step 4.* Assume finally that $r = p \geq 7$.

We take $\Lambda$ to be the set of odd primes $s < \lambda$. Now, $\beta(q,r,1,\lambda,\Lambda)$ is an upper bound for the right hand side of (4.1) for all $m \geq 1$. Since $\log r \geq$

$\beta(r, 1, \lambda, \Lambda)/N(r, \lambda)$, for all $r$, (4.1) is always satisfied, and this completes the proof of Theorem 2.3. □

## 5 Completeness

When considering a finite dimensional Galois extension $E/F$, then $v \in E$ is called *completely free in E over F* [5], if for each intermediate field $K$ of $E/F$ it holds that $\{v^g | g \in G_K\}$ is a $K$-(normal) basis of $E$ over $K$ (where $G_K$ denotes the Galois group of $E/K$). Completely free elements were first studied in 1957 by Faith [Fa]; he proved their existence whenever $F$ has infinite cardinality. For finite fields the corresponding result was only proved in 1986 by Blessenohl and Johnsen [BlJo]: the *complete normal basis theorem* states that for every extension $E/F$ of Galois fields, there exists an element in $E$ which is completely free over $F$. (For an extensive treatment of the structure of completely free elements we refer to the monograph Hachenberger [Ha1].)

It is natural to ask, whether, for a Galois field extension $E/F$, there exists a primitive element of $E$ which is additionally *completely* free over $F$. By means of a computer search, Morgan and Mullen [MoMu] have determined primitive completely free elements for all extensions $\mathrm{GF}(q^n)/\mathrm{GF}(q)$ where $q \leq 97$ and $n \leq 9$; they have therefore conjectured that such elements do always exist (i.e., for all extensions of Galois fields). A considerable step towards proving the *primitive complete normal basis theorem* is provided in Hachenberger [Ha2], where the existence is shown for a (large) class of so-called *regular extensions*. The precise formulation is given in Theorem 5.1.

**Theorem 5.1.** *Consider the extension* $\mathrm{GF}(q^n)/\mathrm{GF}(q)$ *of Galois fields. Let* $n'$ *be the largest divisor of* $n$ *which is relatively prime to* $q$ *and let* $\nu(n')$ *be the square-free part of* $n'$. *Assume that* $\mathrm{ord}_{\nu(n')}(q)$ *and* $n$ *are relatively prime. Assume further that* $q \equiv 1 \bmod 4$ *if* $n'$ *is even. Then there exists a primitive element in* $\mathrm{GF}(q^n)$ *which is completely free over* $\mathrm{GF}(q)$.

Observe that the assumption in Theorem 5.1 is satisfied whenever $n$ is an odd prime power, say $n = r^k$ with $r$ being a prime. If $r \geq 7$, then for the latter class of extensions, Theorem 1.2$^c$ is an improvement of Theorem 5.1. The proof of Theorem 1.2$^c$ is the aim of the following section.

## 6 Proof of Theorem 1.2$^c$

Analogously to Theorem 2.1, the proof of Theorem 1.2$^c$ is established via induction. We first have to formulate a complete version of Definition 2.1.

[5] Again, the terminology is not consistent; such elements are also called *completely normal*.

**Definition 6.1.** A quadruple $(q, k, l, n)$ is called *completely universal*, if for the corresponding quadruple $(F, K, L, E)$ of Galois fields it holds that for **every** $b \in L$ which is primitive in $L$ and for **every** $a \in K$ which is completely free in $K$ over $F$, there exists an element $w = w_{a,b}$ in $E$ which satisfies (1), (3) and (4) in Definition 2.1 as well as

($2^c$) $w$ is completely free in $E$ over $F$. □

As the trace of a completely free element is again completely free, the assumption on $a$ is necessary.

Now, Theorem $1.2^c$ is an immediate consequence of Theorem 6.2.

**Theorem 6.2.** *Assume that $r \geq 7$ is a prime. Let $q > 1$ be any prime power and $m \geq 0$ be any integer. Then the quadruple $(q, r^m, r^m, r^{m+1})$ is completely universal.*

For the proof of Theorem 6.2 we shall use Theorem 2.3 in combination with the characterization of completely free elements in prime power extensions over finite fields as given in Section 17 of Hachenberger [Ha1]. Throughout, let $E$ be the $r^{m+1}$-dimensional extension over the Galois field $F$.

First, the assertion of Theorem 6.2 is valid if $r$ is equal to the characteristic $p$ of $F$, in which case $w \in E$ is completely free over $F$ if and only if $w$ is free over $F$ (see Theorem 5.7 in [Ha1]). We may therefore assume that $r \neq p$. But then the situation is totally different, because for every $q$ there exists a free element in $E$ over $F$ which is not free over $\mathrm{GF}(q^r)$ as long as $m$ exceeds a certain bound depending on $q$ and $r$ (see the proof of Theorem 14.5 in [Ha1] or the discussion in Section 2 of Hachenberger [Ha3]). However, as we will see below, Theorem 2.3 can nevertheless be used, when changing the ground field $F$ to an appropriate intermediate field of $E/F$.

The multiplicative order of $q$ modulo $r^{m+1}$ is of the form

$$\mathrm{ord}_{r^{m+1}}(q) = \mathrm{ord}_r(q) \cdot r^{\alpha}, \tag{6.1}$$

where $\mathrm{ord}_r(q)$ divides $r - 1$ and where $\alpha = \alpha(q, r, m) \leq m$. We define a parameter $\tau = \tau(q, r, m + 1)$ as follows:

$$\tau := \lfloor \frac{\alpha}{2} \rfloor. \tag{6.2}$$

Then $\tau < m$, whence $M := E_{r^\tau}$ is a proper subfield of $E$ containing $F$. Now, in combination with Theorem 2.3, Theorem 6.2 is an immediate consequence of the following proposition.

**Proposition 6.3.** *Let $q \geq 2$ be a prime power, $r$ an odd prime number not dividing $q$, and let $m \geq 0$. Letting $\tau$ be as in (6.2) it holds that $(q, r^m, r^m, r^{m+1})$ is completely universal, if $(q^{r^\tau}, r^{m-\tau}, r^{m-\tau}, r^{m+1-\tau})$ is universal.*

*Proof.* Consider the quadruple $(M, K, L, E)$ corresponding to $(q^{r^\tau}, r^{m-\tau}, r^{m-\tau}, r^{m+1-\tau})$. Let $a \in K = E_{r^m}$ be completely free over $F$ (hence free over $M$) and let $b \in L = K$ be primitive. Assuming the universality of $(q^{r^\tau}, r^{m-\tau}, r^{m-\tau}, r^{m+1-\tau})$, there exists a primitive element $w \in E$ which is free over $M$ such that $\mathrm{Tr}_{E,K}(w) = a$ and $\mathrm{N}_{E,L}(w) = b$. We claim that these conditions already suffice for $w$ to be completely free in $E$ over $F$, whence everything is proved.

Let $\mathcal{T}$ be the kernel of the $(E, K)$-trace mapping. Then, similar to the proof of Lemma 3.3 (with $K = \tilde{K}$), $K \oplus \mathcal{T}$ is a decomposition of the additive group of $E$ as $F[x]$-module with respect to a generator $\sigma$ of the Galois group $G$ of $E$ over $F$. The decomposition likewise respects the action of $M[x]$ with respect to $\psi := \sigma^{r^\tau}$, which is a generator of the Galois group of $E$ over $M$ (in the latter context the scalar multiplication is given by $g \star u := g(\psi)(u)$ for $g \in M[x]$ and $u \in E$). In terms of this decomposition, we have the following characterization of completely free elements for $E$ over $F$ (see Section 17 in [Hal]): $v = v_K + v_\mathcal{T}$ ($v_K \in K$, $v_\mathcal{T} \in \mathcal{T}$) is completely free in $E$ over $F$ if and only if $v_K$ is completely free in $K$ over $F$ and the $M$-order of $v_\mathcal{T}$ (i.e., the monic polynomial $g$ of least degree in $M[x]$ such that $g \star v_\mathcal{T} = 0$) is equal to $\Phi_{r^{m+1-\tau}}$, the $r^{m+1-\tau}$th cyclotomic polynomial.

It therefore remains to verify the latter two conditions for the element $w$ given above. With $w = w_K + w_\mathcal{T}$ we have (by assumption) that $a = \mathrm{Tr}_{E,K}(w) = r w_K$. Hence, $w_K$ is completely free in $K$ over $F$, as $a$ satisfies this property by assumption. We also have chosen $w$ to be free in $E$ over $M$. Consequently, an application of Theorem 8.6 of [Hal] shows that the components $w_K$ and $w_\mathcal{T}$ are generators of the $M[x]$-modules $K$ and $\mathcal{T}$, respectively. But for $w_\mathcal{T}$ this just means that the $M$-order of $w_\mathcal{T}$ is equal to $\Phi_{r^{m+1-\tau}}$, which is the minimal polynomial of $\mathcal{T}$ with respect to $\psi$ over $M$. Thus, everything is proved. □

We finally remark that in general it is not known, whether the intermediate trace of a completely free element can be prescribed, wherefore a complete version of Proposition 2.4 is not (yet) available (see the proof of Proposition 2.4). The latter is however the case, when $E/F$ is of prime power degree (for more details on this problem see Section 26 in [Hal]). We therefore have the following strengthening of Theorem 6.2.

**Theorem 6.2'.** *Assume that $r \geq 7$ is a prime. Let $q > 1$ be any prime power. Then $(q, r^\alpha, r^\beta, r^\gamma)$ is completely universal for all $\alpha, \beta \geq 0$ and all $\gamma > max\{\alpha, \beta\}$.* □

## 7 Concluding Remarks

We would finally like to remark that universal generators might be interesting in view of representing finite fields and their arithmetic in Computer Algebra systems.

It was suggested by J. H. Conway to use norm-compatible primitive generators for describing field extensions. The reason is that, as far as the multiplicative structure is concerned, such presentations include information on the embeddings of the various subfields. This concept is used in the computer algebra system *Magma*.

Analogously, as pointed out in Scheerhorn [Sche] (see also [CiD]), computationally simple embeddings relying on a normal-basis representation consist of trace-compatible free generators. This additive presentation is used in the computer algebra system *Axiom*.

It seems therefore to be of interest to have (complete) universal generators: they yield a *dynamic* data structure for presenting finite fields. It seems also that for finite sets $E_N$ of small degree over small fields $F$ the existence of these objects can only be decided experimentally. When searching for such generators one should assume that the relative degrees of the fields are at least three.

## References

[BlJo] *D. Blessenohl* and *K. Johnsen*, Eine Verschärfung des Satzes von der Normalbasis, J. Algebra **103** (1986), 141-159.

[Ca] *L. Carlitz*, Primitive roots in a finite field. Trans. Am. Math. Soc. **73** (1952), 373-382.

[CiD] Computeralgebra in Deutschland (Bestandsaufnahme, Möglichkeiten, Perspektiven), Herausgegeben von der Fachgruppe Computeralgebra der GI, DMV, GAMM, Passau und Heidelberg (1993).

[Co] *S. D. Cohen*, Gauss sums and a sieve for generators of Galois fields, Publ. Math. Debrecen, **56** (2000), to appear.

[CoHa1] *S. D. Cohen* and *D. Hachenberger*, Primitive normal bases with prescribed trace, Applic. Alg. Engin. Comm. Comp. **9** (1999), 383-403.

[CoHa2] *S. D. Cohen* and *D. Hachenberger*, Primitivity, freeness, norm and trace, Discrete Mathematics **214** (2000), 135-144.

[Da] *H. Davenport*, Bases for finite fields. J. London Math. Soc. **43** (1968), 21-49.

[Fa] *C. C. Faith*, Extensions of normal bases and completely basic fields, Trans. Am. Math. Soc. **85** (1957), 406-427.

[Ha1] *D. Hachenberger*, "Finite Fields: Normal Bases and Completely Free Elements", Kluwer Academic Publishers, Boston, 1997.

[Ha2] *D. Hachenberger*, Primitive normal bases for towers of field extensions, Finite Fields and their Applications **5** (1999), 378-385.

[Ha3] *D. Hachenberger*, Primitive complete normal bases for regular extensions, Glasgow Journal Math., to appear.

[He] *K. Hensel*, Über die Darstellungen der Zahlen eines Gattungsbereiches für einen beliebigen Primdivisor, J. reine angew. Math. **103** (1888), 230-237.

[Ju] *D. Jungnickel*, "Finite Fields. Structure and Arithmetic", Bibliographisches Institut, Mannheim, 1993.

[Le] *H.W. Lenstra, Jr.*, A normal basis theorem for infinite Galois extensions, Nederl. Akad. Wetensch. Indag. Math **47** (1985), no. 2, 221-228.

[LeSc] *H. W. Lenstra, Jr.* and *R. J. Schoof*, Primitive normal bases for finite fields, Math. Comp. **48** (1987), 217-231.

[LiNi] *R. Lidl* and *H. Niederreiter*, "Finite Fields", Addison-Wesley, Reading, Massachusetts, 1983.

[Lü] *H. Lüneburg*, On the early history of Galois fields, in: D. Jungnickel and H. Niederreiter (eds.), Proceedings of the Fifth International Conference on Finite Fields and Applications, Augsburg, August 1999, this volume.

[Sche] *A. Scheerhorn*, Trace- and norm-compatible extensions of finite fields, AAECC **3** (1992), 199-209.

# Irreducible Polynomials Generated by Decimations

Fritz von Haeseler[1] and Wibke Jürgensen*[2]

[1] CeVis
University of Bremen
Universitätsallee 29
28359 Bremen
Germany
e-mail: fritz@cevis.uni-bremen.de

[2] Department of Mathematics and Computer Science
University of Bremen
Postfach 330 440
D-28334 Bremen
Germany
e-mail: wibke@math.uni-bremen.de

**Abstract** Several methods are known, to construct irreducible polynomial over a finite field from known ones. We present a simple approach based on decimations of polynomial, which generalizes an algorithm presented by S. Golomb, and also show its relevance for the generation of primitive polynomials.

## 1 Introduction

In numerous applications, it is necessary to find irreducible polynomials over a finite field $\mathbb{F}_q$ of characteristic $p$. A basic method to generate new irreducible polynomials from known ones is based on the construction of characteristic polynomials.

Starting with a monic irreducible polynomial $P \in \mathbb{F}_q[X]$ of degree $d$ and order $e$, let $\alpha \in \mathbb{F}_{q^d}$ be a root of $P$ in its splitting field. Then the characteristic polynomial $P_m$ of $\alpha^m$ over $\mathbb{F}_q$ is irreducible in $\mathbb{F}_q$ if and only if $d$ is the multiplicative order of $q$ modulo $e/\gcd(e, m)$.

Several different ways are known to calculate $P_m$, see, e.g., [4], and [6,10]:

- $P_m$ is the characteristic polynomial of the $m$th power of the companion matrix of $P$.
- $P_m$ can be calculated from $P$ by

$$P_m(X^m) = (-1)^{d(m-1)} \prod_{j=1}^{m} P(w_j X)$$

where $w_j$ runs through all $m$th roots of unity over $\mathbb{F}_q$.

* Supported by DFG-Graduiertenkolleg "Komplexe Dynamische Systeme"

Other methods to generate irreducible polynomials are presented in [7,10,11].

In [6], Golomb introduces a rational algorithm to calculate $P_m$ for $m = 3$ as well as for $m = 5$ and shows how to generalize it for any prime $m$. This algorithm is also based on the calculation of the characteristic polynomial $P_m$.

The objective of this paper is to present an approach which is more simple and yields the same irreducible polynomials as the methods mentioned above. Surprisingly, our is based on decimations of polynomials, a concept which is an essential part in the theory of automatic sequences (e.g., [1] for an introduction and further references). Our algorithm is not only an extension of the method offered in [6] for $m = 3$ to arbitrary natural numbers, it also provides an explanation why the algorithm in [6] works.

Therefore, we first introduce the relevant notions of decimations and decimation matrices. In Section 3, we consider the $m$-decimation determinant and show that it is related to the construction of irreducible polynomials. Starting with a polynomial $P$ of degree $d$, it is possible to generate all irreducible polynomials whose order divides the order of $P$. Furthermore, we compare the generation of irreducible polynomials by decimation determinants with the generation of irreducible polynomials using companion matrices. We end up with some remarks on the generation of primitive polynomials using the $m$-decimation determinant.

## 2 Decimations of Formal Power Series

For a prime number $p$, we denote by $\mathbb{F}_p$ the Galois field with $p$ elements, and by $\mathbb{F}_q$ the finite field with $q = p^\alpha$ elements. The ring of polynomials over $\mathbb{F}_q$ is denoted by $\mathbb{F}_q[X]$, and with $\mathbb{F}_q[[X]]$ we denote the ring of formal power series.

We begin with the definition of decimation operators, used in the study of automatic sequences and their generating functions, see [1,3,5], and [12].

**Definition 2.1.** Let $m \geq 2$ be a natural number and let $t \in \{0, 1, \ldots, m-1\}$. Then the map $\partial_t^{(m)} : \mathbb{F}_q[[X]] \to \mathbb{F}_q[[X]]$ defined by

$$\sum_{j\in\mathbb{Z}} f_j X^j \mapsto \sum_{j\in\mathbb{Z}} f_{mj+t} X^j$$

is called *$t$-decimation w.r.t. $m$.*

From now on we agree to omit the superscript $m$ if there is no risk of confusion and we also speak of $t$-decimation rather than of $t$-decimation w.r.t. $m$.

Based on the $t$-decimations, we associate with any formal power series $F$ its *$m$-decimation matrix* which is the $m \times m$ matrix $\mathcal{A}_m(F)$ defined as

$$\mathcal{A}_m(F) = (a_{ij})_{ij} \qquad \text{with} \qquad a_{ij} = \begin{cases} \partial_{i-j}^{(m)}(F) & \text{for } i \geq j \\ X\partial_{m+i-j}^{(m)}(F) & \text{for } i < j \end{cases}.$$

*Example 2.2.* Consider the polynomial $1+X+X^4 \in \mathbb{F}_4[X]$ and $m=3$, then the 3-decimation matrix is given by

$$\mathcal{A}_3(1+X+X^4) = \begin{bmatrix} 1 & 0 & X+X^2 \\ 1+X & 1 & 0 \\ 0 & 1+X & 1 \end{bmatrix}.$$

Realizing that the first row of an $m$-decimation matrix gives exactly all $t$-decimations (w.r.t. $m$) of the series under consideration, the notion of $m$-decimation matrices allows an elegant study of properties of decimations, see [8,9].

The most elementary property is the linearity of $m$-decimations which is obvious from the definition.

**Lemma 2.3.** *Let $m > 1$ be a natural number. If $a, b$ are elements in $\mathbb{F}_q$, and $F, G$ are series in $\mathbb{F}_q[[X]]$, then*

$$\mathcal{A}_m(aF + bG) = a\mathcal{A}_m(F) + b\mathcal{A}_m(G).$$

Furthermore, a product rule for decimations can be established.

**Lemma 2.4 ([8]).** *Let $F, G$ be formal power series over $\mathbb{F}_q$. Then*

$$\mathcal{A}_m(FG) = \mathcal{A}_m(F)\mathcal{A}_m(G).$$

The proof relies on the fact that $F(X) = \sum_{j=0}^{m-1} X^j \partial_j(F)(X^m)$ and $G(X) = \sum_{j=0}^{m-1} X^j \partial_j(G)(X^m)$. Multiplying these expressions and rearranging the summands accordingly one obtains the above result.

## 3 Decimation Determinants

In this section, we shall introduce the $m$-determinant. It will be the crucial tool for our algorithm. The $m$-determinant is a map from $\mathbb{F}_q$ to $\mathbb{F}_q$ with surprising properties. Some of them we shall encounter in the course of this and the following section. The importance of the $m$-determinant for the theory of automatic sequences is demonstrated in [8,9].

**Definition 3.1.** The map $\Psi_m : \mathbb{F}_q[[X]] \to \mathbb{F}_q[[X]]$ defined as

$$\Psi_m(F) = \det \mathcal{A}_m(F)$$

is called *$m$-determinant.*

Some properties of $\Psi_m$ are obvious from its definition. By the rule of multiplication of determinants and Lemma 2.4, we obtain $\Psi_m(FG) = \Psi_m(F)\Psi_m(G)$ for series $F, G \in \mathbb{F}_q[[X]]$. In addition, the $m$-decimation matrix $\mathcal{A}_m(F(X^m))$ is of diagonal form with entries equal to $F$. Hence $\Psi_m(F(X^m)) = F(X)^m$.

Amazingly, the following theorem provides a connection between $\Psi_m$ and the generation of irreducible polynomials over $\mathbb{F}_q$.

**Theorem 3.2.** *If $P \in \mathbb{F}_q[X]$ is a polynomial of degree d with leading coefficient $c \in \mathbb{F}_q$ and roots $\xi_1, \dots, \xi_d$ in its splitting field, then*

$$\Psi_m(P)(X) = (-1)^{d(m-1)} c^m \prod_{j=1}^{d} (X - \xi_j^m).$$

*Proof.* We consider $\Psi_m(X - \xi_j)$ for $0 < j \leq d$ over the splitting field of $P(X)$. It is obvious from Definition 3.1 that

$$\Psi_m(X - \xi_j) = \det \begin{bmatrix} -\xi_j & 0 & \cdots & 0 & X \\ 1 & -\xi_j & 0 & \cdots & 0 \\ 0 & 1 & \ddots & \ddots & \vdots \\ \vdots & \ddots & \ddots & -\xi_j & 0 \\ 0 & \cdots & 0 & 1 & -\xi_j \end{bmatrix} = (-1)^{m-1}(X - \xi_j^m).$$

Furthermore, $\Psi_m(c) = c^m$ and hence

$$\Psi_m(P)(X) = \Psi_m\left(c \prod_{j=1}^{d} (X - \xi_j)\right) = (-1)^{d(m-1)} c^m \prod_{j=1}^{d} (X - \xi_j^m).$$

□

In other words, up to a factor, $\Psi_m$ maps a polynomial with roots $\xi_j$ to a polynomial with roots $\xi_j^m$.

*Remark 3.3.* In [2], the authors consider shift register sequences and their decimations to prove the $p$-automaticity of solutions of Mahler equations over the ring of integers modulo $p^\alpha$ for a prime number $p$. An inspection of their proof reveals that their proof is based on the fact that the characteristic polynomial of a decimation of a shift register sequence coincides with the $\Psi_m$-image of the characteristic polynomial of the sequence itself.

At first sight one does not expect that $m$-determinants are related to the roots of polynomials, Lemma 3.2 shows that quite the contrary is true. Moreover, it yields additional conclusions.

**Corollary 3.4.** *Let $P \in \mathbb{F}_q[X]$ be an irreducible polynomial.*

- *If there exists a natural number k with* $\operatorname{ord}(P) = km$ *and if $Q \in \mathbb{F}_q[X]$ is an irreducible factor of $\Psi_m(P)$ then* $\operatorname{ord}(Q) = k$.
- *If* $\gcd(m, \operatorname{ord}(P)) = 1$ *then $\Psi_m(P)$ is irreducible and* $\operatorname{ord}(\Psi_m(P)) = \operatorname{ord}(P)$.

**Corollary 3.5.** *If there are two natural numbers $m_1, m_2$ such that $m = m_1 m_2$, then*

$$\Psi_m(P) = \Psi_{m_1} \circ \Psi_{m_2}(P)$$

*for every polynomial $P \in \mathbb{F}_q[X]$.*

Both corollaries are proved by considering a linear factor of $P$ over the splitting field as done in Theorem 3.2. The assertions then follow from corresponding properties of the roots of $P$.

## 4 Generation of Irreducible Polynomials

In this section, we shall demonstrate how the tools developed so far enable us to construct new irreducible polynomials from polynomials already known to be irreducible.

**Theorem 4.1.** *Let $P \in \mathbb{F}_q[X]$ be an irreducible polynomial of degree $d$ and order $e$. Then $\Psi_m(P)$ is irreducible over $\mathbb{F}_q$ if and only if $d$ is the multiplicative order of $q$ modulo $e/\gcd(e,m)$.*

*Proof.* If $P \in \mathbb{F}_q[X]$ is an irreducible polynomial of degree $d$ and root $\alpha \in \mathbb{F}_{q^d}$ then all roots of $P$ over $\mathbb{F}_{q^d}$ are given by the $d$ distinct elements $\alpha, \alpha^q, \dots, \alpha^{q^{d-1}}$ of order $e$.

According to Theorem 3.2, the $m$-decimation determinant $\Psi_m(P)$ is also of degree $d$ and has roots $\alpha^m, \alpha^{mq}, \dots, \alpha^{mq^{d-1}}$ in $\mathbb{F}_{q^d}$ of order $e/\gcd(e,m)$. Consequently, $\Psi_m(P)$ is irreducible if and only if $d$ is the multiplicative order of $q$ modulo $e/\gcd(e,m)$.

*Example 4.2.* Consider $P(X) = 1 + X^2 + X^5$ of order 31 over $\mathbb{F}_2$ and let $m = 3$. Then

$$\mathcal{A}_3(P) = \begin{bmatrix} 1 & X+X^2 & 0 \\ 0 & 1 & X+X^2 \\ 1+X & 0 & 1 \end{bmatrix},$$

and consequently the 3-decimation determinant of $P$ is the irreducible polynomial given by

$$\Psi_3(P) = 1 + X^2 + X^3 + X^4 + X^5 \qquad \text{(of order 31)}.$$

Using arguments given in [4], it is sufficient for a computation of all irreducible polynomials $Q$ with $\operatorname{ord}(Q) \mid \operatorname{ord}(P)$, where $P$ is a given irreducible polynomial, to compute the value of $\Psi_m(P)$ only for distinct values $m_1, \dots, m_n$. To construct a suitable set $\{m_1, \dots, m_n\}$, let $m_1 = 1$. If $m_1, m_2, \dots, m_{j-1}$ are given, then another element $m_j$ is

$$m_j = \min_{1 \le i < j, a \in \mathbb{N}} \{m \mid m \not\equiv m_i q^a \mod \operatorname{ord}(P)\}.$$

In the example above, one obtains $\{m_1, \dots, m_6\} = \{1, 3, 5, 7, 11, 15\}$ which indeed generates all irreducible polynomials in $\mathbb{F}_2[X]$ of degree 5 and order 31 as shown in the left part of Figure 1.

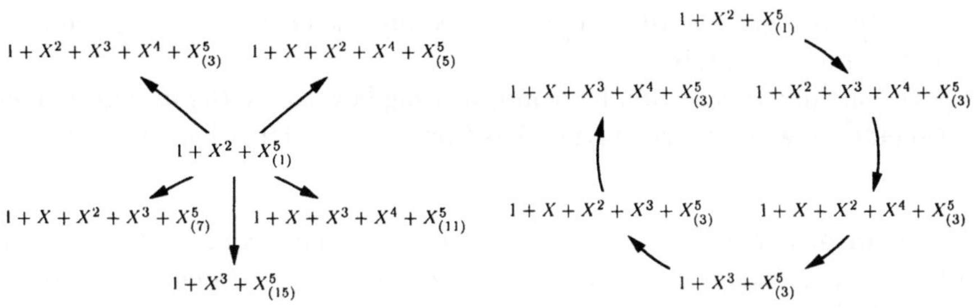


**Figure1.** Visualization of the algorithms for $1 + X^2 + X^5 \in \mathbb{F}_2$ in the usual way and by iterating $\Psi_3$. Both algorithms construct the same irreducible polynomials. However, the left one requires the calculation of determinants of large matrices (size in brackets) whereas the right algorithm only deals with $3 \times 3$ matrices.

Moreover, the calculations can be simplified due to Corollary 3.5. To generate all irreducible polynomials over $F_2$ of degree 5 and order 31, we start with the polynomial $1 + X^2 + X^5 \in \mathbb{F}_2[X]$ and take

$$\{m_1, \ldots, m_6\} = \{1, 3, 3^2, 3^3, 3^4, 3^5\},$$

which means that we iterate $\Psi_3$. The result is shown in the right part of Figure 1. (One can choose $\{m_1, \ldots, m_n\} = \{m^j \mid 0 \le j < n\}$ whenever $\mathrm{ord}(P)$ is a prime number relatively prime to $m$.)

Finally, we make some brief remarks comparing the presented method with the algorithm based on the companion matrix. To this end, let $P \in \mathbb{F}_q[X]$ be a monic irreducible polynomial of degree $d$, $\alpha$ be a root of $P$ over its splitting field and let $P_m \in \mathbb{F}_q[X]$ be the characteristic polynomial of $\alpha^m$.

In our as well as in the companion matrix method determinants play a crucial role. In the later case, the companion matrix $C_P$ is a $d \times d$ matrix with entries in $\mathbb{F}_q$. Firstly one has to calculate the $m$-th power $M = C_P^m$ and then the determinant of $M - X Id$ where $Id$ is the $d \times d$ identity matrix.

To generate $P_m$ via decimation determinants, one first has to find the $m$-decimation matrix, an $m \times m$ matrix with entries in $\mathbb{F}_q[X]$. With regard to operations in $\mathbb{F}_q$, this computation of the matrix $\mathcal{A}_m(P)$ requires no computational effort. The computation of $\Psi_m(P)$, which is a determinant of a $m \times m$ matrix requires less effort than the computation of $\det(M - X Id)$, in particular, if $m < d$.

Especially for polynomials of high degree, it follows that the generation of irreducible polynomials based on $\Psi_m$ requires only the computation of determinants of relatively small matrices.

Furthermore, due to the special structure of $\mathcal{A}_m(P)$ it is reasonable to assume that a special designed algorithm for a computation of $\det(\mathcal{A}_m(P(X)))$ may reduce the amount of computations; a point which has to be investigated further.

In any case our algorithm does not require the computation of the $m$-th power of a $d \times d$-matrix.

We conclude this section by demonstrating how to use the $m$-determinant to generate new primitive polynomials from polynomials which are known to be primitive.

**Theorem 4.3.** *Let $P \in \mathbb{F}_q[X]$ be a primitive polynomial of degree $d$ and order $e$. Then $\Psi_m(P)$ is a primitive polynomial over $\mathbb{F}_q$ if and only if $m$ and $e$ are relatively prime.*

*Proof.* A polynomial $P \in \mathbb{F}_q[X]$ is primitive if and only if it is monic with $P(0) \neq 0$ and $e = q^d - 1$.

Corollary 3.4 yields that $\Psi_m(P)$ is again an irreducible polynomial over $\mathbb{F}_q$ of order $e$. Furthermore, from Theorem 3.2, we have that $\Psi_m(P)$ is monic and $\Psi_m(P)(0) \neq 0$. Hence, $\Psi_m(P)$ is also a primitive polynomial over $\mathbb{F}_q$. □

On the other hand, the same arguments yield that the $\Psi_m$-image of a non-primitive polynomial is not a primitive polynomial.

*Example 4.4.* Since $1 + X^2 + X^5$ is a primitive polynomial over $\mathbb{F}_2$, Fig. 1 shows how to generate all primitive polynomials of degree 5. On the contrary, iterating $\Psi_3$ on $1 + X + X^6$ of order 63 over $\mathbb{F}_2$ gives $1 + X + X^2 + X^4 + X^6$ of order 21 and $(1 + X + X^3)^2$ as well as $(1 + X^2 + X^3)^2$, both squares of primitive polynomials of order 7.

## References


1. J.-P. Allouche. Automates finis en théorie des nombres. *Expo. Math*, 5:239–266, 1987.
2. A. Barbé, H.-O. Peitgen, and G. Skordev. Automaticity of coarse-graining invariant orbits of one-dimensional linear cellular automata. *International Journal of Bifurcation and Chaos*, 9(1), 1999. To appear.
3. A. Cobham. Uniform tag sequences. *Math. Systems Theory*, 6:164–192, 1972.
4. D. E. Daykin. Generation of irreducible polynomials over a finite field. *Amer. Math. Monthly*, 72:646–648, 1965.
5. S. W. Golomb. *Shift Register Sequences.* Holden-Day, San Francisco, 1967.
6. S. W. Golomb. Irreducible polynomials, synchronization codes, primitive necklaces, and the cyclotomic algebra. In *Proc. Conf. Combinatorial Math. and its Appl.*, pages 358–370, Chapel Hill, N.C., 1969. Univ. of North Carolina Press.
7. S. W. Golomb. Obtaining specified irreducible polynomials over finite fields. *SIAM J. Alg. Disc. Meth.*, 1:411–418, 1980.
8. F. von Haeseler and W. Jürgensen. Automaticity of solutions of Mahler equations. In C. Ding, T. Helleseth, and H. Niederreiter, editors, *Sequences and Their Applications*, Discrete Mathematics and Theoretical Computer Science, pages 228–239, London, 1999. Springer-Verlag.
9. W. Jürgensen. Über die Lösungen von Mahlergleichungen. Diploma thesis, University of Bremen, 1997.

10. R. Lidl and H. Niederreiter. *Finite Fields*, volume 20 of *Encyclopedia of Mathematics and its Applications*. Addison-Wesley, 1983.
11. A. J. Menezes, editor. *Applications of Finite Fields*. Kluwer Academic Publishers, 1993.
12. N. Zierler. Linear recurring sequences. *J. Soc. Indust. Appl. Math.*, 7:31–48, 1959.

# Decoding Reed-Muller Codes beyond Half the Minimum Distance

Agnes E. Heydtmann*[1] and Thomas Jakobsen[2]

[1] Department of Mathematics
Building 303
Technical University of Denmark
DK-2800 Lyngby
Denmark
e-mail: agnes@mat.dtu.dk

[2] IO Interactive APS
Store Strandstræde 21, 1.tv.
DK-1255 Copenhagen K
Denmark
e-mail: tj@ioi.dk

**Abstract** Inspired by Sudan's recent algorithm for Reed-Solomon codes we propose an efficient method for decoding $r$-th order Reed-Muller codes of length $2^m$ which can correct errors beyond half the minimum distance.

This procedure involves interpolating $Q \in \mathbb{F}_2[x_1, \dots, x_m, y]$, a polynomial vanishing when evaluated at points in $\mathbb{F}_2^m$ joint with the corresponding received bits. To obtain a list of codewords closest to the received word we need to factor $Q$ considered as an element of the quotient ring of boolean polynomials which is not a unique factorization domain. Therefore we introduce a novel, yet simple polynomial-time factorization algorithm for multivariate boolean polynomials that produces generators for the coset of factors.

Let $p = 2^{-\lambda}$ be the probability of algorithm failure and assume that the weights of a Reed-Muller code are approximately binomially distributed. This assumption is supported by known weight distributions for some short Reed-Muller codes. Then with probability at least $1 - p$, the algorithm corrects

$$\tau \leq \max_{0 \leq \rho \leq m} \min \left\{ 2^m - \sum_{i=0}^{r+\rho} \binom{m}{i} - \lambda, \sum_{i=0}^{\rho} \binom{m}{i} - 1 \right\} \tag{1}$$

independently and uniformly distributed errors.

For the $\mathcal{RM}(2,9)$ code for example, the algorithm corrects up to 122 errors with probability at least 0.99 whereas half the minimum distance is 64. Under the above assumption, we can correct up to half the block length asymptotically for fixed $r$.

---

* During part of this work Agnes E. Heydtmann was supported by a graduate student scholarship (HSPIII) of the German Academic Exchange Service.

# 1 Introduction

Reed-Muller codes (MacWilliams and Sloane [8], Chaps. 13-15) make a class of codes with a simple construction given by Muller ([10]) that allows easy majority logic decoding described by Reed ([12]).

At first glance, the structure seems to be very simple and the minimum distance is in fact also easily derived, but beyond this there are still several unanswered questions. The weight distribution and the covering radius, for example, are known only for specific instances of the codes.

Most known algorithms for decoding Reed-Muller codes do not decode beyond half the minimum distance and those that do so, do not have an efficient running time in the general case. Sudan's recent algorithm (Sudan [15]) allows efficient decoding of Reed-Solomon codes (MacWilliams and Sloane [8], Chap. 10) beyond half their minimum distance. A generalization to some classes of algebraic-geometry codes has been made by Shokrollahi and Wassermann ([14]). As there are several similarities in the construction of Reed-Solomon and Reed-Muller codes, it is natural to speculate whether Sudan-like techniques can be applied to the Reed-Muller case.

This article provides such a generalization. Although the algorithm presented has several similarities to Sudan's algorithm, its requirements and proofs are quite different due to the small size of the underlying field. The algorithm also bears some resemblance to the Welch-Berlekamp algorithm for decoding Reed-Solomon codes (Welch and Berlekamp [17]; Berlekamp [2]). It works best in settings with a low information rate, however the error rate can be correspondingly high. Although when going beyond half the minimum distance, the decoding is not guaranteed to be unique, it is still possible with high probability to correctly determine the codeword sent.

In a sense, decoding of Reed-Muller codes corresponds to the approximation of binary functions by low-degree multivariate polynomials. Consequently, a decoding algorithm which goes beyond half the minimum distance is useful in other applications which are not directly related to error correction. For instance, cryptanalysis of secret-key ciphers (Menezes et al. [9]) often deals with finding approximations, see for example Jakobsen ([4]); here our algorithm could prove to be very useful. In fact, the research resulting in this article was partly initiated because of an observation by Shimoyama and Kaneko ([13]) who realized that a specific quadratic relation of the in- and output of one S-box of the Data Encryption Standard (DES) (Menezes et al. [9]) has the best linear approximation as a factor.

The article is organized as follows: We proceed in Section 2 by defining the Reed-Muller codes and related algebraic concepts. In Sec. 3, the algorithm is presented together with the main results and we give some conjectures about its running time and correctness for random error patterns. The conjectures are related to the generally unknown weight distribution of Reed-Muller codes. They are supported by an implementation of the algorithm and known weight-distributions for short-length Reed-Muller codes. In Sec. 4 we proceed to prove the correctness of the algorithm. Section 5 deals with the error-correcting capabilities and Sec. 6 covers complexity issues. We conclude in Sec. 7.

## 2 Preliminaries

We begin by defining Reed-Muller codes in terms of elements of a quotient ring.

**Definition 2.1.** (i) The *ring of boolean polynomials* in $m$ variables $x_1, \dots,$ $x_m$ is the quotient ring

$$R_{[x_1,\dots,x_m]} = \mathbb{F}_2[x_1,\dots,x_m] / \langle x_1^2 + x_1, \dots, x_m^2 + x_m \rangle \ .$$

Elements of the ring are called *boolean polynomials* and in our notation we naturally identify cosets and polynomials in $\mathbb{F}_2[x_1,\dots,x_m]$ that are linear in each of the $m$ variables. Furthermore, monomials in $\mathbb{F}_2[x_1,\dots,x_m]$ that are linear in each of the $m$ variables and their cosets in $R_{[x_1,\dots,x_m]}$ are called *boolean monomials.*

(ii) The *degree* of a coset $f \in R_{[x_1,\dots,x_m]}$ denoted by $\deg f$, is the degree of the unique element that is linear in each of the $m$ variables.

(iii) The *r-th order Reed-Muller code* of length $2^m$ with $0 \le r \le m$, is the set

$$\mathcal{RM}(r,m) = \{(f(P_1),\dots,f(P_{2^m})) \mid f \in R_{[x_1,\dots,x_m]}, \deg f \le r\}$$

where $P_1,\dots,P_{2^m}$ are the distinct elements of $\mathbb{F}_2^m$. We associate naturally $f \in R_{[x_1,\dots,x_m]}$, $\deg f \le r$ with $\mathcal{RM}(r,m)$ words.

Note that each coset contains exactly one polynomial which is linear in each of the $m$ variables and that all elements of one coset in $R_{[x_1,\dots,x_m]}$ are equal when considered as functions $\mathbb{F}_2^m \to \mathbb{F}_2$. Therefore $\mathcal{RM}(r,m)$ codes are well defined.

The boolean monomials in $R_{[x_1,\dots,x_m]}$ evaluate to linearly independent vectors in $\mathbb{F}_2^{2^m}$ and those of degree $r$ result in codewords of minimal weight. Thus the $\mathcal{RM}(r,m)$ code is a linear $[2^m, \binom{m}{0} + \dots + \binom{m}{r}, 2^{m-r}]$ code. We will also denote the dimension of $\mathcal{RM}(r,m)$ by $k_r = \sum_{i=0}^{r} \binom{m}{i}$.

The ring of boolean polynomials $R_{[x_1,\dots,x_m]}$ is not even an integral domain as $h \cdot (h+1) = 0$ for any $h \in R_{[x_1,\dots,x_m]}$. Therefore the ring is certainly not a unique factorization domain. Nevertheless we define the following:

**Definition 2.2.** Let $R_{[x_1,\dots,x_m]}$ be a ring of boolean polynomials and let $h, f \in R_{[x_1,\dots,x_m]}$. The boolean polynomial $f$ is said to be a *factor* of $h$ or to *divide* $h$, if there exists $g \in R_{[x_1,\dots,x_m]}$ such that

$$h = f \cdot g \ .$$

Certain factors of the $Q \in R_{[x_1,\dots,x_m,y]}$ which we will interpolate in Algorithm 3.1 will correspond to $\mathcal{RM}(r,m)$ codewords closest to the received word. Therefore it is convenient that the following theorem easily characterizes all factors of any boolean polynomial.

**Theorem 2.3.** *A boolean polynomial $f \in R_{[x_1,\dots,x_m]}$ divides a boolean polynomial $h$ if and only if*

$$(f+1) \cdot h = 0 \ .$$

*Proof.* $\Rightarrow$: Assume that there exists a boolean polynomial $g$ such that $h = f \cdot g$. Then

$$(f+1) \cdot h = (f+1) \cdot f \cdot g = (f^2 + f) \cdot g = 0$$

where the last equality holds since $(f^2+f) \in \langle x_1^2 + x_1, \dots, x_m^2 + x_m \rangle$.
$\Leftarrow$: Assume that $(f+1) \cdot h = 0$. Then $h = fh$, i.e. there exists a boolean polynomial $g = h$ such that $h = fg$ which means that $f$ is a factor of $h$ by the definition. □

Theorem 2.3 implies that the factors of a boolean polynomial $h$ form the coset $1 + \{f \in R_{[x_1,\dots,x_m]} | f \cdot h = 0\}$ of the ideal $\{f \in R_{[x_1,\dots,x_m]} | f \cdot h = 0\}$ in $R_{[x_1,\dots,x_m]}$. A factorization algorithm for multivariate, boolean polynomials is immediately obtained: Simply solve the corresponding homogeneous linear system of equations in the coefficients of the factor. In this way it is even straight forward to restrict the solutions to be bounded in their degree.

Let $\tau$ denote the maximum number of errors that should be corrected by the algorithm and let $\rho$ be the smallest integer for which

$$\tau < k_\rho \ .$$

In order to interpolate the boolean polynomial $Q \in R_{[x_1,\dots,x_m,y]}$ from Algorithm 3.1 that should help us to decode up to $\tau$ errors, we need the notion of the order of a boolean polynomial of degree at most $\rho$ in our original ring $R_{[x_1,\dots,x_m]}$ with respect to an arbitrary basis of this subvector space.

**Definition 2.4.** Let $p_0, \dots, p_{k_\rho - 1} \in R_{[x_1,\dots,x_m]}$ denote $k_\rho$ linearly independent, boolean polynomials with $\deg p_j \le \rho$. Let further $0 \ne f \in R_{[x_1,\dots,x_m]}$ of degree at most $\rho$. Then $f$ can be expressed uniquely as

$$f = \sum_{j=0}^{s} c_j p_j$$

where $c_j \in \mathbb{F}_2$, $s \le k_\rho - 1$ and $c_s \ne 0$. Define by the *order* of $f$

$$\operatorname{ord} f = s \ .$$

Note that the above does not define an ordering on $R_{[x_1,\dots,x_m]}$ since the relation is not antisymmetric. Clearly, $\operatorname{ord} p_j = j$ and if $p_j$ is a boolean monomial for all $j$, then the order of a boolean polynomial is simply the maximum order of its boolean monomials.

For Algorithm 3.1 to run correctly the error pattern has to be random in a certain sense. Therefore we introduce the following concepts and notation:

**Definition 2.5.** Let $\mathbf{a} = (a_1, \dots, a_n) \in \mathbb{F}_2^n$ be an $n$-bit word. The *weight* of $\mathbf{a}$ denoted by $\mathrm{w}(\mathbf{a})$ is the number of non-zero positions in $\mathbf{a}$. The *support* of $\mathbf{a}$ is the set of its non-zero positions

$$\operatorname{supp}(\mathbf{a}) = \{i \mid a_i \neq 0\} \ .$$

Clearly, $\mathrm{w}(\mathbf{a}) = |\operatorname{supp}(\mathbf{a})|$. We say that a word $\mathbf{a} \in \mathbb{F}_2^n$ *covers* another word $\mathbf{b} \in \mathbb{F}_2^n$ or that $\mathbf{b}$ is *covered* by $\mathbf{a}$ when $\operatorname{supp}(\mathbf{b}) \subseteq \operatorname{supp}(\mathbf{a})$.

# 3 The Decoding Algorithm

In this section we present the algorithm as well as the main results.

Suppose $\mathbf{y} = (y_1, y_2, \dots, y_{2^m})$ corresponding to a $\mathcal{RM}(r,m)$ word with at most $\tau = k_\rho - 1$ errors is received. Under certain conditions, the following algorithm decodes this received word correctly.

**Algorithm 3.1 (Reed-Muller Decoder).**
*Input:* $\mathbf{y}$, $r$, $\rho$, and $m$
*Output:* All $f \in R_{[x_1,\dots,x_m]}$ corresponding to $\mathcal{RM}(r,m)$ codewords closest to $\mathbf{y}$.

(i) Choose $p_0, \dots, p_{k_\rho - 1} \in R_{[x_1,\dots,x_m]}$ linearly independent with $\deg p_j \leq \rho$.
(ii) Find non-zero $Q = yQ_1 + Q_2 \in R_{[x_1,\dots,x_m,y]}$, $Q_1, Q_2 \in R_{[x_1,\dots,x_m]}$ with $\operatorname{ord} Q_1$ lowest possible and $\deg Q_2 \leq r + \rho$ such that $Q(P_i, y_i) = 0\ \forall i$.
(iii) Find $f \in R_{[x_1,\dots,x_m]}$ with $\deg f \leq r$ such that $f(P_i) = y_i, \forall i : Q_1(P_i) = 1$.
(iv) Return those $f$ that correspond to codewords closest to $\mathbf{y}$.

The conditions for correctness of Algorithm 3.1 is given by Theorem 3.2. It will be proven in Section 4.

**Theorem 3.2 (Correctness).** *Suppose that $f_\mathbf{c} \in R_{[x_1,\dots,x_m]}$ corresponds to a $\mathcal{RM}(r,m)$ codeword $\mathbf{c}$ closest to $\mathbf{y}$. If the error pattern $\mathbf{e} = \mathbf{c} + \mathbf{y}$ does not cover a non-zero $\mathcal{RM}(r+\rho, m)$ codeword and $\mathrm{w}(\mathbf{e}) \leq \tau < k_\rho$, then Algorithm 3.1 returns $f_\mathbf{c}$.*

Of course we have to ask how often the condition that the error pattern $\mathbf{e} = \mathbf{c} + \mathbf{y}$ does not cover a non-zero $\mathcal{RM}(r+\rho, m)$ codeword will be satisfied when errors occur independently and are uniformly distributed. Certainly for $\tau$ less than the code's minimum distance $2^{m-r-\rho}$, the error pattern can never cover a nonzero $\mathcal{RM}(r+\rho, m)$ codeword, but then we are not beyond half the minimum distance either. Conjecture 3.3 addresses the problem for larger $\tau$ and will be discussed further in Sec. 5.

**Conjecture 3.3.** *Let $r+\rho \gg 0$ and $\tau < n-k_{r+\rho}$. Denote the probability that a random error pattern of at most $\tau$ errors covers a non-zero $\mathcal{RM}(r+\rho,m)$ codeword by $\pi_1$. Then*

$$\lim_{m\to\infty} \left|\pi_1 - 2^{k_{r+\rho}-2^m+\tau}\right| = 0 \ .$$

We shall see in Sec. 5 that if it can be proven that for $r+\rho \gg 0$ the weights of $\mathcal{RM}(r+\rho,m)$ are close to being binomially distributed, then we can use $2^{k_{r+\rho}-2^m+\tau}$ as an approximative bound to $\pi_1$. We include the condition $r+\rho \gg 0$, since obviously the weights of $\mathcal{RM}(0,m)$ and $\mathcal{RM}(1,m)$ for example, are far from being binomially distributed.

Theorem 3.4 gives partial insight to the efficiency of the algorithm.

**Theorem 3.4 (Efficiency).** *Suppose that $f_{\mathbf{c}} \in R_{[x_1,\dots,x_m]}$ corresponds to a $\mathcal{RM}(r,m)$ codeword $\mathbf{c}$ closest to $\mathbf{y}$ and that the error pattern $\mathbf{e} = \mathbf{c}+\mathbf{y}$ does not cover a non-zero $\mathcal{RM}(r+\rho,m)$ codeword. In addition, let $\pi_2$ denote the probability that the $\mathcal{RM}(\rho,m)$ codeword corresponding to the boolean polynomial $Q_1$ obtained in Step 2 of Algorithm 3.1 is covered by some $\mathcal{RM}(r,m)$ codeword $\neq (1,1,\dots,1)$. Then with probability at least $1-\pi_2$, the algorithm has a running time of $O(n^3)$.*

The above theorem is proven in Sec. 6. Coupled with Conjecture 3.5 which is further discussed in the same section, we have an efficient algorithm for decoding Reed-Muller codes beyond half the minimum distance.

**Conjecture 3.5.** *Let $I \subseteq \{0,1,\dots,2^m-1\}$ be the set of error positions corresponding to the $\mathcal{RM}(r,m)$ codewords $\mathbf{c}$ closest to $\mathbf{y}$. Further we assume that every error pattern $\mathbf{e} = \mathbf{c}+\mathbf{y}$ has weight $< k_\rho$ and does not cover a non-zero $\mathcal{RM}(r+\rho,m)$ codeword. Let $\mathbf{q}_1$ denote a randomly chosen $\mathcal{RM}(\rho,m)$ codeword such that its components $q_{1i} = 0$ for $i \in I$. Let $\pi_2$ denote the probability that $\mathbf{q}_1$ is covered by some $\mathcal{RM}(r,m)$ codeword $\neq (1,1,\dots,1)$. If $\rho > r$, then $\pi_2 \to 0$ for $m \to \infty$.*

Section 6 should make it clear that the probabilities $\pi_2$ mentioned in Theorem 3.4 and Conjecture 3.5 are indeed the same. Also, we show that $\pi_2$ is the same as the probability that the random boolean polynomial $\mathbf{q}_1$ has factors of low degree. It appears that this probability is negligible.

The following is an example of a Reed-Muller code capable of being decoded beyond half the minimum distance.

*Example 3.6.* Consider $\mathcal{RM}(1,6)$ which has half minimum distance 16 and let $\rho = 2$ which enables us to correct up to 21 errors if there is no non-zero $\mathcal{RM}(3,6)$ codeword covered by the error pattern. Suppose we send the codeword $\mathbf{c}$ corresponding to $f = x_1 + x_2 + x_4$ and $\mathbf{y} = \mathbf{c} + \mathbf{e}$ is received corresponding to the occurrence of 20 errors as indicated by $\mathbf{e}$:

$$\begin{array}{r} \mathbf{c} : \texttt{0110011010011001011001101001100101100110100110010110011010011001} \\ + \, \mathbf{e} : \texttt{1000000001010001100100001100101000001010001011000011000100001001} \\ \hline = \mathbf{y} : \texttt{1110011011001000111101100101001101101100101101010101011110010000} \end{array}$$

Choosing the $p_j$'s as boolean monomials of increasing order, Step 2 of the algorithm interpolates

$$\begin{aligned} Q = & \, y(x_2x_3+x_1x_4+x_3x_4+x_1x_5+x_2x_5+x_3x_5+x_4x_5+x_1x_6+x_2x_6+x_4) \\ & +x_1x_2x_3+x_1x_2x_4+x_1x_3x_4+x_1x_3x_5+x_2x_3x_5+x_3x_4x_5+x_1x_4x_6 \\ & +x_2x_4x_6+x_2x_3+x_1x_4+x_2x_4+x_3x_4+x_1x_5+x_2x_5+x_4x_5+x_1x_6 \\ & +x_2x_6+x_4 \end{aligned}$$

which has the following unique factorization of the form $(y+f)Q_1$ with $\deg f \leq r$:

$$\begin{aligned} Q = & \, (y+x_1+x_2+x_4) \\ & (x_2x_3+x_1x_4+x_3x_4+x_1x_5+x_2x_5+x_3x_5+x_4x_5+x_1x_6+x_2x_6+x_4) \end{aligned}$$

The first factor corresponds to the sent codeword $\mathbf{y}$.

The code $\mathcal{RM}(1,6)$ was chosen to be the smallest example where more errors than half the minimum distance can be corrected with Algorithm 3.1 in practice. Generally, one should not attempt to decode such a small code much beyond half the minimum distance as only few error patterns satisfy the conditions of Theorem 3.2. In Sec. 7 we discuss the decoding of the $\mathcal{RM}(2,9)$ code which is more robust due to its larger size.

## 4 Correctness of the Algorithm

Let boolean polynomials $p_0, \dots, p_{k_\rho - 1}$ be given as in Definition 2.4 and the order of elements of $R_{[x_1,\dots,x_m]}$ of degree at most $\rho$ is thereby defined. Let further $f_\mathbf{c} \in R_{[x_1,\dots,x_m]}$ correspond to an $\mathcal{RM}(r,m)$ codeword $\mathbf{c}$ at distance $t \leq \tau < k_\rho$ to the received word $\mathbf{y} = (y_1, \dots, y_{2^m})$ such that there exists no $\mathcal{RM}(r,m)$ codeword closer to $\mathbf{y}$.

We will investigate two types of boolean polynomials $Q, \tilde{Q} \in R_{[x_1,\dots,x_m,y]}$ interpolating the points $(P_i, y_i)$. Therefore we introduce the following notation: By $Q = yQ_1 + Q_2$, $Q_1, Q_2 \in R_{[x_1,\dots,x_m]}$ we denote a non-zero boolean polynomial that vanishes at $(P_i, y_i)$ with $\operatorname{ord} Q_1$ smallest possible and $\deg Q_2 \leq r + \rho$. By $\tilde{Q} = (y + f_\mathbf{c})\tilde{Q}_1$, $\tilde{Q}_1 \in R_{[x_1,\dots,x_m]}$ we denote another non-zero boolean polynomial that vanishes at $(P_i, y_i)$ with $\operatorname{ord} \tilde{Q}_1$ smallest possible. Note that $\deg f_\mathbf{c}\tilde{Q}_1 \leq r + \rho$ and $\tilde{Q}_1(P_i) = 0$ whenever $f_\mathbf{c}(P_i) \neq y_i$. This means that the existence of $\tilde{Q}$ implies the existence of $Q$ and that the boolean polynomial $\tilde{Q}_1$ may be thought of as an error locator.

Both $Q$ and $\tilde{Q}$ are uniquely determined: Let $Q = yQ_1 + Q_2, Q' = yQ'_1 + Q'_2 \in R_{[x_1,\dots,x_m,y]}$, both vanishing when evaluated in $(P_i, y_i)$ and such that $\operatorname{ord} Q_1 = \operatorname{ord} Q'_1$ lowest possible, $\deg Q_2, \deg Q'_2 \leq r + \rho$. But

$$Q - Q' = y(Q_1 - Q'_1) + (Q_2 - Q'_2)$$

is a new boolean polynomial vanishing when evaluated in $(P_i, y_i)$. Either $\operatorname{ord} Q_1 - Q_1' < \operatorname{ord} Q_1 = \operatorname{ord} Q_1'$ and $\deg Q_2 - Q_2' \leq r + \rho$ which results in a contradiction to the definition of $Q$, or $Q_1 = Q_1'$. In that case we have to conclude that the boolean polynomial $Q_2 - Q_2'$ as an element in $R_{[x_1,\dots,x_m,y]}$ vanishes at all $(P_i, y_i)$. But that means that as an element of $R_{[x_1,\dots,x_m]}$ it vanishes at all points in $\mathbb{F}_2^m$ and must therefore equal the 0-polynomial. That implies $Q_2 = Q_2'$ and we must have had $Q = Q'$ in the first place. Similarly one can prove the uniqueness of $\tilde{Q}$. We will proceed to show that, under some conditions, $Q$ and $\tilde{Q}$ are in fact identical.

**Theorem 4.1.** *If the error pattern* $\mathbf{e} = \mathbf{c} + \mathbf{y}$ *does not cover a* $\mathcal{RM}(m, r+\rho)$ *word, then* $Q = \tilde{Q}$ *and* $\operatorname{ord} Q_1 = \operatorname{ord} \tilde{Q}_1 \leq t$, *where* $t \leq \tau < k_\rho$ *is the number of errors.*

In the proof of the above we will make use of Lemma 4.2. We start by introducing some more notation. Let $l_1, \dots, l_t$ denote the error positions and $I_j = \{l_{j+1}, \dots, l_t\}$ for $0 \leq j \leq t$. Denote by $Q^{(j)} = yQ_1^{(j)} + Q_2^{(j)}$, $Q_1^{(j)}, Q_2^{(j)} \in R_{[x_1,\dots,x_m]}$ the boolean polynomial that vanishes at $(P_i, y_i)$ for $i \notin I_j$ (i.e., at the first $j$ error positions and at every non-erroneous position) with $\operatorname{ord} Q_1^{(j)}$ smallest possible and $\deg Q_2^{(j)} \leq r + \rho$. Analogously, denote by $\tilde{Q}^{(j)} = (y + f_{\mathbf{c}})\tilde{Q}_1^{(j)}$, $\tilde{Q}_1^{(j)} \in R_{[x_1,\dots,x_m]}$ the boolean polynomial that vanishes at $(P_i, y_i)$ for $i \notin I_j$ with $\operatorname{ord} \tilde{Q}_1^{(j)}$ smallest possible. Note that $\deg f_{\mathbf{c}}\tilde{Q}_1^{(j)} \leq r + \rho$ and $\tilde{Q}_1^{(j)}(P_i) = 0$ for $l_1, \dots, l_j$. This means that the existence of $\tilde{Q}^{(j)}$ implies the existence of $Q^{(j)}$ and that the boolean polynomial $\tilde{Q}_1^{(j)}$ is an error locator for the first $j$ error positions.

Under the conditions of Theorem 4.1 both $Q^{(j)}$ and $\tilde{Q}^{(j)}$ are again uniquely determined. Let $Q^{(j)} = yQ_1^{(j)} + Q_2^{(j)}, Q'^{(j)} = yQ'^{(j)}_1 + Q'^{(j)}_2 \in R_{[x_1,\dots,x_m,y]}$, both vanishing when evaluated in $(P_i, y_i)$ for $i \notin I_j$ and such that $\operatorname{ord} Q_1^{(j)} = \operatorname{ord} Q'^{(j)}_1$ lowest possible, $\deg Q_2^{(j)}, \deg Q'^{(j)}_2 \leq r + \rho$. But

$$Q^{(j)} - Q'^{(j)} = y(Q_1^{(j)} - Q'^{(j)}_1) + (Q_2^{(j)} - Q'^{(j)}_2)$$

is a new boolean polynomial vanishing when evaluated in $(P_i, y_i)$ for $i \notin I_j$. Either $\operatorname{ord} Q_1^{(j)} - Q'^{(j)}_1 < \operatorname{ord} Q_1^{(j)} = \operatorname{ord} Q'^{(j)}_1$ and $\deg Q_2^{(j)} - Q'^{(j)}_2 \leq r + \rho$ which results in a contradiction to the definition of $Q^{(j)}$, or $Q_1^{(j)} = Q'^{(j)}_1$. In that case we have to conclude that the boolean polynomial $Q_2^{(j)} - Q'^{(j)}_2$ as element in $R_{[x_1,\dots,x_m,y]}$ vanishes at $(P_i, y_i)$ for $i \notin I_j$. But that means that the $\mathcal{RM}(r + \rho, m)$ codeword corresponding to $Q_2^{(j)} - Q'^{(j)}_2$ is covered by $\mathbf{e}$. We must have had $Q^{(j)} = Q'^{(j)}$ in the first place. Similarly one can prove the uniqueness of $\tilde{Q}^{(j)}$.

**Lemma 4.2.** *Let* $Q^{(j)} = \tilde{Q}^{(j)}$ *with* $\operatorname{ord} Q_1^{(j)} = \operatorname{ord} \tilde{Q}_1^{(j)} = j$.

*(i) If* $Q^{(j)}(P_i, y_i) = 0$ *for all* $i \in I_j$, *then* $Q = Q^{(j)} = \tilde{Q}^{(j)} = \tilde{Q}$.
*(ii) If* $Q^{(j)}(P_i, y_i) \neq 0$ *for some* $i \in I_j$ *(without loss of generality* $i = l_{j+1}$*), then* $Q^{(j+1)} = \tilde{Q}^{(j+1)}$ *with* $\operatorname{ord} Q_1^{(j+1)} = \operatorname{ord} \tilde{Q}_1^{(j+1)} = j + 1$.

*Proof.* Consider the two cases.

(i) If $Q^{(j)}(P_i, y_i) = 0$ for all $i \in I_j$, then $Q^{(j)} = Q$ and by assumption also $Q^{(j)} = \tilde{Q}^{(j)}$ implying $\tilde{Q}^{(j)} = \tilde{Q}$. All in all, we get $Q = Q^{(j)} = \tilde{Q}^{(j)} = \tilde{Q}$.

(ii) If $Q^{(j)}(P_i, y_i) \neq 0$ for an $i \in I_j$, then without loss of generality we may assume that $i = l_{j+1}$.
Consider the boolean polynomials $Q^{(j+1)}$ and $\tilde{Q}^{(j+1)}$ which vanish at $(P_i, y_i)$ for $i \notin I_j$. Note that $\tilde{Q}_1^{(j+1)}$ has to be chosen such that it evaluates to 0 at $P_i$ for $i = l_1, \ldots, l_{j+1}$, but this is possible as a linear combination of the $j+2$ boolean polynomials $p_0, \ldots, p_{j+1}$. Therefore $\operatorname{ord} \tilde{Q}_1^{(j+1)} \leq j+1$. Further, $\operatorname{ord} \tilde{Q}_1^{(j+1)} < j+1$ would contradict the uniqueness of $\tilde{Q}^{(j)}$. Therefore $\operatorname{ord} \tilde{Q}_1^{(j+1)} = j+1$.
On the other hand the existence of $Q^{(j+1)}$ is assured by the existence of $\tilde{Q}^{(j+1)}$. The inequality $\operatorname{ord} Q_1^{(j+1)} < j+1$ would contradict the uniqueness of $Q^{(j)}$. Also $\operatorname{ord} Q_1^{(j+1)} > j+1$ is impossible as in that case $\tilde{Q}^{(j+1)}$ would be of the same type as $Q^{(j+1)}$, but with $\operatorname{ord} \tilde{Q}_1^{(j+1)} < \operatorname{ord} Q^{(j+1)}$. I.e. $\operatorname{ord} Q_1^{(j+1)} = j+1 = \operatorname{ord} \tilde{Q}_1^{(j+1)}$.
Consider now the boolean polynomial $Q' = Q^{(j+1)} - \tilde{Q}^{(j+1)} = yQ_1' + Q_2'$ with $Q_1', Q_2' \in R_{[x_1, \ldots, x_m]}$. Since both $Q_1^{(j+1)}$ and $\tilde{Q}_1^{(j+1)}$ have order $j+1$, $Q_1^{(j+1)} - \tilde{Q}_1^{(j+1)}$ is already a linear combination of $p_0, \ldots, p_j$ and $\operatorname{ord} Q_1' < j+1$. In addition, we have $Q' \neq Q^{(j)}$ as they evaluate to 0 and 1 at $(P_{l_{j+1}}, y_{l_{j+1}})$ respectively, but they both vanish for $(P_i, y_i)$, $i \notin I_j$. This contradicts again the uniqueness of $Q^{(j)}$. Therefore we must have the equality $Q^{(j+1)} = \tilde{Q}^{(j+1)}$ with $\operatorname{ord} Q_1^{(j+1)} = \operatorname{ord} \tilde{Q}_1^{(j+1)} = j+1$. □

*Proof (of Theorem 4.1).* We prove the theorem by induction on the number $j = 0, \ldots, t$ of error positions that are interpolated into $Q^{(j)}$ in addition to the non-erroneous positions.

$j = 0$: Obviously we can choose $\tilde{Q}^{(0)} = (y + f_c) \cdot p_0$ and $Q^{(0)}$ must be of the form $Q^{(0)} = p_0 y + h$ with $h \in R_{[x_1, \ldots, x_m]}$, $\deg h \leq r + \rho$. Both boolean polynomials are of order 0. Now $\tilde{Q}^{(0)} - Q^{(0)} = p_0 \cdot f_c - h$ is a new boolean polynomial vanishing at $i \notin I_0$. Then $(f_{b}c \cdot p_0 + h)(P_i) = 0$ when $i$ is is a non-erroneous position, but $f_{\mathbf{c}} \cdot p_0 + h \neq 0$ would imply that the support of the non-zero $\mathcal{RM}(m, r+\rho)$ word corresponding to $f_{\mathbf{c}} \cdot p_0 + h$ is contained in the support of the error pattern **e**, i.e. that **e** is covered by that word. This contradicts our assumptions and therefore $Q^{(0)} = \tilde{Q}^{(0)}$.

$j \to j+1$: Suppose we have $Q^{(j)} = \tilde{Q}^{(j)}$ with $\operatorname{ord} Q_1^{(j)} = \operatorname{ord} \tilde{Q}_1^{(j)} = j$. By Lemma 4.2 this implies that either $Q = Q^{(j)} = \tilde{Q}^{(j)} = \tilde{Q}$ or after renaming the remaining error positions such that $Q^{(j)}(P_{l_{j+1}}, y_{l_{j+1}}) = 1$, we have $Q^{(j+1)} = \tilde{Q}^{(j+1)}$ with $\operatorname{ord} Q_1^{(j+1)} = \operatorname{ord} \tilde{Q}_1^{(j+1)} = j+1$.

After having interpolated up to $(P_{l_t}, y_{l_t})$, we are done and obtain $Q = \tilde{Q}$ with $\operatorname{ord} Q_1 \leq t$. Note that the order in which we interpolate error positions has no relevance. □

We can now prove that Algorithm 3.1 works in the sense of Theorem 3.2.

*Proof (of Theorem 3.2).* The conditions of Theorem 3.2 and Theorem 4.1 are the same. According to the latter, we have

$$Q = yQ_1 + Q_2 = (y + f_{\mathbf{c}})Q_1$$

after Step 2 in Algorithm 3.1. As mentioned earlier, the boolean polynomial $Q_1$ may be thought of as an error locator since $Q_1(P_i) = 1$ implies that position $i$ is error-free. Consequently, $f_{\mathbf{c}}$ can be obtained when constructing $f \in R_{[x_1,\ldots,x_m]}$ with $\deg f \leq r$ such that $f(P_i) = y_i$ for those $i$ where $Q_1(P_i) = 1$. In other words, $f_{\mathbf{c}}$ is a solution to Step 3 and by its definition also to Step 4. □

Note that Theorem 3.2 implies that if $\mathbf{c}$ and $\mathbf{c}'$ are distinct $\mathcal{RM}(r,m)$ codewords closest to $\mathbf{y}$, then

$$Q = (y + f_{\mathbf{c}})Q_1 = (y + f_{\mathbf{c}'})Q_1 \ ,$$

and both $f_{\mathbf{c}}$ and $f_{\mathbf{c}'}$ are solutions to Step 4.

Further, Step 3 in Algorithm 3.1 could be formulated in two different ways: One can apply Theorem 2.3 and find factors of the form $y + f$ of $Q$ by solving $(y + f + 1)Q = 0$. This is similar to the approach taken in Sudan's algorithm (Sudan [15]). On the other hand $Q = yQ_1 + Q_2 = (y + f_{\mathbf{c}})Q_1$ implies $Q_2 = f_{\mathbf{c}}Q_1$ and a simpler system of linear equations can be obtained. This is similar to the equation solved by the Welch-Berlekamp algorithm (Welch and Berlekamp [17]).

## 5 Error Correction Capability

In this section we will discuss the probability that a random error pattern $\mathbf{e} = \mathbf{c} + \mathbf{y}$ with weight at most $\tau < k_\rho$ will cover a non-zero $\mathcal{RM}(r + \rho, m)$ word. It is essential for the error correction capability of our algorithm that this probability is small.

In the following let $A_0, \ldots, A_n$ be the weight distribution of an arbitrary binary linear code $C$ with parameters $[n, k, d]$. From MacWilliams and Sloane ([8], Chap. 9,§10) we know that a good rule of thumb is

$$A_w \approx 2^{k-n}\binom{n}{w} \ . \tag{2}$$

This rather vague statement is made more precise by Sidelnikov's theorem which shows that the cumulative distribution function associated with the code $C$

$$A(x) = \frac{1}{2^k} \sum_{0 \leq w \leq x}^{n} A_w$$

is in some sense close to the normal cumulative distribution function. But this means that in this sense the function $A$ is also close to the function

$$B(x) = \frac{1}{2^n} \sum_{0 \le w \le x}^{n} \binom{n}{w}$$

which is the cumulative distribution function of the trivial $[n, n, 1]$ code. This justifies (2) somewhat.

The weight distribution of $\mathcal{RM}(5,9)$ is known due to Sugita et al. ([16]) and we compare its cumulative distribution function to $B$ in Fig. 1 (a). The closeness of their course supports the view that Reed-Muller codes are like random binary codes and random binary linear codes among others close to what we will loosely call ***binomially distributed.***

For such a close to binomially distributed $[n, k, d]$ code $C$ we can estimate the probability $\pi_1$ that a random error pattern of weight at most $\tau$ covers a non-zero codeword:

$$\begin{aligned}
\pi_1 &= \frac{|\{\mathbf{e} \in \mathbb{F}_2^n | \mathrm{w}(\mathbf{e}) \le \tau, \exists \mathbf{c} \in C, \mathbf{c} \neq 0 : \mathrm{supp}(\mathbf{c}) \subseteq \mathrm{supp}(\mathbf{e})\}|}{|\{\mathbf{e} \in \mathbb{F}_2^n | \mathrm{w}(\mathbf{e}) \le \tau\}|} \\
&\le \left( \sum_{t=1}^{\tau} \sum_{w=d}^{t} A_w \binom{n-w}{t-w} \right) \cdot \left( \sum_{t=0}^{\tau} \binom{n}{t} \right)^{-1} && (3) \\
&\overset{(2)}{\approx} \left( \sum_{t=1}^{\tau} \sum_{w=d}^{t} 2^{k-n} \binom{n}{w} \binom{n-w}{t-w} \right) \cdot \left( \sum_{t=0}^{\tau} \binom{n}{t} \right)^{-1} \\
&\le 2^{k-n} \left( \sum_{t=0}^{\tau} \sum_{w=0}^{t} \frac{n!}{w!(n-w)!} \frac{(n-w)!}{(t-w)!(n-t)!} \right) \cdot \left( \sum_{t=0}^{\tau} \binom{n}{t} \right)^{-1} \\
&= 2^{k-n} \left( \sum_{t=0}^{\tau} \frac{n!}{t!(n-t)!} \sum_{w=0}^{t} \frac{t!}{w!(t-w)!} \right) \cdot \left( \sum_{t=0}^{\tau} \binom{n}{t} \right)^{-1} \\
&= 2^{k-n} \left( \sum_{t=0}^{\tau} \binom{n}{t} \sum_{w=0}^{t} \binom{t}{w} \right) \cdot \left( \sum_{t=0}^{\tau} \binom{n}{t} \right)^{-1} \\
&= 2^{k-n} \left( \sum_{t=0}^{\tau} \binom{n}{t} 2^t \right) \cdot \left( \sum_{t=0}^{\tau} \binom{n}{t} \right)^{-1} \\
&\le 2^{k-n+\tau} \, . && (4)
\end{aligned}$$

The inequality in (3) is given by counting error patterns with the (un-)desired property by choosing codewords and complementing the support by adding additional non-zero bits.

In Fig. 1 (b) we compare the accurate upper bound (3) for $\pi_1$ in the case of the $\mathcal{RM}(5,9)$ code that we can calculate due to the known weights $A_w$ and the approximate bound (4). This plot demonstrates that the two bounds are close to each other up to $\tau = 130 = 2^9 - k_5$ in an absolute sense, but not relatively. For $\tau > 130$ both bounds rise quickly to large numbers. The reason for this is that the estimate in (3) turns out to be extremely rough for big $\tau$, as different codewords from the considered code can be complemented to give the same error pattern of some weight.

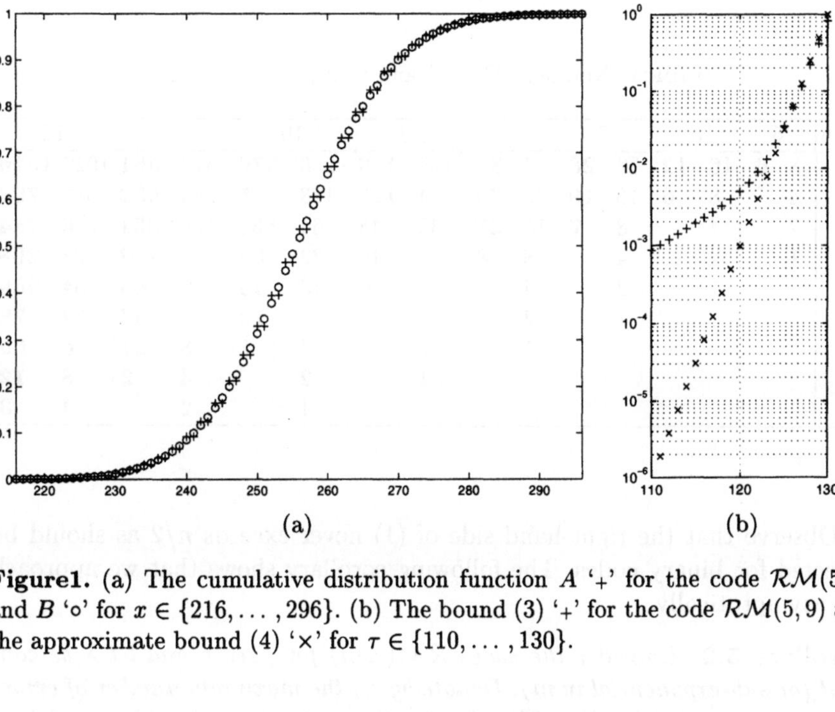

(a) (b)

**Figure1.** (a) The cumulative distribution function $A$ '+' for the code $\mathcal{RM}(5,9)$ and $B$ 'o' for $x \in \{216, \ldots, 296\}$. (b) The bound (3) '+' for the code $\mathcal{RM}(5,9)$ and the approximate bound (4) '×' for $\tau \in \{110, \ldots, 130\}$.

Note that for $r = 0, 1, 2$ the entire weight distributions of $\mathcal{RM}(r,m)$ and their duals are known (MacWilliams and Sloane [8], Chap. 15). For general $\mathcal{RM}(r,m)$ the numbers $A_0, \ldots, A_{5 \cdot 2^{m-r-1}}$ are known (Kasami et al. [6]).

Consequently, if we can prove that Reed-Muller codes of a certain size are approximately binomially distributed which Sidelnikov's theorem is unfortunately too weak for, then the two bounds (3) and (4) are close to each other and the latter can be used as an approximate bound on $\pi_1$ as claimed in Conjecture 3.3.

We can now state the following sufficient conditions that the algorithm decodes correctly.

**Proposition 5.1.** *Consider the Reed-Muller code $\mathcal{RM}(r,m)$ and let*

$$\tau = \max_{0 \le \rho \le m} \min \{2^m - k_{r+\rho} - \lambda, k_\rho - 1\}, \quad \lambda \in \mathbb{N}_0 \ . \tag{5}$$

*If $\pi_1 \le 2^{k_{r+\rho} - 2^m + \tau}$, then Algorithm 3.1 correctly decodes up to $\tau$ random errors that are independently and uniformly distributed, with probability of failure at most $2^{-\lambda}$.*

*Proof.* Obviously if $\pi_1 \le 2^{k_{r+\rho}-2^m+\tau}$ and $\tau \le 2^m - k_{r+\rho} - \lambda$, then $\pi_1 \le 2^{-\lambda}$. But $\tau < k_\rho$ is also required by Theorem 3.2. □

From (5) we immediately get (1). Check Table 1 for a comparison of half the minimum distance $2^{m-r-1}$ and the numbers $\tau$. For small $m$ or high $r$, the number $\tau$ is lower than half the minimum distance, because the probability that the error pattern covers a higher order Reed-Muller word is too large.

**Table1.** Numbers $2^{m-r-1}$ and $\tau$ in (5) for $\lambda = 10$.

| $r \backslash m$ | 5 | | 6 | | 7 | | 8 | | 9 | | 10 | | 11 | | 12 | |
|---|---|---|---|---|---|---|---|---|---|---|---|---|---|---|---|---|
| 1 | 8 | **6** | 16 | **12** | 32 | **28** | 64 | **83** | 128 | **129** | 256 | **376** | 512 | **561** | 1024 | **1576** |
| 2 | 4 | | 8 | **6** | 16 | **19** | 32 | **36** | 64 | **120** | 128 | **175** | 256 | **552** | 512 | **793** |
| 3 | 2 | | 4 | | 8 | **7** | 16 | **27** | 32 | **45** | 64 | **166** | 128 | **231** | 256 | **784** |
| 4 | 1 | | 2 | | 4 | | 8 | **8** | 16 | **36** | 32 | **55** | 64 | **222** | 128 | **298** |
| 5 | | | 1 | | 2 | | 4 | | 8 | **9** | 16 | **46** | 32 | **66** | 64 | **289** |
| 6 | | | | | 1 | | 2 | | 4 | | 8 | **10** | 16 | **57** | 32 | **78** |
| 7 | | | | | | | 1 | | 2 | | 4 | **1** | 8 | **11** | 16 | **69** |
| 8 | | | | | | | | | 1 | | 2 | | 4 | **2** | 8 | **12** |
| 9 | | | | | | | | | | | 1 | | 2 | | 4 | **3** |

Observe that the right-hand side of (1) never exceeds $n/2$ as should be expected for binary codes. The following corollary shows that we approach $n/2$ asymptotically.

**Corollary 5.2.** *Consider the codes $\mathcal{RM}(r,m)$ for fixed $r$ and let $\lambda$ be constant (or sub-exponential in $m$). Denote by $\tau_m$ the maximum number of errors that we can correct by Algorithm 3.1 for the particular value of $m$ and $\lambda$ as in (5). Define the maximum error rate $\varepsilon_m = \tau_m/2^m$. If $\pi_1 \le 2^{k_{r+\rho}-2^m+\tau}$, then*

$$\lim_{m\to\infty} \varepsilon_m = \frac{1}{2}\ .$$

*Proof.* Choose $\rho = \lceil \frac{m-2}{2} \rceil - r$. We show that in this case each of the two expressions in $\min\{2^m - k_{r+\rho} - \lambda, k_\rho - 1\}$ normalized by $2^m$ converges to $1/2$ when taking $m$ to infinity. The first is

$$2^m - \sum_{i=0}^{r+\rho} \binom{m}{i} - \lambda = 2^m - \sum_{i=0}^{\lceil \frac{m-2}{2} \rceil} \binom{m}{i} - \lambda \ge 2^{m-1} - \lambda$$

and the second expression is

$$\sum_{i=0}^{\rho} \binom{m}{i} - 1 \ge \sum_{i=0}^{\lceil \frac{m}{2} \rceil} \binom{m}{i} - \sum_{i=\lceil \frac{m-2}{2} \rceil - r+1}^{\lceil \frac{m}{2} \rceil} \binom{m}{i} \ge 2^{m-1} - r \binom{m}{\lceil \frac{m}{2} \rceil}\ .$$

Both lower bounds converge to 1/2 when normalizing by $2^m$ and $m \to \infty$, the latter by the estimate on binomial coefficients (16) in MacWilliams and Sloane ([8]), Chap. 10. Since one of the values is always below 1/2 for the different $\rho$, the result follows. □

Corollary 5.2 shows that in this setting the decoder is asymptotically capable of correcting up to the maximum number of errors possible.

## 6 Complexity of the Algorithm

We would like to estimate the running time of Algorithm 3.1 as a function in the block length $n = 2^m$ of the Reed-Muller code $\mathcal{RM}(r,m)$. In the following we will assume that solving a system of linear equations takes time $O(a^3)$ with $a$ being the maximum of the number of unknowns and the number of equations (using for example Gaussian elimination).

The complexity of *Step 1* is negligible and can be done once and for all.

A straight-forward implementation of *Step 2* which finds the boolean polynomial $Q$, would repeatedly solve a homogeneous system of equations for increasing order of $Q_1$. The unknowns are the coefficients of $Q_1$ and $Q_2$ which are chosen as linear combinations of the $p_j$ and boolean monomials of degree $\leq r + \rho$ respectively. This gives a complexity of $O(n^4)$. However, as increasing the order of $Q_1$ by 1 simply involves adding another equation and another unknown, a more efficient approach would be to solve the progressively larger system of equations by reusing old knowledge. This gives a complexity of $O(n^3)$ for Step 2. Approaches that exploit the structure of the system of equations might lower this complexity. In Sudan's algorithm for Reed-Solomon codes such an approach is possible where an $O(sn^2)$ interpolation algorithm is described for low values of $s$ (Nielsen and Høholdt [11]).

*Step 3* should find $f \in R_{[x_1,\dots,x_m]}$ such that $\deg f \leq r$, $f(P_i) = y_i$ for those $i$ where $Q_1(P_i) = 1$. In particular if $Q_1 = 1$, then $Q_2(P_i) = y_i$ for all $i$. This is desirable to happen when no errors occurred such that $Q = y + f$ where $f$ corresponds to the sent codeword. Therefore the choice $p_0 = 1$ is advantageous. In general Step 3 defines a system of linear equations and under the conditions of Theorem 3.2 there is a solution. Finding one of those as well as a basis of the solution space of the corresponding homogeneous system of linear equations will then represent all solutions to Step 3. Let $\mathbf{q}_1$ denote the $\mathcal{RM}(\rho,m)$ word that corresponds to $Q_1$. As each non-zero bit in $\mathbf{q}_1$ results in an equation, the complexity of this step depends on the weight of $\mathbf{q}_1$. In the worst case this gives a complexity of $O(n^3)$. In the best case, the first $k_r$ non-zero bits of $\mathbf{q}_1$ treated correspond to information bits of a $\mathcal{RM}(r,m)$ word. Then $k_r$ equations suffice to compute the $k_r$ unknown coefficients of $f$. If the evaluation of $Q_1$ is derived by a fast Fourier-like transform, the overall complexity becomes $O(k_r^3 + n \log n)$. Under all circumstances, however, the complexity of Step 3 is dominated by Step 2.

The complexity of *Step 4* depends on the number of solutions to the equation solved in Step 3. Let $\ell$ be the dimension of the solution space of the homogeneous system from Step 3. Evaluating every solution of which there are $2^\ell$, in all $n = 2^m$ points using a fast Fourier approach then gives a complexity of $O(2^\ell n \log n)$. The overall running time is now $O(2^\ell n \log n + n^3)$. As this is exponential in $\ell$, even moderate values of $\ell$ can give a high running time. However if there is only one solution in Step 3, i.e. $\ell = 0$, then the running time is cubic in $n$. There are strong indications that this is often the case if $p_0 = 1$ and $p_1, \dots, p_{k_\rho - 1}$ are chosen randomly. The conditions for $\ell$ to be zero follow.

**Proposition 6.1.** *Let $\ell$ be the dimension of the solution space of the corresponding homogenous system of equations in Step 3. Furthermore, let $\mathbf{q}_1$ be the $\mathcal{RM}(\rho, m)$ codeword corresponding to $Q_1$. The following three cases are then equivalent*

*(i) $\ell = 0$, i.e., there is exactly one solution to the equation in Step 3;*
*(ii) $\mathbf{q}_1$ is not covered by any $\mathcal{RM}(r, m)$ codeword $\neq (1, 1, \dots, 1)$;*
*(iii) $Q_1$ has no factors $\neq 1$ of degree at most $r$ (possibly including $Q_1$) itself.*

*Proof.* We show each of the following implications by contradiction.

$1 \Rightarrow 2$: Let $f$ be a solution to Step 3, i.e. $(y + f)Q_1 = Q$. Assume that there exists a $\mathcal{RM}(r, m)$ codeword $\mathbf{c} \neq (1, 1, \dots, 1)$ that covers $\mathbf{q}_1$ and the corresponding boolean polynomial $f_\mathbf{c}$ of degree $\leq r$. Then $(f_\mathbf{c} + 1)(P_i) = 0$ when $Q_1(P_i) = 1$. Since $f + f_\mathbf{c} + 1$ now evaluates to $y_i$ at those $P_i$ where $Q_1(P_i) = 1$ and is of degree $\leq r$, this implies that $f + f_\mathbf{c} + 1$ is also a solution to Step 3. The fact that $\mathbf{c} \neq (1, 1, \dots, 1)$ implies $f \neq f + f_\mathbf{c} + 1$ and we get a contradiction to the assumption.

$2 \Rightarrow 3$: Assume that there is a factor $f_\mathbf{c} \neq 1$ of $Q_1$ with $\deg f_\mathbf{c} \leq r$. This implies that $(f_\mathbf{c} + 1)Q_1 = 0$ by Theorem 2.3. Let $\mathbf{c} \neq (1, 1, \dots, 1)$ be the $\mathcal{RM}(r, m)$ codeword corresponding to $f_\mathbf{c}$. Then $(f_\mathbf{c} + 1)$ evaluates to 0 at those $P_i$ where $Q_1(P_i) = 1$ and $\mathbf{c} \neq (1, 1, \dots, 1)$ must cover $q_1$. This contradicts the conditions.

$3 \Rightarrow 1$: Assume that $\ell > 0$, i.e. the equation in Step 3 has two distinct solutions $f_1$ and $f_2$. Then by Theorem 4.1 $(y + f_1)Q_1 = Q_2$ and $(y + f_2)Q_1 = Q_2$. Therefore $(f_1 + f_2)Q_1 = 0$, i.e., $f_1 + f_2 + 1 \neq 1$ is a factor of $Q$ by Theorem 2.3 which contradicts our assumption. □

Proposition 6.1 makes it apparent that $2^\ell$ is the number of factors $\neq 1$ in $Q_1$ of degree at most $r$ and we are able to prove Theorem 3.4.

*Proof (of Theorem 3.4).* Follows directly from Proposition 6.1 and the running time considerations for Step 4. □

Of course the question is, how often does $Q_1$ have factors of degree $\leq r$? Can we choose $Q$ such that $Q_1$ has as few factors of that kind as possible?

When there are more than one codeword **c** closest to the received word **y**, then it is desired that $Q_1$ has factors $f \neq 1$ with $\deg f \leq r$. By Theorem 4.1 this will also happen.

In any case there will definitely be multiple solutions if $r \geq \rho$ and $Q_1 \neq 1$, since then $Q_1$ has itself as a factor of degree $\rho$. If one uses for example $p_0 = 1$, $p_1 = x_1$, $p_2 = x_2, \ldots, p_{k_\rho - 1} = x_{m-\rho+1} \cdots x_{m-1}x_m$ and there are only few errors, then $Q_1$ will be of small degree which might be $\leq r$. As we have seen this is not desirable. Using boolean monomials therefore actually results in a counter intuitive situation where the decoder does not work properly if there are too few errors! By injecting additional errors, one can "fix" this.

The nicer approach however is to choose $r < \rho$, $p_0 = 1$ and $p_1, \ldots, p_{k_\rho - 1}$ at random. The boolean polynomial $Q_1$ will then also be random most likely of degree $\rho$ constrained only by the requirement that $Q_1(P_i) = 0$ for those $i$ that are error positions for some **c** closest to **y**. When choosing the $p_j$ randomly as above in our simulation of Algorithm 3.1 with higher-order Reed-Muller codes of length up to 512, we practically always got only one solution to Step 3 and therefore to Step 4. In other words, a random $Q_1$ with $\deg Q_1 \leq \rho$ and $Q_1(P_i) = 0$ when $i$ is an error position, seems to have very slim chances of having factors of degree $\leq r$. The correct probability does not seem easy to establish or even to estimate although it appears to be very low. Asking when a completely random pattern covers a codeword or vice versa is easier to answer using arguments like those in Sec. 5. For self-dual codes like $\mathcal{RM}(\frac{m-1}{2}, m)$, $m$ odd, one can even upper-bound the probability that a randomly chosen codeword covers another randomly chosen codeword by using generalizations of Gleason's theorem to biweight enumerators (MacWilliams et al. [7]). All these probabilities are asymptotically low. This leads us to Conjecture 3.5 which states that the related probability that we seek also goes to zero as the code length increases.

Even if $\ell > 0$ is too large to search through all the solutions for the correct boolean polynomials, one might choose other $p_0, \ldots, p_{k_\rho - 1}$ resulting in another solution to Step 2, but with the same desired factors by Theorem 4.1. Similarly one can just choose another order of the previous $p_j$. In fact a solution with respect to the original $p_j$ with higher order $Q_1$ will correspond to solutions to Step 2 with respect to a different ordering of the $p_j$. Depending on how Step 2 is implemented one might obtain several boolean polynomials $Q$ automatically. A new $Q \in R_{[x_1, \ldots, x_m, y]}$ will result in a new system of equations in Step 3. Considering this new system together with the old one, should hopefully decrease the number of solutions. The procedure can of course be iterated to further lower the number of solutions and this approach works well in practice. Notice that if one can bound the probability $\pi_2$ of having many solutions just non-negligibly away from 1, then one might also be able to construct a polynomial-time algorithm which fails with negligible probability by simply repeating Steps 1 and 2 with new orderings in the above manner until the combined system of equations in Step 3 has only the closest codewords as solutions.

Furthermore, the algorithm does not have to process all the points. One can stop as soon as Step 3 has an adequate number of solutions. In a situation with side information, a good approach would be to process the most reliable points first. Erasures are dealt with by simply omitting the points in question from the interpolation.

The memory requirement of the algorithm is $O(n^2)$ for storing the various systems of equations.

# 7 Conclusions

Algorithm 3.1 has been implemented in the computer algebra system Singular (Greuel et al. [3]). Among other codes, it successfully decoded $\mathcal{RM}(2,9)$ with half the minimum distance being 64 using $\rho = 3$. Due to the known weight distribution of $\mathcal{RM}(2+3,9)$, we know by Fig. 1 (b) that the failure probability $\pi_1$ for up to 122 random errors is at most 0.01. In each of 10 tries with 120 errors, Step 2 of the algorithm produced only one candidate so no time was wasted filtering away wrong boolean polynomials. The time spent decoding one word was approximately one minute for an unoptimized implementation running on a Pentium 200 MHz under Linux.

Our algorithm improves the results given by Ar et al. ([1]) concerning the reconstruction of "noisy multivariate polynomials" in the binary case. It might be possible to extend our method to work also with generalized Reed-Muller codes which have arbitrary finite fields as alphabets. Further Ar et al. ([1]) present a more general version of Sudan's algorithm not directly related to coding theory which finds relations $f(x,y) = 0$ that hold with high probability. The same sort of generalization ought to be possible in the Reed-Muller setting. It would allow the discovery of probabilistic relations of the type $f'(x_1,\dots,x_{m_1},y_1,\dots,y_{m_2}) = 0$ as opposed to $y = f(x_1,\dots,x_{m_1})$.

The algorithm can be used to generalize the cryptanalytical results of Jakobsen ([4]) and Jakobsen and Knudsen ([5]) to a kind of probabilistic interpolation attack for bit-oriented block ciphers. Such work is currently in progress. For these purposes it sometimes suffices to know whether there exists a low-degree approximation without actually obtaining it. In this case, the first two steps of the algorithm are enough. If the resulting boolean polynomial $Q$ has a low degree, then there must exist a low degree approximation. The algorithm might also be applicable to areas like hardware optimization and construction of binary decision trees.

Finally, to verify the conjectures in practice for longer codes of higher order, more simulation results are needed. To achieve this, an optimized, low-level implementation of the algorithm using direct bit manipulations would be useful.

## Acknowledgments

The authors wish to thank Tom Høholdt and Jørn Justesen for invaluable discussions.

## References

1. Ar, S., R. J. Lipton, and M. Sudan (1999). Reconstructing Algebraic Functions from Erroneous Data. *SIAM Journal on Computing 28*(2), 487–510.
2. Berlekamp, E. R. (1996). Bounded Distance +1 Soft-Decision Reed-Solomon Decoding. *IEEE Transactions on Information Theory 42*(3), 704–720.
3. Greuel, G.-M., G. Pfister, and H. Schönemann (1998). *Singular Version 1.2 User Manual.* Centre for Computer Algebra, University of Kaiserslautern. In *Reports On Computer Algebra*, number 21. `http://www.mathematik.uni-kl.de/~zca/Singular`.
4. Jakobsen, T. (1998). Cryptanalysis of Block Ciphers with Probabilistic Non-linear Relations of Low Degree. In H. Krawczyk (Ed.), *Advances in Cryptology - CRYPTO '98*, Volume 1462 of *Lecture Notes in Computer Science*, pp. 212–222. Springer.
5. Jakobsen, T. and L. Knudsen (1999). Attacks on Block Ciphers of Low Algebraic Degree. Submitted to *Journal of Cryptology.* Preliminary version: The Interpolation Attack on Block Ciphers. In E. Biham (Ed.) (1997), *Fast Software Encryption*, Volume 1267 of *Lecture Notes in Computer Science*, pp. 212–222. Springer.
6. Kasami, T., T. Nobuki, and S. Azumi (1976). On the Weight Enumeration of Weights Less than 2.5$d$ of Reed-Muller Codes. *Information and Control 30*, 380–395.
7. MacWilliams, F. J., C. L. Mallows, and N. J. A. Sloane (1972). Generalizations of Gleason's Theorem on Weight Enumerators of Self-Dual Codes. *IEEE Transactions on Information Theory 18*(6), 794–805.
8. MacWilliams, F. J. and N. J. A. Sloane (1977). *The Theory of Error-Correcting Codes.* North-Holland Mathematical Library.
9. Menezes, A. J., P. C. van Oorschot, and S. A. Vanstone (1996). *Handbook of Applied Cryptography.* Discrete Mathematics and its Applications. CRC Press.
10. Muller, D. E. (1954). Application of Boolean Algebra to Switching Circuit Design and to Error Detection. *IEEE Transactions on Computers 3*, 6–12.
11. Nielsen, R. R. and T. Høholdt (1999). Decoding Reed-Solomon Codes beyond Half the Minimum Distance. In J. Buchmann, T. Høholdt, H. Stichtenoth, and H. Tapia-Recillas (Eds.), *Coding Theory, Cryptography, and Related Areas.* Springer.
12. Reed, I. S. (1954). A Class of Multiple-Error-Correcting Codes and the Decoding Scheme. *IEEE Transactions on Information Theory 4*, 38–49.
13. Shimoyama, T. and T. Kaneko (1998). Quadratic Relation of S-box and Its Application to the Linear Attack of Full Round DES. In H. Krawczyk (Ed.), *Advances in Cryptology - CRYPTO '98*, Volume 1462 of *Lecture Notes in Computer Science*, pp. 200–211. Springer.
14. Shokrollahi, M. A. and H. Wassermann (1998). Decoding of Algebraic-Geometric Codes beyond the Error-Correction Bound. In *Proceedings of the Thirtieth Annual ACM Symposium on Theory of Computing*, pp. 241–248.

15. Sudan, M. (1997). Decoding of Reed-Solomon Codes beyond the Error-Correction Bound. *Journal of Complexity 13*, 180–193.
16. Sugita, T., T. Kasami, and T. Fujiwara (1996). Weight Distributions of the Third and Fifth Order Reed-Muller Codes of Length 512. Technical Report NAIST-IS-TR96006, Nara Institute of Science and Technology. `http://isw3.aist-nara.ac.jp/IS/TechReport2/report/96006.ps`.
17. Welch, L. R. and E. R. Berlekamp (1986). Error Correction of Algebraic Block Codes. U.S. Patent 4 633 470, issued Dec. 30.

# On Binary Cyclic Codes With Few Weights

Henk D. L. Hollmann[1] and Qing Xiang[2]

[1] Philips Research Laboratories
Prof. Holstlaan 4
5656 AA Eindhoven
The Netherlands
e-mail: hollmann@natlab.research.philips.com

[2] Department of Mathematical Sciences
University of Delaware
Newark DE 19716
USA
e-mail: xiang@math.udel.edu

**Abstract** Let $C_{t_0,t_1,\ldots,t_r}$ denote the binary cyclic code of length $n = 2^m - 1$ with defining zeros $\alpha^{t_0}, \alpha^{t_1}, \ldots, \alpha^{t_r}$, where $\alpha$ is a primitive element of $\mathrm{GF}(2^m)$. Using the method in [8], we determine the weight distribution of the following cyclic codes.
(i) $C_{1,t_1,t_2}$, where $m = 2r + 1$, $t_1 = 2^r + 1$, $t_2 = 2^{r-1} + 1$. (This code appeared in Research Problem 9.7 of MacWilliams and Sloane [14].)
(ii) $C_{1,t,t^2}$, where $m = 2r + 1$, $t = 1 + 2^{r+1}$. (This code appeared in a conjecture of Chang, Gaal, Golomb, Gong, and Kumar [5].)
(iii) Several cyclic codes in the paper of Van Lint and Wilson [12].
(iv) $C_{1,t}$, where $m = 2r$, $t = \sum_{i=0}^{r} 2^{ik}$, $\gcd(m, k) = 1$.

## 1 Introduction

In this paper, we continue our investigations of binary cyclic codes of length $n = 2^m - 1$ with only a few defining zeros. Unless otherwise stated, we will always use $\alpha$ to denote a fixed primitive element of the finite field $\mathrm{GF}(2^m)$. The minimal polynomial of $\alpha^i$ over $\mathrm{GF}(2)$ will be denoted by $m_i(X)$. Let $C$ be a binary cyclic code of length $n = 2^m - 1$. As is customary in coding theory, we identify $C$ with the corresponding ideal in $\mathrm{GF}(2)[X]/(X^n - 1)$. We shall say that $C$ has *defining zeros* $\alpha^{t_0}, \alpha^{t_1}, \ldots, \alpha^{t_r}$ if $C$ has generator polynomial $m_{t_0}(X)m_{t_1}(X)\cdots m_{t_r}(X)$; in this case we write $C = C_{t_0,t_1,\ldots,t_r}$. (Note that if we use the terminology in the paper by Van Lint and Wilson [12], then $\{\alpha^{t_0}, \alpha^{t_1}, \ldots, \alpha^{t_r}\}$ should be called a *minimal* defining set for $C$.) There are many cases in which the dual code $C^\perp$ of such a cyclic code $C$ has only a few nonzero weights. The most notable examples of such codes $C$ are the Hamming codes and the binary 2-error-correcting BCH codes. In [8], we have developed a general method to prove that $C^\perp$ has only a few nonzero weights, and applied this method to several classes of cyclic codes with two defining zeros. The cyclic codes that we treated uniformly using the method in [8] include binary cyclic codes $C_{1,t}$ of length $n = 2^m - 1$, where

(a) (the Gold case) $t = 1 + 2^r$, $m/(m, r)$ odd;
(b) (the Kasami case) $t = 2^{2r} - 2^r + 1$, $m/(m, r)$ odd;
(c) (the Welch case) $m = 2r + 1$ odd, $t = 2^r + 3$; and
(d) (the Niho case) $m$ odd, $4r \equiv -1 \bmod m$, $t = 2^{2r} + 2^r - 1$.

each of the above cases, we determined the weight distribution of the cyclic code $C_{1,t}$. We remark that the problem of determining the weight distribution of $C_{1,t}$ is the same as that of determining the crosscorrelation spectrum of two binary m-sequences. Therefore we have solved the problem of determining the crosscorrelation spectrum in the Welch and Niho cases. We note that Canteaut, Charpin and Dobbertin [3] also solved the Welch case.

In this paper, we shall use the same method as in [8] to treat several classes of cyclic codes of length $n = 2^m - 1$ with two or three defining zeros. These include the following codes.
(i) $C_{1,t_1,t_2}$, where $m = 2r+1$, $t_1 = 2^r+1$, $t_2 = 2^{r-1}+1$. (This code appeared in Research Problem 9.7 of MacWilliams and Sloane [14].)
(ii) $C_{1,t,t^2}$, where $m = 2r+1$, $t = 1+2^{r+1}$. (This code appeared in a conjecture of Chang, Gaal, Golomb, Gong, and Kumar [5].)
(iii) Several cyclic codes in the paper of Van Lint and Wilson [12].
(iv) $C_{1,t}$, where $m = 2r$, $t = \sum_{i=0}^{r} 2^{ik}$, $\gcd(m, k) = 1$.

In each of the above cases, we determine the weight distribution of the cyclic codes under consideration.

## 2 Preliminaries

In this section, we shall recall the main ingredients of the method in [8] for showing that certain binary cyclic codes $C^{\perp}_{t_0,t_1,\ldots,t_r}$ of length $n = 2^m - 1$ have only a few nonzero weights. To apply this method to show that $C^{\perp}_{t_0,t_1,\ldots,t_r}$ is a $2s$–weight code, we need to know the first few terms $A_i$, $i = 0, 1, \ldots, 2s$, of the weight enumerator of the code $C = C_{t_0,t_1,\ldots,t_r}$ and the highest power of 2 dividing all weights of $C^{\perp} = C^{\perp}_{t_0,t_1,\ldots,t_r}$.

Briefly, this method works as follows. For given integers $w_1, \ldots, w_{2s}$ with $0 < w_1 \leq w_2 \leq \cdots \leq w_{2s} \leq n$, define

$$g(x) = (x - w_1)(x - w_2)\cdots(x - w_{2s})$$

and consider the expression

$$E = E(w_1, \ldots, w_{2s}) = \sum_{c \in C^{\perp} \setminus \{0\}} g(w(c)) = \sum_{w=1}^{n} g(w) B_w,$$

where $w(c)$ is the Hamming weight of a codeword $c$ and $B_0, B_1, \ldots, B_n$ denote the weight distribution of the code $C^{\perp}$.

On the one hand, since $g(x)$ is a polynomial of degree $2s$, we can use the Pless power moment identities [14] to express $E$ in terms of $n$, $k =$

$\dim(C^{\perp}_{t_0,t_1,\dots,t_r})$, and $A_1, \dots, A_{2s}$. Of course we have $E = 0$ if the $w_i$ include all nonzero weights of the code $C^{\perp}_{t_0,t_1,\dots,t_r}$. On the other hand, suppose that we know or can prove that the code $C^{\perp}$ has no weights in the intervals $(w_1, w_2), (w_3, w_4), \dots, (w_{2s-1}, w_{2s})$, then obviously $E \geq 0$, with equality if and only if all nonzero weights of $C^{\perp}$ are contained in the set $\{w_1, \dots, w_{2s}\}$.

In the following four subsections, we briefly review the techniques used in [8] to obtain the information required to apply this method.

## 2.1 The Largest Power of 2 Dividing all Weights of $C^{\perp}$

As in [8], we may use the following theorem of McEliece to obtain information on the largest power of 2 dividing all weights of $C^{\perp}_{1,t_1,\dots t_r}$.

**Theorem 2.1.** [13] *Let $C$ be a binary cyclic code, and let $\ell$ be the smallest positive integer such that $\ell$ nonzeros of $C$ (with repetitions allowed) have product 1. Then the weight of every codeword in $C$ is divisible by $2^{\ell-1}$, and there is at least one weight which is not divisible by $2^{\ell}$.*

Let $m$ be a positive integer. For a nonnegative integer $a$ with binary representation $a = \sum_{i=0}^{m-1} a_i 2^i$, $a_i \in \{0, 1\}$ for all $i$, we define the (binary) weight $w(a)$ of $a$ as

$$w(a) = \sum_{i=0}^{m-1} a_i.$$

For $t_1, \dots, t_r \in \mathbf{Z}_{2^m-1}$, we let

$$M(m; t_1, \dots, t_r) = \max\left(w(s) - \sum_{i=1}^{r} w(a^{(i)})\right),$$

where the maximum is over all integers $s, a^{(1)}, \dots, a^{(r)}$ for which

$$0 \leq s, a^{(1)}, \dots, a^{(r)} \leq 2^m - 1,$$

and

$$s \equiv \sum_{i=1}^{r} t_i a^{(i)} \bmod 2^m - 1, a^{(i)} \not\equiv 0 \bmod 2^m - 1 \text{ for some } i.$$

The following is a straightforward generalization of Theorem 12 in [8].

**Theorem 2.2.** *Let $C_{1,t_1,\dots,t_r}$ be the binary cyclic code of length $n = 2^m - 1$ with defining zeros $\alpha, \alpha^{t_1}, \dots, \alpha^{t_r}$, where $\alpha$ is a primitive element of* $\mathrm{GF}(2^m)$ *and $t_i \in \mathbf{Z}_n \setminus \{0\}$. Let $M(m; t_1, t_2, \dots, t_r)$ be defined as above. Then the weight of every codeword of $C^{\perp}_{1,t_1,\dots,t_r}$ is divisible by $2^{m-M(m;t_1,t_2,\dots,t_r)-1}$, and there is at least one weight which is not divisible by $2^{m-M(m;t_1,\dots,t_r)}$.*

*Proof.* Let $t_0 = 1$. By definition, the zeros of $C_{1,t_1,\ldots,t_r}$ are $\alpha^{2^j t_i}$, $0 \le j \le m-1$, $0 \le i \le r$, so the nonzeros of $C^{\perp}_{1,t_1,\ldots,t_r}$ are precisely the members of the set $\{\alpha^{-2^j t_i} \mid 0 \le j \le m-1, 0 \le i \le r\}$. Let

$$N = \{-2^j t_i \mid 0 \le j \le m-1, 0 \le i \le r\}.$$

We want to find the smallest $\ell \ge 1$ such that $i_1, i_2, \ldots, i_\ell$ come from the set $N$ and

$$\alpha^{i_1+i_2+\cdots+i_\ell} = 1. \tag{1}$$

Since $N \subseteq \mathbf{Z}_{2^m-1}$ is closed under multiplication by 2, we may assume without loss of generality that all $i_j$ are distinct. Write

$$a^{(i)} = \sum 2^j,$$

where the sum is over all $j$ such that $-2^j t_i \in \{i_1, \ldots, i_\ell\}$. Then (1) holds with $\ell \ge 1$ if and only if

$$\sum_{i=0}^{r} t_i a^{(i)} \equiv 0 \bmod 2^m - 1 \tag{2}$$

and

$$\ell = \sum_{i=0}^{r} w(a^{(i)}) \ge 1. \tag{3}$$

Write

$$s = 2^m - 1 - a^{(0)}.$$

Then $w(s) = m - w(a^{(0)})$, and (2), (3) are equivalent to

$$s \equiv \sum_{i=1}^{r} t_i a^{(i)} \bmod 2^m - 1, \tag{4}$$

$$\ell = m - w(s) + \sum_{i=1}^{r} w(a^{(i)}), \tag{5}$$

and

$$a^{(i)} \not\equiv 0 \bmod 2^m - 1 \text{ for some } i \text{ or } s \ne 2^m - 1. \tag{6}$$

From the above, we conclude that the smallest $\ell \ge 1$ for which (1) holds is equal to

$$m - \max\left( w(s) - \sum_{i=1}^{r} w(a^{(i)}) \right), \tag{7}$$

where $s, a^{(1)}, \ldots, a^{(r)}$ satisfy (4) and (6). To complete the proof, observe that, under the conditions (6), the maximum in (7) is never attained when $a^{(i)} \equiv 0 \bmod 2^m - 1$ for all $i$. □

## 2.2 Modular Add-with-carry Algorithms in $\mathbf{Z}_{2^m-1}$

For the determination of $M(m; t_1, \ldots, t_r)$, we often use the following add-with-carry algorithm for integers modulo $2^m - 1$.

Let $a^{(i)}$, $1 \leq i \leq r$, be nonnegative integers with $0 \leq a^{(i)} \leq 2^m - 1$, and let $a^{(i)}$ have binary representation $a^{(i)} = a_{m-1}^{(i)} \cdots a_0^{(i)}$. Furthermore, let $t_1, t_2 \ldots, t_r$ be nonzero integers. Define $t_+ = \sum_{i,\, t_i > 0} t_i$ and $t_- = \sum_{i,\, t_i < 0} t_i$ so that

$$\sum_{i=1}^{r} t_i = t_+ + t_-, \qquad t_+ \geq 0, \quad t_- \leq 0.$$

Let $s$ be an integer with $0 \leq s \leq 2^m - 1$, with binary representation $s = s_{m-1} \cdots s_0$, say, and suppose that

$$s \equiv \sum_{i=1}^{r} t_i a^{(i)} \bmod 2^m - 1.$$

**Theorem 2.3.** [8] *With the above definitions and assumptions, we have the following.*
*(i) There exist unique integers $c_{-1}, c_0, \ldots, c_{m-1}$ with $c_{-1} = c_{m-1}$ such that*

$$2c_j + s_j = \sum_{i=1}^{k} t_i a_j^{(i)} + c_{j-1}, \quad 0 \leq j \leq m-1. \tag{8}$$

*Moreover, with $w(c) = \sum_{i=0}^{m-1} c_i$, we have that*

$$w(c) = \sum_{i=0}^{r} t_i w(a^{(i)}) - w(s). \tag{9}$$

*The numbers $c_j$ satisfy $t_- - 1 \leq c_j \leq t_+$, and furthermore*

$$t_- \leq c_j < t_+ \tag{10}$$

*for all $j$ if $a^{(i)} \not\equiv 0 \bmod 2^m - 1$ holds for some $i$.*
*The numbers $s_j$ and $c_j$ will be referred to as the* digits *and* carries, *respectively, for the computation modulo $2^m - 1$ of the number $s$.*
*(ii) If $\bar{s}_j = 1 - s_j$, $\bar{a}_j^{(i)} = 1 - a_j^{(i)}$, and $\tilde{c}_j = t_+ + t_- - 1 - c_j$ for all $i$ and $j$, then the $\bar{s}_j$ and $\tilde{c}_j$ are the digits and carries for the modular computation of $\bar{s} \equiv -s \equiv \sum_{j=1}^{k} t_j \bar{a}^{(j)} \bmod 2^m - 1$.*

For the proof of this theorem, we refer the reader to [8]. To stress the periodic nature of this *modular* add-with-carry algorithm, we will often consider all indices of the digits and carries as numbers from $\mathbf{Z}_m$.

### 2.3 Counting Low-weight Codewords

Let $C$ be a binary cyclic code of length $n = 2^m - 1$. Let $A_i$ denote the number of codewords in $C$ of weight $i$. Here we briefly discuss a technique to determine $A_i$ for small $i$ as developed in [11], [12], and [8]. First, we fix some notation.

Let $c(X) = \sum_{i=1}^{k} X^{u_i}$ with $0 \le u_1 < u_2 < \cdots < u_k < n$ be a codeword of weight $k$ in $C$. Denote by $\mathcal{Z}_\rfloor$, the set of zero exponents $\{i \in \mathbf{Z}_n \mid c(\alpha^i) = 0\}$ of the word $c$. For each $a \in \mathbf{Z}_n$, define the vector $v(a)$ in $\mathrm{GF}(2^m)^k$ by

$$v(a) = (\alpha^{au_1}, \alpha^{au_2}, \dots, \alpha^{au_k}).$$

Also we define $p_c(X) = \prod_{i=1}^{k}(X - \alpha^{u_i})$, and use $\mathcal{R}$ to denote the ring $\mathrm{GF}(2^m)[X]/(p_c(X))$. Note that by the Chinese Remainder Theorem, the ring $\mathcal{R}$ is isomorphic to the vector space $\mathrm{GF}(2^m)^k$ with coordinatewise multiplication. Let $A$ be any subset of $\mathbf{Z}_n$. We say that $A$ is *independent* if the corresponding set of vectors $V(A) = \{v(a) \mid a \in A\}$ is independent over $\mathrm{GF}(2^m)$, or, equivalently, if the set of residue classes $\{X^a \mid a \in A\}$ in $\mathcal{R}$ is independent over $\mathrm{GF}(2^m)$. Note that an independent set has size at most $k$.

With the above notation, we now have the following.

(i) For all $r, s \in \mathbf{Z}_n$, we have that $v(r)v(s)^\top = c(\alpha^{r+s})$. So if $A$ is independent, $A \subseteq \mathcal{Z}_\rfloor$, and $s \notin \mathcal{Z}_\rfloor$, then $A \cup \{s\}$ is independent.
Indeed, assume that $\sum_{a\in A} f_a v(a) + f_s v(s) = 0$, where $f_a, f_s \in \mathrm{GF}(2^m)$. We have $0 = \sum_{a\in A} f_a v(a)v(0)^\top + f_s v(s)v(0)^\top = \sum_{a\in A} f_a c(\alpha^a) + f_s c(\alpha^s)$. Since $A \subseteq \mathcal{Z}_\rfloor$ and $s \notin \mathcal{Z}_\rfloor$, the previous equation becomes $f_s = 0$. Since $A$ is independent, we also have $f_a = 0$ for all $a \in A$. This proves that $A \cup \{s\}$ is independent.

(ii) If $A$ is independent and $s \in \mathbf{Z}_n$, then the set $s + A = \{s + a \mid a \in A\}$ is independent.
This statement is obvious when considered in terms of residue classes $X^a \bmod p_c(X)$.

(iii) If $A$ is independent and $\ell$ is any integer, then the set $2^\ell A = \{2^\ell a \mid a \in A\}$ is independent.
Essentially, this statement is a direct consequence of the fact that for each $\ell$, the map $z \mapsto z^{2^\ell}$ from $\mathrm{GF}(2^m)$ to $\mathrm{GF}(2^m)$ is linear.

(iv) (The BCH bound) If $A = \{a + ib \mid i = i_0, i_0 + 1, \dots, i_0 + k - 1\}$, then $A$ is independent.
More generally, suppose that for some $a, \tau \in \mathbf{Z}_n$, $\tau = 2^i$, the sequence $\{v_j\}_{j\ge 0}$ in $\mathbf{Z}_n$ satisfies $v_{j+1} = a + \tau v_j$ for all $j \ge 0$ and the set $\mathcal{Z}_\rfloor$ contains $v_j$ for $j = 0, \dots, k-1$. Then $v_j \in \mathcal{Z}_\rfloor$ for all $j$.
To see this, suppose that $v_j \in \mathcal{Z}_\rfloor$ for $j = 0, \dots, u-1$ and $v_u \notin \mathcal{Z}_\rfloor$. By Step 1 with $A = \emptyset$, the set $\{v_u\}$ is independent, and then by a sequence of applications of Step 2 (with $s = -a$), Step 3 (with $\ell = -i$), and Step 1 (with $s = v_u$), we conclude that the sets $\{v_{u-1}, v_u\}$ , $\{v_{u-2}, v_{u-1}, v_u\}$, $\dots$ , $\{v_0, v_1, \cdots, v_u\}$ are independent, which is only possible if $u + 1 \le k$. (In this paper, we only use the BCH-bound.)

As will be illustrated in the later sections, this approach is very useful for finding the minimum distance of some cyclic codes.

## 2.4 The Power Moment Method

Let $C$ be a binary $[n, n-k]$ linear code, and let $C^\perp$ be the dual code of $C$. Denote by $A_i$ and $B_i$ the number of codewords of weight $i$ in $C$ and $C^\perp$ respectively. Let $P_l = \sum_{w=0}^{n} w^l B_w$ denote the $l$th power moment of the $B_w$'s. Then using the Pless power moment identities ([14, p. 131]), we may express $P_l$ in terms of the length $n$ and dimension $k$ of the code $C^\perp$ and the numbers $A_0, A_1, \ldots, A_l$. In the case where $A_1 = A_2 = 0$, the first five power moments $P_0, P_1, \ldots, P_4$ are given by

$$\begin{aligned}
P_0 &= 2^k, \\
P_1 &= 2^{k-1} n, \\
P_2 &= 2^{k-2} n(n+1), \\
P_3 &= 2^{k-3}(n^2(n+3) - 3!A_3), \\
P_4 &= 2^{k-4}(n(n+1)(n^2+5n-2) + 4!(A_4 - nA_3)).
\end{aligned}$$

In addition to the assumption $A_1 = A_2 = 0$, if $A_3 = A_4 = A_5 = A_6 = 0$, then the next two power moments $P_5, P_6$ are

$$\begin{aligned}
P_5 &= 2^{k-5} n^2 (n^3 + 10n^2 + 15n - 10), \\
P_6 &= 2^{k-6} n(n+1)(n^4 + 14n^3 + 31n^2 - 46n + 16).
\end{aligned}$$

Suppose that we want to prove that the only nonzero weights in $C^\perp$ are $w_1, w_2, \ldots, w_{2s}$, where $0 < w_1 \le w_2 \le \ldots \le w_{2s} \le n$. We may define

$$g(x) = \prod_{i=1}^{2s} (x - w_i),$$

and

$$E(w_1, \ldots, w_{2s}) = \sum_{w=1}^{n} g(w) B_w.$$

Then we have the following easy generalization of Theorem 1 in [8].

**Theorem 2.4.** *With the above notation, if $C^\perp$ has no codewords of weight $w$ in the intervals $(w_1, w_2), (w_3, w_4), \ldots, (w_{2s-1}, w_{2s})$, then*

$$E(w_1, w_2, \ldots, w_{2s}) \ge 0,$$

*with equality if and only if $C^\perp$ has no nonzero weights except $w_1, w_2, \ldots, w_{2s}$.*

*Proof.* We note that $g(x)$ is positive outside the interval $[w_1, w_{2s}]$ and inside the intervals $(w_2, w_3), (w_4, w_5), \ldots, (w_{2s-2}, w_{2s-1})$. Also $g(x)$ is negative in the intervals $(w_1, w_2), (w_3, w_4), \ldots, (w_{2s-1}, w_{2s})$. If $C^\perp$ has no codewords of weight $w$ in the intervals $(w_1, w_2), (w_3, w_4), \ldots, (w_{2s-1}, w_{2s})$, then clearly we have $E(w_1, w_2, \ldots, w_{2s}) \geq 0$, and $E(w_1, w_2, \ldots, w_{2s}) = 0$ if and only if $B_w = 0$ for all nonzero $w$ not equal to $w_i$, $i = 1, 2, \ldots, 2s$. This completes the proof. □

*Remarks*

(i) Using the power moments, $E(w_1, \ldots, w_{2s})$ can be evaluated in terms of $n$, $k$, and $A_1, A_2, \ldots, A_{2s}$.
(ii) When we apply Theorem 2.4, we will usually use the weight divisibility properties (Theorem 2.2) to verify the assumption that $C^\perp$ has no codewords of weight $w$ in the intervals $(w_1, w_2), (w_3, w_4), \ldots, (w_{2s-1}, w_{2s})$.

## 3 On Research Problem 9.7 of MacWilliams and Sloane

We will follow the notation introduced in Section 2. Binary triple-error-correcting BCH codes of length $2^m - 1$, BCH(3), have generator polynomial $m_1(X)m_3(X)m_5(X)$, i.e., have defining zeros $\alpha^1, \alpha^3, \alpha^5$. By the MacWilliams identity, the weight distribution of BCH(3) is known because that of the dual code BCH(3)$^\perp$ is known (cf. [14, p. 669]). Let $C$ be the binary cyclic code of length $n = 2^m - 1$ with defining zeros $\alpha, \alpha^{t_1}, \alpha^{t_2}$, where $m = 2r + 1$, $t_1 = 2^r + 1$, $t_2 = 2^{r-1} + 1$, and $r > 1$. The first part of Research Problem 9.7 in MacWilliams and Sloane [14, p. 291] asks whether the code $C$ has the same weight distribution as BCH(3). In 1996, M. van der Vlugt [16] gave an affirmative answer to this question by using the theory of quadratic forms in characteristic 2. In this section, we give a new proof by using the techniques introduced in Section 2.

First we state the following well-known lemma on the minimum distance of $C$.

**Lemma 3.1.** *The code $C = C_{1,t_1,t_2}$ has minimum distance at least 7. So for the code $C$, we have $A_1 = A_2 = \cdots = A_6 = 0$.*

This lemma was proved several times, for example, see [14, p. 290], [15], [1]. For convenience of the reader, we give a quick proof using the techniques in Section 2.3.

*Proof.* Clearly the code $C$ has minimum distance at least 3. Let $\sigma = 2^{r-1}$. Then we have $8\sigma^2 \equiv 1 (\bmod n)$ since $m = 2r + 1$. Let $c(X) = \sum_{i=1}^{k} X^{u_i}$, $0 \leq u_1 < u_2 < \cdots < u_k < n$, be a nonzero codeword of $C$. Then the set of zero exponents $\mathcal{Z}_{\rfloor} = \{\rangle \in \mathbf{Z}_{\setminus} \mid \rfloor(\alpha^{\rangle}) = \imath\}$ of the word $c(X)$ contains

$1, 2, \sigma+1, 2\sigma+2, 2\sigma+1, 4\sigma+2$. In addition, $\mathcal{Z}_\rfloor$ contains $8\sigma+8$ and $8\sigma+4$, hence (multiply by $\sigma$) also $1+8\sigma, 1+4\sigma$, and (multiply by 2) $2+8\sigma$.

Let $A = \{1,2\}$ and $B = \{0, 2\sigma, 4\sigma\}$. Then $\mathcal{Z}_\rfloor$ includes $A+B = \{a+b \mid a \in A, b \in B\}$. Hence the two sets of vectors $V(A) = \{v(a) \mid a \in A\}$ and $V(B) = \{v(b) \mid b \in B\}$ are orthogonal in $\mathcal{R} = \mathrm{GF}(2^m)[X]/(p_c(X)) \cong (\mathrm{GF}(2^m))^k$, where $p_c(X) = \prod_{i=1}^k (X - \alpha^{u_i})$. Since $k \geq 3$, by the BCH bound, we know that $\{0,1,2\}$ is independent. Therefore both $A$ and $B$ are independent. Noting that $V(A)$ and $V(B)$ are orthogonal in $\mathcal{R}$, we have $k \geq 5$. Next, let $B' = B \cup \{8\sigma\}$. Then by the BCH bound, $B' \cup \{6\sigma\}$, and hence also $B'$, is independent. Now $A + B'$ is included in $\mathcal{Z}_\rfloor$, and $A$, $B'$ both are independent, so $k \geq 6$.

Finally, note that $k$ is even, that is, $c(X)$ has even weight, if and only if $0 \in \mathcal{Z}_\rfloor$. However, if $0 \in \mathcal{Z}_\rfloor$, then with $A' = A \cup \{0\}$, we have that $A' + B'$ is contained in $\mathcal{Z}_\rfloor$, which implies that $k \geq 7$. >From the above we conclude that $C$ has minimum distance at least seven. □

Next, we determine the largest power of 2 dividing all the weights of $C^\perp$. By Theorem 2.2, it suffices to determine $M(m; t_1, t_2)$. We will show that $M(m; t_1, t_2) = (m-1)/2$. In fact, we will prove a more general result from which this claim follows directly.

**Theorem 3.2.** *Let $p, q \in \mathbf{Z}$, and $e = \gcd(m, p, q)$. Then $M(m; 2^p+1, 2^q+1) = (m - \epsilon e)/2$, where $\epsilon = 1$ if $m/e$ is odd and $\epsilon = 0$ otherwise.*

*Proof.* Let $s, a, b$ be integers with $0 \leq s, a, b \leq 2^m - 1$, $s \equiv (2^p+1)a + (2^q+1)b \bmod 2^m - 1$, and assume that at least one of $a, b$ is nonzero modulo $2^m - 1$. Let $s = s_{m-1} \cdots s_0$, $a = a_{m-1}a_{m-2} \cdots a_0$, and $b = b_{m-1}b_{m-2} \cdots b_0$ be the binary representations of $s$, $a$, and $b$, respectively. We wish to show that $w(s) - w(a) - w(b) \leq (m - \epsilon e)/2$. By the modular add-with-carry algorithm in Theorem 2.3, there are carries $c_i \in \{0,1,2,3,4\}$ such that

$$2c_i + s_i = a_i + a_{i-p} + b_i + b_{i-q} + c_{i-1}, \tag{11}$$

for all $i \in \mathbf{Z}_m$, with

$$w(c) + w(s) = 2w(a) + 2w(b). \tag{12}$$

Put $\omega_i = a_i + b_i - c_i$. By (12), we have that $w(\omega) = w(s) - w(a) - w(b)$ and we are done if we can prove the bound $w(\omega) \leq (m - \epsilon e)/2$. From (11), we obtain that

$$2\omega_i = s_i - c_{i-1} + (a_i - a_{i-p}) + (b_i - b_{i-q}) \tag{13}$$

Let $I(i) = \{i - \lambda p - \mu q \mid \lambda, \mu \in \mathbf{Z}_m\}$. Noting that $I(i) = \{i - \nu e \mid \nu \in \mathbf{Z}_m\}$, we see that $|I(i)| = m/e$. Now, summing (13) over all $j \in I(i)$, we have that

$$2 \sum_{j \in I(i)} \omega_j = \sum_{j \in I(i)} s_j - \sum_{j \in I(i)} c_{j-1} \leq |I(i)| = m/e. \tag{14}$$

Hence

$$\sum_{j\in I(i)} \omega_j \le (m/e-\epsilon)/2, \tag{15}$$

where $\epsilon = 1$ if $m/e$ is odd and $\epsilon = 0$ otherwise. Since $\mathbf{Z}_m$ is the disjoint union of $e$ sets $I(i)$, we have that

$$w(\omega) \le e(m/e-\epsilon)/2 = (m-\epsilon e)/2.$$

We remark that our proof above actually shows how to construct $a$ and $b$ such that equality holds in the above inequality. We leave the details to the reader. □

**Corollary 3.3.** *Let* $m = 2r+1$, $t_1 = 2^r+1$, *and* $t_2 = 2^{r-1}+1$. *Then* $M(m;t_1,t_2) = (m-1)/2$.

*Proof.* Take $m = 2r+1$, $p = r$, $q = r-1$ and apply Theorem 3.2. □

**Corollary 3.4.** *[the Gold case] Let* $e = \gcd(m,e)$. *Then* $M(m;2^r+1) = (m-\epsilon e)/2$, *where* $\epsilon = 1$ *if* $m/e$ *is odd and* $\epsilon = 0$ *otherwise.*

*Proof.* This follows from Theorem 3.2 by taking $p = q = r$. (In the proof of Theorem 3.2, take $a = 0$.) □

**Corollary 3.5.** *The weight of every codeword of* $C_{1,t_1,t_2}^{\perp}$ *is divisible by* $2^{(m-1)/2}$, *and there is a codeword of* $C_{1,t_1,t_2}^{\perp}$ *with weight not divisible by* $2^{(m+1)/2}$.

*Proof.* This follows from Theorem 2.2 and Corollary 3.3. □

Now, in order to prove that the code $C_{1,t_1,t_2}^{\perp}$ has the same weight distribution as the binary triple-error-correcting BCH code of the same length, we proceed as follows. We use the expression $E(w_1, w_2, \dots, w_{2s})$ as defined in Section 2.4 with $s = 3$ and weights

$$\begin{aligned}
w_1 &= 2^{m-1} - 2^{(m+1)/2},\\
w_2 &= 2^{m-1} - 2^{(m-1)/2},\\
w_3 &= w_4 = 2^{m-1},\\
w_5 &= 2^{m-1} + 2^{(m-1)/2},\\
w_6 &= 2^{m-1} + 2^{(m+1)/2}.
\end{aligned}$$

We note that these are all the nonzero weights of $\mathrm{BCH}(3)^{\perp}$ of length $2^m - 1$, where $m$ is odd (cf. [14, p. 669]).

By Corollary 3.5, all weights in $C_{1,t_1,t_2}^{\perp}$ are divisible by $2^{(m-1)/2}$. Hence the code $C_{1,t_1,t_2}^{\perp}$ has no weights in the intervals $(w_{2i-1}, w_{2i})$ for $i = 1,2,3$.

Therefore $E(w_1, w_2, \ldots, w_6) \geq 0$. Now we may evaluate $E(w_1, w_2, \ldots, w_6)$ with the aid of the power moment expressions. We find that

$$E(w_1, w_2, \ldots, w_6) = P_6 - P_5 \sum_{i=1}^{6} w_i + P_4 \sum w_i w_j - P_3 \sum w_i w_j w_k$$
$$+P_2 \sum w_i w_j w_k w_l - P_1 \sum w_i w_j w_k w_l w_m$$
$$+(P_0 - 1) w_1 w_2 w_3 w_4 w_5 w_6.$$

Using the expressions for $P_0, P_1, \ldots, P_6$ in Section 2.4 with $n = 2^m - 1$, $k = 3m$, by routine computations using MAPLE, we find that

$$E(w_1, w_2, \ldots, w_6) = 0.$$

Hence the weights $w_i$, $i = 1, 2, \ldots, 6$ are the only nonzero weights of the code $C^{\perp}_{1,t_1,t_2}$. Finally, the numbers $B_{w_i}$, $i = 1, 2, \ldots, 6$, can be found from the seven equations for $P_i$, $i = 0, 1, \ldots, 6$. Therefore we have proved the following theorem.

**Theorem 3.6.** *The code $C_{1,t_1,t_2}$ has the same weight distribution as the binary triple-error-correcting BCH code of the same length.*

*Proof.* We have just shown that $C^{\perp}_{1,t_1,t_2}$ has the same weight distribution as BCH(3)$^{\perp}$ of the same length. The remaining part of the theorem now follows from the MacWilliams identity. □

## 4 Some Codes Having the Same Weight Distribution as a Primitive Binary t-error-correcting BCH Code

In the previous section, we essentially proved the following.

**Theorem 4.1.** *Let $m$ be odd, and let $C$ be a binary cyclic code of length $n = 2^m - 1$, dimension $n - 3m$, and minimum distance at least seven. If all weights in $C^{\perp}$ are divisible by $2^{(m-1)/2}$, then $C$ has the same weight distribution as the binary triple-error-correcting BCH code of the same length.*

We can do a similar thing for a binary double-error-correcting BCH code, BCH(2), of length $n = 2^m - 1$. The dual BCH(2)$^{\perp}$ of such a code has only three non-zero weights

$$2^{m-1}, \qquad 2^{m-1} \pm 2^{(m-1)/2},$$

if $m$ is odd (cf. [14, p. 451]); and it has five non-zero weights

$$2^{m-1}, \qquad 2^{m-1} \pm 2^{m/2-1}, \qquad 2^{m-1} \pm 2^{m/2},$$

if $m$ is even (cf. [14, p. 452]). In exactly the same way, the following result can be proved.

**Theorem 4.2.** *Let $m$ be odd, and let $C$ be a binary cyclic code of length $n = 2^m - 1$, dimension $n - 2m$, and minimum distance at least five. If all weights in $C^{\perp}$ are divisible by $2^{(m-1)/2}$, then $C$ has the same weight distribution as the binary double-error-correcting BCH code of the same length.*

*Remark.* Let $m$ be even, and let $C$ be a binary cyclic code of length $2^m - 1$, dimension $n - 2m$, and minimum distance at least five. In order to prove that $C$ has the same weight distribution as BCH(2), we would need to show that all nonzero weights of $C^{\perp}$ are divisible by $2^{m/2-1}$. Moreover, we need to show that $A_5$ and $A_6$ of $C$ are equal to the number of codewords of weight five and six in the BCH code.

There are several cases where Theorems 4.1 and 4.2 apply. First we handle a number of codes from [12].

**Corollary 4.3.** *(i) Let $n = 2^m - 1$, $\gcd(r, m) = 1$, $\sigma = 2^r$ and $t = \sigma + 1$. If $m$ is odd, then the binary cyclic code $C_{1,t}$ of length $n$ with generator polynomial $m_1(X)m_t(X)$ has the same weight distribution as BCH(2) of length $n$.*

*(ii) Let $n = 2^m - 1$, $m > 3$ odd. Let $r = \lfloor m/4 \rfloor$ and $\sigma = 2^r$. Then the binary cyclic code $C_{\sigma+1,2\sigma+1}$ of length $n$ with generator polynomial $m_{\sigma+1}(X)m_{2\sigma+1}(X)$ has the same weight distribution as BCH(2) of length $n$.*

*(iii) Let $n$, $r$, and $\sigma$ be as in (ii). Then the binary cyclic code $C_{1,\sigma+1,2\sigma+1}$ of length $n$ with generator polynomial $m_1(X)m_{\sigma+1}(X)m_{2\sigma+1}(X)$ has the same weight distribution as BCH(3) of length $n$.*

*(iv) Let $n = 2^m - 1$, $m$ odd. Let $0 < r < m$, $\gcd(r, m) = 1$, and $\sigma = 2^r$. Then the binary cyclic code $C_{\sigma+1,\sigma^3+1}$ of length $n$ with generator polynomial $m_{\sigma+1}(X)m_{\sigma^3+1}(X)$ has the same weight distribution as BCH(2) of length $n$.*

*(v) Let $n$, $r$, and $\sigma$ be the same as in (iv). Then the binary cyclic code $C_{1,\sigma+1,\sigma^3+1}$ of length $n$ with generator polynomial $m_1(X)m_{\sigma+1}(X)m_{\sigma^3+1}(X)$ has the same weight distribution as BCH(3) of length $n$.*

*Proof.* (i). This is the Gold case. Here the cyclic code $C_{1,t}$ has minimum distance 5 because $\gcd(r, m) = 1$ (for details of the proof, see [8]), and the weights of its dual code are all divisible by $2^{(m-1)/2}$ since $M(m;t) = \frac{(m-1)}{2}$. Hence the conclusion in (i) follows from Theorem 4.2.

(ii). The cyclic code $C_{\sigma+1,2\sigma+1}$ is equivalent to $C_{1,t}$ where $t = \sigma^2 - \sigma + 1$ if $m \equiv 1 \bmod 4$ and $t = 4\sigma^2 - 2\sigma + 1$ if $m \equiv 3 \bmod 4$. These codes $C_{1,t}$ (with $t$ as above) are the Kasami codes. It is known that these Kasami codes have the same weight distribution as BCH(2) of the same length (for example, see [8]).

(iii). By Theorem 16 in [12], the cyclic code $C_{1,\sigma+1,2\sigma+1}$ has minimum distance 7. By Theorem 3.2, we have $M(m; \sigma+1, 2\sigma+1) = (m-1)/2$. Hence by Theorem 2.2, all weights in $C^{\perp}_{1,\sigma+1,2\sigma+1}$ are divisible by $2^{m-M(m;\sigma+1,2\sigma+1)-1} = 2^{(m-1)/2}$. Now part (iii) follows from Theorem 4.2.

(iv). The cyclic code $C_{\sigma+1,\sigma^3+1}$ is equivalent to $C_{1,\sigma^2-\sigma+1}$. This is the Kasami case again.

(v). By Theorem 17 in [12], the cyclic code $C_{1,\sigma+1,\sigma^3+1}$ has minimum distance 7. By Theorem 3.2, we have $M(m;\sigma+1,\sigma^3+1) = (m-1)/2$. Hence by Theorem 2.2, all weights in $C^{\perp}_{1,\sigma+1,\sigma^3+1}$ are divisible by $2^{m-M(m;\sigma+1,\sigma^3+1)-1} = 2^{(m-1)/2}$. Now part (v) follows from Theorem 4.1. □

Based on some numerical evidence, we have the following conjecture.

**Conjecture 4.4.** *Let $n = 2^m-1$, $\sigma = 2^r$, $t = \sigma^2-\sigma+1$, where $\gcd(r,m) = 1$ and $m$ is even. Then the code $C_{1,t}$ of length $n$ has the same weight distribution as BCH(2) of length $n$.*

We have checked that the above conjecture is correct for $m = 4, 6, 8, 10$, and all $r$ which are relatively prime to $m$. At present, we have no idea how to prove this by our method without counting $A_5$ and $A_6$; this counting problem seems difficult.

The list of codes of length 31 in Table I ([12], see also Example 16 in [12]) suggests that the cyclic code with $n = 31$ and generator polynomial $m_1(X)m_5(X)m_7(X)$ (which has minimum distance seven) could also have the same weight distribution as the binary triple-error-correcting BCH code of that length. In a recent paper [5], the following result is conjectured.

**Theorem 4.5.** *Let $n = 2^m-1$, $m = 2r+1$, $\sigma = 2^{r+1}$, and $t = 1+\sigma$. Then the cyclic code $C_{1,t,t^2}$ of length $n$ with generator polynomial $m_1(X)m_t(X)m_{t^2}(X)$ has the same weight distribution as BCH(3) of length $n$.*

When $m = 5$, the cyclic code in the theorem above is equivalent to the cyclic code with generator polynomial $m_1(X)m_5(X)m_7(X)$ in Example 16 of [12]. Theorem 4.5 follows directly from Theorem 4.1 and the results from [5] that these codes have minimum distance seven and that all weights in the dual codes are divisible by $2^{(m-1)/2}$. In what follows, we offer direct proofs of these last two facts. Our methods are different from those used in [5].

**Lemma 4.6.** *The code $C_{1,t,t^2}$ in Theorem 4.5 has minimum distance $d$ at least 7.*

*Proof.* First, observe that $\sigma^2 \equiv 2 \bmod n$, hence $(\sigma+1)^{-1} \equiv \sigma - 1 \bmod n$. So the code $C_{1,\sigma+1,(\sigma+1)^2}$ is equivalent to the code $C_{\sigma-1,1,\sigma+1}$. The set of zeros of the latter code contains $\{\alpha^{\sigma-1}, \alpha^{\sigma}, \alpha^{\sigma+1}, \alpha^{\sigma(\sigma+1)} = \alpha^{\sigma+2}\}$. By the BCH bound, the minimum distance of $C_{\sigma-1,1,\sigma+1}$ is at least 5.

Let $c(X)$ be a nonzero codeword in $C_{\sigma-1,1,\sigma+1}$. Let $\mathcal{Z}_\rfloor$ denote the set of zero exponents of $c$. Then $\mathcal{Z}_\rfloor$ contains $1, \sigma-1, \sigma+1$, hence also $\sigma, \sigma+2$ and $2-\sigma$. We know that $c$ has weight at least five. In what follows, we will use the notation and terminology introduced in Section 2.3. In addition, if $A, B$ are two subsets of $\mathbf{Z}_n$, we write $A \sim B$ when $A$ is independent iff $B$ is independent. Also we use rank$(A)$ to denote the rank of the set of vectors $V(A) = \{v(a) \mid a \in A\}$ over GF$(2^m)$.

Now let $A = \{1, \sigma, \sigma+1\}$, $A' = \{1, \sigma-1, \sigma, sigma+1\}$ and $B = \{0, 1, 1-\sigma\}$. We observe that the set $\mathcal{Z}_\mathrm{J}$ of zero exponents of $c$ contains $A+B = \{a+b \mid a \in A, b \in B\}$.

Claim 1: If $\sigma + 4 \notin \mathcal{Z}_\mathrm{J}$, then both $A$ and $B$ are independent. Indeed, suppose that $\sigma + 4 \notin \mathcal{Z}_\mathrm{J}$. By adding $\sigma$ to each element of $B$, multiplying the resulting set by $\sigma$ and then adding 2 (see Section 2.3), we have

$$B \sim \{\sigma, \sigma+1, 1\} \sim \{2, 2+\sigma, \sigma\} \sim \{4, 4+\sigma, \sigma+2\}.$$

Since $4, \sigma+2$ are in $\mathcal{Z}_\mathrm{J}$, and $\sigma+4 \notin \mathcal{Z}_\mathrm{J}$, we have $\mathrm{rank}(B) = 1+\mathrm{rank}(\{4, \sigma+2\})$. Now $\{\sigma+2, 4\} \sim \{\sigma-2, 0\}$ and $\mathrm{rank}(\{\sigma-2, 0\}) = 2$ by the BCH bound. This shows that $B$ is independent. Also, multiplying each element of $A'$ by $\sigma$ and then adding $\sigma+2$, one has

$$A' \sim \{\sigma, 2-\sigma, 2, 2+\sigma\} \sim \{2\sigma+2, 4, 4+\sigma, 4+2\sigma\}.$$

Since $2\sigma+2, 4, 4+2\sigma$ are all contained in $\mathcal{Z}_\mathrm{J}$, and $\sigma+4 \notin \mathcal{Z}_\mathrm{J}$, we have that $\mathrm{rank}(A') = 1+\mathrm{rank}(\{2\sigma+2, 4, 4+2\sigma\})$. Now $\{2\sigma+2, 4, 4+2\sigma\} \sim \{\sigma+1, 2, 2+\sigma\} \sim \{2+\sigma, 2\sigma, 2\sigma+2\} \sim \{4+\sigma, 2\sigma+2, 2\sigma+4\}$ and $2\sigma+2, 2\sigma+4 \in \mathcal{Z}_\mathrm{J}$, hence $\mathrm{rank}(A') = 2+\mathrm{rank}(\{2\sigma+2, 2\sigma+4\}) = 2+2 = 4$ by the BCH bound. This shows that $A'$ is independent, and so is $A$.

Claim 2: If $\sigma + 4 \notin \mathcal{Z}_\mathrm{J}$, then $d \geq 7$. First, by Claim 1 we have that $\mathrm{rank}(A) = 3$, $\mathrm{rank}(B) = 3$. Noting that $A+B \subset \mathcal{Z}_\mathrm{J}$, we know that the two sets of vectors $V(A) = \{v(a) \mid a \in A\}$ and $V(B) = \{v(b) \mid b \in A\}$ are orthogonal in $\mathcal{R}$. Therefore, the minimum distance $d$ of $C_{\sigma-1,1,\sigma+1}$ is at least 6. Also, if $d = 6$, then $0 \in \mathcal{Z}_\mathrm{J}$. In that case, $A'+B$ is contained in $\mathcal{Z}_\mathrm{J}$, so that $d \geq 7$. This is a contradiction. Hence $d \geq 7$.

Claim 3: If $\sigma + 4 \in \mathcal{Z}_\mathrm{J}$, then also $d \geq 7$. If $\sigma+4$ is in $\mathcal{Z}_\mathrm{J}$, then $(\sigma+4)\sigma/2 = 1+2\sigma = 1+\sigma^3$ is in $\mathcal{Z}_\mathrm{J}$. So $\mathcal{Z}_\mathrm{J}$ contains $1, \sigma+1, \sigma^3+1$ and the claim follows from part (v) of Corollary 4.3.

This completes the proof of the lemma. □

**Lemma 4.7.** *Let $m = 2r+1, \sigma = 2^{r+1}$. Then $M(m; \sigma-1, \sigma+1) \leq (m-1)/2$.*

Since the proof of this lemma is long and tedious, we put it in the Appendix. Note that by Lemma 4.7, the weight of every codeword in $C^{\perp}_{1,t,t^2}$ is divisible by $2^{(m-1)/2}$. Combining this with Lemma 4.6, Theorem 4.5 is an immediate corollary of Theorem 4.1.

# 5 A Cyclic Code Related to Some Highly Nonlinear Power Function

Because of their applications in cryptography, highly nonlinear power functions on $\mathrm{GF}(2^m)$ recently received a lot of attention (see [2], [6]). Let $f(x) =$

$x^t$ be a power function from $\mathrm{GF}(2^m)$ to itself. The component functions of $f(x)$ are $f_\gamma(x) = \mathrm{tr}(\gamma x^t)$, where $\gamma$ runs through all nonzero elements of $\mathrm{GF}(2^m)$, and tr is the trace from $\mathrm{GF}(2^m)$ to $\mathrm{GF}(2)$. The nonlinearity of $f(x)$ is measured by the Walsh spectrum of its component functions, i.e., by $\sum_{x\in \mathrm{GF}(2^m)}(-1)^{\mathrm{tr}(\gamma x^t+\beta x)}$, $\gamma \in \mathrm{GF}(2^m)^*$, $\beta \in \mathrm{GF}(2^m)$. Therefore it is an important problem to determine the values of these character sums.

By Delsarte's theorem on trace codes and subfield subcodes ([14, p. 234]), the problem of determining the values of the character sums

$$\sum_{x\in \mathrm{GF}(2^m)} (-1)^{\mathrm{tr}(\gamma x^t+\beta x)}$$

can be translated into a problem of determining the weight distribution of the cyclic code $C_{1,t}^{\perp}$. Therefore we can study these character sums from a coding theoretic point of view.

Let $m = 2r$, $\gcd(m,k) = 1$, and $t = \sum_{i=0}^{r} 2^{ik}$. Let $C_{1,t}$ denote the binary cyclic code of length $n = 2^m - 1$ with generator polynomial $m_1(X)m_t(X)$. This code arises from the highly nonlinear power function $f(x) = x^t$ on $\mathrm{GF}(2^m)$, where $m, t$ are the same as above. In this section we use the method developed in previous sections to show that $C_{1,t}^{\perp}$ has only four nonzero weights $w_1 = 2^{m-1} - 2^{m/2}$, $w_2 = 2^{m-1} - 2^{m/2-1}$, $w_3 = 2^{m-1}$, and $w_4 = 2^{m-1} + 2^{m/2-1}$. This result was previously proved in [6] by evaluating some character sums. We present this coding theoretic proof here because in the process of the proof we discover some cyclic MDS codes which are also related to the Kasami code $C_{1,2^{2k}-2^k+1}$.

We note that all cyclic codes we treated in Sections 3 and 4 have length $n = 2^m - 1$ with $m$ odd. In this section, the cyclic code $C_{1,t}$ defined above has length $n = 2^m - 1$ with $m = 2r$ even. In this case it is usually more difficult to compute the number of low weight codewords in $C_{1,t}$.

Let $A_i$ denote the number of codewords of weight $i$ in $C_{1,t}$. We will first determine $A_3$ and $A_4$ for $C_{1,t}$. It is easy to see that $A_3 \geq (2^{r-1}-1)(2^m-1)/3$, and $A_4 \geq (2^{r-2}-1)(2^{r-1}-1)(2^m-1)/3$. Indeed, let $\alpha$ be a primitive element of $\mathrm{GF}(2^m)$, where $m = 2r$, $\sigma = 2^r$. Then $\beta = \alpha^{\sigma+1}$ is a primitive element of $\mathrm{GF}(2^r)$. For each $e$, $1 \leq e \leq 2^r - 2$, there exists a unique $f$, $1 \leq f \leq 2^r - 2$, such that $1 + \beta^e + \beta^f = 0$. Define

$$c(X) = 1 + X^{e(\sigma+1)} + X^{f(\sigma+1)}.$$

Then $c(\alpha) = 1 + \beta^e + \beta^f = 0$, and $c(\alpha^t) = 1 + \beta^{et} + \beta^{ft} = 0$ since $t = \sum_{i=0}^{r} 2^{ik} \equiv 1 (\mathrm{mod} 2^r - 1)$. Hence $c(X)$ is a codeword of weight 3 in $C_{1,t}$. So are $c(X)X^s$, $0 \leq s \leq 2^m - 2$. This shows that $A_3 \geq (2^r - 2)(2^m - 1)/3! = (2^{r-1}-1)(2^m-1)/3$. Similarly one can also show that $A_4 \geq (2^{r-2}-1)(2^{r-1}-1)(2^m - 1)/3$.

In order to get the exact values of $A_3$ and $A_4$, we need the following theorem.

**Theorem 5.1.** *Let $\alpha$ be a primitive element of* $\mathrm{GF}(2^m)$*, where* $m = 2r$*. Let* $\gcd(k, m) = 1$*,* $\sigma = 2^r$*, and* $\gamma = \alpha^{\sigma-1}$*. Then the cyclic code $C''$ of length $\sigma + 1$ over* $\mathrm{GF}(2^r)$ *with generator polynomial* $(X - \gamma^{2^k-1})(X - \gamma^{-2^k+1})(X - \gamma^{2^k+1})(X - \gamma^{-2^k-1})$ *has minimum distance 5. That is, $C''$ is an MDS code.*

The proof of this theorem is postponed until the end of this section.

**Theorem 5.2.** *Let $C_{1,t}$ be defined as above. Then* $A_3 = (2^{r-1} - 1)(2^m - 1)/3$*, and* $A_4 = (2^{r-2} - 1)(2^{r-1} - 1)(2^m - 1)/3$.

*Proof.* Let $\alpha$ be a primitive element of $\mathrm{GF}(2^m)$, where $m = 2r$. Let $\sigma = 2^r$, $\beta = \alpha^{\sigma+1}$ and $\gamma = \alpha^{\sigma-1}$. Note that $\beta$ is a primitive element of $\mathrm{GF}(2^r)$ and $\gamma$ is a $(\sigma + 1)$th root of unity in $\mathrm{GF}(2^m)$. Let $c(Z) = \sum_{i=0}^{\sigma^2-2} c_i Z^i$ be a codeword in $C_{1,t}$. Since $\gcd(\sigma + 1, \sigma - 1) = 1$, by the Chinese Remainder Theorem, there is a ring isomorphism between $\mathbf{Z}_{\sigma^2-1}$ and $\mathbf{Z}_{\sigma-1} \oplus \mathbf{Z}_{\sigma+1}$, under which each $i \in \mathbf{Z}_{\sigma^2-1}$ is mapped to a pair $(j, k)$, $j \in \mathbf{Z}_{\sigma-1}$, $k \in \mathbf{Z}_{\sigma+1}$, such that

$$i \equiv j \bmod \sigma - 1, \qquad i \equiv k \bmod \sigma + 1. \tag{16}$$

We will identify $i \in \mathbf{Z}_{\sigma^2-1}$ with $(j, k) \in \mathbf{Z}_{\sigma-1} \oplus \mathbf{Z}_{\sigma+1}$, where $(j, k)$ is defined by (16). Such an identification will be denoted by $i \equiv (j, k)$. For $i \equiv (j, k)$, write

$$c_i = c_{j,k}.$$

Since $\gcd(\sigma + 1, \sigma - 1) = 1$, there are powers of $Z$, $X = Z^{(\sigma+1)u}$ and $Y = Z^{(\sigma-1)v}$ such that $Z = XY$. Then we may write $c(Z) = c(X, Y)$, where

$$c(X,Y) = \sum_{j=0}^{\sigma-2} \sum_{k=0}^{\sigma} c_{j,k} X^j Y^k = \sum_{k=0}^{\sigma} c_k(X) Y^k, \quad c_k(X) = \sum_{j=0}^{\sigma-2} c_{j,k} X^j.$$

We say that $c(X, Y)$ has a zero $(\beta^e, \gamma^f)$ if $c(\beta^e, \gamma^f) = c(\alpha^g) = 0$, where $g \equiv (e, f)$. By definition, $t = 1 + \frac{2^k(2^{rk}-1)}{2^k-1} \equiv 1 (\bmod \sigma - 1)$. Since $k$ is odd, we have $t \equiv -(2^k + 1)/(2^k - 1) (\bmod \sigma + 1)$. So the codeword $c(Z) = c(X, Y)$ has zeros $(1, 1)$ and $(1, -(2^k + 1)/(2^k - 1))$. Put $\eta_k = c_k(\beta)$, and let $c'(Y) = \sum_{k=0}^{\sigma} \eta_k Y^k$. From the above we conclude that $c'(Y)$ is a word in the cyclic code $C'$ of length $\sigma + 1$ over $\mathrm{GF}(\sigma)$ with defining zeros $\gamma, \gamma^{-(2^k+1)/(2^k-1)}$. We separate two cases.

(i). If $c'(Y)$ is the all-zero word, then for each $k$, $0 \le k \le \sigma$, the word $c_k(X)$ has zero $\beta$. Hence $c_k(X)$ is contained in the cyclic code of length $\sigma - 1$ over $\mathrm{GF}(2)$ with defining zero $\beta$ (i.e., the Hamming code). So either $c_k(X) = 0$ or $c_k(X)$ has weight at least three. Therefore if the original codeword $c(Z)$ has weight at most five, then $c_k(X) = 0$ for all but at most one $k$. As a consequence, $c(Z)$ is of the desired form $c(Z) = Z^s(Z^{(\sigma+1)i_1} + Z^{(\sigma+1)i_2} + \cdots + Z^{(\sigma+1)i_w})$, where $s$ is some integer and $w = w(c)$.

(ii). Now suppose that $c'(Y)$ is nonzero. We claim that $c'(Y)$, and hence $c(Z)$, has weight at least five. Since $\gcd(k,m)=1$, the cyclic code $C'$ of length $\sigma+1$ over GF$(2^r)$ defined above is equivalent to the cyclic code $C''$ of length $\sigma+1$ over GF$(2^r)$ with generator polynomial $(X-\gamma^{2^k-1})(X-\gamma^{-2^k+1})(X-\gamma^{2^k+1})(X-\gamma^{-2^k-1})$. Now the claim follows from Theorem 5.1.

Combining these with the lower bounds on $A_3$ and $A_4$, we see that $A_3=(2^{r-1}-1)(2^m-1)/3$, and $A_4=(2^{r-2}-1)(2^{r-1}-1)(2^m-1)/3$. This completes the proof. □

Next we shall determine the value of $M(m;t)$.

**Proposition 5.3.** *Let $m=2r$ be even, $\gcd(m,k)=1$, and $t=\sum_{i=0}^{r}2^{ik}$. Then $M(m;t)=m/2$.*

*Proof.* As usual, we apply the add-with-carry algorithm in Theorem 2.3. So let $s,a$ be integers with $0\le s,a\le 2^m-1$, $s\equiv ta \bmod 2^m-1$, and assume that $a$ is nonzero modulo $2^m-1$. Let $s=s_{m-1}\cdots s_0$ and $a=a_{m-1}a_{m-2}\cdots a_0$ be the binary representations of $s$ and $a$, respectively. There are carries $c_i\in\{-1,0,1,2\}$, for all $i\in\mathbf{Z}_m$, such that

$$2c_i+s_i=\sum_{j=0}^{r}a_{i-kj}+c_{i-1}, \tag{17}$$

$w(c)=(r+1)w(a)-w(s)$. We have to prove that $w(s)-w(a)\le m/2$. To this end, we define

$$\gamma_i=c_i+c_{i-rk},\qquad \sigma_i=s_i+s_{i-kr},\qquad \omega_i=w(a)-\gamma_i.$$

Note that $w(\gamma)=2w(c)$ and $w(\omega)=mw(a)-2w(c)=2(w(s)-w(a))$, so we have to show that $w(\omega)\le m$.

Add the equation (17) and the equation obtained from (17) by replacing $i$ by $i-kr$, since $(k,m)=1$, we find that

$$2\gamma_i+\sigma_i=a_i+a_{i-kr}+w(a)+\gamma_{i-1} \tag{18}$$

for all $i\in\mathbf{Z}_m$. Using $\gamma_i=w(a)-\omega_i$, we obtain that

$$2\omega_i=\sigma_i-(a_i+a_{i-kr})+\omega_{i-1} \tag{19}$$

for all $i\in\mathbf{Z}_m$. Since $\sigma_i\le 2$, we conclude that

$$\omega_i\le 1+\lfloor\omega_{i-1}/2\rfloor, \tag{20}$$

for all $i\in\mathbf{Z}_m$. Due to the cyclic nature of this inequality, we first of all conclude that $\omega_i\le 2$ for all $i$. Moreover, we conclude that either $\omega_i\le 1$ for all $i$ (and hence $w(\omega)\le m$, what was to be shown), or $\omega_i=2$ holds for all $i$. In the latter case, from (19) we see that $s_i=s_{i-kr}=1$ and $a_i=a_{i-kr}=0$ for all $i$, that is $a\equiv 0 \bmod n$, contradicting the assumption that $a$ is nonzero modulo $2^m-1$. □

Now we are in the position to prove the main result in this section.

**Theorem 5.4.** *Let $m = 2r$, $\gcd(m,k) = 1$, and $t = \sum_{i=0}^{r} 2^{ik}$. Then the only nonzero weights in the code $C_{1,t}^{\perp}$ are $w_1 = 2^{m-1} - 2^{m/2}$, $w_2 = 2^{m-1} - 2^{m/2-1}$, $w_3 = 2^{m-1}$, and $w_4 = 2^{m-1} + 2^{m/2-1}$.*

*Proof.* First we note that the dimension of the code $C_{1,t}^{\perp}$ is $2m$ since $\gcd(m,k) = 1$, and $2^{jk}(1+2^k+\cdots+2^{kr})$, $j = 0, \ldots, m-1$, are all distinct mod $2^m - 1$.

Second, since $M(m;t) = m/2$, we know by Theorem 2.2 that the weight of every codeword in $C_{1,t}^{\perp}$ is divisible by $2^{m/2-1}$. Therefore there are no codewords with weight in the intervals $(w_i, w_{i+1})$, $i = 1, 2, 3$.

Now we use the power moments to evaluate $E = E(w_1, w_2, w_3, w_4) = \sum_{w=1}^{n} \prod_{i=1}^{4} (w - w_i) B_w$ in terms of $n = 2^m - 1$, $k = 2m$, and $A_3, A_4$. A routine computation shows that

$$E = 2^{2(m-2)}(2^r(2^r - 2)(2^m - 1) - 12(2^r - 2)A_3 + 24A_4). \tag{21}$$

By Theorem 5.2, we have

$$A_3 = (2^{r-1} - 1)(2^m - 1)/3, \qquad A_4 = (2^{r-2} - 1)A_3 \tag{22}$$

Substituting $A_3$, $A_4$ in (21) by the expressions in (22) we immediately find that $E = 0$. Now the theorem follows from Theorem 2.4. □

Finally we give a proof for Theorem 5.1. We will proceed in a number of steps, most of which are from [12]. For convenience, we introduce some notation. Let $\mathbf{F}$ be a finite field of characteristic 2, let $\alpha$ be an $n$th root of 1 in some extension of GF(2), and let $E \subseteq \mathbf{Z}_n$. We define

$$C(\mathbf{F}, n, E, \alpha) = \{c(x) \in \mathbf{F}[x] \bmod (x^n - 1) \mid c(\alpha^e) = 0, e \in E\}$$

to be the cyclic code of length $n$ over $\mathbf{F}$ with zeros $\alpha^e$, $e \in E$. We say that a collection $\mathcal{E} \subseteq \mathrm{GF}(q^m)$ is *complete* over $\mathrm{GF}(q)$ if $a^q \in \mathcal{E}$ for all $a \in \mathcal{E}$.

Step 1. (Theorem 9 in [12]) Let $\mathcal{E} = \{\alpha^e \mid e \in E\} \subseteq \mathrm{GF}(q^m)$ be a set of $n$th roots of 1 which is complete over $\mathrm{GF}(q)$. Then the codes $C(\mathrm{GF}(q), n, E, \alpha)$ and $C(\mathrm{GF}(q^m), n, E, \alpha)$ have the same minimum distance.

Step 2. (Essentially Lemma 3 in [12]) Let $\alpha$ be a primitive $n$th root of 1 in $\mathrm{GF}(2^m)$, and let $\tau = 2^k$. Suppose that $\tau$ and 2 generate the same group modulo $n$, that is, $\tau^j \equiv 2 \bmod n$ for some $j$. Then $C(\mathrm{GF}(2^m), n, \{0, 1, \tau\}, \alpha)$ has minimum distance (at least) four.

Step 3. Let $E \subseteq \mathbf{Z}_n$, $a \in \mathbf{Z}_n^*$, $b \in \mathbf{Z}_n$. Let $\alpha$ be a primitive $n$th root of 1 in $\mathrm{GF}(2^m)$. Then the codes $C(\mathrm{GF}(2^m), n, E, \alpha)$ and $C(\mathrm{GF}(2^m), n, aE + b, \alpha)$ are equivalent, where $aE + b = \{ae + b \mid e \in E\}$.

*Proof.* For $c(x) = \sum_{i=0}^{n-1} c_i x^i \Leftrightarrow (c_0, \ldots, c_{n-1})$, let

$$\phi_b(c) = (c_0, c_1\alpha^{-b}, c_2\alpha^{-2b}, \ldots, c_{n-1}\alpha^{-(n-1)b}),$$

and

$$\theta_a(c) = (c_0, c_a, c_{2a}, \dots, c_{(n-1)a}).$$

Observe that both $\phi_b$ and $\theta_a$ are linear, invertible and monomial, and

$$\phi_b(c)(\alpha^{b+e}) = \sum_i \phi_b(c)_i \alpha^{(e+b)i} = \sum_i c_i \alpha^{-bi} \alpha^{(e+b)i} = c(\alpha^e)$$

and

$$\theta_a(c)(\alpha^{ae}) = \sum_i \theta_a(c)_i \alpha^{aei} = \sum_i c_{ai} \alpha^{aei} = c(\alpha^e).$$

So we see that $\phi_b$ maps $C(\mathrm{GF}(2^m), n, E, \alpha)$ to $C(\mathrm{GF}(2^m), n, E+b, \alpha)$ and $\theta_a$ maps $(\mathrm{GF}(2^m), n, E, \alpha)$ to $C(\mathrm{GF}(2^m), n, aE, \alpha)$. The claim in Step 3 now follows. □

Step 4. Let $\alpha$ be a primitive $n$th root of 1 in $\mathrm{GF}(q^m)$ and $\beta$ be a primitive element of $\mathrm{GF}(q^m)$. Let $E \subseteq \mathbf{Z}_n$ have the property that if $e \in E$, then either $e = q^j$ for some $j$, or $e = q^j + q^k$ for some $j, k$ such that $q^j, q^k \in E$. Then the code $C = C(\mathrm{GF}(q), n, E \cup \{0\}, \alpha)$ has minimum distance at least $d+1$, where $d$ is the minimum distance of the code $C(\mathrm{GF}(q^m), q^m - 1, E, \beta)$.

*Proof.* Let $c(X) = \sum_{i=0}^{n-1} c_i X^i \in C$. Then $c(1) = 0$. For any given $\eta \in \mathrm{GF}(q^m)$, put $X_i = \alpha^i$ and $Y_i = X_i + \eta$. Note that $c(\alpha^e) = \sum_i c_i X_i^e$, also $\sum_i c_i Y_i^e = \sum_i c_i (X_i + \eta)^e$. If $e = q^j \in E$, then

$$\sum_i c_i (X_i + \eta)^e = c(\alpha^e) + \eta^e c(1) = 0,$$

and if $e = q^j + q^k \in E$ for some $q^j, q^k \in E$, we have

$$\sum_i c_i (X_i + \eta)^e = c(\alpha^e) + \eta^{q^k} c(\alpha^{q^j}) + \eta^{q^j} c(\alpha^{q^k}) + \eta^e c(1) = 0.$$

Now for each $i$, $0 \le i \le n-1$, either $Y_i = 0$ or $Y_i = \beta^{\pi_i}$ for some $\pi_i \in \mathbf{Z}_{q^m-1}$. Let $J = \{j \in \mathbf{Z}_{q^m-1} \mid \beta^j = X_i + \eta, \text{ for some } i, 0 \le i \le n-1\}$. We now define a codeword $c^*(X) = \sum_{j=0}^{q^m-2} c_j^* X^j$ in $\mathrm{GF}(q^m)[X]/(X^{q^m-1} - 1)$ by letting $c_j^* = 0$ for $j \notin J$ and $c_j^* = c_i$ if $j \in J$ and $j = \pi_i$. Then for any $e \in E$, we have

$$c^*(\beta^e) = \sum_{j \in J} c_j^* \beta^{je} = \sum_i c_i (X_i + \eta)^e = 0.$$

This shows that $c^*$ is contained in $C(\mathrm{GF}(q^m), q^m - 1, E, \beta)$. Now taking $\eta = -X_\ell$, where $0 \le \ell \le n-1$, and $c_\ell \ne 0$, we obtain a codeword $c^*$ in $C(\mathrm{GF}(q^m), q^m - 1, E, \beta)$ of weight $w(c) - 1$. So $d \le w(c^*) = w(c) - 1$, the claim follows. □

Using Step 1–4, we can now quickly prove the following:

**Theorem 5.5.** *Let* $\alpha$ *be a primitive element in* $\mathrm{GF}(2^m)$, *and let* $\tau = 2^k$ *with* $(k, m) = 1$. *If* $m = 2r$ *is even, we let* $\gamma = \alpha^{2^r-1}$.

*(i) The code* $C(\mathrm{GF}(2^m), 2^m - 1, \{0, 1, \tau\}, \alpha)$ *has minimum distance 4, hence is MDS.*

*(ii) If* $m = 2r$ *is even, then the code* $C(\mathrm{GF}(2^m), 2^r + 1, \{0, 1, \tau\}, \gamma)$ *has minimum distance 4.*

*(iii) The code* $C(\mathrm{GF}(2^m), 2^m - 1, \{0, 1, \tau, \tau+1\}, \alpha)$ *has minimum distance 5, hence is MDS.*

*(iv) If* $m = 2r$ *is even, then the code* $C(\mathrm{GF}(2^m), 2^r + 1, \{0, 1, \tau, \tau + 1\}, \gamma)$ *has minimum distance 5.*

*(v) If* $m = 2r$ *is even, then the code* $C = C(\mathrm{GF}(2^r), 2^r + 1, \{-\tau - 1, \tau - 1\}, \gamma)$ *has minimum distance 5, hence is MDS.*

*Proof.* The Cases (i) and (ii) follow from Step 2. From Step 3 (with $a = -1, b = \tau + 1$), we see that $C(\mathrm{GF}(2^m), 2^m - 1, \{1, \tau, \tau + 1\}, \alpha)$ is equivalent to $C(\mathrm{GF}(2^m), 2^m - 1, \{\tau, 1, 0\}, \alpha)$. Now Case (iii) follows from Case (i) and Step 4. Similarly, Case (iv) follows from Case (ii) and Step 4.

To prove Case (v), first observe that the complete set of zero exponents with respect to $\mathrm{GF}(2^r)$ of this code $C$ is $\{-\tau - 1, -\tau + 1, \tau - 1, \tau + 1\}$. Then, by Step 1, $C$ has the same minimum distance as that of the code $C' = C(\mathrm{GF}(2^m), 2^r + 1, \{-\tau - 1, -\tau + 1, \tau - 1, \tau + 1\}, \gamma)$. By use of Step 3 with $a = 1/2, b = (\tau + 1)/2$, we know that $C(\mathrm{GF}(2^m), 2^m - 1, \{-\tau - 1, -\tau + 1, \tau - 1, \tau + 1\}, \alpha)$ is equivalent to $C(\mathrm{GF}(2^m), 2^m - 1, \{0, 1, \tau, \tau + 1\}, \alpha)$, so the claim now follows from Case (iii). □

*Remarks.*

(i) Note that Case (v) of Theorem 5.5 is exactly Theorem 5.1. Therefore Theorem 5.1 is now proven.

(ii) We note that Case (iii) of Theorem 5.5 also follows from a result on maximal arcs in $PG(3, 2^m)$. A set of $n$ points in $PG(3, q)$, $q$ a prime power, is called an *arc* if no four of which are coplanar. A theorem in finite geometry ([4], see also [7]) says that in $PG(3, 2^m)$, the set of points

$$\mathcal{C}(\tau) = \{(1, x, x^\tau, x^{\tau+1}) \mid x \in \mathrm{GF}(q)\} \cup \{(0, 0, 0, 1)\}$$

is a $(2^m + 1)$-arc if and only if $\gcd(m, k) = 1$. By the equivalence of $n$-arcs of $PG(k - 1, q)$ and linear MDS codes of dimension $k$ and length $n$ over $\mathrm{GF}(q)$, this theorem implies Case (iii) of Theorem 5.5. We remark that Casse and Glynn ([4]) also proved that in $PG(3, q)$, $q = 2^m$, $m > 2$, every $(q + 1)$-arc $\mathcal{C}$ is projectively equivalent to some $\mathcal{C}(\tau)$ defined as above, where $\tau = 2^k$, $\gcd(k, m) = 1$. For more details of this remark, we refer the reader to [9].

Finally, we note that the fact that the Kasami codes are 2-error-correcting also follows directly from Theorem 5.5.

**Theorem 5.6.** *Let $\alpha$ be primitive in* $\mathrm{GF}(2^m)$*, and let $\tau = 2^k$ with $(k,m) = 1$. The code $C(\mathrm{GF}(2^m), 2^m-1, \{1, \tau, \tau^2, \tau^2-\tau+1\}, \alpha)$ has minimum distance 5, hence is MDS.*

*Proof.* By repeated applications of Step 3, since $(\tau-1, 2^m-1) = 1$, we have $\{1, \tau, \tau^2, \tau^2-\tau+1\} \sim \{0, \tau-1, \tau^2-1, \tau^2-\tau\} \sim \{0, 1, \tau+1, \tau\}$, so the theorem follows from Theorem 5.5. □

**Corollary 5.7.** *The binary cyclic code $C_{1,\tau^2-\tau+1}$ of length $n = 2^m-1$ has minimum distance (at least) five.*

*Proof.* The code in this corollary is a subfield subcode of the code in Theorem 5.6, so its minimum distance is at least as large as that code. □

## Appendix I

In this appendix, we give a proof of Lemma 17.

*Proof of Lemma 17:* Let $s, a, b$ be integers with $0 \le s, a, b \le 2^m-1$, $s \equiv (\sigma-1)a + (\sigma+1)b \bmod 2^m-1$, and assume that at least one of $a, b$ is nonzero modulo $2^m-1$. Let $s = s_{m-1}\cdots s_0$, $a = a_{m-1}a_{m-2}\cdots a_0$, and $b = b_{m-1}b_{m-2}\cdots b_0$ be the binary representations of $s, a$, and $b$, respectively. We must show that $w(s) - w(a) - w(b) \le (m-1)/2$. If $a = 0$ or $b = 0$, then this follows from the Gold case (if $b = 0$, replace $a$ by $(\sigma+1)a'$ and use $\sigma^2 \equiv 2 \bmod 2^m-1$). So we may assume that $a, b \ne 0$.

By the modular add-with-carry algorithm in Theorem 2.3, there are carries $c_i \in \{-1, 0, 1, 2\}$, for all $i \in \mathbf{Z}_m$, such that

$$2c_i + s_i = a_{i-r-1} - a_i + b_{i-r-1} + b_i + c_{i-1} \tag{23}$$

for all $i \in \mathbf{Z}_m$, with

$$w(c) + w(s) = 2w(b). \tag{24}$$

Put $\xi_i = s_i - a_i - b_i$. We must show that $2w(\xi) \le m$.

By (24), $w(\xi) = w(s) - w(a) - w(b) = w(b) - w(c) - w(a)$. So if we put $\eta_i = b_i - c_i - a_i$, then it suffices to show that $2w(\eta) \le m$.

Inspired by (23) and the fact that $m = 2r+1$, we consider the following generalisation. Let $s, a, b, s', a', b'$ be integers with $0 \le s, s', a, b, a', b' \le 2^m-1$, $s \equiv a' - a + b' - b \bmod 2^m-1$, $s' \equiv 2a - a' + 2b - b' \bmod 2^m-1$, and assume that at least one of $a, b, a', b'$ is nonzero modulo $2^m-1$. Let $s = s_{m-1}\cdots s_0$, $a = a_{m-1}a_{m-2}\cdots a_0$, $b = b_{m-1}b_{m-2}\cdots b_0$, $a' = a'_{m-1}a'_{m-2}\cdots a'_0$, and $b' = b'_{m-1}b'_{m-2}\cdots b'_0$ be the binary representations of $s, a$, $b$, $a'$, and $b'$, respectively. Again by Theorem 2.3, there are carries $c_i, c'_i \in \{-1, 0, 1, 2\}$ for all $i \in \mathbf{Z}_m$ such that

$$2c_i + s_i = a'_i - a_i + b'_i + b_i + c_{i-1}, \tag{25}$$

$$2c'_i + s'_i = a_{i-1} - a'_i + b_{i-1} + b'_i + c'_{i-1}, \tag{26}$$

for all $i \in \mathbf{Z}_m$, with

$$w(c) + w(s) = w(a') - w(a) + w(b') + w(b), \tag{27}$$

$$w(c') + w(s') = w(a) - w(a') + w(b) + w(b'). \tag{28}$$

Note that if we choose $a' = \sigma a$ and $b' = \sigma b$, then $a'_i = a_{i-r-1}$, $b'_i = b_{i-r-1}$, and (25) and (26) reduce respectively to (23) and the equation obtained from (23) by replacing $i$ by $i - r - 1$. Our goal now is to prove

$$w(s) + w(s') - w(a) - w(b) - w(a') - w(b') \leq m. \tag{29}$$

Note that if $a' = \sigma a$, $b' = \sigma b$, and $m$ is odd, then (29) is equivalent to the statement that $w(s) - w(a) - w(b) \leq (m-1)/2$, which is the claim in the lemma. In what follows, we will call a solution to (25) and (26) *trivial* if one of $a, b, a', b'$ is congruent to zero modulo $n$.

To prove that (29) holds for nontrivial solutions, we proceed as follows. Let us define

$$\eta_i = b_i - c_i - a_i, \qquad \eta'_i = b'_i - c'_i - a'_i, \qquad \omega_i = \eta_i + \eta'_i.$$

Using (27) and (28), we see that (29) follows if we can show that

$$w(\omega) \leq m. \tag{30}$$

In order to prove this, there are several ways to proceed. Often the easiest way to solve this type of problems is to investigate a weighted directed graph associated with the problem. In this case, the graph, which we call $G$, will have as vertices all six-tuples $(a, b, a', b', c, c')$ with $a, b, a', b' \in \{0, 1\}$ and $c, c' \in \{-1, 0, 1, 2\}$. We think of these vertices as all possible six-tuples $(a_i, b_i, a'_i, b'_i, c_i, c'_i)$ corresponding to solutions of (25) and (26). Furthermore, $G$ will have a directed weighted arc

$$(a, b, a', b', c, c') \longrightarrow (A, B, A', B', C, C'),$$

of weight $W = (B - C - A) + (B' - C' - A')$, whenever both

$$S = A' - A + B' + B + c - 2C \in \{0, 1\}$$

and

$$S' = a - A' + b + B' + c' - 2C' \in \{0, 1\}.$$

We think of these arcs as arcs

$$(a_{i-1}, b_{i-1}, a'_{i-1}, b'_{i-1}, c_{i-1}, c'_{i-1}) \longrightarrow (a_i, b_i, a'_i, b'_i, c_i, c'_i)$$

corresponding to solutions of (25) and (26). Given the above motivation for the definition of $G$, the reader will be able to verify without difficulty that solutions of (25) and (26) are in 1-1 correspondence with directed cycles of (odd) length $m$ in the graph; moreover, the number $w(\omega)$ associated with the solution evidently equals the total arcweight of the corresponding directed cycle. As a consequence, (29) will follow from the following.
(Let us call a directed cycle in $G$ *trivial* if the corresponding solution of (25) and (26) is trivial, and nontrivial otherwise.)
Claim 0: For all $m \geq 1$, a non-trivial directed cycle of length $m$ in $G$ has total arcweight at most $m$.

There are now several methods to analyse the graph $G$ in order to verify this claim. One method is to investigate the ***strongly connected components*** of $G$. (Here, call two vertices *equivalent* if there exists a directed cycle in $G$ containing both these vertices; then the strongly connected components of $G$ are the equivalence classes of this equivalence relation.) An analysis of $G$ by computer reveals that $G$ has 133 strongly connected components: one component $G_0$ of size 124, and a further 132 components, each of size one. Of these 132 "small" components, 128 have no arcs at all and four consist of a single loop; moreover, the only loop with arcweight at least 2 is the loop

$$(0,0,0,0,-1,-1) \longrightarrow (0,0,0,0,-1,-1)$$

of arcweight 2 corresponding to the trivial solution $a = b = a' = b' = 0$.

It turns out that all arcweights are at most 2, and in fact the "big" component $G_0$ has only five arcs of weight 2. Also, further inspection of $G_0$ reveals the following facts.
Fact (i): The five arcs in $G_0$ of weight 2 end either in state $v_1 = (0,1,0,0,0,-1)$ or in state $v_2 = (0,1,0,1,0,0)$.
Fact (ii): All arcs with weight at least 1 and starting in either $v_1$ or $v_2$ have weight exactly 1 and end in one of $v_3 = (0,1,0,1,1,0)$, $v_4 = (1,1,0,1,0,0)$, $v_5 = (0,1,0,0,0,0)$, or $v_6 = (0,0,0,1,0,0)$.
Fact (iii): The states $v_3$ and $v_4$ have the property that all arcs leaving these states have arcweight less than 1;
Fact (iv): The only arc with weight at least 1 leaving the state $v_5$ is a loop at that state, of weight 1.
Fact (v): All arcs with weight at least 1 leaving the state $v_6$ have weight exactly 1 and end in one of $v_3, v_4, v_5$, or $v_6$.
The facts (i)–(v) immediately imply that the (cyclic) sequence $w_0, \ldots, w_{m-1}$ of arcweights of a directed cycle of length $m$ in $G_0$ can be partitioned into blocks $B$ of the form $w_i, w_{i+1}, \ldots, w_{i+k-1}$ with $\sum_{j=i}^{i+k-1} w_j \leq k$. As a consequence, Claim 0 follows in $G_0$, and by our earlier remarks on the other strongly connected components, also in $G$.

It is also possible to show directly that Facts (ii)–(v) hold even in $G$. (This can be done easily by a computer or by hand.) Then we are done using similar arguments as before if we can show that the following replacement for Fact (i) holds.

Fact (i'): The only arcs in $G$ that are contained in a nontrivial directed cycle and have weight at least 2 in fact have weight exactly 2 and end either in $v_1$ or in $v_2$.
In order to verify this fact, we will use the following claims.
Claim 1: If $b_i = 1$ and $a_i = 0$, then $c_i \geq 0$. Hence $\eta_i \leq 1$ for all $i \in \mathbf{Z}_m$.
Claim 1': If $b_i' = 1$ and $a_i' = 0$, then $c_i' \geq 0$. Hence $\eta_i' \leq 1$ for all $i \in \mathbf{Z}_m$.
Claim 2: If $a_i = b_i$ and $c_i = -1$, then $c_{i-1} = -1$, $a_i' = b_i' = 0$, and $\eta_i' = -c_i'$.
Claim 2': If $a_i' = b_i'$ and $c_i' = -1$, then $c_{i-1}' = -1$, $a_{i-1} = b_{i-1} = 0$, and $\eta_{i-1} = -c_{i-1}$.
All four of these claims follow immediately from (25) and (26).
Claim 3: If $a_i = b_i$ and $c_i = c_i' = -1$ or if $a_i' = b_i'$ and $c_i' = c_{i-1} = -1$, then $a_j = b_j = a_j' = b_j' = 0$ for all $j \in \mathbf{Z}_m$.
Indeed, if $a_i = b_i$ and $c_i = c_i' = -1$, then by Claim 2 we have that $c_{i-1} = -1$ and $a_i' = b_i' = 0$, so $a_i' = b_i' = 0$ and $c_i' = c_{i-1} = -1$. Similarly, if $a_i' = b_i'$ and $c_i' = c_{i-1} = -1$, then by Claim 2' we have that $c_{i-1}' = -1$ and $a_{i-1} = b_{i-1} = 0$, so $a_{i-1} = b_{i-1} = 0$ and $c_{i-1} = c_{i-1}' = 1$. Now the claim follows by induction on $i$.
Claim 4: If $\eta_i = 1$, then either
(i) $a_i = 0$, $b_i = 1$, $c_i = 0$, and $a_i' + b_i' + c_{i-1} \leq 0$, or
(ii) $a_i = b_i$, $c_i = -1$, $c_{i-1} = -1$, and $a_i' = b_i' = 0$.
Claim 4': If $\eta_i' = 1$, then either
(i') $a_i' = 0$, $b_i' = 1$, $c_i' = 0$, and $a_{i-1} + b_{i-1} + c_{i-1}' \leq 0$, or
(ii')$a_i' = b_i'$, $c_i' = -1$, $c_{i-1}' = -1$, and $a_{i-1} = b_{i-1} = 0$.
Then by Claims 3 and 3', we see that for nontrivial solutions with $\eta_i = 1$ (or with $\eta_i' = 1$), we have in Claim 4, case (ii) (or in Claim 4', case (ii')), in addition also that $c_i' \leq 0$ (or $c_{i-1} \leq 0$). Now by combining Claims 4 and 4' and these observations, we find that the following claim holds.
Claim 5: For a nontrivial solution to (25) and (26) with $\eta_i = \eta_i' = 1$, we have $a_i = 0$, $b_i = 1$, $c_i = 0$, $a_i' + b_i' + c_{i-1} \leq 0$, and either
(i) $a_i' = b_i' = 0$, $c_{i-1} = 0$, $c_i' = -1$, $c_{i-1}' = -1$, and $a_{i-1} = b_{i-1} = 0$, or
(ii) $a_i' = 0$, $b_i' = 1$, $c_i' = 0$, and $a_{i-1} + b_{i-1} + c_{i-1}' \leq 0$.

Now Claim 5, when translated into the graph theoretic terminology introduced earlier, implies Fact (i'). This completes the proof. □

## References


1. R. D. Baker, J. H. van Lint and R. M. Wilson, *On the Preparata and Goethals codes*, IEEE Trans. Inform. Theory, **29** (1983), 342–345.
2. C. Carlet, P. Charpin, V. Zinoviev, *Codes, bent functions and permutations suitable for DES-like cryptosystems*, Designs, Codes and Cryptography **15** (1998), 125–156.
3. C. Canteaut, P. Charpin, H. Dobbertin, *Binary m-sequences with three-valued crosscorrelations: a proof of the Welch conjecture*, IEEE Trans. Inform. Theory, **46** (2000), 4–8.
4. L. R. A. Casse, D. G. Glynn, *The solution to Beniamino Segre's problem* $T_{r,q}, r = 3, q = 2^h$, Geom. Dedicata, **13** (1982), 157–163.

5. A. Chang, P. Gaal, S. W. Golomb, G. Gong, P. V. Kumar, *On a sequence conjectured to have ideal 2-level autocorrelation function*, ISIT 1998, Cambridge, MA, USA.
6. H. Dobbertin, *One-to-one highly nonlinear power functions on* GF($2^n$), AAECC **9**, 139–152 (1998).
7. J. W. P. Hirschfeld, *Finite Projective Spaces of Three Dimensions*, Oxford: Clarendon Press 1985.
8. H. D. L. Hollmann, Q. Xiang, *A proof of the Welch and Niho conjectures on crosscorrelations of binary m-sequences*, accepted for publication in Finite Fields and Their Applications.
9. H. D. L. Hollmann, Q. Xiang, *Maximal arcs in projective three-spaces and double-error-correcting cyclic codes*, J. Combin. Theory (A), **90** (2000).
10. T. Kasami, *Weight distribution of Bose-Chaudhuri-Hocquenghem codes*, in *Proc. Conf. Combinatorial Mathematics and its Applications*,R. C. Bose and T. A. Dowling, Eds. Chapel Hill, NC: University of North Carolina Press, 1969, 335–357.
11. T. Kasami, *The weight enumerator for several classes of subcodes of the 2nd order binary Reed-Muller codes*, Inform. and Control, **18** (1971), 369–394.
12. J. H. van Lint, R. M. Wilson, *On the minimum distance of cyclic codes*, IEEE Trans. Inform. Theory, **32** (1986), 23–40.
13. R. J. McEliece, *On periodic sequences from* GF($q$), J. Combin. Theory (A), **10** (1971), 80–91.
14. F. J. MacWilliams and N. J. A. Sloane, *The Theory of Error-Correcting Codes*, North-Holland, 1977.
15. C. Roos, *A new lower bound for the minimum distance of a cyclic code*, IEEE Trans. Inform. Theory, **29** (1983), 330–332.
16. M. van der Vlugt, *Non-BCH triple-error-correcting codes*, IEEE Trans. Inform. Theory, **42** (1996), 1612–1614.

# MacWilliams Identities for Linear Codes over Finite Frobenius Rings

Thomas Honold[1] and Ivan Landjev[2]

[1] Zentrum Mathematik
Technische Universität München
D-80290 München, Germany
e-mail: honold@mathematik.tu-muenchen.de

[2] Institute of Mathematics and Informatics
Bulgarian Academy of Sciences
8 Acad. G. Bonchev str.
1113 Sofia, Bulgaria
e-mail: ivan@moi.math.bas.bg

**Abstract** The purpose of this paper is a unified treatment of MacWilliams type identities for several kinds of weight distributions for linear codes over a finite FROBENIUS ring $R$. The concept of a $\mathcal{W}$-admissible pair of partitions of the ambient space $R^n$ is introduced, and MacWilliams type identities are proved for the corresponding weight spectra of linear codes of length $n$ over $R$. Several examples—some well-known, others apparently new—are given.

## 1 Introduction

The classical MacWilliams identities relate the Hamming weight enumerators of a linear code $C \subseteq \mathbb{F}_q^n$ and its dual $C^\perp = \{\mathbf{y} \in \mathbb{F}_q^n \mid \forall \mathbf{x} \in C \colon \mathbf{x} \cdot \mathbf{y} = 0\}$. Using the two-variable enumerator $A_C(X,Y) = \sum_{i=0}^n A_i X^{n-i} Y^i \in \mathbb{Q}[X,Y]$, where $A_i = |\{\mathbf{x} \in C \mid w_{\mathrm{Ham}}(\mathbf{x}) = i\}|$, they take the following form:

**Theorem 1.1 (MacWilliams [18]).** *If $C \subseteq \mathbb{F}_q^n$ is linear then*

$$A_{C^\perp}(X,Y) = \frac{1}{|C|} \cdot A_C\bigl(X + (q-1)Y, X - Y\bigr). \tag{1}$$

These identities have been generalized in various ways to other kinds of weight distributions and also to linear codes over finite rings. Our goal is a rational[1] presentation of these identities including several hitherto unknown cases. Since our approach is probably best explained by way of an example, we shall use it now for a proof of the classical MacWilliams identities: Theorem 1.1 is equivalent to the fact that $\alpha \in \operatorname{Aut} \mathbb{Q}[X,Y]$ defined by $(\alpha f)(X,Y) = f\bigl(X + (q-1)Y, X - Y\bigr)$ sends $A_C(X,Y)$ to $|C| A_{C^\perp}(X,Y)$ for

[1] In contrast with the usual character theoretic approach involving roots of unity [19, Chap. 5].

every linear code $C \subseteq \mathbb{F}_q^n$. Obviously it does this for linear codes of length one:

$$\begin{aligned}\alpha(A_{\{0\}}) &= \alpha(X) = X + (q-1)Y = 1 \cdot A_{\mathbb{F}_q},\\ \alpha(A_{\mathbb{F}_q}) &= \alpha\big(X + (q-1)Y\big) = X + (q-1)Y + (q-1)(X-Y)\\ &= qX = q \cdot A_{\{0\}}.\end{aligned}$$

If $C = C_1 \times C_2$ is decomposable into linear codes $C_1$, $C_2$ of smaller length, then $C^\perp = C_1^\perp \times C_2^\perp$ is also decomposable, and hence by induction

$$\alpha(A_C) = \alpha(A_{C_1}A_{C_2}) = \alpha(A_{C_1})\alpha(A_{C_2}) = |C_1||C_2|A_{C_1^\perp}A_{C_2^\perp} = |C|A_{C^\perp}. \quad (2)$$

Now consider the $n+1$ (completely) decomposable linear codes

$$C_k = \underbrace{\mathbb{F}_q \times \cdots \times \mathbb{F}_q}_{k} \times \underbrace{\{0\} \times \cdots \times \{0\}}_{n-k} \qquad (0 \le k \le n). \quad (3)$$

Their weight enumerators $A_{C_k} = \big(X + (q-1)Y\big)^k X^{n-k}$ are linearly independent, hence generate the subspace $\mathbb{Q}[X,Y]_n$ of homogeneous polynomials of degree $n$. Since $\mathbb{Q}[X,Y] = \bigoplus_{n\ge 0} \mathbb{Q}[X,Y]_n$ we are finished once we have proved the existence of a $\mathbb{Q}$-linear endomorphism of $\mathbb{Q}[X,Y]$ which sends $A_C$ to $|C|A_{C^\perp}$. This is done by constructing a linear map $\mathcal{W}_n \colon \mathbb{Q}\mathbb{F}_q^n \to \mathbb{Q}\mathbb{F}_q^n$ which sends the characteristic function $\chi_C$ of $C$ to $|C|\chi_{C^\perp}$, and then identifying $\mathbb{Q}[X,Y]_n$ with the $\mathcal{W}_n$-invariant subspace of $\mathbb{F}_q^n$ spanned by the characteristic functions of the Hamming spheres around $\mathbf{0}$.

In the classical case, however, it is easier to use the following argument of [10] (cf. also [8, p. 265]): Let $\pi \colon \mathbb{F}_q^n \to \mathbb{F}_q^{n-1}$ denote the projection onto the first $n-1$ coordinates and $C_0 = \{\mathbf{x} \in C \mid x_n = 0\}$. By induction we may assume that $C$ is indecomposable. In this case a short calculation reveals

$$\begin{aligned}A_C &= (X-Y)A_{\pi(C_0)} + YA_{\pi(C)},\\ A_{C^\perp} &= (X-Y)A_{\pi(C)^\perp} + YA_{\pi(C_0)^\perp}.\end{aligned}$$

By induction,

$$\begin{aligned}\alpha(A_C) &= \alpha(X-Y)\alpha(A_{\pi(C_0)}) + \alpha(Y)\alpha(A_{\pi(C)})\\ &= qY|C_0|A_{\pi(C_0)^\perp} + (X-Y)|C|A_{\pi(C)^\perp}\\ &= |C|A_{C^\perp}.^2\end{aligned}$$

□

A suitable MacWilliams transform $\mathcal{W}_n$ can be defined for an arbitrary finite Frobenius ring $R$. This is done in Sect. 2, after reviewing basic properties of finite Frobenius rings. So-called regular and $\mathcal{W}$-admissible pairs of partitions of $R^n$ which generalize the Hamming partition are studied in Sect. 3. The

[2] An interesting variant relating weight enumerators of subcodes of $C$ to partial derivatives of $A_C(X,Y)$ can be found in [27].

MacWilliams identities for such partitions are proved in Sect. 4. Finally, in Sect. 5 we discuss several examples.

Regular pairs of partitions are in some sense a semigroup analog of abelian association schemes. Regular partitions of $\mathbb{F}_q^n$ have been considered in [17]. $\mathcal{W}$-admissible pairs are related to F-partitions of finite abelian groups as introduced in [4,5] (cf. also Ex. 5.7). A different set of MacWilliams identities for finite Frobenius rings can be found in [28] (cf. also Ex. 5.3). The related concept of weight functions on finite rings is investigated in [13,29].

The authors wish to thank the referee for some corrections/improvements.

## 2 Linear Codes over Frobenius rings

A finite (associative) ring $R$ (with identity $1 \neq 0$) is called a FROBENIUS ring if it satisfies one of the following equivalent conditions:

(F1) There exists a (necessarily unique) function $\omega\colon R \to \mathbb{Q}$ satisfying $\omega(0) = 1$, $\omega(x) = \omega(uxv)$ for $x \in R$, $u, v \in R^*$, and $\sum_{x\in I} \omega(x) = 0$ for every nonzero left or right ideal $I$ of $R$.[3]

(F2) There exists an additive character $\psi\colon R \to \mathbb{C}^*$ whose kernel contains no nonzero left or right ideal of $R$.

(F3) ${}_R(R/\operatorname{Rad} R) \cong \operatorname{Soc}({}_R R)$, and ${}^\perp(I^\perp) = I$, $({}^\perp J)^\perp = J$ for every left ideal $I$ and every right ideal $J$ of $R$.[3]

Condition (F3) is essentially the original definition of Frobenius rings due to Nakayama [21], (F2) was first studied in [15,16], and (F1) in [13]. The equivalence (F3)$\Longleftrightarrow$(F2) is proved in [7,28], (F3)$\Longleftrightarrow$(F1) in [13]. All these conditions can be weakened; cf. e. g. [9].

Examples of finite Frobenius rings are the rings $\mathbb{Z}_m$ of integers modulo $m$, finite fields and more generally finite chain rings [2,20,22]. Furthermore, the class of Frobenius rings is closed with respect to taking (finite) direct products, matrix rings, and group algebras (of finite groups).

Henceforth we shall assume that $R$ is a finite Frobenius ring and $\omega$ has the same meaning as in (F1).

Let $n \geq 0$ be an integer. With respect to componentwise addition and scalar multiplication the set $R^n$ has the structure of an $(R, R)$-bimodule.[4] A (block) code $C \subseteq R^n$ is said to be *left linear* (resp., *right linear*) if $C$ is an $R$-submodule of ${}_R R^n$ (resp., $R^n_R$). A *linear* code is one which is either left or right linear. We shall formulate our results with a bias towards left modules, often omitting obvious right module counterparts.

[3] Here $R^*$ denotes the group of units of $R$, $\operatorname{Rad} R$ the JACOBSON radical of $R$, $\operatorname{Soc}({}_R R)$ the sum of the minimal left ideals of $R$, and ${}^\perp S = \{x \in R \mid xS = 0\}$, $S^\perp = \{x \in R \mid Sx = 0\}$ for $S \subseteq R$. If $R$ only satisfies the double annihilator conditions ${}^\perp(I^\perp) = I$, $({}^\perp J)^\perp = J$, $R$ is called a quasi-Frobenius ring.

[4] By convention, $R^0 = \{\epsilon\}$ is the zero module having the empty word $\epsilon$ as its single element.

For a code $C \subseteq R^n$ we define

$$\begin{aligned} C^\perp &= \{\mathbf{x} \in R^n \mid \forall \mathbf{y} \in C\colon \mathbf{y}\cdot\mathbf{x} = 0\}, \\ {}^\perp C &= \{\mathbf{x} \in R^n \mid \forall \mathbf{y} \in C\colon \mathbf{x}\cdot\mathbf{y} = 0\}, \end{aligned} \tag{4}$$

where $\mathbf{x}\cdot\mathbf{y} = x_1y_1+\cdots+x_ny_n$. The linear code $C^\perp \leq R^n_R$ (resp., ${}^\perp C \leq {}_RR^n$) is called the *right* (resp., *left*) *orthogonal code* of $C$. As shown in [6], the double annihilator conditions imply ${}^\perp(C^\perp) = C$ for every left linear code $C \leq {}_RR^n$, and similarly $({}^\perp D)^\perp = D$ for every right linear code $D \leq R^n_R$. Hence the submodule lattices $\mathrm{S}({}_RR^n)$ and $\mathrm{S}(R^n_R)$ are antiisomorphic:

$$(C_1 + C_2)^\perp = C_1^\perp \cap C_2^\perp, \quad (C_1 \cap C_2)^\perp = C_1^\perp + C_2^\perp \tag{5}$$

for $C_1, C_2 \in \mathrm{S}({}_RR^n)$. Let $\mathbb{Q}R^n$ denote the $\mathbb{Q}$-space of all functions $f\colon R^n \to \mathbb{Q}$, and $\mathbb{Q}R^\diamond = \bigoplus_{n\geq 0}\mathbb{Q}R^n$.[5] We think of $\mathbb{Q}R^\diamond$ as the semigroup algebra (over $\mathbb{Q}$) of the free monoid $R^\diamond = \bigcup_{n\geq 0} R^n$ over the alphabet $R$. Thus the elements of $\mathbb{Q}R^\diamond$ are functions $f\colon R^\diamond \to \mathbb{Q}$ with finite support, and juxtaposition $\times$ of such functions is defined by $(f \times g)(\mathbf{z}) = \sum_{\mathbf{xy}=\mathbf{z}} f(\mathbf{x})g(\mathbf{y})$.[6] Note that $(\mathbb{Q}R^\diamond, \times)$ is graded w. r. t. to the wordlength on $R^\diamond$.

Furthermore, the space $\mathbb{Q}R^\diamond$ forms an (associative) $\mathbb{Q}$-algebra with respect to the HADAMARD (point-wise) product and the group algebra product (convolution) $*$ defined by $(f * g)(\mathbf{x}) = \sum_{\mathbf{y}\in R^n} f(\mathbf{y})g(\mathbf{x}-\mathbf{y})$ if $f, g \in \mathbb{Q}R^n$ and $f * g = 0$ if $f \in \mathbb{Q}R^n$, $g \in \mathbb{Q}R^m$ with $n \neq m$.

A function $f \in \mathbb{Q}R^\diamond$ is said to be *left invariant* if $f(x) = f(ux)$ for any word $x \in R^\diamond$ and $u \in R^*$. Let $\mathbb{Q}\mathcal{S}^\diamond$ (resp., $\mathbb{Q}\mathcal{T}^\diamond$) denote the spaces of left (resp., right) invariant functions in $\mathbb{Q}R^\diamond$.[7] These spaces are graded, i. e. $\mathbb{Q}\mathcal{S}^\diamond = \bigoplus_{n\geq 0}\mathbb{Q}\mathcal{S}^n$ with $\mathbb{Q}\mathcal{S}^n = \mathbb{Q}\mathcal{S}^\diamond \cap \mathbb{Q}R^n$, closed w. r. t. $\cdot$, $*$, $\times$, and thus in particular graded subalgebras of $(\mathbb{Q}R^\diamond, \times)$.

Obviously, the characteristic function $\chi_C$ of a left linear code $C$ is left invariant.

**Proposition 2.1.** *The space $\mathbb{Q}\mathcal{S}^\diamond$ is generated by the characteristic functions of left linear codes over $R$.*

*Proof.* We must show that every left invariant function $f\colon R^n \to \mathbb{Q}$ belongs to the space $W = \langle \chi_C \mid C \leq {}_RR^n\rangle_{\mathbb{Q}}$. For $C \leq {}_RR^n$ we have

$$\chi_C = \sum_{U\leq C}\Big(\sum_{\substack{\mathbf{y}\in R^n \\ U=R\mathbf{y}}} \chi_{\mathbf{y}}\Big) = \sum_{U\leq C} \chi_{\mathrm{g}(U)}, \tag{6}$$

where $\mathrm{g}(U)$ is the set of generators of the left $R$-module $U$. Möbius inversion on $\mathrm{S}({}_RR^n)$ shows $\chi_{\mathrm{g}(C)} \in W$ for every submodule $C \leq {}_RR^n$. By [28, Prop. 5.1] we have $R^*\mathbf{x} = \mathrm{g}(R\mathbf{x})$ for $\mathbf{x} \in R^n$, and hence $f \in W$.

[5] The symbol '$\diamond$' is pronounced like „Hirsch", the German word for "deer".

[6] If $\chi_{\mathbf{x}}, \chi_{\mathbf{y}}$ are characteristic functions of single words $\mathbf{x}$, $\mathbf{y}$ then $\chi_{\mathbf{x}} \times \chi_{\mathbf{y}} = \chi_{\mathbf{xy}}$ corresponds indeed to the juxtaposition of $\mathbf{x}$ and $\mathbf{y}$.

[7] The reason for using the symbols '$\mathcal{S}^\diamond$', '$\mathcal{T}^\diamond$' is explained in Sect. 3.

**Definition 2.2.** The MacWilliams *transform* of a finite Frobenius ring $R$ is the linear map $\mathcal{W}\colon \mathbb{Q}R^{\diamond} \to \mathbb{Q}R^{\diamond}$ defined by

$$(\mathcal{W}f)(\mathbf{y}) = \sum_{\mathbf{x}\in R^n} f(\mathbf{x})\omega(\mathbf{x}\cdot\mathbf{y}) \quad \text{for } f \in \mathbb{Q}R^n \qquad (n \geq 0). \tag{7}$$

The MacWilliams transform for the opposite ring $R^{\circ}$ is denoted by $\mathcal{W}^{\circ}$, i. e. $\mathcal{W}^{\circ}\colon \mathbb{Q}R^{\diamond} \to \mathbb{Q}R^{\diamond}$ is defined by $(\mathcal{W}^{\circ}f)(\mathbf{y}) = \sum_{\mathbf{x}\in R^n} f(\mathbf{x})\omega(\mathbf{y}\cdot\mathbf{x})$ for $f \in \mathbb{Q}R^n$.

**Theorem 2.3.** *For every left linear code $C \leq {}_RR^n$ we have*

$$\mathcal{W}\chi_C = |C| \cdot \chi_{C^{\perp}}. \tag{8}$$

*Similarly, $\mathcal{W}^{\circ}\chi_D = |D| \cdot \chi_{{}^{\perp}D}$ for every right linear code $D \leq R^n_R$.*

*Proof.* Using (F1) we have

$$(\mathcal{W}\chi_C)(\mathbf{y}) = \sum_{\mathbf{x}\in C} \omega(\mathbf{x}\cdot\mathbf{y}) = \begin{cases} |C| & \text{if } \mathbf{y} \in C^{\perp}, \\ 0 & \text{if } \mathbf{y} \notin C^{\perp}, \end{cases} \tag{9}$$

since the set $I = \{\mathbf{x}\cdot\mathbf{y} \mid \mathbf{x} \in C\}$ is a left ideal of $R$ which is zero iff $\mathbf{y} \in C^{\perp}$. □

**Corollary 2.4.** *We have $|C||C^{\perp}| = |R|^n$ for every left linear code $C \leq {}_RR^n$, and similarly $|D||{}^{\perp}D| = |R|^n$ for every right linear code $D \leq R^n_R$.*

*Proof.* Again using (F1) we have

$$\sum_{\mathbf{x}\in C}\sum_{\mathbf{y}\in R^n} \omega(\mathbf{x}\cdot\mathbf{y}) = |R|^n + \sum_{\mathbf{x}\in C\setminus\{\mathbf{0}\}}\sum_{\mathbf{y}\in R^n} \omega(\mathbf{x}\cdot\mathbf{y}) = |R|^n.$$

This is equal to

$$\sum_{\mathbf{y}\in R^n}\sum_{\mathbf{x}\in C} \omega(\mathbf{x}\cdot\mathbf{y}) = \sum_{\mathbf{y}\in C^{\perp}} |C| = |C||C^{\perp}|.$$ □

Since $\omega$ is right invariant, we have $\operatorname{Im}\mathcal{W} \subseteq \mathbb{Q}\mathcal{T}^{\diamond}$. For $f \in \mathbb{Q}R^{\diamond}$, $u \in R^*$ define $f_u \in \mathbb{Q}R^{\diamond}$ by $f_u(\mathbf{x}) = f(u\mathbf{x})$. Since $\omega$ is left invariant, we have $\mathcal{W}f = \mathcal{W}f_u$ and hence $\mathcal{W}f = \mathcal{W}\overline{f}$ where $\overline{f} \in \mathbb{Q}\mathcal{S}^{\diamond}$ is defined by $\overline{f}(\mathbf{x}) = |R^*|^{-1} \times \sum_{u\in R^*} f(u\mathbf{x})$. Thus we need only consider the restriction $\mathcal{W}\colon \mathbb{Q}\mathcal{S}^{\diamond} \to \mathbb{Q}\mathcal{T}^{\diamond}$ which in fact is one-to-one:

**Theorem 2.5.** *We have $\mathcal{W}^{\circ}(\mathcal{W}f) = |R|^n \cdot f$ for $f \in \mathbb{Q}\mathcal{S}^n$, $n \geq 0$; furthermore, $\mathcal{W}(f \times g) = \mathcal{W}f \times \mathcal{W}g$ and $\mathcal{W}(f * g) = (\mathcal{W}f)(\mathcal{W}g)$ for $f, g \in \mathbb{Q}\mathcal{S}^{\diamond}$. In particular, $\mathcal{W}$ induces isomorphisms $(\mathbb{Q}\mathcal{S}^{\diamond}, \times) \to (\mathbb{Q}\mathcal{T}^{\diamond}, \times)$ and $(\mathbb{Q}\mathcal{S}^{\diamond}, *) \to (\mathbb{Q}\mathcal{T}^{\diamond}, \cdot)$.*

*Proof.* Since $\mathbb{Q}S^\diamond$ is generated by the characteristic functions of left-linear codes (cf. Prop. 2.1), we may assume $f = \chi_C$, $g = \chi_D$ with $C \leq {}_RR^n$, $D \leq {}_RR^m$. By Th. 2.3 and Cor. 2.4,

$$\mathcal{W}^\circ(\mathcal{W}\chi_C) = |C||C^\perp| \cdot \chi_{\perp(C^\perp)} = |R|^n \cdot \chi_C \tag{10}$$

proving the first assertion. Further,

$$\begin{aligned}\mathcal{W}(\chi_C \times \chi_D) &= \mathcal{W}(\chi_{C\times D}) = \\ &= |C \times D| \cdot \chi_{(C\times D)^\perp} = |C||D| \cdot \chi_{C^\perp \times D^\perp} = \mathcal{W}\chi_C \times \mathcal{W}\chi_D\end{aligned} \tag{11}$$

and

$$\begin{aligned}&\mathcal{W}(\chi_C * \chi_D) = \mathcal{W}(|C \cap D| \cdot \chi_{C+D}) = \\ &= |C \cap D||C + D| \cdot \chi_{(C+D)^\perp} = |C||D| \cdot \chi_{C^\perp \cap D^\perp} = (\mathcal{W}\chi_C)(\mathcal{W}\chi_D),\end{aligned} \tag{12}$$

as required. □

*Remark 2.6.* The MacWilliams transform of a Frobenius ring $R$ can be viewed as a rational version of the FOURIER transform of $(R^n, +)$: Let $\psi$ be a character of $(R, +)$ as in (F2). Such a character is called a generating character of $R$. Then

$$\omega(x) = \frac{1}{|R^*|} \cdot \sum_{u\in R^*} \psi(ux) \qquad (x \in R), \tag{13}$$

since the right-hand side of (13) also satisfies (F1). This implies

$$\mathcal{W}f = \mathcal{F}\overline{f} \quad \text{for } f \in \mathbb{Q}R^n, \tag{14}$$

where $\mathcal{F}\colon \mathbb{C}R^n \to \mathbb{C}R^n$ is defined by $(\mathcal{F}f)(\mathbf{y}) = \sum_{\mathbf{x}\in R^n} f(\mathbf{x})\psi(\mathbf{x}\cdot\mathbf{y})$. In particular $\mathcal{W}f = \mathcal{F}f$ iff $f$ is left-invariant. The identities $\mathcal{W}^\circ(\mathcal{W}f) = |R|^n \cdot f$, $\mathcal{W}(f * g) = (\mathcal{W}f)(\mathcal{W}g)$ of Th. 2.5 can be derived from the Fourier inversion formula and the Plancherel-Parseval formula, respectively.

## 3 Regular and $\mathcal{W}$-Admissible Pairs of Partitions

For our purposes a *partition of* $R^n$ is a family $\mathcal{P} = (P_i \mid i \in I)$ of nonempty subsets of $R^n$ satisfying the following conditions:

(i) $R^n = \bigcup_{i\in I} P_i$;
(ii) $P_i \cap P_j = \emptyset$ for all pairs $i \neq j$.

The set $I$ is called the *weight set* of $\mathcal{P}$.[8] If $\mathbf{x} \in P_i$ we say that $\mathbf{x}$ has $\mathcal{P}$*-type* $i$.[9]

[8] Sometimes it is necessary to work with equivalence classes of partitions relative to reindexing. Thus an indexed partition is probably best viewed as an ordinary partition with some extra structure.

[9] These and subsequent definitions have natural extensions to partitions $\mathcal{P}$ of $R^\diamond$ which are *graded* in the sense that for every $n$ the set $R^n$ is a union of members of $\mathcal{P}$.

The set of partitions of $R^n$ forms a lattice with respect to the usual partial order defined by $\mathcal{P}_1 \leq \mathcal{P}_2$ if '$\subseteq$' holds for the corresponding equivalence relations. The *cartesian product* of two partitions $\mathcal{P} = (P_i \mid i \in I)$ and $\mathcal{Q} = (Q_j \mid j \in J)$ of $R^n$ and $R^m$, respectively, is the partition $\mathcal{P} \times \mathcal{Q} = (P_i \times Q_j \mid (i,j) \in I \times J)$ of $R^{m+n}$. Using this we define the $n$th *cartesian power* of a partition $\mathcal{P}$ of $R$ as $\mathcal{P}^{\times n} = \underbrace{\mathcal{P} \times \cdots \times \mathcal{P}}_{n}$ which is a partition of $R^n$, and $\mathcal{P}^{\times\Diamond} = \bigcup_{n\geq 0} \mathcal{P}^{\times n}$.

**Definition 3.1.** Let $\mathcal{L} = (L_i \mid i \in I)$, $\mathcal{R} = (R_j \mid j \in J)$ be two partitions of $R^n$. The (ordered) pair $(\mathcal{L}, \mathcal{R})$ is called a *regular pair of partitions of* $R^n$ if for any two elements $i \in I$, $j \in J$ and any $\alpha \in R$ there exist constants $\lambda_{ij}^{\alpha}, \rho_{ij}^{\alpha}$ such that for each $\mathbf{x} \in L_i$ there are exactly $\lambda_{ij}^{\alpha}$ members $\mathbf{y} \in R_j$ with $\mathbf{x} \cdot \mathbf{y} = \alpha$, and for each $\mathbf{y} \in R_j$ exactly $\rho_{ij}^{\alpha}$ members $\mathbf{x} \in L_i$ with $\mathbf{x} \cdot \mathbf{y} = \alpha$. The single partition $\mathcal{P}$ is called *regular* if $(\mathcal{P}, \mathcal{P})$ is a regular pair.

For a partition $\mathcal{P}$ of $R^n$ denote by $\mathbb{Q}\mathcal{P}$ the $\mathbb{Q}$-linear span of the characteristic functions of the members of $\mathcal{P}$. The space $\mathbb{Q}\mathcal{P}$ consists of all functions $f\colon R^n \to \mathbb{Q}$ which are constant on the members of $\mathcal{P}$. We have

$$\begin{aligned} \mathbb{Q}(\mathcal{P}_1 \vee \mathcal{P}_2) &= \mathbb{Q}(\mathcal{P}_1) \cap \mathbb{Q}(\mathcal{P}_2), \\ \mathbb{Q}(\mathcal{P}_1 \wedge \mathcal{P}_2) &= \mathbb{Q}(\mathcal{P}_1)\mathbb{Q}(\mathcal{P}_2). \end{aligned} \tag{15}$$

**Definition 3.2.** A pair $(\mathcal{L}, \mathcal{R})$ of partitions of $R^n$ is said to be $\mathcal{W}$*-admissible* if both $\mathcal{W}(\mathbb{Q}\mathcal{L}) \subseteq \mathbb{Q}\mathcal{R}$ and $\mathcal{W}^\circ(\mathbb{Q}\mathcal{R}) \subseteq \mathbb{Q}\mathcal{L}$. A single partition $\mathcal{P}$ is said to be $\mathcal{W}$*-admissible* if the pair $(\mathcal{P}, \mathcal{P})$ is $\mathcal{W}$-admissible.

Clearly, $(\mathcal{L}, \mathcal{R})$ is $\mathcal{W}$-admissible iff for every $i, j$ the values

$$\sum_{\mathbf{x}\in L_i} \omega(\mathbf{x}\cdot\mathbf{y}) \quad \text{and} \quad \sum_{\mathbf{y}\in R_j} \omega(\mathbf{x}\cdot\mathbf{y}) \tag{16}$$

depend only on the sets $R_j$ resp., $L_i$ to which $\mathbf{y}$ resp., $\mathbf{x}$ belong, but not on the particular choice of $\mathbf{y}$ within $R_j$ resp., $\mathbf{x}$ within $L_i$. For a $\mathcal{W}$-admissible pair $(\mathcal{L}, \mathcal{R})$ we denote these values by $k_{ji}^\circ$ resp., $k_{ij}$ and call $\mathbf{K} = (k_{ij}) \in \mathbb{Q}^{I\times J}$ (resp., $\mathbf{K}^\circ = (k_{ji}^\circ) \in \mathbb{Q}^{J\times I}$) the *first* (resp., *second*) *intersection matrix of* $(\mathcal{L}, \mathcal{R})$.

Our next proposition gives a criterion for the $\mathcal{W}$-admissibility of $(\mathcal{L}, \mathcal{R})$ which will be used in Sect. 4. Let $\mathcal{A}\colon \mathbb{Q}R^n \to \mathbb{Q}\mathcal{L}$ denote the linear map sending $\chi_{\mathbf{x}}$ to $\chi_{L_i}$ where $i$ is determined by $\mathbf{x} \in L_i$, and define $\mathcal{B}\colon \mathbb{Q}R^n \to \mathbb{Q}\mathcal{R}$ similarly.

**Proposition 3.3.** *The pair* $(\mathcal{L}, \mathcal{R})$ *is* $\mathcal{W}$*-admissible iff there exist (linear) maps* $\mathcal{K}\colon \mathbb{Q}\mathcal{L} \to \mathbb{Q}\mathcal{R}$, $\mathcal{K}^\circ\colon \mathbb{Q}\mathcal{R} \to \mathbb{Q}\mathcal{L}$ *such that* $\mathcal{B}\mathcal{W} = \mathcal{K}\mathcal{A}$ *and* $\mathcal{A}\mathcal{W}^\circ = \mathcal{K}^\circ\mathcal{B}$. *If* $(\mathcal{L}, \mathcal{R})$ *is* $\mathcal{W}$*-admissible, we have* $\mathcal{K}\colon \chi_{L_i} \mapsto \sum_{j\in J} k_{ij}\chi_{R_j}$ *and* $\mathcal{K}^\circ\colon \chi_{R_j} \mapsto \sum_{i\in I} k_{ji}^\circ \chi_{L_i}$.

The proof is straightforward and omitted.

Observe that every regular pair is $\mathcal{W}$-admissible, but not conversely. For instance, $(\{\mathbf{0}\}, R^n \setminus \{\mathbf{0}\})$ is always a $\mathcal{W}$-admissible partition of $R^n$, but is regular only if $R$ is a field. The special case $n = 1$, $R = \mathbb{F}_q$ further illustrates the difference between regular and $\mathcal{W}$-admissible partitions: Every partition of $\mathbb{F}_q$ which has $\{0\}$ as a member is $\mathcal{W}$-admissible. On the other hand, a regular partition of $\mathbb{F}_q$ either contains only singletons or has no singleton besides $\{0\}$.

Let $\mathcal{S}^\diamond$ denote the set of orbits of $R^*$ on $R^\diamond$ under the left action and $\mathcal{S}^n = \mathcal{S}^\diamond \cap R^n$. Define $\mathcal{T}^\diamond$, $\mathcal{T}^n$ similarly.[10]

**Definition 3.4.** A pair $(\mathcal{L}, \mathcal{R})$ of partitions of $R^n$ is said to be *invariant* if $(\mathcal{S}^n, \mathcal{T}^n) \le (\mathcal{L}, \mathcal{R})$.[11] A single partition $\mathcal{P}$ is said to be invariant if $(\mathcal{P}, \mathcal{P})$ is an invariant pair.

Thus $\mathcal{P}$ is invariant iff $\mathcal{P}$ is both left and right invariant in the sense that $\mathcal{S}^n \le \mathcal{P}$ (resp., $\mathcal{T}^n \le \mathcal{P}$).

**Theorem 3.5.** *An invariant pair $(\mathcal{L}, \mathcal{R})$ satisfying $\mathcal{W}(\mathbb{Q}\mathcal{L}) \subseteq \mathbb{Q}\mathcal{R}$ is $\mathcal{W}$-admissible iff $|\mathcal{L}| = |\mathcal{R}|$.*

*Proof.* Since $(\mathcal{L}, \mathcal{R})$ is invariant, we have $\mathbb{Q}\mathcal{L} \subseteq \mathbb{Q}\mathcal{S}^n$ and $\mathbb{Q}\mathcal{R} \subseteq \mathbb{Q}\mathcal{T}^n$. By Th. 2.5, $\mathcal{W}^\circ(\mathbb{Q}\mathcal{R}) \subseteq \mathbb{Q}\mathcal{L}$ iff $\mathcal{W}(\mathbb{Q}\mathcal{L}) = \mathbb{Q}\mathcal{R}$ iff $|\mathcal{L}| = \dim \mathbb{Q}\mathcal{L} = \dim \mathbb{Q}\mathcal{R} = |\mathcal{R}|$. □

**Theorem 3.6.** *If $(\mathcal{L}_1, \mathcal{R}_1)$ and $(\mathcal{L}_2, \mathcal{R}_2)$ are both $\mathcal{W}$-admissible resp., regular pairs of partitions of $R^n$, then so is $(\mathcal{L}_1 \vee \mathcal{L}_2, \mathcal{R}_1 \vee \mathcal{R}_2)$.*

*Proof.* If $(\mathcal{L}_1, \mathcal{R}_1)$ and $(\mathcal{L}_2, \mathcal{R}_2)$ are $\mathcal{W}$-admissible, we have $\mathcal{W}(\mathbb{Q}(\mathcal{L}_1 \vee \mathcal{L}_2)) = \mathcal{W}(\mathbb{Q}\mathcal{L}_1 \cap \mathbb{Q}\mathcal{L}_2) \subseteq \mathbb{Q}\mathcal{R}_1 \cap \mathbb{Q}\mathcal{R}_2 = \mathbb{Q}(\mathcal{R}_1 \vee \mathcal{R}_2)$ proving the $\mathcal{W}$-admissibility of $(\mathcal{L}_1 \vee \mathcal{L}_2, \mathcal{R}_1 \vee \mathcal{R}_2)$. Let now $(\mathcal{L}_1, \mathcal{R}_1)$ and $(\mathcal{L}_2, \mathcal{R}_2)$ be regular and $\alpha \in R$. The relation $\mho \subseteq R^n \times R^n$ defined by $\mathbf{x} \mho \mathbf{x}'$ iff $|\{\mathbf{y} \in T \mid \mathbf{x} \cdot \mathbf{y} = \alpha\}| = |\{\mathbf{y} \in T \mid \mathbf{x}' \cdot \mathbf{y} = \alpha\}|$ for every member $T \in \mathcal{R}_1 \vee \mathcal{R}_2$ contains both $\mathcal{L}_1$ and $\mathcal{L}_2$, since $T$ is a union of members of $\mathcal{R}_1$ resp., of $\mathcal{R}_2$. Hence $\mho$ contains $\mathcal{L}_1 \vee \mathcal{L}_2$ proving the regularity of $(\mathcal{L}_1 \vee \mathcal{L}_2, \mathcal{R}_1 \vee \mathcal{R}_2)$. □

**Theorem 3.7.** *Let $(\mathcal{L}_1, \mathcal{R}_1)$ and $(\mathcal{L}_2, \mathcal{R}_2)$ be regular resp., invariant $\mathcal{W}$-admissible pairs of partitions of $R^n$ resp., $R^m$. Then $(\mathcal{L}_1 \times \mathcal{L}_2, \mathcal{R}_1 \times \mathcal{R}_2)$ is also regular resp., $\mathcal{W}$-admissible.*[12]

---

[10] Thus $\mathbb{Q}\mathcal{S}^\diamond$, $\mathbb{Q}\mathcal{S}^n$, $\mathbb{Q}\mathcal{T}^\diamond$, $\mathbb{Q}\mathcal{T}^n$ are indeed the spaces introduced in Sect. 2.

[11] Here '$\le$' denotes the product order.

[12] The corresponding assertion for $\mathcal{W}$-admissible pairs is false as the example $R = \mathbb{F}_q$, $q$ odd, $q > 3$, $n_1 = n_2 = 1$, $\mathcal{P}_1 = \mathcal{P}_2 = (\{0\}, \{1\}, \mathbb{F}_q \setminus \{0, 1\})$ shows: For $\mathbf{x} = (\alpha, \alpha)$, $\mathbf{x}' = (\alpha, -\alpha)$ with $\alpha \in \mathbb{F}_q \setminus \{0, \pm 1\}$, $\mathbf{y} = (1, 1)$ we have $\omega(\mathbf{x} \cdot \mathbf{y}) = -1/(q-1) \ne 1 = \omega(\mathbf{x}' \cdot \mathbf{y})$. But $\mathbf{x}, \mathbf{x}'$ belong to the same member of $\mathcal{P}_1 \times \mathcal{P}_2$, and the singleton $\{\mathbf{y}\}$ is a member of $\mathcal{P}_1 \times \mathcal{P}_2$, showing that $\mathcal{P}_1 \times \mathcal{P}_2$ is not $\mathcal{W}$-admissible.

*Proof.* Suppose first that $(\mathcal{L}_1, \mathcal{R}_1)$ and $(\mathcal{L}_2, \mathcal{R}_2)$ are regular. For $\mathbf{x} = \mathbf{x}_1\mathbf{x}_2 \in R^{n+m}$, $R_1 \times R_2 \in \mathcal{R}_1 \times \mathcal{R}_2$ and $\alpha \in R$ we have

$$\begin{aligned} &|\{\mathbf{y} \in R_1 \times R_2 \mid \mathbf{x}\cdot\mathbf{y} = \alpha\}| = \\ &\quad = \sum_{\substack{\alpha_1,\alpha_2\in R\\ \alpha_1+\alpha_2=\alpha}} |\{\mathbf{y}_1 \in R_1 \mid \mathbf{x}_1\cdot\mathbf{y}_1 = \alpha_1\}| \cdot |\{\mathbf{y}_2 \in R_2 \mid \mathbf{x}_2\cdot\mathbf{y}_2 = \alpha_2\}|, \end{aligned} \tag{17}$$

which by the regularity of $(\mathcal{L}_1, \mathcal{R}_1)$ and $(\mathcal{L}_2, \mathcal{R}_2)$ is independent of $\alpha \in R$ and the choices of $\mathbf{x}_1 \in L_1$, $\mathbf{x}_2 \in L_2$, i.e. independent of the choice of $\mathbf{x} = \mathbf{x}_1\mathbf{x}_2 \in L_1 \times L_2$. Thus $(\mathcal{L}_1 \times \mathcal{L}_2, \mathcal{R}_1 \times \mathcal{R}_2)$ is regular.

Suppose now that $(\mathcal{L}_1, \mathcal{R}_1)$ and $(\mathcal{L}_2, \mathcal{R}_2)$ are both invariant and $\mathcal{W}$-admissible. The space $\mathbb{Q}(\mathcal{L}_1 \times \mathcal{L}_2) = \mathbb{Q}\mathcal{L}_1 \times \mathbb{Q}\mathcal{L}_2$ is generated by the products $f \times g$ with $f \in \mathbb{Q}\mathcal{L}_1$, $g \in \mathbb{Q}\mathcal{L}_2$. By assumption, $f, g$ are left-invariant. Hence by Th. 2.5

$$\begin{aligned} \mathcal{W}(\mathbb{Q}(\mathcal{L}_1 \times \mathcal{L}_2)) &= \mathcal{W}(\mathbb{Q}\mathcal{L}_1) \times \mathcal{W}(\mathbb{Q}\mathcal{L}_2) \\ &\subseteq \mathbb{Q}\mathcal{R}_1 \times \mathbb{Q}\mathcal{R}_2 = \mathbb{Q}(\mathcal{R}_1 \times \mathcal{R}_2) \end{aligned} \tag{18}$$

proving the $\mathcal{W}$-admissibility of $(\mathcal{L}_1 \times \mathcal{L}_2, \mathcal{R}_1 \times \mathcal{R}_2)$. □

*Constructions from groups.* The general semilinear group $\Gamma\mathrm{L}(n, R)$ is the set $\mathrm{GL}(n, R) \times \mathrm{Aut}\, R$ with multiplication rule $(\mathbf{B}, \tau)(\mathbf{A}, \sigma) = (\mathbf{B}\tau(\mathbf{A}), \tau\sigma)$. The group $\Gamma\mathrm{L}(n, R)$ acts on both sides of $R^n$ via

$$(\mathbf{A}, \sigma)\mathbf{x} = \mathbf{A}\sigma(\mathbf{x}), \quad \mathbf{x}(\mathbf{A}, \sigma) = \sigma^{-1}(\mathbf{x})\sigma^{-1}(\mathbf{A}) \qquad (\mathbf{x} \in R^n). \tag{19}$$

The right action is essentially the adjoint of the left action:

$$\mathbf{x} \cdot (\mathbf{A}, \sigma)\mathbf{y} = \sigma\big(\mathbf{x}(\mathbf{A}, \sigma) \cdot \mathbf{y}\big). \tag{20}$$

For a subgroup $G$ of $\Gamma\mathrm{L}(n, R)$ denote by $\mathcal{L}_G$ (resp., $\mathcal{R}_G$) the set of orbits of $G$ on $R^n$ with respect to the *right* (resp., *left*) action.

**Definition 3.8.** Let $(\mathcal{L}, \mathcal{R})$ be a $G$-invariant pair (i.e. $\mathcal{L}G = \mathcal{L}$ and $G\mathcal{R} = \mathcal{R}$) of partitions of $R^n$. The pair $(\mathcal{L} \vee \mathcal{L}_G, \mathcal{R} \vee \mathcal{R}_G)$ whose members are the sets $\bigcup_{g\in G} Lg$ $(L \in \mathcal{L})$ and $\bigcup_{g\in G} gR$ $(R \in \mathcal{R})$, respectively, is called the *fusion of* $(\mathcal{L}, \mathcal{R})$ *with respect to* $G$.

**Theorem 3.9.** *For any subgroup $G$ of $\Gamma\mathrm{L}(n, R)$ the pair $(\mathcal{L}_G, \mathcal{R}_G)$ is $\mathcal{W}$-admissible. If $G \leq \mathrm{GL}(n, R)$ then $(\mathcal{L}_G, \mathcal{R}_G)$ is regular.*

*Proof.* Theorem 3.9 is proved using (20) and $\omega(\sigma(x)) = \omega(x)$ for $\sigma \in \mathrm{Aut}\, R$. We omit the details.

**Corollary 3.10.** *Let $G$ be a subgroup of $\Gamma\mathrm{L}(n, R)$. If $(\mathcal{L}, \mathcal{R})$ is a $G$-invariant $\mathcal{W}$-admissible pair of partitions of $R^n$, then so is the fusion $(\mathcal{L} \vee \mathcal{L}_G, \mathcal{R} \vee \mathcal{R}_G)$ with respect to $G$. Furthermore, if $(\mathcal{L}, \mathcal{R})$ is regular and $G \leq \mathrm{GL}(n, R)$, then $(\mathcal{L} \vee \mathcal{L}_G, \mathcal{R} \vee \mathcal{R}_G)$ is regular.*

The $n$th power $(\mathcal{L}^{\times n}, \mathcal{R}^{\times n})$ of any pair $(\mathcal{L}, \mathcal{R})$ of partitions of the ring $R$ is obviously invariant under the symmetric group $\mathrm{Sym}(n)$.

**Definition 3.11.** The fusion of $(\mathcal{L}^{\times n}, \mathcal{R}^{\times n})$ with respect to $\mathrm{Sym}(n)$ is called the *nth extension* of $(\mathcal{L}, \mathcal{R})$, and is denoted by $(\mathcal{L}^n, \mathcal{R}^n)$.

We take $I^n = \{w \in \mathbb{N}_0^I \mid \sum_{i \in I} w(i) = n\}$ as the weight set of $\mathcal{L}^n$, the indexing being defined by $\mathbf{x} \in L_w$ iff $\mathbf{x}$ has exactly $w(i)$ entries belonging to $L_i$ ($i \in I$). Let further $\mathcal{L}^\diamond = \bigcup_{n \geq 0} \mathcal{L}^n$, $\mathcal{R}^\diamond = \bigcup_{n \geq 0} \mathcal{R}^n$, $I^\diamond = \bigcup_{n \geq 0} I^n = \mathbb{N}_0^I$.

By Th. 3.7 and Cor. 3.10, if $(\mathcal{L}, \mathcal{R})$ is regular resp., invariant $\mathcal{W}$-admissible, then so is $(\mathcal{L}^n, \mathcal{R}^n)$. The maps $\mathcal{A}$, $\mathcal{B}$, $\mathcal{K}$, $\mathcal{K}^\circ$ can be extended to graded maps satisfying the conditions of Prop. 3.3 and such that e.g. $\mathcal{A}(\chi_{\mathbf{x}}) = \chi_L$ if $\mathbf{x} \in L \in \mathcal{L}^\diamond$. The corresponding maps for $(\mathcal{L}^{\times n}, \mathcal{R}^{\times n})$, will be distinguished, if necessary, by an additional superscript (i.e. $\mathcal{A}^\times$, $\mathcal{B}^\times$).

*Constructions from submodule lattices.* Let $\mathfrak{S}$ be a complete sublattice of $\mathrm{S}({}_RR^n)$, i.e. a set of submodules of ${}_RR^n$ containing $\{0\}$ and $R^n$, and such that $N_1, N_2 \in \mathfrak{S}$ implies $N_1 \cap N_2 \in \mathfrak{S}$ and $N_1 + N_2 \in \mathfrak{S}$. Let $\mathbb{Q}\mathfrak{S} = \langle \chi_N \mid N \in \mathfrak{S} \rangle_{\mathbb{Q}}$. It is readily seen that $\mathbb{Q}\mathfrak{S} = \mathbb{Q}\mathcal{L}_{\mathfrak{S}}$ for a unique partition $\mathcal{L}_{\mathfrak{S}}$ of $R^n$. Two elements $\mathbf{x}, \mathbf{y} \in R^n$ belong to the same member of $\mathcal{L}_{\mathfrak{S}}$ iff $\{N \in \mathfrak{S} \mid \mathbf{x} \in N\} = \{N \in \mathfrak{S} \mid \mathbf{y} \in N\}$. It follows from (5) that $\mathfrak{S}^\perp = \{N^\perp \mid N \in \mathfrak{S}\}$ is a complete sublattice of $\mathrm{S}(R^n_R)$. Let $\mathcal{R}_{\mathfrak{S}}$ be the partition of $R^n$ with $\mathbb{Q}(\mathfrak{S}^\perp) = \mathbb{Q}\mathcal{R}_{\mathfrak{S}}$.

**Theorem 3.12.** *For every complete sublattice $\mathfrak{S}$ of $\mathrm{S}({}_RR^n)$ the pair $(\mathcal{L}_{\mathfrak{S}}, \mathcal{R}_{\mathfrak{S}})$ is $\mathcal{W}$-admissible.*

*Proof.* By Th. 2.3, $\mathcal{W}(\mathbb{Q}\mathcal{L}_{\mathfrak{S}}) = \mathcal{W}(\mathbb{Q}\mathfrak{S}) \subseteq \mathbb{Q}(\mathfrak{S}^\perp) = \mathbb{Q}\mathcal{R}_{\mathfrak{S}}$. The assertion now follows from Th. 3.5. □

## 4 Weight Spectrum and MacWilliams Identities

Let $\mathcal{P}$ be a partition of $R^n$ and $C \subseteq R^n$ a (not necessarily linear) code.

**Definition 4.1.** The function $\big(|C \cap P_i| \mid i \in I\big)$ is called the *weight spectrum of $C$ with respect to the partition $\mathcal{P}$.*

Let $(\mathcal{L}, \mathcal{R})$ be a $\mathcal{W}$-admissible pair of partitions of $R^n$. Considering $\mathcal{L} = (L_i)_{i \in I}$, $\mathcal{R} = (R_j)_{j \in J}$ as fixed, we denote the weight spectrum of a code $C \subseteq R^n$ with respect to $\mathcal{L}$ by $A(C) = \big(A_i(C) \mid i \in I\big)$ and the weight spectrum of $C$ with respect to $\mathcal{R}$ by $B(C) = \big(B_j(C) \mid j \in J\big)$.

**Theorem 4.2 (MacWilliams identities).** *Let $(\mathcal{L}, \mathcal{R})$ be a $\mathcal{W}$-admissible pair of partitions of $R^n$ with intersection matrices $\mathbf{K}, \mathbf{K}^\circ$, and $C \leq {}_RR^n$ be a linear code. The weight spectrum of $C$ with respect to $\mathcal{L}$ and the weight spectrum of the orthogonal code $C^\perp$ with respect to $\mathcal{R}$ are related by*

$$B(C^\perp) = \frac{1}{|C|} A(C) \cdot \mathbf{K}, \qquad A(C) = \frac{1}{|C^\perp|} B(C^\perp) \cdot \mathbf{K}^\circ. \tag{21}$$

*Proof.* Observe that $\mathcal{A}(\chi_C) = \sum_{i\in I} A_i(C)\chi_{L_i}$, $\mathcal{B}(\chi_{C^\perp}) = \sum_{j\in J} B_j(C^\perp)\chi_{R_j}$. By Th. 2.3 and Prop. 3.3,

$$|C| \cdot \mathcal{B}(\chi_{C^\perp}) = \mathcal{B}(|C| \cdot \chi_{C^\perp}) = \mathcal{B}(\mathcal{W}\chi_C) = \mathcal{K}(\mathcal{A}\chi_C) \tag{22}$$

proving the first identity, since $\mathbf{K}$ represents $\mathcal{K}$ with respect to the bases $(\chi_{L_i} \mid i \in I)$ and $(\chi_{R_j} \mid j \in J)$. The second identity then follows, e.g., by taking $R^\circ$ instead of $R$. □

Theorem 4.2 reduces the problem of determining, for any linear code $C \leq {}_RR^n$, the weight spectrum $B(C^\perp)$ from the weight spectrum $A(C)$ to the computation of the $|I| \times |J|$-matrix $\mathbf{K} = (k_{ij})$. Instead of computing the constants $k_{ij} = \sum_{\mathbf{y}\in R_j} \omega(\mathbf{x}\cdot\mathbf{y})$ directly, which can be very tedious, we use the following approach: Find $|I|$ linear codes $C \leq {}_RR^n$ whose weight enumerators $\mathcal{A}(\chi_C) \in \mathbb{Q}\mathcal{L}$ are linearly independent and such that both $A(C)$ and the weight distribution $B(C^\perp)$ are known. Then we can solve (21) for $\mathbf{K}$. Of course, this method can only be applied if $\mathbb{Q}\mathcal{L}$ is generated by the weight enumerators of left linear codes, and similarly for $\mathbb{Q}\mathcal{R}$. Invariant pairs have this property, and the method works particularly well for the $n$th cartesian power resp., extension of an invariant $\mathcal{W}$-admissible pair of partitions of the ring $R$.

First let $\mathcal{P}$ be an arbitrary partition of $R$. The space $\mathbb{Q}\mathcal{P}^{\times\diamond}$ forms a subalgebra of $(\mathbb{Q}R^\diamond, \times)$ isomorphic to the semigroup algebra of the free monoid over $\mathcal{P}$. The associated map $\mathcal{A}^\times : \mathbb{Q}R^\diamond \to \mathbb{Q}\mathcal{P}^{\times\diamond}$ is defined for words $\mathbf{x} \in R^\diamond$ by $\mathcal{A}^\times(\chi_\mathbf{x}) = \chi_P$ if $P \in \mathcal{P}^{\times\diamond}$ contains $\mathbf{x}$. Since $\mathcal{A}^\times(\chi_\mathbf{x} \times \chi_\mathbf{y}) = \mathcal{A}^\times(\chi_{\mathbf{xy}}) = \chi_{L_1\times L_2} = \chi_{L_1} \times \chi_{L_2}$ if $\mathbf{x} \in L_1$, $\mathbf{y} \in L_2$, the map $\mathcal{A}^\times$ is a (graded) homomorphism from $(\mathbb{Q}R^\diamond, \times)$ onto $(\mathbb{Q}\mathcal{P}^{\times\diamond}, \times)$.

In general, the space $\mathbb{Q}\mathcal{P}^\diamond$ is not closed with respect to $\times$, but it can be given a different multiplication using its weight set $\mathbb{N}_0^I$. Defining $\chi_{P_w} \odot \chi_{P_{w'}} = \chi_{P_{w+w'}}$ we see that $(\mathbb{Q}\mathcal{P}^\diamond, \odot) \cong \mathbb{Q}(\mathbb{N}_0^I, +) = \mathbb{Q}[X_i; i \in I]$ as graded $\mathbb{Q}$-algebras.[13] For example, the weight enumerator of $C \subseteq R^n$ with respect to the $n$th extension $\mathcal{P}^n$ may be written as either $\mathcal{A}(\chi_C) = \sum A_w(C)\chi_{P_w}$ or $A_C = \sum A_w(C) \prod_{i\in I} X_i^{w(i)}$, just as we please. Since $\mathbf{x} \in P_w$, $\mathbf{y} \in P_{w'}$ implies $\mathbf{xy} \in P_{w+w'}$, we have $\mathcal{A}(\chi_\mathbf{x} \times \chi_\mathbf{y}) = \mathcal{A}(\chi_{\mathbf{xy}}) = \chi_{P_{w+w'}} = \chi_{P_w} \odot \chi_{P_{w'}} = \mathcal{A}(\chi_\mathbf{x}) \odot \mathcal{A}(\chi_\mathbf{y})$, i. e., $\mathcal{A}\colon (\mathbb{Q}R^\diamond, \times) \to (\mathbb{Q}\mathcal{L}^\diamond, \odot)$ is a (graded) epimorphism.

**Theorem 4.3.** *Let $(\mathcal{L}, \mathcal{R})$ be an invariant $\mathcal{W}$-admissible pair of partitions of the ring $R$. Then both $\mathcal{K}^\times : (\mathbb{Q}\mathcal{L}^{\times\diamond}, \times) \to (\mathbb{Q}\mathcal{R}^{\times\diamond}, \times)$ and $\mathcal{K}\colon (\mathbb{Q}\mathcal{L}^\diamond, \odot) \to (\mathbb{Q}\mathcal{R}^\diamond, \odot)$ are graded isomorphisms. In particular, if $C \leq {}_RR^n$ is a left linear*

[13] By definition (see e. g. [1, Chap. III, §2.9]), the polynomial ring $\mathbb{Q}[X_i; i \in I] = \mathbb{Q}(\mathbb{N}_0^I, +)$ consists of all functions $f\colon \mathbb{N}_0^I \to \mathbb{Q}$ with finite support which are multiplied according to the rule $\chi_w\chi_{w'} = \chi_{w+w'}$ $(w, w' \in \mathbb{N}_0^I)$. The symbol '$X_i$' is used as an abbreviation for $\chi_w$ where $w(i) = 1$ and $w(i') = 0$ if $i' \neq i$.

*code, then*

$$\sum_{\mathbf{j}\in J^n} B_{\mathbf{j}} T_{j_1} T_{j_2} \cdots T_{j_n} = \frac{1}{|C|} \cdot \sum_{\mathbf{i}\in I^n} A_{\mathbf{i}} (\mathcal{K}S_{i_1})(\mathcal{K}S_{i_2}) \cdots (\mathcal{K}S_{i_n}), \tag{23}$$

$$B_{C^\perp}(Y_j; j \in J) = \frac{1}{|C|} \cdot A_C(\mathcal{K}X_i; i \in I), \tag{24}$$

*where* $A_{\mathbf{i}} = A_{\mathbf{i}}(C)$, $B_{\mathbf{j}} = B_{\mathbf{j}}(C^\perp)$, $\mathcal{K}S_i = \sum_{j\in J} k_{ij} T_j$, $\mathcal{K}X_i = \sum_{j\in J} k_{ij} Y_j$.[14] *Similar statements hold for* $\mathcal{K}^{\times\circ}$, $\mathcal{K}^\circ$ *and right linear codes* $D \le R_R^n$.

*Proof.* We only give the proof for $\mathcal{K}$. As remarked in Sect. 3, each pair $(\mathcal{L}^n, \mathcal{R}^n)$ is $\mathcal{W}$-admissible. For $f \in \mathbb{Q}S^n$, $g \in \mathbb{Q}S^m$ we have

$$\begin{aligned}\mathcal{K}(\mathcal{A}f \odot \mathcal{A}g) &= \mathcal{K}\mathcal{A}(f \times g) = \mathcal{B}\mathcal{W}(f \times g) = \\ &= \mathcal{B}(\mathcal{W}f \times \mathcal{W}g) = \mathcal{B}(\mathcal{W}f) \odot \mathcal{B}(\mathcal{W}g) = \mathcal{K}(\mathcal{A}f) \odot \mathcal{K}(\mathcal{A}g).\end{aligned} \tag{25}$$

Since $\mathbb{Q}\mathcal{L}^\Diamond = \mathcal{A}(\mathbb{Q}S^\Diamond)$ (due to the invariance of $\mathcal{L}^n$), this already proves that $\mathcal{K}$ is a homomorphism. By Th. 2.5, $\mathcal{K}$ is one-to-one.

Theorem 4.3 gives an "easy" method for computing $B(C^\perp)$ from $A(C)$. Since $\mathcal{L}$ is invariant, we can express e. g. the indeterminate $X_i$ as a linear combination $X_i = \sum_{I \le_R R} a_I A_I$ of the weight enumerators of left ideals of $R$. By Th. 4.2, $\mathcal{K}X_i = \sum_{I \le_R R} a_I |I| B_{I^\perp}$, which in view of (24) gives $B(C^\perp)$. This will be illustrated in the examples below.

## 5 Examples

*Example 5.1.* Since $\mathcal{H} = \big(\{0\}, R\backslash\{0\}\big)$ is invariant and $\mathcal{W}$-admissible (though not, in general, regular), $\mathcal{H}$ gives rise to $\mathcal{W}$-admissible partitions $\mathcal{H}^{\times n}$, $\mathcal{H}^n$ of $R^n$. The partition $\mathcal{H}^{\times n}$ is called the ***support partition*** of $R^n$, since its members $H_1 \times \cdots \times H_n$ consist of the words $\mathbf{x} \in R^n$ with fixed support $S = \big\{i \in \overline{1,n} \mid H_i = R \setminus \{0\}\big\}$.[15] With this indexing we obtain

$$A_T(C^\perp) = \frac{1}{|C|} \cdot \sum_{S \subseteq \overline{1,n}} (-1)^{|T\cap S|} (|R|-1)^{|T\backslash S|} A_S(C) \qquad (T \subseteq \overline{1,n}) \tag{26}$$

from Th. 4.3.[16] For, $A_T(C^\perp)$ is the coefficient of $S_{j_1} S_{j_2} \cdots S_{j_n}$ on the right-hand side of (23), where $j_k = 1$ if $k \in T$ and $j_k = 0$ if $k \notin T$.[16] Since $\mathcal{K}S_0 = S_0 + (|R|-1)S_1$, $\mathcal{K}S_1 = S_0 - S_1$, the coefficient of $S_{j_1} S_{j_2} \cdots S_{j_n}$ in

---

[14] We have used further abbreviations: $S_i = \chi_{L_i} \in \mathbb{Q}\mathcal{L}^{\times\Diamond}$, $T_j = \chi_{R_j} \in \mathbb{Q}\mathcal{R}^{\times\Diamond}$, $Y_j = \chi_{R_w}$ where $w \in \mathbb{N}_0^J$ is defined by $w(j) = 1$ and $w(j') = 0$ if $j' \ne j$, and omitted '$\times$', '$\odot$'.

[15] Here $\overline{1,n}$ denotes the set $\{1, 2, \ldots, n\}$.

[16] Note that $\mathcal{A} = \mathcal{B}$, $T_i = S_i$, $Y_i = X_i$ for pairs of the special form $(\mathcal{P}, \mathcal{P})$.

$(\mathcal{K}S_{i_1})(\mathcal{K}S_{i_2})\cdots(\mathcal{K}S_{i_n})$ is $(|R|-1)^{|T\setminus S|}(-1)^{|T\cap S|}$ where $S=\{k\in\overline{1,n}\mid i_k=1\}$.

The partition $\mathcal{H}^n$ is called the *Hamming partition* of $R^n$. Its members are $H_i^n=\{\mathbf{x}\in R^n\mid w_{\mathrm{Ham}}(\mathbf{x})=i\}$ $(0\le i\le n)$. Writing $X=X_0$, $Y=X_1$ we have $\mathcal{K}X=X+(|R|-1)Y$, $\mathcal{K}Y=X-Y$ and thus from (24),

$$A_{C^\perp}(X,Y)=\frac{1}{|C|}\cdot A_C\bigl(X+(|R|-1)Y,X-Y\bigr).^{17} \tag{27}$$

*Example 5.2.* Let $\mathcal{C}=(\{x\}\mid x\in R)$ be the partition of $R$ into singletons. Its $n$th power $\mathcal{C}^{\times n}$ is the family of singleton subsets of $R^n$ which is trivially regular. The weight spectrum of a code $C\subseteq R^n$ with respect to $\mathcal{C}^{\times n}$ is known as the *exact weight distribution of* $C$. The $n$th extension $\mathcal{C}^n$ is regular by the results of Sect. 3. Weight spectra with respect to $\mathcal{C}^n$ are known as *complete weight spectra.* They satisfy the MacWilliams identities (21), but since $\mathcal{C}$ is not invariant, (21) can't be put into the simple form (24).

*Example 5.3.* Let $\mathfrak{I}=\mathrm{S}({}_RR)$, $\mathfrak{J}=\mathrm{S}(R_R)$, and denote by $\mathfrak{I}_1\subseteq\mathfrak{I}$ (resp., $\mathfrak{J}_1\subseteq\mathfrak{J}$) the subset of principal left (resp., principal right) ideals. By [28, Prop. 5.1] the partition $\mathcal{I}=\mathcal{L}_{\mathfrak{I}}$ of $R$ generated by $\mathfrak{I}$ is the set $\mathcal{S}^1$ of orbits of $R^*$ under the left action and admits the indexing $Rx\mapsto R^*x$ with weight set $\mathfrak{I}_1$. In view of $ux=(uxu^{-1})u$ we have $(\mathcal{I},\mathcal{J})=(\mathcal{L}_G,\mathcal{R}_G)$ where $G\le\Gamma\mathrm{L}(1,R)$ is isomorphic to $R^*$.[18] In order to apply Th. 4.3 we now compute the weight enumerators of the left ideals of $R$. Let $\mu$, $\ni$ denote the Möbius and Euler function for finite left $R$-modules defined by $\mu(\{0\})=1$, $\sum_{N\le M}\mu(N)=0$ if $M\ne\{0\}$, $\ni(M)=|\{x\mid M=Rx\}|$, and similarly for right $R$-modules.[19] >From $A_I=\sum_{K\le I}\ni(K)S_K$ we have, by Möbius inversion, $S_I=\ni(I)^{-1}\times\sum_{K\le I}\mu(I/K)A_K$, whence

$$\begin{aligned}\mathcal{K}S_I&=\frac{1}{\ni(I)}\cdot\sum_{K\le I}\mu(I/K)|K|A_{K^\perp}\\&=\frac{1}{\ni(I)}\cdot\sum_{K\le I}\sum_{J\le K^\perp}\mu(I/K)|K|\ni(J)T_J\\&=\frac{1}{\ni(I)}\cdot\sum_{J\le R_R}\ni(J)\left(\sum_{K\le^\perp J\cap I}\mu(I/K)|K|\right)T_J.\end{aligned}\tag{28}$$

We give the explicit form of (28) in two special cases:

[17] Equation (27) is also (in a more general form) in [28], in [14] for the case $R=\mathbb{Z}_m$, and in [23] for modules over commutative Frobenius rings.

[18] In general, $G\not\le\mathrm{GL}(1,R)$ and $(\mathcal{I},\mathcal{J})$ is not regular.

[19] Note that by [13, Th. 3] the function $\mu$ is well-defined for (quotients of) two-sided ideals of $R$. The same is true of $\ni$, since in view of $|Rx|=|xR|$ (cf. e. g. [9]) we have $I=Rx$ iff $I=xR$ for any two-sided ideal $I$ of $R$ and $x\in R$.

*Chain rings.* Let $R$ be a chain ring of order $q^m$ with maximal ideal $N$ and residue field $R/N$ of order $q$. In this case $\mathfrak{I} = \mathfrak{I}_1 = \mathfrak{J} = \mathfrak{J}_1 = \{N^i \mid 0 \leq i \leq m\}$, so we take $\mathcal{I} = (R_i \mid 0 \leq i \leq m)$ with $R_i = N^i \setminus N^{i+1}$. Further, $|N^i| = q^{m-i}$, $(N^i)^\perp = N^{m-i}$, $\mu(N^{m-1}) = -1$, $\mu(N^i) = 0$ if $i \leq m-2$, $\ni(N^i) = q^{m-i} - q^{m-i-1}$ if $i < m$. Using (24) and (28) we obtain

$$A_{C^\perp}(X_0, X_1, \ldots, X_m) = \frac{1}{|C|} \cdot A_C\big(A_m - X_{m-1}, A_{m-1} - qX_{m-2}, \ldots, A_1 - q^{m-1}X_0, A_0\big) \quad (29)$$

where $A_i = X_m + (q-1)X_{m-1} + \cdots + (q^{m-i} - q^{m-i-1})X_i$ is the weight enumerator of $N^i$. A slightly different version of the MacWilliams identities for chain rings is proved in [12]. In the case $R = \mathbb{Z}_{p^m}$ the matrix $\mathbf{K}$ can be found in [14]. For chain rings of prime characteristic[20] it corresponds to an identity in [3].

*Matrix rings of size two.* Left (right) ideals of $\mathrm{M}_m(\mathbb{F}_q)$ of order $q^{mk}$ correspond to subspaces of $\mathbb{F}_q^m$ of rank $k$ with $J = I^\perp$ iff $W = V^\perp$ for the corresponding subspaces.[21] Thus $R = \mathrm{M}_2(\mathbb{F}_q)$ has (besides $\{0\}$, $R$) exactly $q+1$ left and right ideals $I_s$, $J_s$ ($s \in \overline{1, q+1}$) with $|I_s| = |J_s| = q^2$ and $J_s = I_s^\perp$. Using $\mu(I_s) = \mu(J_s) = -1$, $\mu(R) = q$, $\ni(I_s) = \ni(J_s) = q^2 - 1$, $\ni(R) = |\mathrm{GL}(2, R)| = (q^2-1)(q^2-q)$ we obtain after a short calculation

$$\begin{aligned} &B_{C^\perp}(Y_0, \ldots, Y_t, \ldots, Y_R) = \\ &= \frac{1}{|C|} \cdot A_C\Big(Y_0 + (q^2-1)\sum_{s=1}^{q+1} Y_s + (q^2-1)(q^2-q)Y_R, \ldots, \\ &\quad Y_0 + q^2 Y_t - \sum_{s=1}^{q+1} Y_s - (q^2-q)Y_R, \ldots, Y_0 - \sum_{s=1}^{q+1} Y_s + qY_R\Big).^{22} \quad (30) \end{aligned}$$

*Remark 5.4.* In [28] the so-called *symmetrized weight composition with respect to a subgroup $U$ of $R^*$* was defined as $\mathrm{swc}_r(\mathbf{x}) = |\{t \mid 1 \leq t \leq n \text{ and } x_t \in rU\}|$. The $\mathcal{J}$-type of $\mathbf{x}$ is thus essentially the symmetrized weight composition of $\mathbf{x}$ with respect to the full group of units of $R$. The MacWilliams identities for $\mathrm{swc}_r$ proved in [28] are similar to (24), but they involve characters of $(R, +)$.

---

[20] These are precisely the truncated skew polynomial rings $\mathbb{F}_q[X; \sigma]/(X^m)$ with $\sigma \in \mathrm{Aut}\,\mathbb{F}_q$; cf. e. g. [11].

[21] The space associated with a left (resp., right) ideal of $\mathrm{M}_m(\mathbb{F}_q)$ is the set of row (resp., column) vectors of its elements or, equivalently, the row (resp., column) space of each of its generators.

[22] The abbreviations $X_s = X_{I_s}$, $Y_s = Y_{J_s}$ for $s \in \overline{1, q+1}$ were used.

*Example 5.5.* Let $R = \mathbb{F}_q$ be a finite field and $\mathrm{Mon}(n,q) \leq \mathrm{GL}(n,q)$ the group of all monomial matrices. Choose a partition $\mathcal{T} = \{T_1, \dots, T_r\}$ of the set $\{1, 2, \dots, n\}$ of positions and let $G$ be the subgroup of $\mathrm{Mon}(n,q)$ consisting of all monomial matrices whose permutation part preserves the sets $T_1, \dots, T_r$. Let $w_{\mathcal{T}}(\mathbf{x}) = \big(|\mathrm{supp}\,\mathbf{x} \cap T_1|, \dots, |\mathrm{supp}\,\mathbf{x} \cap T_r|\big)$ be the so-called $\mathcal{T}$-weight of $\mathbf{x} \in \mathbb{F}_q^n$ defined in [24]. Two vectors $\mathbf{x}, \mathbf{y} \in \mathbb{F}_q^n$ belong to the same orbit of $G$ iff $w_{\mathcal{T}}(\mathbf{x}) = w_{\mathcal{T}}(\mathbf{y})$. Hence the orbits of $G$ form an invariant regular partition $\mathcal{P} = \big(P_{\mathbf{i}} \mid \mathbf{i} \in \mathrm{W}(\mathcal{T})\big)$ with weight set $\mathrm{W}(\mathcal{T}) = \{\mathbf{i} \in \mathbb{N}^r \mid 0 \leq i_s \leq |T_s| \text{ for } 1 \leq s \leq r\}$. The weight spectrum of codes with respect to $\mathcal{P}$ was investigated in [24], including a proof of the MacWilliams identities.

*Example 5.6.* The *rank partition* of $\mathrm{M}_m(\mathbb{F}_q)$ is $\mathcal{D} = (D_0, D_1, \dots, D_m)$ where $D_i = \{X \in \mathrm{M}_m(\mathbb{F}_q) \mid \mathrm{rk}\, X = i\}$. Since $(\mathcal{D}, \mathcal{D})$ is the fusion of $(\mathcal{I}, \mathcal{J})$ (cf. Ex. 5.3) with respect to $\mathrm{GL}(1, R) = R^*$, the partition $\mathcal{D}$ is invariant and $\mathcal{W}$-admissible, but $\mathcal{D}$ is not regular for $m \geq 2$. It is not difficult to see that the MacWilliams identity for $\mathcal{D}^n$ is obtained from that for $(\mathcal{I}^n, \mathcal{J}^n)$ by substituting e. g. $X_k$ for $X_I$ in (30) if $|I| = q^{mk}$. Thus

$$
\begin{aligned}
A_{C^\perp}(X_0, X_1, X_2) &= \\
&= \frac{1}{|C|} \cdot A_C\Big(X_0 + (q^2-1)(q+1)X_1 + (q^2-1)(q^2-q)X_2, \\
&\qquad X_0 + (q^2-q-1)X_1 - (q^2-q)X_2, X_0 - (q+1)X_1 + qX_2\Big). \quad (31)
\end{aligned}
$$

For example, in the binary case $A_{C^\perp}(X_0, X_1, X_2) = 4^{-k} A_C(X_0 + 9X_1 + 6X_2, X_0 + X_1 - 2X_2, X_0 - 3X_1 + 2X_2)$ if $C$ is a left linear code over $\mathrm{M}_2(\mathbb{F}_2)$ of length $n$ and composition length $k$, i. e. $|C| = 4^k$ and $|C^\perp| = 4^{2n-k}$.

*Example 5.7.* In [4,5] a partition $\mathcal{P}$ of a finite abelian group $A$ was called an *F-partition* if the space $\mathbb{C}\mathcal{P}$ is invariant under the FOURIER transform $\mathcal{F} \colon \mathbb{C}A \to \mathbb{C}A$ defined by $(\mathcal{F}f)(x) = \sum_{y \in A} \psi_x(y) f(y)$, where $x \mapsto \psi_x$ is a fixed isomorphism from $A$ onto its character group $\widehat{A}$. This definition certainly depends on the particular choice of the map $x \mapsto \psi_x$.

The group $A$ can be given (non-canonically) the structure of a commutative Frobenius ring as follows: Write $A = A_1 \oplus \cdots \oplus A_r$ with each $A_j$ cyclic, say $A_j = \langle e_j \rangle$ where $e_j$ has order $q_j$, and define multiplication by $e_i^2 = e_i$, $e_i e_j = 0$ if $i \neq j$. Let $\psi \in \widehat{A}$ be defined by $\psi(e_j) = \exp(2\pi\mathrm{i}/q_j)$ for $1 \leq j \leq r$. Then $\psi$ is a generating character of the ring $A$. For $x \in A$ let $\psi_x(y) = \psi(xy)$. Then it is immediate that every regular partition of the ring $A$ is an F-partition. Conversely, it can be shown that an F-partition $\mathcal{P}$ is a $\mathcal{W}$-admissible partition of the ring $A$ if $\mathcal{P}$ is invariant under every automorphism of $A$ which fixes each $A_j$. F-partitions admit a MacWilliams identity of the form (21); cf. [5].

## References


1. N. Bourbaki. *Algebra I, Chapters 1–3 (Algebraic Structures; Linear Algebra; Tensor Algebras, Exterior Algebras, Symmetric Algebras)*. Elements of Mathematics. Addison-Wesley, 1974.
2. W. E. Clark and D. A. Drake. Finite chain rings. *Abh. Math. Sem. Univ. Hamburg*, 39:147–153, 1974.
3. S. T. Dougherty and M. M. Skriganov. MacWilliams-type theorems for a non-Hamming metric. Preprint, Sept. 1999.
4. T. Ericson, J. Simonis, H. Tarnanen, and V. A. Zinoviev. $F$-partitions of cyclic groups. *AAECC*, 8:387–393, 1997.
5. T. Ericson and V. A. Zinoviev. On Fourier-invariant partitions of finite abelian groups and the MacWilliams identity for group codes. *Probl. Inform. Trans.*, 32(1):117–122, 1996. English translation of [26].
6. M. Hall, Jr. A type of algebraic closure. *Ann. of Math. (2)*, 40:360–369, 1939.
7. Y. Hirano. On admissible rings. *Indag. Math. (N.S.)*, 8(1):55–59, 1997.
8. W. Heise and P. Quattrocchi. *Informations- und Codierungstheorie*. Springer-Verlag, 3. Auflage, 1995.
9. T. Honold. Characterization of finite Frobenius rings. Submitted for Publication.
10. T. Honold. A Proof of MacWilliams' Identity. *J. Geometry*, 57:120–122, 1996.
11. T. Honold and I. Landjev. Linearly representable codes over chain rings. *Abh. Math. Sem. Univ. Hamburg*, 69:187–203, 1999.
12. T. Honold and I. Landjev. Linear codes over finite chain rings. *Electronic Journal of Combinatorics*, 7(#R11), 2000.
13. T. Honold and A. A. Nechaev. Weighted modules and representations of codes. *Probl. Inform. Trans.*, 35(3):205–223, 1999. English translation of [25].
14. M. Klemm. Über die Identität von MacWilliams für die Gewichtsfunktion von Codes. *Arch. Math. (Basel)*, 49:400–406, 1987.
15. E. Lamprecht. Allgemeine Theorie der Gaußchen Summen in endlichen kommutativen Ringen. *Math. Nachr.*, 9:149–196, 1953.
16. E. Lamprecht. Über $I$-reguläre Ringe, reguläre Ideale und Erklärungsmoduln. I. *Math. Nachr.*, 10:353–382, 1953.
17. I. Landjev. A note on the MacWilliams identities. *Compt. Rend. Bulg. Acad. Sci.*, 52, 1999.
18. F. J. MacWilliams. A theorem on the distribution of weights in a systematic code. *Bell System Tech. J.*, 42(1):79–94, 1963.
19. F. J. MacWilliams and N. J. A. Sloane. *The Theory of Error-Correcting Codes*. North-Holland, 1977.
20. B. R. McDonald. *Finite Rings with Identity*. Marcel Dekker, New York, 1974.
21. T. Nakayama. On Frobeniusean algebras. II. *Ann. of Math. (2)*, 42:1–21, 1941.
22. A. A. Nechaev. Finite principal ideal rings. *Math. USSR Sbornik*, 20:364–382, 1973.
23. A. A. Nechaev. Linear codes over modules and over spaces. MacWilliams' identity. In *Proceedings of the 1996 IEEE Int. Symp. Inf. Theory and Appl.*, pages 35–38, Victoria B.C., Canada, 1996.
24. J. Simonis. MacWilliams identities and coordinate partitions. *Linear Algebra Appl.*, 216:81–91, 1995.

25. Томас Хонольд, Александр А. Нечаев. Полновесные модули и представления кодов. Проблемы передачи информации, 35(3):18–39, July–September 1999. English translation in [13].
26. В. А. Зиновиев, Т. Ериксон. О фурые-инвариантных разбиениях конечных абелевых групп и тождестве мак-бильямс для групповых кодов. Проблемы передачи информации, 32(1):137–143, 1996. Translated in [5].
27. H. N. Ward. The MacWilliams identity. *Arch. Math. (Basel)*, 74:95–96, 2000.
28. J. A. Wood. Duality for modules over finite rings and applications to coding theory. *Amer. J. Math.*, 121(3):555–575, 1999.
29. J. A. Wood. Weight functions and the extension theorem for linear codes over finite rings. In G. L. M. Ronald C. Mullin, editor, *Finite Fields: Theory, Applications, and Algorithms*, vol. 225 of *Contemp. Math.*, pages 231–243. Amer. Math. Soc., 1999.

# Cyclotomic Function Fields with Many Rational Places

Alice Keller

Fachbereich Mathematik
Universität des Saarlandes
Postfach 15 11 50
66041 Saarbrücken
Germany
e-mail: liz@math.uni-sb.de

**Abstract** Let $A = \mathbb{F}_q[T]$ be the polynomial ring in the variable $T$ and $K = \mathbb{F}_q(T)$ the rational function field over $\mathbb{F}_q$ (the finite field with $q$ elements), and let $K_\infty$ be the completion of $K$ at the place $\infty := \frac{1}{T}$. Furthermore let $C$ be the completion of a fixed algebraic closure of $K_\infty$.
We aim to construct extensions $K \subset K' \subset C$ with many rational places relative to the genus $g(K')$ of $K'$.
As a first step we consider the cyclotomic fields $K(n)/K$ with $n \in A$, which are generated analogously to the classical cyclotomic fields over $\mathbb{Q}$. Then we consider certain decomposition fields and their intersections. Here we know a lower bound for the number of rational places. We get explicit formulas to calculate the genus, but they depend on the relative position of some subgroups of the multiplicative group $(A/(n))^*$ of the ring $A/(n)$. So the concrete calculation of examples must be done by computer. With a special program we made a systematic search for $q = 2$ and found three improved lower bounds.

## 1 Cyclotomic Extensions of $K$

First we introduce some notations which are used in the whole article.

**Definition 1.1.** Let $n \in A$ be a monic polynomial of positive degree. Then there exists a unique factorisation of $n$ in irreducible polynomials, namely:

$$n = \prod_{\nu=1}^{s} p_\nu^{r_\nu} ,$$

with $s, r_\nu$ in $\mathbb{N}$ and $p_\nu \in A$ monic, irreducible and pairwise different polynomials of degree $\geq 1$. We write for short

$$d_\nu := deg(p_\nu) , \qquad q_\nu := q^{d_\nu} ,$$
$$n_\nu := p_\nu^{r_\nu} , \qquad m_\nu := \frac{n}{n_\nu} \qquad \text{and} \quad \varphi(n) := |(A/(n))^*| .$$

The extensions $K(n)$ we are looking for are just the splitting fields of the polynomials $\rho_n$ in $C$ which are defined by the following conditions:

(i) $\rho_T(X) = TX + X^q$,
(ii) $\rho_{T^i}(X) = \rho_T(\rho_{T^{i-1}}(X))$,
(iii) $\rho_{f+g}(X) = \rho_f(X) + \rho_g(X)$ for all $f, g \in \mathbb{F}_q[T]$
(iv) $\rho_{cf}(X) = c \cdot \rho_f(X)$ for all $c \in \mathbb{F}_q$ and $f \in A$.

In other words, $K(n)$ is obtained from $K$ by adjoining the $n$-torsion of the Carlitz–module $\rho$ (cf. [1, chapter 3]).

**Theorem 1.2.** *(i) The extension $K(n)/K$ is galois and abelian with Galois group $(A/(n))^*$.*
*(ii) Let $n = p^r$ be a primary polynomial, then the extension $K(n)/K$ is totally ramified in $\mathfrak{p} = (p)$ and unramified in all other finite places $\mathfrak{q} \neq \mathfrak{p}$.*
*(iii) Let $K_+(n)$ be the fixed field of the embedding $\mathbb{F}_q^* \hookrightarrow (A/(n))^*$. Then the place $\infty$ is totally split in $K_+(n)$ and any place of $K_+(n)$ over $\infty$ is totally ramified in the extension $K(n)/K_+(n)$.*
*(iv) Let $n$ be the product of $s$ primary factors in $A$. Then $K(n)$ is the compositum of the fields $K(p_\nu^{r_\nu})$, $\nu = 1, \ldots, s$, and all these $K(p_\nu^{r_\nu})$ are linearly disjoint.*
*(v) Let $\mathcal{O}(n)$ be the integral closure of $A$ in $K(n)$ and $\lambda$ a primitive root of $\rho_n$. Then*

$$\mathcal{O}(n) = A[\lambda] \ .$$

*Proof.* In [2] as well as in [3].

**Corollary 1.3.** *The field $\mathbb{F}_q$ is algebraically closed in $K(n)$.*

*Remark 1.4.* The theorem shows that the splitting fields of $\rho_n$ have many of the properties of the cyclotomic fields over $\mathbb{Q}$. Therefore they are called *cyclotomic extensions of the rational function field.* The field $K_+(n)$ is the analogue of the maximal real extension of $\mathbb{Q}$ which is contained in the cyclotomic extension. The integral closure $\mathcal{O}(n)$ could be compared with the ring of integers $\mathbb{Z}[\zeta_m]$ for some primitive $m$-th root of unity $\zeta_m$.

## 2 The Genus of $K(n)$

Our aim is now to find a closed formula for the genus $g(K(n))$. Such a formula has been known to several people since about the appearance of [2]. Surprisingly, that formula for the case of a general $n$ has never been published. For special cases, see [7].
First we consider the field $K(p^r)$ and then we can derive the genus of $K(n)$ via induction.
We need some notations:

**Definition 2.1.** For a field extension $L$ of $K$ let $S(L)$ be the set of places. Let $\mathcal{O}$ be the integral closure of $A$ in $L$. Then for any $\mathfrak{q} \in S(L)$ we write $\mathcal{O}_{(\mathfrak{q})}$

for the localization of $\mathcal{O}$ at $\mathfrak{q}$. With $\mathcal{D}(L/K)$ we denote the different of $L/K$. Let $\mathfrak{p}_\nu := (p_\nu)$ for all $\nu = 1, \dots, s$ and for $1 \le \alpha \le r_\nu$

$$G(n_\nu)^\alpha := \{b \in (A/(n_\nu))^* \,|\, b \equiv 1 \bmod p_\nu^\alpha\} .$$

Furthermore we fix an algebraic closure $\overline{\mathbb{F}}_q$ of $\mathbb{F}_q$ and denote the constant field extension by $K'(n) := \overline{\mathbb{F}}_q\, K(n)$.
Later we consider some subextension $K_*(n)$ of $K(n)/K$ for which we will write $K_*'(n) := \overline{\mathbb{F}}_q\, K_*(n)$.
Let $K \subset L \subset K(n)$. For any place $\mathfrak{q} \in S(L)$ and $S(L') \ni \mathfrak{q}'|\mathfrak{q}$ with $L' := \overline{\mathbb{F}}_q\, L$, we write

$$g_i(\mathfrak{q}') := |(G(K'(n)/L'))_i(\mathfrak{q}')| = |(G(K(n)/L))_i(\mathfrak{q})| =: g_i(\mathfrak{q})$$

for the cardinality of the $i$-th ramification group in the extension $K'(n)/L'$.

Since $\mathbb{F}_q$ is algebraically closed in $K(n)$ neither the genus $g(K(n))$ nor the ramification index $e_\mathfrak{p}(K(n)/K)$ change by a constant field extension. The cardinalities of the higher ramification groups are invariant too. So we get:

$$\begin{aligned} [K'(n):K'] &= [K(n):K] = \varphi(n) \quad \text{and} \\ e_{\mathfrak{p}'}(K'(n)/K') &= e_\mathfrak{p}(K(n)/K) \quad \text{as well as} \quad g(K'(n)) = g(K(n)) , \end{aligned}$$

where $\mathfrak{p}' \in S(K')$ with $\mathfrak{p}'|\mathfrak{p}$.

Let $p \in A$ be a prime polynomial of degree $d > 0$ and $r > 0$ an integer. The genus of the field $K(p^r)$ can be calculated by the Riemann–Hurwitz–formula

$$2g(K(p^r)) - 2 = [K(p^r):K](2g(K) - 2) + \deg \mathcal{D}(K(p^r)/K)).$$

The degree of the different is the sum of degrees of the local differents. For calculating them we can use the cardinalities of the ramification groups (cf. [4, chap. 4, §2]).
We know from 1.2 that $\infty$ is tamely ramified, so the higher ramification groups are trivial and the local different $\mathcal{D}(K(p^r)_\mathfrak{Q}/K_\infty)$ at a place $\mathfrak{Q}$ with $\mathfrak{Q}|\infty$ has degree $q-2$.
Another fact from theorem 1.2 is that $\mathfrak{p} = (p)$ is the only finite ramified place. Since $\mathfrak{p}$ is totally ramified, the ramification index is

$$e_\mathfrak{p}(K(p^r)/K) = |(A/(p^r))^*| = \varphi(p^r) = q^{d(r-1)}(q^d - 1) .$$

For $\mathfrak{P}|\mathfrak{p}$ let $K_\mathfrak{p}$ (resp. $K(p^r)_\mathfrak{P}$) the completion of $K$ (resp. $K(p^r)$) with respect to $\mathfrak{p}$ (resp. $\mathfrak{P}$). We write $G := G(K(p^r)/K)$ for the Galois group of the extension $K(p^r)/K$.
Then we can identify $G$ with the Galois group $G(K(p^r)_\mathfrak{P}/K_\mathfrak{p})$, which is equal to the decomposition and inertia group.
For all $i \in \mathbb{Z}$ and $i \ge -1$ the $i$-th ramification group of $\mathfrak{P}$ in $K(p^r)_\mathfrak{P}/K_\mathfrak{p}$ is

$$G_i := \{\sigma \in G \,|\, v_\mathfrak{P}(\sigma(a) - a) \ge i + 1 \text{ for all } a \in \mathcal{O}_{(\mathfrak{P})}\} .$$

The next result and its proof are similar to Prop. 18 in [4, chap. 4].

**Proposition 2.2.** *The ramification groups of $K(p^r)_{\mathfrak{P}}/K_{\mathfrak{p}}$ are*
$G_0 = G(K(p^r)_{\mathfrak{P}}/K_{\mathfrak{p}})$ ,
$G_i \cong G(p^r)^\alpha$ *for* $q^{d(\alpha-1)} \leq i \leq q^{d\alpha} - 1$ *and* $1 \leq \alpha \leq r-1$ ,
*a group of cardinality* $q^{d(r-\alpha)}$,
$G_i \cong G(p^r)^r = \{1\}$ *for* $q^{d(r-1)} \leq i$ .

*Remark 2.3.* The filtration $\{G(p^r)^\alpha\}$ corresponds to the filtration in upper numbering via the Herbrand $\varphi$-function (cf. [4]).

We write $G(n)$ for $(A/(n))^*$ and identify the Galois group $G(K(n)/K)$ with $G(n)$.
Now we use the above results to calculate the genus via the Riemann–Hurwitz–formula and get

**Theorem 2.4.** *(i) The genus of the field $K(p^r)$ is*

$$g(K(p^r)) = 1 + \frac{1}{2}\varphi(p^r)\left(-2 + \frac{q-2}{q-1} + d\frac{rq^d - r - 1}{q^d - 1}\right) .$$

*(ii) More generally, for any monic polynomial $n \in A$ we get*

$$g(K(n)) = 1 + \frac{1}{2}\varphi(n)\left(-2 + \frac{q-2}{q-1} + \sum_{\nu=1}^{s} d_\nu \frac{r_\nu q_\nu - r_\nu - 1}{q_\nu - 1}\right) .$$

*Proof.* The first part is shown. The second part of the theorem can be proved by induction under using some well–known facts:

(i) The Riemann–Hurwitz–formula gives a recursion:

$$2g(K(n)) - 2 = [K(n) : K(m_\nu)](2g(K(m_\nu)) - 2) + \deg \mathcal{D}(K(n)/K(m_\nu))$$

(ii) $K(n_\nu)$ and $K(m_\nu)$ are linearly disjoint over $K$ and $\mathfrak{p}_\nu$ is the only finite ramified place in $K(n_\nu)/K$, it is unramified in $K(m_\nu)/K$ and totally split in $K'(m_\nu)/K'$ of degree $d_\nu \cdot \varphi(m_\nu)$ (cf. 1.2).
(iii) For $s \geq 2$ the ramification indices of $\infty$ in $K(n)/K$ and $K(m_\nu)/K$ are equal.
(iv) Let $\mathfrak{P} \in S(K(n))$ with $\mathfrak{P}|\mathfrak{p}_\nu$, and let $H_i(\mathfrak{P})$ be the $i$-th ramification group of $\mathfrak{P}$ in $K(n)/K(m_\nu)$. Then we have

$$H_i(\mathfrak{P}) = G(K(n)/K(m_\nu)) \cap G_i(\mathfrak{P}) = G_i(\mathfrak{P}) ,$$

because of $G(K(n)/K(m_\nu)) = G(K(n_\nu)/K) = G(n_\nu)$.
(v) So for $s \geq 2$ we have

$$2g(K(n)) - 2 = \varphi(n_\nu)(2g(K(m_\nu)) - 2) + d_\nu\varphi(m_\nu)\sum_{i=0}^{\infty}(g_i(\mathfrak{p}_\nu) - 1) .$$

□

## 3 The Decomposition Field $K_+(n)$

Now we calculate the genus of the decomposition field of the place $\infty$. We do this from the top and consider the extension $K(n)/K_+(n)$. Since $|\mathbb{F}_q^*| = q-1$, the ramification is tame for all places and all higher ramification groups are trivial. So we have only to calculate the inertia group. For a (fixed) finite place $\mathfrak{P}$ of $K(n)$ with $\mathfrak{P}|\mathfrak{p}_\nu$ for any $\nu = 1, \dots, s$ the inertia group is

$$T_\mathfrak{P}(K(n)/K_+(n)) = T_\mathfrak{P}(K(n)/K) \cap G(K(n)/K_+(n)) .$$

Because of $T_\mathfrak{P}(K(n)/K) \cong G(n_\nu)$ we get

$$T_\mathfrak{P}(K(n)/K) \cong \{b \in G(n) \mid b \equiv 1 \bmod m_\nu\}$$

and therefore

$$T_\mathfrak{P}(K(n)/K) \cap G(K(n)/K_+(n)) \cong \begin{cases} \mathbb{F}_q^*, & \text{for } s = 1 \\ \{1\}, & \text{for } s > 1 \end{cases} .$$

All $\frac{\varphi(n)}{q-1}$ different places lying over $\infty$ have the ramification index $q-1$. This proves

**Theorem 3.1.** *The genus of $K_+(n)$ is*

$$g(K_+(n)) = 1 + \frac{1}{q-1}\left(g(K(n)) - 1 - \frac{1}{2}\left(\varphi(n)\frac{q-2}{q-1} + d_1(q-2)\right)\right)$$

*for $s = 1$ and*

$$g(K_+(n)) = 1 + \frac{1}{q-1}\left(g(K(n)) - 1 - \frac{1}{2}\varphi(n)\frac{q-2}{q-1}\right) \quad \textit{for } s > 1 .$$

## 4 Decomposition Fields and Their Intersections

Let the monic polynomial $n \in A$ now be relatively prime to $T$ and $T-1$. So we can consider the decomposition field of $(T)$ and $(T-1)$, respectively, in $K(n)$. This is no restriction, since for any two rational places $(a)$ and $(b)$ in $K$ there exists a linear transformation which maps these places to $(T)$ and $(T-1)$, respectively, and lets $\infty$ invariant.
We define $K_T(n)$ and $K_{T-1}(n)$ as the fixed field of the decomposition groups $Z_{(T)}(K(n)/K)$ and $Z_{(T-1)}(K(n)/K)$, respectively. We further put

$$\begin{aligned} K_{T,+}(n) &:= K_T(n) \cap K_+(n) , \\ K_{T,T-1}(n) &:= K_T(n) \cap K_{T-1}(n) , \\ K_{T,T-1,+}(n) &:= K_{T,T-1}(n) \cap K_+(n) . \end{aligned}$$

Now we will calculate the genus for each of these fields.

## 4.1 The Decomposition Group of a Finite Place

To do that we need an explicit description of the decomposition group of a place $\mathfrak{a} \in S(K)$ of degree $\alpha$. It is well–known that this group is cyclic and the generator is the Frobenius automorphism, which is described by

$$\sigma(x) \equiv x^{q^\alpha} \bmod \mathfrak{A} \quad \text{for all } x \in \mathcal{O} .$$

Here $\mathfrak{A}$ is a prime divisor of $\mathfrak{a}$ in the integral closure $\mathcal{O}$ of $A$ in $K(n)$. From [2, cor. 2.5.] follows

**Proposition 4.1.** *Let $a \in A$ be a monic, irreducible polynomial prime to $n$. The Frobenius automorphism of $(a)$ is $\sigma_{\overline{a}}$ with $\sigma_{\overline{a}}(\lambda) = \rho_a(\lambda)$ and $\lambda$ a primitive root of $\rho_n$.*

**Definition 4.2.** Let $a \in G(n)$. Then $\langle a \rangle_{G(n)}$ denotes the subgroup of $G(n)$ generated by $a$.

In view of our identifications we have for the decomposition group of the places $(T)$ and $(T-1)$:

$$\begin{aligned} & Z_{(T)}(K(n)/K) = \langle \sigma_{\overline{T}} \rangle = \langle T \rangle_{G(n)} \\ \text{and} \quad & Z_{(T-1)}(K(n)/K) = \langle \sigma_{\overline{T-1}} \rangle = \langle T-1 \rangle_{G(n)} . \end{aligned}$$

Now we can describe the Galois groups of the extensions.

We have

$$\begin{aligned} G(K(n)/K_{T,+}(n)) &= \langle T \rangle_{G(n)} \mathbb{F}_q^* , \\ G(K(n)/K_{T,T-1}(n)) &= \langle T, T-1 \rangle_{G(n)} , \\ G(K(n)/K_{T,T-1,+}(n)) &= \langle T, T-1 \rangle_{G(n)} \mathbb{F}_q^* . \end{aligned}$$

## 4.2 The Genus of the Subextensions

**Definition 4.3.** For $\alpha = 1, \ldots, r_\nu$ we write $H(n_\nu)^\alpha$ for the embedding of $G(n_\nu)^\alpha$ in $G(n)$ and $F_n$ for $\mathbb{F}_q^*$ regarded as a subgroup of $G(n)$.

With this notation we have a description of the ramification groups for the different extensions $K(n)/L$, where $L$ is one of the fields $K_T(n)$, $K_{T,+}(n)$, $K_{T,T-1}(n)$ and $K_{T,T-1,+}(n)$. For any $\nu \in \{1, \ldots, s\}$ let $\mathfrak{P} | \mathfrak{p}_\nu$ be a place of $L$ and $q_\nu^{\alpha-1} \leq i \leq q_\nu^\alpha - 1$, then we have

$$\begin{aligned} (G(K(n)/K_T(n))_i(\mathfrak{P}) &= \langle T \rangle_{G(n)} \cap H(n_\nu)^\alpha , \\ (G(K(n)/K_{T,+}(n))_i(\mathfrak{P}) &= \langle T \rangle_{G(n)} F_n \cap H(n_\nu)^\alpha , \\ (G(K(n)/K_{T,T-1}(n))_i(\mathfrak{P}) &= \langle T, T-1 \rangle_{G(n)} \cap H(n_\nu)^\alpha , \\ (G(K(n)/K_{T,T-1,+}(n))_i(\mathfrak{P}) &= \langle T, T-1 \rangle_{G(n)} F_n \cap H(n_\nu)^\alpha . \end{aligned}$$

Now it follows that the sequences

$$
\begin{array}{lllllll}
1 \to & F_n \cap H(n_\nu)^\alpha & \to & F_n & \to & F_{m_\nu p_\nu^\alpha} & \to 1\,, \\
1 \to & \langle T \rangle_{G(n)} \cap H(n_\nu)^\alpha & \to & \langle T \rangle_{G(n)} & \to & \langle T \rangle_{G(m_\nu p_\nu^\alpha)} & \to 1\,, \\
1 \to & \langle T \rangle_{G(n)} F_n \cap H(n_\nu)^\alpha & \to & \langle T \rangle_{G(n)} F_n & \to & \langle T \rangle_{G(m_\nu p_\nu^\alpha)} F_{m_\nu p_\nu^\alpha} & \to 1\,, \\
1 \to & \langle T, T-1 \rangle_{G(n)} \cap H(n_\nu)^\alpha & \to & \langle T, T-1 \rangle_{G(n)} & \to & \langle T, T-1 \rangle_{G(m_\nu p_\nu^\alpha)} & \to 1
\end{array}
$$

are exact. The maps between the groups are the usual injections and projections.

So it is possible to calculate the cardinality of a higher ramification group as the quotient of the cardinalities of two known groups. For a concrete example it is more or less necessary to build the different groups explicitly and count the elements. So we get explicit formulas which depend on the relative position of $\langle T \rangle_{G(n)}$ and $\langle T-1 \rangle_{G(n)}$, and are therefore difficult to evaluate.

We give two examples.

Let $f \in A$ be a monic polynomial prime to $T$ and $T-1$. Then we define

$$\tilde{e}_{T,T-1}(f) := |\langle T \rangle_{G(f)} \cap \langle T-1 \rangle_{G(f)}|\,, \quad e_{T,T-1}(f) := |\langle T, T-1 \rangle_{G(f)}|,$$

$$\tilde{e}_{T,T-1,+}(f) := |\langle T, T-1 \rangle_{G(f)} \cap F_f|\,, \quad e_{T,T-1,+}(f) := |\langle T, T-1 \rangle_{G(f)} F_f|\,,$$

and for $s = 1$ we put $\tilde{e}_*(m_\nu) := 1$ and $e_*(m_\nu) := 1$.

**Formula 4.4.** *The field $K_{T,T-1}(n)$ has genus*

$$g(K_{T,T-1}(n)) = 1 + \frac{1}{e_{T,T-1}(n)} \left( g(K(n)) - 1 - \frac{1}{2} \left[ \frac{\varphi(n)}{q-1} \left( \tilde{e}_{T,T-1,+}(n) - 1 \right) + \sum_{\nu=1}^{s} \varphi(m_\nu) d_\nu \cdot a_\nu(K'_{T,T-1}(n)) \right] \right)\,,$$

$$a_\nu(K'_{T,T-1}(n)) = \frac{e_{T,T-1}(n)}{e_{T,T-1}(m_\nu)} - q_\nu^{r_\nu - 1} + (q_\nu - 1) \sum_{\alpha=1}^{r_\nu - 1} q_\nu^{\alpha-1} \frac{e_{T,T-1}(n)}{e_{T,T-1}(m_\nu p_\nu^\alpha)}\,.$$

**Formula 4.5.** *The genus of $K_{T,T-1,+}(n)$ is*

$$g(K_{T,T-1,+}(n)) = 1 + \frac{1}{e_{T,T-1,+}(n)} \left( g(K(n)) - 1 - \frac{1}{2} \left[ \frac{\varphi(n)}{q-1} (q-2) + \sum_{\nu=1}^{s} \varphi(m_\nu) d_\nu \cdot a_\nu(K'_{T,T-1,+}(n)) \right] \right)\,,$$

$$a_\nu(K'_{T,T-1,+}(n)) = \frac{e_{T,T-1,+}(n)}{e_{T,T-1,+}(m_\nu)} - q_\nu^{r_\nu - 1} + (q_\nu - 1) \sum_{\alpha=1}^{r_\nu - 1} \frac{q_\nu^{\alpha-1} e_{T,T-1,+}(n)}{e_{T,T-1,+}(m_\nu p_\nu^\alpha)}\,.$$

*Remark 4.6.* It is clear that this method can also be used to calculate the ramification groups of the intersection of more than three decomposition fields.

## 5 Number of Rational Places

In the subextensions from above we get lower bounds for the number of rational places.
Let $\mathfrak{p} \in S(K)$ be a rational place. In its decomposition field $\tilde{K}$ in $K(n)/K$, every place over $\mathfrak{p}$ is again rational. So this field has degree of $\tilde{K}/K$ many rational places. Therefore the field $L$ defined as above has – depending on the situation – at least one, two or three times $[L : K]$ many rational places.

For $q = 2$ we made explicit calculations for the fields $K_*(n)$ with $deg\,(n) \leq 15$. Here the place $\infty$ is totally split in $K(n)$ and in any subextension because the inertia group is trivial. Furthermore $\mathbb{F}_2(T)$ has only three rational places and so we got through our systematic search the exact number of rational places in the intersection and not only a lower bound. Comparing our results with the tables published in [8] and [6], this yields three new lower bounds for the number of rational places for fields with fixed genus.

*Example 5.1.* For the polynomial $n = (T^4+T^3+1)(T^4+T^3+T^2+T^1+1)$ with $\varphi(n) = 225$ and $e_T(n) = 15$ the field $K_T(n)$ has the genus 38 and 30 rational places. A theoretical upper bound is 33 (cf. [5]).

*Example 5.2.* The field $K_{T,T-1}(n)$ with $n = (T^3 + T^2 + 1)^3(T^2 + T + 1)^2$, where $\varphi(n) = 5376$ and $e_{T,T-1}(n) = 336$, has the genus 66 and 48 rational places. Here we have 50 as an upper bound.

*Example 5.3.* Let $n = (T^3+T^2+1)^2(T^3+T+1)(T^{11}+T^{10}+T^9+T^7+1)$, then the field $K_T(n)$ has the genus 81 and 56 rational places. For this genus 59 is an upper bound for the number of rational places.

The following table gives the best results we obtain with our method. In the first column the genus is written, in the second an upper bound which is taken from [8]. The third column gives the best results which are cited in the tables of [8] and [6]. (In the extensive bibliographies of these two papers, the reader will find many related articles.) The last column contains our best results for each of these genera.
*Acknowledgements.* This paper comprises the results of my Diplomarbeit, and I would like to thank my teacher, Prof. Dr. E.-U. Gekeler, for his support and Bodo Wack for writing the programs for the systematic search.

## References

1. Goss, D.: Basic Structures of Function Field Arithmetic. Springer, Berlin (1996)

**Table1.** Best results

| $g$ | upper bound | lit. | here |
|---|---|---|---|
| 2 | 6 | 6 | 6 |
| 5 | 9 | 9 | 8 |
| 6 | 10 | 10 | 9 |
| 8 | 11 | 11 | 9 |
| 9 | 12 | 12 | 12 |
| 11 | 14 | 14 | 12 |
| 20 | 21 | 19 | 18 |
| 21 | 21 | 21 | 21 |
| 22 | 22 | 21 | 18 |
| 26 | 25 | 24 | 24 |
| 28 | 26 | 24 | 24 |
| 29 | 27 | 25 | 24 |
| 31 | 28 | 27 | 24 |
| 32 | 29 | 26 | 24 |
| 34 | 30 | 27 | 27 |
| 37 | 32 | 28 | 28 |
| 38 | 33 | 28 | 30 |
| 41 | 35 | 32 | 32 |
| 42 | 35 | 30 | 30 |
| 48 | 39 | 34 | 32 |
| 51 | 41 | 36 | 36 |
| 53 | 42 | 40 | 36 |
| 54 | 43 | 42 | 42 |
| 55 | 43 | 36 | 36 |
| 62 | 48 | 44 | 42 |

| $g$ | upper bound | lit. | here |
|---|---|---|---|
| 65 | 50 | 48 | 48 |
| 66 | 50 | 42 | 48 |
| 67 | 51 | 44 | 42 |
| 68 | 51 | 45 | 42 |
| 69 | 52 | 49 | 48 |
| 70 | 53 | 46 | 45 |
| 75 | 56 | 48 | 48 |
| 76 | 56 | 50 | 45 |
| 81 | 59 | 48 | 56 |
| 105 | 73 | | 64 |
| 108 | 75 | | 63 |
| 128 | 87 | | 72 |
| 135 | 91 | | 84 |
| 149 | 99 | | 96 |
| 154 | 101 | | 90 |
| 161 | 105 | | 96 |
| 167 | 109 | | 96 |
| 172 | 112 | | 96 |
| 173 | 112 | | 96 |
| 185 | 119 | | 96 |
| 186 | 119 | | 96 |
| 238 | 148 | | 135 |
| 244 | 151 | | 126 |
| 357 | 212 | | 192 |
| 521 | 298 | | 240 |

2. Hayes, D.R.: Explicit class field theory for rational function fields. Trans. Am. Math. Soc. 189 (1974) 77–91
3. Rust, I. and Scheja, O.: A guide to explicit class field theory in global function fields. Proceedings in Drinfeld Modules, Modular Schemes and Applications. World Scientific Publishing (1997) 44–65
4. Serre, J-P.: Local Fields. Springer, New York (1979)
5. Serre, J-P.: Sur le nombre des points rationnels d'une courbe algébrique sur un corps fini. C. R. Acad. Sci. Paris Sér. I Math 296 (1983) 397–402
6. Van der Geer, G. and Van der Vlugt, M.: Tables of curves with many points. Math. Comp., electronically published (1999)
7. Niederreiter, H. and Xing, C.: Explicit global function fields over the binary field with many rational places. Acta Arith. 75 (1996) 383–396
8. Xing, C. and Niederreiter, H.: Drinfeld modules of rank 1 and algebraic curves with many rational points. Mh. Math. 127 (1999) 219–241

# On the Siamese Twin Designs

H. Kharaghani*

Department of Mathematics & Computer Science
University of Lethbridge
Lethbridge
Alberta, T1K 3M4
Canada
e-mail: hadi@cs.uleth.ca

**Abstract** Let $4n^2$ be the order of a Bush-type Hadamard matrix with $q = (2n+1)^2$ a prime power. It is shown that there is a weighing matrix

$$W(4(q^m + q^{m-1} + \cdots + q + 1)n^2, 4q^m n^2)$$

which can be used to construct a pair of symmetric designs with the parameters

$$\nu = 4(q^m + q^{m-1} + \cdots + q + 1)n^2, \quad \kappa = q^m(2n^2 + n), \quad \lambda = q^m(n^2 + n)$$

for every positive integer $m$. As a corollary we get a new class of symmetric designs with parameters

$$\nu = 16(q^m + q^{m-1} + \cdots + q + 1)n^2, \quad \kappa = q^m(8n^2 + 2n), \quad \lambda = q^m(4n^2 + 2n)$$

for all positive integers $m$ and $n$, where $4n$ is the order a Hadamard matrix.

## 1 Introduction

A Bush-type Hadamard matrix is a block matrix $H = [H_{ij}]$ of order $4n^2$ with block size $2n$, $H_{ii} = J_{2n}$ and $H_{ij}J_{2n} = J_{2n}H_{ij} = 0$, $i \neq j$, $1 \leq i \leq 2n$, $1 \leq j \leq 2n$, where $J_{2n}$ is the $2n$ by $2n$ matrix of all entries 1. It is known that symmetric Bush–type Hadamard matrices exist for all orders $16n^2$ for all values of $n$ for which there is a Hadamard matrix of order $4n$, see [8] for details.

A balanced generalized weighing matrix $BGW(\nu, \kappa, \lambda)$ over a group $G$ is a matrix $W = [w_{ij}]$ of order $\nu$ , with $w_{ij} \in G \cup \{0\}$ such that each row and column of $W$ has $\kappa$ non–zero entries and for each $k \neq l$, the multiset $\{w_{kj}w_{lj}^{-1} : 1 \leq j \leq \nu, w_{kj} \neq 0, w_{lj} \neq 0\}$ contains $\lambda/|G|$ copies of every element of $G$.

In [7] the author used a Bush–type Hadamard matrix of order $4n^2$, with $q = (2n-1)^2$ a prime power, in a class of balanced generalized weighing matrices over a cyclic group to construct a new class of symmetric designs with parameters $\nu = 4(q^m + q^{m-1} + \cdots + q + 1)n^2$, $\kappa = q^m(2n^2 - n)$, $\lambda =$

* Thanks to W. Holzmann for his help and useful conversation.

$q^m(n^2 - n)$, for every positive integers $m$. These symmetric designs were shown to be generated in pairs by equating certain entries of a weighing matrix to zero. The first class of symmetric designs with these parameters was first constructed by Ionin in [6] for $n = 3$. More precisely, Ionin [6] used a very special regular Hadamard matrix of order 36 in a class of balanced generalized weighing matrices $BGW(q^m + q^{m-1} + \cdots + q + 1, q^m, q^m - q^{m-1})$ over a cyclic group of order $t$, where $q$ is a prime power, $m$ is a positive integer and $t$ is a divisor of $q - 1$, to construct his symmetric designs.

In this paper we again use a Bush–type Hadamard matrix of order $4n^2$, but this time with $q = (2n + 1)^2$ a prime power to construct the sister class of symmetric designs with parameters $\nu = 4(q^m + q^{m-1} + \cdots + q + 1)n^2$, $\kappa = q^m(2n^2 + n)$, $\lambda = q^m(n^2 + n)$ for every positive integer $m$. We will show that these designs again are generated in pairs, but unlike the designs in [7], the blocks of the designs here have some "common intersections". We use the term "Siamese twin design" to draw the attention to this property.

The problem of investigating the existence of Bush–type Hadamard matrices of order $4n^2$, $n$ an odd integer, is a tough one. It is quite interesting to note that if such matrices exist, then the construction method given in this paper simplifies Ionin's method significantly. Moreover, the method used in this paper produces an infinite class of new symmetric designs from any Bush–type Hadamard matrix of new order $4n^2$, $n$ odd such that the number $(2n + 1)^2$ is a prime power, see [5]. On the other hand, the nonexistence of a symmetric Bush–type Hadamard matrix of order $4n^2$ would imply the nonexistence of projective plane of order $2n$. The nonexistence of a symmetric Bush–type Hadamard matrix of order 36 has been established in [2], though in a completely different setting. The smallest order for which the existence of Bush–type Hadamard matrices is unknown is $196 = 14^2$. We conjecture here, as in [7], that Bush-type Hadamard matrices exist for all orders $4n^2$, $n$ an odd integer.

For a $(0, \pm 1)$-matrix $K$, let $K = K^+ - K^-$, where $K^+$ and $K^-$ are $(0, 1)$-matrices. The Kronecker product of two matrices $A = [a_{ij}]$ and $B$, denoted $A \otimes B$, is defined, as usual, by $A \otimes B = [a_{ij}B]$. For a matrix $A = [a_{ij}]$, denote by $|A|$ the matrix $[|a_{ij}|]$. Throughout the paper, $-$ represents $-1$.

## 2 Bush-type Hadamard Matrices and Siamese Twin Designs

K. A. Bush [1] proved that if there exists a projective plane of order $2n$, then there is a Hadamard matrix $H$ of order $4n^2$, such that:

(i) $H$ is symmetric.
(ii) $H = [H_{ij}]$, where $H_{ij}$ are blocks of order $2n$, $H_{ii} = J_{2n}$ and $H_{ij}J_{2n} = J_{2n}H_{ij} = 0$, for $i \neq j$, $1 \leq i \leq 2n$, $1 \leq j \leq 2n$.

Bush's interest was mainly in the nonexistence of such matrices. Whereas there are different methods to construct matrices of order $16n^2$ of the above type, (see [8]), we are not aware of nonexistence of matrices of this form of order $4n^2$ for any odd value of $n$, $n > 3$.

Please note that in this paper by a *Bush-type Hadamard matrix* we mean an Hadamard matrix satisfying only condition 2 above, as we do not need to assume that $H$ is symmetric for our construction. For completeness we include the following result of the author [8].

**Theorem 2.1.** *Let $4n$ be the order of a Hadamard matrix, then there is a Bush-type Hadamard matrix of order $16n^2$.*

*Proof.* Let $K$ be a normalized Hadamard matrix of order $4n$. Let $r_1, r_2, \dots, r_{4n}$ be the row vectors of $K$. Let $C_i = r_i^t r_i$, $i = 1, 2, \dots, 4n$. Then it is easy to see that:

(i) $C_i^t = C_i$, for $i = 1, 2, \dots, 4n$.
(ii) $C_1 = J_{4n}$, $C_i J_{4n} = J_{4n} C_i = 0$, for $i = 2, \dots, 4n$.
(iii) $C_i C_j^t = 0$, for $i \neq j$, $1 \leq i, j \leq 4n$.
(iv) $\sum_{i=1}^{4n} C_i C_i^t = 16n^2 I_{4n}$.

Now let $H = \text{circ}(C_1, C_2, \dots, C_{4n})$, the block circulant matrix with first row $C_1 C_2 \dots C_{4n}$. Then $H$ is a Bush-type Hadamard matrix of order $16n^2$.

*Example 2.2.* Let

$$K = \begin{pmatrix} 1 & 1 & 1 & 1 \\ 1 & 1 & - & - \\ 1 & - & 1 & - \\ 1 & - & - & 1 \end{pmatrix}$$

Then,

$$r_1 = (1\,1\,1\,1)$$

$$r_2 = (1\,1\,-\,-)$$

$$r_3 = (1\,-\,1\,-)$$

$$r_4 = (1\,-\,-\,1)$$

$$C_1 = r_1^t r_1 = \begin{pmatrix} 1\,1\,1\,1 \\ 1\,1\,1\,1 \\ 1\,1\,1\,1 \\ 1\,1\,1\,1 \end{pmatrix}$$

$$C_2 = r_2^t r_2 = \begin{pmatrix} 1 & 1 & - & - \\ 1 & 1 & - & - \\ - & - & 1 & 1 \\ - & - & 1 & 1 \end{pmatrix}$$

$$C_3 = r_3^t r_3 = \begin{pmatrix} 1 & - & 1 & - \\ - & 1 & - & 1 \\ 1 & - & 1 & - \\ - & 1 & - & 1 \end{pmatrix}$$

$$C_4 = r_4^t r_4 = \begin{pmatrix} 1 & - & - & 1 \\ - & 1 & 1 & - \\ - & 1 & 1 & - \\ 1 & - & - & 1 \end{pmatrix}$$

Then $H = \mathrm{circ}(C_1, C_2, C_3, C_4)$ is the following matrix,

$$\begin{bmatrix}
1 & 1 & 1 & 1 & 1 & 1 & - & - & 1 & - & 1 & - & 1 & - & - & 1 \\
1 & 1 & 1 & 1 & 1 & 1 & - & - & - & 1 & - & 1 & - & 1 & 1 & - \\
1 & 1 & 1 & 1 & - & - & 1 & 1 & 1 & - & 1 & - & - & 1 & 1 & - \\
1 & 1 & 1 & 1 & - & - & 1 & 1 & - & 1 & - & 1 & 1 & - & - & 1 \\
1 & - & - & 1 & 1 & 1 & 1 & 1 & 1 & 1 & - & - & 1 & - & 1 & - \\
- & 1 & 1 & - & 1 & 1 & 1 & 1 & 1 & 1 & - & - & - & 1 & - & 1 \\
- & 1 & 1 & - & 1 & 1 & 1 & 1 & - & - & 1 & 1 & 1 & - & 1 & - \\
1 & - & - & 1 & 1 & 1 & 1 & 1 & - & - & 1 & 1 & - & 1 & - & 1 \\
1 & - & 1 & - & 1 & - & - & 1 & 1 & 1 & 1 & 1 & 1 & 1 & - & - \\
- & 1 & - & 1 & - & 1 & 1 & - & 1 & 1 & 1 & 1 & 1 & 1 & - & - \\
1 & - & 1 & - & - & 1 & 1 & - & 1 & 1 & 1 & 1 & - & - & 1 & 1 \\
- & 1 & - & 1 & 1 & - & - & 1 & 1 & 1 & 1 & 1 & - & - & 1 & 1 \\
1 & 1 & - & - & 1 & - & 1 & - & 1 & - & - & 1 & 1 & 1 & 1 & 1 \\
1 & 1 & - & - & - & 1 & - & 1 & - & 1 & 1 & - & 1 & 1 & 1 & 1 \\
- & - & 1 & 1 & 1 & - & 1 & - & - & 1 & 1 & - & 1 & 1 & 1 & 1 \\
- & - & 1 & 1 & - & 1 & - & 1 & 1 & - & - & 1 & 1 & 1 & 1 & 1
\end{bmatrix}$$

**Lemma 2.3.** *Let $H$ be a Bush-type Hadamard matrix of order $4n^2$. Let $H = H^+ - H^-$. Then each of $H^+$ and $H^- + I_{2n} \otimes J_{2n}$ is a symmetric $(4n^2, 2n^2 + n, n^2 + n)$–design. Furthermore, the two designs share the diagonal block entries $I_{2n} \otimes J_{2n}$.*

*Proof.* The proof is simple, but we present an unusual proof here which inspires a similar proof for Theorem 8; the main result of the paper.

First note that,

$$H^+ + H^- = J_{2n} \otimes J_{2n}.$$

So, $2H^+ = H + J_{2n} \otimes J_{2n}$. Now,

$$\begin{aligned} 4H^+H^{+t} &= (H + J_{2n} \otimes J_{2n})(H^t + J_{2n} \otimes J_{2n}) \\ &= HH^t + 4n^2 J_{2n} \otimes J_{2n} + 4(n^2 + n))J_{2n} \otimes J_{2n} \\ &= 4n^2 I_{4n^2} + 4(n^2 + n)J_{2n} \otimes J_{2n}. \end{aligned}$$

Therefore, $H^+H^{+t} = n^2 I_{4n^2} + (n^2 + n)J_{2n} \otimes J_{2n}$. This means that $H^+$ is a symmetric $(4n^2, 2n^2 + n, n^2 + n)$–design. It is similarly simple to see that $H^- + I_{2n} \otimes J_{2n}$ is also a symmetric $(4n^2, 2n^2 + n, n^2 + n)$–design. Note that Bush–typeness is essential for $H$ to be a twin design. Computer computation shows that the two designs are different in general.

*Example 2.4.* Let $H$ be the matrix of example 2, then, $H^+$ is the following matrix,

$$\begin{bmatrix} 1&1&1&1&1&1&0&0&1&0&1&0&1&0&0&1 \\ 1&1&1&1&1&1&0&0&0&1&0&1&0&1&1&0 \\ 1&1&1&1&0&0&1&1&1&0&1&0&0&1&1&0 \\ 1&1&1&1&0&0&1&1&0&1&0&1&1&0&0&1 \\ 1&0&0&1&1&1&1&1&1&1&0&0&1&0&1&0 \\ 0&1&1&0&1&1&1&1&1&1&0&0&0&1&0&1 \\ 0&1&1&0&1&1&1&1&0&0&1&1&1&0&1&0 \\ 1&0&0&1&1&1&1&1&0&0&1&1&0&1&0&1 \\ 1&0&1&0&1&0&0&1&1&1&1&1&1&1&0&0 \\ 0&1&0&1&0&1&1&0&1&1&1&1&1&1&0&0 \\ 1&0&1&0&0&1&1&0&1&1&1&1&0&0&1&1 \\ 0&1&0&1&1&0&0&1&1&1&1&1&0&0&1&1 \\ 1&1&0&0&1&0&1&0&1&0&0&1&1&1&1&1 \\ 1&1&0&0&0&1&0&1&0&1&1&0&1&1&1&1 \\ 0&0&1&1&1&0&1&0&0&1&1&0&1&1&1&1 \\ 0&0&1&1&0&1&0&1&1&0&0&1&1&1&1&1 \end{bmatrix}$$

and the matrix $H^- + I_{2n} \otimes J_{2n}$ is

$$\begin{bmatrix}
1&1&1&1&0&0&1&1&0&1&0&1&0&1&1&0\\
1&1&1&1&0&0&1&1&1&0&1&0&1&0&0&1\\
1&1&1&1&1&1&0&0&0&1&0&1&1&0&0&1\\
1&1&1&1&1&1&0&0&1&0&1&0&0&1&1&0\\
0&1&1&0&1&1&1&1&0&0&1&1&0&-&0&1\\
1&0&0&1&1&1&1&1&0&0&1&1&1&0&1&0\\
1&0&0&1&1&1&1&1&1&1&0&0&0&1&0&1\\
0&1&1&0&1&1&1&1&1&1&0&0&1&0&1&0\\
0&1&0&1&0&1&1&0&1&1&1&1&0&0&1&1\\
1&0&1&0&1&0&0&1&1&1&1&1&0&0&1&1\\
0&1&0&1&1&0&0&1&1&1&1&1&1&1&0&0\\
1&0&1&0&0&1&1&1&1&1&1&1&1&1&0&0\\
0&0&1&1&0&1&0&1&0&1&1&0&1&1&1&1\\
0&0&1&1&1&0&1&0&1&0&0&1&1&1&1&1\\
1&1&0&0&0&1&0&1&1&0&0&1&1&1&1&1\\
1&1&0&0&1&0&1&0&0&1&1&0&1&1&1&1
\end{bmatrix}$$

Note that the two design share only the diagonal block entries.

## 3 A Cyclic Subgroup of Signed Permutation Matrices

Let $SP_m$ be the set of all signed permutation matrices of order $m$. Let $U = \text{circ}(0,1,0,\dots,0)$ be the circulant shift permutation matrix of order $2n$ (this is a circulant matrix of order $2n$ with first row $010\dots0$) and $N = \text{diag}(-,1,1,\dots,1)$ be the diagonal matrix of order $2n$ with $-1$ at the $(1,1)$-position and 1 elsewhere on the diagonal. Let $E = UN$, then $E$ is in $SP_{2n}$. Let $G_{4n} = \{\gamma^i = E^i \otimes I_{2n} \colon i = 1,2,\dots,4n\} = \prec \gamma \succ$.

**Lemma 3.1.** *$G_{4n}$ is a cyclic subgroup of $SP_{4n^2}$, of order $4n$.*

*Proof.* For $1 \leq r \leq 2n$, $(UN)^r$ is $U^r$ with its first $r$ columns negated. Thus $\gamma^{2n} = E^{2n} \otimes I_{2n} = -I_{2n} \otimes I_{2n} = -I_{4n^2}$. It now follows that $G_{4n}$ is a cyclic subgroup of $SP_{4n^2}$, of order $4n$.

Note that $G_{4n}$ is a (signed) group subgroup of $SP_{4n^2}$ and $\sum_{\gamma \in G_{4n}} \gamma = 0$.

*Example 3.2.* Let $n = 2$ in lemma 5, then

$$\gamma = E \otimes I_4 = \begin{pmatrix} 0 & I_4 & 0 & 0 \\ 0 & 0 & I_4 & 0 \\ 0 & 0 & 0 & I_4 \\ -I_4 & 0 & 0 & 0 \end{pmatrix}$$

$$\gamma^2 = E^2 \otimes I_4 = \begin{pmatrix} 0 & 0 & I_4 & 0 \\ 0 & 0 & 0 & I_4 \\ -I_4 & 0 & 0 & 0 \\ 0 & -I_4 & 0 & 0 \end{pmatrix}$$

$$\gamma^3 = E^3 \otimes I_4 = \begin{pmatrix} 0 & 0 & 0 & I_4 \\ -I_4 & 0 & 0 & 0 \\ 0 & -I_4 & 0 & 0 \\ 0 & 0 & -I_4 & 0 \end{pmatrix}$$

$$\gamma^4 = E^4 \otimes I_4 = \begin{pmatrix} -I_4 & 0 & 0 & 0 \\ 0 & -I_4 & 0 & 0 \\ 0 & 0 & -I_4 & 0 \\ 0 & 0 & 0 & -I_4 \end{pmatrix}$$

$$\gamma^{4+j} = E^{(4+j)} \otimes I_4 = -E^j \otimes I_4 = -\gamma^j, j = 1, 2, 3, 4.$$

So for this example, $G_8 = \{\gamma^i = E^i \otimes I_4 \colon i = 1, 2, \dots, 8\}$ is the cyclic subgroup of $SP_{16}$, of order 8.

**Lemma 3.3.** *Let $q = (2n+1)^2$ be a prime power. Then there is a balanced weighing matrix $BGW(q^m + q^{m-1} + \cdots + q + 1, q^m, q^m - q^{m-1})$ over the cyclic group $G_{4n}$ for each positive integer $m$.*

*Proof.* Note that $4n$ is a divisor of $q-1$ and apply [3], IV.4.22, or [4], Theorem 2.2.

## 4 The Siamese Twin Designs

Ionin [6] was the first to investigate symmetric designs with parameters $\nu = 4(q^m + q^{m-1} + \cdots + q + 1)n^2$, $\kappa = q^m(2n^2 + n)$, $\lambda = q^m(n^2 + n)$ for every positive integer $m$ and $n = 3$. In [7], the author considered the case where $q = (2n-1)^2$ was a prime power. We now consider the case where $q = (2n+1)^2$ is a prime power.

The following is the main result of this paper.

**Theorem 4.1.** *Let $H$ be a Bush-type Hadamard matrix of order $4n^2$ with $q = (2n+1)^2$ a prime power. Then there is a weighing matrix*

$$W(4(q^m + q^{m-1} + \cdots + q + 1)n^2, 4q^m n^2)$$

*which can be used to construct a Siamese twin design with parameters*

$$\nu_m = 4(q^m + q^{m-1} + \cdots + q + 1)n^2, \quad \kappa_m = q^m(2n^2 + n), \quad \lambda_m = q^m(n^2 + n),$$

*for each positive integer $m$.*

*Proof.* Let $m$ be a positive integer. Let $W = [w_{ij}]$ be the balanced generalized weighing matrix $BGW(\nu, \kappa, \lambda)$ of Lemma 7, where $\nu = q^m + q^{m-1} + \cdots + q + 1$, $\kappa = q^m$, $\lambda = q^m - q^{m-1}$. Consider the block matrix $A = [Hw_{ij}]$ of order $4\nu n^2$. Let $AA^t = [B_{kl}]$.
For $k \neq l$,

$$\begin{aligned}B_{kl} &= \sum_{j=1}^{\nu} Hw_{kj}(Hw_{jl})^t \\ &= \sum_{j=1}^{\nu} H(w_{kj}{w_{lj}}^t)H^t \\ &= H(\sum_{j=1}^{\nu} w_{kj}{w_{lj}}^t)\, H^t \\ &= H(\sum_{\gamma \in G_{4n}} \frac{\lambda}{4n}\gamma)\, H^t = 0.\end{aligned}$$

For $k = l$,

$$\begin{aligned}B_{kl} &= H\sum_{j=1}^{\nu}(w_{kj}{w_{lj}}^t)\, H^t \\ &= \kappa HH^t \\ &= 4n^2\kappa I_{2n}.\end{aligned}$$

So $A$ is a weighing matrix $W(4n^2\nu, 4n^2\kappa)$. Now, let

$$D = [Mw_{ij} + (I_{2n} \otimes J_{2n})|w_{ij}|],$$

where $M = H - I_{2n} \otimes J_{2n}$. This matrix (which is obtained from $A$ by negating some entries) contains a symmetric designs with the parameters, $\nu_m = 4(q^m + q^{m-1} + \cdots + q + 1)n^2$, $\kappa_m = q^m(2n^2 + n)$, $\lambda_m = q^m(n^2 + n)$ for every positive integer $m$. To see this, let

$$D = D^+ - D^- = [Mw_{ij} + (I_{2n} \otimes J_{2n})|w_{ij}|].$$

Now observe that

$$D^+ + D^- = [(J_{2n} \otimes J_{2n})|w_{ij}|].$$

Therefore,

$$2D^+ = [Mw_{ij} + ((J_{2n} + I_{2n}) \otimes J_{2n})|w_{ij}|].$$

First note that for all $i, j, k, l$,

$$((J_{2n} + I_{2n}) \otimes J_{2n})|w_{ij}|)(Mw_{kl})^t = (Mw_{kl})((J_{2n} + I_{2n}) \otimes J_{2n})|w_{ij}|)^t = 0.$$

Also note that a calculation on $[Mw_{ij}]$ akin of that on $A$ above would give,

$$[Mw_{ij}][Mw_{ij}]^t = \kappa(I_\nu \otimes MM^t).$$

So, we have,

$$\begin{aligned}
&4D^+D^{+t} \\
&= [((J_{2n}+I_{2n})\otimes J_{2n})|w_{ij}|][((J_{2n}+I_{2n})\otimes J_{2n})|w_{ij}|]^t + [Mw_{ij}][Mw_{ij}]^t \\
&= [((J_{2n}+I_{2n})\otimes J_{2n})|w_{ij}|][((J_{2n}+I_{2n})\otimes J_{2n})|w_{ij}|]^t + \kappa(I_\nu \otimes MM^t).
\end{aligned}$$

For $k \neq l$,

$$\begin{aligned}
&\sum_{j=1}^{\nu} ((J_{2n}+I_{2n})\otimes J_{2n})|w_{kj}|(((J_{2n}+I_{2n})\otimes J_{2n})|w_{jl}|)^t \\
&\quad = ((J_{2n}+I_{2n})\otimes J_{2n})\Big(\sum_{j=1}^{\nu}(|w_{kj}||w_{lj}{}^t|\Big)((J_{2n}+I_{2n})\otimes J_{2n}) \\
&\quad = \frac{\lambda}{2n}((J_{2n}+I_{2n})\otimes J_{2n})(J_{2n}\otimes I_{2n})((J_{2n}+I_{2n})\otimes J_{2n}) \\
&\quad = \frac{\lambda}{2n}((J_{2n}+I_{2n})\otimes J_{2n})(J_{2n}\otimes I_{2n})((J_{2n}+I_{2n})\otimes J_{2n}) \\
&\quad = \frac{\lambda}{2n}(2n+1)^2 J_{2n}\otimes 2nJ_{2n} \\
&\quad = \lambda(2n+1)^2 J_{2n}\otimes J_{2n} \\
&\quad = 4n(n+1)q^m J_{2n}\otimes J_{2n}.
\end{aligned}$$

Therefore, all the $(k,l)$, $1 \leq k \neq l \leq \nu$ blocks of the matrix $D^+D^{+t}$ consist of the matrix $n(n+1)q^m J_{2n}\otimes J_{2n}$.
For $k = l$ we have,

$$\begin{aligned}
&\sum_{j=1}^{\nu} (((J_{2n}+I_{2n})\otimes J_{2n})|w_{kj}|)\big(((J_{2n}+I_{2n})\otimes J_{2n})|w_{kj}|\big)^t + \\
&\sum_{j=1}^{\nu} Mw_{kj}(Mw_{kj})^t \\
&\quad = \kappa\big((J_{2n}+I_{2n})\otimes J_{2n}\big)\big((J_{2n}+I_{2n})\otimes J_{2n}\big)^t + MM^t\big) \\
&\quad = q^m\big(4n^2 I_{4n^2} + 4n(n+1)J_{2n}\otimes J_{2n}\big).
\end{aligned}$$

>From this we conclude that all the $(k,k), 1 \leq k \leq \nu$ blocks of the matrix $D^+D^{+t}$ consist of the matrix $q^m\big(n^2 I_{4n^2} + n(n+1)J_{2n}\otimes J_{2n}\big)$. Therefore, $D^+$ is a symmetric $(\nu_m, \kappa_m, \lambda_m)$–design.

By a similar argument we can show that

$$\frac{1}{2}[-Mw_{ij} + ((J_{2n}+I_{2n})\otimes J_{2n})|w_{ij}|]$$

is also a symmetric $(\nu_m, \kappa_m, \lambda_m)$–design. Please note that the two designs share the part $(I_{2n} \otimes J_{2n})|w_{ij}|$ and both could be obtained from the weighing matrix $A$.

**Corollary 4.2.** *Let $4n$ be the order of a Hadamard matrix with $q = (4n+1)^2$ a prime power. Then there is a weighing matrix*

$$W(16(q^m + q^{m-1} + \cdots + q + 1)n^2, 16q^m n^2)$$

*from which two symmetric designs are generated with the parameters*

$$\nu_m = 16(q^m + q^{m-1} + \cdots + q + 1)n^2, \quad \kappa_m = q^m(8n^2 + 2n),$$
$$\lambda_m = q^m(4n^2 + 2n)$$

*for every positive integer $m$.*

*Proof.* This follows from Theorems 1 and 8.

*Remark 4.3.* Corollary 9 provides a new class of symmetric designs.

**Acknowledgment.** The research is supported by an NSERC operating grant.

## References

1. K.A. Bush: Unbalanced Hadamard matrices and finite projective planes of even order, JCT, 11(1971), pp. 38–44.
2. Frans C. Bussemaker, Willem Haemers and Edward Spence: The search for pseudo orthogonal Latin squares of order six, preprint.
3. Warwick de Launey: Section on "Bhaskar Rao designs," CRC Handbook of Combinatorial Designs, edited by Charles J. Colbourn and Jeffrey H. Dinitz, Kluwer Academic Press, 1996.
4. Dieter Jungnickel and Vladimir Tonchev: Perfect Codes and Balanced Weighing Matrices, To appear (1999).
5. Zvonimir Janko, Hadi Kharaghani and Vladimir D. Tonchev, The existence of a Bush–type Hadamard matrix of order 324 and two new infinite classes of symmetric designs, preprint.
6. Y.J. Ionin: New symmetric designs from regular Hadamard matrices, The Electronic Journal of Combinatorics, Vol. 5, No.1 (1998), R1.
7. H. Kharaghani: On the twin designs with the Ionin–type parameters, The Electronic Journal of Combinatorics, (2000), #R1.
8. H. Kharaghani: New classes of weighing matrices, ARS comb., 19(1985), pp. 69–72.

# On the Brauer Monoid for Finite Fields

Vladimir V. Kirichenko[1] and Boris V. Novikov[2]

[1] Vozdukhoflotskij prospekt 20/1, apt.47
Kiev, 252049
Ukraine
e-mail: vkir@mechmat.univ.kiev.ua

[2] Saltovskoye shosse 258, apt.20
Kharkov, 310178
Ukraine
e-mail: boris.v.novikov@univer.kharkov.ua

**Abstract** The definition of the Brauer monoid $M(G, L)$ ($G$ is the Galois group of a field $L$) was given in [3]. In this article it is studied by the notions of modifications [6] and 0-cohomology [5]. Namely, $M(G, L)$ is a semilattice of Abelian groups each of which is isomorphic to $H_0^2(S, L)$, the second 0-cohomology of a modification of $G$. We show that if the subgroup $U$ of the invertible elements of $S$ is normal in $S$ then $\mathrm{Ker}(H_0^2(S, L) \to H^2(U, L))$ is isomorphic to the second 0-cohomology of $S/U$. In particular, for finite fields the influence of invertible elements of modifications on the Brauer monoid may be eliminated in some sense.

## 1 Introduction

It is well-known that the Brauer group of any finite field is trivial [2]. Therefore the so called Brauer monoid proposed in [3] is of interest. This monoid generalizes the Brauer group and isn't trivial for any non-trivial field extension. One can hope that the studying of its properties will be useful for the investigation of algebras over finite fields.

The description of the Brauer monoid by modifications and their 0-cohomology proposed in [6] is more convenient in our opinion than original one [3]. We give in Section 2 this description and a necessary information about semigroup 0-cohomology too.

Section 3 is devoted to the proof of a theorem, which facilitates essentially the calculation of a Brauer monoid for finite fields by elimination of the invertible elements.

Finally we note that throughout this article relative Brauer monoids (and relative Brauer groups) are considered only, so the adjective "relative" will be omitted.

## 2 Preliminary: 0-Cohomology and Modifications

Semigroup 0-cohomology is a specific case of partial cohomologies which were built in [5]; it had appeared in the investigation of the projective representations of semigroups.

Let $S$ be an arbitrary semigroup with a zero. An Abelian group $A$ is called *a 0-module* over $S$, if an action $(S \setminus 0) \times A \to A$ is defined which satisfies for all $s, t \in S \setminus 0,\ a, b \in A$ the following conditions:

$$s(a+b) = sa + sb,$$

$$st \neq 0 \Rightarrow s(ta) = (st)a.$$

*A n-dimensional 0-cochain* is a partial $n$-place mapping out of $S$ to $A$ which is defined for all $n$-tuples $(s_1, \dots, s_n)$, such that $s_1 \cdot \dots \cdot s_n \neq 0$. The coboundary operator is given like for the usual cohomology by the formula

$$\partial^n f(s_1, \dots, s_{n+1}) = s_1 f(s_2, \dots, s_{n+1}) + \sum_{i=1}^{n} (-1)^i f(s_1, \dots, s_i s_{i+1}, \dots, s_{n+1}) + (-1)^{n+1} f(s_1, \dots, s_n)$$

The equality $\partial^2 = 0$ is valid too; obtained cocycles (cohomology) are called *0-cocycles (0-cohomology)* and their groups are denoted by $Z_0^n(S, A)$ (resp. $H_0^n(S, A)$).

Note that for a semigroup $T^0 = T \cup 0$ with an adjointed zero

$$H_0^n(T^0, A) \cong H^n(T, A),$$

so 0-cohomology may be considered as a generalization of Eilenberg–MacLane cohomology.

Let $L$ be a finite-dimensional normal extension of a field $K$ with the Galois group $G$, $L^\times$ be the multiplicative group of $L$.

By *a modification* $G(\star)$ of the group $G$ we mean a semigroup on the set $G^0 = G \cup 0$ with operation $\star$ such that $x \star y$ is equal either to $xy$ or to $0$, while

$$0 \star x = x \star 0 = 0 \star 0 = 0$$

and the identity of $G$ is the same for the semigroup $G(\star)$.

In other words, to obtain a modification, one must erase the contents of some inputs in the multiplication table of $G$ and insert there zeros so that the new operation would be associative.

Note some general properties of modifications. Firstly, a modification of $G$ satisfies the weak cancellation condition: from $x \star z = y \star z \neq 0$ it follows $x = y$ and analogously for left cancellation. Secondly, let $U$ be the subgroup of invertible elements in $G(\star)$. Then its complement $I = G(\star) \setminus U$ is a two-sided ideal. It follows from finiteness of $G$ that $I$ is nilpotent[6].

$L^\times$ is a 0-module over every modification, where elements of the modification act on $L^\times$ as automorphisms of the field. 0-cohomology groups $H_0^2(G(\star), L^\times)$ will be called *components of the Brauer monoid.* In the case when operation $\star$ is defined by such a way that $x \star y = xy$ for $x, y \neq 0$, the component of the Brauer monoid turns out the Brauer group.

Let $S = G(\star)$ and $T = G(*)$ be modifications of $G$. We write $S \prec T$, if $x*y = 0$ implies $x\star y = 0$ for all $x, y \in G$. Clearly, in this case a homomorphism is defined

$$\varepsilon_{T,S} : H_0^2(T, L^\times) \longrightarrow H_0^2(S, L^\times)$$

Since for $S \prec T \prec U$ these homomorphisms yield the equalities

$$\varepsilon_{U,T}\varepsilon_{T,S} = \varepsilon_{U,S}, \qquad \varepsilon_{S,S} = id,$$

one can build in the standard way [1] the semilattice of groups $H_0^2(S, L^\times)$ (where $S$ runs over all modifications of $G$), which is called *a (relative) Brauer monoid* $M(G, L)$.

## 3 Invertible Elements in Modifications

As above let $S = G(\star)$ be a modification of the Galois group $G$, $U$ the subgroup of invertible elements of $S$, $I = S \setminus U$. We shall write in this section $xy$ instead of $x \star y$, inasmuch as the operation of the group $G$ will not be used. Besides let us agree to employ additive notation for the $G$-module $L^\times$.

The inclusion $U \hookrightarrow G$ induces a homomorphism

$$\varphi : H_0^2(S, L^\times) \to H^2(U, L^\times)$$

We study this homomorphism in the situation when $U$ is a normal subgroup of $G$. Then $U$ turns out normal in $S$ too (in the meaning that $xU = Ux$ for all $x \in S$). The partition into cosets of $U$ (together with $\{0\}$) is a congruence on $S$. The quotient-semigroup on this congruence will be denoted by $S/U$.

Further, if $U \triangleleft S$ then the multiplicative group of the subfield $P$ of all $U$-fixed elements is a $S/U$-module. The inclusion $P^\times \hookrightarrow L^\times$ and the epimorphism $S \to S/U$ induce a homomorphism

$$\psi : H_0^2(S/U, P^\times) \to H_0^2(S, L^\times)$$

**Theorem 3.1.** *Let $G$ be the Galois group of a field extension $L/K$ as above, $S$ a modification of $G$, $U$ the subgroup of invertible elements of $S$, $P$ the subfield of all $U$-fixed elements from $L$. If $U \triangleleft S$ then the sequence*

$$0 \longrightarrow H_0^2(S/U, P^\times) \xrightarrow{\psi} H_0^2(S, L^\times) \xrightarrow{\varphi} H^2(U, L^\times)$$

*is exact.*

*Proof.* We fix a system $T$ of representatives of the cosets with respect to $U$: $S = \{0\} \cup (\bigcup_{t \in T} Ut)$ (as usually the representative for $U$ is the identity). In what follows we denote by $a, b, c$ elements of $U$, by $x, y, z$ elements of $I \setminus 0$, by $r, s, t$ elements of $T$. If $f$ is a (0)-cocycle then the respective element of the (0)-cohomology group is denoted by $[f]$.

A remark must yet be made. It follows out of normality of $U$ that for every $a \in U$, $x \in I \setminus 0$ there is such $a_x \in U$ that $ax = xa_x$. In addition $a_x$ is defined uniquely because $xa_x \neq 0$. Next, if $a$ runs over $U$ then $a_x$ runs over it too, and $(ab)_x = a_x b_x$ for $a, b \in U$.

Let $[f] \in \operatorname{Ker}\varphi$ where $f$ is a 0-cocycle of $Z_0^2(S, L^\times)$. One can assume that $f(a,b) = 0$ for all $a, b \in U$. Consider the cochain of the group $U$

$$\pi_x(a) = f(a,x) - f(x,a_x).$$

It is a cocycle because

$$\begin{aligned}\partial\pi_x(a,b) &= af(b,x) - af(x,b_x) - f(ab,x) + f(x,a_x b_x) + f(a,x) - f(x,a_x)\\ &= \partial f(a,b,x) - \partial f(a,x,b_x) - f(ax,b_x) + f(x,a_x b_x) - f(x,a_x)\\ &= -f(xa_x,b_x) + f(x,a_x b_x) - f(x,a_x) = \partial f(x,a_x,b_x) = 0\end{aligned}$$

Since $H^1(U, L^\times) = 0$ (see, e. g., [4]), it follows from this that

$$\pi_x(a) = (a-1)\lambda(x),$$

where $\lambda(x) \in L^\times$ for all $x \in I \setminus 0$. Set $\lambda(a) = 0$ and $g = f + \partial\lambda$. Then $g(a,b) = 0$ and besides

$$\begin{aligned}g(a,x) &= f(a,x) - a\lambda(x) + \lambda(ax)\\ &= \pi_x(a) + f(x,a-x) - a\lambda(x) + \lambda(xa_x)\\ &= f(x,a-x) - \lambda(x) + \lambda(xa_x) = g(x,a_x).\end{aligned}$$

Next let us set $\rho(at) = g(a,t)$ for $t \in T$ and consider the 0-cocycle $h = g + \partial\rho$. Then

$$h(a,bt) = g(a,bt) + ag(b,t) - g(ab,t) = \partial g(a,b,t) + g(a,b) = 0 \qquad (1)$$
$$h(at,b) = g(at,b) - g(t,a_x b) + g(a,t) = \partial g(t,a_x,b) - tg(a_x,b) = 0 \qquad (2)$$

>From here we obtain for $xy \neq 0$:

$$\begin{aligned}ah(x,y) &= h(ax,y) - h(a,xy) + h(a,x) = h(xa_x,y)\\ &= xh(a_x,y) + h(x,a_x y) - h(x,a_x) = h(x,a_x y) = h(x,y(a_x)_y)\\ &= -xh(y,(a_x)_y) + h(xy,(a_x)_y) + h(x,y) = h(x,y)\end{aligned}$$

Hence $h(x,y) \in P^\times$. Besides the last calculation implies

$$h(ax,y) = h(x,a_x y) = h(x,y)$$

Since $a$ and $a_x$ both run over all group $U$ we have:

$$h(ax,y) = h(x,ay) = h(x,y)$$

This means that $h$ defines a 0-cocycle $\overline{h} \in Z_0^2(S/U, P^\times)$ in the following way:

$$\overline{h}(Us, Ut) = h(s,t) \quad \text{for } st \neq 0.$$

So for a given 0-cocycle $f \in Z_0^2(S, L^\times)$ by the 0-cocycles $g$ and $h$ which are cohomological to it, we construct the 0-cocycle $\overline{h} \in Z_0^2(S/U, P^\times)$.

We show that the correspondence $h \to \overline{h}$ extends to the cohomology mapping. Let $h = \partial\sigma$ for some 0-cochain $\sigma \in C_0^1(S, L^\times)$. Since $\partial\sigma(a,b) = h(a,b) = 0$, one has $\sigma(a) = (a-1)\mu$ for the restriction of $\sigma$ on $U$, where $\mu \in L$. Further, it follows out of (1) and (2) $\partial\sigma(a,x) = \partial\sigma(x,a) = 0$. Therefore

$$\begin{aligned} a[\sigma(x) - (x-1)\mu] &= \partial\sigma(a,x) + \sigma(ax) - \sigma(a) - xa_x\mu + a\mu \\ &= \sigma(xa_x) - xa_x\mu + \mu \\ &= \partial\sigma(x, a_x) - x\sigma(a_x) + \sigma(x) - xa_x\mu + \mu \\ &= \sigma(x) - (x-1)\mu, \end{aligned}$$

so $\sigma(x) - (x-1)\mu \in P^\times$. Let $\tau(g) = \sigma(g) - (g-1)\mu$ for any $g \in S \setminus 0$. Then $\partial\tau = \partial\sigma = h$, $\tau(g) \in P^\times$ and in addition

$$\begin{aligned} \tau(ax) &= -\partial\sigma(a,x) + a\sigma(x) + \sigma(a) - (ax-1)\mu \\ &= a[\tau(x) - (x-1)\mu] + (a-1)\mu - (ax-1)\mu \\ &= a\tau(x) = \tau(x), \end{aligned}$$

i.e. 0-cochain $\tau$ is constant on cosets of $U$. Setting $\overline{\tau}(Ut) = \tau(t)$ we get $\overline{h} = \partial\overline{\tau}$. Thus a homomorphism $\mathrm{Ker}\varphi \to H_0^2(S/U, P^\times)$ is defined.

Now we construct an inverse map. Let $\overline{h} \in Z_0^2(S/U, P^\times)$. Set $h(at,b) = h(a,bt) = 0$ and $h(as,bt) = \overline{h}(Us,Ut)$ for $st \neq 0$. Then one can verify straightforward that $\partial h = 0$, and since the restriction of $h$ on $U$ equals zero, $[h] \in \mathrm{Ker}\varphi$.

Let $\overline{h} = \partial\overline{\gamma}$ for some 0-cochain $\overline{\gamma} \in C_0^1(S/U, P^\times)$. Setting $\gamma(at) = \overline{\gamma}(Ut)$ we get $h = \partial\gamma$. Thus we constructed the sought mapping

$$\mathrm{Ker}\varphi \to H_0^2(S/U, P^\times)$$

and proved that these groups are isomorphic. □

*Remark 3.2.* The proved theorem generalizes the results from [3] (where one assumed that $I^2 = 0$) and [6] (where one assumed that the modification $S$ is commutative).

*Remark 3.3.* Indeed $S/U$ is a modification of the group $G/U$, the Galois group of the extension $P/K$, so $H_0^2(S/U, P^\times)$ is a component of the respective Brauer monoid $M(G/U, P)$.

**Corollary 3.4.** *If the field L is finite then*

$$H_0^2(S, L^\times) \cong H_0^2(S/U, P^\times)$$

*Proof.* Since in this case the group $G$ is Abelian (even cyclic) so $U \triangleleft S$. Then $U$ is the Galois group of $L/P$ and $H^2(U, L^\times)$ is trivial as a Brauer group of a finite field. □

## References

1. A. H. Clifford, G. B. Preston. *Algebraic Theory of Semigroups,*, Amer. Math. Soc., Providence, 1964.
2. Ju. A. Drozd, V. V. Kirichenko. *Finite Dimensional Algebras.* Springer-Verlag, Berlin-Heidelberg-New-York, 1994.
3. D. E. Haile, R. G. Larson, M. E. Sweedler. *A new invariant for* **C** *over* **R**. Amer. J. Math., **105**(1983), N3, p.689-814.
4. S. Lang. *Algebra,* Addison-Wesley Publ. Comp., 1965.
5. B. V. Novikov. *On partial cohomologies of semigroups.* Semigroup Forum, **28**(1984), N1-3, pp.355-364.
6. B. V. Novikov. *The Brauer monoid.* Matem. zametki, **57**(1995), No. 4, pp.633-636 (Russian). Translated in: Math. Notes **57**(1995), No. 3-4, pp.440-442.

# Algorithms for Factoring Polynomials over Finite Fields

Tanja Lange[1] and Arne Winterhof[2]

[1] Institut für Geometrie
TU Braunschweig
Pockelsstr. 14
D-38106 Braunschweig
Germany
e-mail: ta.lange@tu-bs.de

[2] Institut für Diskrete Mathematik
Österreichische Akademie der Wissenschaften
Sonnenfelsgasse 19
A-1010 Wien
Austria
e-mail: arne.winterhof@oeaw.ac.at

**Abstract** Results on the worst case behavior of the authors' extension (Theor. Comput. Sci. 234 (2000), 301-308) of Shoup's algorithm for factoring polynomials over finite prime fields (Inf. Process. Lett. 33 (1990), 261-267) are improved. Moreover, the consequences of the average case behavior of the extended algorithm for multivariate algorithms are described, and an extension of Lenstra's algorithm (Lond. Math. Soc, Lect. Note Ser. 154 (1990), 76-85) for root finding over finite prime fields is presented.

## 1 Introduction

In [4] the authors analyze an extension of Shoup's [7] deterministic algorithm for factoring polynomials over finite prime fields $\mathbf{F}_p$ to arbitrary finite fields $\mathbf{F}_q$. In particular, for a polynomial of degree $n$ over $\mathbf{F}_q$, $q$ odd, they show that the fraction of all monic polynomials of degree $n$ for which the extended algorithm fails to halt in $O\left(n^{2+\varepsilon}\log^2 q\right)$ $\mathbf{F}_q$-operations is $O\left(n^2\log^2 q/q\right)$. This is the first average case analysis of a deterministic algorithm over arbitrary $\mathbf{F}_q$ of this quality. Unfortunately the worst case running time of the algorithm is $O\left(q^{1/2}n^{2+\varepsilon}\log q\right)$ $\mathbf{F}_q$-operations.

In Section 2 we recall the extended algorithm and its worst case analysis. In Section 3 we show that the worst case bound can be improved essentially under some certain restrictions. We show that the fraction of all monic polynomials of degree $n$ for which the algorithm fails to halt in $O\left(n^{2+\varepsilon}r^2p^{1/2}\log p\right)$ is $O\left(n^2/q\right)$ if $r$ is an odd integer satisfying $r < p^{1/2}/4$, where $q = p^r$ is an odd prime power. Therefore, for a large fraction of polynomials the algorithm is slightly better than Shoup's generalization of his algorithm to arbitrary finite fields ([8], [2]). (Shoup did not prove any bounds on the average case complexity. In the worst case his algorithm has a complexity of $O\left((nr)^{2+\varepsilon}p^{1/2}\log p\right)$.)

In Section 4 we describe the consequences of the average case bound of [4] for multivariate factorization.

Moreover, in Section 5 we extend Lenstra's algorithm for root finding ([5]). Although the running time of the latter algorithm is not better, the interesting algebraic structure of this algorithm for arbitrary finite fields warrants attention.

In Section 2 - 4 we restrict ourselves to the case that $q = p^r$ is an odd prime power. The case of even characteristic can be handled similarly (see [4]). In Section 5 we consider arbitrary prime powers $q$.

## 2 Extension of Shoup's Algorithm

In this section we give a short description of the factoring algorithm of [4]

We expect the reader to be familiar with the basic results on factoring methods. For a recent survey see Naudin/Quitté [6]. We may restrict ourselves to the case of equal-degree factorization, i.e. we want to factor $g = g_1 \cdots g_k \in \mathbf{F}_q[X]$ where the $g_i$'s are distinct monic irreducible polynomials of degree $d$.

Let

$$R := \mathbf{F}_q[X]/(g) \cong \mathbf{F}_q[X]/(g_1) \oplus \ldots \oplus \mathbf{F}_q[X]/(g_k),$$

$$x = X \bmod g \text{ and } \alpha_i = \alpha \bmod g_i \in \mathbf{F}_q[X]/(g_i) \cong \mathbf{F}_{q^d} \text{ for } \alpha \in R.$$

$$B = \{\alpha \in R | \alpha^q = \alpha\} = \{\alpha \in R | \alpha_i \in \mathbf{F}_q, i = 1, \ldots, k\}$$

is called the *(relative) Berlekamp subalgebra*, and a subset $S \subseteq B$ is called a *(relative) separating set* if for all $1 \le i,j \le k$, $i \ne j$, there exists an element $\alpha \in S$ such that $\alpha_i \ne \alpha_j$.

The algorithm constructs a separating set in the following way. We compute a polynomial $h(Y) = h_0 + \ldots + h_{d-1}Y^{d-1} + Y^d \in R[Y]$ where

$$h(Y) = (Y - x)(Y - x^q) \cdots \left(Y - x^{q^{d-1}}\right).$$

$x \bmod g_i \in \mathbf{F}_{q^d}$ is a zero of the irreducible polynomial $g_i \in \mathbf{F}_q[Y]$ and thus

$$h(Y) \bmod g_i = (h_0 \bmod g_i) + \ldots + (h_{d-1} \bmod g_i)Y^{d-1} + Y^d = g_i(Y). \quad (1)$$

For $i \ne j$ there exists an $l$ such that the coefficients $h_l \bmod g_i$ of $g_i(Y)$ and $h_l \bmod g_j$ of $g_j(Y)$ are different. Hence, $S = \{h_0, \ldots, h_{d-1}\}$ is a separating set.

For $U \subset \mathbf{F}_q[X]$ and $v \in \mathbf{F}_q[X]$ let

$$\text{Refine}(U, v) = \bigcup_{u \in U} \left\{\gcd(u,v), \frac{u}{\gcd(u,v)}\right\} \setminus \{1\}.$$

Let $(x_0, \ldots, x_{q-1})$ be an ordering of the elements of $\mathbf{F}_q$. We can factor $g$ as follows.

**Algorithm 2.1.** Initially, let $U = \{g\}$ and $z = 0$.
While $|U| \neq k$ do the following refinement step
for $l = 0, \dots, d-1$ do
1. replace $U$ with Refine$(U, h_l + x_z)$.
2. replace $U$ with Refine$(U, (h_l + x_z)^{(q-1)/2} - 1)$.

increment $z$.

The correctness of the algorithm follows from the fact, that $\{h_0, \dots, h_{d-1}\}$ is a separating set.

The running time depends on the choice of the ordering of $\mathbf{F}_q$. We choose the following special ordering.

Let $\{b_0, b_1, \dots, b_{r-1}\}$ be a basis of $\mathbf{F}_q$ over $\mathbf{F}_p$. If

$$z = z_0 + z_1 p + \cdots + z_{r-1} p^{r-1}, \quad 0 \leq z_i < p, \quad i = 0, \dots, r-1$$

is the $p$-adic expansion of $z$, then let

$$x_z = z_0 b_0 + z_1 b_1 + \cdots + z_{r-1} b_{r-1}. \tag{2}$$

To determine the complexity of the algorithm, we have to estimate the number of times the refinement step is executed. Therefore we need the following lemma.

**Lemma 2.2.** *Let $\chi$ be the quadratic character of $\mathbf{F}_q$ and $f \in \mathbf{F}_q[X]$ a monic polynomial which is not a square and has $w$ distinct zeros in its splitting field over $\mathbf{F}_q$. Let $L = L(q, f)$ denote the largest number of consecutive elements $x_0, x_1, \dots, x_{L-1}$ in (2) satisfying*

$$\chi(f(x_0)) = \chi(f(x_1)) = \ldots = \chi(f(x_{L-1})). \tag{3}$$

*Then*

$$L(q, f) \leq 3wq^{1/2} - 1.$$

*Proof.* [4, Lemma 1]. □

For prime fields and $f \in \mathbf{F}_p[X]$ a nonsquare it can be proved that

$$L(p, f) \leq 2\deg f p^{1/2} - 1 \tag{4}$$

(see the remark after Lemma 1 in [4]).

**Theorem 2.3.** *Let $f \in \mathbf{F}_q[X]$ be a monic polynomial of degree $n$.*

*Then the algorithm completely factors $f$ in $O\left(q^{1/2} n^{2+\varepsilon} \log q\right)$ operations in $\mathbf{F}_q$.*

*Proof.* The proof is a straightforward extension of Shoup's proof for prime fields based on the previous lemma. □

## 3 Improvement of the Worst Case Analysis

The estimation of the number of executions of the refinement step characterizes the worst case bound. The following lemma shows that under some restrictions the bound for the number of executions of the refinement step can be improved essentially.

**Lemma 3.1.** *Let $r < \frac{p^{1/2}}{4}$ be an odd integer, $\chi$ be the quadratic character of $\mathbf{F}_q$, $a, b \in \mathbf{F}_q$, $a \neq b$, such that $a$ and $b$ are not conjugates over $\mathbf{F}_p$, $f(X) = (X-a)(X-b) \in \mathbf{F}_q[X]$ and $b_0 = 1$. Then*

$$L(q,f) \leq 4rp^{1/2} - 1.$$

*Proof.* $\chi$ can be seen as obtained by lifting the quadratic character $\chi'$ of the prime field $\mathbf{F}_p$.

$$\chi(x) = \chi'(N(x)), \; x \in \mathbf{F}_q,$$

where $N(x) = x^{(q-1)/(p-1)}$ denotes the norm of $\mathbf{F}_q$ onto $\mathbf{F}_p$. Consider the polynomial

$$F(X) = (X-a)(X-a^p)\cdots(X-a^{p^{r-1}})(X-b)(X-b^p)\cdots(X-b^{p^{r-1}})$$

over $\mathbf{F}_p$ of degree $2r$. For $x \in \mathbf{F}_p$ we have

$$F(x) = N(f(x)).$$

Since $b_0 = 1$ we have $x_0, x_1, \ldots, x_{p-1} \in \mathbf{F}_p$ and thus

$$\chi'(F(x_0)) = \chi'(F(x_1)) = \ldots = \chi'(F(x_{L-1})),$$

provided $L < p$.

Hence, $L(q,f) \leq L(p,F)$. Since $a$ and $b$ are not conjugates and $r$ is odd $F(X)$ is not a square and (4) yields $L(p,F) \leq 4rp^{1/2} - 1 < p$. □

Now we establish an upper bound on the number of polynomials for which this better bound does not hold. So in this section we restrict ourselves to the case $r < p^{1/2}/4$.

Let $f_1 \cdots f_n$ be the distinct degree factorization of the square-free part of a polynomial $f$, where $f_d$ is the product of irreducible polynomials of degree $d$, and $K_d$ the number of times the refinement step is executed to factor $f_d$. Let $P$ be the probability that a polynomial $f$ is *bad* in the sense that $K_d > 4rp^{1/2} - 1$ for some $1 \leq d \leq n$. We will prove

$$P = O(n^2 q^{-1}).$$

The assertion follows easily from the following lemmas.

**Lemma 3.2.** *Let $N$ be the number of pairs $(g_1, g_2)$ of monic polynomials of degree $d$ such that each pair of corresponding coefficients of $g_1$ and $g_2$ is conjugated over $\mathbf{F}_p$. Then*

$$N < \frac{q^{d(1+1/p)}}{p}.$$

*Proof.* Let $g_1 = X^d + a_{d-1}X^{d-1} + \ldots + a_0$ and $g_2 = X^d + b_{d-1}X^{d-1} + \ldots + b_0$ such that $a_i$ and $b_i$ are conjugates for $i = 0, 1, \ldots, d-1$. Then

$$\operatorname{Tr}(g_1(x) - g_2(x)) = \operatorname{Tr}(a_{d-1} - b_{d-1})x^{d-1} + \ldots + \operatorname{Tr}(a_0 - b_0) = 0 \text{ for all } x \in \mathbf{F}_p$$

and thus

$$g_1(x) = g_2(x) + h(x)^p - h(x) \text{ for all } x \in \mathbf{F}_p$$

with a polynomial $h \in \mathbf{F}_q[X]$ of degree $\leq \left\lfloor \frac{d-1}{p} \right\rfloor$. The number of such pairs $(g_1, g_2)$ is bounded by $q^d \cdot \frac{q^{d/p}}{p}$. □

The *factorization pattern* of a polynomial $f$ of degree $n$ is the $n$-tuple $\pi = (k_1, \ldots, k_n)$ where $k_d$ is the number of irreducible factors (counting multiplicity) of degree $d$ that divide $f$.

Let $P_\pi$ be the conditional probability that $f$ is *bad* given its factorization pattern $\pi$. We can write

$$P_\pi \leq P_{\pi,1} + \cdots + P_{\pi,n},$$

where $P_{\pi,d}$ is the probability that $K_d > 4rp^{1/2} - 1$ given the factorization pattern $\pi$.

**Lemma 3.3.** *For fixed* $1 \leq d \leq n$ *we have*

$$P_{\pi,d} = O\left(k_d^2 q^{-1}\right).$$

*Proof.* Two polynomials $g_i, g_j$ of equal degree are called *resembling* if each pair of corresponding remainders of the $d$ polynomials $h_l \in S$ modulo $g_i$ and $g_j$ is conjugated. Equivalently, by (1) $g_i$ and $g_j$ are resembling if each pair of corresponding coefficients is conjugated.

Let $\tilde{f}^{(d)}$ be the product of all irreducible factors of $f$ of degree $d$ (including multiplicity). $P_{\pi,d}$ is not larger than the probability that $\tilde{f}^{(d)}$ is divisible by two resembling monic polynomials of degree $d$ by Lemma 3.1. As $\tilde{f}^{(d)}$ contains $k_d$ irreducible factors, this probability is no greater than $k_d^2$ times the probability $P'$ that a randomly chosen pair of irreducible polynomials of degree $d$ is resembling.

For the number $I(d)$ of resembling pairs of monic irreducible polynomials of degree $d$ we have

$$I(d) < q^{d(1+1/p)}/p$$

by Lemma 3.2.

For the number $N(d)$ of all monic irreducible polynomials of degree $d$ we have

$$N(d) = \Theta(q^d/d).$$

Hence,

$$P' = I(d)/N(d)^2 = O\left(\frac{q^{d(1+1/p)}d^2}{pq^{2d}}\right) = O\left(\frac{q^{1/p}}{pq}\right) = O\left(q^{-1}\right)$$

since $r < p$. Thus

$$P_{\pi,d} \leq k_d^2 P' = O\left(k_d^2 q^{-1}\right).$$

□

We have proved the following theorem.

**Theorem 3.4.** *Let $r < p^{1/2}/4$ be an odd integer. Then the fraction of all monic polynomials of degree $n$ for which the algorithm fails to halt in*

$$O\left(n^{2+\varepsilon} r^2 p^{1/2} \log p\right)$$

*is*

$$O\left(n^2 q^{-1}\right).$$

## 4 Multivariate Factoring

It is well-known that multivariate factorization can be reduced to bivariate factorization (see [1], [3], [10]). We restrict ourselves to bivariate factorization and sufficiently large $q$ with respect to $n$, namely $q > n^3$.

We make use of the average case analysis of [4] showing that the refinement step is executed at most $O(\log q)$ times for all but at most a fraction of $O\left((n \log q)^2/q\right)$ monic polynomials of degree $n$ over $\mathbf{F}_q[X]$ hence resulting in an average case running time of $O\left(n^{2+\varepsilon} \log^2 q\right)$.

Let

$$f(X,Y) = a(f_n(X,Y) + f_{n-1}(X,Y) + \cdots + f_0(X,Y)),$$

where $a \in \mathbf{F}_q^*$, $f_k(X,Y)$ is the homogeneous part of degree $k$ in $f(X,Y)$ for $k = 0, \ldots, n$ and $f_n(X,1)$ is monic. To ease the analysis of the algorithm we suppose $f_n(X,1)$ to be squarefree. (Note that otherwise some preprocessing is necessary that might skew the distribution.) Given the factorization of $f_n(X,1)$ over $\mathbf{F}_q[X]$ then by homogenization with respect to the total degree $n$ we obtain

$$f_n(X,Y) = p_1(X,Y) \cdots p_t(X,Y) = g_v(X,Y) \cdot h_s(X,Y),$$

where $g_v$ and $h_s$ are homogeneous polynomials of degree $v$ and $s$, respectively. Comparing the homogeneous parts of degree $k$ of

$$f = f_n + \cdots + f_0 = (g_v + \cdots + g_0) \cdot (h_s + \cdots + h_0) = g \cdot h$$

we get

$$f_{n-k} - \sum_{i=1}^{k-1} g_{v-i} h_{s-k+i} = g_{v-k} h_s + h_{s-k} g_v \text{ for } k = 0, 1, \ldots, n,$$

where $g_j, h_j = 0$ if $j < 0$. As $g_v$ and $h_s$ are relatively prime (because $f_n$ is squarefree), $g_{v-k}$ and $h_{s-k}$ can be computed by the Euclidean algorithm and there are $2^t$ possible choices for $g_v$ once the factorization of $f_n$ is known. If during the computation a contradiction appears then we have to choose a different $g_v$. If there is no further choice then $f(X,Y)$ is irreducible over $\mathbf{F}_q[X,Y]$.

Wan [10] shows that for factoring polynomials $f$ with less than $c \ln n$ irreducible factors in the part $f_n$ his factoring algorithm takes

$$O\left(n^{3+c\ln 2} \log^2 n\right) + O\left(\tau(n,q)\right) \ \mathbf{F}_q\text{-operations},$$

where $O\left(\tau(n,q)\right)$ denotes the costs for the univariate factoring of a monic polynomial of degree $n$. We only use Wan's average case results.

Using the algorithm in Section 2 we have

$$O\left(\tau(n,q)\right) = O\left(n^{2+\varepsilon} \log^2 q\right) \text{ for all except a fraction of } O\left((n \log q)^2 / q\right)$$

polynomials. By Shparlinski [9, Corollary to Theorem 3.10], the number of distinct irreducible factors of monic polynomials of degree $n$ is less than

$$\ln n + \Delta (\ln n)^{1/2}$$

for all but at most $O\left(q^n \Delta^{-2}\right)$ polynomials. Choosing

$$\Delta = \varepsilon \log\log q / (\ln n)^{1/2}$$

(i.e. $c = 1 + \varepsilon \log\log q / \ln n$) we are now able to prove an extension of Shparlinski [9, Theorem 1.8].

**Theorem 4.1.** *There exists a deterministic algorithm that factors all but at most*

$$O\left(q^{(n+1)(n+2)/2} \ln n / (\varepsilon \log\log q)^2\right)$$

*polynomials* $f(X,Y) \in \mathbf{F}_q[X,Y]$ *of total degree* $n$ *in time*

$$O\left(n^{3.7} \log^{\varepsilon} q + n^{2+\varepsilon} \log^2 q\right).$$

*Proof.* For a given $f_n(X,Y)$ there are less than $q^{n(n+1)/2+1}$ polynomials $f(X,Y) \in \mathbf{F}_q[X,Y]$ of total degree $n$ with $f_n$ as homogeneous part of degree $n$ with $f_n(X,1)$ monic. The theorem follows by excluding the

$$O\left(q^{(n+1)(n+2)/2-1}\right)$$

polynomials $f(X,Y) \in \mathbf{F}_q[X,Y]$ for which $f_n(X,1)$ is not squarefree, and the

$$O\left(q^{(n+1)(n+2)/2-1}(n \log q)^2\right)$$

polynomials for which $f_n(X,1)$ is in the excluded set of the univariate factoring algorithm. Furthermore, by the above estimates at most

$$O\left(q^{n+n(n+1)/2+1}\frac{\ln n}{(\varepsilon \log\log q)^2}\right)$$

polynomials have too many factors. Using $q > n^3$ the result follows. □

## 5 An Extension of Lenstra's Algorithm

We extend H. W. Lenstra's algorithm for root finding over prime fields to a factoring algorithm over arbitrary finite fields. As in the algorithm of Section 2 we may restrict ourselves to the case of equal-degree factorization. After computing the separating set $S$ as in the first algorithm we reduce the number of elements to be put in the refinement step:

We divide $\mathbf{F}_q$ into sets of cardinality $p^{\lfloor r/2 \rfloor} \le q^{1/2}$ and test these field elements together on $a$ or $b$ being in this set. Only then these elements are used in the Refine procedure.

Ordering (2) provides an additive property like in $\mathbf{Z}_p$:

$$x_{jp^i} + x_\nu = x_{jp^i+\nu} \text{ for } \nu < p^i.$$

Let $h_i \in S$, $u \in U$, $t = \lfloor r/2 \rfloor$ and

$$d(Y) = (Y - x_0)(Y - x_1)\cdots(Y - x_{p^t-1}).$$

Then $\gcd(u, d(h_i - x_{jp^t})) \ne 1$ if and only if $\gcd(u, h_i - x_l) \ne 1$ for $x_l \in \{x_{jp^t}, x_{jp^t+1}, \ldots, x_{(j+1)p^t-1}\}$.

In order to assure that the inner loop is not repeated too often we introduce the routine *Test*$(U, s)$ for $s \in S$ stating if $s$ is of any further use in the factorization. This is tested by computing the remainder of $s$ by all $u_l \in U$ with $\deg u_l > d$. Only if this remainder is nonconstant for at least one such $u_l$ then $s$ can lead to a further factorization.

**Algorithm 5.1.**
Compute $d(Y) = (Y - x_0)(Y - x_1) \cdots (Y - x_{p^t - 1})$
$i = 0, U = \{g\}$.
repeat until $|U| = k$
1.if $Test(U, h_i) \neq 0$:
2. for $j = 0, \ldots, p^t - 1$
3. if gcd( $\prod_{\substack{l \\ \deg u_l > d}} u_l, d(h_i - x_{jp^t})$) is nontrivial
4. for $\nu = 0, \ldots, p^t - 1$
Refine$(U, h_i - x_{jp^t + \nu})$
5. $i = i + 1$
If $r$ is odd we add after step 4:
4'. if $Test(U, h_i) \neq 0$:
5'.for $j = p^t, \ldots, p^{t+1} - 1$
6'. if gcd( $\prod_{\substack{l \\ \deg u_l > d}} u_l, d(h_i - x_{jp^t})$) is nontrivial
7'. for $\nu = 0, \ldots, p^t - 1$
Refine$(U, h_i - x_{jp^t + \nu})$

**Theorem 5.2.** *The algorithm factors the product of k distinct monic irreducible polynomials of degree d in*

$$O\left((dm)^{1+\varepsilon} \log q + \min\{d, k\}(mp^t)^{1+\varepsilon}\right)$$

$\mathbf{F}_q$*-operations, where* $m = kd$.

*Proof.* The computation of the separating set takes $O\left((dm)^{1+\varepsilon} \log q\right)$ $\mathbf{F}_q$-operations (see [4]), $d(Y)$ can be computed using $O\left((p^t)^{1+\varepsilon}\right)$, the procedure *Test* is used at most $d$ times and each execution takes $O\left(m^{1+\varepsilon}\right)$.

The following steps are iterated at most $\min\{d, k\}$ times:

Using multipoint-evaluation at $p^t$ points of the degree $p^t$ polynomial $d(Y)$ modulo $g$ needs $O\left((mp^t)^{1+\varepsilon}\right)$ $\mathbf{F}_q$-operations. Step 3 takes $p^t \cdot O\left(m^{1+\varepsilon}\right)$ and for each useful $x_{jp^t}$ we get an extra $p^t \cdot O\left(m^{1+\varepsilon}\right)$ for the Refine procedures, but only at most $\min\{k, d\}$ elements $x_{jp^t}$ can be useful.

If $r$ is odd, i. e. $p^t < q^{1/2}$ then additional costs arise for the test ($\min\{d, k\}$ times $O\left(m^{1+\varepsilon}\right)$) and for at most $\min\{k, d\}$ times $p$ gcd's and for further Refine computations. □

## References


1. J. von zur Gathen and E. Kaltofen, Factorization of multivariate polynomials over finite fields, *Math. Comput.* **45** (1985), 251-261.
2. J. von zur Gathen and V. Shoup, Computing Frobenius maps and factoring polynomials, *Comput. Complex.* **2** (1992), 187-224.
3. E. Kaltofen, Polynomial-time reductions from multivariate to bi- and univariate integral polynomial factorization, *SIAM J. Comput.* **14** (1985), 469-489.

4. T. Lange and A. Winterhof, Factoring polynomials over arbitrary finite fields, *Theor. Comput. Sci.* **234** (2000), 301-308.
5. H. W. Lenstra jun., Algorithms for finite fields, in: J. H. Loxton (ed.) "Number theory and cryptography", *Lond. Math. Soc., Lect. Note Ser.* **154**, Cambridge University Press, Cambridge, 1990, 76-85.
6. P. Naudin and C. Quitté, Univariate polynomial factorization over finite fields, *Theor. Comput. Sci.* **191** (1998), 1-36.
7. V. Shoup, On the deterministic complexity of factoring polynomials over finite fields, *Inf. Process. Lett.* **33** (1990), 261-267.
8. V. Shoup, A fast deterministic algorithm for factoring polynomials over finite fields of small characteristic, in: Proc. Int. Symp. Symbolic and Algebraic Comput. (ISSAC) (1991), 14-21.
9. I.E. Shparlinski, "Finite Fields: Theory and Computation", Kluwer, Dordrecht/Boston/London, 1999.
10. D. Wan, Factoring multivariate polynomials over large finite fields, *Math. Comput.* **54** (1990), 755-770.

# Gauss Sums over Quasi-Frobenius Rings

Philippe Langevin and Patrick Solé

[1] Groupe d'Etude de Codage de Toulon
Université de Toulon et du Var
83957 La Garde Cedex, B.P. 132
France
e-mail: langevin@univ-tln.fr

[2] CNRS-13S, ESSI
route des Colles
06903 Sophia Antipolis
France
e-mail: Patrick.Sole@agat.univ-lille1.fr

**Abstract** Quasi-Frobenius (QF) rings comprise finite fields, Galois rings and enjoy remarkable character-theoretic properties. We define two types of Gauss sums over QF rings : complete and incomplete. We compute the modulus of the former. We estimate the modulus of the latter from above and give an application to Gauss sums over Galois rings. We derive the analogue of Stickelberger theorem for Gauss sums indexed by the Teichmüller set.

## 1 Gauss Sums Over Finite Fields

Let $K$ be a finite field of characteristic $p$ and order $q = p^f$. Let $\chi$ be a multiplicative character of $K$ and let $\psi$ be an additive character of $K$. The Gauss sum related to the pair $(\chi, \psi)$ is defined by :

$$G_K(\chi, \psi) = \sum_{x \in K^\times} \chi(x)\psi(x). \tag{1}$$

If both $\chi$ are $\psi$ not trivial, one uses the orthogonality relations of characters to establish that $G_K(\chi, \psi)$ has absolute value $\sqrt{q}$ and that

$$G_K(1,1) = q-1, \quad G_K(\chi, 1) = 0, \quad G_K(1, \psi) = -1. \tag{2}$$

Let us denote respectively by $\zeta$ and $\xi$ the principal $p$-root and principal $(q-1)$-root of the unity in the complex numbers field. Let $\wp$ be any prime ideal above $p$ lying in the cyclotomic ring $\mathbb{Z}[\zeta, \xi]$. The identification between $\mathbb{Z}[\zeta, \xi]/\wp$ and the field $K$ suggests the definition of two special characters : an additive *canonical* character $\mu_K$ which sends $x$ on $\mu_K(x) = \zeta^{\mathrm{tr}_{K/\mathbb{F}_p}(x)}$ and a multiplicative *Teichmüller* character which sends the class of $\xi$ on $\xi$ character of $K$, denoted by $\omega_K$, it has order $q-1$. For any positive integer $a < q-1$, the $\wp$-adic valuation of the Gauss sum $G_K(\bar{\omega}^a, \mu)$ is given by the Stickelberger 's congruences [20] :

$$G_K(\bar{\omega}_K^a, \mu_K) \equiv -\frac{(\zeta-1)^{S(a)}}{R(a)} \pmod{\wp^{S(a)+1}} \tag{3}$$

where $S(a)$ is the sum of the $p$-digit of $a$ and $R(a)$ the product of their factorial. The factorization of the principal ideal generated by $G_K(\bar{\omega}_K^a, \mu_K)$ follows, see e.g. [12].

## 2 Gauss Sums Over Finite Rings

Let $A$ be finite commutative ring. The multiplicative characters group is denoted by $\widehat{A^\times}$ and the additive one by $\widehat{A^+}$. As in the case of fields, we define the Gauss sum related to the characters $\chi \in A^\times$ and $\psi \in A^+$ in the same manner :

$$G_A(\chi, \psi) = \sum_{x \in A^\times} \chi(x)\psi(x) \tag{4}$$

It is a *complete Gauss sum.* For any subgroup $X$ of $A^\times$, we also define the *incomplete Gauss sum*

$$G_X(\chi, \psi) = \sum_{x \in X} \chi(x)\psi(x) \tag{5}$$

In this paper, we study the elementary properties of these Gauss sums. The *quadratic Gauss sums* have been studied by [11]. Some works about the non-commutative case can be found in [7] [8], see also the references pages 55–56 of [1].

All along this paper, we will assume commutativity. A commutative ring decomposes as a direct sum of local ring and we can focus our attention on Gauss sum over commutative, local ring. In order to get general results, it is necessary to put restrictions on the ring $A$. We will work with quasi-Frobenius rings which is a general framework since most of the usual finite rings are quasi-Frobenius : finite fields, the quotient rings $\mathbb{Z}/m\mathbb{Z}$, and the Galois rings are quasi-Frobenius. Moreover the class of quasi-Frobenius rings is stable with the classical constructors : a product of quasi-Frobenius rings, group rings and matrix rings based over quasi-Frobenius ring are quasi-Frobenius rings...

Up to the proposition 4.2, the results of the first sections are known, thought not readily avaible, see for example [10] [21]. For completion, important results on Gauss sums are collected in sections 5-6, with an application to the trivial incomplete Gauss sums. A generalization of the Stickelberger's congruences is proposed in the last section.

$A$ commutative local ring of characteristic $p^\ell$
$M$ maximal ideal of $A$
$K$ residue field, $A/M$ of order $q$
$A^\times$ units
$S$ strong units, $1 + \text{ann}(M)$
$T$ Teichmüller units, $\sim K^\times$
$U$ principal units, $1 + M$, note that $A^\times \sim T \times U$
$\mu$ generating character when $A$ is a quasi-Frobenius ring

**Figure1.** notations for a local commutative ring

## 3 Gauss Sums Over Local Rings

In this section, we settle notations and definitions summarized by fig.1. For details on the basic properties of finite rings see the monography of McDonald [14]. A finite commutative ring $A$ of characteristic $\gamma$ decomposes as a direct sum of rings

$$A = \bigoplus_{\substack{p|\gamma \\ p,\text{ prime}}} A_p$$

where $A_p$ is the $p$-ring $\{a \in A \mid \exists n,\, p^n a = 0\}$. The $p$-ring $A_p$ decomposes as a direct sum of local rings of characteristic a power of $p$. It follows that the Gauss sum $G_A(\chi, \psi)$ is a product of Gauss sums over finite local rings, without a loss of generality, we may assume that $A$ is local of characteristic $p^m$.

Let $M$ the maximal ideal of $A$. The residue field $K := A/M$ is a finite extension of degree $f$ of the prime field $\mathbb{F}_p$, of order $q = p^f$. For any ideal $I \neq A$ of $A$ the set $1 + I$ is a subgroup of $A^\times$, we will distinguish the group $U := 1 + M$ and the group of *strong units* $S := 1 + \text{ann}(M)$ assuming $A$ is not a field ! Since $M$ is a $p$-group, we can use the exact sequence coming from the reduction map :

$$1 \longrightarrow U \longrightarrow A^\times \longrightarrow K^\times \longrightarrow 1 \tag{6}$$

to decompose $A^\times$ as a direct product of $T \times U$ where $T$ is a cyclic group of order $q - 1$ that we call the *Teichmüller units* of $A$. The monoïd $T \cup \{0\}$ is isomorphic to $(K, \times)$. We call $T \cup \{0\}$ the *Teichmüller set* of $A$. Each $a \in A$ admits a unique decomposition as a sum :

$$a = t + m, \qquad t \in T \cup \{0\}, \quad m \in M, \tag{7}$$

where $t$ is the *Teichmüller representative* of $a$. For any $x$, $y$ in $T \cup \{0\}$, we write $x \oplus y$ the Teichmüller representative of $x + y$. The Teichmüller set equipped with the laws $\times$ and $\oplus$ becomes a field isomorphic to $K$, of course.

The exact values of complete or incomplete Gauss sums are not easy to evaluate. We begin by an elementary case i.e. the case of complete Gauss sum associated to a multiplicative character trivial on the group $U$.

**Proposition 3.1.** *Let $\psi$ be an additive character and let $\chi$ be a multiplicative character. If $\chi$ is trivial on $U$ then*

$$G_A(\chi,\psi) = \begin{cases} G_K(\chi,\psi)\,|M|, & \psi \text{ trivial over } M; \\ 0, & \text{else.} \end{cases} \tag{8}$$

*Proof.* Indeed,

$$G_A(\chi,\psi) = \sum_{t\in T}\sum_{m\in M} \chi(t+m)\psi(t+m) = \sum_{t\in T}\chi(t)\psi(t)\sum_{m\in M}\psi(m). \tag{9}$$

# 4 Quasi-Frobenius Rings

Let $A$ a finite ring, may be not local. The group $\widehat{A^+}$ becomes a $A$-module with the external law $(a,\psi)\mapsto\psi_a$, where $\psi_a(x) = \psi(ax)$. It is an isomorphism if and only if there exists a *generating character* $\mu$ i.e. a character such that the kernel does not contain a proper ideal. In this case, the ring is called *quasi-Frobenius* or equivalentely *self-injective.* Note that a direct sum of quasi-Frobenius rings is quasi-Frobenius and conversely. Of course, finite fields are quasi-Frobenius. The group rings and the matrix rings based over a quasi-Frobenius ring are quasi-Frobenius. Indeed, it is easy to construct a generating character for these rings starting from the one of the basic ring.

**Theorem 4.1.** *Let $A$ be a finite ring. The following assertions are equivalent :*

*(1) The ring $A$ is quasi-Frobenius.*
*(2) The $A$-module $A$ is injective.*
*(3) The map $I\mapsto ann(I)$ is an involution of the set of ideals of $A$.*

*Proof.* For the the equivalence between (1) and (2) see [14], for those between (2) and (3) see [3].

**Proposition 4.2.** *Let $A$ be a local ring of maximal ideal $M$ and residual field $K$. The ring $A$ is quasi-Frobenius if and only if it satisfies one of the two following properties :*

*(1) there is only one minimal ideal.*
*(2) the order of $ann(M)$ is equal to the order of $K$.*

For example, the ring $A = \mathbb{F}_2[X,Y]/(X^2,XY,Y^2)$ is not quasi-Frobenius. Indeed, let $x$ be the class of $X$, and let $y$ be the class of $Y$. The ring $A$ is local of maximal ideal $M = (x,y)$. The annihilator of $M$ is $M$ of cardinality 4, and the residual field has only two elements. En passant, remark that :

$$\mathrm{ann}\big(\mathrm{ann}(A.x)\big) = \mathrm{ann}(M) = M \neq A.x \tag{10}$$

*Proof.* First point. Let $I$ be the minimal ideal of $A$, and consider a non-trivial character over $I$. We can extend it to a character of $A$, see e.g. [18] which is a generating character. Conversely, let $I$ be a minimal ideal of $A$. We have $IM = I$ or $IM = (0)$ and from Nakayama lemma [12], we deduce $IM = (0)$ and thus $\mathrm{ann}(I) = M$ since the ring is quasi-Frobenius $I = \mathrm{ann}(M)$.

Second point. $\mathrm{ann}(M)$ is a $K$-vector space. If its dimension is not one 1 then it contains a subspace of dimension 1 which is necessary an ideal. Conversely, if $I$ is a minimal ideal then it is included in $\mathrm{ann}(M)$ but that last one is a $K$-vector space of dimension 1.

**Proposition 4.3.** *Let $\psi$ be a generating character of $A$ and let $\chi$ be a multiplicative character. The modulus of a complete Gauss sum is completely determined :*

$$|G_A(\chi,\psi)|^2 = \begin{cases} |A|, & \chi \not\perp S \\ 0, & \chi \perp S \end{cases} \tag{11}$$

*Proof.*

$$\begin{aligned} |G_A(\chi,\psi)|^2 &= \sum_{x\in A^\times}\sum_{y\in A^\times} \chi(x/y)\psi(x-y) \\ &= \sum_{z\in A^\times}\sum_{y\in A^\times} \chi(z)\psi((z-1)y) \\ &= |A| - |M| \sum_{z\in S} \chi(z) \end{aligned} \tag{12}$$

>From this result, we deduce an estimation of for incomplete Gauss sum.

**Proposition 4.4.** *Let $X$ be a subgroup of $A^\times$, let $\chi$ be a multiplicative character and let $\psi$ be an generating character of $A$. If $\chi \perp S$ that is if $\chi$ is trivial over $S$ then*

$$|G_X(\chi,\psi)| \leq \frac{q-\epsilon}{q}\sqrt{|A|} \tag{13}$$

*where $\epsilon$ is the order of the intersection of $X$ and $S$.*

*Proof.*

$$\begin{aligned} G_X(\chi,\psi) &= \frac{1}{|X^\perp|}\sum_{\theta\in X^\perp} G_A(\chi\theta,\psi) \\ &= \frac{1}{|X^\perp|}\sum_{\theta\perp X,\chi\theta\not\perp S} G_A(\chi\theta,\psi) \\ &\leq \frac{1}{|X^\perp|}|\{\theta\in X^\perp \mid \theta \not\perp XS\}|\sqrt{|A|} \end{aligned} \tag{14}$$

The order of $SX$ is $\frac{q|S|}{\epsilon}$, whence

$$|X^\perp \cap S^\perp| = |(XS)^\perp| = \frac{|A^\times|}{|XS|} = \frac{|X^\perp||X\cap S|}{|S|} = \frac{\epsilon|X^\perp|}{q}$$

# 5 Galois Rings

Let $\ell$ be a positive integer. Up to isomorphism , there exists one and only one Galois extension of degree $f$ of the ring $\mathbb{Z}/p^\ell\mathbb{Z}$. It is the Galois ring $GR(p^\ell, f)$ :

$$A := GR(p^\ell, f) \sim \mathbb{Z}[X]/(g(X), p^\ell) \tag{15}$$

where $g(X)$ is an unitary polynomial of degree $f$ in $\mathbb{Z}[X]$ dividing $X^{p^f} - X$ modulo $p^\ell$ whose the reduction modulo $p$ is irreducible over $\mathbb{F}_p$. It is a local ring of characteristic $p^\ell$ whose maximal ideal is principal generated by $p$. Any element $z$ of $A$ decomposes as :

$$z = \sum_{0 \le i < \ell} t_i p^i, \tag{16}$$

where the $t_i$ are in the Teichmüller set. The Galois group of $A$ is cyclic of order $f$ generated by the Frobenius automorphism which sends $z$ to $\sigma(z) = \sum_{0 \le i < \ell} t_i^p p^i$. The trace of $A$ sends $z$ on $\mathrm{tr}_A(z) = \sum_{k=1}^{f} \sigma^k(z)$, and one can check that the character $\mu_A(z) = \exp(2i\pi \mathrm{tr}_A(z)/p^\ell)$ is a generating character i.e. the Galois ring $A$ is a quasi-Frobenius ring.

They are three ways to handle the Galois rings : Witt vectors, Hensel lift and extension of $p$-adic fields. Let us say a few words about these different approaches.

## 5.1 Witt's Vectors

One uses the map

$$x_0 + px_1 + \cdots + x_{\ell-1}p^{\ell-1} \mapsto (x_0, x_1^p, \ldots, x_{\ell-1}^{p^{\ell-1}}) \tag{17}$$

to carry on $K^f$ the structure of $A$ [17]. For any index $j$, there exists polynomials $S_j$ and $P_j$ in $\mathbb{Z}[X_0, X_1, \ldots X_j, Y_0, Y_1, \ldots, Y_j]$ such that :

$$\begin{aligned} (x+y)_j &= S_j(x_0, x_1, \ldots, x_j, y_0, y_1, \ldots, y_j), \\ (xy)_j &= P_j(x_0, x_1, \ldots, x_j, y_0, y_1, \ldots, y_j). \end{aligned} \tag{18}$$

The polynomial $S_j$ and $P_j$ do not depend on $\ell$.

$$S_0(x_0, y_0) = x_0 + y_0 \quad S_1(x_0, x_1, y_0, y_1) = x_1 + y_1 + \frac{x_0^p + y_0^p - (x_0 + y_0)^p}{p}$$

$$P_0(x_0, y_0) = x_0 . y_0 \qquad P_1(x_0, x_1, y_0, y_1) = y_0^p x_1 + y_1 x_0^p$$

### 5.2 Hensel Lifts

This approach is based on the following proposition which is a special case of Hensel's Lemma.

**Proposition 5.1.** *Let $P(X)$ and $Q(X)$ two monic polynomial of $\mathbb{Z}[X]$ such that $P(X)Q(X) = X^q - X$ modulo $p^{\ell-1}$. There exist two monic polynomials $P_\ell$ and $Q_\ell$ with integer coefficients satisfying the modulo $p^\ell$ congruences :*

$$P_\ell(X) \equiv P(X) \quad Q_\ell(X) \equiv Q(X), \quad P_\ell(X) \equiv Q_\ell(X) \tag{19}$$

Modulo $p^\ell$, $P_\ell$ is completely determined, it is the *Hensel lift* of $P$. Let $F_\ell$ the polynomial obtained after $\ell - 1$ lifts of a unitary polynomial $F_1$ whose the reduction modulo $p$ is primitive then the Galois ring $A$ is isomorphic to the quotient ring $\mathbb{Z}[X]/(p^\ell, F_\ell(X))$, moreover the class of $X$ generates the group of Teichmüller units.

### 5.3 Extensions of $\mathbb{Q}_p$.

Let $\alpha$ a root of the unity of order $q - 1$ in an algebraic extension of the field of $p$-adic numbers. The extension $\mathbb{Q}_p(\alpha)$ of $\mathbb{Q}_p$ is unramified of degree $f$. The ideal generated by $p$ in the ring $\mathbb{Z}_p[\alpha]$ is maximal, the Galois ring $A$ is isomorphic to

$$\mathbb{Z}_p[\alpha]/(p^\ell) \tag{20}$$

## 6 Character Sums Over Galois Rings

Let $F(X) = f_0(X) + p^1 f_1(X) + \cdots + p^{\ell-1} f_{\ell-1}(X)$, where the $f_i(X)$'s are polynomials whose the coefficients are in the Teichmüller set. Let $\mathrm{Deg}(F)$ denote the *weighted degree* of $F$ i.e.

$$\mathrm{Deg}(F) = \max\{p^{\ell-i} \deg(f_i) \mid 0 \leq i < \ell\} \tag{21}$$

For any positive integer $s$, let $A_s$ be a Galois extension of degree $s$ of $A$. For any additive character $\psi \in \widehat{A^+}$ , let $\psi^{(s)} \in \widehat{A_s^+}$ the lift of $\psi$ by the trace from $A_s$ onto $A$ :

$$\psi^{(s)} = \psi \circ \mathrm{tr}_{A_s/A} \tag{22}$$

In [2], is proved that the set of solutions $(z, x) \in A \times T$ of the equation $\sigma(z) - z = F(x)$ deduces from a curve of genus $\frac{(\mathrm{Deg}(F)-1)(q-1)}{2}$ obtained by a tower of $\ell$ Artin-Schreier extensions.

**Theorem 6.1.** *Let $\psi$ be a non-trivial character of $A$. There exists* $\mathrm{Deg}(F)-1$ *algebraic integers $\omega_i$ of modulus exactly $\sqrt{q}$ such that*

$$-\sum_{x\in T(A_s)} \psi^{(s)}(F(x)) = \sum_{i=1}^{\mathrm{Deg}(F)-1} \omega_i^s \tag{23}$$

*Proof.* See [2].

Using a variation of the Stepanov's method [13], one can prove a analog in the multiplicative case. More precisely, for any multiplicative character $\chi \in \widehat{A^\times}$ let us denote by $\chi^{(s)}$ the lift of $\chi$ by the norm of $A_s$ onto $A$ :

$$\chi^{(s)} = \chi \circ \mathrm{N}_{A_s/A} \tag{24}$$

**Theorem 6.2.** *Let $\psi$ be a non-trivial additive character of $A$. Let $\chi$ be a multiplicative character of $A$ not trivial over $T$. There exists* $\mathrm{Deg}(F)$ *algebraic integers $\omega_1, \omega_2, \dots, \omega_{\mathrm{Deg}(F)}$, of modulus at most $\sqrt{q}$ such that*

$$-\sum_{x\in T(A_s)} \chi^{(s)}(x)\psi^{(s)}(F(x)) = \sum_{i=1}^{D(f)} \omega_i^s \tag{25}$$

*Proof.* See [6].

These two theorems give bounds for Gauss sums under Teichmüller units.

$$|G_T(1,\mu_A)+1| \leq (p^{\ell-1}-1)\sqrt{q} \tag{26}$$

and if $\chi \in A^\times$ is not trivial over $T$ then

$$|G_T(\chi,\mu_A)| \leq p^{\ell-1}\sqrt{q} \tag{27}$$

when $f = \ell = 2$ the estimate of the proposition (4.4) is slightly more precise.

## 7 Trivial Incomplete Gauss Sum

Let $x$ be a Teichmüller element of the Galois ring $A$. Let us consider the $p$-adic developement of the trace of $x$ :

$$\mathrm{tr}_A(x) = \sum_{j=0}^{\ell-1} t_j(x)p^j. \tag{28}$$

where the functions $t_j$ can be seen as functions from $K$ into $\mathbb{F}_p$. We say that the $t_j$ are *tracic functions.* The first tracic function $t_0$ is the trace of the field $K$. We determine the polynomial expression of the $t_1$ by mean of Witt vectors. Let us continue assuming that $p = 2$. The second tracic function is equal to :

$$t_1(x) = \sum_{0 \le i < j \le f-1} x^{2^i + 2^j} \tag{29}$$

The case of $t_2$ is more difficult, but we can use the Hensel lift to get it. Indeed, let $\beta \in K$, for any integer $\ell$, let $P_\ell(X)$ be the $\ell$-lifting of $P$ of the polynomial $P(X) = \prod_{i=0}^{f-1}(X - \beta^{p^i})$. Let us set

$$P_\ell(X) = \sum_{i=0}^{f} S_i^{(\ell)}(\beta) X^{f-i}. \tag{30}$$

Using the Graeffe's root-squaring method [19], one can see that $P_{\ell+1}(X^2)$ is equal to $(-1)^f P_\ell(X) P_\ell(-X)$. It follows

$$\begin{aligned} S_1^{(\ell+1)}(\beta) &= 2S_2^{(\ell)}(\beta) - S_1^{(\ell)}(\beta)^2, \\ S_2^{(\ell+1)}(\beta) &= 2S_4^{(\ell)}(\beta) - 2S_1^{(\ell)}(\beta)S_3^{(\ell)}(\beta) + S_2^{(\ell)}(\beta)^2 \end{aligned} \tag{31}$$

In particular, we recover $t_1(\beta) = \sum_{0 \le i < j < f} \beta^{2^i + 2^j}$. Hence, $t_1$ is a quadratic form and one can check that $t_2$ is a boolean function of degree 4. The tracic function $t_1$ is a quadratic form whose associated bilinear form is :

$$\phi(x, y) = \mathrm{tr}(x)\mathrm{tr}(y) + \mathrm{tr}(xy) \tag{32}$$

In particular, $t_1$ is a quadratic form with maximal rank. The trivial Gauss sum depends on the Arf invariant ( see e.g [4], [5] ) its hyperplane intersection with $t_0(x) = 0$. For that approach, one has to determine of symplectic basis, it is not easy but done by Netchaev and Kuzmin [15], see also [16]. But, there exists a very simple way based on Theorem (6.1) which gives :

$$S_A(\mu_A) = 1 + G_T(1, \mu_A) = -(-1)^m (1+i)^m \tag{33}$$

We finish this section with a nice formula that links the trivial Incomplete Gauss sums $G_T(1, \mu_\theta)$ and $G_T(1, \mu_A)$ where $\theta \in A$ and $\mu_\theta : x \mapsto \mu(\theta x)$.

**Proposition 7.1.** *[9] Let $A$ be a Galois ring of characteristic 4. Let $\theta = a + 2b \in A$ where $a$ and $b$ are Teichmüller elements then*

$$S_A(\mu_\theta) = \begin{cases} \mu_A(\frac{b}{a}) S_A(\mu_A), & a \neq 0; \\ q, & a = b = 0; \\ 0, & a = 0 \neq b. \end{cases} \tag{34}$$

*Proof.* The proof is based on the well known relation

$$x \oplus y = x + y + 2\sqrt{xy} \tag{35}$$

that gives the Teichmüller representative of the sum two elements of the Teichmüller set. In particular, $\mu_A(x \oplus y) = \mu_A(x + y + 2xy)$, and :

$$1 + G_T(1, \mu_\theta) = \sum_{x \in T \cup \{0\}} \mu_A\big((a + 2b)x\big) = \sum_{x \in T \cup \{0\}} \mu_A\big((1 + 2\frac{b}{a})x\big)$$

$$= \sum_{x \in T \cup \{0\}} \mu_A(x + 2\frac{bx}{a}) = \mu_A(b/a) \sum_{x \in T \cup \{0\}} \mu_A(x \oplus b/a)$$

## 8 Stickelberger Theorem for Incomplete Gauss Sums

In this section, we show that Gauss sums over Galois ring of characteristic 4 satisfy analogues of Stickelberger congruences. The notation are unchanged : $\mu_A$ and $\mu_K$ are the canonical additive characters of $A := GR(4, f)$ and its residue field is denoted by $K$. We start by a formula linking Gauss sums with the character sum

$$H(\chi, \mu) = \sum_{s \in T} \mu_A(s)\chi(s/s + 1), \tag{36}$$

we get the relation

$$G_T(\chi, \mu_A)^2 = G_K(\chi, \mu_K)\big[1 + G_T(1, \mu_A) + H(\chi, \mu_A)\big] \tag{37}$$

*Proof.* This is a straightforward calculation. Let $\chi$ be a multiplicative character and let us square $G_T(\chi, \mu_A)$.

$$\begin{aligned}
G_T(\chi, \mu_A)^2 &= \sum_{x \in T} \sum_{y \in T} \chi(xy)\mu_A(x + y) \\
&= \sum_{x \in T, y \in T} \chi(xy)\mu_A(x \oplus y + 2xy) \quad \text{by 35,} \\
&= \sum_{\substack{x \oplus y = u \\ xy = v}} \chi(v)\mu_A(u + 2v) \\
&= \sum_{x \in K^\times} \chi(x^2)\mu_K(x^2) + 2 \sum_{\substack{u \in T, v \in T \\ \mathrm{tr}_{K/\mathbb{F}_2}(v/u^2) = 0}} \chi(v)\mu_A(u + 2v) \\
&= G_{K^\times}(\chi, \mu_K) + 2 \sum_{\substack{u \in T, w \in T \\ \mathrm{tr}_{K/\mathbb{F}_2}(w) = 0}} \chi(u^2 w)\mu_A(u + 2u^2 w)
\end{aligned}$$

changing $u$ by $u^2$,

$$= G_{K^\times}(\chi, \mu_K) + 2 \sum_{\substack{u \in T, w \in T \\ \mathrm{tr}_{K/\mathbb{F}_2}(w)=0}} \chi(uw)\mu_A(u + 2uw)$$

$$= G_{K^\times}(\chi, \mu_K) + 2 \sum_{u \in T} \mu_A(u)\chi(u) \sum_{\substack{w \in T \\ \mathrm{tr}_{K/\mathbb{F}_2}(w)=0}} \chi(w)\mu_K(uw).$$

One can check, for example with the Poisson formula, that

$$\sum_{\mathrm{tr}_{K/\mathbb{F}_2}(w)=0} \chi(w)\mu_K(uw) = \frac{1}{2}[G_K(\chi, \mu_{K_{u+1}}) + G_K(\chi, \mu_{K_u})]$$

$$= \frac{1}{2} \begin{cases} [\bar{\chi}(u+1) + \bar{\chi}(u)] G_K(\chi, \mu_K), & u \neq 1; \\ G_K(\chi, \mu_K), & u = 1. \end{cases}$$

and we can continue,

$$G_T(\chi, \mu_A)^2 = G_{K^\times}(\chi, \mu_K) + \mu_A(1) G_K(\chi, \mu_K) + G_K(\chi, \mu_K) \sum_{u \notin \{0,1\}} \mu_A(u)\chi(u)[\bar{\chi}(u+1) + \bar{\chi}(u)]$$

$$= G_K(\chi, \mu_K)[1 + G_T(1, \mu_A) + H(\chi, \mu_A)]$$

**Proposition 8.1.** *Let $\wp$ be a prime ideal above 2 in the ring $\mathbb{Z}[i, \xi]$. Let $\omega$ be the Teichmüller character of the Galois rings that occurs in the Stickelberger congruences of the residue field. Then*

$$\mathrm{val}_\wp(G_T(\omega^a, \mu_A)) = S(a). \tag{38}$$

*Proof.* It suffices to show that the sum $H(\chi)$ is invertible modulo $\wp$.

$$\sum_{u \notin \{0,1\}} \omega^a(u+1)\bar{\omega}^a(u)\mu(u) \equiv \sum_{u \notin \{0,1\}} \omega^a(u+1)\bar{\omega}^a(u) \pmod{\mathcal{P}}$$

$$\equiv \sum_{u \notin \{0,1\}} (u+1)^a u^{q-1-a} \equiv -1$$

## References


1. B. C. Berndt, R. J. Evans, and K. S. Williams. *Gauss and Jacobi sums*, volume 21. Wiley & sons, 1998.
2. A. R. Calderbank, T. Helleseth, and P. V. Kumar. An upper bound for weil exponential sums over galois rings and applications. *IEEE transactions on Information Theory*, 41(2):456–468, 1995.

3. C. W. Curtis and I. Reiner. *Representation Theory of Finite Groups and Associative Algebras.* Wiley Classics Library. Wiley & sons, 1988.
4. J. Dieudonné. *La géométrie des groupes classiques*, volume 5. Springer-Verlag, 1991.
5. R. H. Dye. On the arf invariant. *Journal of Algebra*, 53:36–39, 1978.
6. T. Helleseth, P. V. Kumar, and A. G. Shanbag. Exponential sums over galois rings and their applications. volume 233, pages 109–128. 3rd International Conference of Glasgow, Cambridge Press University, 1996.
7. D. S. Kim. Gauss sums for $o^-(2n,q)$. *Acta Arithmetica*, LXXX.4:343–365, 1997.
8. D. S. Kim and In Sok Lee. Gauss sums for $o^+(2n,q)$. *Acta Arithmetica*, LXXVIII.1:75–89, 1996.
9. P. V. Kumar, T. Helleseth, A. G. Shanbag, and K. Yang. On the weight hierarchy of kerdock codes. *IEEE transactions on Information Theory*, 42, 1996.
10. E. Lamprecht. Algemeine theorie der gauschen summen in endlichen kommutativen ringen. *Mathematische Nachrichten (MR:14,942c)*, 9:149–196, 1953.
11. E. Lamprecht. Calculation of general gauss sums and quadratic gauss sums in finite rings. In *C.R. Conf. Int. Quebec/Can*, pages 561–573, 1987.
12. S. Lang. *Cyclotomic fields.* Springer, 1990.
13. R. Lidl and H. Niederreiter. *Finite Fields*, volume 20 of *Encyclopedia of Mathematics and its Applications.* Addison-Wesley, 1983.
14. B. R. McDonald. *Finite Rings with Identity*, volume 28 of *Pure and Applied Math.* M. Dekker, 1974.
15. A. A. Netchaev and A. S. Kuzmin. Trace-function on a galois ring in coding theory. In *AAECC-12*, pages 277–290.
16. Langevin Ph. Sommes de gauss sur un anneau local. Technical Report RR 98-26, laboratoire I3S, 1998.
17. J.-P. Serre. *Corps locaux*, volume 1296 of *Actualités Scientifiques et Industrielles.* Hermann, 1968.
18. J.-P. Serre. *Représentation des groupes finis.* Hermann, troisième edition, 1978.
19. P. Solé. A quaternary cyclic code and a family of quadriphase sequence with low correlation properties. In *Lecture Notes in Computer Sciences*, volume 388, pages 193–201, 1989.
20. J. Stickelberger. Uber eine verallgemeinerung der kreistheilung. *Math. Ann.*, 37:321–367, 1890.
21. J. Wang. On the jacobi sums modulo $p^n$. *Journal of Number Theory*, 39:50–64, 1991.

# On the Early History of Galois Fields

Heinz Lüneburg

FB Mathematik
Universität Kaiserslautern
PO Box 3049
67653 Kaiserslautern
Germany
e-mail: luene@mathematik.uni-kl.de

**Abstract** In this paper, I narrate the development of Galois field theory in the $19^{\text{th}}$ Century. This includes, of course, the description of the situation Galois met when he introduced his Galois fields. The first applications of Galois field theory in group theory and in the foundations of geometry are also coped with. The narration ends with a few spectacular results in this area obtained in the $20^{\text{th}}$ century.

The first step towards Galois field theory was done in antiquity. For one of the fundamental theorems of this theory is Proposition 24 of Book VII of Euclid's Elements. In our language, it reads as follows:

**Theorem 0.1.** *If $a$, $b$, $n \in \mathbb{N}$ and if $a$ und $n$ as well as $b$ and $n$ are relatively prime, then $ab$ and $n$ are relatively prime.*

This shows that the set of integers which are relatively prime to $n$ is closed under multiplication. Closer analysis shows that this multiplication is compatible with taking residues modulo $n$. Thus we get in the end the group of units of the ring $\mathbb{Z}/n\mathbb{Z}$ and, in particular, that $\mathbb{Z}/p\mathbb{Z}$ is a field if $p$ is a prime, the *Galois field* $\mathrm{GF}(p)$ with $p$ elements.

The next spectacular result in our field is an algorithm due to Bachet (1624) (Claude Gaspar Bachet de Meziriac 1581–1638). Given relatively prime natural numbers $a$ and $b$, it computes non-negative integers $x$ and $y$ such that $x < b$ and $y < a$ and

$$ax = by + 1.$$

He used this algorithm within another algorithm solving the Chinese remainder problem. It is interesting to note that Bachet does not use negative numbers in the formulation and verification of his algorithm. Subtraction means partial subtraction for him, i.e., subtracting a smaller number from a larger one. To establish the algorithm, he uses the Euclidean algorithm first and then, substituting backwards, he finds $x$ und $y$. Therefore, he has to store at least the remainder of the division at each stage. Nowadays, one uses an algorithm for this purpose based on Joseph Louis Lagrange's (1736–1813) theory of continued fractions (Lagrange (1770)). It needs only a small number of storage addresses which does not depend on the number of steps the algorithm takes. In either algorithm, the intermediate results are smaller than the numbers one started with.

Bachet's and Lagrange's algorithms allow to compute inverses in GF($p$) and, hence, to perform division in GF($p$). As these algorithms work in any Euclidean domain, they will allow to compute inverses in the end also in GF($p^n$).

In a letter to Bernard Frenicle de Bessy (~1605–1675), dated October 18th, 1640, Pierre de Fermat (1601–1665) stated that

$$x^{p-1} \equiv 1 \bmod p$$

whenever $p$ is a prime that does not divide $x$ (Dickson (1971), p. 59). This letter I believe is the birth certificate of Galois field theory. The first proof of this result was published by Leonhard Euler (1707–1783) in 1741. He published two more proofs, the second one in 1750 and the third one in 1761 (see the bibliography for details). Two of the three proofs are inductive and use the binomial formula

$$(x+1)^p = \sum_{i:=0}^{p} \binom{p}{i} x^i \equiv x^p + 1 \bmod p.$$

The third one he liked best, because he thought that the binomial formula did not suit well the particular purpose (Euler (1750), § 53, p. 510). The third one is remarkable, indeed, because it is an early piece of implicit group theory which is still alive. I shall reproduce it here.

*Proof.* First, Euler argues quoting Euklid VII.24 that none of the powers of $x$ is divisible by $p$. Then he shows using the pigeon hole principle that there is a natural number $d$ such that $x^d \equiv 1 \bmod p$. Pick $d$ minimal. Then

$$1, x, x^2, \ldots, x^{d-1}$$

are pairwise incongruent modulo $p$. Moreover, $x^d \equiv 1 \bmod p$. If $d = p-1$, then we are done. If $d < p-1$, then there is a residue $y$ which is distinct from $1, x, \ldots, x^{d-1}$. It turns out that $y, xy, x^2y, \ldots, x^{d-1}y$ are also pairwise incongruent modulo $p$ and also incongruent to $1, x, x^2, \ldots, x^{d-1}$. Thus we have now $2d$ different residues modulo $p$. If $2d = p-1$ then $d$ is a divisor of $p-1$. If $2d < p-1$, then there is a residue $z$ which is incongruent to all of $1, x, \ldots, x^{d-1}, y, xy, \ldots, x^{d-1}y$. Continuing this way, one sees that $d$ is a divisor of $p-1$. Hence $p-1 = ed$ with $e \in \mathbb{N}$. It follows

$$x^{p-1} - 1 = (x^d - 1) \sum_{i:=0}^{e-1} x^{di} \equiv 0 \bmod p.$$

This proves the assertion.

Euler saw that his argument yields more. He used it to prove that

$$x^{\varphi(n)} \equiv 1 \bmod n,$$

if $x$ and $n$ are relatively prime and where $\varphi$ denotes as usual Euler's totient function (Euler (1763)). He also proved in this paper that $\varphi(ab) = \varphi(a)\varphi(b)$ provided $a$ and $b$ are relatively prime. The letter $\varphi$ for Euler's totient function was introduced by Gauß (1801), Art. 38, and used ever since.

We still use Euler's argument, as Camille Jordan did (Jordan (1870/1957), p. 25), to show what we call Lagrange's theorem that the order of a subgroup of a finite group divides the order of the group. Actually, Lagrange does not speak of groups. He investigates polynomial functions invariant under certain permutations of the arguments and he gets that their numbers divide $n!$. His argument is similar to Euler's, but one has to abstract a lot to see Lagrange's theorem in his writing (Lagrange (1770/71), §§ 63, 64 and also 83). We shall meet Euler's argument again.

In 1769, J. H. Lambert (1728–1777) claimed the existence of a primitive element modulo $p$, if $p$ is a prime (Dickson (1919/1971), p. 181), i.e., the cyclicity of the multiplicative group of GF($p$). Euler wrote that one has to have a proof for that fact (Euler (1774), § 37), as there is no primitive element for most composite numbers. Euler's proof, however, (Euler (1774), § 38) was rejected by Gauß as being lacunary (Gauß (1801), Art. 56). Gauß himself gave two proofs (Gauß (1801), Art. 52, 53). The first one is most beautiful and gives even more, namely:

**Theorem 0.2.** *Let $G$ be a finite group. If the equation $x^d = 1$ has at most $d$ solutions in $G$ for all divisors $d$ of $|G|$, then $G$ is cyclic.*

*Proof.* Denote by $\psi(d)$ the number of elements of order $d$ in $G$. As any element of $G$ has an order that divides $n := |G|$, we have

$$n = \sum_{d|n} \psi(d).$$

Let $g$ be an element of order $d$. Then $1, g, \ldots, g^{d-1}$ are solutions of the equation $x^d = 1$. Therefore, this equation has exactly $d$ solutions in $G$. In particular, all elements of order $d$ belong to the cyclic group generated by $g$. Hence $\psi(d) = \varphi(d)$ provided $\psi(d) \neq 0$. It follows

$$n = \sum_{d|n} \psi(d) \leq \sum_{d|n} \varphi(d) = n,$$

the equation $n = \sum_{d|n} \varphi(d)$ being also established by Gauß (Gauß (1801), Art. 39). It follows $\psi(d) = \varphi(d)$ for all divisors $d$ of $|G|$. In particular, $\psi(|G|) = \varphi(|G|) \geq 1$. This shows that $G$ contains an element of order $|G|$, whence $G$ is cyclic.

It seems that Legendre was the first who proved the existence of a primitive element modulo a prime $p$. He published a proof in Legendre VI, p. 413, and it seems that he had published a proof before.

To apply this theorem to the original situation, one has to know that the polynomial $x^d - 1$ has at most $d$ roots in GF($p$). Euler saw the necessity of proving this, but his attempt failed (Euler (1774), § 37 and § 28). His proof would have been correct if he had stated that any polynomial of degree $n > 0$ has at most $n$ roots. This more general result was proved by Lagrange (Mém. de l'Ac. de Berlin, Année 1768, p. 192. Quoted after Gauß (1801), Art. 44).

According to Dickson (1905), p. 182, Legendre published a proof of the existence of primitive roots modulo a prime $p$ in 1785 (Mém. Ac. R. Sc. Paris 1785. 471–473. This is Dickson's quotation). I do not know whether the proof given then is the same proof he published in his famous *Essai sur la théorie des nombres*. There he characterizes the primitive roots modulo $p$ as the roots of $x^{p-1} - 1$ which are not roots of any of the polynomials

$$x^{\frac{p-1}{q}} - 1,$$

where $q$ is a prime divisor of $p-1$. He also showed that $x^{p-1} - 1$ has exactly $p-1$ roots modulo $p$. Then he claims that there are exactly $\varphi(p-1)$ elements enjoying the above property among the $p-1$ roots of $x^{p-1} - 1$. His proof is a not very well formulated inclusion–exclusion induction on the number of prime divisors of $p-1$ (Legendre (1797/98), p. 413 ff.). Legendre does not use the function symbol $\varphi$, but writes $\varphi(p-1)$ explicitly as a product.

The name primitive element for a generator of a cyclic group is due to Euler (1774).

This was the situation when Galois entered the stage. Évariste Galois is a most remarkable man. He was born on October 25$^{\text{th}}$, 1811 and he died on May 31$^{\text{st}}$, 1832, fatally wounded during a duel the day before. Not yet 21 years of age when he died, he left a few papers and notes that should influence immensely the development of mathematics in the 19$^{\text{th}}$ century. His tying the algebraic equations solvable by radicals to the structure of their Galois groups was one of the sources of the structural thinking in present day mathematics.

Primitive soluble permutation groups have a minimal normal subgroup of order the power of a prime $p$. This normal subgroup is actually an elementary abelian $p$-group and the prototype of a primitive soluble group is the group of all mappings

$$x \to ax^{p^i} + b$$

of GF($p^n$) onto itself with $a, b \in$ GF($p^n$) and $a \neq 0$ and $0 \leq i < n$, the minimal normal subgroup consisting of the set of all mappings of the form $x \to x + b$. Galois introduced Galois fields just for the purpose of constructing these groups. Obviously, he knew what is now known under the name of *Frobenius austomorphism*.

Galois uses irreducible polynomials of degree $n$ to construct what we now call $\mathrm{GF}(p^n)$. In order to define irreducible polynomials, he uses $\mathbb{Z}[x]$ and the prime $p$ saying that the polynomial $F$ with coefficients in $\mathbb{Z}$ is called irreducible modulo $p$, if, and only if, there are no polynomials $\varphi$, $\psi$ and $\chi$ over $\mathbb{Z}$ with

$$\varphi\psi = F + p\chi.$$

He does not mention that the degrees of $\varphi$ and $\psi$ have to be greater than or equal to 1. In the sequel, he does not speak of congruences modulo $p$ anymore. He uses $\mathrm{GF}(p)$ and then $\mathrm{GF}(p^n)$ as we do today.

Given an irreducible polynomial $F$ of degree $n$, he considers general expressions (*expressions générales*) of the form

$$\text{(GF)} \qquad a + a_1 i + a_2 i^2 + \cdots + a_{n-1} i^{n-1},$$

where $i$ is a root of $F$ and the coefficients $a$, $a_1$, ..., $a_{n-1}$ are elements of $\mathrm{GF}(p)$. He says that this root $i$ may be used in some circumstances as advantageously as the complex unity $\sqrt{-1}$. The expression root of $F$ means, as the further developments show, that powers $i^k$ with $k \geq n$ which occur in products of two expressions of the form (GF) are to be reduced modulo $F$. Galois handles his general expressions (GF) in the same way Cauchy handles his imaginary expressions (*expressions imaginaires*) in his paper Cauchy (1821).

Galois states that there are exactly $p^n$ expressions of the form (GF). Let $u$ be one of these expressions distinct from zero. Then there exists an $m \in \mathbb{N}$ such that $u^m = 1$. Choose $m$ minimal with this property then

$$1, u, u^2, \ldots, u^{m-1}$$

are different elements of the form (GF). Using the argument Euler used to prove Fermat's little theorem he gets that $m$ divides $p^n - 1$. He infers that

$$u^{p^n-1} = 1 \quad \text{resp.} \quad u^{p^n} = u$$

where the latter equation is also valid for $u = 0$.

Next Galois states that one proves like in the theory of numbers the existence of primitive elements among the elements of the form (GF). This means, to repeat it in our language, that the multiplicative group of a Galois field is cyclic. Galois does not say anything else about this fact. So it is not clear whether he refers to Gauß who proved that there is a primitive root modulo any prime $p$, or to Legendre who also gave a proof of this fact, or to both.

Concerning the determination of a primitive element, Euler said that there is no rule to find a primitive element modulo $p$ and Gauß repeated it (Gauß (1801), Art. 73). We still find a primitive element of $\mathrm{GF}(p^n)$ only by trial and error. A good strategy is to use aleatoric algorithms (see e. g. Lüneburg (1987)) where one finds the modern version of Gauß (1801), Art. 73.

The group of units of $\mathbb{Z}/p^n\mathbb{Z}$ is cyclic for $p$ odd. Jacobi computed primitive elements for all these rings with $p^n < 1000$. He gave two tables for each such module, one where the powers of the primitive element are given as residues and the other one where the residues are given as powers of the primitive element (Jacobi (1839)). Bussey did the same job for all $\mathrm{GF}(p^n)$ with $p^n < 1000$ (Bussey (1906, 1910)). Jacobi's *canon arithmeticus* contained a lot of mistakes. The tables were computed anew in 1956. Galois comes back to the "remarkable consequence" that

$$u^{p^n} = u$$

for all elements of the form (GF) and he draws the conclusion that all irreducible polynomials of degree $n$ over $\mathrm{GF}(p)$ divide the polynomial

$$x^{p^n} - x.$$

This yields that any irreducible polynomial of degree $n$ will define the same elements (GF).

The question of the existence of irreducible polynomials of degree $n$ is answered in the affirmative by Galois without giving many details. What he says is important for the history of field theory. Therefore, I shall quote it here explicitely.

> Il s'agit maintenant de faire voir que, réciproquement à ce que nous venons de dire, les racines de l'équation ou de la congruence $x^{p^\nu} = x$ dépendront toutes d'une seule congruence du degré $\nu$.
>
> Soit en effet $i$ une racine d'une congruence irréductible, et telle que toutes les racines de la congruence $x^{p^\nu} = x$ soient fonctions rationelles de $i$. (Il est clair qu'ici, comme dans les équations ordinaires, cette propriété a lieu)**.

And the footnote ** reads:

> La proposition générale dont il s'agit ici peut s'énoncer ainsi: étant donnée une équation algébrique, on pourra trouver une fonction rationnelle $\theta$ de toutes ses racines, de telle sorte, que réciproquement chacune des racines s'exprime rationnellement en $\theta$. Ce théorème était connu d'Abel, ainsi qu'on peut le voir par la première partie du mémoire que ce célèbre géomètre a laissé sur les fonctions elliptiques.

This means in our language that any polynomial has a splitting field and that this splitting field has a primitive element. We know since Steinitz (1910) that this is not always true. It is true, however, for perfect fields, i.e., for fields of characteristic 0 and for fields of characteristic $p$ having the further property that the mapping $x \to x^p$ is surjective. Finite fields, of course, enjoy the latter property, and, hence, are perfect.

The paper by Abel that Galois mentions is Abel (1827). Abel showed in this paper that the splitting field of certain polynomials connected with elliptic functions have a primitive element. Of course, the notions of a field, of a splitting field, and of a primitive element of a field extension were not coined yet. The calculations are very cumbersome.

Back to Galois. What he shows explicitely in this context is: Given an irreducible polynomial $G$ of degree $m$ such that all elements of the form (GF) depend rationally on a root of $G$ and if this root, conversely, depends rationally on elements of the form (GF), then $m = n$.

If one deliberates the polynomial $x^{p^n} - x$ from all divisors common to $x^{p^m} - x$ for all $m \leq n$ — actually for all $m$ dividing $n$, but Galois does not say this —, then one gets a polynomial $H$ which ist the product of all irreducible polynomials of degree $n$ and leading coefficient 1 over GF$(p)$. This $H$ could be split into its irreducible factors by the method of Gauß. He does not give a precise reference to Gauß, so I cannot tell which method he means. So far Galois.

Denote by $D(n,p)$ the number of irreducible polynomials of degree $n$ over GF$(p)$ having leading coefficient 1. We have only Galois's word that $D(n,p) \geq 1$. Gauß computed these numbers in the early nineties of the 18[th] Century, but his notes were only published posthumously in Gauß (1863), p. 222). In 1857, however, Dedekind published the following formula Gauß also had in his notes

$$p^n = \sum_{d|n} dD(d,p).$$

This formula stems from the fact that an irreducible polynomial of degree $d$ divides $x^{p^n} - x$ if, and only if, $d$ divides $n$ and that $x^{p^n} - x$ has only simple roots. The latter observation was already made by Galois using the derivative of the polynomial. Using the Möbius inversion formula which is not due to Möbius, but to Dedekind who established it in the same paper, one gets

$$nD(n,p) = \sum_{d|n} \mu(d)p^{\frac{n}{d}}$$

where $\mu$ denotes the Möbius function which was introduced by Möbius in his paper Möbius (1832). If $\sigma$ denotes the product over all primes dividing $n$, then this formula yields

$$nD(n,p) \equiv \mu(\sigma)p^{\frac{n}{\sigma}} \bmod p^{\frac{n}{\sigma}+1}.$$

This shows that $D(n,p) \neq 0$ for all $n$ and $p$.

There is another possibility to prove the existence of GF$(p^n)$. Using Steinitz's theorem on the existence of a splitting field for any polynomial over any commutative field, one gets GF$(p^n)$ as the splitting field of the polynomial $x^{p^n} - x$ over GF$(p)$ (Steinitz (1910)). Before Steinitz, the existence of a splitting field was established only for polynomials with non-vanishing discriminant (Kronecker (1887), Weber (1893)).

There is a strange paper by Eisenstein in the Crelle Journal with title "Lehrsätze" (Eisenstein (1850)). It contains three theorems from three different fields without any motivation or comment or proof. The third one is of interest to us. It reads as follows:

**Theorem 0.3.** *Given an irreducible polynomial $f$ of degree $n$ over* GF$(p)$, *if $\alpha$ is a root of $f$, then there is always a suitable expression $\beta$ of $\alpha$ such that any expression*

$$a_0 + a_1\alpha + a_2\alpha^2 + \cdots + a_{n-1}\alpha^{n-1}$$

*with $a_i \in$ GF$(p)$ can be expressed in one, and only one, way in the form*

$$b_0\beta + b_1\beta^p + b_2\beta^{p^2} + \cdots + b_{n-1}\beta^{p^{n-1}}$$

*with $b_i \in$ GF$(p)$.*

Thus, Eisenstein claims that there is always a *normal basis* for the extension GF$(p^n)$ over GF$(p)$. A proof of this theorem in the case that $p$ does not divide $n$ was published by Schönemann in the following volume of Crelle's Journal (Schönemann (1850)). Hensel was the first to establish the theorem in its full generality (Hensel (1888)). Hensel indicates its usefulness in algebraic number theory and he announces a paper where he will show this more explicitely. I have not found this paper.

Hensel does even more. He also derives a formula for the number of $\beta$'s generating normal bases, namely the formula

$$p^{n-m}\prod_{d|m}(p^{\nu_d}-1)^{\frac{\varphi(d)}{\nu_d}},$$

where $m$ is the $p$-free part of $n$ and $\nu_d$ the order of $d$ modulo $p$ and where $\varphi$ denotes Euler's totient function.

In Jacobson (1975) (first edition (1964)), I found the most beautiful proof for the existence of a normal basis of GF$(q^n)$ over GF$(q)$. It uses the following result, a forerunner of which was given by Dedekind (Dirichlet (1893), pp. 460–461):

*Proof.* Let $G$ be a group and $F$ a commutative field. If $\eta_1, \ldots, \eta_n$ are distinct homomorphisms of $G$ into the multiplicative group of $F$, then $\eta_1, \ldots, \eta_n$ are linearly independent.

If $\phi$ denotes the Frobenius automorphism of GF$(q^n)$ over GF$(q)$, then 1, $\phi$, $\phi^2$, ... , $\phi^{n-1}$ are linearly independent by Dedekind's result, whence the minimal polynomial $\mu$ of $\phi$ has degree at least $n$. As $\phi^n = 1$, it follows that $\mu$ divides $x^n - 1$. Hence $\mu = x^n - 1$. This implies that the GF$(q)$-vector space

GF($q^n$) is a cyclic GF($q$)[$\phi$]-module, i.e., that GF($q^n$) has a normal basis over GF($q$). Using the techniques of linear algebra, it is easy to construct a $\beta$ generating a normal basis (Lüneburg (1987), (1993)).

Jacobson does not indicate to whom this proof is due.

Weber was the first to give an abstract definition of a field (Weber (1893). It is amusing to note that he forgot to postulate associativity of multiplication). The name he used for a field was *Körper*, the name Dedekind introduced for what we now call subfield of the complex number field. Dedekind's definition appeared for the first time in his supplements to the second edition of Dirichlet's number theory (Dirichlet (1871), Suppl. X, § 159, S. 424). Having an abstract definition for the notion of a field, one has to admit finite examples. Some were mentioned immediately by Weber. Now one could ask the question: What do the finite commutative fields look like? Moore asked this question and answered it promptly (Moore (1896)):

**Theorem 0.4.** *Every existent field* $F[s]$ *is the abstract form of a Galois field* $GF[q^n]$; $s = q^n$.

To prove this theorem, it suffices to show after all that Galois already proved that $s$ is a power of a prime $q$. For this token, Moore proved that all elements of $F[s]$ distinct from 0 have order $q$ with respect to addition where $q$ is a prime. He proves that the additive group of $F[s]$ is an $F[q]$-vector space and that $F[q]$ is nothing but GF($q$). He defines *linear independence* over GF($q$) and remarks that $n+1$ elements of an elementary abelian $q$-group of *rank* $n$ are always linearly dependent. I wished the word rank had been used instead of the word dimension for the cardinal of a maximal set of linearly independent vectors. A lot of confusion would have been avoided that arose and still arises from the use of the word dimension for two different numbers attached to the same object, namely, the dimension of the vector space and the dimension of its lattice of subspaces.

Moore also proves that GF($q^d$) is contained in GF($q^n$) if, and only if, $d$ divides $n$.

Moore indicates in a footnote to his paper on page 208 that he uses the term *field* as a synonym for Weber's term *endlicher Körper*. This is the earliest use of the word field I know of for structures we still call fields. The name *Galois field* was also coined in this paper.

The most spectacular result in this area is Wedderburn's theorem (Wedderburn (1905)):

**Theorem 0.5.** *If* $F$ *is a finite skew field, then* $F$ *is a field.*

Observe that one has to have the abstract notion of a skew field even to formulate the theorem. The situation is similar to Moore's when he formulated his result on finite fields. It could not have been formulated before the axiomatic description of a field was given.

Wedderburn gave three proofs of this result in his paper, the first of which being not correct as Artin pointed out (Artin (1927)). The other two use a beautiful result of Zsigmondy's. Let $p$ be a prime and $a$ an integer which is not divisible by $p$. Then $a$ can be interpreted as an element of the multiplicative group of $\mathrm{GF}(p)$. As such it has an order which we shall denote by $\mathrm{ord}_p(a)$. Zsigmondy's result then reads as follows (Zsigmondy (1892)):

**Theorem 0.6.** *Let $a$ and $n$ be natural numbers distinct from 1. Then there exists a prime $p$ with $\mathrm{ord}_p(a) = n$, unless $n = 2$ and $a + 1$ is a power of 2 or $n = 6$ and $a = 2$.*

There are several proofs available for Zsigmondy's theorem: Zsigmondy (1892), Birkhoff and Vandiver (1904), Dickson (1905), Artin (1955), Hering (1974), Lüneburg (1981). All of them use the cyclotomic polynomials.

A prime $p$ with the property that $\mathrm{ord}_p(a) = n$ is also called an *a-primitive prime divisor* of $a^n - 1$.

Let $F$ be a finite non-commutative skew-field and denote by $C$ its centre. Then $|C| = p^m$ and $|F| = p^n = p^{sm}$. Given $a \in F$, then $C[a]$ is a commutative subfield of $F$. Hence $C[a] \neq F$. Moreover, $|C[a]| = p^{tm}$. According to Sylow's theorem (Sylow (1872)), this can only happen if there is no $p$-primitive prime divisor of $p^n - 1$. Hence $n = 2$ and $p + 1$ is a power of 2 or $n = 6$ and $p = 2$. The exceptions have to be handled separately. In the first case $C[a] = F$ for any $a \in F - C$, whence $F$ is commutative. In the second case, one sees immediately that $m = 1$. Then one will get a contradiction using the class equation. Wedderburn uses the term "Sylow's theorem" without a quotation and his argument is as short as reproduced here. So it must have been familiar to his public. The class equation is written down for the particular situation. It does not have a name.

The class equation is to be found implicitely in Sylow (1872). He uses it to show that any $p$-group has a non-trivial centre. When Sylow wrote his paper, group theory was still implicit group theory. Weber's paper, where groups were defined axiomatically for the first time, making group theory explicit, was published in 1893. So Sylow formulated his results for permutation groups, but his arguments did not make really use of the permutation group structure.

Wedderburn's third proof is even more delicate. For $0 \neq a \in F$ he defines mappings $\lambda_a$ and $\rho_a$ of $F$ into itself by

$$\lambda_a(x) := ax$$

and

$$\rho_a(x) := xa^{-1}.$$

The mappings $\lambda$ and $\rho$ are monomorphisms of the multiplicative group $F^*$ of $K$ into the group $\mathrm{GL}_C(F)$ of $C$-linear automorphisms of the $C$-vector space $F$. Call the images $L$ and $R$, resp. Then

$$L| = |R| = |F| - 1 = p^n - 1.$$

The groups $L$ and $R$ centralize each other. Hence $LR$ is a subgroup of $\mathrm{GL}_C(F)$. As this group is a subgroup of $G := \mathrm{GL}_{\mathrm{GF}(p)}(F)$, also $LR$ is a subgroup of the latter group. Hence $|LR|$ divides the order of $G$, i.e.

$$g := p^{\frac{1}{2}n(n-1)} \prod_{i:=1}^{n} (p^i - 1).$$

Now $L \cap R$ is isomorphic to $C^*$. Hence

$$|LR| = \frac{(p^n - 1)^2}{p^m - 1}.$$

Let $q$ be a $p$-primitive prime divisor of $p^n - 1$ and let $q^t$ be the highest power of $q$ dividing $p^n - 1$. Then $q^{t+1}$ does not divide $g$, as $q$ does not divide $p^i - 1$ for $i < n$. Therefore, $q^{t+1}$ does not divide $|LR|$. Therefore, $m = n$ and hence $C = F$ in this case. Thus $n = 2$ and $p + 1$ is a power of 2 or $n = 6$ and $p = 2$, etc.

Wedderburn's paper ends with the remark that also L. E. Dickson has proved that finite skew fields are commutative using the theory of canonical forms. Many proofs have been given in the meantime. The one by Witt is the one reproduced most often (Witt (1931)).

Let $R$ be a set with two binary operations called addition and multiplicaton such that $(R, +)$ is an abelian group and such that both distributive laws hold. One calls $(R, +, \cdot)$ *alternative ring* if $a(ab) = a^2b$ and $(ba)a = ba^2$ hold for alle $a$, $b \in R$. The following theorem was published by Zorn (Zorn (1931), Satz 5), but it is due to Artin, as Zorn writes. It is usually called the theorem of Artin–Zorn in the literature.

**Theorem 0.7.** *If $R$ is a finite alternative ring without zero-divisors, then $R$ is a finite field.*

This theorem has many implications in the theory of finite projective planes (see Dembowski (1968)).

Another theorem in this family of characterizations of Galois fields is the following. It was proved for characteristic $p > 2$ by G. Glauberman using the techniques given by the theory of Jordan-algebras. (Pasqual Jordan 1902–1980. German physicist.) Günter Heimbeck gave a much simpler proof that works also for characteristic 2 (Glaubermann (1972), Heimbeck (1984)).

**Theorem 0.8.** *Let $V$ be a non-trivial finite-dimensional vector space over a finite field. If $S$ is a subset of the group of all regular linear mappings of $V$ onto itself enjoying the following properties*

*(i) If $a \in S$, then $a^{-1} \in S$,*
*(ii) $S \cup \{0\}$ is closed under addition,*
*(iii) $1 \in S$,*

*then* $S \cup \{0\}$ *is a field.*

Hensel praised in 1888 the usefulness of normal bases of $\mathrm{GF}(p^n)$ over $\mathrm{GF}(p)$ in establishing theorems in algebraic number theory. But eighteen years before Hensel published his paper, Jordan's *Traité des substitutions et des équations algébriques* appeared in which Jordan made heavily use of Galois fields. The *Traité* is divided into four books, a short first one on Galois fields, the second one on permutation groups in general and linear groups over $\mathrm{GF}(p)$ in particular. Books III and IV are devoted to algebraic equations and their groups. In Book I on Galois fields, Jordan repeats what Galois did in the paper we discussed above. He adds a proof for the existence of irreducible polynomials of any degree over $\mathrm{GF}(p)$. His proof is similar to what Dedekind did. However, he does not count the number of irreducible polynomials explicitly, he shows only that it is not zero.

One expects that Jordan now studies linear groups over arbitrary Galois fields. He does not! He starts to investigate the composition series of the linear group $L(n,m)$ of a free $\mathbb{Z}/m\mathbb{Z}$-module of rank $n$. If $m = ab$ with $a$ and $b$ relatively prime, then $L(n,m)$ is isomorphic to the direct product of $L(n,a) \times L(n,b)$. Thus he is led to study $L(n,p^e)$ where $p$ is a prime. This group has a series of normal $p$-subgroups according to the ideals $p^i\mathbb{Z}/p^e\mathbb{Z}$ of $\mathbb{Z}/p^e\mathbb{Z}$. On top of it sits the group $L(n,p) = \mathrm{GL}(n,p)$ which has to be to investigated next.

Starting at the wrong end, Jordan missed the more general situation of investigating the groups $\mathrm{GL}(n,p^e)$ and all the other interesting subgroups of it. What does he need Galois fields then for? Well, he establishes what is nowadays called the Jordan normal form for matrices over $\mathrm{GF}(p)$. As the characteristic polynomial does not always split into linear factors over $\mathrm{GF}(p)$, he adds a root of an irreducible factor of it to $\mathrm{GF}(p)$ — repeating it, if necessary — in order to obtain the normal form. Then he poses and solves five problems using the normal form:

**Problem I.** Given a linear mapping $A$ and a natural number $\lambda$, find *a priori* the $\lambda^{th}$ power of $A$.

**Problem II.** Find the order of $A \in \mathrm{GL}(n,p)$.

**Problem III.** Determine the centralizer of $A \in \mathrm{GL}(n,p)$.

**Problem IV.** Determine all $A \in \mathrm{GL}(n,p)$, as well as their number, that have the same normal form.

**Problem V.** Determine the structure of all maximal abelian subgroups of $\mathrm{GL}(n,p)$.

Weierstraß published the Jordan normal form for matrices over the field of complex numbers already in 1868. But there is something new in Jordan's approach, as Jordan needs a field extension to obtain the eigenvalues of the given linear mapping, whereas Weierstraß uses that the field of complex numbers is algebraically closed.

In the winter quarter of 1905, O. Veblen lectured on axiomatically defined projective geometry at the University of Chicago (Veblen and Bussey (1906), footnote on page 245). Using the possibility established in that lecture of coordinatizing a projective space of dimension at least 3 by a not necessarily commutative field, Veblen and Bussey obtain the result that any finite projective geometry of dimension $d \geq 3$ has $p^n + 1$ points on each line where $p$ is a prime and $n$ a suitable natural number. Conversely, they show that given $d \geq 3$, a prime $p$ and a natural number $n$, then there exists up to isomorphisms exactly one projective geometry of dimension $d$ having $p^n + 1$ points on a line. Invoking Wedderburn's theorem on finite skew fields, they get that Pappos's theorem holds in all finite projective spaces of dimension $d \geq 3$ and also in all finite desarguesian projective planes. There is no geometrical proof for this result up to now (I do know Tecklenburg (1987)).

One of the purposes they investigate finite projective geometries for is that they "have a geometrical theory identical in most of its general theorems with ordinary projective geometry and thus (to) afford a treatment of finite linear group theory analogous to the ordinary theory of collineations."

## References

1. Abel, Niels Henrik: Rherches sur les fonctions elliptiques. J. reine und angew. Mathematik 2, 101–181, 1827
2. Artin, Emil: Über einen Satz von Herrn J. H. Maclagan Wedderburn. Abh. Math. Sem. Univ. Hamburg 5, 245–250, 1927
3. Artin, Emil: The orders of the linear groups. Comm. Pure & Appl. Math. 8, 355–365, 1955
4. Bachet de Meziriac, Claude Gaspar: Problemes plaisans e delectables qui se font par les nombres. 2nd ed. Lyon 1624
5. Birkhoff, Garret D.; Vandiver, H. S.: On the integral divisors of $a^n - b^n$. Ann. of Math. (ser. 2) 5, 173–180, 1904
6. Bussey, W. H.: Galois field tables for $p^n \leq 169$. Bull. Amer. Math. Soc. 12, 22–38, 1906
7. Bussey, W. H.: Tables of Galois fields of order less than 1,000. Bull. Amer. Math. Soc. 16, 188–206, 1910
8. Cauchy, Augustin-Louis: Cours d'analyse. Paris 1821. Œuvres, Ser. 2, Vol. 3. Paris 1897
9. Dedekind, Richard: Abriß einer Theorie der höheren Kongruenzen in Bezug auf einen reellen Primzahl-Modulus. J. reine und angew. Mathematik 54, 1–26, 1857
10. Dembowski, Peter: Finite Geometries. Ergebnisse der Mathematik und ihrer Grenzgebiete. Bd. 44. Berlin, Heidelberg, New York 1968
11. Dickson, Leonhard Eugene: On the cyclotomic function. Amer. Math. Monthly 12, 86–89, 1905
12. Dickson, Leonhard Eugene: History of the Theory of Numbers. Vol. 1. Reprint of the edition 1919. New York 1971
13. Dirichlet, Peter Gustav Lejeune: Vorlesungen über Zahlentheorie. Herausgegeben und mit Zusätzen versehen von R. Dedekind. Zweite umgearbeitete und vermehrte Auflage. Braunschweig 1871

14. Dirichlet, Peter Gustav Lejeune: Vorlesungen über Zahlentheorie. Herausgegeben und mit Zusätzen versehen von R. Dedekind. 4. Auflage. Braunschweig 1893. Korrigierter Nachdruck New York 1968
15. Eisenstein, Gotthold: Lehrsätze. J. reine und angew. Mathematik 39, 180–182, 1850
16. Euklid: Die Elemente. Nach Heibergs Text aus dem Griechischen übersetzt und herausgegeben von Clemens Thaer. Darmstadt 1980
17. Euler, Leonard: Theorematum quorundam ad numeros primos spectantium demonstratio. Commentarii academiae scientiarum Petropolitanae 8 (1736), 141–146, 1741. Opera omnia I, 33–37
18. Euler, Leonard: Theoremata circa divisores numerorum. Novi commentarii academiae scientiarum Petropolitanae 1 (1747/8), 20–48, 1750. Opera omnia I, 62–85
19. Euler, Leonard: Theoremata circa residua ex divisione potestatum relicta. Novi commentarii academiae scientiarum Petropoltanae 7 (1758/9), 49–82, 1761. Opera omnia I, 493–518
20. Euler, Leonard: Theoremata arithmetica nova methodo demonstrata. Novi commentarii academiae scientiarum Petropolitanae 8 (1760/1), 74–104, 1763. Opera omnia I, 531–555
21. Euler, Leonard: Demonstrationes circa residua ex divisione potestatum per numeros primos resultantia. Novi commentarii academiae scientiarum Petropolitanae 18 (1773), 85–135, 1774. Opera omnia II, 240–281
22. Galois, Évariste: Sur la théorie des nombres. Bull. des sciences mathématiques de Ferussac 13, 428–435, 1830
23. Galois, Évariste: Écrits et mémoires mathématiques d'Évariste Galois. Édition critique intégrale de ses manuscrits et publications par Robert Bourgne et J.-P. Azra. Préface de J. Dieudonné. Paris 1962
24. Gauß, Carl Friedrich: Disquisitiones arithmeticae. Leipzig 1801
25. Gauß, Carl Friedrich: Disquisitiones generales de congruentiis. Works, vol. 2, 212–240. Göttingen 1863
26. Glauberman, G.: A sufficient condition of $p$-stability. Proc. London Math. Soc. (3), 25, 253–287, 1972
27. Heimbeck, Günter: Bemerkungen zu einem Satz von Glaubermann. Elemente der Mathematik 39, 95–98, 1984
28. Hensel, Kurt: Ueber die Darstellung der Zahlen eines Gattungsbereiches für einen beliebigen Primdivisor. J. reine und angew. Mathematik 103, 230–237, 1888
29. Hering, Christoph: Transitive linear groups and linear groups which contain irreducible subgroups of prime order. Geom. Dedicata 2, 425–460, 1974
30. Jacobi, Carl Gustav Jacob: Canon arithmeticus. Berlin 1839.
31. Jacobi, Carl Gustav Jacob: Canon arithmeticus. Nach Berechnungen von Wilhelm Patz in verbesserter und erweiterter Form herausgegeben von Heinrich Brandt. Berlin 1956
32. Jacobson, Nathan: Lectures in Abstract Algebra III. Theory of Fields and Galois Theory. Second Printing. New York, Heidelberg, Berlin 1975
33. Jordan, Camille: Traité des substitutions et des équations algébriques. Reprint of the edition Paris 1870. Paris 1957
34. Kronecker, Leopold: Ein Fundamentalsatz der allgemeinen Arithmetik. J. reine und angew. Mathematik 100, 490–510, 1887

35. Lagrange, Joseph Louis: Nouvelle methode pour résoudre les problèmes indeterminés en nombres entiers. Mémoires de l'Académie royale des Sciences et Belles-Lettres de Berlin. T. XXIV, 1770. Œuvres, vol. 2, 655–726. Paris 1869. Reprint Hildesheim 1973
36. Lagrange, Joseph Louis: Réflexions sur la résolution algébrique des équations. Nouv. Mémoires de l'Académie royale des Sciences et Belle-Lettres de Berlin 1770/71. Œuvres, vol. 3, 205–421. Paris 1869. Reprint Hildesheim 1973
37. Legendre, Adrien-Marie: Essai sur la théorie des nombres. Paris, An VI (i. e. 1797/98)
38. Lüneburg, Heinz: Ein einfacher Beweis für den Satz von Zsigmondy über primitive Primteiler von $a^n - 1$. In: Geometries and Groups. Proceedings of a Colloquium Held at the Freie Universität Berlin, May 1881. Ed. by Martin Aigner and D. Jungnickel. Springer LNM 969, 251–257. Berlin, Heidelberg, New York 1981
39. Lüneburg, Heinz: On the Rational Normal Form of Endomorphisms. A Primer to Constructive Algebra. Mannheim 1987
40. Lüneburg, Heinz: Vorlesungen über Lineare Algebra. Mannheim 1993
41. Möbius, August Ferdinand: Über eine besondere Art der Umkehrung der Reihen. J. reine und angew. Mathematik 9, 105–123, 1832
42. Moore, Eliakim Hastings: A doubly-infinite system of simple groups. Mathematical Papers Read at the International Mathematical Congress Chicago 1893, 208–242. New York 1896
43. Schönemann, Theodor: Über einige von Herrn Dr. Eisenstein aufgestellte Lehrsätze, irreductible Congruenzen betreffend (S. 182 Bd. 39 dieses Journals). J. reine und angew. Mathematik 40, 185–186, 1850
44. Steinitz, Ernst: Algebraische Theorie der Körper. J. reine und angew. Mathematik 137, 167–309, 1910
45. Sylow, Peter Ludwig Mejdell: Théorèmes sur les group des substitutions. Math. Ann. 5, 584–594, 1872
46. Tecklenburg, Helga: A proof of the theorem of Pappus in finite desarguesian affine planes. J. of Geometry 30, 172–181, 1987
47. Veblen, Oswald; Bussey, W. H.: Finite projective geometries. Trans. Amer. Math. Soc. 7, 241–259, 1906
48. Weber, Heinrich: Die allgemeinen Grundlagen der Galois'schen Gleichungstheorie. Math. Ann. 43, 521–549, 1893
49. Wedderburn, J. H. Maclagan: A theorem on finite algebras. Trans. Am. Math. Soc. 6, 349–352, 1905
50. Weierstraß, Karl: Zur Theorie der bilinearen und quadratischen Formen. Monatsbericht der Beliner Akademie der Wissenschaften vom 18. Mai 1868. Math. Werke, vol. II, 19–44.
51. Witt, Ernst: Über die Kommutativität endlicher Schiefkörper. Abh. Math. Sem. Hamburg 8, 413, 1931
52. Zorn, Max: Theorie der alternativen Ringe. Abh. Sem. Univ. Hamburg 8, 123–147,1931
53. Zsigmondy, K.: Zur Theorie der Potenzreste. Monatshefte für Math. und Physik 3, 265–284, 1892

# Linear Blocking Sets: a Survey

Guglielmo Lunardon[1] and Olga Polverino[2]

[1] Dip. di Matematica e Applicazioni Università di Napoli
Complesso di Monte S. Angelo
Edificio T- Via Cintia
80126 Napoli
Italy
e-mail: Lunardon@Guglielmo Lunardon
[2] Dip. di Matematica Seconda Università di Napoli
Complesso Universitario Via Vivaldi
81100 Caserta
Italy
e-mail: Polverino@Olga Polverino

**Abstract** A class of small minimal blocking sets, called *linear*, was introduced in [9]. Using a result of [2], it has been proven that, with few exceptions, all small minimal blocking sets of Rédei type in the finite desarguesian projective planes are linear [9], and that there are examples of non-Rédei type in $PG(2,q^t)$, $t \geq 4$, [14]. The aim of this paper is to collect the known results and to list all the known examples of linear blocking sets of minimum and maximum size.

## 1 Introduction

A *blocking set* in a projective plane is a set of points which intersects every line. A blocking set is *trivial* if it contains a line, and it is *minimal* if no proper subset of it is a blocking set. Two blocking sets are *equivalent* if there is a collineation of the plane which maps one to the other. In the projective plane over the Galois field of order $q$, a blocking set $B$ is *small* if its size is less than $3(q+1)/2$, and is called of *Rédei type* if there is a line $l$ such that $|B \setminus l| = q$; the line $l$ is a *Rédei line* of $B$.

A class of small minimal blocking sets, called *linear*, was introduced in [9]. They have been successfully used for proving the existence of blocking sets of minimum size in $PG(3,q^2)$ (see [9]), and for constructing the first examples of small minimal non-Rédei type blocking sets in $PG(2,q^t)$, $t \geq 4$ (see [14]). The possibility of defining a linear blocking set in three different and equivalent ways has introduced methods of linear algebra, projective geometry, and finite field theory into the investigation of these structures. Moreover, the "linear representation" of desarguesian planes was used by Ball, Blokhuis and Lavrauw ([1]) to prove the existence of $(q+1)$-fold blocking sets in $PG(2,q^4)$ (they are called *linear $(q+1)$-fold blocking sets*), which are not union of $(q+1)$ disjoint Baer subplanes. Presently, we know many examples of linear non-Rédei type blocking sets and some new classes of Rédei type.

The aim of this paper is to collect the known results, also emphasizing some applications to the Rédei blocking sets case. Moreover, we describe the first examples of small minimal non-Rédei type blocking sets of maximum size, and we discuss some open problems.

## 2 Linear Blocking Sets

Let $V$ be a vector space over a field $F$. Denote by $PG(V,F)$ the projective geometry of the $F$–linear subspaces of $V$. Accordingly, for any subfield $K$ of $F$, the symbol $PG(V,K)$ stands for the projective geometry of the $K$–linear subspaces of $V$. As usual, if $F = GF(q)$ and $V$ has finite rank $n+1$, we write $PG(V,F) = PG(n,q)$, and we denote by $\langle v \rangle$ the point of $PG(V,F)$ defined by the non-zero vector $v$ of $V$.

A $(t-1)$-spread $\mathcal{S}$ of $PG(3t-1,q)$ is a set of $(t-1)$-dimensional subspaces such that each point of $PG(3t-1,q)$ is contained in exactly one element of $\mathcal{S}$. Such a $(t-1)$-spread is called *normal* if for every pair $(X,Y)$ of distinct elements of $\mathcal{S}$, the set $\mathcal{S}_T = \{Z \in \mathcal{S} \mid Z \cap T \neq \emptyset\}$ is a $(t-1)$-spread of $T = < X, Y >$. Let $\mathcal{S}$ be a normal $(t-1)$-spread of $PG(3t-1,q)$, and set

$$\mathcal{L} = \{\mathcal{S}_T \mid T = < X, Y >,\ X, Y \in \mathcal{S} \text{ and } X \neq Y\}.$$

Let $\Pi$ be the plane whose pointset is $\mathcal{S}$ and whose lineset is $\mathcal{L}$. Then, $\Pi$ is a desarguesian plane of order $q^t$ (see [18]). If $L$ is a $t$-dimensional subspace of $PG(3t-1,q)$, the subset of $\mathcal{S}$ given by

$$B_L = \{X \in \mathcal{S} \mid X \cap L \neq \emptyset\}$$

is a small minimal blocking set of the plane $\Pi$, called *$GF(q)$-linear* (see [9]). $B_L$ is a line if and only if $L$ is contained in some subspace $T$ joining two distinct elements of $\mathcal{S}$. A $GF(q)$-linear blocking set in $PG(2,q^t)$ contains at most $q^t + q^{t-1} + \cdots + q + 1$ points and, if the intersection of $L$ with an element of $\mathcal{S}$ is either empty or a point, then $B_L$ has exactly $q^t + q^{t-1} + \cdots + q + 1$ points. Moreover, if $L$ intersects an element of $\mathcal{S}$ in a subspace of dimension $t-2$ and $B_L$ is not a line, then the number of points of $B_L$ is $q^t + q^{t-1} + 1$.

Now, suppose that the plane $\Pi = PG(2,q^t)$ is given by a vector space $V$ of rank 3 over $GF(q^t)$. Let $W$ be a $GF(q)$–vector subspace of $V$ of rank $t+1$, and define

$$B_W = \{\langle w \rangle \,:\, w \in W \setminus \{0\}\}.$$

**Theorem 2.1 ([10]).** *$B_W$ is a $GF(q)$-linear blocking set of $PG(2,q^t)$. Moreover, every $GF(q)$-linear blocking set of $PG(2,q^t)$ is equivalent to some $B_W$.*

Note that the subspace $W$ is not uniquely defined, as proven in [10].

Finally, it is possible to obtain linear blocking sets by projecting, in a suitable way, canonical subgeometries of a projective space. Let $\Sigma = PG(t,q)$, $t \geq 3$, be a canonical subgeometry of $\Sigma^* = PG(t,q^t)$. Let $\pi^*$ be a $(t-$

3)−dimensional subspace of $\Sigma^*$ disjoint from $\Sigma$, and let $\pi$ be a plane of $\Sigma^*$ disjoint from $\pi^*$. Define the map $p_{\pi^*,\pi,\Sigma}$ from $\Sigma$ to $\pi$ by $x \mapsto \langle x, \pi^* \rangle \cap \pi$ for each point $x$ of $\Sigma$.

**Theorem 2.2 ([10]).** *$B = p_{\pi^*,\pi,\Sigma}(\Sigma)$ is a $GF(q)-$linear blocking set of $\pi = PG(2, q^t)$.*

Also any $GF(q)-$linear blocking set of $PG(2, q^t)$ can be constructed as a projection of a suitable canonical subgeometry of $PG(t, q^t)$ (see [11]).

Finally, we recall that the $GF(q)-$linear blocking sets of $PG(r-1, q^t)$, $r > 2$, have been studied in [10].

## 3 Blocking Sets of Rédei Type

Let $B$ be a minimal blocking set of Rédei type in the plane $PG(2, s)$. Let $L$ be a Rédei line of $B$ and let $P$ be a point of $L \setminus B$. Suppose that $L$ has equation $X_2 = 0$, that $P$ has coordinates $(0, 1, 0)$, and that $(0, 0, 1) \in B$. Let

$$B \setminus L = \{(a_i, b_i, 1) : i = 1, \ldots, s\}$$

and define the following map

$$f\colon a_i \in GF(s) \longrightarrow b_i \in GF(s).$$

So, $B \setminus L$ is the graph $G_f$ of the map $f$ in the affine plane $AG(2, s) = PG(2, s) \setminus L$ and the points of $L \setminus B$ are all the points determined by the secants of the graph of $f$.

**Theorem 3.1 ([2]).** *Let $B$ be a small minimal blocking set in $PG(2, s)$ ($s = p^n$, $p$ any prime) of Rédei type and let $L$ be a Rédei line of $B$. Then, there exists a positive integer $e$, with $e|n$, such that every line intersects $B$ in a number of points congruent to 1 modulo $p^e$. Moreover, if $p^e > 3$ or $p^e = 3$ and $|L \cap B| = s/3 + 1$, then the map $f$, with $G_f = B \setminus L$, is linear over the subfield $GF(p^e)$.*

In [9] it has been proven that a blocking set defined by a linear function is a linear blocking set. So, by Theorem 3, it follows:

**Theorem 3.2 ([9]).** *In the hypotheses of Theorem 3, if $p^e > 3$ or, $p^e = 3$ and $|L \cap B| = s/3 + 1$, then $B$ is a $GF(p^e)-$linear blocking set.*

Therefore, with few exceptions, small blocking sets of Rédei type of $PG(2, s)$ are linear.

# 4 Blocking Sets of Minimum Size

It has been proven by Bruen in [5] that, if $B$ is a non-trivial blocking set in a projective plane of order $q$, then $|B| \geq q + \sqrt{q} + 1$, and equality holds if and only if $B$ is a Baer subplane. Therefore, in a desarguesian plane of square order, all the blocking sets of minimum size are equivalent.

Let $GF(q) \subset GF(q^t)$, and let $tr(x) = x + x^q + \cdots + x^{q^{t-1}}$ be the trace of $GF(q^t)$ over $GF(q)$. If $(X_0, X_1, X_2)$ are the homogeneous coordinates of a point of $PG(2, q^t)$, then $R = \{(\alpha, tr(\alpha), b) \mid \alpha \in GF(q^t), b \in GF(q)\}$ is a small minimal blocking set of $PG(2, q^t)$ of size $q^t + q^{t-1} + 1$ such that $X_2 = 0$ is a Rédei line. It is easy to prove that all secant lines incident with the point $(1, 0, 0)$ are Rédei lines, i.e. $R$ has at least $q + 1$ Rédei lines. For $t = 2$, $R$ is a Baer subplane of $PG(2, q^2)$. If $GF(q)$ is the largest subfield of $GF(q^t)$, it has been conjectured in [3] that a small minimal blocking set in $PG(2, q^t)$ has at least $q^t + q^{t-1} + 1$ points, and it has exactly this number of points if and only if $B$ is equivalent to $R$.

The proof of the conjecture for $t = 2$ follows from Bruen's characterisation of the Baer subplane as the blocking set of minimum size. Moreover, it has been proven in [16] and [17] that the conjecture is also true for $t = 3$ and $q = p$ any prime. Hence, the first open case is $PG(2, p^5)$. In [12], the authors construct two linear blocking sets of size $q^t + q^{t-1} + 1$ in $PG(2, q^t)$, one of which is not of Rédei type (here $t > 4$) while the other one has exactly one Rédei line (and $t > 3$). If $GF(q)$ is the largest subfield of $GF(q^t)$, their results disprove the above conjecture.

In the sequel of this section, we present some examples of linear blocking sets of size $q^t + q^{t-1} + 1$ constructed in [12].

### Reguli and Blocking Sets in $PG(2, q^5)$

Let $\Sigma^* = PG(V^*, GF(q^5)) = PG(5, q^5)$. Fix a basis $\{e_0, e_1, e_2, e_3, e_4, e_5\}$ of $V^*$ and denote by $(x_0, x_1, x_2, x_3, x_4, x_5)$ the homogeneous coordinates of the point $\langle x_0e_0 + x_1e_1 + x_2e_2 + x_3e_3 + x_4e_4 + x_5e_5 \rangle$ of $\Sigma^*$. Let $\sigma$ be the semilinear map from $V^*$ to itself defined by

$$\sigma : \sum_{i=0}^{5} x_i e_i \mapsto \sum_{i=0}^{5} x_i^q e_i.$$

Then $V = \{v \in V^* \mid v^\sigma = v\}$ is a 6–dimensional $GF(q)$–vector space and $\{e_0, e_1, e_2, e_3, e_4, e_5\}$ is also a basis of $V$ over $GF(q)$. Hence $V^* = GF(q^t) \otimes V$, i.e. $\Sigma = PG(V, GF(q))$ is a canonical subgeometry of $\Sigma^*$. If $\mathbf{x} = (x_0, x_1, x_2)$ and $\mathbf{y} = (x_3, x_4, x_5)$, put $(\mathbf{x}, \mathbf{y}) = (x_0, x_1, x_2, x_3, x_4, x_5)$. Let

$$F(\infty) = \{(\mathbf{0}, \mathbf{y}) \mid \mathbf{y} = (a, b, c); a, b, c \in GF(q)\},$$
$$F(\lambda) = \{(\mathbf{x}, \lambda\mathbf{x}) \mid \mathbf{x} = (a, b, c); a, b, c \in GF(q)\}, \quad \lambda \in GF(q),$$
$$\mathcal{R} = \{F(i) \mid i = \infty \text{ or } i \in GF(q)\}.$$

Then, $\mathcal{R}$ is a regulus of $\Sigma$. Denote by $\mathcal{R}^*$ the regulus of $\Sigma^*$ defined by $\mathcal{R}$, i.e.

$$\mathcal{R}^* = \{F^*(i) \mid i = \infty \text{ or } i \in GF(q^5)\},$$

where

$$F^*(\infty) = \{(\mathbf{0}, \mathbf{y}) \mid \mathbf{y} = (a, b, c); a, b, c \in GF(q^5)\},$$
$$F^*(\lambda) = \{(\mathbf{x}, \lambda\mathbf{x}) \mid \mathbf{x} = (a, b, c); a, b, c \in GF(q^5)\}, \quad \lambda \in GF(q^5).$$

Note that, if $i = \infty$ or $i \in GF(q)$, then $F^*(i) \cap \Sigma = F(i)$, and if $\lambda \in GF(q^5) \setminus GF(q)$, then $F^*(\lambda) \cap \Sigma = \emptyset$.

**Theorem 4.1 ([12]).** *Fix $\lambda \in GF(q^5) \setminus GF(q)$. Let $\pi^* = F^*(\lambda)$ and $\pi = F^*(\infty)$. Then $B = p_{\pi^*,\pi,\Sigma}(\Sigma)$ is a non-Rédei type blocking set of size $q^5 + q^4 + 1$ in $\pi = PG(2, q^5)$.*

### Two Examples of Blocking Sets

If $\pi = PG(2, q^t) = PG(V, GF(q^t))$ and $w_0, w_1, w_2$ is a $GF(q^t)$–basis of $V$, denote by $(X_0, X_1, X_2)$ the homogeneous coordinates of the point $\langle X_0w_0 + X_1w_1 + X_2w_2\rangle$ of $\pi$. Let $\xi \in GF(q^t)$ be a root of an irreducible polynomial $f(x) \in GF(q)[x]$ of degree $t$, so that $\{1, \xi, \xi^2, \ldots, \xi^{t-1}\}$ is a $GF(q)$-basis of $V$. Let $W$ be the set of all vectors of $V$ with coordinates

$$(x_0 + x_1\xi, y_0 + y_1\xi + \cdots + y_{t-3}\xi^{t-3}, z_0),$$

where $x_0, x_1, y_0, y_1, \ldots, y_{t-3}, z_0 \in GF(q)$. It is easy to see that $W$ is a $GF(q)$-vector subspace of $V$ of rank $t + 1$, so it defines a $GF(q)$-linear blocking set of $PG(2, q^t)$.

**Theorem 4.2 ([12]).** *If $t > 3$, then $B_W$ is a $GF(q)$–linear blocking set of size $q^t + q^{t-1} + 1$, and $X_2 = 0$ is the unique Rédei line of $B_W$.*

Now, let $W'$ be the set of all vectors of $V$ with coordinates

$$(x_0 + x_1\xi + \cdots + x_{t-4}\xi^{t-4}, y_0 + y_1\xi, z_0 + z_1\xi)$$

where $x_0, x_1, \ldots, x_{t-4}, y_0, y_1, z_0, z_1 \in GF(q)$.

**Theorem 4.3 ([12]).** *If $t \geq 5$, then $B_{W'}$ is a $GF(q)$–linear blocking set of size $q^t + q^{t-1} + 1$ not of Rédei type.*

## 5 Linear blocking Sets of Maximum Size

A $GF(q)$-linear blocking set $B$ in $PG(2, q^t)$ has at most $q^t + q^{t-1} + \cdots + q + 1$ points. If $B$ has such a number of points and is of Rédei type, then every secant different from the Rédei line intersects it in $q+1$ points. Such a blocking set defines a derivable net of $PG(2,q^t)$ and, if $(t,q-1)=1$, then the derived plane is a generalized André plane ([13]).

Let $\pi = PG(2,q^t) = PG(V,q^t)$ and let $(X_0, X_1, X_2)$ be the homogeneous coordinates of the points of $\pi$. If $f : GF(q^t) \longrightarrow GF(q^t)$ is a $GF(q)$-linear map, then the set $B = \{(x, f(x), a) : x \in GF(q^t), a \in GF(q)\}$ is a $GF(q)$-linear Rédei blocking set of $\pi$ and the line $X_2 = 0$ is a Rédei line (see Section 3). If $f$ is the map $x \longrightarrow x^q$, then the blocking set $\mathcal{B}$ defined by $f$ is a $GF(q)$-linear blocking set of maximum size (see [6]).

Let $\lambda$ be a fixed element of $GF(q^t)$ different from 0, and denote by $N$ the norm function of $GF(q^t)$ over $GF(q)$, i.e. $N(x) = x^{q^{t-1}+\cdots+q+1}$, for $x \in GF(q^t)$. Define

$$B_\lambda = \{(x, \lambda x^q + x^{q^{t-1}}, a) \mid x \in GF(q^t), a \in GF(q)\}.$$

Since $x \longrightarrow \lambda x^q + x^{q^{t-1}}$ is a $GF(q)$-linear map, $B_\lambda$ is a $GF(q)$-linear blocking set of $PG(2,q^t)$ and it is of a Rédei type with respect to the line $X_2 = 0$.

**Theorem 5.1 ([13]).** *If $N(\lambda) \neq 1$, then $B_\lambda$ is a blocking set of size $q^t + q^{t-1} + \cdots + q + 1$. Moreover, if $q > 3$ and $t \geq 4$, then $B_\lambda$ and $\mathcal{B}$ are not equivalent.*

Finally, we describe the first examples of linear blocking sets of maximum size not of Rédei type. Let $\xi$ be, as before, an element of $GF(q^t)$ root of an irreducible polynomial of $GF(q)[x]$ of degree $t$. Let $r$ be a positive integer and let $W_r$ be the set of vectors of $V$ with coordinates $(x_0, x_0^q, x_2)$ where

$$x_0 = a_0 + a_1\xi + \cdots + a_{t-r}\xi^{t-r} + b_0\xi^{t-r+1} + \cdots + b_{r-2}\xi^{t-1}$$

and $x_2 = b_0 + b_1\xi + \cdots + b_{r-1}\xi^{r-1}$, with $a_i, b_j \in GF(q)$ for $i = 0, \ldots, t-r$, $j = 0, \ldots, r-1$. $W_r$ is a $GF(q)$-vector subspace of $V$ of rank $t+1$. Therefore, it defines a $GF(q)$-linear blocking set $B_r$.

**Theorem 5.2 ([17]).** *If $2 \leq r \leq t-2$, then $B_r$ has size $q^t + q^{t-1} + \cdots + q + 1$, and it is not of Rédei type. Moreover, if $2 \leq r, r' < t/2$ with $r \neq r'$, then $B_r$ and $B_{r'}$ are not equivalent.*

## 6 Some Open Problems

The main problem in the theory of small minimal blocking sets in desarguesian planes is to decide whether they are linear or not. The second question is to count the number of non equivalent $GF(q)$-linear blocking sets in $PG(2,q^t)$. This problem is equivalent to calculating the number of the orbits of the group $P\Gamma L(t+1,q)$ on the $(t-3)$-dimensional subspaces of $\Sigma^* = PG(t,q^t)$ disjoint from the canonical subgeometry $\Sigma = PG(t,q)$.

## References


1. Ball, S., Blokhuis, A., Lavrauw, M.: Linear $(q+1)$-fold blocking sets in $PG(2,q^4)$. Finite Fields and their Applications (to appear).
2. Ball, S., Blokhuis, A., Brouwer, A.E., Storme, L., Szőnyi, T.: On the number of slopes of the graph of a function defined on a finite field. J. Comb. Theory (A) **86** (1999) 187–196.
3. Blokhuis, A.: Blocking sets in projective and affine planes. Galois Geometry and Generalized Polygons, Gent 1998.
4. Brouwer, A.E., Wilbrink, H.A.: Blocking sets in translation planes. J. Geom **19** (1982) 200.
5. Bruen,A.A.: Baer subplanes and blocking sets. Bull. Amer. Math. Soc. **76** (1970) 342–344.
6. Bruen, A.: Blocking sets in finite projective planes. Siam J. Appl. Math. Vol. 21 **3** (1971) 380–392.
7. Dembowski, P.: Finite geometries. Springer-Verlag (1968).
8. Lüneburg, H.: Translation planes. Springer-Verlag (1980).
9. Lunardon, G.: Normal spreads. Geom. Dedicata **75** (1999), 245–261.
10. Lunardon, G.: Linear $k$-blocking sets. Submitted.
11. Lunardon, G., Polito, P., Polverino, O.: A geometric characterisation of $k$-linear blocking sets. Preprint.
12. Lunardon, G., Polverino, O.: Blocking sets of size $q^t + q^{t-1} + 1$. J. Comb. Theory (A) (to appear).
13. Lunardon, G., Polverino, O.: Blocking sets and derivable nets. Submitted.
14. Polito, P., Polverino,O.: On small blocking sets. Combinatorica **18** (1998) 133–137.
15. Polito, P., Polverino,O.: Blocking sets in André planes. Geom. Dedicata **75** (1999) 199–207.
16. Polverino, O.: Small blocking sets in $PG(2,p^3)$. Designs, Codes and Cryptography (to appear).
17. Polverino, O.: Blocking set nei piani proiettivi. Ph.D. Thesis, University of Naples "Federico II", (1998).
18. Segre, B.: Teoria di Galois, fibrazioni proiettive e geometrie non desarguesiane. Ann. Mat. Pura e Appl. **64** (1964) 1–76.
19. Szőnyi, T.: Blocking sets in desarguesian affine and projective planes. Finite Fields Appl. **3** (1997) 187–202

# Comparison of Techniques on Divisibility Properties of Exponential Sums and Applications

Oscar Moreno and Francis N. Castro

University of Puerto Rico
Rio Piedras Campus
Mathematics Department
PO Box 23355, SJ
PR 00931-3355

**Abstract** We give an exposition of the present day methods to obtain the divisibility properties of exponential sums. Also we compare the methods and evaluate their performance with respect to the to bound the exponential sums and the minimum distance of cyclic codes. Finally, we give some new results applying the packing method in order to improve the divisibility properties of certain families of exponential sums and cyclic codes.

## 1 Introduction

The object of the present paper is to give an overview of the techniques for obtaining the divisibility properties of exponential sums. We also illustrate the concepts with examples, to make the subject easier to understand. In order to emphasize the importance of exponential sums in the applications, we begin the paper (in section 1) with an exposition of the close relation between coding theory and exponential sums (for further reading about this see [4],[10],[16], [12] ,and [30]).

In section 2 we deal with the classical results by Ax-Katz ( [3], [8]), Moreno-Moreno ( [20] , [21]) and Adolphson-Sperber ( [1]). In section 3 we introduce the method of Moreno-Castro-Cáceres(see [17]) that combines Adolphson-Sperber method with the reduction to the ground field method. In section 4 we review the method of covers ([22])and notice that this method works only in characteristic 2.

In the next three sections we discuss the methods of Moreno et al ( [19]) and we illustrate its power by giving some new results in exponential sums which are of use in coding theory. Finally, we compare (in the last section) the different methods presenting many examples.

We want to point out that divisibility properties of exponential sums are very useful, in number theory, in the sense that the divisibility improvement gives also an improvement on the absolute value of exponential sums (see section 1 and also [10], [18], [22], [23]).

## 2 Coding Theory and a Question of MacWilliams and Sloane

The result of Moreno-Moreno of the next section arose from coding theory applications. In this section we want to point out the close relation between BCH codes and exponential sums in one variable. Also note how the divisibility leads to improved Serre type bounds.

Let $C_f$ be a binary BCH code of length $q-1 = 2^f - 1$ with designed distance $\delta = 2t+1$. Let $C_f^\perp$ be the dual of $C_f$. Estimates for exponential sums obtained by Carlitz-Uchiyama yield that the weight $w$ of any non-zero code word of $C_f^\perp$ satisfies

$$2^{f-1} - (t-1)2^{f/2} \leq w \leq 2^{f-1} + (t-1)2^{f/2}.$$

On the basis of known weight distribution of codes, in Research Problem 9.5, MacWilliams and Sloane (see [11]) have posed the following question:

Let $w$ be the weight of any non-zero code word of $C_f^\perp$. Can the above inequality be strengthened to

$$2^{f-1} - (t-1)2^{[f/2]} \leq w \leq 2^{f-1} + (t-1)2^{[f/2]}\,?$$

(Here $[f/2]$ is the integral part of $f/2$).

The relation between the weight of a code word in $C_f^\perp$ and exponential sums is the following. A code word in $C_f^\perp$ has the form (see [11])

$$c = \left(Tr(F(\alpha))\right)_{\alpha\in\mathbf{F}_q^\times}$$

where $Tr(x) = x + x^2 + \cdots + x^{2^{f-1}}$ for $x \in \mathbf{F}_q$, and where $F(x)$ is a polynomial over $\mathbf{F}_q$ of degree $(2t-1)$ and the weight of $c$ is the number of non-zero components, i.e. the number of $\alpha \in \mathbf{F}_q^\times$ with $Tr(F(\alpha)) = 1$. If we define an exponential sum by

$$S(F) = \sum_{x\in\mathbf{F}_q} (-1)^{Tr(F(x))},$$

then, whenever $F(0) = 0$, clearly

$$S(F) = 1 + \left(\left(q - 1 - w(c)\right) - w(c)\right),$$

that is

$$2w(c) = q - S(F).$$

**Definition 2.1.** For each integer $n$ with binary expansion

$$n = 2^{a_1} + 2^{a_2} + \cdots + 2^{a_\ell},$$

we denote its binary weight by $\sigma(n) = \ell$.

In a series of papers ( [10], [22], [13], [12] and [23]) improved bounds for exponential sums were developed and we will describe some in the following theorems.

**Theorem 2.2.** . *Let $q = 2^f$, and let*

$$F(x) = \sum_j a_j x^{d_j}$$

*be a polynomial with coefficients in $\mathbf{F}_q$ of odd degree d. Let*

$$S(F) = \sum_{x \in \mathbf{F}_q} (-1)^{Tr(F(x))}.$$

*We then have*

$$|S(F)| \leq \left(\frac{d-1}{2}\right)\left[2\sqrt{q}\right],$$

*where $[a]$ is the integer part of a. Furthermore, if*

$$s = \max_j \{\sigma(d_j)\}$$

*is the maximum of the binary weights of the exponents of $F$ and $\mu$ is the smallest positive integer $\geq f/s$, then $2^\mu \mid S(F)$ and*

$$|S(F)| \leq 2^\mu \left[\frac{\left(\frac{d-1}{2}\right)\left[2\sqrt{q}\right]}{2^\mu}\right].$$

**Theorem 2.3.** *Let $F(x_1, \ldots, x_n)$ be a polynomial in n variables with coefficients in $\mathbf{F}_q$, $q = 2^f$. Let*

$$s = \max_{(d_1, \ldots, d_n)} \{\sigma(d_1) + \cdots + \sigma(d_n)\},$$

*where the maximum is taken over the degrees of all the monomials in F. We then have that the exponential sum*

$$S(F) = \sum_{x_1, \ldots, x_n \in \mathbf{F}_q} (-1)^{Tr(F(x_1, \ldots, x_n))}$$

*is divisible by $2^\mu$, where $\mu = \lceil nf/s \rceil$ is the smallest integer $\geq fn/s$.*

For the proof of the above theorems see [10].

In [22], Moreno-Moreno improved Theorem 1 for polynomials of degree $\leq 9$. Now, we state their improvement.

**Theorem 2.4.** *Let $q = 2^{3f}$, $f$ odd and let $F(x)$ be a polynomial, with coefficients in $\mathbf{F}_q$, of degree r, with r equal to 7 or 9. We then have*

$$\left|\sum_{x \in \mathbf{F}_q} (-1)^{Tr(F(x))}\right| \leq (r-1) \cdot 2^{f-1}\left[2^{1-f}\sqrt{q}\right].$$

*Remark 2.5.* The inequality in the above theorem is tight for $q = 2^9$ and $q = 2^{39}$ (see [4], [25]).

In [23], Moreno-Moreno also improved Theorem 1 and furthermore generalized Theorem 3 to any polynomial. Their improvement is the following:

**Theorem 2.6.** *Let $q = 2^f$, and let*

$$F(x) = \sum_j a_j x^{d_j}$$

*be a polynomial with coefficients in $\mathbf{F}_q$. Suppose the maximum binary weight of the exponents is*

$$s = \max_j\{\sigma(d_j)\}.$$

*Let $\mu = [f/s]$. We then have*

$$\left|\sum_{x\in\mathbf{F}_q}(-1)^{Tr(F(x))}\right| \le \frac{(\deg F - 1)}{2}2^{\mu}\left[2^{1-\mu}\sqrt{q}\right].$$

Even though the question of MacWilliams and Sloane led to the just mentioned result, unfortunaly the full question is too ambitious, in fact the following theorem gave a doubly infinite family of counter examples to the MacWilliams and Sloane question(see [22]).

**Theorem 2.7.** *For each prime $p$ for which 2 is of order odd exactly $\frac{p-1}{2}$, let $q = 2^{((p-1)/2)f}$. Let $\epsilon > 0$. Then for infinitely many odd $f$, as well as for infinitely many even $f$, we have*

$$\left|\sum_{x\in\mathbf{F}_q}(-1)^{Tr(x^p)}\right| \ge (p-1)\sqrt{q}\,(1-\epsilon).$$

*Remark 2.8.* Papers [15] and [24] independently were the first to find counter examples to Problem 9.5. Rodier in [25] was the first to find an infinite family, the one corresponding to the single prime $p = 7$.

## 3 Results of Ax-Katz, Moreno-Moreno and Adolphson-Sperber

In 1935, Chevalley proved a conjectured by E. Artin: If $F(X_1, \ldots, X_n)$ is a homogeneous polynomial of total degree $d$ over a finite field $\mathbf{F}_q$ having $q = p^f$ elements and $n > d$, then $F$ has a nontrivial zero. C. Chevalley showed that it is true even with the homogeneity hypothesis replaced by a weaker assumption of no constant term. Immediately afterwards, E. Warning showed that the characteristic of the field divides the number of zeros without the last assumption. A proof of the Chevalley-Warning theorem can be found in J. Ax's paper [3]. In the same paper, Ax following Dwork's ideas [6] developed a $p$-adic technique in which previously unseen solutions became visible in characteristic zero. He proved that if $b$ is equal to $\lceil n/d \rceil - 1$, where $\lceil a \rceil$ is the smallest integer larger or equal to $a$, then the number of zeros of $F$ is divisible by $q^b$. Using the principle of inclusion and exclusion, Ax's result is extended to a system of polynomials: Let $F_i(X_1, \ldots, X_n)$, $(i = 1, \ldots, r)$ be polynomials over $\mathbf{F}_q$ of degree $d_i$ respectively, and $\lambda$ be

$$\lambda = \left\lceil \frac{n - \sum_{i=1}^{r} d_i}{\sum_{i=1}^{r} d_i} \right\rceil .$$

Then the number of zeros of the system of polynomials is divisible by $q^\lambda$.

In 1971, N. Katz( [8]) improved Ax's theorem as follows:

**Theorem 3.1.** *Let $F_1, \ldots, F_t$ be polynomials in $n$ variables with coefficients in $F_q$ of total degrees $d_1, \ldots, d_t$. Let $\mu$ be the least integer that satisfies*

$$\mu \geq \frac{n - \sum_{i=1}^{t} d_i}{max_i\, d_i} .$$

*Then $q^\mu$ divides $N(F_1, \ldots, F_t)$, where $N(F_1, \ldots, F_t)$ is the number of simultaneous solutions of $F_1 = 0, \ldots, F_t = 0$.*

D. Wan gave a simpler proof of Katz's Theorem in [29].

Let $p$ be a prime number. For any integer $a$, define $\operatorname{ord}_p(a)$ as follows:

$$\operatorname{ord}_p(a) = \max\{\, k \mid p^k \text{ divides } a\} .$$

*Example 3.2.* Let $I_1, \ldots, I_m$ be mutually disjoint subsets of $\{1, \ldots, n\}$. Let $F(x_1, \ldots, x_n) = \sum_{j=1}^{m} \prod_{i \in I_j} x_i$ be a polynomial of degree $d$ over $\mathbf{F}_q$. Ax's theorem implies that $\operatorname{ord}_q(N(F)) \geq \frac{n-d}{d}$. If the sets $I_j$'s have same cardinality and $d$ divides $n$, Ax's theorem is tight, i.e., $\operatorname{ord}_q(N(F)) = \frac{n-d}{d}$.

Moreno and Moreno gave in [20] an improvement to the Ax-Katz's theorem. Before we state the Moreno and Moreno result, we need to give a definition.

**Definition 3.3.** For each integer $n$ with $p$-expansion

$$n = a_0 + a_1 p \cdots + a_s p^s \text{ where } 0 \leq a_i < p ,$$

we denote its $p$-weight by $\sigma_p(n) = \sum_{i=0}^{s} a_i$. The $p$-weight degree of a monomial $x^d = x_1^{d_1} \cdots x_n^{d_n}$ is $w_p(x^d) = \sigma_p(d_1) + \cdots + \sigma_p(d_n)$. The $p$-weight degree of a polynomial $F(x_1, \ldots, x_n) = \sum_d a_d x^d$ is $w_p(F) = max_{x^d,\, a_d \neq 0} w_p(x^d)$

The Moreno-Moreno's result is the following:

**Theorem 3.4.** *Let $F_1, \dots, F_t$ be polynomials in $n$ variables with coefficients in $\mathbf{F}_q$, a finite field with $q = p^f$ elements. Let $w_p(F_i)$ be the $p$-weight degree of $F_i$ and let $\mu$ be the smallest integer such that*

$$\mu \geq f\left(\frac{n - \sum_{i=1}^{t} w_p(F_i)}{max_i\, w_p(F_i)}\right).$$

*Then $p^\mu$ divides $N(F_1, \dots, F_t)$.*

*Remark 3.5.* Theorem 7 improves Theorem 6 whenever the characteristic is small with comparison to the degrees, i.e., we need, say $p < \max_i d_i$ in order for an improvement to occur.

Now, combining the Ax-Katz's and Moreno-Moreno's results we get the following corollary.

**Corollary 3.6.** *Let $F_1, \dots, F_t$ be polynomials in $n$ variables with coefficients in $\mathbf{F}_q$ of total degrees $d_1, \dots, d_t$. Let $\mu$ be the least integer that satisfies*

$$\mu \geq \max\{f\lceil\frac{n - \sum_{i=1}^{t} d_i}{max_i\, d_i}\rceil, \lceil\frac{f(n - \sum_{i=1}^{t} w_p(F_i))}{max_i\, w_p(F_i)}\rceil\}.$$

*Then $p^\mu$ divides $N(F_1, \dots, F_t)$.*

Let $\psi$ be an additive character of $\mathbf{F}_q$, then the exponential sums associated to a polynomial $F$ is defined by

$$S(F) = \sum_{X_1, \dots, X_n \in \mathbf{F}_q} \psi(F(X_1, \dots, X_n)).$$

Now we state the relation of exponential sums and solutions of polynomial equations.

$$N(F_1, \dots, F_t) = \frac{1}{q^t} \sum_{\substack{X_1, \dots, X_n \in \mathbf{F}_q \\ Y_1, \dots, Y_t \in \mathbf{F}_q}} \psi(\sum_{i=1}^{t} Y_i F_i).$$

Hence $\mathrm{ord}_p(N(F_1, \dots, F_t)) = \mathrm{ord}_p(S(\sum_{i=1}^{t} Y_i F_i(X_1, \dots, X_n))) - tf$.

Adolphson and Sperber in [1] gave a lower bound of the $p$-divisibility of exponential sums. Their result is an improvement of the Ax-Katz's theorem. They use the theory of Newton Polyhedra in the proof of their theorem.

Let $F(X) = F(X_1, \dots, X_n)$ be a polynomial over $\mathbf{F}_q$ and let $D$ be the set of all the integral vectors $d$ corresponding to the monomials, where $aX^d = a_{d_1, \dots, d_n} X_1^{d_1} \cdots X_n^{d_n}$ is a monomial of $F$. The Newton polyhedron $\Delta(F)$ is defined to be convex hull in $\mathbf{R}^n$ of the set $D \bigcup \{(0, \dots, 0)\}$. Let $\omega(F)$ be the smallest positive rational number such that $\omega(F)\Delta(F)$ contains at least one point with positive integral coordinates. Now, we state (see [1] and [2]) the Adolphson and Sperber's theorem:

**Theorem 3.7.** *With the above notation and assumptions, we have*

$$ord_q(S(F)) \geq \omega(F).$$

Let us illustrate Theorem 8, computing $\omega(F)$ for a polynomial. We will use this in example 4 to give an instance where Theorem 9 improves Theorem 8.

*Example 3.8.* Let $F(X,Y,Z,V) = V(X^4 + Y^4 + Z^4 + XY + XZ)$ be a polynomial over $\mathbf{F}_q$. The convex hull associated to $F$ is the convex set generated by the following points: $v_1 = (4,0,0,1), v_2 = (0,4,0,1), v_3 = (0,0,4,1), v_4 = (1,1,0,1), v_5 = (1,0,1,1), v_6 = (0,0,0,0)$. Then $\omega(F) = 1$, i.e., $\frac{1}{8}v_1 + \frac{1}{8}v_2 + \frac{1}{4}v_3 + \frac{1}{2}v_4 = (1,1,1,1)$. We do not get any information about $N(X^4 + Y^4 + Z^4 + XY + XZ)$.

In the following two examples we illustrate Theorems 7 and 8.

*Example 3.9.* Let $F_1(x_1, \ldots, x_6) = x_1 + \cdots + x_5 + \beta_1 x_6$, $F_2(x_1, \ldots, x_6) = x_1^3 + \cdots + x_5^3 + \beta_2 x_6^3$ and $F_3(x_1, \ldots, x_6) = x_1^5 + \cdots + x_5^5 + \beta_3 x_6^5$ be polynomials over $\mathbf{F}_{2^f}$. Applying Theorem 8, we obtain that $\omega(\sum_{i=1}^3 y_i F_i) = 3$. Hence we do not get any information about $N(F_1, F_2, F_3)$. Applying Theorem 7, we get that $2^{\lceil f/2 \rceil}$ divides $N(F_1, F_2, F_3)$.

*Example 3.10.* Let $F_1(x_1, x_2) = x_1^3 + x_2^5 + x_1 x_2$ be polynomials over $\mathbf{F}_q$. Applying Theorem 8, we obtain that $\omega(F) = 1$. Hence $q$ divides $S(F)$. Applying Theorem 7, we get

$$\text{ord}_p(S(F)) \geq \begin{cases} f \text{ if } p = 2 \\ \frac{2f}{3} \text{ if } p = 2 \text{ or } 5 \\ \frac{2f}{5} \text{ otherwise.} \end{cases}$$

Consequently in this case Theorem 8 gives a better result than Theorem 7, whenever $p \neq 2$.

## 4 Improvement to the Adolphson and Sperber's Theorem

In this section following [17], we will combine the reduction to the ground field method and the Newton Polyhedra method of Adolphson and Sperber to give an improvement to the Moreno-Moreno's result and Adolphson-Sperber's result.

In the proof of theorem 7, Moreno and Moreno used the reduction to ground field method. It consists in transforming a polynomial in $n$ variables over $\mathbf{F}_{p^f}$ to a polynomial in $fn$ variables over $\mathbf{F}_p$ by applying the trace map. We are going to illustrate the reduction to the ground field method for a polynomial over a finite field of characteristic 2. Given a monomial $X_1^{d_1}\cdots X_n^{d_n}$ over $\mathbf{F}_{2^f}$, we can choose a $\mathbf{F}_2$-basis $\alpha_1,\ldots,\alpha_f$ of $\mathbf{F}_{2^f}$ such that

$$X_j=\sum_{i=1}^{f} x_{ji}\alpha_i \text{ for } j=1,\ldots,n\,.$$

If we substitute the above in $X_1^{d_1}\cdots X_n^{d_n}$, we get

$$\begin{aligned}X_1^{d_1}\cdots X_n^{d_n} &= (\textstyle\sum_{i=1}^f x_{1i}\alpha_i)^{d_1}\cdots(\sum_{i=1}^f x_{ni}\alpha_i)^{d_n}\\ &= (\textstyle\sum_{i=1}^f x_{1i}\alpha_i)^{2^{j_{11}}}(\sum_{i=1}^f x_{1i}\alpha_i)^{2^{j_{12}}}\cdots\\ &= (\textstyle\sum_{i=1}^f x_{1i}\alpha_i^{2^{j_{11}}})(\sum_{i=1}^f x_{ji}\alpha_i^{2^{j_{21}}})\cdots \end{aligned}\tag{1}$$

where $d_s=2^{j_{s1}}+2^{j_{s2}}+\cdots$. Applying to (1) the trace map from $\mathbf{F}_{2^f}$ to $\mathbf{F}_2$, we get a polynomial in $nf$ variables over $\mathbf{F}_2$. We can transform any polynomial over $\mathbf{F}_{p^f}$ by using the ground field method to a polynomial over $\mathbf{F}_p$ since any polynomial is a sum of monomials and the trace map is linear. Now we give an example of the reduction to the ground field method.

*Example 4.1.* Let $F(X,Y,X,V)=V(X^{p+1}+Y^{p+1}+Z^{p+1}+XY+XZ)$ over $\mathbf{F}_{p^2}$. Let $1,\alpha$ be a $\mathbf{F}_p$-basis of $\mathbf{F}_{p^2}$. We assume that $\alpha^2=-1$ and $Tr(\alpha)=0$. Note that this is true for $p=3,7,11$. We have the following:

$$Tr_{\mathbf{F}_{p^2}/\mathbf{F}_p}(VX^{p+1})=2v_1x_1^2+2v_1x_2^2$$

$$Tr_{\mathbf{F}_{p^2}/\mathbf{F}_p}(VY^{p+1})=2v_1y_1^2+2v_1y_2^2$$

$$Tr_{\mathbf{F}_{p^2}/\mathbf{F}_p}(VZ^{p+1})=2v_1z_1^2+2v_1z_2^2$$

$$Tr_{\mathbf{F}_{p^2}/\mathbf{F}_p}(VXY)=2(v_1x_1y_1-v_1x_2y_2-v_2x_1y_2-v_2y_1x_2)$$

$$Tr_{\mathbf{F}_{p^2}/\mathbf{F}_p}(VXZ)=2(v_1x_1z_1-v_1x_2z_2-v_2x_1z_2-v_2x_2z_1)\,.$$

The polynomial $F'$ corresponding to $F$ over $\mathbf{F}_p$ is the following.

$$\begin{aligned}&F'(x_1,x_2,y_1,y_2,z_1,z_2,w_1,w_2,v_1,v_2)\\ &=Tr_{\mathbf{F}_{p^2}/\mathbf{F}_p}(F(X,Y,Z,V))\\ &=2(v_1x_1^2+v_1x_2^2+v_1y_1^2+v_1y_2^2+v_1z_1^2+v_1z_2^2+v_1x_1y_1\\ &\qquad -v_1x_2y_2-v_2x_1y_2-v_2y_1x_2-v_1x_1z_1-v_1x_2z_2-v_2x_1z_2-v_2x_2z_1).\end{aligned}$$

Let $F'$ be the polynomial corresponding to $F$ after we apply the reduction to ground field method to $F$. Let $\psi(x)=e^{2\pi\imath Tr(x)/p}$, then

$$\begin{aligned}S(F)&=\sum_{(X_1,\ldots,X_n)\in\mathbf{F}_{p^f}^n}\\ &=e^{2\pi\imath Tr(F(X_1,\ldots,X_n))/p}\sum_{(x_1,\ldots,x_{nf})\in\mathbf{F}_p^{nf}} e^{2\pi\imath F'(x_1,\ldots,x_{nf})/p}=S(F')\,.\end{aligned}$$

The above implies that any estimate of $S(F')$ gives an estimate of $S(F)$.

Let $\Delta(F')$ be the Newton polyhedron associated to $F'$. Recall that $\Delta(F')$ is a convex subset of $\mathbf{R}^{fn}$. Let $\omega(F')$ be the smallest positive rational number such that $\omega(F')\Delta(F')$ contains at least one point with positive integral coordinates.

*Example 4.2.* Let $F(X,Y,Z,V)$ be the polynomial defined in example 5. Note that $\Delta(F')$ is a convex subset of $\mathbf{R}^8$. The Newton polyhedron of $F'$ over $\mathbf{F}_p$ is the convex hull of the following set of points:

$$\{ (2,0,0,0,,0,0,1,0),(0,2,0,0,0,0,0,1,0),(0,0,2,0,0,0,1,0),$$
$$(0,0,0,2,0,0,1,0),(0,0,0,0,2,0,1,0),(0,0,0,0,0,2,1,0),$$
$$(1,0,1,0,0,0,1,0),(0,1,0,1,0,0,1,0),(1,0,0,1,0,0,0,1)$$
$$(0,1,1,0,0,0,0,1),(1,0,0,0,0,1,0,1),(0,1,0,0,1,0,0,1)$$
$$(1,0,0,0,1,0,1,0),(0,1,0,0,0,1,1,0),(0,\dots,0)\}.$$

In this case we have that $\omega(F') = 3$, i.e.,

$$(\tfrac{1}{3},\tfrac{1}{3},\tfrac{1}{3},\tfrac{1}{3},\tfrac{1}{3},\tfrac{1}{3},\tfrac{1}{3},\tfrac{1}{3}) = \tfrac{1}{6}(0,0,2,0,0,0,1,0)+\tfrac{1}{6}(0,0,0,2,0,0,1,0)+$$
$$\tfrac{1}{3}(1,0,0,0,0,1,0,1)+\tfrac{1}{3}(0,1,0,0,1,0,0,1).$$

The next result is a combination of Theorem 8 of [17] and Theorem 8.

**Theorem 4.3.** *Let $F_1,\dots,F_t$ be polynomials in $n$ variables with coefficients in $F_q$, a finite field with $q=p^f$ elements. Let $(y_iF_i)'$ be the polynomial coresponding to $y_iF_i$ after we apply to it the reduction to the ground field method. Let*

$$\omega = max\{f\omega(\sum_{i=1}^{t} y_iF_i),\omega(\sum_{i=1}^{t}(y_iF_i)')\},$$

*then*

$$ord_p(N(F_1,\dots,F_t)) \geq \omega - tf.$$

*Remark 4.4.* We want to point out that we have two Newton Polyhedra, $\Delta(F)$ and $\Delta(F')$ associated to $S$, where $S=S(F)=S(F')$. $\Delta(\sum_i^t(y_iF_i)')$ is the Newton polyhedron associated to $\sum_i^t(y_iF_i)'$, where $\sum_i^t(y_iF_i)'$ is the polynomial over $\mathbf{F}_p$ corresponding to $\sum_i^t y_iF_i$. Also, $\omega(\sum_i^t(y_iF_i)')$ is the smallest positive rational number such that $\omega(\sum_i^t(y_iF_i)')\Delta(\sum_i^t(y_iF_i)')$ contains at least one point with positive integral coordinates.

*Proof.* We apply Theorem 8 to the Newton polyhedron $\Delta(\sum_{i=1}^t(y_iF_i)')$ and $\Delta(\sum_{i=1}^t y_iF_i)$. □

As a corollary (this is Corollary 1 of section 2) we obtain Theorem 1.1 of [23].

**Corollary 4.5.** *Let $F_1, \dots, F_t$ be polynomials in $n$ variables with coefficients in $F_q$ ( $q = p^f$ ) with total degrees $d_1, \dots, d_t$. Let*

$$\mu = max\{f\lceil\frac{n - \sum_{i=1}^t d_i}{\max_i d_i}\rceil, \lceil\frac{f(n - \sum_i^t w_p(F_i))}{\max_i w_p(F_i)}\rceil\},$$

*where $\lceil a \rceil$ denotes the the least integer $\geq a$. Then*

$$ord_p(N(F_1, \dots, F_t)) \geq \mu .$$

*Proof.* In [1, section 5], it was proved that $\omega(\sum_{i=1}^t y_i F_i) - t \geq \lceil\frac{n-\sum_{i=1}^t d_i}{\max_i d_i}\rceil$. Substituting $F$ by $F'$ and $\mathbf{F}_q$ by $\mathbf{F}_p$ in the proof of Adolphson and Sperber mentioned in the last sentence, we prove that $\omega(\sum_{i=1}^t (y_i F)') - tf \geq \lceil\frac{f(n-\sum_i^t w_p(F_i))}{\max_i w_p(F_i)}\rceil$. This completes the proof of the corollary since $\mu = \max\{f\omega(\sum_{i=1}^t y_i F_i), \omega(\sum_{i=1}^t (y_i F_i)')\}$. □

*Example 4.6.* Let $F$ be the polynomial defined in examples 5 and 6. Applying Theorem 9 to this polynomial, we obtain that $p$ divides $N(X^{p+1} + Y^{p+1} + Z^{p+1} + XY + XZ)$. Note that when $p = 3$, $F$ is the polynomial defined in example 2 and that we improve upon Theorem 8.

*Example 4.7.* Let $\epsilon_i, \beta_i \in \mathbf{F}_{2^f}$. for $i = 1, \dots, 5$. Let $F(X_1, \dots, X_5) = \epsilon_1 X_1^7 + \epsilon_2 X_2^7 + \epsilon_3 X_3^7 + \epsilon_4 X_4^7 + \epsilon_5 X_5^3 + \beta_1 X_1^5 + \beta_2 X_2^5 + \beta_3 X_3^5 + \beta_4 X_4^5 + \beta_5 X_5^5 + G(X_1, \dots, X_5)$ over $\mathbf{F}_{2^f}$, where $G(X_1, \dots, X_5)$ is a polynomial of 2-weight degree $\leq 2$. We have $\omega((yF)') > \frac{f(5-3)}{3} + f = \frac{5f}{3}$ and $\omega(yF) = 1$. Applying Theorem 9 ($\omega = \max\{\frac{5f}{3}, f\}$) we have that $\mathrm{ord}_2(N(F)) > \frac{2f}{3}$. Note that in this case the new constant given by Theorem 9 is better than the one given by Theorem 8.

*Remark 4.8.* Given polynomials $F_1, \dots, F_t$ over $\mathbf{F}_q$, in the method of Adolphson and Sperber, the Newton polyhedron $\Delta(\sum_{i=1}^t y_i F_i)$ does not change as we vary the characteristic of the field or as we extend the field $\mathbf{F}_q$. On the other hand, the method we introduced, the Newton polyhedron $\Delta(\sum_{i=1}^t (y_i F_i)')$ ($\sum_{i=1}^t (y_i F_i)'$ is the polynomial associated to $\sum_{i=1}^t y_i F_i$ over $\mathbf{F}_p$) changes with the characteristic of the field (Note that the $p$-weight of a polynomial changes with the characteristic of the field). This implies that the constant given by the result of Adolphson and Sperber does not change if it changes $\mathbf{F}_q$ and the new constant given by Theorem 9 changes if it changes $\mathbf{F}_q$ . Probably, the constant $\omega(F')$ also changes when we extend the field $\mathbf{F}_q$ . We can think of $\Delta(F)$ as the 'global' Newton polyhedron associated to number of zeros of $F$ and $\Delta(F')$ as the local Newton polyhedron associated to number of zeros of $F$.

# 5 On 2-Divisibility of Exponential Sums; Method of Coverings

In this section we review the method of covers introduced in [22]. Let $\mathbf{F}_{2^f}$ be the binary field and $F(X) = F(X_1, \ldots, X_n)$ be a polynomial in $n$ variables over $\mathbf{F}_{2^f}$. By using the reduction to ground field method of the previous section we reduce a polynomial $F$ to another polynomial $F'$ over the binary field. In other words, we will write $F'(x) = F'(x_1, \ldots, x_{nf})$ as sum of monomials of the form $a(d)x^d$ where $d = (d_1, \ldots, d_n)$ is a vector of nonnegative integers such that $x^d = x_1^{d_1} \cdots x_n^{d_n}$ $(d_i, a(d_i) \in \{0, 1\})$. Let $\mathcal{C}$ be a minimal set of the monomials of $F$ covering all variables, that is, every variable $x_i$ is in at least one monomial in $\mathcal{C}$ and $\mathcal{C}$ is minimal with that property. We call this set $\mathcal{C}$ a *minimal covering of the variables of* $F$ and we assume that its cardinality is $r$.

The following lemma (see [15] and [22]) establishes divisibility properties of an exponential sum which will determine the divisibility of the number of zeros of a polynomial equation. Lemma 1 is an improvement to the Ax's theorem ([22]), as we will illustrate in a simple example below.

**Lemma 5.1.** *Let $F(x) = F(x_1, \ldots, x_n)$ be a polynomial in $n$ variables over $\mathbf{F}_2$. Let $\mathcal{C}$ be a minimal set of monomials of $F(x)$ covering all the variables and $r$ be the cardinality of $\mathcal{C}$. Then $2^r$ divides $S(F)$.*

*Proof.* Throughout the proof, we will use the identities $(-1)^{a(d)x^d} = 1 - 2a(d)x^d$ and $x_i^l = x_i$ for $l > 0$, which hold only on binary variables. Therefore, if $F(x_1, \ldots, x_n) = \sum_{d \in D} a(d)x^d$ then

$$S(F) = \sum_{x_1, \ldots, x_n \in \mathbf{F}_2^n} \prod_{d \in D} (1 - 2a(d)x^d) .$$

If we expand this equation, we get

$$\begin{aligned} S(F) &= \sum_{x_1, \ldots, x_n} (1 + \sum_{\lambda} 2^{n(\lambda)} g_\lambda(x_1, \ldots, x_n)) \\ &= 2^n + \sum_{\lambda} 2^{n(\lambda)} \sum_{(x_1, \ldots, x_n) \in \mathbf{F}_2^n} g_\lambda(x_1, \ldots, x_n) \qquad (2) \end{aligned}$$

where the $g_\lambda$'s are monomials.

We claim that the minimum power of 2 that divides

$$2^{n(\lambda)} \sum_{(x_1, \ldots, x_n) \in \mathbf{F}_2^n} g_\lambda(x_1, \ldots, x_n) ,$$

for all the $\lambda$ in (2), happens when $n(\lambda) = r$ and $g_\lambda(x_1, \ldots, x_n) = \prod_{x^d \in \mathcal{C}} x^d$. Recall that if $g_\lambda(x_1, \ldots, x_n)$ lacks $l$ variables, then

$$2^{n(\lambda)} \sum_{(x_1,\dots,x_n)\in \mathbf{F}_2^n} g_\lambda(x_1,\dots,x_n) = 2^{n(\lambda)+l}.$$

We assume that there is a term $2^{n(\lambda)} \sum_{(x_1,\dots,x_n)\in \mathbf{F}_2^n} g_\lambda(x_1,\dots,x_n) = 2^a$ where $a < r$. Hence $a = n(\lambda) + l$, where $l$ is the number of variables that does not appear in $g_\lambda$. This is a contradiction to the minimality of the set $C$. That ends the proof. □

**Corollary 5.2.** *Let $F(x) = F(x_1,\dots,x_n)$ be a polynomial in $n$ variables over $\mathbf{F}_2$. Let $C$ be a minimal set of monomials of $F(x)$ covering all the variables and $r$ be the cardinality of $C$. Then $2^{r-1}$ divides $N(F)$.*

*Example 5.3.* Let $F(x_1,\dots,x_n) = x_1x_2x_3 + x_3x_4 + x_4x_5 + x_5x_6 + \cdots + x_{n-1}x_n + x_1x_2x_n$ be a polynomial over $\mathbf{F}_2$. Applying corollary 3, we obtain that $2^{[n/2]-1}$ divides $N(F)$ since the cardinality of a minimal covering $C$ of the variables of $F$ is $[n/2]$. Applying Ax's theorem, we obtain that $2^{\lceil n/3\rceil-1}$ divides $N(F)$.

*Remark 5.4.* Lemma 1 has applications to the number of zeros of a system of polynomial equations over $\mathbf{F}_2$ . Let $F_1,\dots,F_t$ be a system of polynomial equations over $\mathbf{F}_2$ and let $r$ be minimal number of monomials of $F_1,\dots,F_t$ that covers $x_1,\dots,x_n$. Then $2^{r-t}$ divides $N(F_1,\dots,F_t)$, where $N(F_1,\dots,F_t)$ is number of simultaneous solutions of $F_1,\dots,F_t$.

The following example illustrate how Lemma 1 can be applied to estimate the number of solutions of a system of polynomial equations.

*Example 5.5.* Let $F_1(x_1,\dots,x_n) = x_1x_2 + x_2x_3 + \cdots + x_{n-1}x_n + x_1x_n$ and $F_2(x_1,\dots,x_n) = x_1x_2x_3 + x_2x_3x_4 + \cdots + x_{n-2}x_{n-1}x_n + x_{n-1}x_nx_1$ over $\mathbf{F}_2$. In this case $r = [n/2]$, hence $2^{[n/2]-2}$ divides $N(F_1,F_2)$. The Ax-Katz's and Moreno-Moreno's theorems (see theorems 6 and 7) imply that

$$\mathrm{ord}_2(N(F_1,F_2)) \geq \frac{n-5}{3}.$$

Note that $[n/2] - 2 > \frac{n-5}{3}$ for $n > 8$.

## 6 Method of Modular Equations

First some notation and relevant concepts. Throughout this paper we use mostly finite extensions of the binary field $\mathbf{F}_2$ and for these we define exponential sums whose divisibility properties are of interest in coding theory. The proper setting for a precise discussion of these questions is the theory of $p$-adic fields and their finite algebraic extensions. To facilitate the understanding of the statements in the theorems we give a brief description of the basic ideas about the field of dyadic numbers $\mathbb{Q}_2$.

Given a nonzero rational number $r = \frac{a}{b}$ we can write it in the form $r = 2^e \cdot \frac{a'}{b'}$, where the numerator $a'$ and the denominator $b'$ are odd. Using this representation we define the dyadic absolute value by

$$|r|_2 = \frac{1}{2^e},$$

and postulate that $|0|_2 = 0$. It is also useful to consider the exponential valuation defined by $\mathrm{ord}_2(r) = e$. The pair $(\mathbb{Q}, |\,|_2)$ consisting of the rational numbers together with the dyadic absolute value form a metric space and its completion $\mathbb{Q}_2$ is the dyadic analogue of the real numbers $\mathbf{R}$ after completing the rationals with respect to the ordinary absolute value. The subset of elements in $\mathbb{Q}_2$ which satisfy the inequality $|\alpha|_2 \leq 1$ form the so-called dyadic integers, denoted by $\mathbf{Z}_2$; $\mathbb{Q}_2$ is in fact the quotient field of $\mathbf{Z}_2$. The dyadic integers is a local ring whose maximal ideal is the set of all multiples of 2, i.e., $2 \cdot \mathbf{Z}_2 = (2)$. Divisibility properties of numbers in $\mathbb{Q}_2$ are measured in terms of this ideal, more precisely, to say that a number $s$ in $\mathbb{Q}_2$ is divisible by $2^f$, means that in terms of the ideal generated by $s$, we have

$$(s) = (2)^f.$$

An even more concrete way of viewing the concept of divisibility is to observe that any element in $\mathbb{Q}_2$ can be written as a Laurent series in (the local uniformizing parameter) 2:

$$\mathbb{Q}_2 = \left\{ \alpha = \sum_{j=-e}^{\infty} a_j 2^j \;\middle|\; a_j \in \{0,1\} \right\}.$$

Of course, in such a representation, the integer $e$ depends on $\alpha$.

If $q = 2^f$, we obtain a copy of the finite set $\mathbf{F}_q$ by solving the equation $t^q - t = 0$ in $\mathbb{Q}_2$ (and we will denote this set as $T = \{0, 1, t, \ldots, t^{q-2}\}$). Aside from $t = 0$, the other non-zero roots satisfy $t^{q-1} - 1 = 0$ and these give rise to an algebraic extension $\mathbf{K} = \mathbb{Q}_2(t)$ obtained from $\mathbb{Q}_2$ by adjoining a primitive root $t$. In the following we shall use the set $T \setminus \{0\}$ interchangeably as an (isomorphic) image of the multiplicative group of $\mathbf{F}_q \setminus \{0\}$ inside of the multiplicative group of the field $\mathbf{K} \setminus \{0\}$. This field has an absolute value which is an extension of $|\,|_2$. The elements in $\mathbf{K}$ with absolute value $\leq 1$ are those in $\mathbf{Z}_2[t]$, which is again a local ring.

An element $x \in \mathbf{F}_q$ has a unique representative $t_x$ in $T$ (and they are called Teichmuller representatives) and the trace taken in $\mathbf{F}_q$

$$Tr(x) = x + x^2 + \cdots + x^{2^{f-1}}$$

can be compared with the trace taken in $\mathbf{K}$

$$Tr(t_x) = t_x + t_x^2 + \cdots + t_x^{2^{f-1}}$$

via the relation

$$Tr(x) \equiv Tr(t_x) \pmod{2\mathbf{Z}_2}.$$

The map $x \mapsto (-1)^{Tr(x)} = (-1)^{Tr(t_x)}$ defines a non-trivial additive character of $\mathbf{F}_q$ with values in $\mathbf{Z}_2$. Similarly, any multiplicative character of $\mathbf{F}_q^\times$ with values in $\mathbf{K}$ is given by the map

$$x \mapsto \chi(x) = t_x^{-a},$$

for some integer $a$, $0 \le a < q-1$.

**Definition 6.1.** To each integer $j, 0 \le j \le q-1$, there corresponds a Gauss sum

$$g(j) = \sum_{t \in T^\times} t^{-j}(-1)^{Tr(t)}.$$

We will use frequently the following formula:

$$(-1)^{Tr(t)} = \begin{cases} \frac{1}{q-1}\sum_{1\le j\le q-1} g(j)t^j & \text{if } t \neq 0 \\ 1 & \text{if } t = 0. \end{cases}$$

(see [3, p. 257]). If we define coefficients $C(j)$ by $C(j)(q-1) = g(j)$ for $0 < j < q-1$ and $C(0) = 1$, $C(q-1)(q-1) = -q$, then we can write

$$(-1)^{Tr(t)} = \sum_{0\le j\le q-1} C(j)t^j. \tag{3}$$

The main tool for the study of congruences of exponential sums is the following result.

**Theorem 6.2 (Stickelberger).** *With notation as above we have*

$$C(j) \equiv 0 \pmod{2^{\sigma(j)}}, Quad0 \le j \le q-1.$$

First we will deal with polynomials in one variable. In section 7 we will handle the general case.

Now we are going to show how to associate a modular equation to an exponetial sum in one variable over a finite field of characteristic 2. For simplicity, we are going to consider the following polynomial $F(x) = \epsilon_1 x^{11} + \epsilon_2 x^9 + \epsilon_3 x^5 + \epsilon_4 x^3 + \epsilon_5 x$ over $\mathbf{F}_q$. Choosing the Teichmuller representatives so that $\epsilon_i \to a_i$ for $i = 1, \dots, 5$ and we get

$$\begin{aligned} S(F) &= \sum_{x\in\mathbf{F}_q} (-1)^{Tr(\epsilon_1 x^{11} + \epsilon_{2r} x^9 + \epsilon_3 x^5 + \epsilon_4 x^3 + \epsilon_5 x)} \\ &= \sum_{t\in T} (-1)^{Tr(a_1 t^{11} + a_2 t^9 + a_3 t^5 + a_4 t^3 + a_5 t)}. \end{aligned}$$

Using the identity (3) we can expand $S(F)$ as

$$S(F) = \sum_{t\in T}\sum_{j_1=0}^{q-1}\cdots\sum_{j_5=1}^{q-1} a_1^{j_1}\cdots a_5^{j_5} C(j_1)\cdots C(j_5) t^{11j_1+\cdots+j_5}.$$

We split the multiple sums into three distinct parts

$$S(F) = \sum_{t\in T}\overbrace{C(0)\cdots C(0)}^{5-times} +$$

$$\sum_{t\in T}\overbrace{C(q-1)+\cdots C(q-1)}^{5-times} +$$

$$\sum_{t\in T}\sum_{(j_1,\dots,j_5)\in A} a_1^{j_1}\cdots a_5^{j_5} C(j_1)\cdots C(j_5) t^{11j_1+\cdots+j_5},$$

where $A = [0, q-1]^5 - \{(0,\dots,0), (q-1,\dots,q-1)\}$. Since $C(0) = 1$ and $C(q-1) = q$, we obtain the following:

$$\sum_{t\in T}\overbrace{C(0)\cdots C(0)}^{5-times} = \sum_{t\in T} 1 = q$$

and

$$\sum_{t\in T}\overbrace{C(q-1)\cdots C(q-1)}^{5-times} = (-q)^5 \sum_{t\in T^\times} 1 = (-q)^5(q-1).$$

Finally, the third part

$$\sum_{(j_1,\dots,j_5)\in A} a_1^{j_1}\cdots a_5^{j_5} C(j_1)\cdots C(j_5) \sum_{t\in T} t^{11j_1+\cdots+j_5}$$

has

$$\sum_{t\in T} t^{11j_1+9j_2+5j_3+3j_4+j_5} \tag{4}$$

as a factor, which is zero unless

$$11j_1 + 9j_2 + 5j_3 + 3j_4 + j_5 \equiv 0 \bmod q-1.$$

The actual value of (4) is $q-1$. By Stickelberger's theorem, we know that the power of 2 that divides $C(j_1)\cdots C(j_5)$ is

$$\sigma(j_1) + \sigma(j_2) + \sigma(j_3) + \sigma(j_4) + \sigma(j_5). \tag{5}$$

Hence, the actual divisor of $S(F)$ is determined by the minimum value of (5).

This method can be done for any polynomial over a finite field of characteristic $p$ (not necessarily equal to 2). In section 7, we review the results of [19], where we also associate a system of modular equations to exponential sums in several variables and for a field of arbitrary characteristic.

## 7 Packing Method and Some New Applications

In the previous section we associated a modular equation to $S(F)$. In [16] and [18], we have introduced a new technique, that we call the packing method, which is a combinatorial technique that estimates the solutions of the modular equation; and the corresponding divisibility. Using it in this section, we are able to improve the 2-divisibility for certain families of exponential sums.

For the proof of our main result of this section we need the following technical lemma whose proof appears in [22, sec. 4, claim 3].

**Lemma 7.1.** *If $w < 2^k$, then $\sigma((w+1)(2^k-1)) = k$.*

Now we illustrate the packing method with an example

*Example 7.2.* Let $S = \sum_{x \in \mathbf{F}_{2^{3f}}} (-1)^{Tr(\epsilon_1 x^9 + \epsilon_2 x^7 + \epsilon_3 x^5 + \epsilon_4 x^3 + \epsilon_5 x)}$. The modular equation associated to $S$ is

$$9j_1 + 7j_2 + 5j_3 + 3j_4 + j_5 \equiv 0 \bmod 2^{3f} - 1 \quad (0 \leq j_i \leq 2^{3f} - 1)\,.$$

This modular equation determines the 2-divisibility of $S$. The modular equation is equivalent to the following equation

$$9j_1 + 7j_2 + 5j_3 + 3j_4 + j_5 = c(2^{3f} - 1) \quad (1 \leq c \leq 25)\,.$$

A important point is that $\sigma_2(2^{3f} - 1) = 3f$ for $1 \leq c \leq 25$. For any $c$ in the range, we need to reproduce the 2-expansion of $c(2^{3f} - 1)$ by adding the two expansion of 9(1001), 7(111), 5(101), 3(11) and 1(1). For simplicity, we assume that $c = 1$, therefore

$$2^{3f} - 1 = \overbrace{111\cdots111}^{3f}\,.$$

We want to cover a total of $3f$ positions with 1's using 9(1001), 7(111), 5(101), 3(11) and 1(1). Note that the more economical way to cover the $3f$ positions is using 7, i.e.,

$$\overbrace{111}^{7}\cdots\overbrace{111}^{7}\,.$$

We used $f$ of the above ($\overbrace{111}^{7}$'s) to cover the $3f$ positions with 1's. Therefore

$$\min_{\substack{(j_1,\cdots,j_5) \\ 9j_1+7j_2+5j_3+3j_4+j_5 \equiv 0 \bmod 2^{3f}-1}} \{\sigma_2(j_1) + \cdots + \sigma_2(j_5)\} = f\,.$$

We have produced a solution $j_1 = 7(1+2^3+2^6+\cdots+2^{3f-3})$, $j_2 = \cdots = j_5 = 0$ of $9j_1 + 7j_2 + 5j_3 + 3j_4 + j_5 \equiv 0 \bmod q - 1$ that is minimal.

Our attention goes now to degree 11 polynomials over $\mathbf{F}_q$, $q = 2^{5f}$ of the form

$$f(x) = \epsilon_1 x^{11} + \epsilon_2 x^9 + \epsilon_3 x^5 + \epsilon_4 x^3 + \epsilon_5 x \,.$$

For these polynomials we determine the 2-divisibility of their Weil sums.

**Theorem 7.3.** *Let $F(x) = \epsilon_1 x^{11} + \epsilon_2 x^9 + \epsilon_3 x^5 + \epsilon_4 x^3 + \epsilon_5 x$ be a polynomial over $\mathbf{F}_q (q = 2^{5f})$, then*

$$2^{2f} \textit{ divides } S(F)\,.$$

*Proof.* Following section 5, the modular equation associated to $F(x)$ is $11j_1 + 9j_2 + 5j_3 + 3j_4 + j_5 \equiv 0 \bmod q-1$. We need to prove that

$$\sigma(j_1) + \sigma(j_2) + \sigma(j_3) + \sigma(j_4) + \sigma(j_5) \geq 2f\,.$$

We will prove the theorem in two different claims:
**claim 1.** Let $r_1, r_2$ be nonnegative integers. For $|r_1 - r_2| \leq 3$ and $l_i \in \{11, 9, 5, 3, 1\}$, we have $\sigma(11 \times 2^{r_1} + l_i \times 2^{r_2}) \leq 5$.
**Proof.** Multiplication by a power of 2 does not change the binary weight. Without of loss of generality we can assume $r_1 \geq r_2$. If $r_1 \geq r_2$, then $\sigma(2^{r_1 - r_2}(11 + l_i \times 2^{r_1 - r_2})) = \sigma(11 + l_i \times 2^{r_1 - r_2})$. But when adding, $\sigma(m+n) \leq \sigma(m) + \sigma(n)$ and equality holds only when not two 1's of the 2-expansions of $m$ and $n$ share the same positional value. The hypothesis $|r_1 - r_2| \leq 3$ and the fact that the $l_i$'s are odd ensure that in each sum both summands have at least one 1 in the same positional value. That proves the claim.

We now introduce the "no loss" concept when adding binary numbers. We say that the sum of two positive integers $m$ and $n$ *does not have any loss* if $\sigma(m+n) = \sigma(m) + \sigma(n)$. The only two cases where Claim 2 is tight are $\sigma(11 \times 2^r + 11 \times 2^{r+2}) = \sigma(11 \times 2^r + 5 \times 2^{r+2}) = 5$. Note that if $|r_1 - r_2| \leq 3$, we have $\sigma(11 \times 2^{r_1} + 11 \times 11^{r_2}) < \sigma(11 \times 2^{r_1}) + \sigma(11 \times 2^{r_1}) = 6$. The case $\sigma(11 \times 2^r + 5 \times 2^{r+2}) = \sigma(11 \times 2^r) + \sigma(5 \times 2^{r+1}) = 5$ is the only case with "no loss".
**claim 2.** A lower bound for $\sigma(j_1) + \cdots + \sigma(j_5)$, where $11j_1 + 9j_2 + 5j_3 + 3j_4 + j_5 = c(q-1)$, is $2f$.
**Proof.** For any set of values $0 \leq j_1, j_2, \ldots, j_5 \leq 2^{5f} - 1$, we have

$$11j_1 + 9j_2 + 5j_3 + 3j_4 + j_5 < (11 + \cdots + 1)(2^{5f} - 1) = 29(2^{5f} - 1).$$

Let us assume that we have a solution $\{j_1, j_2, \ldots j_5, c\}$ of $11j_1 + \cdots + j_5 = c(q-1)$ where $0 < c < 29$. The cases $c = 0$ and $c = q-1$ are trivial. For any such $c$ we need to know how we can re-produce the 2-expansion of $c(q-1)$ by adding the 2-expansion of 11 (1101), 9 (1001), 5(101), 3(11) and 1(1). By Lemma 2, $\sigma(11j_1 + \cdots + j_5) = \sigma(c(q-1)) = 5f$, i.e., we need to *cover* the $5f$ 1's in the 2-expansion of $c(q-1)$ with $2^\mu \cdot 11 = \overbrace{0\cdots0}^{\mu}1101$, $2^\mu \cdot 9 = \overbrace{0\cdots0}^{\mu}1001$, $2^\mu \cdot 5 = \overbrace{0\cdots0}^{\mu}101$, $2^\mu \cdot 3 = \overbrace{0\cdots0}^{\mu}11$ and $2^\mu \cdot 1 = \overbrace{0\cdots0}^{\mu}1$.

Note that $\sigma(l_i) \le 2$, for $l_i \in \{9,5,3,2\}$. This is the "digits packing" process. For $c$ in the prescribed range, $0 < c < 29$, let us call by $c_0, c_1, \ldots, c_{k-1}, c_k$, $k \le 5f + 5$ ($\lfloor \log_2 29 \rfloor = 5$), the binary digits of $c(q-1)$. Actually $k = 5f +$ the number of digits of $c$ in the binary expansion of $c$. By *covering* $c(q-1) = c_0 + c_1 \cdot 2 + \cdots + c_k \cdot 2^k$ we mean to find integers $j_1, j_2, \ldots, j_5$ such that the 2-expansions of $11j_1 = a_{01} + a_{11} \cdot 2 + \cdots + a_{s_1 1} \cdot 2^{s_1}$ added to the 2-expansions of $l_i j_i = a_{0i} + a_{1i} \cdot 2 + \cdots a_{s_i i} \cdot 2^{s_i}$, $2 \le i \le 5$ with $l_i \in \{9,5,3,1\}$, produce the 2-expansion of $c(q-1)$, i.e.,

$$c_1 \cdots c_l = (a_{01} \cdots a_{s_1 1}) + \cdots + (a_{0r} \cdots a_{s_r r}) .$$

By Lemma 2, we have $\sum_{i=1}^{k} c_i = 5f$. The minimum of $\sigma(j_1) + \cdots + \sigma(j_r)$ happens when we use the minimum of $2^\mu \times 11 = 0 \cdots 0\overbrace{1101}^{11}$'s and $2^s l_i$'s for $l_i \in \{1001, 101, 11, 1\}$. Since 11(11001) has larger binary weight than the $l_i$'s, we "cover more ground" using $2^\mu(1101)$ with " no loss" as often as possible than using $2^\nu l_i$ of weight $\le 2$.

Note that if $c = 2^\mu$, then $c(q-1)$ has $5f$ consecutive 1's and its length is $5f + \mu$, i.e.,

$$\overbrace{0 \cdots 0}^{\mu-times}\ \overbrace{1 \cdots 1}^{5f} . \tag{6}$$

Note that $2^\mu((11 + 2^2 \times 5) + 2^5(11 + 2^2 \times 5) + \cdots + 2^{5(f-1)}(11 + 2^2 \times 5)$ is a solution of $11j_1 + \cdots + j_5 = c(q-1)$. More especific the solution is $j_1 = 2^\mu \times 11(1 + 2^5 + \cdots + 2^{5(f-1)})$, $j_2 = 2^m \times 5 \times 2^2(1 + 2^5 + \cdots + 2^{5(f-1)})$, $j_3 = j_4 = j_5 = 0$. Recall that in our case $m < 5$. By claim 2, this is the more economical way to cover the $5f + \mu$ positions with 1's in (6). Hence

$$\sigma(j_1) + \cdots + \sigma(j_5) \ge 2f .$$

Without loss of generality, we may assume that $c$ is odd and $c > 1$. Let $\lambda =$ number of digits of $c$ in its binary expansion. Note $\lambda \le 5$. Then the 2-expansion of $c(q-1)$ has a sequence of 0's and 1's as their first $\lambda$ entries, followed by a sequence of $5f - \lambda$ consecutive 1's, and a sequence of 0's and 1's as their last $\lambda$ entries. The first $\lambda$ entries and last $\lambda$ entries of $c(q-1)$ are complements of each other, i.e., if $c(q-1) = c_0 \cdots c_\lambda \overbrace{1 \cdots 1}^{5f-2\lambda} c_{5f+1} \cdots c_{5f+\lambda}$ then $c_i = 1 \Leftrightarrow c_{5f+i} = 0$ for $0 \le i \le \lambda$.

Suppose that $c(q-1)$ has exactly $s$ consecutive 1's. Note that we want to fill $s$ consecutive 1's with 11 and the $l_i$'s where $l_i \in \{9(1001), 5(101), 3(11), 1(1)\}$. If $s$ is a multiple of 5, then $s = 5(f-1)$. This happens when $c \in$ { 17, 19, 21, 23, 25, 27 }. We fill the $s$ positions with $(2f-1)$ 11111's ($2^\mu(11 + 5 \times 2^2)$). Now we have to analyze the first 5 entries of $c(q-1)$ and the last 5 entries of the 2-expansion of $c(q-1)$. We need at least two $l_i$'s to fill the first and last 5 entries. Hence

$$\sigma_{j_1} + \cdots + \sigma(j_5) \ge 2f - 2 + 2 = 2f .$$

This proves this case, i.e., $s$ is multiple of 5. If $s$ is not a multiple of 5, then $s = 5k + r$, where $1 \leq r \leq 4$. We have $k = f - 1$ since $1 \leq r \leq 4$ and the total number of 1's in the first $\lambda$ entries and in the last $\lambda$ entries of the 2-expansion of $c(q-1)$ is $\leq 5$. Then

$$c(q-1) = c_0 \cdots c_\lambda \overbrace{1 \cdots 1}^{5(f-1)} \overbrace{1 \cdots 1}^{r} c_{5f+1} \cdots c_{5f+\lambda}.$$

We cover the $5(f-1)$ consecutive 1's as we did in the case when $s$ was a multiple of 5. The last two sentences imply that we need to use $2(f-1)$ 11111's to cover the $5(f-1)$ consecutive 1's. We need to analyze the first $\lambda$ entries $c_0 \cdots c_\lambda$ and the last $r + \lambda$ entries $\overbrace{1 \cdots 1}^{r} c_{5f+1} \cdots c_{5f+\lambda}$. To cover these entries we need at least two $l_i$'s. Therefore

$$\sigma(j_1) + \cdots + \sigma(j_5) \geq 2(f-1) + 2 = 2f .$$

This ends the proof.

Now, we apply Theorem 11 to coding theory (see section 1).

**Corollary 7.4.** *Let $\mathbf{C}_{5f}$ be the cyclic code with zeros $\alpha^{11}, \alpha^9, \alpha^5, \alpha^3, \alpha$, where $\alpha$ is primitive over $\mathbf{F}_{2^{5f}}$, then $2^{2f-1}$ divides the weights of the elements of $\mathbf{C}_{5f}^{\perp}$.*

The following theorems can be proved in a similar way to previous theorem.

**Theorem 7.5.** *Let $F(x) = \epsilon_1 x^{13} + \epsilon_2 x^9 + \epsilon_3 x^5 + \epsilon_4 x^3 + \epsilon_5 x$ be a polynomial over $\mathbf{F}_{2^{5f}}$, then*

$$2^{2f} \text{ divides } S(F) .$$

**Corollary 7.6.** *Let $\mathbf{C}_{5f}$ be the cyclic code with zeros $\alpha^{13}, \alpha^9, \alpha^5, \alpha^3, \alpha$, where $\alpha$ is primitive over $\mathbf{F}_{2^{5f}}$, then $2^{2f-1}$ divides the weights of the elements of $\mathbf{C}_{5f}^{\perp}$.*

**Theorem 7.7.** *Let $F(x) = \epsilon_1 x^{23} + \epsilon_2 x^{13} + \epsilon_3 x^9 + \epsilon_4 x^5 + \epsilon_5 x^3 + \epsilon_6 x$ be a polynomial over $\mathbf{F}_{2^{7f}}$, then*

$$2^{2f} \text{ divides } S(F) .$$

**Corollary 7.8.** *Let $\mathbf{C}_{7f}$ be the cyclic code with zeros $\alpha^{23}, \alpha^{13}, \alpha^9, \alpha^5, \alpha^3, \alpha$, where $\alpha$ is primitive over $\mathbf{F}_{2^{7f}}$, then $2^{2f-1}$ divides the weights of the elements of $C_{7f}^{\perp}$.*

Applying Moreno-Moreno's Theorem of section 2 to the polynomials of Theorems 11 and 12 we obtain that $2^{\lceil \frac{5f}{3} \rceil}$ divides $S(F)$. Our results give a better 2-divisibility since $2f > \lceil \frac{5f}{3} \rceil$ for $f$ a multiple of 3. Applying also to the polynomials of Theorem 13 we obtain $2^{\lceil \frac{7f}{4} \rceil}$ divides $S(F)$. Theorem 13 gives a better 2-divisibility since $2f > \lceil \frac{7f}{4} \rceil$ for $f$ a multiple of 4.

# 8 The Case of Exponential Sums in Several Variables

In this section we review the results of [19]. Following the work of Chevalley-Warning, J. Ax obtained a result that bounds the order of $p$ divisibility of exponential sums where $p$ is the characteristic of the corresponding finite field, and his bound is tight for the class of polynomials of fixed total degree $d$ (see example 1). On the other hand consider a set of $m$ multi-variable monomials, and the $q^m$ multi-variable polynomials obtained by taking coefficients in a finite field $\mathbf{F}_q$ with $q = p^f$. In recent years Moreno-Moreno and Adolphson-Sperber (see Theorems 8 and 9 of section 3) have given improvements to the result of Ax for the set of polynomials we just mentioned in the last sentence. In this section we present two of the main results obtained in [19] that provides the best order of the $p$ divisibility for exponential sums. We give a bound, which is tight for a choice of the coefficients. Even though to obtain the exact value of our theorem we have to solve a system of modular equations, our methods give approximates general solutions for many classes of examples, improving results of Ax-Katz, Moreno-Moreno and Adolphson-Sperber. Please look at [19] for a full treatment of the techniques with many more applications like that of Theorems 14 and 15.

Let $F$ be a polynomial in $n$ variables with $m$ terms, namely,

$$F(x_1, \dots, x_n) = \sum_{j=1}^{m} a_j x_1^{e_{1j}} \cdots x_n^{e_{nj}}, \quad a_j \in \mathbf{F}_q .$$

Let $\zeta$ be a $p$-root of unity and $\lambda = (1-\zeta)$ be the ideal generated by $1-\lambda$. Let $a \in \mathbb{Q}_p(\zeta)$, then we define the order $\mathrm{ord}_\lambda(a)$ of $a$ at $\lambda$ by $\mathrm{ord}_\lambda(a) = k \Leftrightarrow (a) = (\lambda)^k$.

The system of modular equations associated to $F$ is the following:

$$\begin{aligned} e_{11}j_1 + e_{12}j_2 + \cdots + e_{1m}j_m &\equiv 0 \bmod q-1 \\ e_{21}j_1 + e_{22}j_2 + \cdots + e_{2m}j_m &\equiv 0 \bmod q-1 \\ \vdots \quad \cdots \qquad\qquad \cdots \vdots \quad & \qquad \vdots \\ e_{n1}j_1 + e_{n2}j_2 + \cdots + e_{nm}j_m &\equiv 0 \bmod q-1 \end{aligned} \tag{7}$$

Let

$$L = \min_{\substack{(j_1,\dots,j_m)\text{ is a} \\ \text{solution of (7)}}} \{\sigma_p(j_1) + \cdots + \sigma_p(j_m)\} .$$

Now we state two of the main theorems of [19].

**Theorem 8.1.** *Let $F$ be a polynomial in $n$ variables with $m$ terms, namely,*

$$F(x_1, \dots, x_n) = \sum_{j=1}^{m} a_j x_1^{e_{1j}} \cdots x_n^{e_{nj}}, \quad a_j \in \mathbf{F}_q .$$

*and $\psi$ be an additive character of $\mathbf{F}_q$. Let $S(F)$ be the exponential sum associated to $F$ over $\mathbf{F}_q$,*

$$S(F) = \sum_{x_1,\dots,x_n \in \mathbf{F}_q} \psi(F(x_1,\dots,x_n)),$$

*then*

$$ord_\lambda(S(F)) = L,$$

*for a choice of coefficients $a_1,\dots,a_m \in \mathbf{F}_q$.*

Note that the exponential sum associated to $N(F)$ is $S(yF)$. Therefore the system of modular equations associated to $N(F)$ is the following:

$$\begin{aligned} e_{11}j_1 + e_{12}j_2 + \cdots + e_{1m}j_m &\equiv 0 \bmod q-1 \\ e_{21}j_1 + e_{22}j_2 + \cdots + e_{2m}j_m &\equiv 0 \bmod q-1 \\ \vdots \quad \cdots \qquad\qquad \cdots \vdots \quad & \qquad \vdots \\ e_{n1}j_1 + e_{n2}j_2 + \cdots + e_{nm}j_m &\equiv 0 \bmod q-1 \\ j_1 + j_2 + j_3 + \cdots + j_{m-1} + j_m &\equiv 0 \bmod q-1 \end{aligned} \tag{8}$$

Let

$$L = \lceil \min_{\substack{(j_1\dots,j_m)\text{ is a} \\ \text{solution of (8)}}} \{\frac{\sigma_p(j_1) + \cdots + \sigma_p(j_m)}{p-1}\}\rceil .$$

**Theorem 8.2.** *Let $F$ be a polynomial in $n$ variables with $m$ terms, namely,*

$$F(x_1,\dots,x_n) = \sum_{j=1}^{m} a_j x_1^{e_{1j}} \cdots x_n^{e_{nj}}, \quad a_j \in \mathbf{F}_q,$$

*then*

$$ord_p(N(F)) = L,$$

*for a choice of coefficients $a_1,\dots,a_m \in \mathbf{F}_q$.*

*Example 8.3.* Let $F(x_1,x_2,x_3) = a_1x_1^7 + a_2x_2^7 + a_3x_3^7 + a_4x_1x_2x_3 + a_5x_1 + a_6x_2 + a_7x_3 + a_8x_1^3 + a_9x_2^3 + a_{10}x_3^3 a_{11}x_1x_2 + a_{12}x_2x_3 + a_{13}x_1x_3$ be a polynomial over $\mathbf{F}_{2^f}$. The system of modular equation associated to $F$ is the following:

$$\begin{aligned} 7j_1 + j_4 + j_5 + 3j_8 + j_{11} + j_{13} &\equiv 0 \bmod 2^f - 1 \\ 7j_2 + j_4 + j_6 + 3j_9 + j_{11} + j_{12} &\equiv 0 \bmod 2^f - 1 \\ 7j_3 + j_4 + j_7 + 3j_{10} + j_{12} + j_{13} &\equiv 0 \bmod 2^f - 1 \end{aligned}$$

In this case $L = f$. Applying Theorem 14, we know that exists $a_1,\dots,a_{13} \in \mathbf{F}_{2^f}$ such that $\text{ord}_2(S(F)) = f$. In particular, we have $\text{ord}_2(S(F)) = f$, if $a_1 = a_2 = a_3 = 0$ and $a_4 \neq 0$

## 9 Comparison of the Methods of Previous Sections

The following theorem shows that the covering method (of section 4) gives better results than the method of section 3.

**Theorem 9.1.** *Let $F(X_1, \dots, X_n)$ be a polynomial in $n$-variables over $\mathbf{F}_{2^f}$. Let $F'(x_1, \dots, x_{nf})$ be the polynomial associated to $F$ over $\mathbf{F}_2$ and $r$ be the cardinality of a minimal set of monomials of $F'$ covering all variables, $x_1, \dots, x_{nf}$. Then*

$$r \geq \omega(F').$$

*Moreover, if $F(X_1, \dots, X_n)$ has 2-weight degree $l$, then $r \geq \frac{fn}{l}$.*

*Proof.* We have $r \geq \omega(F')$, since

$$r\Big(\sum_{x^d \in C} \frac{1}{r} d\Big)$$

has nonnegative integral coordinates. The second part follows from the fact that the monomials of $F'$ have 2-weight degree $\leq l$. This completes the proof. □

Theorem 16 is a generalization of Theorem 6 in [17]. We want to point out that the second part of the theorem was obtained by Litsyn et al in [10]. The following is an example where $r > \omega(F')$.

*Example 9.2.* Let $F(x_1, x_2, x_3, x_4) = x_1x_2x_3 + x_2x_3x_4 + x_1x_4 + x_1 + x_2 + x_3$ be polynomial over $\mathbf{F}_2$. In this case the cardinality of $C$ is 2, i.e., $C = \{x_1x_2x_3,\ x_1x_4\}$. Hence 2 divides $N(F)$. In this case $\omega(F') = \frac{3}{2}$ since

$$\frac{1}{3}(1,1,1,,0) + \frac{1}{3}(0,1,1,1) + \frac{1}{3}(1,0,0,1) = (\frac{2}{3}, \frac{2}{3}, \frac{2}{3}, \frac{2}{3}, \frac{2}{3}).$$

This implies that 2 divides $N(F)$, since we take the smallest positive integer $\geq \omega(F')$.

Let $a_1, \dots a_m \in \mathbf{F}_q$ $(q = p^f)$. Let $\mathcal{F}$ be the set of all the polynomials generated by the monomials $x_1^{e_{11}} \cdots x_n^{e_{n1}}, \dots, x_1^{e_{1m}} \cdots x_n^{e_{nm}}$.

$$\mathcal{F} = \{a_1x_1^{e_{11}} \cdots x_n^{e_{n1}} + \cdots + a_mx_1^{e_{1m}} \cdots x_n^{e_{nm}} \mid a_i \in \mathbf{F}_q\}$$

Applying Theorem 14, there exists $F_0 \in \mathcal{F}$ such that

$$\mathrm{ord}_\lambda(S(F_0)) = L.$$

We assume that $F_0(x_1, \dots, x_n) = a_{10}x_1^{e_{11}} \cdots x_n^{e_{n1}} + \cdots + a_{r0}x_1^{e_{1r}} \cdots x_n^{e_{nr}}$ where $a_{i0} \in \mathbf{F}_q^\times$. Let $\mathcal{G}$ be generated by the monomials in $F_0$, i.e.,

$$\mathcal{G} = \{a_1x_1^{e_{11}} \cdots x_n^{e_{n1}} + \cdots + a_rx_1^{e_{1r}} \cdots x_n^{e_{nr}} + a_{r+1}x_1^{e_{1r+1}} \cdots x_n^{e_{nr+1}} + \cdots + \cdots + a_mx_1^{e_{1m}} \cdots x_n^{e_{nm}} \mid a_1, \dots, a_r \in \mathbf{F}_q^\times \text{ and } a_{r+1}, \dots, a_m \in \mathbf{F}_q\}.$$

Note that $\mathcal{G}$ contains all the polynomials for which the monomials of $F_0$ appear( in other words their coefficients are non-zero). Let

$$\mathcal{G}' = \{a_1 x_1^{e_{11}} \cdots x_n^{e_{n1}} + \cdots + a_r x_1^{e_{1r}} \cdots x_n^{e_{nr}} \mid a_i \in \mathbf{F}_q^{\times}\}.$$

Note $|\mathcal{G}| = (q-1)^r + q^{n-r}$, $|\mathcal{G}'| = (q-1)^r$ and $\mathcal{G}' \subseteq \mathcal{G}$.

**Theorem 9.3.** *Let $\mathcal{G}$ and $\mathcal{G}'$ as defined above, then for every $F \in \mathcal{G}$ we have*

$$L \geq f(p-1)\omega(F).$$

*Proof.* Following the above notation we have $F_0 \in \mathcal{G}$ such that $\mathrm{ord}_\lambda(S(F_0)) = L$. We have $\omega(G) \geq \omega(F)$ for all $G \in \mathcal{G}'$ and $F \in \mathcal{G}$. This is true since $\mathcal{G}' \subseteq \mathcal{G}$. We have that $\mathrm{ord}_\lambda(S(F_0)) \geq f(p-1)\omega(F)$ for all $F \in \mathcal{G}'$ using Theorem 8 and using that $F_0 \in \mathcal{G}$ and the definition of $\omega(F)$. Hence $\mathrm{ord}_\lambda(S(F_0)) \geq f(p-1)\omega(F)$ for all $F \in \mathcal{G}$. □

Now, let us illustrate the procedure of Theorem 17 in the following example.

*Example 9.4.* Let $F(x_1,x_2,x_3) = a_1x_1^7 + a_2x_2^7 + a_3x_3^7 + a_4x_1x_2x_3 + a_5x_1 + a_6x_2 + a_7x_3 + a_8x_1^3 + a_9x_2^3 + a_{10}x_3^3 + a_{11}x_1x_2 + a_{12}x_2x_3 + a_{13}x_1x_3$ be a polynomial over $\mathbf{F}_{2^f}$. In example 12 we said that the minimum binary weight for any solution of the modular system associated to $F$ is $L = f$. This value is attained for the solutions $(0,0,0,q-1,0\ldots,0)$ and $(2^i \cdot \frac{q-1}{7}, 2^i \cdot \frac{q-1}{7}, 2^i \cdot \frac{q-1}{7}, 0, \ldots, 0)$ for $i = 0,1,2$. If we take $a_1 = a_2 = a_3 = 0$ and $a_4 \neq 0$, we have $\mathrm{ord}_2(S(F)) = f$. Now consider the set $\mathcal{G} = \{a_1x_1^7 + a_2x_2^7 + a_3x_3^7 + a_4x_1x_2x_3 + a_5x_1 + a_6x_2 + a_7x_3 + a_8x_1^3 + a_9x_2^3 + a_{10}x_3^3 + a_{11}x_1x_2 + a_{12}x_2x_3 + a_{13}x_1x_3 \mid a_i \in \mathbf{F}_q$ and $a_4 \neq 0\}$. If we consider $\Delta(F)$ for any polynomial in $\mathcal{G}$, we have $f \geq f\omega(F)$ .

In the following examples we compare the methods described of sections 3, 4 and 5. At the end, after example 19, we give a table that summarizes our comparison results.

*Example 9.5.* Let $F(x) = x^3 + x$ be a polynomial over $\mathbf{F}_{16}$. The modular equation associated to $S(F)$ over $\mathbf{F}_{16}$ is $3j_1 + j_2 \equiv 0 \bmod 15$, where $0 \leq j_i \leq 15$. We obtain $\min\{\sigma(j_1) + \sigma(j_2) \mid 3j_1 + j_2 \equiv 0 \bmod 15\} = 2$. This minimum is attained for the solutions $(5,0), (10,0)$. Hence 4 divides $S(F)$. Now applying the reduction to ground field method to $F$, we obtain $Tr(F(x)) = F'(x_1,x_2,x_3,x_4) = x_2 + x_2x_4 + x_4x_3 + x_4$. Note that the variable $x_1$ is missing of $F'$. A minimal covering of the variables $x_2, x_3, x_4$ of $F'$ is $\mathcal{C} = \{x_2x_4, x_4x_3\}$. Applying Lemma 1 and the fact that in $F'$ is missing the variable $x_1$, we obtain that 8 divides $S(F)$. Applying Theorem 9 to $F$ and $F'$, we obtain that $\omega = \max\{\omega(F), \omega(F')\} = \max\{\frac{4}{3}, 3\} = 3$. Hence 8 divides $S(F)$.

*Example 9.6.* Let $F(x) = x^5 + x$ be a polynomial over $\mathbf{F}_{16}$. The modular equation associated to $S(F)$ over $\mathbf{F}_{16}$ is $5j_1 + j_2 \equiv 0 \bmod 15$, where $0 \leq j_i \leq 15$. We obtain $\min\{\sigma(j_1) + \sigma(j_2) \mid 3j_1 + j_2 \equiv 0 \bmod 15\} = 2$. This minimum is attained for the solutions $(3,0)$, $(6,0)$, $(12,0)$ . Hence 4 divides $S(F)$. Now applying the reduction to ground field method to $F$, we obtain $Tr(F(x)) = F'(x_1, x_2, x_3, x_4) = x_3$. Note that the variables $x_1, x_2, x_4$ are missing of $F'$. Therefore 16 divides $S(F)$.

*Example 9.7.* Let $F(x) = x^{11} + x^5$ be a polynomial over $\mathbf{F}_{1024}$. The modular equation associated to $S(F)$ over $\mathbf{F}_{1024}$ is $11j_1 + 5j_2 \equiv 0 \bmod 1023$, where $0 \leq j_i \leq 1023$. The minimum is $\min\{\sigma(j_1) + \sigma(j_2) \mid 11j_1 + 5j_2 \equiv 0 \bmod 1023\} = 4$. Hence 16 divides $S(F)$. Now applying the reduction to ground field method to $F$, we obtain a polynomial in ten variables, i.e., $Tr(F(x)) = F'(x_1, \dots, x_{10})$. A minimal covering of the variables of $F'$ is $\mathcal{C} = \{x_1x_2x_8, x_5x_7x_9, x_1x_6x_3, x_4x_6x_{10}\}$. Applying Lemma 1 we obtain that 16 divides $S(F)$. Applying Theorem 8 to $F$, we obtain that $\omega(F) = \frac{10}{11}$. Hence 2 divides $S(F)$.

*Example 9.8.* Let $F(x) = x^{23} + x$ be a polynomial over $\mathbf{F}_{64}$. The modular equation associated to $S(F)$ over $\mathbf{F}_{64}$ is $23j_1 + j_2 \equiv 0 \bmod 63$, where $0 \leq j_i \leq 63$.The minimum is $\min\{\sigma(j_1) + \sigma(j_2) \mid 23j_1 + j_2 \equiv 0 \bmod 63\} = 3$. Hence 8 divides $S(F)$. Now applying the reduction to ground field method to $F$, we obtain a polynomial in six variables, i.e., $Tr(F(x)) = F'(x_1, \dots, x_6)$. A minimal covering of the variables of $F'$ has cardinality 2. Applying Lemma 1 we obtain that 4 divides $S(F)$. Applying Theorem 8 to $F$, we obtain that $\omega(F) = \frac{6}{23}$. Hence 2 divides $S(F)$.

*Example 9.9.* Let $F(x) = x^{23} + x^7$ be a polynomial over $\mathbf{F}_{64}$. The modular equation associated to $S(F)$ over $\mathbf{F}_{64}$ is $23j_1 + 7j_2 \equiv 0 \bmod 63$, where $0 \leq j_i \leq 63$. The minimum is $\min\{\sigma(j_1) + \sigma(j_2) \mid 23j_1 + j_2 \equiv 0 \bmod 63\} = 2$. Hence 8 divides $S(F)$. Now applying the reduction to ground field method to $F$, we obtain a polynomial in six variables, i.e., $Tr(F(x)) = F'(x_1, \dots, x_6)$. A minimal covering of the variables of $F'$ has cardinality 2, i.e., $\mathcal{C} = \{x_1x_5x_6x_2,\ x_4x_6x_5\}$. Applying Lemma 1 we obtain that 4 divides $S(F)$. Applying Theorem 8 to $F$, we obtain that $\omega(F) = \frac{6}{23}$. Hence 2 divides $S(F)$.

| Example | Method 2 | Method 3 | Method 4 | Method 5 |
|---|---|---|---|---|
| 15 | $\frac{4}{3}$ | 2 | 3 | 3 |
| 16 | $\frac{4}{5}$ | 2 | 4 | 4 |
| 17 | $\frac{10}{11}$ | 4 | 4 | 4 |
| 18 | $\frac{6}{23}$ | 3 | 2 | 2 |
| 19 | $\frac{6}{23}$ | 2 | 2 | 2 |

Now, we compare the methods, which are :

(i) The method of Ax-Katz and Moreno-Moreno described in section 2 (depends only on the total degrees or total $p$- weight degrees of the polynomials).

(ii) The Newton polyhedra method (of Adolphson-Sperber), described in section 2 (depends on the degree of each monomial and each variable).

(iii) The method of modular equations, described in sections 5, 6 and 7 (depends on the degree of each monomial and each variable, and also on the field $\mathbf{F}_q$).

(iv) The combination of the Newton polyhedra method and the reduction to the ground field method, described in section 3 ( depends on the $p$-weight degree of each monomial and each variable and also on the coefficients of the polynomials).

(v) The method of covering described in section 4 (same as the last one above).

Following the idea presented in Remark 5, we can therefore see that we have enumerated the methods in descending order of globality. In other words as you go down in our list the methods depend upon more parameters, become so to speak more local. Note that usually as you go down in the list the divisibility improves (this is true in the table with the exception of example 18). From a practical point of view if we are studying divisibility properties of exponential sums for a particular polynomial then perhaps we should use one of the methods down on the list. If, on the other hand, we are studying a whole class of polynomials then we use the methods at the beginning of the list.

Note that Theorem 16, loosely speaking, says that method 5 gives better divisibility than method 4 and Theorems 14 and 15 say that method 3 improves on method 2. Now let us explain example 18 above that seems to contradict our philosophy. In other words locally for certain coefficients method 3 can be improved, but in certain cases Theorem 17 asserts, you can not improve method 3. We want to remark that method 3 also gives global solutions of the type belonging to category 1 (like Ax-Katz, etc.), and for this see [19]. Finally for application of divisibility to $p$-adic Serre type bounds methods 1 or 2 are the useful ones (see [19], [22]).

## References


1. A. Adolphson and S. Sperber, $p$-adic Estimates for Exponential Sums and the Theorem of Chevalley-Warning, *Ann. Sci. Ec. Norm. Super.*, 4$^e$ série, vol **20**, pp. 545-556, 1987.
2. A. Adolphson and S. Sperber, $p$-adic Estimates for Exponential Sums, *Lectures Notes in Mathematics*, **1454**, pp. 11-22, Springer 1990.
3. J. Ax, Zeros of Polynomials over Finite Fields, *Am. J. Math*, vol. **86**, pp. 255-261, 1964.
4. A. Cacéres and O. Moreno, On the Estimation of the Minimum Distance of Duals of BCH Codes, *Cong. Numer.*, vol. **81**, pp. 205-208, 1991.
5. P. Deligne, La Conjecture de Weil, I, *Inst. Hautes Études Sci. Publ. Math.* **43** (1974), 273–308.

6. B. Dwork, On the Zeta Function of a Hypersurfase, *Inst. Hautes Études Sci. Publ. Math.* **12** (1962), 5–68.
7. F. Q. Gouvêa, *p- adic Numbers*, Universitext, Springer, Second Edition 1997.
8. N. M. Katz, On a Theorem of Ax, *Amer. J. Math.*, **93** (1971),485-499.
9. R. Lidl and H. Niederreiter, *Finite Fields, (Encyclopedia of Mathematics and its Applications)* **20**, Addison-Wesley, Reading, Mass., 1983.
10. S. Litsyn, C.J. Moreno, O. Moreno, Divisibility Properties and New Bounds for Cyclic Codes and Exponential Sums in One and Several Variables, *Applicable Algebra in Engineering, Communication and Computing*, **5**, 105-116(1994).
11. F. J. MacWilliams and N. J. A. Sloane, *The Theory of Error Correcting Codes*, North Holland, Amsterdam, 2nd reprint, 1983.
12. C. J. Moreno and O. Moreno, Exponential Sums and Goppa Codes I, *Proc. AMS* vol **111**, no. 2 pp. 523-531, 1991.
13. C. J. Moreno and O. Moreno, Exponential Sums and Goppa Codes II, *IEEE Trans. Inform. Theory*, vol. **38** no. 4 pp. 1222-1229, July 1992.
14. C. J. Moreno and O. Moreno, An Improved Bombieri-Weil Bound and Applications to Coding Theory, *J. Number Theory*, no. 42, pp. 32-46, 1992.
15. O. Moreno, C. Cáceres and M. Alonso, An Improved and Simplified Binary Ax Theorem, *Proc. 1994 IEEE International Symposium on Information Theory*, Trondheim, Norway June-July 1994.
16. O. Moreno, F.N. Castro, A. Cacéres, On the Tightness of Carlitz-Uchiyama Bound and an Improvement of Ax's Theorem for Binary Exponential Sums, *Proc. of the 29-th Southeastern International Conference on Combinatorics, Graph Theory and Computing, Congr. Numer.* **134**(1998), 147-161.
17. O. Moreno, F.N. Castro, A. Caceres, Exponential Sums in Several Variables over Finite Fields, *Coding Theory, Cryptography and Related Areas*, 209-220, Springer, 2000.
18. O. Moreno, F.N. Castro, A. Cacéres, Improvement of the Weil-Carlitz-Uchiyama Bound and Exact Distance of some Binary Cyclic Codes (submitted).
19. O. Moreno, W. Shum, F.N. Castro, P. Vijay Kumar, An Improved Tight Bound for the Divisibility of Exponential Sums (submitted).
20. O. Moreno and C.J. Moreno, Improvement of the Chevalley-Warning and the Ax-Katz theorems, *Amer. J. Math.* **117**:1 (1995),241-244.
21. O. Moreno and C.J. Moreno, An Elementary Proof of a Partial Improvement to the Ax-Katz Theorem, *Proc. of Applied Algebra, Algebraic Algorithms an Error Correcting Codes*-AAECC-10, (*Lectures Notes In Comput. Sci.* **673**), 257-268, Springer-Verlag, Berlin, 1993.
22. O. Moreno and C. J. Moreno, The MacWilliams-Sloane Conjecture on the Tightness of the Carlitz-Uchiyama Bound and the Weights of Dual of BCH Codes, *IEEE Trans. Inform. Theory* **40:6** (1994), 1894-1907.
23. O. Moreno and C. J. Moreno, A $p$-adic Serre Bound, *Finite Fields and their Applications*, vol. **4** pp. 201-217, 1998.
24. O. Moreno and N. Santiago, An Extension of a Theorem of Welch and Uniform Cyclotomy, *Congress Numerantium* **49** (1985) 75-78.
25. F. Rodier, On the Spectra of the Duals of the Binary BCH Codes of Designed Distance $\delta = 9$, *IEEE Trans. Inform. Theory*, vol. **38:2** (1992) pp. 478-479.
26. J. P. Serre, Sur le Nombre des Points Rationnels d'une Courbe Algébrique sur un Corps Fini, *C. R. Acad. Sci. Paris*, vol. **296**, série I, pp. 397-402, 1983.

27. J. P. Serre, Nombres de Points de Courbes Algebriques sur $\mathbf{F}_q$, *Sémin. Théorie Nombres de Bordeaux*, exposé **22**, pp. 1-8, 1983.
28. J. P. Serre, Resumé des Cours des 1983-1984, *Annu. College de France*, pp. 79-83, 1984 [Collected Papers III, New York: Springer-Verlag].
29. D. Wan, An Elementary Proof of a Theorem of Katz, *Amer. J. Math*, **111** (1989), pp. 1-8.
30. H. N. Ward, A Bound for Divisible Codes, *IEEE Trans. Inform. Theory*, vol. **38**(1), pp. 191-194 (1992).
31. J. Wolfmann, New Bounds on Cyclic Codes from Algebraic Curves, *Lectures Notes in Computer Science*, vol. **388**, pp. 47-62, Berlin, Heidelberg, New York: Springer 1980.

# On the Rank of Appearance and the Number of Zeros of the Lucas Sequences over $\mathbb{F}_q$

Siguna Müller*

University of Klagenfurt
Dept. of Math.
A-9020 Klagenfurt
Austria
e-mail: siguna.mueller@uni-klu.ac.at

**Abstract** Let $D_n(P,Q)$ respectively $E_n(P,Q)$ denote the Dickson polynomial of the first, respectively second kind of degree $n$ in the indeterminate $P$ and with parameter $Q$. It is well known that these polynomials are periodic over $\mathbb{F}_q$, the finite field consisting of $q$ elements. That is, there exist some positive integers $t_1, t_2$ such that $D_{n+t_1}(P,Q) = D_n(P,Q)$, respectively $E_{n+t_2}(P,Q) = E_n(P,Q)$ for all $n \geq 0$.

In this paper we will be dealing with the periodicity of the occurrence of 0. This special type of periodicity is known as the rank of appearance of the corresponding Lucas sequence (respectively Dickson polynomial). For any possible value of the rank of appearance we will give the exact number of parameters with this specific rank over $\mathbb{F}_q$. We will firstly adopt the ideas of [12] in keeping $Q = Q_0$ fixed and considering $D_n$ and $E_n$ as polynomials in $P$. Conversely then we will also be dealing with the opposite case, where we keep $P = P_0$ fixed and regard $Q$ as the indeterminate. Our main results will extend previous results in [8] and [9] concerning parameters with the same rank of appearance over $\mathbb{F}_p$, $p$ an odd prime. We will show that similar results can be obtained over any finite field $\mathbb{F}_q$.

In the second part of the paper we show how this number of parameters with the same rank can efficiently be utilized into establishing the number of zeros of the Dickson polynomials. That is, for any fixed $Q = Q_0$ we will establish the number of zeros $P$ of both of the Dickson polynomials $D_n(P,Q_0)$ and $E_n(P,Q_0)$ over $\mathbb{F}_q$ and give distinct formulas which are specified by certain attributes of the parameter $Q_0$. Again, we then also reverse the role of $P$ and $Q$ in that we keep $P = P_0$ fixed, and regarding $D_n(P_0,Q)$ and $E_n(P_0,Q)$ as polynomials in the variable $Q$, we will establish the number of zeros $Q$ over $\mathbb{F}_q$. Most interestingly, in this case, the formulas will not depend on $P_0$.

## 1 Motivation

Throughout the entire paper, $q$ is a power of a prime $p$. It is known that the Dickson polynomials can be described by the second order linear recurrences

* Research supported by the Österreichischen Fonds zur Förderung der wissenschaftlichen Forschung, FWF-Project no. P 13088-MAT

$(U)$, $(V)$, which are, for all integers $n$, defined by

$$U_{n+2} = PU_{n+1} - QU_n, \quad U_0 = 0, U_1 = 1,$$
$$V_{n+2} = PV_{n+1} - QV_n, \quad V_0 = 2, V_1 = P,$$

where $P \in \mathbb{F}_q$, $Q, D \in \mathbb{F}_q^*$ for $D = P^2 - 4Q$. These sequences are usually referred to as the Lucas sequences of the first and second kind, respectively (cf. [10]). Due to the standard notation given in the literature, we have the relationship between the Dickson polynomials of the first and second kind, respectively, with the Lucas sequences of the second and first kind, respectively, $D_n = V_n$ and $E_{n-1} = U_n$. The shift of the degree in the latter relation is simply caused by the numbering of the initial terms. Traditionally this has been done differently for the Lucas sequences than for the Dickson polynomials [5,10,20].

Among many others, one of the most interesting features of these polynomials/sequences is that of periodicity. Just as a matter of nomenclature, since most of the periodicity properties have been extensively studied in relation to recurrence sequences (cf., e.g., [4,17,18]) we will be referring to Lucas sequences rather than Dickson polynomials. A number of results relating to recurrence sequences over $\mathbb{F}_q$ have been obtained in [6,12,13,14]. In particular, necessary and sufficient conditions on the periodicity, and in some cases, the number of parameters which correspond to a specific period, are given.

A particularly interesting type of periodicity is the occurrence of 0 in the $(U)$ sequence. This is known as the ***rank of appearance*** (rank of apparition) of $(U)$ [10]. The principal goal of this paper is to analyze properties of the rank of appearance over $\mathbb{F}_q$.

One of the reasons why this concept is so important is that it can be seen as *the analogue of the multiplicative order of a group element in terms of Lucas sequences.* As an example of this relation the author has investigated the characterization of pseudoprimes (cf. [7]). This can be achieved by utilizing either the order of an element (when dealing with Fermat based pseudoprimes), or alternatively, the rank of appearance (when dealing with Lucas based pseudoprimes). For both types the results are then analogous.

In this paper we will prove an even more interesting result, namely *the analogue between the order and the rank when counting the number of parameters with the corresponding value.* In terms of the order, we have the following well-known fact.

**Proposition 1.1.** *Let $q$ be a power of a prime. If $d \nmid (q-1)$, there is no element $a \in \mathbb{F}_q$ that has order $d$. If $d|(q-1)$, the number of $a \in \mathbb{F}_q^*$ with $ord(a) = d$ is $\phi(d)$, where $\phi(d)$ denotes Euler's $\phi$ function.*

As a first main result of this paper we will give analogous results for the rank of appearance over $\mathbb{F}_q$. These will generalize the ones obtained in [9] where $q = p$, an odd prime.

In the second part of the paper we will give an application of these results. As an easy consequence, we will show how to establish ***the number of zeros***

***of the Lucas sequences***. That is, we will give a characterization and an exact formula for the number of zeros of both the $(U)$ and the $(V)$ sequence.

We note that there are certainly other means of counting the zeros. As an example one could rely on the factorization of the Dickson polynomials given in [1,2]. Nonetheless, this approach seems to be complicated by the structure of the irreducible factors which involve sums of primitive roots of unity. In particular, for the case of $Q = Q_0$ and the $(V)$ sequence the number of zeros can be derived from properties of the value sets of the Dickson polynomials in [3].

However, since the rank of appearance is literally defined by means of the occurrence of 0, we believe that our method in counting the zeros actually represents the most natural and easiest way.

## 2 Lucas Sequences over $\mathbb{F}_q$

### 2.1 Some Basic Properties

Associated with both $(U)$ and $(V)$ is the characteristic polynomial $f(x) = f(x, P, Q) = x^2 - Px + Q$ with characteristic roots $r$, $\overline{r}$ that can be expressed as $\frac{P \pm \sqrt{P^2-4Q}}{2}$ when $q$ is odd.

Alternatively, the Lucas sequences of degree $k$ can be described by means of the Binet forms $U_k(P,Q) = \frac{r^k - \overline{r}^k}{r - \overline{r}}$, $V_k(P,Q) = r^k + \overline{r}^k$. If the roots $r, \overline{r}$, are considered over any field $\mathbb{F}_q$, then they are clearly either both elements of $\mathbb{F}_q$, or $\mathbb{F}_{q^2}$.

As an important proposition, we need the following well-known fact, that asserts, that even when the roots themselves are in the extension field $\mathbb{F}_{q^2}$, specific powers of them are in $\mathbb{F}_q$.

**Proposition 2.1.** – *If $f(x)$ splits over $\mathbb{F}_q$, then any root $r$ of $f(x)$ satisfies $r^{q-1} = 1$ in $\mathbb{F}_q$.*
– *If $f(x)$ is irreducible over $\mathbb{F}_q$, and if $r$, $\overline{r}$ denote the roots of $f(x)$ in $\mathbb{F}_{q^2}$ then $r^{q+1} = \overline{r}^{q+1} = Q \in \mathbb{F}_q$.*

*Proof.* If $f(x)$ is reducible, then it can certainly be written as $f(x) = (x - r)(x - \overline{r})$ for $r$, $\overline{r} \in \mathbb{F}_q$, where $r, \overline{r}$ are both different from 0, since $Q = r\overline{r} \in \mathbb{F}_q^*$. The assertion now immediately follows, since $r^q = r$. If $f(x)$ is not reducible in $\mathbb{F}_q[x]$ then it is known (cf. e.g. Theorem 9.3.11 of [20]) that $r$, $r^q \in \mathbb{F}_{q^2}$ are the two different roots of $f(x)$. Consequently, $r^q$ equals the conjugate root $\overline{r}$ of $r$, and therefore, $r^{q+1} = r\overline{r} = Q$. □

Throughout this paper *we will denote $\epsilon = 1$ if $f(x)$ is reducible in $\mathbb{F}_q[x]$, and $\epsilon = -1$ otherwise.*

For $2 \nmid q$ necessary and sufficient criteria for reducibility/irreducibility can be obtained by means of the quadratic character of the discriminant $D = P^2 - 4Q$. For $q$ odd let $\eta$ denote the *quadratic character* of $\mathbb{F}_q$, that is the real-valued function on $\mathbb{F}_q^*$ with $\eta(c) = 1$ if $c$ is a square of an element of $\mathbb{F}_q^*$ and $\eta(c) = -1$ otherwise.

It is known then that the polynomial $f(x) \in \mathbb{F}_q[x]$, $q$ odd, is irreducible in $\mathbb{F}_q[x]$ if and only if $\eta(D) = -1$.

### 2.2 Arithmetical Properties of the Lucas Sequences over $\mathbb{F}_q$

For $q = p$, an odd prime, a number of identities involving modular properties of the Lucas sequences modulo $p$ are known (cf., e.g. [4,5,10,20]). Here, we generalize these results over any finite field $\mathbb{F}_q$.

**Lemma 2.2.** *For any $P, Q \in \mathbb{F}_q$, $Q \neq 0$, the following identities hold in $\mathbb{F}_q$.*

$$V_q(P,Q) = P, \tag{1}$$

$$U_q(P,Q) = \epsilon, \tag{2}$$

$$U_{q-\epsilon}(P,Q) = 0, \tag{3}$$

$$V_{q-\epsilon}(P,Q) = \begin{cases} 2, & \text{if } \epsilon = 1, \\ 2Q, & \text{if } \epsilon = -1. \end{cases} \tag{4}$$

*Proof.* Properties (3) and (4) are immediate consequences of Proposition 2.1. Since the equations (1)-(4) are linearly dependent (cf. [10]) the other two assertions follow. □

Consider now the case that $q$ is odd. Since $\mathbb{F}_q^*$ is cyclic we can generalize Euler's criterion over $\mathbb{F}_q$. That is, for $a \neq 0 \in \mathbb{F}_q$ the equation $x^2 = a$ is solvable in $\mathbb{F}_q$ if and only if $a^{(q-1)/2} = 1$ in $\mathbb{F}_q$. (This can easily be derived from various sources, e.g. [16] p. 165.) An equivalent statement is $\eta(a) = a^{(q-1)/2}$.

**Lemma 2.3.** *If $q$ is odd and $Q, D \in \mathbb{F}_q^*$, then, in $\mathbb{F}_q$,*

$$\begin{cases} V_{(q-\eta(D))/2}(P,Q) = 0 & \text{when} \quad \eta(Q) = -1, \\ U_{(q-\eta(D))/2}(P,Q) = 0 & \text{when} \quad \eta(Q) = 1. \end{cases}$$

*Proof.* By applying formulas (4.2.7) and (4.3.4) of [20] we obtain for $\epsilon = \eta(D)$, $V_{q-\epsilon} = V^2_{\frac{q-\epsilon}{2}} - 2Q^{(q-\epsilon)/2}$, and, therefore, by Lemma 2.2, $V^2_{\frac{q-\epsilon}{2}} = 2Q^{(q-\epsilon)/2} + 2Q^{(1-\epsilon)/2} = 2Q^{(1-\epsilon)/2}(1 + Q^{(q-1)/2})$. If $\eta(Q) = -1$ then clearly $V_{(q-\epsilon)/2} = 0$. If $\eta(Q) = 1$ then the result follows from the identity $U_{q-\epsilon} = U_{(q-\epsilon)/2}V_{(q-\epsilon)/2}$, because $V_{\frac{q-\epsilon}{2}} \neq 0$. □

## 3 The Rank of Appearance of $(U)$ over $\mathbb{F}_q$

### 3.1 Definition and Fundamental Properties

**Definition 3.1.** Let $U = U(P,Q)$ be the Lucas sequence of the first kind. The rank of appearance $\rho = \rho(P,Q)$ in $\mathbb{F}_q$ is the least positive integer $l$ such that, in $\mathbb{F}_q$, $U_l(P,Q) = 0$.

The rank of appearance is sometimes called the rank of apparition. We refer to [4,10,11,12,17,18,20] for more details and historical developments.

**Lemma 3.2.** *The rank of appearance has the following properties.*

$$\rho(P,Q)|(q-\epsilon), \tag{5}$$

$$U_m(P,Q)=0 \text{ in } \mathbb{F}_q \quad \text{if and only if} \quad \rho(P,Q)|m, \tag{6}$$

$$\text{if } 2\nmid q \text{ then} \quad \rho(P,Q)\,\Big|\,\frac{q-\epsilon}{2} \quad \text{if and only if} \quad \eta(Q)=1. \tag{7}$$

*Proof.* Assertion (6) is well-known. Property (5) follows from Lemma 2.2, property (7) is an immediate consequence of Lemma 2.3. □

*Remark 3.3.* In [12,14] L. Somer has shown that for any divisor $d$ of $q-\epsilon$ when $2|q$, respectively for any divisor $d$ of $\frac{q-\eta(D)}{2}$, or $q-\eta(D)$ according as $\eta(Q_0)=1$ or $-1$ when $2\nmid q$ there do exist parameters $P$ for which $\rho(P,Q_0)=d$ when $Q=Q_0$ is fixed. Below we will actually establish the exact number of parameters that have the same rank.

The rank of appearance is closely connected to the period of linear recurrence sequences (cf. [4,12,13,14]).

In any finite field $\mathbb{F}_q$, the *period* $\mu$ of $(U)$ is defined to be the least positive integer $t$ such that $U_{n+t}=U_n$ for all $n\geq 0$. The *restricted period* $\alpha$ of $(U)$ is the least positive integer $t$ such that $U_{n+t}=sU_n$ for all non-negative integers $n$ and some non-zero element $s$, called the multiplier of $(U)$.

It is well-known that for second order recurrences, $\alpha=\rho$. Moreover, if the multiplicative order of the multiplier $s$ is denoted by $\beta$, then

$$\beta=\frac{\mu}{\alpha}. \tag{8}$$

## 3.2 The Rank of Appearance for a Fixed Parameter $Q$

### The case that $f(x)$ is irreducible in $\mathbb{F}_q[x]$.

In the following, denote by ord$(c)$ the multiplicative order of any $c\in\mathbb{F}_q^*$. Let $\nu_2(a)$ denote the highest power of 2 in the prime decomposition of $a$ and let $e=(\frac{q-1}{ord(Q)},q+1)$. We note that then $e\in\{1,2\}$ if $2\nmid q$. The following result can easily be derived from [12] for second order recurrences.

**Proposition 3.4.** *(i) For any $P\in\mathbb{F}_q$, $Q\in\mathbb{F}_q^*$ the rank of appearance has the form $\rho(P,Q)=\frac{q+1}{h}$, where $h|(q+1)$, $\gcd(\frac{h}{e},\text{ord}(Q))=1$ and*

$$\begin{cases} e=2 & \text{iff } \nu_2(\text{ord}(Q))<\nu_2(q-1), \quad \text{when } 2\nmid q,\\ e=1, & \text{when } 2|q.\end{cases}$$

*(ii) If $2 \nmid q$ then*

$$\begin{cases} \operatorname{ord}(Q) | \beta | 2\operatorname{ord}(Q), & \text{if } 2 \nmid \operatorname{ord}(Q), \\ \beta = 2\operatorname{ord}(Q), & \text{if } \left(\nu_2(\operatorname{ord}(Q)) < \nu_2(q-1), \text{ and } 2 | \operatorname{ord}(Q)\right), \\ \beta = \operatorname{ord}(Q), & \text{otherwise, i.e. when } \nu_2(\operatorname{ord}(Q)) = \nu_2(q-1). \end{cases}$$

*(iii) If $2|q$ then $\beta = \operatorname{ord}(Q)$.*

**The case that $f(x)$ is reducible in $\mathbb{F}_q[x]$.**

If $\eta$ is the quadratic character of $\mathbb{F}_q$ when $2 \nmid q$, then $\eta$ is defined for all nonzero elements of $\mathbb{F}_q$. It is convenient to extend this definition by setting $\eta(0) = 0$. Then one obtains (cf. [6]) $\sum_{c \in \mathbb{F}_q} \eta(c) = 0$ and, moreover, the property $\eta(a_1 a_2) = \eta(a_1)\eta(a_2)$ holds then for all $a_1, a_2 \in \mathbb{F}_q$. These properties will be fundamental in allowing us to apply Theorem 5.48 of [6] in the proof to the following lemma.

**Lemma 3.5.** *Let $Q_0 \in \mathbb{F}_q^*$ be fixed where $2 \nmid q$. Then the number of elements $P \in \mathbb{F}_q$ with $\eta(P^2 - 4Q_0) = \epsilon \in \{-1, 1\}$ is equal to*

$$\frac{q-\epsilon}{2} - \frac{1}{2}\left(\eta(Q_0) + 1\right).$$

*Proof.* Let $D(P) = P^2 - 4Q_0$. Then Theorem 5.48, p. 230 of [6] implies

$$\sum_{P \in \mathbb{F}_q} \eta(D(P)) = -\eta(1) = -1$$

because of $Q_0 \neq 0$. Hence, the number of $P \in \mathbb{F}_q$ with $\eta(D(P)) = 1$ equals

$$\sum_{\substack{P \in \mathbb{F}_q \\ P^2 \neq 4Q_0}} \frac{1}{2}(1 + \eta(D(P))) = \sum_{P \in \mathbb{F}_q} \frac{1}{2}(1 + \eta(D(P))) - \frac{1}{2}\left(\eta(4Q_0) + 1\right) =$$

$$= \frac{1}{2} \sum_{P \in \mathbb{F}_q} 1 + \frac{1}{2} \sum_{P \in \mathbb{F}_q} \eta(D(P)) - \frac{1}{2}\left(\eta(Q_0) + 1\right) = \frac{q-1}{2} - \frac{1}{2}\left(\eta(Q_0) + 1\right).$$

Consequently, the number of $P \in \mathbb{F}_q$ with $\eta(D(P)) = -1$ becomes

$$\sum_{P \in \mathbb{F}_q} 1 - \frac{q-1}{2} - \frac{1}{2}\left(\eta(Q_0) + 1\right). \qquad \square$$

**Lemma 3.6.** *Suppose that the characteristic polynomial $f(x) = x^2 - Px + Q$ splits over $\mathbb{F}_q$ with roots $r_1$ and $r_2$. Denote $G \in \mathbb{F}_q$ the quotient $\frac{r_1}{r_2}$, respectively $\frac{r_2}{r_1}$ where $Q = Q_0 \in \mathbb{F}_q^*$ is fixed. Let $P \in \mathbb{F}_q$ be varied. Then the number of such different $G$'s obtained equals*

$$\begin{cases} \frac{q-1}{2} - \frac{1}{2}\left(\eta(Q_0) + 1\right), & \text{if } 2 \nmid q, \\ q - 1, & \text{if } 2 | q. \end{cases}$$

*Proof.* Consider firstly the case $2 \nmid q$. Assume that $Q = Q_0$ is kept fixed and $P$ is being varied. According to Lemma 3.5 there are exactly $\frac{q-1}{2} - \frac{1}{2}(\eta(Q_0)+1)$ parameters $P \in \mathbb{F}_q$ for which $\eta(P^2 - 4Q_0) = \eta(D(P)) = 1$. Each of those $P$ determines the roots $r_1 = r_1(P)$ and $r_2 = r_2(P)$, and consequently, the corresponding value $G$.

For $2 \nmid q$ it can now be shown that the number of $G'$s in $\mathbb{F}_q$ equals the number of $P'$s obtained for a fixed $Q = Q_0$. Since $\mathbb{F}_q$ is a field, this follows exactly in the same manner as in the proof to Lemma 3.5 of [9] where the analogous statement was proved over $\mathbb{F}_q$ in the case that $q = p$, an odd prime.

Now, let $2|q$. Then $2q' = q$ for some integer $q'$. Consequently, $\left(Q^{q'}\right)^2 = Q$ for any $Q \in \mathbb{F}_q^*$. Moreover, $Q^{q'}$ is the uniquely determined element in $\mathbb{F}_q^*$ with this property. For, suppose to the contrary that there exists some $b \in \mathbb{F}_q^*$ such that $\left(Q^{q'}+b\right)^2 = Q = \left(Q^{q'}\right)^2$. Because $char(\mathbb{F}_q) = 2$, this yields, $\left(Q^{q'}\right)^2 + b^2 = \left(Q^{q'}\right)^2$, and therefore, $b = 0$, a contradiction.

In the following, let $Q = Q_0 \in \mathbb{F}_q^*$. We now determine the number of $P \in \mathbb{F}_q$ such that the characteristic polynomial w.r.t. $P$, $f(x) = f(x, P, Q_0)$ splits in $\mathbb{F}_q[x]$. Clearly, the reducibility of $f(x)$ in $\mathbb{F}_q[x]$ is equivalent to the existence to the roots $r_1, r_2 \in \mathbb{F}_q$ with $f(x) = (x - r_1)(x - r_2)$.

We firstly show that $P = 0$ if and only if $r_1 = r_2 = Q^{q'}$. Obviously, $P = r_1 + r_2 = 0$ exactly when $r_1 = r_2$ because $char(\mathbb{F}_q) = 2$. Moreover, since $r_2 = \frac{Q}{r_1}$, and $\left(Q^{q'}\right)^2 = Q$ we obtain $r_1 = Q^{q'}$ if and only if $r_2 = r_1$. Hence, if $r_1 \in \mathbb{F}_q^*$, $r_1 \neq Q^{q'}$, then the corresponding root $r_2$ is different from $r_1$.

Now, consider the case that $P \neq 0$. We show that exactly two roots $r_1, r_2 \in \mathbb{F}_q^*$ distinguish one parameter $P \in \mathbb{F}_q^*$. This follows, since $r_1, r_2$ can be listed in either order and $P = r_1 + r_2 = r_2 + r_1$. Also, if $r_1' \neq r_1$, $r_1' \neq r_2 \in \mathbb{F}_q^*$, then the $P'$ corresponding to $r_1'$ is different to the $P$ corresponding to $r_1$. For, assume otherwise that $r_1' + r_2' = r_1 + r_2$, then, since $char(\mathbb{F}_q) = 2$, we obtain $r_1' + r_1 = r_2' + r_2 = \frac{Q_0}{r_1'} + \frac{Q_0}{r_1} = \frac{Q_0(r_1'+r_1)}{r_1 r_1'}$. By hypothesis, $r_1' + r_1 \neq 0$ in $\mathbb{F}_q$, so that $r_1 r_1' = Q_0 = r_1 r_2$ and thus, $r_1' = r_2$, a contradiction.

In summarizing, we have shown that any two $r_1, r_2 \in \mathbb{F}_q^*$, $r_1 \neq Q^{q'}$ correspond to one $P \in \mathbb{F}_q$, $P \neq 0$. Consequently, there are $\frac{q-2}{2}$ parameters $P \in \mathbb{F}_q^*$ obtained this way. Similarly as in the proof to Lemma 3.5 in [9] we obtain the number of $G$'s by means of the number of the $P$'s. It can easily be shown that, for any $P \neq 0$ one obtains two distinct $G$'s. Additionally, since $2|q$, it can be shown that if $P'$ is different from $P$ then the $G$'s obtained for $P'$ are distinct from the $G$'s obtained for $P$. Since, additionally, for $r_1 = r_2 = Q^{q'}$ we have $P = 0$, one more $G$ is obtained, namely $G = 1$, since $r_1 = r_2$. This yields a total of $2\frac{q-2}{2} + 1$ of such values $G \in \mathbb{F}_q$ as claimed. □

By using the notation of Lemma 3.6, we have $s = r_1^\rho = r_2^\rho = G^\rho r_1^\rho$ in $\mathbb{F}_q$, and therefore (cf. [14])

**Corollary 3.7.** *If the characteristic polynomial $x^2 - Px + Q$ splits over $\mathbb{F}_q$, then for the value $G$ of Lemma 3.6 one has that $\rho(P, Q_0) = \mathrm{ord}(G)$.*

We have seen that the parameter $Q$ plays a unique role. In detail, the rank of appearance intimately depends on the quadratic character of $Q$ (compare Lemma 3.2). Accordingly, it makes sense to explicitly consider the case that $Q$ is a fixed value. We now establish the number of parameters $P$ with a specific rank. In section 5 below we then reverse the role of $P$ and $Q$ by keeping $P$ fixed and varying $Q$.

# 4 The Number of $P$ With the Same Rank When $Q = Q_0$

We now determine the number of parameters $P \in \mathbb{F}_q$ with the same rank of appearance $\rho(P, Q_0)$ for a fixed $Q = Q_0 \in \mathbb{F}_q^*$. The techniques employed will be generalizations of those in [9].

## 4.1 The cases $2|q$, respectively $2 \nmid q$ and $\eta(Q_0) = -1$.

If $2 \nmid q$, Lemma 3.2 asserts that $U_{(q-\epsilon)/2}(P, Q_0) \neq 0$, since, by hypothesis, $\eta(Q_0) = -1$. Therefore, it suffices to count the parameters with rank equal to $d$, where $d$ is a divisor of $q - \eta(D)$ with $\nu_2(d) = \nu_2(q - \eta(D))$.

Consider firstly the case of irreducible characteristic polynomials. By hypothesis, we have $\nu_2(\mathrm{ord}(Q_0)) = \nu_2(q-1)$ when $2 \nmid q$. Hence, for both $2 \nmid q$ and $2|q$, it follows from Proposition 3.4 that every $\alpha = \rho$ has the form $\frac{q+1}{t}$, where $t$ is any integer relatively prime to $\mathrm{ord}(Q)$ such that $t|(q+1)$.

Now, let $d = \frac{q+1}{t}$ arbitrarily. We wish to evaluate the number of sequences $U = U(P)$ over $\mathbb{F}_q$ with $\rho(P, Q_0) = d$. By Proposition 3.4, $\beta = \mathrm{ord}(Q_0)$. This follows immediately for the case $2|q$, while if $2 \nmid q$, then $\nu_2(\mathrm{ord}(Q)) = \nu_2(q-1)$ by hypothesis. Therefore, as $Q = Q_0$ is fixed, we are by property (8) counting the sequences with $d' := \mu = \mathrm{ord}(Q_0)\frac{q+1}{t}$. However, the number of sequences over $\mathbb{F}_q$ with irreducible characteristic polynomial and fixed $Q_0$ such that $\mu = d'$, is according to Theorem 2 of [12] equal to $\frac{\phi(d')}{2\phi(ord(Q_0))}$. By extracting the powers of 2 and applying the multiplicative property of the Euler $\phi$-function, this quantity becomes $\phi(\frac{q+1}{t}) = \phi(d)$.

We thus have shown

**Theorem 4.1.** *Let $Q_0 \in \mathbb{F}_q^*$ be fixed and $\eta(D) = -1$ where $D = P^2 - 4Q_0$.*

- *If $2 \nmid q$ let $\eta(Q_0) = -1$. Then, for any divisor $d$ of $q+1$ with $\nu_2(d) = \nu_2(q+1)$ there are $\phi(d)$ parameters $P \in \mathbb{F}_q$ for which $\rho(P, Q_0) = d$.*
- *If $2|q$, then for any divisor $d > 1$ of $q+1$ there are $\phi(d)$ parameters $P \in \mathbb{F}_q$ for which $\rho(P, Q_0) = d$.*

Observe that, if $2 \nmid q$, the condition $\nu_2(\mathrm{ord}(Q_0)) = \nu_2(q-1)$ is essential for allowing us to apply Proposition 3.4. Consequently, the case $\eta(Q_0) = 1$ will have to be dealt with in a different way.

The corresponding result is also true when the characteristic polynomial splits in $\mathbb{F}_q[x]$.

**Theorem 4.2.** *Let $Q_0 \in \mathbb{F}_q^*$ be fixed and $\eta(D) = 1$ where $D = P^2 - 4Q_0$.*

- *If $2 \nmid q$ let $\eta(Q_0) = -1$. Then, for any divisor $d$ of $q-1$ with $\nu_2(d) = \nu_2(q-1)$ there are $\phi(d)$ parameters $P \in \mathbb{F}_q$ for which $\rho(P, Q_0) = d$.*
- *If $2|q$, then for any divisor $d > 1$ of $q-1$ there are $\phi(d)$ parameters $P \in \mathbb{F}_q$ for which $\rho(P, Q_0) = d$.*

*Proof.* The desired number can by Lemma 3.6 and Corollary 3.7 be determined as the number of $G \in \mathbb{F}_q^*$ for which $\mathrm{ord}(G) = d$ and, if $2 \nmid q$, additionally, $\eta(G) = -1$ since $Q_0^{\frac{q-1}{2}} = G^{\frac{q-1}{2}}$. Let firstly $2 \nmid q$. Then, obviously the number of $a \in \mathbb{F}_q^*$ with $\eta(a) = -1$ and $\mathrm{ord}(a) = d$ equals the number of all $a \in \mathbb{F}_q^*$ with $\mathrm{ord}(a) = d$, when $\nu_2(d) = \nu_2(q-1)$. Consequently, for $2 \nmid q$, and trivially also for the case $2|q$, Proposition 1.1 establishes the result. □

## 4.2 The case $2 \nmid q$ and $\eta(Q_0) = 1$.

It turns out that the methods developed above do not carry over to the case that $2 \nmid q$, $\eta(Q_0) = 1$. We therefore have to introduce some new concepts. Below, we let $\eta(Q_0) = \pm 1$ and $q$ arbitrarily unless it is explicitly stated that $\eta(Q_0) = 1$, respectively $2 \nmid q$.

**Definition 4.3.** Let $Q_0 \in \mathbb{F}_q^*$, $\epsilon \in \{-1, 1\}$ be fixed, and $d > 1$ be any divisor of $q - \epsilon$. The function $\bar{\psi}(d, Q_0, \epsilon)$ is defined to be the number of distinct values of $P \in \mathbb{F}_q$ for which $\rho(P, Q_0) = d$.

*Remark 4.4.* Note that, if $2 \nmid q$, then $\bar{\psi}(d, Q_0, \epsilon)$ counts the values of $P \in \mathbb{F}_q$ with both $\eta(P^2 - 4Q_0) = \epsilon$ and $\rho(P, Q_0) = d|(q-\epsilon)$ if $d > 1$. If $2|q$, then $\bar{\psi}(d, Q_0, \epsilon)$ counts the corresponding $P$'s for which $f(x) = f(x, P, Q_0)$ is reducible, respectively irreducible over $\mathbb{F}_q$, according as $\epsilon = 1$ or $-1$.

In the case under inspection we have $2 \nmid q$ and $\eta(Q_0) = 1$. Therefore, over $\mathbb{F}_q$, $U_{q-\epsilon}(P, Q_0) = 0$ if and only if $U_{(q-\epsilon)/2}(P, Q_0) = 0$.

For any divisor $d > 2$ of $\delta = \frac{q-\epsilon}{2}$ we now count the number of solutions $P \in \mathbb{F}_q$ of the polynomial $U_d(P) = 0$. Since $d|\delta$, we have $U_\delta(P) = U_d(P) W_{\delta,d}(P)$. The degrees of both of the polynomials $U_k$ and $W_{\delta,d}$ can be derived by means of the formulas of the Dickson polynomials $D_k$ and $E_{k-1}$ (cf. [19,9]). One then can conclude exactly in the same manner as in the proof of Theorem 3.12 of [9] and establish the exact number of solutions $U_d(P)$, and consequently, of $U_{q-\epsilon}(P)$. This is possible because the number of $P$ with $\eta(P^2 - 4Q_0) =$

$\left(\frac{P^2-4Q_0}{p}\right) = \epsilon$ over $\mathbb{F}_p$, $p$ an odd prime, that is essential in establishing the proof in [9] can, according to Lemma 3.5, analogously be obtained over $\mathbb{F}_q$, where $q$ is an odd prime power.

As a generalization of the results in [9] one then obtains the analogous result to $\mathbb{F}_p$ for any $\mathbb{F}_q$.

**Theorem 4.5.** *Let $2 \nmid q$ and $Q = Q_0 \in \mathbb{F}_q^*$ be fixed such that $\eta(Q_0) = 1$. If $d > 1$ is a divisor of $q - \eta(D)$ with $\nu_2(d) < \nu_2(q - \eta(D))$, then there are $\phi(d)$ parameters $P \in \mathbb{F}_q$ for which $\rho(P, Q_0) = d$.*

In summarizing the above results and noting that $U_2(P, Q_0) = 0$ over $\mathbb{F}_q$ exactly for $P = 0$, we therefore obtain the main result concerning the number of $P$ with the same rank when $Q = Q_0$ is fixed. As indicated above, this number turns out to be the exact analogue to the result of Proposition 1.1 concerning the number of parameters with the same multiplicative order.

**Corollary 4.6.** *Number of parameters P with same rank of appearance*

*Let $d \geq 1$ and suppose $d$ is a divisor of $q - \epsilon$.*

*(i) If $2 \nmid q$ and $\eta(Q_0) = -1$ then*

$$\bar{\psi}(d, Q_0, \epsilon) = \begin{cases} \phi(d), & \text{if } \nu_2(d) = \nu_2(q-\epsilon), \\ 0, & \text{otherwise.} \end{cases}$$

*(ii) Further if $2 \nmid q$ and $\eta(Q_0) = 1$ then*

$$\bar{\psi}(d, Q_0, \epsilon) = \begin{cases} \phi(d), & \text{if } \nu_2(d) < \nu_2(q-\epsilon), \\ 0, & \text{otherwise.} \end{cases}$$

*(iii) If $2 | q$, then*

$$\bar{\psi}(d, Q_0, \epsilon) = \phi(d).$$

## 5 The Number of $Q$ With the Same Rank When $P = P_0$

We now reverse the role of $P$ and $Q$ by keeping $P = P_0$ fixed. Our goal now consists of determining the number of $Q$ with the same rank. It turns out that the above method developed for $2 \nmid q$ and $\eta(Q_0) = 1$ carries over to this case. The analogous definition to above is

**Definition 5.1.** Let $P_0 \in \mathbb{F}_q^*$, $\epsilon \in \{-1, 1\}$ be fixed, and $d > 1$ be any divisor of $q - \epsilon$. Denote by $\underline{\psi}(d, P_0, \epsilon)$ the number of distinct values of $Q \in \mathbb{F}_q^*$ for which $\rho(P_0, Q) = d$.

*Remark 5.2.* Obviously, if $2 \nmid q$, we now count the number of $Q$'s such that $\eta(P_0^2-4Q)=\epsilon$ and $\rho(P_0,Q)=d|(q-\epsilon)$ if $d>2$. If $2|q$, then $\bar{\psi}(d,Q_0,\epsilon)$ counts the corresponding $Q$'s for which $f(x)=f(x,P_0,Q)$ is reducible, respectively irreducible over $\mathbb{F}_q$, according as $\epsilon=1$ or $-1$. We exclude $P_0=0$, because otherwise trivially $U_2(P_0,Q)=0$, so that the rank would be 2 for all $Q$. Consequently, if $P_0\in\mathbb{F}_q^*$, $\underline{\psi}(d,P_0,\epsilon)=0$ when $d=2$.

Let firstly $2\nmid q$. Then the method above can be employed in an analogous way. We need to establish the number of $Q\in\mathbb{F}_q^*$ such that $\eta(P_0^2-4Q)=\epsilon$ when $P_0$ is fixed and $\epsilon\in\{-1,1\}$. The desired number becomes, firstly in $\mathbb{F}_q$

for $\epsilon=1$,
$$\sum_{\substack{Q\in\mathbb{F}_q\\ P_0^2\neq 4Q}}\frac{1}{2}\left(1+\eta(P_0^2-4Q)\right)=\frac{q}{2}-\frac{1}{2}+\frac{1}{2}\sum_{Q\in\mathbb{F}_q}\eta(P_0^2-4Q)=$$
$$=\frac{q-1}{2}+\frac{1}{2}\sum_{Q\in\mathbb{F}_q}\eta(Q)=\frac{q-1}{2},$$
and thus, for $\epsilon=-1$,
$$\sum_{\substack{Q\in\mathbb{F}_q\\ P_0^2\neq 4Q}}1-\frac{q-1}{2}=\frac{q-1}{2}.$$

Now, the number of parameters $Q\in\mathbb{F}_q$, $Q\neq 0$, follows by observing that, if $Q=0$, we have $\eta(P_0^2)=1$, so that for $\epsilon=1$ we obtain the desired number as $\frac{q-1}{2}-1=\frac{q-3}{2}$. Consequently, the above number of $Q$ in $\mathbb{F}_q^*$ becomes
$$\frac{q-2-\epsilon}{2}.$$

This is the exact generalization of the analogous number over $\mathbb{F}_p$, which is required in the proof [9]. In exactly the same manner as in [9] we therefore obtain for $2\nmid q$ the main result in Theorem 5.3 stated below.

However, if $2|q$, we cannot directly employ the techniques of [9]. This is due to the functions
$$2^{k-1}U_k(P,Q_0)=\tilde{T}_k(Q)=\sum_{r=0}^{[k/2]-\gamma}\binom{k}{2r+1}P_0^{k-2r-1}(P_0^2-4Q)^r$$
$$2^{k-1}V_k(P,Q_0)=\tilde{S}_k(Q)=\sum_{r=0}^{[k/2]}\binom{k}{2r}P_0^{k-2r}(P_0^2-4Q)^r$$
that constituted the underlying part in determining the number of zeros of $U_k$ and $V_k$ in [9].

However, as $2|q$ and $|\epsilon|=1$, we only need to deal with formulae for $U_k$ and $V_k$, where $2\nmid k$, since, in this case, $k=q-\epsilon$, respectively $k|(q-\epsilon)$ is odd.

We can thus replace the above function $\tilde{T}$ by the following function $\bar{T}$, where $2 \nmid d$ (cf. [19]),

$$(-1)^r U_{(2r+1)d}(P_0, Q) = \bar{T}_{(2r+1)d}(Q) =$$
$$= U_d(P_0, Q) \sum_{j=0}^{r} (-1)^j \binom{r+j}{r-j} Q^{d(r-j)} V_d^{2j}(P_0, Q). \qquad (9)$$

For a fixed $P = P_0$ the Lucas function $U_k(P_0, Q)$ of odd degree $k = (2r+1)d$, $\sum_{i=0}^{(k-1)/2} \binom{k-1-i}{i} (-Q)^i P_0^{k-1-2i}$ is a monic polynomial of degree $\frac{k-1}{2}$, so that also $\bar{T}_k(Q)$ has degree $\frac{k-1}{2}$.

Similarly, the function $V_d(P_0, Q)$ for $2 \nmid d$, $\sum_{i=0}^{[d/2]} \frac{d}{d-i} \binom{d-i}{i} (-Q)^i P_0^{d-2i}$ has degree $\frac{d-1}{2}$ with leading coefficient $dP_0$.

If we denote $\mathcal{R}$ the sum occurring in (9), then it follows that $\mathcal{R} = \mathcal{R}_{k,d} = \mathcal{R}_{(2r+1)d,d}(Q)$ is a monic polynomial of degree $\frac{k-d}{2} = \frac{(2r+1)d-d}{2} = dr$. This follows, since the exponent of $Q$ is $d(r-j) + j(d-1)$ which is maximal for $j = 0$.

We can now adapt the proof of [9]. Thus, we count the number of solutions of $\bar{T}_d(Q) = 0$ over $\mathbb{F}_q$, where $d|(q-\epsilon)$.

If, firstly $d = \delta = q - \epsilon$, then

$$\bar{T}_\delta(Q) = 0 \text{ if and only if } U_\delta(P_0, Q) = 0.$$

The solutions are obviously those $Q \in \mathbb{F}_q^*$ for which $f(x, P_0, Q)$ is reducible, respectively, irreducible, according as $\epsilon = 1$ or $-1$. We claim that this number of $Q$ equals

$$\frac{q - \epsilon - 1}{2} = \frac{\delta - 1}{2}. \qquad (10)$$

For $\epsilon = 1$ this claim follows by counting the number of characteristic roots $r, \bar{r} \in \mathbb{F}_q^*$ of $f(x)$. There are obviously $q-1$ choices for $r$. Since $char(\mathbb{F}_q) = 2$ we need to have $r \neq P_0$ since otherwise $\bar{r} = 0$ and thus $Q = 0$. For $r \neq P_0 \in \mathbb{F}_q^*$ it follows then that $\bar{r} \neq r$ and $\bar{r}$ is uniquely determined from $r$. We therefore have $q-2$ different pairs $(r, \bar{r})$. Since $r\bar{r} = \bar{r}r$, there are $\frac{q-2}{2}$ parameters $Q$ obtained this way. These are all different. For suppose to the contrary that there exists some $r'$ with $r' \neq r$, $r' \neq \bar{r}$, but $r\bar{r} = r'\overline{r'}$. Then, as $P_0$ is fixed, $r(r + P_0) = r'(r' + P_0)$, so that, since $char(\mathbb{F}_q) = 2$, $r^2 + r'^2 = P_0(r + r')$, and thus, $r + r' = P_0$, and consequently, $r' = \bar{r}$, a contradiction.

For $\epsilon = -1$ the assertion (10) follows from the number of all $Q \in \mathbb{F}_q^*$ minus the number of those where $\epsilon = 1$ since $P_0^2 - 4Q = P_0^2 = 0$ is impossible by the choice of $P_0$. This proves the number of zeros of $\bar{T}_{\delta(Q)} = 0$ stated in (10).

Now, for a proper divisor $d$ of $(q-\epsilon)$ let

$$\bar{T}_d(Q) = 0 \text{ in } \mathbb{F}_q \qquad (11)$$

have $j$ solutions. Then the identity

$$\bar{T}_\delta(Q) = \bar{T}_d(Q)\mathcal{R}_{\delta,d}(Q) \tag{12}$$

shows that

$$\mathcal{R}_{\delta,d}(Q) = 0 \tag{13}$$

has $\frac{\delta-1}{2} - j$ solutions. On the other hand, if we put $\delta = (2r+1)d$, then we know from the degree of $\mathcal{R}_{\delta,d}(Q)$ that the number of solutions $Q$ of (13) is at most $\frac{\delta-d}{2}$. We note that $j \leq \frac{d-1}{2}$ since $\bar{T}_d(Q)$ has degree $\frac{d-1}{2}$ with leading coefficient 1. If we assume that $j < \frac{d-1}{2}$ then (13) has $\frac{\delta-1}{2} - j > \frac{\delta-1}{2} - \frac{d-1}{2} = \frac{\delta-d}{2}$ solutions which cannot occur.

Consequently, for all divisors $d$ of $q - \epsilon$, we have shown that $j = \frac{d-1}{2}$.

By introducing the function $\chi(h) = \underline{\psi}(h, P_0, \epsilon)$ for $h \neq 1$, and $\chi(1) = 1$ otherwise, we therefore can argue as in [9] and obtain, since $2 \nmid d$,

$$\sum_{h|d} \chi(h) = \frac{d-1}{2}. \tag{14}$$

Finally, Möbius inversion formula gives

$$\chi(d) = \sum_{h|d} \mu(h) \frac{d-h}{2h} = \frac{1}{2}\left( d \sum_{h|d} \frac{\mu(h)}{h} - \sum_{h|d} \mu(h) \right) = \frac{\phi(d)}{2},$$

since for $d > 2$ the last sum reduces to zero. By definition, for $d > 2$ this now is the desired number of $Q$ with the same rank $d$.

We thus have established our main result of this section. Recall, that if $d = 2$ then there are no elements $Q$ with rank equal to $d$.

**Theorem 5.3.** *Number of parameters Q with same rank of appearance*

*If $d > 2$, $d|(q - \epsilon)$, with $|\epsilon| = 1$, then*

$$\underline{\psi}(d, P_0, \epsilon) = \frac{\phi(d)}{2}.$$

*That is, if $P = P_0 \in \mathbb{F}_q^*$ is fixed, then for any divisor $d > 2$ of $q - \epsilon$ there exist exactly $\frac{\phi(d)}{2}$ elements $Q \in \mathbb{F}_q^*$ with rank of appearance equal to $d$.*

In summarizing the results of the last two sections, we have shown another attribute of the rank of appearance that demonstrates its similarity to the multiplicative order of any group element. By considering the two above cases of counting the parameters $P$ and $Q$ we have proven the analogous results to Proposition 1.1 for the number of parameters with the same rank.

# 6 The Number of Zeros of the Lucas Sequences

In this section we show that the number of zeros can very easily be obtained from the number of parameters with the same rank of appearance. We need some preliminary facts.

**Proposition 6.1.** *If $k, n, i, j$ are positive integers, then the following conditions hold over any field $\mathbb{F}_q$.*

$$U_k(P,Q) = 0 \text{ iff } V_{2^i k}(P,Q) \neq 0 \text{ for every } i \geq 0 \tag{15}$$
$$V_{2^i k}(P,Q) = 0 \text{ iff } V_{2^j k}(P,Q) \neq 0 \text{ whenever } i \neq j \tag{16}$$
$$U_{2k}(P,Q) = 0 \text{ iff either } U_k(P,Q) \text{ or } V_k(P,Q) = 0 \tag{17}$$

*Proof.* The first two assertions are obvious. So is the third when considering the well known identity $U_{2n}(P,Q) = U_n(P,Q)V_n(P,Q)$ and the fact that $U_n(P,Q)$ and $V_n(P,Q)$ cannot simultaneously be zero in $\mathbb{F}_q$. □

We firstly state the number of zeros of the Lucas sequences in dependence on the value $\epsilon$ and then give the general result. Recall that $\epsilon$ is defined to be 1 if the characteristic polynomial $f(x,P,Q)$ of the Lucas sequences is reducible, and $\epsilon = -1$ if $f(x,P,Q)$ is irreducible over $\mathbb{F}_q[x]$. We will denote the *signature of* $(P,Q)$ *over* $\mathbb{F}_q$ the corresponding value $\epsilon$ (cf. also [15]).

## 6.1 The Number of Zeros $Q$ when $P = P_0$ is Fixed

**Lemma 6.2.** *Let $k$ be a positive integer and $\epsilon \in \{-1, 1\}$ a constant. For a fixed value of $P_0 \in \mathbb{F}_q^*$, the number of $Q \in \mathbb{F}_q^*$ such that $(P_0, Q)$ has signature $\epsilon$ and $U_k(P_0, Q) = 0$ in $\mathbb{F}_q$, is*

$$\begin{cases} \frac{(k,q-\epsilon)}{2} - 1, & \text{if } 2|k \text{ and } 2 \nmid q, \\ \frac{(k,q-\epsilon)-1}{2}, & \text{else.} \end{cases}$$

*Proof.* Lemma 3.2 asserts that $U_k(Q) = 0$ in $\mathbb{F}_q$ iff $\rho(P_0,Q)|k$. Moreover, since $\rho(P_0,Q)$ always divides $q-\epsilon$, we obtain the condition

$$\rho(P_0,Q)|(k,q-\epsilon). \tag{18}$$

By definition, $\underline{\psi}(t,P_0,\epsilon)$ counts the number of $Q \in \mathbb{F}_q^*$ with $\rho(P_0,Q) = t|(q-\epsilon)$. By aid of Theorem 5.3 we thus obtain the number we are looking for as

$$\sum_{\substack{d|(k,q-\epsilon) \\ d>2}} \frac{\phi(d)}{2} = \begin{cases} \frac{(k,q-\epsilon)-2}{2}, & \text{if } 2|k \text{ and } 2 \nmid q, \\ \frac{(k,q-\epsilon)-1}{2}, & \text{else.} \end{cases}$$

□

We therefore have our main theorem concerning the number of zeros $Q$ w.r.t. the $(U)$ sequence when $P = P_0$ is fixed.

**Theorem 6.3.** *Zeros of $U_k(P_0, Q)$*

*Let $k$ be a positive integer. For a fixed value of $P_0 \in \mathbb{F}_q^*$, the number of $Q \in \mathbb{F}_q^*$ such that $P_0^2 - 4Q = D(Q) \neq 0$ and $U_k(P_0, Q) = 0$ in $\mathbb{F}_q$, is*

$$\begin{cases} \frac{(k,q-1)+(k,q+1)}{2} - 2, & \text{if } 2|k \text{ and } 2 \nmid q, \\ \frac{(k,q-1)+(k,q+1)-2}{2}, & \text{else.} \end{cases}$$

Most interestingly, for the companion $(V)$ sequence in many cases there don't exist any zeros at all.

**Lemma 6.4.** *For any integer $k$ and any fixed $P_0 \in \mathbb{F}_q^*$ the number of $Q$ with signature $\epsilon$ and $V_k(P_0, Q) = 0$ in $\mathbb{F}_q$, is given as*

$$\begin{cases} \frac{1}{2}(k, q-\epsilon), & \text{for } 2|k \text{ and } \nu_2(k) < \nu_2(q-\epsilon), \\ \frac{(k,q-\epsilon)-1}{2}, & \text{for } 2 \nmid k \text{ and } 2 \nmid q, \\ 0, & \text{otherwise.} \end{cases}$$

*Proof.* By Proposition 6.1 we need to count those $Q$ for which $U_{2k}(P_0, Q) = 0$ and additionally $U_k(P_0, Q) \neq 0$. As a consequence of Lemma 6.2 we obtain the desired number as $\frac{1}{2}(2k, q-\epsilon) - 1 - \frac{1}{2}(k, q-\epsilon) + 1$, when $2|k$ and $2 \nmid q$, as $\frac{(2k,q-\epsilon)}{2} - 1 - \frac{(k,q-\epsilon)-1}{2}$, when $2 \nmid k, 2 \nmid q$, and as $\frac{(2k,q-\epsilon)-1}{2} - \frac{(k,q-\epsilon)-1}{2} = 0$, when $2|q$. The desired assertions then can easily be verified. □

Let now $2 \nmid q$. Then, if $2|k$ the number of zeros $Q$ does not vanish only when $\nu_2(k) < \nu_2(q-\epsilon)$. This is only possible for $4|(q-\epsilon)$. Consequently, if $\epsilon$ is replaced by $-\epsilon$ then there are no zeros $Q$ of $V_k(P_0, Q)$. This shows that for even $k$ *there is at most one signature $\epsilon$ (that allows zeros). If this $\epsilon \in \{-1, 1\}$ with $\nu_2(k) < \nu_2(q-\epsilon)$ exists, then we will denote this unique signature by $\tilde{\epsilon}$.*

We obtain therefore the formula for the number of zeros $Q$ w.r.t. the $(V)$ sequence as follows.

**Theorem 6.5.** *Zeros of $V_k(P_0, Q)$*

*Let $k$ be a positive integer and $2 \nmid q$. For a fixed value of $P_0 \in \mathbb{F}_q^*$, the number of $Q \in \mathbb{F}_q^*$ such that $V_k(P_0, Q) = 0$ in $\mathbb{F}_q$, is*

$$\begin{cases} 0, & \text{for } 2|k \text{ and } \nu_2(k) \geq \nu_2(q-1),\ \nu_2(k) \geq \nu_2(q+1), \\ \frac{1}{2}(k, q-\tilde{\epsilon}), & \text{for } 2|k \text{ when } \nu_2(k) < \nu_2(q-\tilde{\epsilon}), \\ \frac{(k,q-1)+(k,q+1)-2}{2}, & \text{for } 2 \nmid k. \end{cases}$$

*If $2|q$ then there are no zeros $Q$ of $V_k(P_0, Q)$ in $\mathbb{F}_q$.*

### 6.2 The Number of Zeros $P$ When $Q = Q_0$ is Fixed

As above, we firstly establish the number of zeros $P$ w.r.t. a specific signature $\epsilon$. Since $Q_0$ is fixed, the zeros of $U_k$, respectively $V_k$ heavily depend on the quadratic character of $Q_0$. We will consider these cases separately.

**Lemma 6.6.** *Let $k$ be a positive integer and $\epsilon \in \{-1, 1\}$ a constant. For a fixed value of $Q_0 \in \mathbb{F}_q^*$, the number of $P \in \mathbb{F}_q^*$ such that $(P, Q_0)$ has signature $\epsilon$ and $U_k(P, Q_0) = 0$ in $\mathbb{F}_q$, is, when $2 \nmid q$, given in the following way.*

*(i) For* $\eta(Q_0) = -1$ *as* $\begin{cases} \frac{(k,q-\epsilon)}{2}, & \text{when } \nu_2(k) \geq \nu_2(q-\epsilon), \\ 0, & \text{otherwise,} \end{cases}$

*(ii) for* $\eta(Q_0) = 1$ *as* $\begin{cases} \frac{(k,q-\epsilon)}{2} - 1, & \text{when } \nu_2(k) \geq \nu_2(q-\epsilon), \\ (k, q-\epsilon) - 1, & \text{otherwise.} \end{cases}$

*If $2|q$, then the corresponding number of zeros equals* $(k, q-\epsilon) - 1$.

*Proof.* We now count the parameters $P \in \mathbb{F}_q$ with $\rho(P, Q_0) = t|(q-\epsilon)$. This number was defined to be $\overline{\psi}(t, Q_0, \epsilon)$. By Theorem 4.6 the number of the $P$'s depends on the quadratic residue symbol $\eta(Q_0)$ when $2 \nmid q$. Since the proof is very similar to the one for Lemma 6.2, we omit the details. □

Obviously, also in this case, if $2 \nmid q$, the signature depends on $k$. To obtain a general formula of the zeros of $U_k(P, Q_0)$ in $\mathbb{F}_q$ the corresponding quantities occurring in Lemma 6.6 need to be added with respect to the values of $\epsilon = 1$ or $-1$ and in dependence of $\nu_2(k)$, when necessary.

We then obtain the main result concerning the zeros of the $(U)$ sequence.

**Theorem 6.7.** *Zeros of $U_k(P, Q_0)$*

*Let $k$ be a positive integer and $2 \nmid q$. For a fixed value of $Q_0 \in \mathbb{F}_q^*$, the number of $P \in \mathbb{F}_q^*$ such that $U_k(P, Q_0) = 0$ in $\mathbb{F}_q$, is, when $2 \nmid q$,*

- *for $\eta(Q_0) = -1$ given as*

$$\begin{cases} \gamma_1 \frac{(k,q+1)}{2} + \gamma_2 \frac{(k,q-1)}{2}, & \text{if } \nu_2(k) > 1, \text{ where} \\ & \begin{cases} \gamma_1 = 1, \text{ if } \nu_2(k) \geq \nu_2(q+1), \text{ and 0 otherwise,} \\ \gamma_2 = 1, \text{ if } \nu_2(k) \geq \nu_2(q-1), \text{ and 0 otherwise,} \end{cases} \\ \frac{(k,q-\check{\epsilon})}{2}, & \text{if } \nu_2(k) = 1 \text{ and } \check{\epsilon} \equiv -q \bmod 4, |\check{\epsilon}| = 1, \\ 0, & \text{if } \nu_2(k) = 0, \end{cases}$$

- *for $\eta(Q_0) = 1$ given as*

$$\begin{cases} \frac{(k,q+1)}{2^{\gamma_1}} + \frac{(k,q-1)}{2^{\gamma_2}} - 2, & \text{for } \nu_2(k) > 1, \text{ where } \gamma_1, \gamma_2 \text{ are as above,} \\ \frac{(k,q-\check{\epsilon})}{2} + (k, q+\check{\epsilon}) - 2, & \text{for } \nu_2(k) = 1, \text{ where } \check{\epsilon} \text{ is as above,} \\ (k, q-1) + (k, q+1) - 2, & \text{for } \nu_2(k) = 0. \end{cases}$$

*If* $2|q$*, then the corresponding number of zeros equals* $(k,q+1)+(k,q-1)-2$.

For completeness, we establish the corresponding number for the $(V)$ sequence. We note that for the case that $Q=Q_0$ is fixed this number already follows from a special case of the value set problem of the Dickson polynomials [3].

By means of the rank of appearance we very easily obtain an alternative proof of this result.

**Lemma 6.8.** *Under the hypotheses of Lemma 6.6, the number of* $P \in \mathbb{F}_q$ *with* $V_k(P,Q_0)=0$ *for a fixed* $Q_0 \in \mathbb{F}_q^*$ *is, for* $2 \nmid q$

– *when* $\eta(Q_0)=-1$, *given as* $\begin{cases} (k, \frac{q-\epsilon}{2}), & \text{for } \nu_2(k)+1=\nu_2(q-\epsilon), \\ 0, & \text{otherwise,} \end{cases}$

– *when* $\eta(Q_0)=1$, *given as* $\begin{cases} (k, q-\epsilon), & \text{for } \nu_2(k)+1<\nu_2(q-\epsilon), \\ 0, & \text{otherwise.} \end{cases}$

*If* $2|q$ *then there are no zeros* $P$ *of* $V_k(P,Q_0)$.

*Proof.* We again omit the details since the proof runs exactly along the same lines as above. □

We note that, w.r.t. the zeros of $V_k(P,Q_0)$ when $2 \nmid q$, there is also at most one signature such that the number of parameters does not vanish. *If it exists, we denote this signature* $\epsilon$ *by* $\hat{\epsilon}$*, that is, let* $\hat{\epsilon} \in \{-1,1\}$ *be such that* $\nu_2(2k)=\nu_2(q-\hat{\epsilon})$*, respectively* $\nu_2(2k)<\nu_2(q-\hat{\epsilon})$*, according as* $\eta(Q_0)=-1$ *or* $1$.

Consequently, we can formulate the precise number of zeros $P$ of $V_k(P,Q_0)$ as follows.

**Theorem 6.9.** *Zeros of* $V_k(P,Q_0)$

*Let* $k$ *be a positive integer. For a fixed value of* $Q_0 \in \mathbb{F}_q^*$*, the number of* $P \in \mathbb{F}_q^*$ *such that* $V_k(P,Q_0)=0$ *in* $\mathbb{F}_q$*, for* $2 \nmid q$*, is*

- *when* $\hat{\epsilon}$ *exists, equal to*

$$\begin{cases} (k, \frac{q-\hat{\epsilon}}{2}), & \text{if } \eta(Q_0)=-1, \\ (k, q-\hat{\epsilon}), & \text{if } \eta(Q_0)=1. \end{cases}$$

- *equal to* $0$ *otherwise.*

*If* $2|q$ *then there are no zeros* $P$ *of* $V_k(P,Q_0)$ *in* $\mathbb{F}_q$.

In summarizing this section, we have demonstrated how the number of parameters with the same rank can very efficiently be applied to establishing a detailed formula for the number of zeros of the Lucas sequences, respectively Dickson polynomials.

**Acknowledgements**

I am grateful to Professor Winfried B. Müller for numerous helpful and stimulating discussions, and to Professor Lawrence Somer for his qualifying advice. Also, I am indebted to Professor Gary Mullen for his interest in my work, his many insightful remarks, and for sending me reference [3]. Thanks are also due to Professor Wun-Seng Chou for sending me references [1,2], and to the referee for his careful reading of the manuscript.

## References

1. Bhargava, M., Zieve, M. E., Factoring Dickson polynomials over finite fields, *Finite Fields and Their Applications* **5**, (1999), 103-111.
2. Chou, W.S., The factorization of Dickson polynomials over finite fields, *Finite Fields and Their Applications* **3**, (1997), 84-96.
3. Chou, W.S., Gomez-Calderon, J., Mullen, G.L., Value sets of Dickson polynomials over finite fields, *J. of Number Theory* **30**, (1988), 334-344.
4. Carmichael, R.D., On sequences of integers defined by recurrence relations, *Quart. J. Pure Appl. Math.* **48**, (1920), 343-372.
5. Lidl, R., Mullen, G.L., Turnwald, G., *Dickson Polynomials.* Pitman Monographs and Surveys in Pure and Applied Mathematics, Vol. 65. Longman, London, 1993.
6. Lidl, R., Niederreiter, H., *Finite Fields*, Encyclopedia of Mathematics and its Applications, Vol. 20, Addison-Wesley, Reading, Massachusetts, 1983.
7. Müller, S., On strong Lucas pseudoprimes. In: Dorninger, D., Eigenthaler, G., Kaiser, H.K., Kautschitsch, H., More, W., Müller, W.B. (eds.) *Contribution to General Algebra,* **10** (1998), 237-249.
8. Müller, S., Carmichael numbers and Lucas tests. In: Mullin, R.C., Mullen, G. (eds.) *Contemporary Mathematics* **225** (1999), 193-202.
9. Müller, S., On the rank of appearance of Lucas sequences. In: Howard, F. (ed.) *Applications of Fibonacci Numbers*, Vol. **8**, Kluwer (1999), pp. 259-276.
10. Ribenboim, P., *The Book of Prime Number Records.* Berlin: Springer 1996.
11. Selmer, E., *Linear Recurrence Relations Over Finite Field*, Lecture notes, Dept. of Math., Univ. of Bergen, Norway, 1966.
12. Somer, L., Periodicity properties of $k$th order linear recurrences with irreducible characteristic polynomial over a finite field, In: Gary L. Mullen and Peter Jau-Shyong Shiue (eds.), *Finite Fields, Coding Theory and Advances in Communications and Computing*, Marcel Dekker Inc., (1993), 195-207.
13. Somer, L., Periodicity properties of $k$th order linear recurrences whose characteristic polynomial splits completely over a finite field, I, *Contemporary Mathematics* **168** (1994), 327-339.
14. Somer, L., Periodicity properties of $k$th order linear recurrences whose characteristic polynomial splits completely over a finite field, II, In: S. Cohen, H. Niederreiter (eds.), *Finite Fields and Applications*, Cambridge University Press (1996), 333-347.
15. Somer, L., On Lucas $d$-pseudoprimes. In: Bergum, G., Philippou, A., Horadam, A. (eds.) *Applications of Fibonacci Numbers*, Vol. 7, Kluwer (1998), 369-375.

16. Spindler, K.H., *Abstract Algebra with Applications*, Volume II, Marcel Dekker, Inc., New York, 1994.
17. Ward, M., The arithmetical theory of linear recurring series, *Trans. Amer. Math. Soc.* **35**, (1933), 600-628.
18. Ward, M., The laws of apparition and repetition of primes in a cubic recurrence, *Trans. Amer. Math. Soc.* **79**, (1955), 72-90.
19. Williams, H. C., On numbers analogous to the Carmichael - numbers. *Canad. Math. Bull.* **20** (1977), 133-143.
20. Williams, H. C., *Éduard Lucas and Primality Testing*, John Wiley & Sons, 1998.

# A Permutation Problem for Finite Fields

Alan R. Prince

Department of Mathematics
Heriot-Watt University
Edinburgh
Scotland

**Abstract** We consider the block structure of the incidence matrix of a projective plane of order $p+1$ which admits a collineation of order $p$ with three fixed points, where $p$ is a prime. We show that, if $p \equiv 3 \pmod 4$, a $p^2 \times p^2$ block in the incidence matrix can always be completed. The construction utilises the square root map on the quadratic residues mod $p$. The problem leads to a much more general question about the existence of a certain type of permutation of the nonzero elements of a finite field $\mathrm{GF}(q)$. The existence of a permutation with the required properties would lead to a construction of a projective plane of order $q+1$.

## 1 Introduction

Janko and van Trung [3] showed that the collineation group of a possible projective plane of order 12 is a $\{2,3\}$-group. In particular, they showed that there is no projective plane of order 12 admitting a collineation of order 11 with three fixed points. One subcase of this problem led, by a computer search, to the construction of a block of size $121 \times 121$ in a possible incidence matrix of a plane of order 12. In this paper, we analyse the structure of this block and show that, remarkably, there is a general construction that gives a $p^2 \times p^2$ block in the incidence matrix of a possible projective plane of order $p+1$ admitting a collineation of order $p$ with 3 fixed points, where $p$ is any prime satisfying $p \equiv 3 \pmod 4$. To construct the block we utilise the square root map on the quadratic residues mod $p$.

For large $p$, this means that we can construct most of the incidence matrix of a projective plane of order $p+1$. This is interesting in itself but it may also have some application to finding mutually orthogonal Latin squares (MOLS). The existence of a projective plane of order $n$ is well known to be equivalent to the existence of $n-1$ mutually orthogonal Latin squares of order $n$. Using the $p^2 \times p^2$ block in the incidence matrix of a plane of order $p+1$, $p$ incomplete mutually orthogonal Latin squares of order $p+1$ can be constructed. For known results on MOLS and incomplete MOLS, see [1].

The square root map is a permutation of the quadratic residues with a certain property (see (A1) of Section 3) and, in fact, any permutation of the quadratic residues with this property could be used to construct the $p^2 \times p^2$ block. The problem of extending to a full incidence matrix leads to a more general question about the existence of a certain type of permutation of the nonzero elements of a finite field GF($q$). The existence of a permutation with the required properties would lead to a construction of a projective plane of order $q+1$ (see Theorem 3).

## 2 The Background

We consider the problem of the existence of a projective plane of order $p+1$ admitting a collineation $\sigma$ of order $p$ with 3 fixed points. Since the number of fixed points of any collineation of a projective plane is always the same as the number of fixed lines, $\sigma$ also fixes 3 lines. The orbits of $\sigma$ on points consist of 3 orbits of length 1 (the fixed points), 3 orbits of length $p$ consisting of points on one of the 3 fixed lines, and $p$ orbits of length $p$ consisting of points which are not incident with any fixed line. Similarly, the orbits of lines consist of 3 orbits of length 1 and $p+3$ orbits of length $p$, precisely $p$ of which consist of lines not incident with any fixed point. Thus, we can order the rows and columns so that the incidence matrix of the plane has a $p^2 \times p^2$ submatrix $A$ corresponding to the points and lines not incident with any fixed line or point respectively. This submatrix consists of a $p \times p$ array of circulant $p \times p$ matrices and gives rise to an orbit matrix $M$, which is a $p \times p$ matrix with $(i,j)$-entry being the number of 1's in any row or column of the $(i,j)$-circulant matrix of $A$. The matrix $M$ consists of non-negative integer entries and satisfies the equation

$$\boxed{MM^t = (p+1)I + (p-3)J}$$

where $J$ is the $p \times p$ matrix with all entries equal to 1.

For example, when $p = 11$, we are in the situation investigated by Janko and van Trung [1]. The matrix $M$ is an $11 \times 11$ matrix satisfying $MM^t = 12I + 8J$. They showed that there are three possibilities for $M$ up to permutation of rows and columns:

$$\begin{pmatrix}
0&2&0&2&2&2&0&0&0&2&0\\
0&0&2&0&2&2&2&0&0&0&2\\
2&0&0&2&0&2&2&2&0&0&0\\
0&2&0&0&2&0&2&2&2&0&0\\
0&0&2&0&0&2&0&2&2&2&0\\
0&0&0&2&0&0&2&0&2&2&2\\
2&0&0&0&2&0&0&2&0&2&2\\
2&2&0&0&0&2&0&0&2&0&2\\
2&2&2&0&0&0&2&0&0&2&0\\
0&2&2&2&0&0&0&2&0&0&2\\
2&0&2&2&2&0&0&0&2&0&0
\end{pmatrix},
\begin{pmatrix}
2&0&0&0&2&0&0&2&2&0&2\\
2&0&0&0&0&2&2&0&0&2&2\\
0&2&2&0&0&0&0&2&0&2&2\\
0&2&0&2&0&2&0&0&2&0&2\\
0&0&2&2&2&0&2&0&0&0&2\\
3&2&2&1&1&1&0&0&0&0&0\\
1&0&0&3&1&1&0&2&0&2&0\\
0&2&0&0&3&1&1&0&1&2&0\\
0&0&2&0&1&3&1&2&1&0&0\\
1&2&0&1&0&0&3&2&1&0&0\\
1&0&2&1&0&0&1&0&3&2&0
\end{pmatrix},
\begin{pmatrix}
0&2&0&2&2&0&2&0&0&0&2\\
0&3&0&0&0&2&0&1&2&1&1\\
1&0&0&3&0&2&1&0&1&2&0\\
1&0&0&1&3&1&0&2&2&0&0\\
2&1&0&0&0&0&3&2&1&1&0\\
2&0&0&0&1&1&0&1&0&2&3\\
3&2&2&1&1&1&0&0&0&0&0\\
0&0&2&0&1&3&2&1&0&0&1\\
0&1&2&2&0&0&0&3&0&1&1\\
1&0&2&1&0&0&1&0&3&0&2\\
0&1&2&0&2&0&1&0&1&3&0
\end{pmatrix}.$$

By a computer search, Janko and van Trung found that the construction of the submatrices $A$ corresponding to these orbit matrices is possible only for the first orbit matrix above and that there is an essentially unique solution for $A$ in this case. In the next section we describe their solution in a conceptual way and generalise it to give a large block in the incidence matrix of a possible projective plane of order $p+1$ admitting a collineation of order $p$ with three fixed points, where $p$ is any prime satisfying $p \equiv 3 \pmod 4$.

It would be of interest to know all the possibilities for the orbit matrices in the case of a possible projective plane of order 18 admitting a collineation of order 17 with 3 fixed points. Here, the orbit matrix equation is $MM^t = 18I + 14J$, where $M$ is a $17 \times 17$ matrix.

## 3 The Construction of a $p^2 \times p^2$ Block in the Incidence Matrix of a Projective Plane of Order $p + 1$ for Any Prime $p \equiv 3 \pmod 4$.

The first possibility for $M$ in the previous section is actually the incidence matrix of the (11,5,2)-biplane with 1's replaced by 2's. In the first row of $M$, the 2's occur in positions corresponding to the quadratic residues mod 11 (if the columns are labelled $0, 1, 2, ..., 10$). The submatrix $A$ is obtained by replacing each 2 in $M$ by an $11 \times 11$ circulant matrix with two 1's in each row. The circulant matrix that replaces a 2 in the first row of $M$ has its two 1's in its first row in positions corresponding to the square roots of the appropriate quadratic residue (again the columns are labelled $0, 1, 2, ..., 10$). The other 2's in $M$ are similarly replaced: each 2 is associated to a quadratic residue, namely the position in the first row to which it corresponds. The resulting $A$ in block form is the following:

$$\begin{pmatrix}
0&1&0&5&2&4&0&0&0&3&0\\
0&0&1&0&5&2&4&0&0&0&3\\
3&0&0&1&0&5&2&4&0&0&0\\
0&3&0&0&1&0&5&2&4&0&0\\
0&0&3&0&0&1&0&5&2&4&0\\
0&0&0&3&0&0&1&0&5&2&4\\
4&0&0&0&3&0&0&1&0&5&2\\
2&4&0&0&0&3&0&0&1&0&5\\
5&2&4&0&0&0&3&0&0&1&0\\
0&5&2&4&0&0&0&3&0&0&1\\
1&0&5&2&4&0&0&0&3&0&0
\end{pmatrix}$$

where $i$ represents the $11 \times 11$ circulant matrix with two 1's in each row, the entries in the first row being in the columns labelled $\pm i$. For example, 3 is the matrix

$$\begin{pmatrix}
0&0&0&1&0&0&0&0&1&0&0\\
0&0&0&0&1&0&0&0&0&1&0\\
0&0&0&0&0&1&0&0&0&0&1\\
1&0&0&0&0&0&1&0&0&0&0\\
0&1&0&0&0&0&0&1&0&0&0\\
0&0&1&0&0&0&0&0&1&0&0\\
0&0&0&1&0&0&0&0&0&1&0\\
0&0&0&0&1&0&0&0&0&0&1\\
1&0&0&0&0&1&0&0&0&0&0\\
0&1&0&0&0&0&1&0&0&0&0\\
0&0&1&0&0&0&0&1&0&0&0
\end{pmatrix}$$

It is possible to generalise this to construct a $p^2 \times p^2$ block in the incidence matrix of a possible plane of order $p + 1$ where $p$ is any prime $\equiv 3 \pmod 4$. If $p \equiv 3 \pmod 4$, the quadratic residues form a $(p, \frac{p-1}{2}, \frac{p-3}{4})$-difference set. In the incidence matrix of the corresponding symmetric design, replacing the 1's by 2's gives a matrix $M$ satisfying the orbit matrix equation:

$$\boxed{MM^t = (p+1)I + (p-3)J}$$

corresponding to a collineation of order $p$ with 3 fixed points. We then replace the 2's in the first row by the $p \times p$ circulant matrices determined by the square roots of the corresponding quadratic residues and cycle these through the matrix as before.

The property we require of the submatrix $A$ is that any two rows (or columns) have inner product at most 1. Denote by $Q$ the quadratic residues mod $p$, where $p \equiv 3 \pmod 4$. Let $\phi$ be a permutation of $Q$ with the property:

(A1): for fixed $d \neq 0$, as $(a, b)$ range over the pairs with $a - b = d$, the values $\pm a\phi \pm b\phi$ are distinct.

We can use $\phi$ to construct a submatrix $A$ corresponding to the orbit matrix $M$. The circulant $p \times p$ block corresponding to the quadratic residue $a$ has 1's in columns $\pm a\phi$ in the first row. The property (A1) is the critical property required for any two rows to have inner product at most 1. The square root map on $Q$ is one such permutation $\phi$.

**Lemma 3.1.** *The map $\phi : a \mapsto \sqrt{a}$ is a permutation of $Q$ satisfying property (A1).*

*Proof.* Fix $d \neq 0$. Assume $x^2 - y^2 = d$. Then, $(x-y)(x+y) = d$. For any nonzero $u$, $d = uv$ where $v = u^{-1}d$ and the equations $x - y = u$ and $x + y = v$ have a unique solution $x = \frac{u+v}{2}, y = \frac{v-u}{2}$ (since $p$ is odd). We have $x = 0$ if and only if $u = -v$ which occurs if and only if $u^2 = -d$, while $y = 0$ if and only if $u = v$ which occurs if and only if $u^2 = d$. Since $p \equiv 3 \pmod 4$, precisely one of $d, -d$ is a quadratic residue. Thus, if $d$ is a quadratic residue there are two values of $u$ (namely $\pm\sqrt{d}$) which give $x = 0$, while if $d$ is a quadratic non-residue then there are two values of $u$ (namely $\pm\sqrt{-d}$) which give $y = 0$ and these are the only occurences of zero values for $x$ or $y$. Thus, as $u$ ranges over the nonzero elements of GF($p$), all except two values give a solution $(x, y)$ of $x^2 - y^2 = d$ with neither $x$ nor $y$ zero. These solutions $(x, y)$ can be partitioned into quadruples $(\pm a, \pm b)$ which give the same quadratic residue pairs $(a^2, b^2)$ showing that each $d$ can be written as a difference of quadratic residues in precisely $\frac{p-3}{4}$ ways. The quadruples correspond to the different factorisations $d = uv = vu = (-u)(-v) = (-v)(-u)$. The four values of $x - y = u$ corresponding to the quadruple $(\pm a, \pm b)$ are $\pm a \pm b$ where $(a, b)$ may be assumed to be the quadratic residue pair amongst the quadruple. It follows that as $(a^2, b^2)$ range over the $\frac{p-3}{4}$ quadratic residue pairs with $a^2 - b^2 = d$ then $\pm(a^2)\phi \pm (b^2)\phi$ range over all the nonzero values of GF($p$) other than either $\pm\sqrt{d}$ or $\pm\sqrt{-d}$, depending on whether $d$ is a quadratic residue or not. □

For any permutation $\phi$ of $Q$ satisfying property (A1), the values $\pm a\phi \pm b\phi$ range over $p - 3$ nonzero values, for fixed $d \neq 0$, and there are two omitted nonzero values. In the above proof, we showed that for the square root map, these omitted nonzero values are $\pm\sqrt{d}$ or $\pm\sqrt{-d}$ depending on whether $d$ is a quadratic residue mod $p$ or not. It turns out that this is an obstacle to extending the submatrix $A$ to a full incidence matrix. If a permutation $\phi$ satisfies not only property (A1) but also the property:

(A2): the two omitted values are $\pm d$.

then it would be possible to extend the submatrix $A$, constructed using $\phi$ as described above, to a full incidence matrix of a projective plane of order $p+1$. In the case $p = 7$, the permutation $\phi = (1,4,2)$ of $Q$ satisfies the conditions (A1) and (A2), while the square root map $\phi = (2,4)$ satisfies (A1) but not (A2).

Although the $p^2 \times p^2$ block obtained using the square root map does not extend to a full incidence matrix, it is the incidence matrix of some new finite geometry with $p^2$ points and $p^2$ lines, which is of some interest in its own right. We can describe this geometry directly. If we take the points to be ordered pairs $(a, b)$ of elements of GF$(p)$ and the lines to be ordered pairs $[c, d]$, then $(a, b)$ is incident with $[c, d]$ if and only if $(d - b)^2 = c - a \neq 0$.

**Lemma 3.2.** *Suppose that a permutation $\phi$ of $Q$ exists satisfying (A1),(A2). Define a map $\sigma : GF(p)^* \to GF(p)$ by*

$$(a\phi)\sigma = \frac{a + a\phi}{2}, \qquad (-a\phi)\sigma = \frac{a - a\phi}{2}$$

*where $a$ ranges over $Q$. Then:*
*(i) $\sigma$ is a permutation of $GF(p)^*$*
*(ii) if $(i, j) \neq (k, \ell)$ and $i - j = k - \ell \neq 0$ then $i\sigma - j\sigma \neq k\sigma - \ell\sigma$*
*(iii) if $i \neq j$ then $i - j \neq i\sigma - j\sigma$.*

*Proof.* (i) Let $a, b \in Q$. If $(a\phi)\sigma = (b\phi)\sigma$, then $a + a\phi = b + b\phi$ so that $b\phi - a\phi = a - b$. If $(a\phi)\sigma = (-b\phi)\sigma$, then $a + a\phi = b - b\phi$ so that $-a\phi - b\phi = a - b$. If $(-a\phi)\sigma = (-b\phi)\sigma$, then $a - a\phi = b - b\phi$ so that $a\phi - b\phi = a - b$. These conclusions contradict the fact that $\phi$ satisfies (A2). Thus, $\sigma : GF(p)^* \to GF(p)$ is 1:1. Since $\sum_{a \in Q} (a\phi)\sigma + (-a\phi)\sigma = \sum_{a \in Q} a = 0$, the missing value in the range of $\sigma$ must be 0. Hence, $\sigma$ is a permutation of GF$(p)^*$.
(ii) Let $a, b, c, d \in Q$. If $a\phi - b\phi = c\phi - d\phi \neq 0$ and $(a + a\phi) - (b + b\phi) = (c + c\phi) - (d + d\phi)$, then $a - b = c - d \neq 0$ which contradicts (A1). The other cases are similar, establishing (ii).
(iii) Let $a, b \in Q$. If $a\phi - b\phi = \frac{(a+a\phi)-(b+b\phi)}{2}$, then $a\phi - b\phi = a - b$ contradicting (A2). The other cases are similar, establishing (iii). □

## 4 The Related Permutation Problem

Let GF$(q)$ be any finite field. We consider the following two properties relating to a permutation $\sigma$ of GF$(q)^*$:

(P1) if $(i,j) \neq (k,\ell)$ and $i-j = k-\ell \neq 0$ then $i\sigma - j\sigma \neq k\sigma - \ell\sigma$;

(P2) if $i \neq j$ then $i - j \neq i\sigma - j\sigma$.

The properties (P1) and (P2) are equivalent to the single property:

(P3) $(i,j) \mapsto (i-j, i\sigma - j\sigma)$ is a bijection of $\Omega = \mathrm{GF}(q)^* \times \mathrm{GF}(q)^* - D$, where $D = \{(a,a) | a \in \mathrm{GF}(q)^*\}$.

**Theorem 4.1.** *Suppose that there exists a permutation $\sigma$ of $GF(q)^*$ satisfying (P1),(P2). Let $X$ and $\infty$ be new symbols. Define a point-line incidence structure as follows. The points are the ordered pairs $(a,b),(a,X),(X,b)$, $(X,X)$ together with $(a),(X),(\infty)$, where $a,b$ range over $GF(q)$. The lines are defined as follows, where $a,b,c$ range over $GF(q)$:*

$$\begin{gathered}\{(i+a, i\sigma+b) \mid i \in \mathrm{GF}(q)^*\} \cup \{(a,X),(X,b),(b-a)\} \\ \{(c,i) \mid i \in \mathrm{GF}(q)\} \cup \{(c,X),(\infty)\} \\ \{(i,c) \mid i \in \mathrm{GF}(q)\} \cup \{(X,c),(X)\} \\ \{(i,i+c) \mid i \in \mathrm{GF}(q)\} \cup \{(X,X),(c)\} \\ \{(i,X) \mid i \in \mathrm{GF}(q)\} \cup \{(X,X),(c)\} \\ \{(X,i) \mid i \in \mathrm{GF}(q)\} \cup \{(X,X),(\infty)\} \\ \{(i) \mid i \in \mathrm{GF}(q)\} \cup \{(X),(\infty)\}\end{gathered}$$

*Then, this incidence structure is a projective plane of order $q+1$.*

*Proof.* Denote by $\ell_{ab}$ the first line defined in statement of the theorem. If $\ell_{cd}$ is also a line of this type with $(a,b) \neq (c,d)$, then: (i) if $a = c$ the unique point of intersection is $(a,X)$; (ii) if $b = d$ the unique point of intersection is $(X,b)$; (iii) if $b-a = d-c$ the unique point of intersection is $(b-a)$ since $(i+a, i\sigma+b) = (j+c, j\sigma+d)$ implies $i\sigma - i = j\sigma - j$, contradicting (P2); (iv) if $a \neq c$, $b \neq d$ and $b-a \neq d-c$ then, since $(i+a, i\sigma+b) = (j+c, j\sigma+d)$ if and only if $i-j = c-a$ and $i\sigma - j\sigma = d-b$, there is a unique point of intersection by property (P3). In a similar way, it can be checked that any two distinct lines meet in a unique point. Thus, the incidence structure is a projective plane of order $q+1$. □

We can define an algebraic system $(R,+,\cdot)$ associated to a permutation $\sigma$ of $\mathrm{GF}(q)^*$ satisfying (P1) and (P2) in the following way. Let $\theta$ be a symbol and let $R = \{0\} \cup \{\theta^a \mid a \in \mathrm{GF}(q)\}$. Multiplication in $R$ is given by $\theta^a \cdot \theta^b = \theta^{a+b}$ together with $0 \cdot \theta^a = 0 = \theta^a \cdot 0$. We define addition on $R$ by $\theta^a + \theta^b = \theta^{a+(a-b)\sigma}$, if $a \neq b$, and by $\theta^a + \theta^a = 0$ together with the condition that 0 is the additive identity.

The algebraic system $(R,+,\cdot)$ is a *division neo-ring*, that is it satisfies the axioms:

(i) $(R,+)$ is a loop with identity element 0;
(ii) $(R^*,\cdot)$ is a loop with identity element $\theta^0$;
(iii) $0 \cdot x = 0 = x \cdot 0$ for all $x \in R$;
(iv) both distributive laws hold.

The system $(R, +, \cdot)$ gives the associated linear planar ternary ring of the projective plane described in the theorem, so is an example of a *planar* division neo-ring. Planar division neo-rings have been investigated by Hughes [2].

**Lemma 4.2.** *If $p = 2^n - 1$ is a Mersenne prime and $\theta$ is a primitive element of $GF(2^n)$, then the Jacobi log map on $GF(2^n)$ gives a permutation $\sigma$ of $GF(p)^*$, defined by $1 + \theta^i = \theta^{i\sigma}$, which satisfies conditions (P1) and (P2). The projective plane determined by $\sigma$ is the Desarguesian plane of order $2^n$.*

*Proof.* If $i - j = k - \ell = d \neq 0$, then $\theta^i = \theta^j\theta^d$ and $\theta^k = \theta^\ell\theta^d$. If we have $i\sigma - j\sigma = k\sigma - \ell\sigma = e$ (say), then $\theta^i + 1 = (\theta^j + 1)\theta^e$ and $\theta^k + 1 = (\theta^\ell + 1)\theta^e$. Thus:

$$\frac{\theta^j\theta^d + 1}{\theta^\ell\theta^d + 1} = \frac{\theta^j + 1}{\theta^\ell + 1}.$$

Hence:

$$\theta^j\theta^d + \theta^\ell = \theta^\ell\theta^d + \theta^j.$$

So $\theta^j(\theta^d - 1) = \theta^\ell(\theta^d - 1)$. Since $d \neq 0$, $\theta^d \neq 1$ and hence $\theta^j = \theta^\ell$. Thus, $j = \ell$ and $(i, j) = (k, \ell)$. This establishes that property (P1) holds.
If $i\sigma - j\sigma = i - j$, then

$$\frac{\theta^i}{\theta^j} = \frac{\theta^i + 1}{\theta^j + 1}$$

which implies that $\theta^i = \theta^j$ and so $i = j$. Thus, property (P2) holds.

The planar division neo-ring determined by $\sigma$ is GF$(2^n)$ since the addition defined above is just the addition of the field. Thus, the corresponding projective plane is the Desarguesian plane of order $2^n$. □

The author has checked, by a computer search, that the only permutations $\sigma$ of GF(31)* satisfying (P1) and (P2) are the 6 Jacobi log maps corresponding to the 6 primitive polynomials of degree 5 over GF(2). For reference, we list these 6 permutations:

(1, 18)(2, 5)(3, 29)(4, 10)(6, 27)(7, 22)(8, 20)(9, 16)(11, 19)(12, 23)(13, 14)(15, 24)(17, 30)(21, 25)(26, 28)

(1, 14)(2, 28)(3, 5)(4, 25)(6, 10)(7, 16)(8, 19)(9, 24)(11, 23)(12, 20)(13, 30)(15, 22)(17, 18)(21, 27)(26, 29)

(1, 12)(2, 24)(3, 8)(4, 17)(5, 28)(6, 16)(7, 9)(10, 25)(11, 30)(13, 27)(14, 18)(15, 21)(19, 20)(22, 29)(23, 26)

(1, 19)(2, 7)(3, 11)(4, 14)(5, 29)(6, 22)(8, 28)(9, 15)(10, 27)(12, 13)(16, 25)(17, 21)(18, 30)(20, 23)(24, 26)

(1, 13)(2, 26)(3, 23)(4, 21)(5, 7)(6, 15)(8, 11)(9, 25)(10, 24)(12, 30)(16, 22)(17, 27)(18, 19)(20, 28)(24, 29)

(1, 20)(2, 9)(3, 26)(4, 18)(5, 8)(6, 21)(7, 29)(10, 16)(11, 12)(13, 17)(14, 27)(15, 25)(19, 30)(22, 24)(23, 28)

These correspond, respectively, to a primitive root $\theta$ satisfying $\theta^5 = \theta^2 + 1$, $\theta^5 = \theta^3 + 1$, $\theta^5 = \theta^3 + \theta^2 = \theta + 1$, $\theta^5 = \theta^4 + \theta^2 + \theta + 1$, $\theta^5 = \theta^4 + \theta^3 + \theta + 1$, $\theta^5 = \theta^4 + \theta^3 + \theta^2 + 1$. None of these arises from a permutation $\phi$ of the quadratic residues mod 31 satisfying conditions (A1) and (A2), as in Lemma 3.2. This follows since $x\sigma + (-x)\sigma$ is not always a quadratic residue as $x$ ranges over GF(31)*.

**Lemma 4.3.** *Let $\sigma$ be a permutation of GF(q)* satisfying (P1),(P2). Then:*
*(a) $\sigma$ is fixed point-free;*
*(b) $a\sigma - a = (-a)\sigma$ for all $a \in GF(q)^*$;*
*(c) $a\sigma + (-a)\sigma \neq b\sigma + (-b)\sigma$ unless $a = \pm b$.*

*Proof.* (a) The property $i\sigma - j\sigma \neq i - j$ if $i \neq j$ (in GF$(q)^*$) is equivalent to the property that $i\sigma - i$ takes distinct values in GF$(q)$ as $i$ ranges over GF$(q)^*$. If $a$ is the missing value of $i\sigma - i$ then $\sum(i\sigma - i) = -a$. However, $\sum(i\sigma - i) = \sum i\sigma - \sum i = 0$. In these sums, $i$ ranges over the elements of GF$(q)^*$. Hence, $a = 0$, in other words $i\sigma - i \neq 0$ for all $i \neq 0$.
(b) Consider all $(i, j)$ with $i - j = d \neq 0$, $i, j \in$ GF$(q)^*$. Then, $i\sigma - j\sigma$ takes distinct values. Note that $j = -d$ is excluded here (since the corresponding $i$ would be 0). So there are $q - 2$ values in GF$(q)^*$ taken by $(j + d)\sigma - j\sigma$ as $j$ ranges over GF$(q)^* - \{-d\}$. The missing value is $d$ since $\sigma$ satisfies condition (P2). Hence,

$$\sum_{j \neq -d} (j + d)\sigma - j\sigma = -d.$$

However, $\sum(j+d)\sigma = -d\sigma$ (since $j \neq 0$) and $\sum j\sigma = -(-d)\sigma$ (since $j \neq -d$). This gives $-d\sigma + (-d)\sigma = -d$, that is $d\sigma - d = (-d)\sigma$.
(c) $a\sigma + (-a)\sigma = b\sigma + (-b)\sigma$ implies $a\sigma - b\sigma = (-b)\sigma - (-a)\sigma$, which implies $a = \pm b$ □

The following is a deeper result on the existence of a multiplier.

**Lemma 4.4.** *Let $\sigma$ be a permutation of GF(q)* satisfying (P1),(P2). If $r$ is a prime dividing $q + 1$, then $(ra)\sigma = r(a\sigma)$ for all $a \in$ GF$(q)^*$.*

*Proof.* Let $(R, +, \cdot)$ be the planar division neo-ring associated to $\sigma$ (see the discussion following the proof of Theorem 3). The multiplication in $R$ is both commutative and associative and the abelian group $(R^*, \cdot)$ is isomorphic to (GF$(q)$,+). Let $G$ be the direct product $(R^*, \cdot) \times (R^*, \cdot)$ and let $H_1, H_2, H_3$ denote the subgroups of $G$ consisting of all elements of the form $(\theta^0, a), (a, \theta^0), (a, a)$, respectively. Denote by $D$ the subset $\{(\theta^a, \theta^{a\sigma}) \mid a \in$ GF$(q)^*\}$ of $G$. Then, elements of $G$ outside $H_1, H_2, H_3$ have a unique representation as $d_1 d_2^{-1}$, where $d_1, d_2$ are elements of $D$, while non-identity elements of $H_1, H_2, H_3$ have no such representation. Thus, $D$ is a *difference set* in $G$ in the sense of Hughes [2, page 515]. By equation (9) on page 518 of [2], $D\phi = D$, where $\phi$ is the $r$th-power map on $G$, for $r$ any prime dividing $q + 1$. Hence, $(\theta^{ra}, \theta^{r(a\sigma)}) \in D$, for all $a \in$ GF$(q)^*$, so that $(ra)\sigma = r(a\sigma)$. □

## 5 The Desarguesian Case, $p$ Mersenne

We conclude with an observation on the Desarguesian plane of order $p + 1$ admitting a collineation of order $p$ with 3 fixed points, in the case that $p$ is a Mersenne prime. Let $\theta$ be a primitive element of GF$(2^n)$, where $2^n - 1$ is a

Mersenne prime $p$. Then, $\theta$ has order $p$ and the matrix diag$(1, \theta, \theta^{-1})$ defines a collineation of order $p$ of the Desarguesian plane of order $2^n$ with three fixed points and lines. The points not on any fixed line are the points $(1, a, b)$ with both $a, b$ nonzero. The orbits are parametrized by $\lambda \neq 0$ and consist of the points $(1, a, b)$ with $ab = \lambda$. This gives $p$ orbits of length $p$. Similarly, the lines not containing any fixed point are the lines $x + cy + dz = 0$ with both $c, d \neq 0$. Two lines are in the same orbit if and only if the product $cd$ is the same for both. Thus, we can take representatives of the $p$ orbits to be the lines $x + y + \mu z = 0$. The line $x + y + \mu z = 0$ contains the point $(1, a, a^{-1}\lambda)$ $(a \neq 0)$ if and only if $1 + a + a^{-1}\mu\lambda = 0$, that is $a^2 + a + \mu\lambda = 0$. Thus, there is a solution if and only if trace$(\mu\lambda)$=0 and then the solutions are given by the two solutions for $a$ of the quadratic equation $a^2 + a + \mu\lambda = 0$. This describes the points $(1, a, a^{-1}\lambda)$ which lie on the line $x+y+\mu z = 0$ and shows that there are two points from those orbits parametrised by $\lambda$, where $\mu\lambda$ has trace zero. Here, the trace map is the trace from GF$(2^n)$ into GF(2). The corresponding orbit matrix is thus twice the incidence matrix of the 2-design corresponding to the difference set of trace zero elements of GF$(2^n)$.

## References


1. Colbourn, C.J., Dinitz, J.H. (editors): *CRC Handbookof Combinatorial Designs*, CRC Press, Boca Raton, 1996
2. Hughes, D.R.: Planar division neo-rings. Trans. Amer. Math. Soc. **80** (1955) 502–527
3. Janko, Z., Tran van Trung: The full collineation group of any projective plane of order 12 is a $\{2, 3\}$-group. Geom. Dedicata **12** (1982) 101–110

# The $L$-Function of Gold Exponential Sum

Kenneth Shum[1], P. Vijay Kumar[1], and Tor Helleseth[2]

[1] University of Southern California
Department of Electrical Engineering – Systems
EEB 500
Los Angeles, CA 90089-2565
USA.
e-mail: keshum@milly.usc.edu, vijayk@ceng.usc.edu

[2] University of Bergen
Department of Informatics
Høyteknologisenteret
N-5020 Bergen
NORWAY.
em-ail: torh@ii.uib.no.

**Abstract** In this paper, we show that the Gold exponential sum is maximal in some extension field of the ground field. The smallest cyclotomic field which contains the roots of the $L$-function is identified. This simplifies the explicit calculation of the roots.

## 1 Introduction

*Notation:* $\mathbb{F}_q$ is a finite field of characteristic 2 and $q = 2^n$; $\mathsf{T}^n_m$ is the trace function from $\mathbb{F}_{2^n}$ to $\mathbb{F}_{2^m}$.

In [4], Gold discusses the cross-correlation of two pseudo-noise sequences and studies the following exponential sum,

$$\sum_{x\in\mathbb{F}_q} (-1)^{\mathsf{T}^n_1(x^{2^k+1}+ax)} \tag{1}$$

where $a$ is an element in $\mathbb{F}_q$. This exponential sum will be called the Gold exponential sum. It is well known that the Gold exponential sum takes on 3 values and the corresponding Gold sequences have good cross-correlation property. [4] [5] [7]. For example, when $n$ is odd, they are optimal and are much better than the Weil bound. The family of Gold sequences are employed in European third-generation code-division multiple-access (CDMA) systems and global positioning system (GPS). In this paper, we will study the $L$-function of the Gold exponential sum and determine when the Weil bound is met.

An exponential sum with polynomial argument is said to be *maximal* if its magnitude meets the Weil bound. It is related to function field with maximal or minimal number of rational points [3]. The corresponding Artin-Schreier curve is maximal if the Gold exponential sum is maximal.

For a polynomial $f(x) \in \mathbb{F}_q[x]$ of degree $d$, the exponential sums over extension fields of $\mathbb{F}_q$,

$$T_s(f) = \sum_{x \in \mathbb{F}_{q^s}} (-1)^{\mathrm{T}_1^{ns}(f(x))}$$

have a generating function of complex variable, called the $L$-function, equal to,

$$L(f,z) = \exp\left(\sum_{s=1}^{\infty} T_s(f)\frac{z^s}{s}\right).$$

If $(d,q) = 1$, then $L(f,z)$ is a polynomial of degree $d-1$ [8, p.45] [6]. $L(f,z)$ can be factored as,

$$L(f,z) = (1-\eta_1 z)(1-\eta_2 z)\cdots(1-\eta_{d-1} z).$$

$T_s$ can be expressed in terms of the reciprocal of the zeros of $L$-function,

$$T_s = -\eta_1^s - \eta_2^s - \ldots - \eta_{d-1}^s.$$

The celebrated Weil's theorem determines the absolute value of the $\eta_j$'s.

**Theorem 1.1 (Weil).** *If $f(x)$ is a polynomial over $\mathbb{F}_q$ of degree $d \geq 2$, $(d,q) = 1$, then there exist $d-1$ complex numbers $\eta_j$, each of absolute value $\sqrt{q}$ such that for all $s$*

$$\sum_{x \in \mathbb{F}_{q^s}} (-1)^{\mathrm{T}_1^{ns}(f(x))} = -\sum_{j=1}^{d-1} \eta_j^s.$$

As a direct corollary of Weil's theorem, we have the following upper bound on the absolute value of exponential sum, which is also known as Carlitz-Uchiyama bound [1] [6, p.223],

**Theorem 1.2 (Weil bound).** *Under the conditions stated in the previous theorem,*

$$\left|\sum_{x \in \mathbb{F}_{q^s}} (-1)^{\mathrm{T}_1^{ns}(f(x))}\right| \leq (d-1)\sqrt{q^s}.$$

The normalized values $\eta_j/\sqrt{q}$ are roots of unity. The main theorem of this paper determines the smallest cyclotomic field which contains the normalized reciprocal roots of $L$-function when the function $f(x)$ has the special form $x^{2^k+1} + ax$.

**Theorem 1.3.** *Let $q = 2^n$ and $a \in \mathbb{F}_q$. There exists $2^k$ complex number $\eta_j$, $j = 1, 2, \ldots, 2^k$, such that for all $s$,*

$$\sum_{x \in \mathbb{F}_{q^s}} (-1)^{\mathsf{T}_1^{ns}(x^{2^k+1}+ax)} = -\sum_{j=1}^{2^k} \eta_j^s$$

*Let $m$ be the smallest integer such that, $2k | mn$ and $\mathsf{T}_{2k}^{nm}(a) = 0$. Let $\beta$ be a root of $x^{2^k} + x^{2^{-k}} + a$ in $\mathbb{F}_{q^m}$. If*

$$\mathsf{T}_1^{nm}(\beta^{2^k+1}) = mn/2k$$

*then $\eta_j/\sqrt{2^n}$ are $m$th roots of unity. Otherwise, they are $2m$th roots of unity but not $m$th roots of unity.*

Note that the value of $m$ in the main theorem is no greater than $4k$. That implies that the normalized $\eta_j$'s belong to the $8k$th cyclotomic field.

Related results can be found in [2].

## 2 Squaring Method

A tighter bound is possible in the case of the Gold exponential sum.

**Proposition 2.1.** *Let $m = (2k, ns)$. Then for all $a \in \mathbb{F}_q$,*

$$\left| \sum_{x \in \mathbb{F}_{q^s}} (-1)^{\mathsf{T}_1^{ns}(x^{2^k+1}+ax)} \right| \leq 2^{m/2}\sqrt{q^s}.$$

*Hence, the Weil bound can be achieved only if $m = 2k$, i.e. only if $2k | ns$.*

*Proof.* Squaring the exponential sum and making change of variables $y \mapsto x + z$,

$$\begin{aligned}
\left[ \sum_{x \in \mathbb{F}_{q^s}} (-1)^{\mathsf{T}_1^{ns}(x^{2^k+1}+ax)} \right]^2 &= \sum_{x,y \in \mathbb{F}_{q^s}} (-1)^{\mathsf{T}_1^{ns}(x^{2^k+1}+ax+y^{2^k+1}+ay)} \\
&= \sum_{x,z} (-1)^{\mathsf{T}_1^{ns}(x^{2^k+1}+ax+(z+x)^{2^k+1}+a(z+x))} \\
&= \sum_{z} (-1)^{\mathsf{T}_1^{ns}(z^{2^k+1}+az)} \sum_{x} (-1)^{\mathsf{T}_1^{ns}(x^{2^k}z+xz^{2^k})} \\
&= \sum_{z} (-1)^{\mathsf{T}_1^{ns}(z^{2^k+1}+az)} \sum_{x} (-1)^{\mathsf{T}_1^{ns}(x(z^{2^{-k}}+z^{2^k}))}
\end{aligned}$$

The inner summation is equal to $q^s$ if $z^{2^{-k}} = z^{2^k}$. An element $z \in \mathbb{F}_{q^s}$ satisfies $z^{2^{-k}} = z^{2^k}$ iff $z \in \mathbb{F}_{q^m}$.

$$\left[\sum_{x\in\mathbb{F}_{q^s}} (-1)^{\mathrm{T}_1^{ns}(x^{2^k+1}+ax)}\right]^2 = q^s \sum_{z\in\mathbb{F}_{2^m}} (-1)^{\mathrm{T}_1^{ns}(z^{2^k+1}+az)} \le 2^m q^s. \tag{2}$$

Since the Weil bound for this exponential sum gives

$$\left|\sum_{x\in\mathbb{F}_{q^s}} (-1)^{\mathrm{T}_1^{ns}(x^{2^k+1}+ax)}\right| \le 2^k\sqrt{q^s},$$

the exponential sum can be maximal only if $m = 2k$. □

We ask the question when the Weil bound is achieved. From the previous lemma, it is sufficient to consider the case when $\mathbb{F}_{q^s}$ is an extension of $\mathbb{F}_{2^{2k}}$, i.e. $2k|ns$. The first case is treated in the following proposition.

**Proposition 2.2.** *For $a \in \mathbb{F}_{2^{2k}}$, we have,*

$$\sum_{x\in\mathbb{F}_{2^{2k}}} (-1)^{\mathrm{T}_1^{2k}(x^{2^k+1}+ax)} = \begin{cases} 0 & \text{if } a \neq 0, \\ 2^{2k} & \text{otherwise.} \end{cases}$$

*Proof.* Because $(x^{2^k+1})^{2^k-1} = x^{2^{2k}-1} = 1$, $x^{2^k+1}$ lies in the subfield $\mathbb{F}_{2^k}$ and $\mathrm{T}_1^{2k}(x^{2^k+1}) = \mathrm{T}_1^k(x^{2^k+1}\mathrm{T}_k^{2k}(1)) = 0$. Therefore,

$$\sum_{x\in\mathbb{F}_{2^{2k}}} (-1)^{\mathrm{T}_1^{2k}(x^{2^k+1}+ax)} = \sum_{x\in\mathbb{F}_{2^{2k}}} (-1)^{\mathrm{T}_1^{2k}(ax)}.$$

The last sum is zero unless $a = 0$. □

In general if $\mathbb{F}_{2^{\mu k}}$ is an extension field of $\mathbb{F}_{2^{2k}}$, we can determine the magnitude of the Gold exponential sum by borrowing some argument used in the proof of Proposition 2.1.

**Proposition 2.3.** *Let $\mu$ be even. For $a \in \mathbb{F}_{2^{\mu k}}$,*

$$\left|\sum_{x\in\mathbb{F}_{2^{\mu k}}} (-1)^{\mathrm{T}_1^{\mu k}(x^{2^k+1}+ax)}\right| = \begin{cases} 2^k\sqrt{2^{\mu k}} & \text{if } \mathrm{T}_{2k}^{\mu k}(a) = 0, \\ 0 & \text{otherwise,} \end{cases}$$

*i.e. the exponential sum is maximal iff $\mathrm{T}_{2k}^{\mu k}(a) = 0$.*

*Proof.* If $\mu$ is a multiple of 2, then $m = (2k, \mu k) = 2k$ in the proof of Proposition 2.1. Put $m = 2k$ in equation (2),

$$\left[ \sum_{x \in \mathbb{F}_{2^{\mu k}}} (-1)^{\mathsf{T}_1^{\mu k}(x^{2^k+1}+ax)} \right]^2 = 2^{\mu k} \sum_{u \in \mathbb{F}_{2^{2k}}} (-1)^{\mathsf{T}_1^{\mu k}(u^{2^k+1}+au)}$$

The proof of Proposition 2.2 shows that $u^{2^k+1}$ belongs to a subfield and $\mathsf{T}_1^{\mu k}(u^{2^k+1})$ is equal to zero. Therefore the square of the exponential sum is,

$$2^{\mu k} \sum_{u \in \mathbb{F}_{2^{2k}}} (-1)^{\mathsf{T}_1^{2k}(u\mathsf{T}_{2k}^{\mu k}(a))} = \begin{cases} 2^{\mu k}2^{2k} & \text{if } \mathsf{T}_{2k}^{\mu k}(a) = 0, \\ 0 & \text{if } \mathsf{T}_{2k}^{\mu k}(a) \neq 0. \end{cases}$$

Thus, Weil bound is achieved if $\mathsf{T}_{2k}^{\mu k}(a) = 0$. □

## 3 Proof of Main Theorem

If an exponential sum achieves the Weil bound, the normalized reciprocal roots $\eta_j/\sqrt{q}$ raised to some power are either 1 or $-1$. To find the smallest cyclotomic fields containing the normalized $\eta_j$'s, we need to determine the sign of the exponential sum in Proposition 2.3.

**Proposition 3.1.** *Suppose $\mu$ is an even number. For any $a \in \mathbb{F}_{2^{\mu k}}$,*

$$\sum_{x \in \mathbb{F}_{2^{\mu k}}} (-1)^{\mathsf{T}_1^{\mu k}(x^{2^k+1}+ax)}$$
$$= \begin{cases} [(-1)^{\mathsf{T}_1^{\mu k}(\beta^{2^k+1})}(-1)^{(\mu/2)+1}]2^k\sqrt{2^{\mu k}} & \textit{if } \mathsf{T}_{2k}^{\mu k}(a) = 0, \\ 0 & \textit{otherwise,} \end{cases}$$

*where $\beta$ is a root of $x^{2^k} + x^{2^{-k}} + a = 0$ in $\mathbb{F}_{2^{\mu k}}$.*

*Proof.* We follow the approach in Gold's paper [4]. For any $\beta \in \mathbb{F}_{2^{\mu k}}$, we can make a change of variable $x \mapsto x + \beta$,

$$\begin{aligned} x^{2^k+1} + ax &\mapsto (x+\beta)^{2^k+1} + a(x+\beta) \\ &= x^{2^k+1} + \beta^{2^k+1} + x^{2^k}\beta + x\beta^{2^k} + ax + a\beta. \end{aligned}$$

Thus,

$$\sum_{x \in \mathbb{F}_{2^{\mu k}}} (-1)^{\mathsf{T}_1^{\mu k}(x^{2^k+1}+ax)} = \sum_x (-1)^{\mathsf{T}_1^{\mu k}(x^{2^k+1}+\beta^{2^k+1}+a\beta+x(\beta^{2^{-k}}+\beta^{2^k}+a))}. \quad (3)$$

In Lemma 3.2 we will determine,

$$\sum_{x \in \mathbb{F}_{2^{\mu k}}} (-1)^{\mathsf{T}_1^{\mu k}(x^{2^k+1})} = (-1)^{\mu/2+1}2^k\sqrt{2^{\mu k}}.$$

Also, in Lemma 3.3, we will show that if $\mathsf{T}_{2k}^{\mu k}(a) = 0$, we can choose $\beta \in \mathbb{F}_{2^{\mu k}}$ such that $\beta^{2^{-k}} + \beta^{2^k} + a = 0$. Furthermore for such choice of $\beta$, $\mathsf{T}(\beta^{2^k+1} + a\beta) = \mathsf{T}(\beta^{2^{-k}+1}) = \mathsf{T}(\beta^{1+2^k})$.

These results simplify equation (3). When $\mathsf{T}_{2k}^{\mu k}(a) = 0$,

$$\begin{aligned}\sum_{x\in\mathbb{F}_{2^{\mu k}}} (-1)^{\mathsf{T}_1^{\mu k}(x^{2^k+1}+ax)} &= (-1)^{\mathsf{T}_1^{\mu k}(\beta^{2^k+1})} \sum_x (-1)^{\mathsf{T}_1^{\mu k}(x^{2^k+1})} \\ &= (-1)^{\mathsf{T}_1^{\mu k}(\beta^{2^k+1})} (-1)^{\mu/2+1} 2^k \sqrt{2^k}.\end{aligned}$$

□

**Lemma 3.2.** *Consider $\mathbb{F}_{2^{\mu k}}$ as extension field over $\mathbb{F}_{2^k}$.*

$$\sum_{x\in\mathbb{F}_{2^{\mu k}}} (-1)^{\mathsf{T}_1^{\mu k}(x^{2^k+1})} = \begin{cases} 0 & \text{if } \mu \text{ is odd,} \\ (-1)^{(\mu/2)+1} 2^k \sqrt{2^{\mu k}} & \text{if } \mu \text{ is even.} \end{cases}$$

*Proof.* Apply Weil's theorem, there exist $2^k$ complex numbers $\eta_j$'s, $|\eta_j| = \sqrt{2^k}$, that satisfy,

$$\sum_{x\in\mathbb{F}_{2^{\mu k}}} (-1)^{\mathsf{T}_1^{\mu k}(x^{2^k+1})} = -\sum_{j=1}^{2^k} \eta_j^{\mu} \qquad \text{for all integer } \mu.$$

When $\mu = 1$,

$$\mathsf{T}_1^k(x^{2^k+1}) = \mathsf{T}_1^k(x \cdot x) = \mathsf{T}_1^k(x)^2 = \mathsf{T}_1^k(x).$$

Therefore,

$$\sum_{x\in\mathbb{F}_{2^k}} (-1)^{\mathsf{T}_1^k(x^{2^k+1})} = \sum_{x\in\mathbb{F}_{2^k}} (-1)^{\mathsf{T}_1^k(x)} = 0.$$

The case $\mu = 2$ is included in Proposition 2.2, and hence the first two cases are known,

$$-\sum_{j=1}^{2^k} \eta_j = 0 \qquad \text{and} \qquad -\sum_{j=1}^{2^k} \eta_j^2 = 2^{2k}. \tag{4}$$

Cauchy-Schwarz inequality shows that $\eta_j^2 = -2^k$ for $j = 1, 2, \ldots, n$. Taking square root, we know that $\eta_j = \pm i\sqrt{2^k}$. But the sum of $\eta_j$'s equals zero. It is possible only if half of them are positive and half of them negative. We can assume that $\eta_1 = \eta_2 = \ldots = \eta_{2^{k-1}} = i\sqrt{2^k}$, and $\eta_{2^{k-1}+1} = \eta_{2^{k-1}+2} = \ldots = \eta_{2^k} = -i\sqrt{2^k}$.

$$\sum_{x\in\mathbb{F}_{2^{\mu k}}} (-1)^{\mathsf{T}_1^{\mu k}(x^{2^k+1})} = -\sum_{j=1}^{2^{k-1}} (i\sqrt{2^k})^\mu - \sum_{j=2^{k-1}+1}^{2^k} (-i\sqrt{2^k})^\mu$$
$$= -i^\mu 2^{k-1}\sqrt{2^{\mu k}}(1+(-1)^\mu)$$
$$= \begin{cases} 0 & \text{if } \mu \text{ is odd,} \\ -i^\mu 2^k\sqrt{2^{\mu k}} & \text{if } \mu \text{ is even.} \end{cases}$$

□

The next lemma is essentially the proposition in appendix of [4].

**Lemma 3.3.** *Let $\ell$, $k$ be integers, and $a \in \mathbb{F}_{2^{\mu k}}$. Suppose $g = (\ell, 2k)$. Then $x^{2^k} + x^{2^{-k}} + a$ has roots in $\mathbb{F}_{2^\ell}$ iff $\mathsf{T}_g^\ell(a) = 0$.*

*Proof.* ($\Rightarrow$) If $\beta \in \mathbb{F}_{2^\ell}$ satisfies $\beta^{2^k} + \beta^{2^{-k}} + a = 0$, then by taking the trace $\mathsf{T}_g^\ell$ on both sides we know that $\mathsf{T}_g^\ell(a) = 0$.

($\Leftarrow$) The map $\phi : x \mapsto x^{2^k} + x^{2^{-k}}$ is a linear transformation over $\mathbb{F}_2$. The kernel of $\phi = \mathbb{F}_{2^g}$; indeed, if $x^{2^k} + x^{2^{-k}} = 0$, then $x^{2^{2k}} = x$ and hence $x \in \{x \in \mathbb{F}_{2^\ell} : x^{2^{2k}} = x\} = \mathbb{F}_{2^g}$. The range of $\phi \subseteq R := \{x \in \mathbb{F}_{2^\ell} : \mathsf{T}_g^\ell(x) = 0\}$. As vector spaces over $\mathbb{F}_2$, $\dim(ker(\phi)) = \dim(\mathbb{F}_{2^g}) = g$, and $\dim(range(\phi)) \le \dim(R) = \ell - g$. By comparing dimensions, $\phi$ is surjective onto the set of trace-zero elements. □

*Proof (of theorem 1.3).* We want to find the smallest extension field of $\mathbb{F}_{2^n}$ such that equality in the Weil bound holds. By Lemma 3.2, it is sufficient to consider $\mathbb{F}_{2^{ns}}$ when $ns$ is divisible by $2k$. If $2k|ns$, then the exponential sum in the theorem is either 0 or maximal by Proposition 2.3. The Weil bound is achieved if and only if $\mathsf{T}_{2k}^{ns}(a) = 0$. Therefore, the $m$ defined in Theorem 1.3 is the smallest integer such that

$$\left|\sum_{x\in\mathbb{F}_{2^{nm}}} (-1)^{\mathsf{T}_1^{nm}(x^{2^k+1}+ax)}\right| = 2^k\sqrt{2^{mn}}$$

The Cauchy-Schwarz inequality implies that this is possible only when the $\eta_i^m$'s are real and have the same sign, i.e. either $\eta_j^m = (\sqrt{2^n})^m$ for all $j$, or $\eta_j^m = -(\sqrt{2^n})^m$ for all $j$.

The sign of them is determined by applying Proposition 3.1. Let $\mu = mn/k$. Then $\mu$ is even and,

$$sign(\sum_{x\in\mathbb{F}_{2^{nm}}} (-1)^{\mathsf{T}_1^{nm}(x^{2^k+1}+ax)}) = sign(\sum_{x\in\mathbb{F}_{2^{\mu k}}} (-1)^{\mathsf{T}_1^{\mu k}(x^{2^k+1}+ax)})$$
$$= (-1)^{\mathsf{T}_1^{nm}(\beta^{2^k+1})}(-1)^{(\mu/2)+1}.$$

The sign in the above equation is $-1$ when $\mathsf{T}_1^{nm}(\beta^{2^k+1}) = nm/2k$. Then $\eta_j^m = (\sqrt{2^n})^m$ for all $j$ and $(\eta_j/\sqrt{2^n})^m = 1$; they are $m$th roots of unity. Otherwise if the sign is 1, $\eta_j^m = -(\sqrt{2^n})^m$ for all $j$ and $(\eta_j/\sqrt{2^n})^{2m} = 1$. □

## 4 Numerical Examples

Consider the simplest case $a = 1$, $n = 1$,

$$\sum_{x \in \mathbb{F}_{2^s}} (-1)^{\mathsf{T}_1^s(x^{2^k+1}+x)} = -\sum_{j=1}^{2^k} \eta_j^s$$

The condition in Theorem 1.3 is satisfied if $m = 4k$. Also, if $\beta^{2^k} + \beta^{2^{-k}} + 1 = 0$, then $1 = \beta^{2^{2k}} + \beta$. So $\beta^{2^{3k}} + \beta^{2^k}$ is also equal to 1 and

$$\begin{aligned} 1 &= (\beta^{2^{3k}} + \beta^{2^k})(\beta^{2^{2k}} + \beta) \\ &= \beta^{2^{3k}+2^{2k}} + \beta^{2^{3k}+1} + \beta^{2^{2k}+2^k} + \beta^{2^k+1} \\ &= \beta^{(2^k+1)2^{2k}} + \beta^{(2^k+1)2^{3k}} + \beta^{(2^k+1)2^k} + \beta^{2^k+1} \\ &= \mathsf{T}_k^{4k}(\beta^{2^k+1}). \end{aligned}$$

Hence, $\mathsf{T}_1^{4k}(\beta^{2^k+1}) = \mathsf{T}_1^k(\mathsf{T}_k^{4k}(\beta^{2^k+1})) = \mathsf{T}_1^k(1) = k$. We have $4k$th roots of unity when $k$ is even and $8k$th roots of unity when $k$ odd. The multiplicies can be found using linear algebra.

*Example 4.1 (n=1, a=1, k=5).* Let

$$E_5(s) = \sum_{x \in \mathbb{F}_{2^s}} (-1)^{\mathsf{T}_1^s(x^{33}+x)}.$$

The following table lists the values of the $E_5(s)$, $1 \le s \le 19$.

| $s$ | 1 | 2 | 3 | 4 | 5 | 6 | 7 | 8 | 9 | 10 | 11 | 12 | 13 | 14 | 15 | 16 | 17 | 18 | 19 |
|---|---|---|---|---|---|---|---|---|---|---|---|---|---|---|---|---|---|---|---|
| $E_5(s)$ | 2 | 0 | $-4$ | 8 | 32 | 0 | 16 | $-32$ | 32 | 0 | $-64$ | 128 | $-128$ | 0 | $-2^{10}$ | $-512$ | 512 | 0 | $-2^{10}$ |

The normalized $\eta_j$'s belong to the 40th cyclotomic field but not the 20th cyclotomic field. Some of them are repeated with multiplicity two.

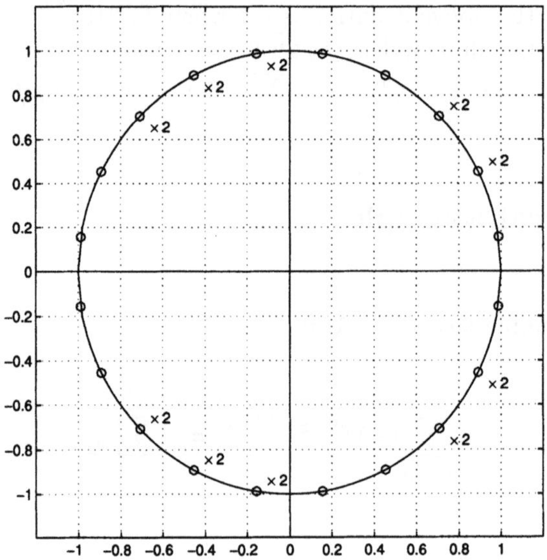


*Example 4.2 (n=1, a=1, k=6).* Let

$$E_6(s) = \sum_{x \in \mathbb{F}_{2^s}} (-1)^{\mathrm{T}_1^s(x^{65}+x)}.$$

The values of $E_6(s)$ are calculated for $1 \leq s \leq 19$.

| $s$ | 1 | 2 | 3 | 4 | 5 | 6 | 7 | 8 | 9 | 10 | 11 | 12 | 13 | 14 | 15 | 16 | 17 | 18 | 19 |
|---|---|---|---|---|---|---|---|---|---|---|---|---|---|---|---|---|---|---|---|
| $E_6(s)$ | 2 | 4 | 8 | 0 | $-8$ | 64 | 16 | $-64$ | $-64$ | 64 | $-64$ | 0 | $-2^7$ | $2^8$ | $-2^9$ | $-2^{10}$ | $2^9$ | $2^{12}$ | $-2^{10}$ |

The normalized $\eta_j$'s are 24th roots of unity.

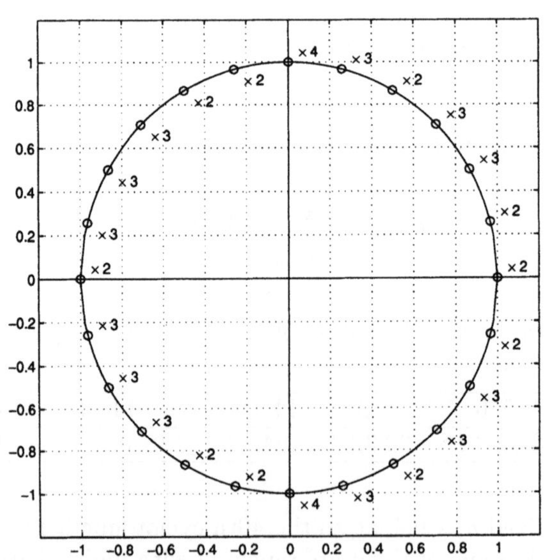

Note that the normalized reciprocal roots are "evenly" distributed on the unit circle. The cancellations explains why the Gold exponential sums are (usually) small.

## References


1. L. Carlitz and S. Uchiyama, Bounds on Exponential Sums, *Duke Math. J.*, vol 24, pp37-41, 1957.
2. R. Coulter, On the Evaluation of a Class of Weil Sums in Characteristic 2, *New Zealand J. of Math*, vol 28, pp1–14, 1999.
3. I. Duursma, H. Stichtenoth and C. Voss, Generalized Hamming weights for duals of BCH codes, and maximal algebraic function fields, *Arithmetic, Geometry and Coding Theory*, Pellikaan eds, pp 53–66, Walter de Gruyter, Berlin, 1996.
4. R. Gold, Maximal Recursive Sequences with 3-valued Recursive Cross-Correlation Functions, *IEEE Trans of Info. Theory*, vol 14, pp154–156, Jan, 1968.
5. P. V. Kumar and T. Helleseth, Sequences with Low Correlation, *Handbook in Coding Theory*, V. Pless eds, 1999, North-Holland.
6. R. Lidl and H. Niederreiter, *Finite Fields*, Second Edition, Cambridge University Press, 1997.
7. D. Sarwate and M. Pursley, Crosscorrelation Properties of Pseudorandom and Related Sequences, *Proc. of IEEE*, vol 68, pp593–619, no. 5, May 1980.
8. S. Stepanov, *Arithmetic of Algebraic Curves*, Monographs in contemporary Mathematics, Plenum, New York, 1994.

# On Permutation Polynomials and Derivable Translation Planes

Rita Vincenti

Dipartimento di Matematica e Informatica
Università di Perugia
Via Vanvitelli, 1
06100 Perugia
Italy

**Abstract** Permutation polynomials are involved in various different situations, for instance, in the resolution of singularities or in the construction of cryptographic systems for the secure transmission of data. Here we show that permutation polynomials occur also in the classification of translation planes.

## 1 Introduction

The classification of translation planes has always played a central role in finite geometries and it is at the basis of interest in new combinatorial structures. André, Bruck and Bose (see [1], [3] , [4], [5]) proved that it is equivalent to classifying translation planes or to classifying spreads in projective spaces.

In recent years many results on the classification have been due to the definition of flock of quadratic sets of $PG(3,q)$ (see Thas [17], Johnson [13]).

Examples of flock planes include the Walker regular nearfield planes, the irregular nearfield planes, the Roo–Rodabaugh–Wilke–Zemmer planes of order $p^2$. On the other hand, Dempwolff and Reifart [10], Czerwinski and Oakden [8] used the computer to classify translation planes of order 16 and 25 respectively. The translation planes $\Pi$ of order $q^2$ and kernel $K = GF(q)$ can be represented by spreads of lines of $PG(3,q)$ or, equivalently, by spread sets $\mathcal{M}$ of $2 \times 2$ matrices over $K$ (we refer to [9], pp. 131–133, 219–221, to [16], pp. 1–10, 15–29 for definitions and representations). To this large family belong Desarguesian planes, Hall planes, flock planes and at least 13 among the 21 planes of order 25. We prove that the existence and the classification of such sets $\mathcal{M}$ (that is, the existence of non-isomorphic classes of planes) depends on the permutation polynomials of $K[x]$ involved in the entries of the elements of $\mathcal{M}$. For the order 25 a class of planes, called *non-Walker*, is exhibited (see Theorem 14).

## 2 Permutation Polynomials and Spread Sets

Let $K = GF(q)$ be the Galois field of order $q = p^n$ and let $\Sigma = PG(4,q)$ be the 4-dimensional projective geometry over $K$. Let $\Pi$ be a derivable translation plane of order $q^2$ and with kernel containing $K$.

Let us fix a hyperplane $\Sigma' = PG(3,q)$ in $\Sigma$ and assume that

$$S = \{L_\infty\} \cup \{L_M \mid M \in \mathcal{M}\} \tag{1}$$

is the spread of lines of $\Sigma'$ representing $\Pi$ where $\mathcal{M}$ is a spread set of $2 \times 2$ matrices (see [9], pp. 131–133, 219–221, [16], pp. 1–10).

From Johnson [13] we know that $S$ must contain the set

$$\mathcal{D} = \{L_\infty\} \cup \{L_k \mid k \in K\} \tag{2}$$

where

$$L_k = \{(x_1, x_2, x_1 k, x_2 k^\sigma) \mid x = (x_1, x_2) \in L_\infty,\ \sigma \in Aut\,K\}.$$

Note that $\mathcal{D}$ represents a derivation set for $\Pi$ and it is a regulus precisely when $\sigma = id$.

From the above it follows that the spread set $\mathcal{M}$ contains matrices of the type

$$kI^\sigma = \begin{pmatrix} k & 0 \\ 0 & k^\sigma \end{pmatrix}, \quad \sigma \in Aut\,K,\ \forall k \in K. \tag{3}$$

The set $\mathcal{M}_k = \{L_\infty\} \cup \{kI^\sigma \mid k \in K\}$ represents the partial spread corresponding to $\mathcal{D}$ . Denote $\bar{\mathcal{M}} = \mathcal{M} - \mathcal{M}_k$, that is, any matrix $M \in \bar{\mathcal{M}}$ is the generic element of the spread set $\mathcal{M}$. Set $M = (m_{ij})$, $m_{ij} \in K$, $i,j = 1,2$.

**Lemma 2.1.** *If $M = (m_{ij}) \in \bar{\mathcal{M}}$ then $m_{12} \neq 0 \neq m_{21}$. Moreover any function $h : K \times K^* \to K$ such that for each $u \in K^*$ the map $h_u : K \times K^* \to K$, $t \longmapsto h(u,t)$ is bijective defines unique functions $f, g : K \times K^* \to K$ and a bijective labeling $M : K \times K^* \to \bar{\mathcal{M}}$, $(u,t) \longmapsto M(u,t)$ with*

$$M(u,t) = \begin{pmatrix} h(u,t) & u \\ g(u,t) & f(u,t) \end{pmatrix}, \quad u \in K^*,\ t \in K. \tag{4}$$

*Moreover $f_u$ (defined similar as $h_u$) is bijective too. If $h$ is constant as a function in the first variable the map $g_t : K^* \to K$, $u \longmapsto g(u,t)$ is bijective for $t \in K$.*

*Proof.* For any $M \in \bar{\mathcal{M}}$, $M' \in \mathcal{M}_k$ we have $det(M - M') = (m_{11} - k)(m_{22} - k^\sigma) - m_{12}m_{21} \neq 0$. If $m_{12} = 0$ or $m_{21} = 0$, choose $k = m_{11}$ to have a contradiction.

Set $m_{12} = u$ and $m_{11} = h(u,t)$ where $u \in K^*$ and $h : K^* \times K \to K$ is a function. Since the first row uniquely determines the second, there exist functions $f, g : K^* \times K \to K$ such that $m_{21} = f(u,t)$, $m_{22} = g(u,t)$.

As $\mathcal{M}$ is a spread set, $det(M - M') \neq 0$ for any $M, M' \in \mathcal{M}$, $M \neq M'$. Choose $M, M' \in \mathcal{M}$ and write them in the functional form (4) with respect to $(u,t)$, $(u',t') \in K^* \times K$, respectively.

Fixing $u' = u$ then $det(M - M') = (h(u,t') - h(u,t))(f(u,t') - f(u,t))$. If $M \neq M'$ or $t \neq t'$ then $h(u,t') \neq h(u,t)$ and $f(u,t') \neq f(u,t)$, that is, $h_u(t)$ and $f_u(t)$ are bijective functions of $t$.

Let $h(u,t) = h(u',t)$ for any $u, u' \in K^*, u \neq u'$, for any $t \in K$, that is, assume $h$ to be constant in $t$.

Then $det(M - M') = (u,u')(g(u,t) - g(u',t'))$. If $M \neq M'$ or, $u \neq u'$ then $g(u,t) \neq g(u',t)$, that is, $g_t(u)$ is bijective for $t \in K$.

From now on we will refer to (4) as a *functional form* of the spread set $\mathcal{M}$. Now the question is to determine these three functions.

The following result is due to Dickson [11].

**Lemma 2.2.** *Every function from $K$ into $K$ can be represented uniquely by an element of $K[x]$ of degree less than $q$.*

**Lemma 2.3.** *Every function $\gamma : K^* \times K \to K$ can be represented by an element of $K[x,y]$ of degree less than $2q - 3$.*

*Proof.* From Lemma 2 we can write

$$\gamma(x,y) = \gamma_0(x) + \gamma_1(x)y + \gamma_2(x)y^2 + \cdots + \gamma_{q-1}(x)y^{q-1}$$

where $\gamma_i(x)$ is a polynomial of $K[x]$ for $i = 0, 1, \ldots, q-1$.

As $u^{q-1} = 1$ when $u \neq 0$, the degree $deg\, \gamma_i(x)$ is less than $q - 2$ for any $i = 0, 1, \ldots, q-1$. Therefore $deg\, \gamma(x,y) < q - 2 + q - 1 = 2q - 3$.

From Lemma 1 it follows immediately

**Proposition 2.4.** *For any fixed $u \in K^*$ and for any $t \in K$, the functions $h(u,t)$ and $f(u,t)$ are permutation polynomials of $K[t]$; if $h(u,t) = h(u',t)$ for all $u, u' \in K^*$, $t \in K$ then for any fixed $t \in K$ and for any $u \in K^*$, $g(u,t)$ is a permutation polynomial of $K[u]$.*

From Hermite's Criterion (see [15]) and from the definition of spread set we can easily prove:

**Proposition 2.5.** *Each of the degrees of $f(u,t), h(u,t) \in K[t]$ (and of $g(u,t) \in K[u]$ when $h(u,t)$ does not depend on $u$) is non-zero and at most $q - 2$.*

In order to recognize non-isomorphic planes, we need to classify spread sets representing them.

Let $\Pi$ and $\Pi'$ be derivable translation planes of order $q^2$ and with kernels containing $K$. Let $\mathcal{S}$ and $\mathcal{S}'$ be the spreads of $\Sigma'$ representing them, let $\mathcal{D} \subset \mathcal{S}$ and $\mathcal{D}' \subset \mathcal{S}'$ be the derivation sets and $\mathcal{M}$ and $\mathcal{M}'$ the spread sets related to them, respectively.

Assume that the 0-matrix and the $I$-matrix belong to both $\mathcal{M}$ and $\mathcal{M}'$. For any matrix $M = (m_{ij})$ and for any $\alpha \in Aut\, K$ denote by $M^\alpha$ the matrix $(m_{ij}^\alpha)$.

The planes $\Pi$ and $\Pi'$ are isomorphic precisely when there exists a semi-linear automorphism $\gamma$ of $K^4$ such that $\gamma\mathcal{S} = \mathcal{S}'$ (see [16]).

**Theorem 2.6.** *If $\Pi$ and $\Pi'$ are isomorphic via $\gamma$ and $\gamma\mathcal{D} = \mathcal{D}'$, one can choose a labeling $\mathcal{M}'$ for $\Pi'$ such that*

$$\mathcal{M}' = P^{-1}\mathcal{M}^{\alpha}P$$

*where $P$ is non singular and $\alpha \in Aut\, K$.*

*Proof.* Without loss of generality we can require that and $\gamma L_\infty = L'_\infty$, $\gamma L_0 = L'_0$, $\gamma L_I = L'_I$, that is, $\gamma(x,y) = (x^\alpha P, y^\alpha P)$ where $P$ is non-singular and $\alpha$ is the companion automorphism of $\gamma$. It is now easy to end the proof.

Let $K' \subset K$ be the field pointwise fixed by the automorphism $\sigma$ of $K$ (see (3)).

For any $M \in \mathcal{M}$, $k \in K^*$ the matrix $\bar{M} = M - kI^\sigma$ is non-singular or $\bar{M} = 0$ (see [16]), therefore:

(i) $det\, \bar{M} = k^{\sigma+1} - k^\sigma h(u,t) - kf(u,t) + h(u,t)f(u,t) - ug(u,t) \neq 0$ for any $k, t \in K$, $u \in K^*$.
In case $p$ is odd and $k$ ranges over $K'$ (or, $\sigma = id$) from (i) follows:

(ii) $(h(u,t) - f(u,t))^2 + 4ug(u,t) = s$ is a non-square of $K'$ for any $t \in K'$, $u \in K'^*$.

Assume from now on $p$ odd and $\mathcal{D}$ is a regulus.

For a given $\mathcal{M}$ representing $\Pi$ in functional form (4), denote by $S$ the subset of the non-squares of $K$ consisting of the elements $s$ of (ii).

Let $\mathcal{M}'$ be a spread set representing a plane $\Pi'$ isomorphic to $\Pi$.

Because of Theorem 6, $\mathcal{M}'$ contains $\mathcal{M}_k = \{kI \mid k \in K\}$. Assume that for any matrix $M' \in \bar{\mathcal{M}}' = \mathcal{M}' - \mathcal{M}_k$

$$M' = \begin{pmatrix} h'(u',t') & u' \\ g'(u',t') & f'(u',t') \end{pmatrix} \qquad \text{(compare with (4))}.$$

Let $\gamma : \Pi \to \Pi'$ be an isomorphism and let $\mathcal{M}$, $\mathcal{M}'$ be labeled as in Theorem 6. Denote by $S'$ the subset of the non-squares of $K$ analogous to $S$ relatively to $\mathcal{M}'$.

**Theorem 2.7.** (i) *and* (ii) *are equivalent. The sets $S$ and $S'$ coincide.*

*Proof.* To prove that (ii) implies (i), assume (i) is false, that is, $det(M - kI) = 0$ for some $M \in \bar{\mathcal{M}}'$ (or, for some $u \in K^*$, $t \in K$) and for some $k \in K$. As $\mathcal{M}$ is a spread set, it must be $M - kI = 0$, the null matrix. In particular, this means $g(u,t) = 0$, a contradiction to Lemma 1.

To prove that $S$ and $S'$ coincide, note first that for any $M' \in \bar{\mathcal{M}}'$ there exists $M \in \bar{\mathcal{M}}$ such that $M' = P^{-1}M^\alpha P$, $\alpha \in Aut\, K$ (see Theorem 6). We can write $det(M' - k'I) = det(P^{-1}M^\alpha P - k^\alpha I) = (det(M - kI))^\alpha$. The last equality shows that the matrices of $\mathcal{M}$ and $\mathcal{M}'$ have the same sets of eigenvalues (in a suitable extension of $K$) and for that it must be $S = S'$.

From Theorem 7 it follows that we can write

$$g(u,t) = 4^{-1}u^{q-2}(s - (h(u,t) - f(u,t))^2) \tag{5}$$

where $s$ ranges over $S$ and an analogous expression holds for $g'(u',t')$.

**Proposition 2.8.** *$g(u,t)$ is homogeneous of degree 1 if and only if*

$$s - (h(u,t) - f(u,t))^2 = 4\gamma u^2$$

*and $g(u,t) = \gamma u$ for some $\gamma \in K^*$.*

*Proof.* If $s - (h(u,t) - f(u,t))^2 = 4\gamma u^2$, $\gamma \in K^*$ then $g(u,t) = \gamma u$. Assume $g(u,t) = \alpha u + \beta t$, $\alpha, \beta \in K$, $\beta \neq 0$. We obtain $s - (h(u,t) - f(u,t))^2 = 4u(\alpha u + \beta t)$ which becomes zero when $t = -\alpha\beta^{-1}u$, a contradiction.

**Proposition 2.9.** *The matrices $M$ and $M' = P^{-1}M^{\alpha}P$ are related by the followings*

$$u' = au^{\alpha}, \quad h'(u',t') = h(u,t)^{\alpha} + bu^{\alpha}, \quad f'(u',t') = f(u,t)^{\alpha} - bu^{\alpha},$$

$$g'(u',t') = 4^{-1}u'^{-1}(s' - (h(u,t) - f(u,t) + 2bu)^2), \quad a, b \in K, \ \alpha \in Aut\, K.$$

*Proof.* As by assumption $P^{-1}M(u,t_i)^{\alpha}P = M'(u',t_i')$, $i = 1,2$; $t_1 \neq t_2$ we see that $P$ has lower triangular form. After easy computations, one can end the proof by comparing all the entries of $M$ and $M'$, respectively, and using the expressions chosen for both $g(u,t)$ and $g'(u',t')$.

Let $\mathcal{M}$ and $\mathcal{M}'$ be spread sets representing planes $\Pi$ and $\Pi'$, respectively. Assume that for $M \in \mathcal{M}$, $M' \in \mathcal{M}'$ is

$$M = \begin{pmatrix} h(u,t) & u \\ g(u,t) & f(u,t) \end{pmatrix}, \qquad M' = \begin{pmatrix} ah(u',t') & bu' \\ cg(u',t') & df(u',t') \end{pmatrix}$$

for $a, b, c, d \in K^*$.

**Proposition 2.10.** *If $g(u,t) = cg(u',t')$ when $u = bu'$ and $t', t$ satisfy*

$$df^{-1}(u,t)h(u,t) = af^{-1}(u',t')h(u',t'),$$

*then $\Pi$ and $\Pi'$ are isomorphic.*

*Proof.* The hypothesis says that for any matrix $M \in \bar{\mathcal{M}}$, corresponding to given $u, t \in K$, there exists a matrix $M' \in \bar{\mathcal{M}}'$ depending on $u', t' \in K$ such that $M = M'$. That is, $\bar{\mathcal{M}} = \bar{\mathcal{M}}'$ and the planes $\Pi, \Pi'$ are isomorphic.

From the Proposition 10 it follows

**Corollary 2.11.** *We can choose $f(u,t)$ (or $h(u,t)$) as a monic polynomial of $K[t]$ .*

## 3 The Classification

Some first considerations about the classification of derivable translation planes using this permutation polynomials point of view are summarized by the following theorems.

Let $\mathcal{M}$ be a spread set in the functional form (4) representing $\Pi$.

**Theorem 3.1.**
(i) *$\Pi$ is an elliptic flock (= desarguesian) plane precisely when $h(u,t) = f(u,t)$, $g(u,t) = \gamma u$ where $\gamma \in K^*$ and $f^2(u,t) - \gamma u^2$ is irreducible for all $u \in K^*$;*
(ii) *$\Pi$ is a hyperbolic flock plane precisely when $h(u,t) = ut$, $f(u,t) = u\hat{f}(t)$, $g(u,t) = u\hat{g}(t)$ where $t\hat{f}(t) - \hat{g}(t)$ is irreducible;*
(iii) *$\Pi$ is a Hall plane precisely when $h(u,t) = t$, $f(u,t) = -t$, $g(u,t) = -u^{q-2}(t^2 - s)$ where $t^2 - s$ is irreducible.*

*Proof.* $\Pi$ is desarguesian if and only is $\mathcal{M}$ is a field isomorphic to $K^2$ (see [5]). The functions $f$, $g$ and $h$ must be bilinear (therefore $g(u,t) = \gamma u$, $\gamma \in K^*$ for all $u \in K^*$, see Proposition 8) then it is immediate to prove that $h(u,t) = f(u,t)$ and $f^2(u,t) - \gamma u^2 \neq 0$ for all $u \in K^*$.

To prove (ii) and (iii), see Theorem 7, Proposition 8 and [9], p. 233, [13], Theorem B, respectively.

**Theorem 3.2.** *For $q = 3$ there exists only the desarguesian plane and the Hall plane of order* 9.

*Proof.* If $q = 3$ from Proposition 5 it follows that $f(u,t)$ and $h(u,t)$ have both degree 1 in $t$.

Set $h(u,t) = a + bu + t$, $f(u,t) = c + du + t(e + lu)$ (see Corollary 11).

If $f(u,t) = h(u,t)$ for all $(u,t) \in K^* \times K$ then set $t' = a + t$, $g = 2u$ to recognize $\Pi$ is desarguesian.

If $f \neq h$ assume that there is $(\bar{u},\bar{t})$ such that $f(\bar{u},\bar{t}) = h(\bar{u},\bar{t})$ and let us consider the following two matrices of $\mathcal{M}$

$$M = \begin{pmatrix} \bar{f} & \bar{u} \\ 2\bar{u} & \bar{f} \end{pmatrix} \quad \text{and} \quad M' = \begin{pmatrix} h & u \\ u & 2h \end{pmatrix} \quad \text{where } \bar{f} = f(\bar{u},\bar{t}), h = h(u,t).$$

It is $det(M - M') = \bar{f}^2 + 2h^2 - (2\bar{u}^2 + u^2) = 0$, therefore $f(u,t) \neq h(u,t)$ for all $(u,t) \in K^* \times K$, or $f = 2h = -h$, $g(u,t) = u$ and the plane is a Hall plane.

A permutation polynomial $f \in K[x]$ is in *normalized form* if $f$ is monic, $f(0) = 0$ and, when its degree $n$ is not divisible by the characteristic of $K$, the coefficient of $x^{n-1}$ is 0. Over $K = GF(5)$ there are only two normalized polynomials : $x^5 - ax$ where $a \neq 1$ and $x^5 - 2bx^3 + b^2x$ where $b \neq 1, 4$, that is, $x$ and $x^3$, respectively ( see [14], p. 352).

Therefore, for the possible choices of normalized non-monic polynomials we will have to consider the following ones:

$$\alpha x,\ \alpha \in K^*,\ \beta x^3,\ \beta \in \{1,4\}.$$

Let $h(u,t)$ be the polynomials in position $(1,1)$ of the matrices $\bar{\mathcal{M}}$. From (5), Proposition 4 and Proposition 5 it follows

$$h(u,t) = \sum_{i=0}^{3}(h_{0i} + h_{1i}t^j)u^i, \quad j \in \{1,3\}, \quad h_{0,i}, h_{1,i} \in K,\ i = 0,1,2,3. \quad (6)$$

Moreover, if for all $u \in K^*$ $h(u,0) = 0$, then

$$h(u,t) = \sum_{i=0}^{3} h_{1i}t^j u^i, \quad j \in \{1,3\}, \quad h_{1i} \in K. \quad (7)$$

Analogous considerations hold for the polynomials $f(u,t)$ in position $(2,2)$. Note that to exchange the role of $h(u,t)$ and $f(u,t)$ (which is equivalent to conjugation by $\begin{pmatrix} 0 & 1 \\ 1 & 0 \end{pmatrix}$ of the matrices followed by transposition), no non-isomorphic spread sets are added (see Theorem 6).

We can conclude that the classification of the translation planes of order 25 depends on the non-isomorphic occurrence of such kinds (6) and (7) of polynomials in the matrices $\bar{\mathcal{M}}$.

**Theorem 3.3.** *For $q = 5$, the following first five classes of planes are obtained:*

(i) *the elliptic flock plane,*
(ii) *two hyperbolic flock planes, when $h(u,t) = ut$, $f(u,t) = u\hat{f}(t)$ according to the choices $\hat{f}(t) = t$ and $\hat{f}(t) = \alpha t$ or $\hat{f}(t) = \beta t^3$, $\alpha \neq 1 \neq \beta$, respectively;*
(iii) *the Walker (non-linear flock of quadratic cone) plane of order 25 when $h(u,t) = 2u^2 + t$, $f(u,t) = 3u^2 + t$, $g(u,t) = -2u^3$;*
(iv) *the non-Walker plane when $h(u,t) = 2u^2 + t$, $f(u,t) = 3u^2 + t$, $g(u,t) = -u^3$.*

*Proof.* To prove (i) and (ii) assume $h(u,t) = ut$ and $f(u,t) = u\hat{f}(t)$ (see Corollary 11). In this case $g(u,t) = u\hat{g}(t)$ where $\hat{g}(t) = 4(s - (t - \hat{f}(t))^2)$ and $s$ is a fixed non-square of $K$.

Set

$$M_\alpha = \begin{pmatrix} ut & u \\ 4u(s - t^2(1-\alpha)^2) & \alpha ut \end{pmatrix}, \quad M_\beta = \begin{pmatrix} ut & u \\ 4u(s - (t - \beta t^3)^2) & \beta ut^3 \end{pmatrix}$$

so that

$$\mathcal{M}_\gamma = \mathcal{M}_k \cup \{M_\gamma \mid (u,t) \in K^* \times K\}, \quad \gamma \in \{\alpha, \beta\}$$

are the spread sets corresponding to $\hat{f}(t) = \alpha t$ and $\hat{f}(t) = \beta t^3$, respectively, where $\alpha \in K^*$ and $\beta \in \{1,4\}$ (see (6)).

If $\gamma = \alpha$ then it can be proved that $\mathcal{M}_1$ is a desarguesian (or elliptic flock) spread set (see Theorem 12) and, by using Proposition 10, that $\mathcal{M}_\alpha \cong \mathcal{M}_{\alpha'}$, where $\alpha \neq \alpha' \neq 1$.

If $\gamma = \beta$ it is easy to prove that $\mathcal{M}_\beta \cong \mathcal{M}_{\beta'}$, $\beta \neq \beta'$. Compare [13] to conclude that $\mathcal{M}_\alpha$ and $\mathcal{M}_\beta$ represent the two classes of hyperbolic flock planes.

To prove (iii) and (iv), assume $h(u,t) = 2u^2 + t$ and $f(u,t) = -2u^2 + t$. In this case $g(u,t) = 4u^3(s-1)$ where $s$ is a non-square of $K$.

If $s = 3$ then $g(u,t) = 3u^3$ and the Walker planes are obtained (see [16], p.237). If $s = 2$ then $g(u,t) = 4u^3$. Denote by $\mathcal{M}$ and $\mathcal{M}'$ the two corresponding spread sets, respectively. We will prove that it is impossible, by using Theorem 6 and Proposition 10, to obtain a matrix $M' \in \bar{\mathcal{M}}'$ from any matrix $M \in \bar{\mathcal{M}}$. Set $u' = au$, $a \in K^*$, and compare the entries $(1,1)$, $(2,2)$ to obtain $t' = 2u^2(1-a^2) + t$ and then $a^2 = 1$. Hence $t' = t$ and in both cases, corresponding to $a = 1, 4$, we obtain that $4u'^3 \neq 3u^3$.

We can conclude that $\mathcal{M}$ and $\mathcal{M}'$ represent non-isomorphic planes, that is Walker and non-Walker planes, respectively.

## References


1. André, J.: Über nicht-Desarguessche Ebenen mit transitiver Translationsgruppe Math. Z. **80** (1954), 156–186.
2. Bertini, E.: Geometria proietttiva degli iperspazi. Spoerri Ed., Pisa (1907).
3. Bruck, R.H.: Construction Problems of Finite Projective Planes in Combinatorial Mathematics and its Applications. Bose, R.C. and Dowling, T.A., Eds, The Univ. of N.Carolina Press, Chapel Hill, NC (1969), 426–514.
4. Bruck, R.H. & Bose, R.C.: The Construction of Translation Planes from Projective Spaces. J Algebra **1** (1964), 85–102.
5. Bruck, R.H. & Bose, R.C.: Linear Representation of Projective Planes in Projective Spaces. J Algebra **4** (1966), 117–172.
6. Bruen, A.A.: Inverse Geometry and Some New Translation Planes. Geom. Dedicata **7** (1978) 81–88.
7. Bruen, A.A. & Silverman, R.: Switching Sets in $PG(3,q)$. Proc. Amer. Math. Soc. **43** (1974) 176–180.
8. Czerwinski, T. & Oakden, D.: The Translation Planes of Order Twenty-five. J. Comb Theory, Ser. A **59** (1992) 193–217.
9. Dembowski, P.: Finite geometries. Springer_Verlag, Berlin (1968).
10. Dempwolff, U. and Reifart, A.: The Classification of the Translation Planes of Order 16. I. Geom. Dedicata **15** (1983) 137–153.
11. Dickson, L.E.: Linear Group with an Exposition of the Galois Field Theory. Teubner, Leipzig (1901); Dover, N.Y. (1958).
12. Johnson, N.L.: Lezioni sui piani di traslazione. Quaderni Dip. Mat. Univ. Lecce Q. **3** (1986) 1–121.
13. Johnson, N.L.: Flocks of Hyperbolic Quadrics and Translation Planes Admitting Affine Homologies. J. Geometry **34** (1989) 50–73.

14. Lidl, R. & Niederreiter, H.: Finite Fields. Encyclopedia of Math. and its Appl., Vol. **20**, Addison_Wesley Reading (1983).
15. Lidl, R. & Mullen, G.L.: When Does a Polynomial Over a Finite Field Permute the Elements of the Field? Amer. Math. Monthly **95** (1988) 243–246.
16. Lũneburg, H.: Translation Planes. Springer_Verlag, Heidelberg, N.Y. (1979).
17. Thas, J.A.: Flocks of Non Singular Ruled Quadrics in $PG(3,q)$. Rendic. Accad. Taz. Linecei, **59** (1975) 83–85.
18. Thas, J.A.: Recent Results on Flocks, Maximal Exterior Sets and Inversive Planes. Combinatorics 88 **1**, A.Barlotti et al.(eds) Research and Lecture Notes in Mathematics, Mediterranean Press (1991) 95–108.

# Pure $L$-Functions from Algebraic Geometry over Finite Fields

Daqing Wan

Department of Mathematics
University of California
Irvine, CA 92697-3875
e-mail: dwan@math.uci.edu

**Abstract** This survey gives a concrete and intuitively self-contained introduction to the theory of pure L-functions arising from a family of algebraic varieties defined over a finite field of characteristic $p$. The standard fundamental questions in any theory of L-functions include the meromorphic continuation, functional equation, Riemann hypothesis (RH for short), order of zeros at special points and their special values. Our emphasis here will be on the meromorphic continuation and the RH. These two questions can be described in a general setup without introducing highly technical terms.

The construction of the pure L-function depends on the choice of an absolute value of the rational number field $\mathbf{Q}$. In the case that the absolute value is the complex or $\ell$-adic absolute value ($\ell \neq p$), the rationality of the pure L-function requires the full strength of the $\ell$-adic cohomology including Deligne's main theorem on the Weil conjectures. In the case that the absolute value is the $p$-adic absolute value, the pure L-function is no longer rational but conjectured by Dwork to be $p$-adic meromorphic. This conjecture goes beyond all existing $p$-adic cohomology theories. Its truth opens up several new directions including a possible $p$-adic RH for such pure L-functions.

The guiding principle of our exposition in this paper is to describe all theorems and problems as simple as possible, directly in terms of zeta functions and L-functions without using cohomological terms. In the case that this is not easy to do so, we simply give an intuitive discussion and try to convey a little feeling. Along the way, a number of natural open questions and conjectures are raised, some of them may be accessible to certain extent but others may be somewhat wild due to the lack of sufficient evidences.

## 1 Introduction

The most basic question in number theory is to understand the integers. In particular, for a given integer $N$, we need to understand the absolute value $||N||$ for every absolute value $||?||$ on the rational number field $\mathbf{Q}$. For the complex absolute value, this is to determine

$$||N|| = |N| = ?$$

For the $p$-adic absolute value with $p$ being a prime, this is to determine

$$|N|_p = p^{-a_p},\ a_p = \mathrm{ord}_p(N) = ?$$

This last theoretical question is practically the problem of factoring integers which has important applications.

More generally, suppose that we are given a sequence of interesting integers

$$\{N_1, N_2, \cdots,\}$$

In order to understand this sequence of integers, one naturally forms a suitable generating function $Z(\{N_i\}, T)$ which contains all information about the given sequence. The basic question is then to understand the analytic properties of the generating function $Z(\{N_i\}, T)$ with respect to each absolute value $||?||$ of $\mathbf{Q}$. This includes the possible meromorphic continuation $Z(\{N_i\}, T)$ and a suitable RH about its zeros and poles, for both the complex absolute value and the $p$-adic absolute value. If we have a family of such generating functions, then we would like to understand its analytic variation when the parameter varies.

The most interesting type of sequences arises from counting prime ideals in a finitely generated commutative ring or equivalently from counting rational points on an algebraic variety. In the case of counting prime numbers in the ring $\mathbf{Z}$ of integers, the natural generating function is the Riemann zeta function. This first example was studied by Riemann from complex point of view and by Kummer-Kubota-Leopoldt from $p$-adic point of view. It is the motivating example for much of the modern developments on general Hasse-Weil zeta functions of algebraic varieties as well as their conjectural $p$-adic analogues.

Our interesting sequence of integers in this paper arises from counting rational points over various finite extension fields of an algebraic variety $X$ defined over a finite field of characteristic $p$. The resulting generating function is the zeta function of $X$ which is the object of study in the celebrated Weil conjectures. The zeta function is a rational function as proved by Dwork using $p$-adic methods. It satisfies a suitable complex and $\ell$-adic RH as proved by Deligne using $\ell$-adic methods, where $\ell$ is a prime number different from $p$. The $p$-adic RH for the zeta function is more complicated and remains mysterious in general. The variation of the whole zeta function, when the variety moves through an algebraic family, leads to new interesting questions which are understood to certain extent.

The zeta function is however not pure. That is, the zeros and poles have different absolute values. This is especially so from $p$-adic point of view. Thus, the zeta function decomposes as a product of pure pieces defined in terms of the absolute values of the zeros and poles. A finer form of the RH is to understand the purity decomposition. A further question is to understand the variation of each pure piece of the zeta function when the variety moves through an algebraic family. This naturally leads to the construction of pure L-functions arising from algebraic geometry. Our fundamental question here is then to understand the analytic properties of such a pure L-function, notably its meromorphic continuation and RH. Since the zeta function has integer coefficients, there are three different types of absolute values (complex, $\ell$-adic and $p$-adic) that we can choose to work. These lead to different results and different theories.

In the case that the absolute value is the complex or the $\ell$-adic absolute value, Deligne's main theorem shows that the pure L-function from algebraic geometry can be identified with a geometric L-function, that is the L-function of a certain geometric constructible $\ell$-adic étale sheaf. One can then apply the full machinery of $\ell$-adic étale cohomology. In particular, the pure L-function from algebraic geometry is always rational by Grothendieck's rationality theorem. It satisfies a suitable complex and $\ell$-adic RH by Deligne's theorem. The situation is quite algebraic in nature. All the expected finiteness properties hold. The main point is that one does not need to distinguish the subcategory of geometric $\ell$-adic sheaves from the full category of $\ell$-adic sheaves. Much of the relevant theory works for every $\ell$-adic sheaf, whether it is geometric or not.

In the case that the absolute value is the $p$-adic absolute value, the situation is quite different and much more complicated. A pure L-function from algebraic geometry is not rational any more. However, Dwork conjectured that a pure L-function from algebraic geometry is $p$-adic meromorphic. The situation is quite transcendental in nature. Grothendieck's specialization theorem, Katz's isogeny theorem and Berthelot's finiteness theorem on relative crystalline cohomology show that, at least in nice cases, a pure L-function from algebraic geometry can be identified with the L-function of a certain geometric $p$-adic étale sheaf. The trouble is that the L-function of a general $p$-adic étale sheaf does not behave well. The usual trace formula does not hold. Even worse, the L-function is not meromorphic in general, unlike what Katz conjectured. Thus, in order to prove Dwork's conjecture, one must distinguish the subcategory of geometric $p$-adic sheaves from the full category of $p$-adic sheaves. Our recent work shows that the geometric $p$-adic sheaves can be understood by introducing a new category with growth condition. Roughly speaking, this new category consists of infinite nuclear complexes of infinite rank nuclear overconvergent F-isocrystals. The nuclear overconvergent condition insures that the L-function is $p$-adic meromorphic. This then establishes the meromorphic continuation of a pure L-function from algebraic geometry and thus proves Dwork's conjecture. The $p$-adic RH for such a pure L-function is extremely mysterious. A good understanding seems to require entirely new ideas.

We would like to point out that there is a more general and more difficult type of zeta functions arising from counting algebraic cycles on an algebraic variety $X$ defined over a finite field. These zeta functions are called the zeta functions of algebraic cycles. They seem to be out of reach at this time. For zero cycles, they reduce to the zeta functions studied in the Weil conjectures, which are already quite interesting and fruitful. In general, they contain important arithmetic information about algebraic cycles and are related to Tate's conjecture. Under a mild finiteness condition on the effective cone of the Chow group, these zeta functions of algebraic cycles are conjectured to be $p$-adic meromorphic. If this conjecture is true, one could go on to understand its $p$-adic RH, their variation when the variety moves through an algebraic family, the purity decomposition and the resulting pure L-functions of algebraic cycles. Any proof of this meromorphic conjecture would likely have a profound impact on arithmetic and geometry of algebraic cycles.

## 2 Rationality of Zeta Functions

Let $\mathbb{F}_q$ be the finite field of $q$ elements of characteristic $p$. Let $X$ be an algebraic variety defined over $\mathbb{F}_q$, namely, a separated scheme of finite type over $\mathbb{F}_q$. For example, if $X$ is affine, then $X$ is defined by a system of polynomial equations

$$f_1(x_1,\cdots,x_n) = \cdots = f_r(x_1,\cdots,x_n) = 0$$

in some affine $n$-space $\mathbf{A}^n$, where each $f_i$ is a polynomial defined over $\mathbb{F}_q$. For an extension field $\mathbf{F}_{q^k}$ of degree $k$ over $\mathbb{F}_q$, let $X(\mathbf{F}_{q^k})$ denote the set of $\mathbf{F}_{q^k}$-rational points on $X$. The zeta function of $X/\mathbb{F}_q$ is then defined to be the following formal power series

$$\begin{aligned} Z(X/\mathbb{F}_q,T) &= \exp\left(\sum_{k=1}^{\infty} \frac{\#X(\mathbf{F}_{q^k})}{k} T^k\right) \\ &= \prod_{x\in X(\bar{\mathbf{F}}_q)} \frac{1}{(1-T^{\deg(x)})^{1/\deg(x)}} \\ &= \prod_{x\in X_0} \frac{1}{1-T^{\deg(x)}} \in \mathbf{Z}[[T]], \end{aligned}$$

where $X_0$ is the set of closed points on $X/\mathbb{F}_q$ and $\bar{\mathbf{F}}_q$ denotes a fixed algebraic closure of $\mathbb{F}_q$. Recall that a closed point on $X/\mathbb{F}_q$ is simply the orbit of an actual geometric point $x \in X(\bar{\mathbf{F}}_q)$ under the $q$-th power Frobenius map $\sigma : x \to x^q$ and $\deg(x)$ is the smallest positive integer $k$ such that $\sigma^k(x) = x$. The integer $\#X(\mathbf{F}_{q^k})$ is simply the number of fixed points of the $k$-th power $\sigma^k$ acting on $X(\bar{\mathbf{F}}_q)$. We shall write $Z(X,T)$ for $Z(X/\mathbb{F}_q,T)$ when the ground field $\mathbb{F}_q$ is clear.

The zeta function is a generating function for counting rational points on the variety $X$ over various finite extension fields of $\mathbf{F}_q$. Although at each stage $\mathbf{F}_{q^k}$, only finitely many points are counted, the generating function counted all points of the variety $X$ over the algebraic closure $\bar{\mathbf{F}}_q$. This explains why the zeta function should contain a great deal of geometric and arithmetic information about the variety $X$. Thus, our first fundamental question is

*Question 2.1.* Understand the zeta function $Z(X,T)$.

A general principle in analytic arithmetic algebraic geometry is that all zeta functions and L-functions arising naturally from arithmetic algebraic geometry are analytically good functions. Here we are considering algebraic geometry over $\mathbb{F}_q$. Based on earlier results in various special cases such as diagonal hypersurfaces, curves and abelian varieties, Weil conjectured the following rationality result.

**Theorem 2.2.** *The zeta function $Z(X,T)$ is a rational function in $T$.*

This theorem was first proved by Dwork [Dw1] using $p$-adic analysis. His basic idea is to establish a trace formula which expresses the zeta function as a finite alternating product of the Fredholm determinants of several nuclear operators acting on certain $p$-adic Banach spaces. Since the Fredholm determinant of a nuclear operator is entire [Se], it follows that the zeta function is $p$-adic meromorphic. One then concludes the proof of Theorem 2.2 with the following $p$-adic analogue of Borel's rationality criteria.

**Lemma 2.3.** *Let $f(T) \in \mathbf{Z}[[T]]$ be a power series with integer coefficients. Then, $f(T)$ is a rational function if and only if $f(T)$ is analytic at the origin as a complex function and meromorphic as a $p$-adic function for some prime number $p$.*

Dwork's rationality proof pioneered his $p$-adic theory of zeta functions. Although his proof is not cohomological in nature, it can be viewed as a proof on the chain level in a suitable sense. Its refinement by introducing commuting differential operators is cohomological in nature. The relevant cohomology is of De Rham type as shown by Katz [K1]. This motivated later more systematic developments of various $p$-adic cohomology theories such as the formal cohomology [MW] for smooth affine varieties, the liftable cohomology [Lu] for smooth projective liftable varieties, crystalline cohomology for smooth projective varieties and rigid cohomology [Be] for arbitrary varieties. These $p$-adic cohomology theories in their full generality have not been well understood. The relevance of differential operators suggests the existence of a close relationship between zeta functions and differential equations. This is the subject of F-crystals [K3], which we will not discuss here.

Weil's conjectural approach to Theorem 2.2, generalizing his method for curves and abelian varieties, is to construct a suitable cohomology theory for varieties in characteristic $p$ so that the Lefschetz fixed point theorem holds which immediately implies the rationality of the zeta function. This led to the introduction and full development of the $\ell$-adic étale cohomology theory by Grothendieck and others, where $\ell$ is a prime number different from $p$. This theory provides some of the most powerful tools in the study of zeta functions, especially from the complex point of view. For the exceptional prime $p$, the theory of $p$-adic étale cohomology has also been studied by Grothendieck's school but it does not behave well for L-function purpose. Nevertheless, the $p$-adic étale cohomology contains important $p$-adic information about the zeta function. This turns out to be closely related to Dwork's conjecture. We shall however not discuss this cohomological point of view here due to our restricted simple nature of this survey.

The zeta function is a power series with integer coefficients. However, all proofs of Theorem 2.2 use non-archimedian methods, either $\ell$-adic or $p$-adic. No direct "motivic" proof over $\mathbf{Q}$ or $\mathbf{C}$ is known. Although the classical method of Gauss sums and Jacobi sums for the elementary diagonal case can be viewed as a method over $\mathbf{C}$, it seems hopeless to extend this method to the general case. In the case of curves, there is a direct approach over $\mathbf{Z}$ using the Riemann-Roch theorem. It would be of great interest to extend this Riemann-Roch approach to the general case. Such a direct approach would make it possible to attack Tate's conjecture [Ta] relating orders of zeros and ranks of algebraic cycles which so far has been inaccessible. Such a Riemann-Roch approach should also be useful in understanding the harder zeta functions of algebraic cycles introduced in [W1].

The first application of Theorem 2.2 is the existence of a formula for the number of rational points on $X$. Since the zeta function is rational, there are finitely many algebraic integers $\alpha_i$ and $\beta_j$ such that

$$Z(X,T) = \frac{\prod_i (1-\alpha_i T)}{\prod_j (1-\beta_j T)}.$$

Taking logarithmic derivative, one deduces the following well structured formula for every positive integer $k$:

$$\#X(\mathbf{F}_{q^k}) = \beta_1^k + \beta_2^k + \cdots - \alpha_1^k - \alpha_2^k - \cdots .$$

In particular, this provides a fast algorithm to compute the number of rational points on the variety $X$ over a large finite field $\mathbf{F}_{q^k}$ provided $X$ is defined over a small finite field $\mathbf{F}_q$. The theory can be improved to get a fast algorithm, even for $X$ defined over a large finite field $\mathbb{F}_q$ as long as the characteristic $p$ is small ($q$ can be large), see [W5] for a perspective on this algorithmic subject which has important practical applications.

Once we know that the zeta function is a rational function, we can move on to the next fundamental question. What can we say about its zeros and poles? Ideally, we would like to know how many zeros and poles with a given absolute value. This is the RH for the zeta function. Since the reciprocal zeros $\alpha_i$ and $\beta_j$ are algebraic integers, there are several different types of absolute values (complex, $\ell$-adic and $p$-adic) that we can consider. Accordingly, we can talk about the complex RH, the $\ell$-adic RH and the $p$-adic RH for the zeta function $Z(X,T)$. These questions and their family versions are discussed in the following sections.

## 3 Purity Decomposition and RH

The zeta function has rational coefficients. In order to understand its analytic properties, we have to choose an absolute value of the rational number field

$\mathbf{Q}$. For this purpose, we let $||?||$ be a fixed absolute value on $\mathbf{Q}$. Let $\Omega$ be the smallest extension field of $\mathbf{Q}$ such that $\Omega$ is both algebraically closed and topologically complete with respect to $||?||$. If $||?||$ is the complex absolute value $|?|$, $\Omega$ is the field $\mathbf{C}$ of complex numbers. If $||?||$ is the $p$-adic absolute value $|?|_p$ for some prime number $p$, $\Omega$ is the field $\mathbf{C}_p$ of $p$-adic numbers which is the completion of an algebraic closure of the $p$-adic rational numbers $\mathbf{Q}_p$, where the $p$-adic absolute value is normalized by $|p|_p = 1/p$. For a fixed prime power $q$ of $p$, we define the slope function on $\Omega$ by

$$s(\alpha) = \log_q ||\alpha||, \text{ if } ||?|| = |?|$$

and

$$s(\alpha) = -\log_q ||\alpha||, \text{ if } ||?|| \neq |?|.$$

This definition depends on $q$ which is our base of the slope function. For $\alpha \in \Omega$, we can write

$$\alpha = q^{s(\alpha)} u(\alpha),$$

where $u(\alpha)$ is a number with absolute value 1. Note that in the complex absolute value case, our slope function $s(\alpha)$ is twice the weight function $w(\alpha)$ as defined by Deligne. In the $p$-adic case, the slope function is simply the order function

$$s(\alpha) = \mathrm{ord}_q(\alpha).$$

For a given polynomial

$$P(T) = \prod_i (1 - \alpha_i T) \in \Omega[T],$$

and a real number $s$, we define the slope $s$ part of $P(T)$ by

$$P_s(T) = \prod_{s(\alpha_i)=s} (1 - \alpha_i T) \in \Omega[T].$$

This immediately yields the purity (or slope) decomposition of $P(T)$:

$$P(T) = \prod_{s \in \mathbb{R}} P_s(T).$$

This is a finite product since $P(T)$ is a polynomial. This definition easily extends to rational functions in $\Omega(T)$ as well as meromorphic functions in $\Omega((T))$ by the Weierstrass factorization theorem. In the latter case, the purity decomposition is an infinite product in general. Applying the purity decomposition to the zeta function $Z(X,T)$, our notation becomes

$$Z(X,T) = \prod_{s \in \mathbf{R}} Z_s(X,T),$$

where $Z_s(X,T)$ is the slope $s$ part of $Z(X,T)$. This is called the purity (or slope) decomposition of the zeta function $Z(X,T)$. Our next fundamental question is then to understand this purity decomposition. That is,

*Question 3.1.* Understand each pure slope $s$ part $Z_s(X,T)$ of $Z(X,T)$.

The first step is to understand the degree of the rational function $Z_s(X,T)$ for each $s$. To be precise, we recall that the degree of a rational function is the degree of the numerator minus the degree of the denominator. Similarly, the total degree of a rational function is the degree of the numerator plus the degree of the denominator.

**Definition 3.2.** Let $d(X)$ (resp. $D(X)$) denote the degree (resp. the total degree) of the zeta function $Z(X,T)$. Similarly, for each real number $s$, let $d_s(X)$ (resp. $D_s(X)$) denote the degree (resp. the total degree) of the slope $s$ part $Z_s(X,T)$ of the zeta function $Z(X,T)$.

It is clear that we have the purity decomposition for the degrees $d(X)$ and $D(X)$:

$$d(X) = \sum_{s\in\mathbf{R}} d_s(X),$$

$$D(X) = \sum_{s\in\mathbf{R}} D_s(X).$$

The RH for the zeta function is to determine the exact slopes of the zeros and poles. It is easy to see that the number of reciprocal zeros of slope $s$ is given by $(D_s(X) + d_s(X))/2$. Similarly, the number of reciprocal poles of slope $s$ is given by $(D_s(X) - d_s(X))/2$. Thus, the following weaker but more precise form of Question 3.1 is already the RH for $Z(X,T)$.

*Question 3.3.* Understand the pure degree $d_s(X)$ and the pure total degree $D_s(X)$ for all $s$.

Both the degree $d(X)$ and the total degree $D(X)$ of the whole zeta function $Z(X,T)$ can be effectively bounded using $p$-adic methods as shown by Bombieri [Bo]. This is because Dwork's $p$-adic theory is constructive. Thus, the pure degree $d_s(X)$ and the pure total degree $D_s(X)$ are also effectively bounded for all $s$. The integers $d_s(X)$ and $D_s(X)$ depend both on the slope $s$ and on the variety $X$. Of course, they also depend on the choice of the absolute value $\|?\|$ which was built in the definition of the slope decomposition. In this section, we consider the case that $X$ is fixed. In next section, we consider how $d_s(X)$ and $D_s(X)$ vary when $X$ varies.

Deligne's main theorem [De2] on the complex and the $\ell$-adic RH can be stated in our notations as follows.

**Theorem 3.4 (complex case).** *Let $\|?\|$ be the complex absolute value. Let $n$ be the dimension of the variety $X$. If*

$$s \notin \left\{0, \frac{1}{2}, 1, \frac{3}{2}, 2, \cdots, n\right\},$$

*then*

$$d_s(X) = D_s(X) = 0, \ Z_s(X,T) = 1.$$

This shows that for the complex absolute value, the non-trivial slopes are rational numbers in the interval $[0, n]$ with denominators at most 2. Thus, for the complex RH, it remains to determine the $2n+1$ values $d_s(X)$ and $D_s(X)$, where $s$ varies in the above exceptional set of $2n+1$ numbers. These remaining values are in general difficult to determine although some extremal cases such as $s = n, n-\frac{1}{2}$ can be done. They depend on the detailed geometry of the variety $X$. However, in nice situations, they can be determined by the Betti numbers. This includes the smooth projective (more generally smooth proper) case as conjectured by Weil and first proved by Deligne [De1] [De2] using $\ell$-adic cohomology. A similar proof was later given by Faltings [Fa] using crystalline cohomology.

For the $\ell$-adic RH, the answer is much simpler and very clean.

**Theorem 3.5 ($\ell$-adic case).** *Let $||?||$ be the $\ell$-adic absolute value for some prime $\ell \neq p$. If $s \neq 0$, then*

$$d_s(X) = D_s(X) = 0, \ Z_s(X,T) = 1.$$

*In particular,*

$$D_0(X) = D(X), \ d_0(X) = d(X), \ Z_0(X,T) = Z(X,T).$$

*That is, all zeros and poles of the zeta function are $\ell$-adic units.*

Since the $\ell$-adic case is always pure and thus gives no interesting decomposition, from now on, we shall mostly restrict our attention to the complex case and the $p$-adic case.

For the $p$-adic RH, unfortunately, no clean general answer is possible, even in the smooth projective case, even in the case of smooth projective curves. However, one has the following weak but simple general result, which is a consequence of the rationality of the zeta function.

**Theorem 3.6 ($p$-adic case).** *Let $||?||$ be the $p$-adic absolute value. There is an effectively computable positive integer $N$ such that if*

$$s \notin \left\{0, \frac{1}{N}, \frac{2}{N}, \cdots, n\right\},$$

*then*

$$d_s(X) = D_s(X) = 0, \ Z_s(X,T) = 1,$$

*where $n$ is the dimension of $X$.*

This result is far weaker than Theorem 3.4, because the denominator $N$ here is not bounded by 2. In fact, the denominator $N$ cannot be bounded by any finite absolute constant. It depends very much on the variety $X$ and the prime number $p$, not just on the geometry of $X$. This explains why the $p$-adic RH for the zeta function is very complicated. It can be determined in a few special cases such as the elementary diagonal hypersurface case where one can use the Stickelberger theorem for Gauss sums. In nice situations such as the smooth projective case, a good lower bound for the Newton polygon (which determines the $p$-adic RH) is given by the Hodge polygon (constructed using the Hodge numbers of a lifting of $X$), as conjectured by Katz and proved by Mazur [M2]. Strictly speaking, both polygons are defined for each cohomological dimension. For our restricted purpose, one could either fix a cohomological dimension or take the collection over all cohomological dimensions. If the two polygons coincide, the variety $X$ is called ordinary.

For a given smooth projective variety $X$, Mazur's theorem provides a geometric lower bound for the arithmetic Newton polygon but it does not tell if $X$ is ordinary or how far $X$ is from being ordinary. There is no known clean recipe (conjectural or not) to determine the Newton polygon of $X$. A preliminary step might be to look at the size of the endomorphism group of $X$. The larger the endomorphism group for $X$ is, the more relations there would be among the zeros and poles and thus $X$ would be less likely ordinary, as one observes in the diagonal case and the supersingular elliptic curve case. It would be interesting to make such heuristic arguments more precise.

Another interesting question is to consider the limiting behavior of the Newton polygon as $p$ varies. For instance, let $X$ be a smooth projective variety defined over $\mathbf{Q}$ and let $X_p$ be the reduction of $X$ mod $p$ for large prime $p$. Let $NP(X_p)$ denote the Newton polygon of $X_p$ and let $HP(X)$ denote the Hodge polygon of $X$. As $p$ goes to infinity, the Newton polygon $NP(X_p)$ would not have a limit in general. But it always has its lower limit. As this lower limit has nothing to do with any particular prime $p$, it should be a geometric invariant of $X$. Thus, it is tempting to make

**Conjecture 3.7.** *Let $X$ be a smooth projective variety over $\mathbf{Q}$. Then,*

$$\lim_{p\to\infty} \inf NP(X_p) = HP(X).$$

Since there are only finitely many possibilities for the $NP(X_p)$ for a fixed smooth projective $X/\mathbf{Q}$, the above conjecture is equivalent to saying that there are infinitely many ordinary primes $p$ (probably of positive density) for a given smooth projective $X/\mathbf{Q}$. If the endomorphism group of $X$ is sufficiently small, one may further hope that the set of ordinary primes $p$ for $X$ often has density 1. A somewhat related conjecture is given by Serre [Oo] in the case of abelian varieties over number fields. Similarly, one could ask

$$\lim_{p\to\infty} \sup NP(X_p) = ?$$

This sup limit should again be a geometric invariant of $X$.

So far, we have been concerned with the degree of the pure slope $s$ part $Z_s(X,T)$. A finer question is to understand the rational function $Z_s(X,T)$ itself for non-trivial slope $s$. For instance, one could ask about the possible rationality of the coefficients for $Z_s(X,T)$ when $Z_s(X,T)$ is written as the quotient of two relatively prime polynomials with constant term 1. The answer depends on the absolute value $\|?\|$ we choose. If $\|?\|$ is the $\ell$-adic absolute value, then Theorem 3.5 shows that $Z_s(X,T)$ trivially has integer coefficients. The same result holds in the complex absolute value case. This follows from Deligne's main theorem and Galois theory.

**Theorem 3.8.** *Let $\|?\|$ be the complex absolute value. Then, the pure slope $s$ part $Z_s(X,T)$ has integer coefficients.*

In the $p$-adic case, it is no longer true that the coefficients of $Z_s(X,T)$ are integers, because the purity decomposition becomes more substantial, even in nice situations. One has the following $p$-adic rationality result which is a consequence of the rationality of $Z(X,T)$.

**Theorem 3.9.** *Let $\|?\|$ be the $p$-adic absolute value. Then, the coefficients of the pure slope $s$ part $Z_s(X,T)$ are $p$-adic integers in $\mathbf{Z}_p$, which are also algebraic integers.*

Of course, the reciprocal roots of $Z_s(X,T)$ will not be in $\mathbf{Z}_p$ in general.

## 4 Variation of the Pure Degrees

In the previous section, we discussed the pure degrees $d_s(X)$ and $D_s(X)$ for the zeta function of a single variety $X$. In this section, we turn to discussing how the pure degrees $d_s(X)$ and $D_s(X)$ vary when $X$ moves through an algebraic family.

Let $f : Y \to X$ be a family of algebraic varieties over $\mathbb{F}_q$ parametrized by $X$. For each geometric point $x \in X(\mathbb{F}_q^{\deg(x)})$, the fibre $Y_x = f^{-1}(x)$ is an algebraic variety defined over $\mathbb{F}_q^{\deg(x)}$ and thus we have the purity decomposition:

$$Z(Y_x/\mathbb{F}_q^{\deg(x)},T) = \prod_{s\in\mathbf{R}} Z_s(Y_x/\mathbb{F}_q^{\deg(x)},T),$$

where the slope function is defined with respect to $q^{\deg(x)}$ since $\mathbb{F}_q^{\deg(x)}$ is the base field of $Y_x$. For each rational number $s$, let $d_s(Y_x)$ (resp. $D_s(Y_x)$) denote the degree (resp. the total degree) of the slope $s$ part $Z_s(Y_x,T)$ of the zeta function $Z(Y_x,T)$, where we stress again that the slope function is defined with respect to $q^{\deg(x)}$. The general results in Section 3 give information about the nature of the numbers $d_s(Y_x)$ and $D_s(Y_x)$ for a fixed $x$. We would like to understand how these numbers $d_s(Y_x)$ and $D_s(Y_x)$ vary when the geometric point $x$ varies. For a given integer $m$, we define

$$X(d_s,m) = \{x \in X | d_s(Y_x) = m\}, \ X(D_s,m) = \{x \in X | D_s(Y_x) = m\}.$$

These are subsets of $X$. We would like to understand the possible algebraic and geometric structure of these sets.

The theory of $\ell$-adic cohomology and Deligne's main theorem [De2] imply

**Theorem 4.1.** *Let $f : Y \to X$ be a family of algebraic varieties over $\mathbb{F}_q$. Let $||?||$ be the complex or the $\ell$-adic absolute value. Then for every integer $m$ and every slope $s$, the set $X(d_s, m)$ is a constructible subset of $X$.*

The same general statement for the harder $X(D_s, m)$ does not seem to follow from the existing results because the nature of cancellation is not well understood in general. Thus, we have

*Question 4.2.* Let $f : Y \to X$ be a family of algebraic varieties over $\mathbb{F}_q$. Let $||?||$ be the complex or the $\ell$-adic absolute value. Is it true that for every integer $m$ and every slope $s$, the set $X(D_s, m)$ is a constructible subset of $X$?

This is known to be true if $f$ is smooth and proper because there is then no cancellation of zeros and poles by Deligne's theorem. We hope and are slightly inclined to believe that Question 4.2 has a positive answer, at least after replacing the constructible notion by the more general definable notion of logic.

In the $p$-adic case, one can ask a similar question.

**Conjecture 4.3.** *Let $f : Y \to X$ be a family of algebraic varieties over $\mathbb{F}_q$. Let $||?||$ be the p-adic absolute value. Then for every integer $m$ and every slope $s$, the set $X(d_s, m)$ is a constructible subset of $X$.*

*Question 4.4.* Let $f : Y \to X$ be a family of algebraic varieties over $\mathbb{F}_q$. Let $||?||$ be the $p$-adic absolute value. Is it true that for every integer $m$ and every slope $s$, the set $X(D_s, m)$ is a constructible (or more generally definable) subset of $X$?

Conjecture 4.3 and Question 4.4 are known to have a positive answer in nice situations such as when $f : Y \to X$ is proper and smooth over $\mathbb{F}_q$ with a proper smooth lifting to characteristic zero. In such a case, Conjecture 4.3 is a consequence of Grothendieck's specialization theorem [K4] and Berthelot's finiteness theorem [Be] on the relative crystalline cohomology or the relative rigid cohomology in the proper smooth liftable case. The stronger Question 4.4 in such a nice situation also needs Deligne's theorem to avoid the cancellation problem.

The general case of Conjecture 4.3 follows from the conjectural finiteness of the relative rigid cohomology. However, Conjecture 4.3 is already known to be true for $s = 0$ by the congruence formula of Deligne-Katz [De3] [K2] for zeta functions. The general case of Question 4.4 is apparently more difficult. It is not known to be true even for $s = 0$.

In some nice cases such as the universal family of hypersurfaces (or more generally complete intersections), the generic Newton polygon coincides with the Hodge polygon as conjectured by Mazur [M1]. This can be proved in two ways. One approach is to use hyperplane sections to reduce the question to the case of a generic plane curve, as worked out by Illusie [Il] using crystalline cohomology and some ideas of Deligne. Another more flexible approach introduced by the author [W2] is to establish suitable local to global decomposition theorems to reduce the question to the diagonal case where Stickelberger theorem applies, see [W7] for an exposition of this method. This method has other applications such as the more general Adolphson-Sperber conjecture [AS] for the generic Newton polygon of exponential sums. Hence, in such a generic ordinary case, the generic values of $d_s(X)$ and $D_s(X)$ are determined by the Hodge numbers of $X$. Of course, for a given smooth projective hypersurface, there is still no simple recipe to determine if $X$ is ordinary.

It seems very difficult and complicated to have a complete understanding of the stratification of the universal family of hypersurfaces by Newton polygons. This is so, even in the case of curves. However, for the more managable family of abelian varieties, the stratification question by Newton polygons is reasonably well understood by the work of de Jong and Oort [DO].

## 5 Variation of the Zeta Function

Let $f : Y \to X$ be a family of algebraic varieties over $\mathbb{F}_q$ parametrized by $X$. In this section, we consider how the zeta function $Z(Y_x, T)$ varies when the parameter $x$ varies. A standard procedure is to understand all the higher moments of the zeros and poles of the family of rational functions $Z(Y_x, T)$. Write

$$Z(Y_x, T) = \frac{\prod_i (1 - \alpha_i(x)T)}{\prod_j (1 - \beta_j(x)T)},$$

where the $\alpha_i(x)$ and the $\beta_j(x)$ are algebraic integers parametrized by $x$. For a positive integer $k$, the $k$-th moment of the zeros and poles of $Z(Y_x, T)$ for $x \in X(\mathbb{F}_q)$ is given by the sum

$$S_k(f) = \sum_{x \in X(\mathbb{F}_q)} (\alpha_1(x)^k + \alpha_2(x)^k + \cdots - \beta_1(x)^k - \beta_2(x)^k - \cdots).$$

More generally, the $k$-th moment over the $d$-th extension field $\mathbf{F}_{q^d}$ of $\mathbb{F}_q$ is defined by

$$S_{k,d}(f) = \sum_{x \in X(\mathbf{F}_{q^d})} (\alpha_1(x)^{k\frac{d}{\deg(x)}} + \alpha_2(x)^{k\frac{d}{\deg(x)}} + \cdots$$
$$- \beta_1(x)^{k\frac{d}{\deg(x)}} - \beta_2(x)^{k\frac{d}{\deg(x)}} - \cdots).$$

This is an integer by Galois theory. The variation question is then to understand the $k$-th moment sequence $S_{k,d}(f)$ $(d = 1, 2, \cdots)$ for every $k$. In terms of generating functions, we need to understand the $k$-th power L-function of the family $f$ defined by

$$L^{[k]}(f,T) = \exp\left(\sum_{d=1}^{\infty} \frac{S_{k,d}(f)}{d} T^d\right)$$
$$= \prod_{x \in X(\bar{\mathbf{F}}_q)} \frac{1}{Z(Y_x/\mathbf{F}_{q^{k\deg(x)}}, T^{\deg(x)})^{1/\deg(x)}}$$
$$= \prod_{x \in X_0} \frac{1}{Z(Y_x/\mathbf{F}_{q^{k\deg(x)}}, T^{\deg(x)})} \in \mathbf{Z}[[T]].$$

In the last two equations, the degree of $x$ is defined over $\mathbb{F}_q$, but the ground field for the fibre $Y_x$ has been extended from $\mathbb{F}_q^{\deg(x)}$ to its $k$-th extension field $\mathbf{F}^{k\deg(x)}$. Explicitly,

$$L^{[k]}(f,T) = \prod_{x \in X_0} \frac{\prod_j (1 - \beta_j^k(x) T^{\deg(x)})}{\prod_i (1 - \alpha_i^k(x) T^{\deg(x)})} \in \mathbf{Z}[[T]].$$

**Theorem 5.1.** *The $k$-th power L-function $L^{[k]}(f,T)$ of the family $f$ is a rational function for every positive integer $k$.*

This result is a special case of a more general rationality result for certain partial zeta functions studied in [W11]. It can be proved using either Dwork's $p$-adic method or Grothendieck's $\ell$-adic method. The only new ingredient is to use tensor operations and Newton's formula expressing the $k$-th power symmetric functions in terms of elementary symmetric functions. A theorem of Faltings [W11] shows that the partial zeta function is always nearly rational.

By Theorem 5.1, for each $k$, there are finitely many algebraic integers $\gamma_i(k)$ and $\delta_j(k)$ such that

$$L^{[k]}(f,T) = \frac{\prod_i (1 - \gamma_i(k)T)}{\prod_j (1 - \delta_j(k)T)}.$$

Thus, for every positive integer $k$, we have the formula for the $k$-th moment sequence

$$S_{k,d}(f) = \delta_1(k)^d + \delta_2(k)^d + \cdots - \gamma_1(k)^d - \gamma_2(k)^d - \cdots, \quad d = 1, 2, \cdots.$$

Such formulas imply the existence of a suitable equi-distribution theorem for the zeros and poles of the family of rational functions $Z(Y_x, T)$ parametrized by $x \in X$. The simple approach we take here is related to but different from the approach of Deligne-Katz [K5] via monodromy representations.

To be more precise in the application of the equi-distribution problem, one needs to know information about the RH for the $k$-th power L-function $L^{[k]}(f,T)$. Applying the purity decomposition to $L^{[k]}(f,T)$, we can write

$$L^{[k]}(f,T) = \prod_{s \in \mathbf{R}} L_s^{[k]}(f,T),$$

where $L_s^{[k]}(f,T)$ is the slope $s$ part of $L^{[k]}(f,T)$ and the slope is defined with respect to the base $q^k$. Recall that the purity decomposition always depends on our choice of the absolute value $||?||$ on $\mathbf{Q}$.

**Definition 5.2.** Let $d(f,k)$ (resp. $d_s(f,k)$) be the degree of the rational function $L^{[k]}(f,T)$ (resp. $L_s^{[k]}(f,T)$). Similarly, let $D(f,k)$ (resp. $D_s(f,k)$) be the total degree of the rational function $L^{[k]}(f,T)$ (resp. $L_s^{[k]}(f,T)$).

Although the present situation is more complicated than the zeta function case treated in Section 3, it is essentially of the same nature. That is, similar RH holds for $L^{[k]}(f,T)$ in the complex and $\ell$-adic case. One simply keeps track of the simple proof of Theorem 5.1 and applies Deligne's theorem. In particular, the zeros and poles of $L^{[k]}(f,T)$ are always $\ell$-adic units. For the complex absolute value, we make it explicit as follows, see [W11] for a more general result on partial zeta functions.

**Theorem 5.3 (complex case).** *Let $||?||$ be the complex absolute value. Let $n$ be the dimension of the variety $X$ and let $m$ be the relative dimension of the family $f$. For each non-negative integer $k$, if*

$$s \notin \left\{0, \frac{1}{2}, 1, \frac{3}{2}, 2, \cdots, km+n\right\},$$

*then*

$$d_s(f,k) = D_s(f,k) = 0, \ L_s^{[k]}(f,T) = 1.$$

This result shows that it suffices to understand $d_s(f,k)$ and $D_s(f,k)$ for the above $2(km+n)+1$ exceptional values of the slope $s$. These are in general quite complicated to determine except in some extremal cases.

As for the $p$-adic RH, one only has the following very weak general result.

**Theorem 5.4 ($p$-adic case).** *Let $||?||$ be the $p$-adic absolute value. There is an effectively computable positive integer $N(f,k)$ such that if*

$$s \notin \left\{0, \frac{1}{N(f,k)}, \frac{2}{N(f,k)}, \cdots, km+n\right\},$$

*then*

$$d_s(f,k) = D_s(f,k) = 0, \ L_s^{[k]}(f,T) = 1,$$

*where $n$ is the dimension of $X$ and $m$ is the relative dimension of $f$.*

Again, this result for the $p$-adic case is far weaker than Theorem 5.3 for the complex case. The problem is that we know almost nothing about the denominator $N(f,k)$. An important problem is to estimate the size of $N(f,k)$. This is related to the size of the degree $d_s(f,k)$. The extra integer $k$ provides a new dimension of variation. That is, how the degrees $d(f,k)$,

$D(f,k)$, $d_s(f,k)$ and $D_s(f,k)$ vary with $k$. Although the size of the degrees $d(f,k)$ and $D_s(f,k)$ can be effectively bounded, no explicit general bounds are available in the literature. This should be a realistic interesting problem to study.

The first example of great arithmetic interest is the universal family $f$ of elliptic curves parametrized by a modular curve or an Igusa curve. In this case, the sequence of integers $d(f,k)$ is mutually determined by the sequence of dimensions of modular forms of weight $k+2$. The harder sequence $D(f,k)$ is unknown because of possible cancellation of zeros and poles. For the complex absolute value, the easier degree sequence $d_s(f,k)$ is understood by Deligne's theorem on the Ramanujan-Peterson conjecture [De4] but again the harder total degree sequence $D_s(f,k)$ is unknown. For the $p$-adic absolute value, the easier degree sequence $d_s(f,k)$ is already quite mysterious and one conjectures [W6] that $N(f,k)$ is bounded independent of $k$, or even stronger, $d_s(f,k)$ is bounded independent of $k$ and $s$. This is the $p$-adic analogue of the Ramanujan-Peterson conjecture. Very little is known about it. The variation of $d_s(f,k)$ with $k$ is closely related to the Gouvêa-Mazur conjecture [GM], see Coleman [Co] and [W4] for positive results in this direction. The variation of $d_s(f,k)$ with $k$ is also closely related to the geometry of the eigencurve as studied by Coleman-Mazur [CM].

## 6 Variation of the Pure Part of the Zeta Function

Let $f: Y \to X$ be a family of algebraic varieties over $\mathbb{F}_q$ parametrized by $X$. In this section, for each fixed slope $s$, we consider how the pure part $Z_s(Y_x,T)$ varies when the parameter $x$ varies. This is the full aspect of Question 3.1. This question is deeper than just how the degree $d_s(Y_x)$ of $Z_s(Y_x,T)$ varies with $x$. It is also deeper than how the total zeta function $Z(Y_x,T)$ varies with $x$. As in Section 5, the standard procedure is to understand the $k$-th moment sequence associated to the reciprocal zeros and reciprocal poles of $Z_s(Y_x,T)$. Equivalently, we need to understand the $k$-th power L-function associated to the $k$-th moment sequence.

Write
$$Z_s(Y_x/\mathbf{F}_{q^{\deg(x)}},T) = \frac{\prod_i (1-\alpha_i(x,s)T)}{\prod_j (1-\beta_j(x,s)T)}.$$
The associated $k$-th moment over the $d$-th extension field $\mathbf{F}_{q^d}$ of $\mathbb{F}_q$ is defined by
$$S_{k,d}(s,f) = \sum_{x\in X(\mathbf{F}_{q^d})} (\alpha_1(x,s)^{k\frac{d}{\deg(x)}} + \alpha_2(x,s)^{k\frac{d}{\deg(x)}} + \cdots$$
$$- \beta_1(x,s)^{k\frac{d}{\deg(x,s)}} - \beta_2(x,s)^{k\frac{d}{\deg(x)}} - \cdots).$$
This is an integer in the complex absolute value case by a generalization of Theorem 3.8. It is a $p$-adic integer in $\mathbf{Z}_p$ in the $p$-adic case by a generalization of Theorem 3.9. The variation question is then to understand the $k$-th

moment slope $s$ sequence $S_{k,d}(s,f)$ $(d=1,2,\cdots)$ for every $k$. In terms of generating functions, we need to understand the following pure L-function from algebraic geometry.

**Definition 6.1.** Let $k$ be a positive integer. For a given slope $s \in \mathbf{Q}$, the $k$-th power slope $s$ L-function $L^{[k]}(s,f,T)$ attached to the family $f$ is defined to be

$$\begin{aligned} L^{[k]}(s,f,T) &= \exp\left(\sum_{d=1}^{\infty} \frac{S_{k,d}(s,f)}{d} T^d\right) \\ &= \prod_{x \in X(\bar{\mathbf{F}}_q)} \frac{1}{Z_s(Y_x/\mathbf{F}_{q^{k\deg(x)}}, T^{\deg(x)})^{1/\deg(x)}} \\ &= \prod_{x \in X_0} \frac{1}{Z_s(Y_x/\mathbf{F}_{q^{k\deg(x)}}, T^{\deg(x)})} \in \Omega[[T]], \end{aligned}$$

where the slope function is defined with respect to the base $q^k$ (note that we have already replaced $T$ by $T^{\deg(x)}$ in the Euler factors).

In the last two equations, the degree of $x$ is defined over $\mathbb{F}_q$, but the ground field for the fibre $Y_x$ has been extended from $\mathbb{F}_q^{\deg(x)}$ to its $k$-th extension field $\mathbf{F}_q^{k\deg(x)}$. Explicitly,

$$L^{[k]}(s,f,T) = \prod_{x \in X_0} \frac{\prod_j (1-\beta_j^k(x,s)T^{\deg(x)})}{\prod_i (1-\alpha_i^k(x,s)T^{\deg(x)})} \in \Omega[[T]].$$

The $k$-the power slope $s$ L-function arises in a natural way from arithmetic and geometry. Thus, by the general principle, we expect it to be an analytically good function. For the $\ell$-adic absolute value, the situation reduces to the $k$-th power L-function of Section 5, since the purity decomposition is trivial. For the complex absolute value, the situation is deeper than Theorem 5.1. One needs the full strength of Deligne's main theorem [De2] which says that the higher direct image sheaf is mixed. This together with Grothendieck's rationality theorem and Newton's formula implies the following result which can be viewed as the archimedian analogue of Dwork's conjecture.

**Theorem 6.2 (complex case).** *Let $||?||$ be the complex absolute value. For every positive integer $k$ and every rational number $s$, the $k$-th power slope $s$ L-function $L^{[k]}(s,f,T)$ is a rational function.*

In the case that the absolute value is the $p$-adic absolute value, the situation is much more complicated. For slope $s=0$, the pure L-function $L^{[k]}(0,f,T)$ can be obtained as a suitable limit of the easier mixed L-functions treated in the previous section. To see this, one observes that the total degree of the zeta function of each fibre $Y_x$ is uniformly bounded by the results in

[Bo]. This implies that for certain choices of positive integer $M$ depending on $f$, we have the following congruence

$$L^{[k+p^m M]}(f,T) \equiv L^{[k+p^{m+1}M]}(f,T) \ (\mathrm{mod} p^m)$$

for all positive integers $m$. This $p$-adic continuity relation shows that the $p$-adic limit

$$\lim_{m\to\infty} L^{[k+p^m M]}(f,T)$$

exists as a formal $p$-adic power series. In fact, one checks easily from the definition of the pure L-function that the above $p$-adic limit is precisely the pure slope 0 L-function:

$$L^{[k]}(0,f,T) = \lim_{m\to\infty} L^{[k+p^m M]}(f,T).$$

For slope $s > 0$, we do not know any $p$-adic limiting formula for the pure slope L-function $L^{[k]}(s,f,T)$ in terms of the mixed L-functions $L^{[k]}(f,T)$. For slope $s = 0$, although each mixed L-function $L^{[k]}(f,T)$ is rational by Theorem 5.1, we cannot conclude that its limit $L^{[k]}(0,f,T)$ would be rational or even meromorphic. There are geometric examples (the universal family of elliptic curves, for instance) which show that the $k$-th power slope 0 L-function $L^{[k]}(0,f,T)$ is not rational in the $p$-adic case. Dwork, however, conjectured [Dw5] that the pure slope L-function is always $p$-adic meromorphic.

**Conjecture 6.3 (Dwork).** *Let $\|?\|$ be the p-adic absolute value. For every positive integer k and every rational number s, the k-th power slope s L-function $L^{[k]}(s,f,T)$ is a p-adic meromorphic function.*

In [Dw2-5], Dwork showed that Conjecture 6.3 is true for several examples. This includes the universal family of elliptic curves and a certain family of K3-surfaces. His idea is to reduce the problem to the classical overconvergent setting by establishing the existence of the so-called excellent lifting. The excellent lifting however rarely exists and thus one cannot hope that it would work in general. Another approach studied in [DS] and [W3] is to try to relax the overconvergent condition by using the weaker $c\log$-convergent condition. This approach pushes Dwork's trace formula to its full potential but again it cannot work in general by the counter-example in [W3]. The general case of Dwork's conjecture is being proved in our recent series of papers [W8-10] by introducing an entirely new method building on the fundamental Dwork-Monsky trace formula [Mo]. At this point, the abstract version of Dwork's conjecture for F-crystals and $\sigma$-modules, as proved in [W8-10], implies Conjecture 6.3 whenever the finiteness of the relative rigid cohomology with compact support is known such as the smooth proper liftable case. Our strategy to handle singular and open family is to directly work with the infinite rank version. This requires additional work which is in progress.

Thus, the $k$-th power slope $s$ L-function $L^{[k]}(s,f,T)$ is always a good function. An immediate application is the existence of a formula for the $k$-th moment slope $s$ sequence $S_{k,d}(s,f)$. For each $k$ and each $s$, we can write

$$L^{[k]}(s,f,T)=\frac{\prod_i(1-\gamma_i(k,s)T)}{\prod_j(1-\delta_j(k,s)T)},$$

where the product is finite in the complex absolute value case and infinite in the $p$-adic case. We have

$$\lim_{i\to\infty}\gamma_i(k,s)=0=\lim_{j\to\infty}\delta_j(k,s).$$

The formula for the $k$-th moment slope $s$ sequence is

$$S_{k,d}(s,f)=\delta_1(k,s)^d+\delta_2(k,s)^d+\cdots-\gamma_1(k,s)^d-\gamma_2(k,s)^d-\cdots,\ d=1,2,\cdots.$$

Such formulas imply the existence of a suitable equi-distribution theorem for the zeros and poles of the family of pure rational functions $Z_s(Y_x,T)$ parametrized by $x\in X$. In the complex absolute case, this is related to but different from the approach of Deligne-Katz via monodromy representations. In the $p$-adic absolute value case, such a result is completely new.

Once we know that $L^{[k]}(s,f,T)$ is a good function, we can then ask for the various RHs for $L^{[k]}(s,f,T)$. In the case of Theorem 6.2, there are three types of RHs (the complex, the $\ell$-adic and the $p$-adic). The situation is similar to Section 5. The L-function in Theorem 6.2 has integer coefficients in view of Theorem 3.8.

In the case of Conjecture 6.3, one can only ask the $p$-adic RH since the zeros and poles are $p$-adic numbers. This $p$-adic RH for the L-function in Conjecture 6.3 is apparently more difficult than the already mysterious $p$-adic RH for the L-function in Theorem 6.2, because there are now infinitely many zeros and poles in the case of Conjecture 6.3. In fact, we have no ideas what the general conjectural formulation for the $p$-adic RH should be. It seems un-reasonable to expect a finite upper bound for the denominators of the slopes of the zeros and poles of $L^{[k]}(s,f,T)$, although such a finite bound is conjectured to be true [W6] in the simplest elliptic family case.

A harder higher dimensional example is given by the following family of $n$-dimensional Calabi-Yau varieties:

$$X_1^{n+2}+X_2^{n+2}+\cdots+X_{n+2}^{n+2}+\lambda X_1X_2\cdots X_{n+2}=0.$$

In the case $n=1$ (elliptic curves), the $k$-th power slope $s$ ($s=0$) L-function $L^{[k]}(s,f,T)$ in the $p$-adic case already contains a great deal of arithmetic information about classical and $p$-adic modular forms. In the case $n=2$ (K3-surfaces), Conjecture 6.3 was proved by Dwork but its arithmetic consequences have not been explored. In the higher dimensional case $n>2$, Conjecture 6.3 follows from our recent work. It would be of great interest to explore its arithmetic consequences. For instance, the $k$-th power slope $s$ ($s = 0$) L-function $L^{[k]}(s, f, T)$ in the $p$-adic case should be closely related to the arithmetic of the mysterious mirror map in mirror symmetry, see Lian and Yau [LY] for another relation bewteen Dwork's work and the mirror map. In some cases, the special value of the L-function $L^{[k]}(s, f, T)$ at $T = 1$ seems to be related to the conjectural $p$-adic L-functions of algebraic varieties defined over a number field. Very little is known in this direction.

## 7 Zeta Functions of Algebraic Cycles

In this final section, following [W1], we define the zeta functions of algebraic cycles of a projective variety embedded in a given projective space. Several standard conjectures associated with such zeta functions are described.

We first recall the definition of the degree of a projective variety. Let $n$ and $m$ be positive integers with $n \leq m$. Let $X$ be a closed $n$ dimensional subscheme of the $m$ dimensional projective space $\mathbf{P}^m$ over $\mathbf{F}_q$ (the embedding will be fixed). We shall always work over the ground field $\mathbf{F}_q$. Let

$$S(X) = \oplus_{d=0}^{\infty} S_d(X)$$

be the homogeneous coordinate ring of $X$, where $S_d(X)$ consists of the homogeneous elements of degree $d$ and $S(X)$ is finitely generated by $S_1(X)$ as an $\mathbf{F}_q$-algebra ($S_0(X) = \mathbf{F}_q$). Let

$$l(d) = \dim_{\mathbf{F}_q} S_d(X).$$

For large $d$, $l(d)$ is a polynomial in $d$ of degree $n$, called the Hilbert polynomial of $X$. The degree of $X$, denoted by $\deg X$, is defined to be the coefficient of $d^n/n!$ in this polynomial. If $X$ is irreducible, an alternative definition of $\deg X$ is the intersection number $X \cdot H^n$, where $H$ is a hyperplane in $\mathbf{P}^m$. This means that the degree of $X$ is the number of intersection points cut out by the intersection of $n$ sufficiently general hyperplanes.

Fix a closed $n$ dimensional subscheme $X$ of $\mathbf{P}^m$ over $\mathbf{F}_q$. A prime cycle on $X/\mathbf{F}_q$ is a closed integral subscheme of $X$ (i.e., a reduced and irreducible closed subscheme). For each integer $0 \leq r \leq n$, we define the zeta function of $r$-cycles on $X$ to be the following formal power series

$$Z_{r-\text{cycles}}(X, T) = \prod_{P} (1 - T^{\deg P})^{-1},$$

where $P$ runs over all prime cycles of dimension $r$ on $X$ and $\deg P$ is the degree of $P$ viewed as a closed subscheme of $\mathbf{P}^m$. Since the definition of the degree of a variety depends on the embedding, thus the zeta function of $r$-cycles for $0 < r < n$ depends on the embedding of $X$ in the projective space $\mathbf{P}^m$.

Let $N_r(d)$ (resp. $M_r(d)$) be the number of prime $r$-cycles (resp. effective $r$ cycles) of degree $d$ on $X$. By a theorem of Chow and van der Waerden, the set of effective $r$-cycles of degree $d$ is parametrized (one-to-one) by an algebraic set (the Chow variety) in a projective space. Thus, $M_r(d)$ and $N_r(d)$ are all finite. The zeta function of $r$-cycles on $X/\mathbf{F}_q$ is well defined. Since $Z_{r-\text{cycles}}(X,T)$ is a power series with integer coefficients, it is trivially $p$-adic analytic in the open unit disk.

Alternative expressions for $Z_{r-\text{cycles}}(X,T)$ are given by

$$\begin{aligned} Z_{r-\text{cycles}}(X,T) &= \prod_{d=1}^{\infty}(1-T^d)^{-N_r(d)} \\ &= \sum_{d=0}^{\infty} M_r(d)T^d \\ &= \exp\left(\sum_{d=1}^{\infty}\frac{W_r(d)}{d}T^d\right), \end{aligned}$$

where $W_r(d)$ is the following weighted sum (each prime $r$-cycle of degree $k$ is counted $k$ times)

$$W_r(d) = \sum_{k|d} kN_r(k).$$

Geometrically, a point counted in $N_r(d)$ consists of $d$ conjugate "non closed points" of dimension $r$. Thus, $W_r(d)$ can be interpreted as the number of "non closed points" of degree $d$ and dimension $r$. In particular, $W_0(d)$ is just the number of $\mathbf{F}_{q^d}$-rational points on $X$ and $Z_{0-\text{cycles}}(X,T)$ is the classical zeta function of a projective scheme over a finite field. The zeta function $Z_{0-\text{cycles}}(X,T)$ of zero cycles is a rational function by Dwork's theorem. It is easy to show that for $0<r<n$, the zeta function $Z_{r-\text{cycles}}(X,T)$ of $r$-cycles is never rational and in fact never complex meromorphic, as the radius of convergence for $Z_{r-\text{cycles}}(X,T)$ is zero as a complex function. However, some simple examples [W1] suggest that $Z_{r-\text{cycles}}(X,T)$ might be $p$-adic meromorphic in some cases. If $Z_{r-\text{cycles}}(X,T)$ is indeed $p$-adic meromorphic, it would imply that there are two sequences of $p$-adic integers $\alpha_i$ and $\beta_i$ approaching zero such that for all positive integers $d$, we have

$$W_r(d) = \beta_1^d + \beta_2^d + \cdots - \alpha_1^d - \alpha_2^d - \cdots.$$

This would generalize the classical formula for the number $W_0(d)$, in which there are only finitely many terms, because the function $Z_{0-\text{cycles}}(X,T)$ of zero cycles is rational.

To formulate our conjectures, we let $A_r(X)$ be the Chow group of $r$-cycles on $X/\mathbf{F}_q$ modulo rational equivalence. It is conjectured that $A_r(X)$ is a finitely generated abelian group. Let $A_r^+(X)$ be the monoid in $A_r(X)$

generated by the effective $r$-cycles on $X/\mathbf{F}_q$. This is the effective cone in $A_r(X)$. The monoid $A_r^+(X)$ is in general not a finitely generated monoid for $0 < r < n$. Although we do not have a proven counter-example, we feel that the $p$-adic meromorphic continuation of $Z_{r-\text{cycles}}(X,t)$ might be false if the monoid $A_r^+(X)$ is not finitely generated. This is a little inconsistent with the general philosophy that L-functions arising in a natural way from algebraic geometry are analytically good functions. Perhaps, one should not push the general principle too far. Thus, all our conjectures assume that $A_r^+(X)$ is a finitely generated monoid. In fact, we are confident about the truth of our conjectures only under the stronger assumption that $A_r(X)$ is of rank one. However, we will state our conjectures under the weaker assumption that the monoid $A_r^+(X)$ is finitely generated. The meromorphic continuation conjecture is

**Conjecture 7.1 (meromorphic continuation).** *Let $X$ be an $n$ dimensional projective variety in $\mathbf{P}^m$ over $\mathbf{F}_q$. Assume that the monoid $A_r^+(X)$ is a finitely generated monoid. Then $Z_{r-\text{cycles}}(X,T)$ is a $p$-adic meromorphic function.*

Since the zeta function of $r$-cycles is supposed to be a $p$-adic meromorphic function, we can only ask for its $p$-adic RH. Any RH is really a finiteness property about zeros and poles. Thus, the $p$-adic RH in our current situation should be a finitness property on the slopes of the zeros and poles. Since there are infinitely many zeros and poles in general for the zeta function of $r$-cycles, the set of slopes for the zeros and poles is unbounded. Thus, the best we can hope for is to bound the denominators of the slopes. In this direction, we propose

**Conjecture 7.2 ($p$-adic RH).** *Let $X$ be an $n$ dimensional projective variety in $\mathbf{P}^m$ over $\mathbf{F}_q$. Assume that the monoid $A_r^+(X)$ is a finitely generated monoid. Then the denominators of the slopes (which are rational numbers) of the reciprocal zeros and reciprocal poles of $Z_{r-\text{cycles}}(X,T)$ are bounded by a constant which may depend on $X$.*

In connection with Tate's conjecture about the order of poles, we propose

**Conjecture 7.3 (order of pole).** *Let $X$ be an $n$-dimensional smooth projective variety in $\mathbf{P}^m$ over $\mathbf{F}_q$. Assume that the monoid $A_r^+(X)$ is a finitely generated monoid. Then, the rank of $A_r(X)$ is equal to the order of pole of $Z_{r-\text{cycles}}(X,T)$ at $T=1$.*

All three conjectures are true in the two extremal cases $r = 0, n$. The first accessible new case is the case of divisors, i.e., when $r = n-1$. If $r = n-1$, the above three conjectures are known to be true in the case when $A_{n-1}(X)$ has rank one [W1]. In this case, Conjecture 7.2 was not stated and hence not proved in [W1], but its truth follows from the meromorphy proof given there. Furthermore, Conjecture 7.3 is always true in the divisor case $r = n-1$ without the rank one assumption.

Thus, even in the case of divisors, Conjectures 7.1-7.2 are not known to be true in general if $A_{n-1}(X)$ has rank greater than 1. At this point, the most fundamental conjecture seems to be Conjecture 7.1. Once Conjecture 7.1 is proved (if it is true), its proof should give a great deal of information about the other two conjectures. But no single example is known for $1 \leq r \leq n-2$. The simplest substantial example is to consider $Z_{1-\text{cycles}}(\mathbf{P}^3, T)$ (counting space curves of degree $d$ in $\mathbf{P}^3$ as $d$ varies), which seems already sufficiently difficult and requires new mathematics.

Tate's conjecture says that the order of pole at $T = q^{-r}$ of the zeta function of zero cycles is equal to the rank of the group of algebraic $r$-cycles modulo $\ell$-adic homological equivalence. For a variety over a finite field, this latter rank should equal to be the rank of the Chow group $A_r(X)$. With this modification, Tate's conjecture can be reformulated as

**Conjecture 7.4 (Tate).** *Let $X$ be an $n$-dimensional smooth projective variety over $\mathbf{F}_q$. Then, the rank of $A_r(X)$ is equal to the order of pole of the zeta function $Z_{0-\text{cycles}}(X, T)$ at $T = q^{-r}$.*

As indicated in the above, Conjecture 7.3 is known to be true in the divisor case $r = n-1$. This suggests that Tate's conjecture might also be provable in the divisor case $r = n-1$ if the monoid $A_{n-1}^{+}(X)$ is finitely generated. Combining the previous two conjectures together, we obtain the following conjecture relating the zeta function of $r$-cycles to the zeta function of zero cycles.

**Conjecture 7.5.** *Let $X$ be an $n$-dimensional smooth projective variety in $\mathbf{P}^m$ over $\mathbf{F}_q$. Assume that the monoid $A_r^{+}(X)$ is a finitely generated monoid. Then, the order of pole of $Z_{r-\text{cycles}}(X, T)$ at $T = 1$ is equal to the order of pole of $Z_{0-\text{cycles}}(X, T)$ at $T = q^{-r}$.*

Based on the Griffith-Katz counter-example about 1-cycles on certain 3-fold, which applies only to varieties over fields with transcendental elements, we remarked in [W1] that one should not expect the equality between the order of pole at $T = 1$ of $Z_{r-\text{cycles}}(X, T)$ and the order of pole at $T = 1$ of $Z_{(n-r)-\text{cycles}}(X, T)$. This remark is misleading. As Sòule pointed out to me sometime ago, one should expect the equality of those two numbers by Beilinson's conjectures.

The special value at $T = 1$ of the zeta function of $r$-cycles is undoubtedly related to the torsion order of $A_r(X)$ and certain regulator of $A_r(X)$. This is proved in [W1] in the divisor case $r = n-1$. We leave it to interested readers to find a conjectural formula for the special value for other $r$.

## References

[AS] A. Adolphson and S. Sperber, Exponential sums and Newton polyhedra, Ann. Math., 130(1989), 367-406.

[Be] P. Berthelot, Géométrie rigide et cohomologie des variétés algébriques de caractéristique $p$, Mém. Soc. Math. France (N.S.) No. 23(1986), 3, 7–32.

[Bo] E. Bombieri, On exponential sums in finite fields, II, Invent. Math., 47(1978), 29-39.

[Co] R. Coleman, $p$-adic Banach spaces and families of modular forms, Invent. Math., 127(1997), 417-479.

[CM] R. Coleman and B. Mazur, The eigencurve, in Galois Representations in Arithmetic Algebraic Geometry (Durham, 1996), London Math. Soc. Lecture Note Ser., 254, 1998, 1-113.

[De1] P. Deligne, La Conjecture de Weil, I, Publ. Math., IHES, 43(1974), 273-307.

[De2] P. Deligne, La Conjecture de Weil, II, Publ. Math., IHES, 52(1980), 137-252.

[De3] P. Deligne, Fonctions L modulo $\ell^n$ et modulo $p$, in: Cohomologie étale (SGA $4\frac{1}{2}$), 110–128, Lecture Notes in Math. 569, Springer-Verlag, 1977.

[De4] P. Deligne, Formes modulaires et representations $\ell$-adiques, Sém Bourbaki, 355, Lecture Notes in Math., Vol. 179(1971), 136-172.

[DO] A.J. de Jong and F. Oort, Purity of the stratification by Newton polygons, J. Amer. Math. Soc., 13(2000), 209-241.

[Dw1] B. Dwork, On the rationality of the zeta function of an algebraic variety, Amer. J. Math., 82(1960), 631-648.

[Dw2] B. Dwork, $p$-adic cycles, Publ. Math., IHES, 37(1969), 27-115.

[Dw3] B. Dwork, On Hecke polynomials, Invent. Math., 12(1971), 249-256.

[Dw4] B. Dwork, Normalized period matrices, Ann. Math., 94(1971), 337-388.

[Dw5] B. Dwork, Normalized period matrices II, Ann. Math., 98(1973), 1-57.

[DS] B. Dwork and S. Sperber, Logarithmic decay and overconvergence of the unit root and associated zeta functions, Ann. Sci. Ecole Norm. Sup. 24(1991), 575-604.

[Fa] G. Faltings, F-isocrystals on open varieties: results and conjectures, The Grothendieck Festschrift, Vol II(1990), 219-248.

[GM] F. Gouvêa and B. Mazur, Families of modular eigenforms, Math. Comp., 58(1992), 793-806.

[Gr] A. Grothendieck, Formule de Lefschetz et rationalité des fonctions L, Séminaire Bourbaki, exposé 279, 1964/65.

[Il] L. Illusie, Ordinarité des intersections complètes générales, in Grothendieck Festschrift, Volume II (1990), 375-405.

[K1] N. Katz, On the differential equations satisfied by period matrices, Publ. Math., IHES., 35(1968), 223-258.

[K2] N. Katz, Une formule de congruence pour la fonction $\zeta$, in: Groupes de monodromie en géométrie algébrique (SGA 7 II), Exp. XXII, 401-438, Lecture Notes in Math. 340, Springer-Verlag, 1973

[K3] N. Katz, Travaux de Dwork, Séminaire Bourbaki, exposé 409(1971/72), Lecture Notes in Math., No. 317, 1973, 167-200.

[K4] N. Katz, Slope filtration of F-crystals, Astérisque, 63(1979), 113-164.

[K5] N. Katz, Gauss sums, Kloosterman sums, and monodromy groups, Annals of Mathematics Studies, 116, 1988, Princeton University Press.

[LY] B. Lian and S.T. Yau, Integrality of certain power series, Proceedings of the International Conference on Algebra and Geometry, ed, M.-C. Kang, International Press, 1998, 213-225.

[Lu] S. Lubkin, A $p$-adic proof of Weil's conjectures, Ann. Math., 87(1968), 105-225.

[M1] B. Mazur, Frobenius and the Hodge filtration, Bull. Amer. Math. Soc., 78(1972), 653-667.

[M2] B. Mazur, Frobenius and the Hodge filtration (estimates), Ann. of Math., 98(1973), 58-95.

[Mo] P. Monsky, Formal cohomology III, Ann. Math., 93(1971), 315-343.

[MW] P. Monsky and G. Washnitzer, Formal cohomology I, Ann. Math., 88(1968), 181-217.

[Oo] F. Oort, Some questions in algebraic geometry, preprint.

[Se] J.-P. Serre, Endomorphismes complétements continues des espaces de Banach $p$-adiques, Publ. Math., IHES., 12(1962), 69-85.

[Ta] J. Tate, Algebraic cycles and poles of zeta functions, 1965 Arithmetic Algebraic Geometry (Proc. Conf. Purdue Univ., 1963), 93-110.

[W1] D. Wan, Zeta functions of algebraic cycles, Manuscripta Math., 74(1992), 413-444.

[W2] D. Wan, Newton polygons of zeta functions and L-functions, Ann. Math., 137(1993), 247-293.

[W3] D. Wan, Meromorphic continuation of L-functions of $p$-adic representations, Ann. Math., 143(1996), 469-498.

[W4] D. Wan, Dimension variation of classical and $p$-adic modular forms, Invent. Math., 133(1998), 449-463.

[W5] D. Wan, Computing zeta functions over finite fields, Contemporary Mathematics, 225(1999), 131-141.

[W6] D. Wan, A quick introduction to Dwork's conjecture, Contemporary Mathematics, 245(1999), 147-163.

[W7] D. Wan, An introduction to the theory of Newton polygons for L-functions of exponential sums, to appear.

[W8] D. Wan, Dwork's conjecture on unit root zeta functions, Ann. Math., 150(1999), 867-927.

[W9] D. Wan, Rank one case of Dwork's conjecture, J. Amer. Math. Soc., to appear.

[W10] D. Wan, Higher rank case of Dwork's conjecture, J. Amer. Math. Soc., to appear.

[W11] D. Wan, Partial zeta functions of algebraic varieties over finite fields, Finite Fields & Applications, to appear.

# Incomplete Additive Character Sums and Applications

Arne Winterhof

Institut für Diskrete Mathematik
Österreichische Akademie der Wissenschaften
Sonnenfelsgasse 19
A-1010 Wien
Austria
e-mail: arne.winterhof@oeaw.ac.at

Dedicated to Klaus Burde on the occasion of his 65th birthday

**Abstract** We prove some bounds on incomplete additive character sums of polynomials over finite fields. We also apply results on incomplete additive character sums to get a distribution property of irreducible polynomials, we estimate the maximum running digital sums of some dc-constrained codes, and we describe a variant of Waring's problem in finite fields.

## 1 Introduction

In the present article we consider incomplete additive character sums of polynomials $\sum_{x\in S}\psi(g(x))$, where $S$ is a subset of the finite field $\mathbf{F}_q$, $\psi$ the additive canonical character of $\mathbf{F}_q$, and $g$ a polynomial over $\mathbf{F}_q$.

In Section 2 we deduce a lower bound for arbitrary subsets $S$, that is an existence theorem, in Section 3 we prove upper bounds for some special $S$, and in Section 4 we describe some applications of results on incomplete additive character sums. In particular, we present a distribution property of irreducible polynomials, we investigate some dc-constrained codes, and we discuss a variant of Waring's problem for finite fields.

## 2 A Lower Bound

**Theorem 2.1.** *Let $n \geq 2$ and $(n, q-1) = 1$. Then for any subset $S \subset \mathbf{F}_q$ there exist $b \in \mathbf{F}_q^*$ and a monic irreducible polynomial $g \in \mathbf{F}_q[x]$ of degree $n$ such that*

$$\left|\sum_{x\in S}\psi(bg(x))\right| \geq \sqrt{|S|\left(\frac{|S|-1}{(q-1)^2}+1\right)}.$$

We prove the theorem after some preliminary lemmas.

We denote by

$$N_n(a) = \prod_{i=0}^{n-1} a^{q^i} = a^{(q^n-1)/(q-1)}; \; a \in \mathbf{F}_{q^n},$$

the norm of $\mathbf{F}_{q^n}$ onto $\mathbf{F}_q$.

**Lemma 2.2.** *For $d|n$, $(\frac{n}{d}, q-1) = 1$ and $x \neq y$; $x, y \in \mathbf{F}_q$, the number of $a \in \mathbf{F}_{q^d}$ with*

$$N_n(x-a) = N_n(y-a)$$

*is $\frac{q^d-q}{q-1}$.*

*Proof.* For $x \neq y$ we have $N_n(x-a) = N_n(y-a)$ if and only if

$$N_d\left(\frac{x-a}{y-a}\right)^{n/d} = 1; \; a \neq y.$$

As $a$ runs over $\mathbf{F}_{q^d} \setminus \{y\}$, $t = \frac{x-a}{y-a}$ runs over $\mathbf{F}_{q^d} \setminus \{1\}$. Thus the number of $a \in \mathbf{F}_{q^d}$ with $N_n(x-a) = N_n(y-a)$ equals the number of $1 \neq t \in \mathbf{F}_{q^d}$ with $N_d(t)^{n/d} = 1$. Since $(\frac{n}{d}, q-1) = 1$, the mapping $x \mapsto x^{n/d}$ is a bijection on $\mathbf{F}_q$ and thus $N_d(t)^{n/d} = 1$ if and only if $N_d(t) = 1$. Hence the number of $a \in \mathbf{F}_{q^d}$ with $N_n(x-a) = N_n(y-a)$ equals the number of $1 \neq t \in \mathbf{F}_{q^d}$ with $N_d(t) = 1$, which is $\frac{q^d-1}{q-1} - 1 = \frac{q^d-q}{q-1}$. □

**Lemma 2.3.** *For $d|n$, $(\frac{n}{d}, q-1) = 1$ and $x, y \in \mathbf{F}_q$ we have*

$$\sum_{a \in \mathbf{F}_{q^d}} \sum_{b \in \mathbf{F}_q^*} \psi(bN_n(x-a))\overline{\psi}(bN_n(y-a)) = \begin{cases} q^d(q-1) & \text{if } x = y \\ \frac{q^d-q^2}{q-1} & \text{if } x \neq y \end{cases}.$$

*Proof.* For $x = y$ the result is trivial. Remember that

$$\sum_{b \in \mathbf{F}_q^*} \psi(bz) = \begin{cases} -1; & z \neq 0 \\ q-1; & z = 0 \end{cases}; \; z \in \mathbf{F}_q.$$

By Lemma 2.2 we get for $x \neq y$

$$\sum_{a \in \mathbf{F}_{q^d}} \sum_{b \in \mathbf{F}_q^*} \psi(b(N_n(x-a) - N_n(y-a))$$

$$= \frac{q^d-q}{q-1}(q-1) - \left(q^d - \frac{q^d-q}{q-1}\right) = \frac{q^d-q^2}{q-1}. \; □$$

**Lemma 2.4.** *For $d|n$, $(\frac{n}{d}, q-1) = 1$ and any subset $S \subset \mathbf{F}_q$ we have*

$$\sum_{a\in\mathbf{F}_{q^d}} \sum_{b\in\mathbf{F}_q^*} \left|\sum_{x\in S} \psi(bN_n(x-a))\right|^2 = |S|\left((q^d - q^2)\frac{|S|-1}{q-1} + q^d(q-1)\right).$$

*Proof.* We have

$$\begin{aligned}
&\sum_{a\in\mathbf{F}_{q^d}} \sum_{b\in\mathbf{F}_q^*} \left|\sum_{x\in S} \psi(bN_n(x-a))\right|^2 \\
&= \sum_{x,y\in S} \sum_{a\in\mathbf{F}_{q^d}} \sum_{b\in\mathbf{F}_q^*} \psi(bN_n(x-a))\overline{\psi}(bN_n(y-a)) \\
&= \sum_{x\in S} \left((|S|-1)\frac{q^d-q^2}{q-1} + q^d(q-1)\right)
\end{aligned}$$

by Lemma 2.3. □

Let $D_n(q)$ denote the set of defining elements of $\mathbf{F}_{q^n}$ over $\mathbf{F}_q$.

**Lemma 2.5.** *For $n \geq 2$, $(n, q-1) = 1$ and any subset $S \subset \mathbf{F}_q$ we have*

$$\sum_{a\in D_n(q)} \sum_{b\in\mathbf{F}_q^*} \left|\sum_{x\in S} \psi(bN_n(x-a))\right|^2 = |S||D_n(q)|\left(\frac{|S|-1}{q-1} + q - 1\right).$$

*Proof.* We have for $n \geq 2$ and $(n, q-1) = 1$ by Möbius' inversion formula and Lemma 2.4

$$\begin{aligned}
&\sum_{a\in D_n(q)} \sum_{b\in\mathbf{F}_q^*} \left|\sum_{x\in S} \psi(bN_n(x-a))\right|^2 \\
&= \sum_{d|n} \mu\left(\frac{n}{d}\right) \sum_{a\in\mathbf{F}_{q^d}} \sum_{b\in\mathbf{F}_q^*} \left|\sum_{x\in S} \psi(bN_n(x-a))\right|^2 \\
&= \sum_{d|n} \mu(n/d)|S|\left((q^d - q^2)\frac{|S|-1}{q-1} + q^d(q-1)\right) \\
&= |S||D_n(q)|\left(\frac{|S|-1}{q-1} + q - 1\right). \; \square
\end{aligned}$$

*Proof (of Theorem 2.1).* By Lemma 2.5 there exist $a \in D_n(q)$ and $b \in \mathbf{F}_q^*$ with

$$\left|\sum_{x\in S} \psi(bN_n(x-a))\right|^2 \geq |S|\left(\frac{|S|-1}{(q-1)^2} + 1\right).$$

Choosing $g(x) = N_n(x-a)$ we get the assertion. □

*Remark.* For $n = 1$ the method yields that for any $a \in \mathbf{F}_q$ and any subset $S \subset \mathbf{F}_q$ there exists a $b \in \mathbf{F}_q^*$ such that

$$\left|\sum_{x \in S} \psi(b(x-a))\right| \geq \sqrt{|S| \frac{q-|S|}{q-1}}.$$

*Example 2.6.* We consider $\mathbf{F}_9$, choose $S = \mathbf{F}_3$, and verify that

$$\left|\sum_{\nu=0}^{2} \psi(b(N_2(\nu - a))\right| = \sqrt{3} < \sqrt{\frac{33}{32}}\sqrt{3} \text{ for all } b \in \mathbf{F}_3^*,\ a \in D_2(3).$$

Hence in Theorem 2.1 the condition that $(n, q-1) = 1$ is necessary.

For multiplicative character sums we can prove the following result (see [11, Theorem 1]):
Let $n \geq 2$, $\chi$ be a multiplicative character of $\mathbf{F}_q$ and $\mathrm{ord}\chi \not| n$. Then for any subset $S \subset \mathbf{F}_q$ there exists a monic irreducible polynomial $f \in \mathbf{F}_q[x]$ of degree $n$ such that

$$\left|\sum_{x \in S} \chi(f(x))\right| \geq \sqrt{|S|}. \tag{1}$$

Equation (1) and Theorem 2.1 can be extended to hybrid sums. First we extend (1).

**Theorem 2.7.** *Let $n \geq 2$, $\chi$ be a multiplicative character of $\mathbf{F}_q$ and $\mathrm{ord}\chi \not| n$. Then for any $g \in \mathbf{F}_q[x]$ and any subset $S \subset \mathbf{F}_q$ there exists a monic irreducible polynomial $f \in \mathbf{F}_q[x]$ of degree $n$ such that*

$$\left|\sum_{x \in S} \chi(f(x))\psi(g(x))\right| \geq \sqrt{|S|}.$$

*Proof.* By [11, Lemma 1] we have

$$\sum_{a \in \mathbf{F}_{q^d}} \chi(N_n(x-a))\overline{\chi}(N_n(y-a)) = \begin{cases} q^d - 1 & \text{if } x = y \\ -1 & \text{else} \end{cases}.$$

Hence we get for $n \geq 2$ and $s \not| n$

$$\sum_{a \in D_n(q)} \left|\sum_{x \in S} \chi(N_n(x-a))\psi(g(x))\right|^2$$

$$= \sum_{x,y \in S} \sum_{d|n} \mu\left(\frac{n}{d}\right) (\psi(g(x) - g(y)) \sum_{a \in \mathbf{F}_{q^d}} \chi(N_n(x-a))\overline{\chi}(N_n(y-a))$$

$$= \sum_{x \in S} \sum_{d|n} \mu\left(\frac{n}{d}\right) \left(q^d - 1 - \sum_{y \in S \setminus \{x\}} \psi(g(x) - g(y))\right)$$

$$= \sum_{x \in S} \sum_{d|n} \mu\left(\frac{n}{d}\right) q^d = |S||D_n(q)|.$$

Now the result follows analogously to Theorem 2.1. □

Now we extend Theorem 2.1.

**Theorem 2.8.** *Let $n \geq 2$ and $(n, q-1) = 1$. Then for any subset $S \subset \mathbf{F}_q$ and any $f \in \mathbf{F}_q[x]$ with $f(x) \neq 0$; $x \in S$, there exist $b \in \mathbf{F}_q^*$ and a monic irreducible polynomial $g \in \mathbf{F}_q[x]$ of degree $n$ such that*

$$\left|\sum_{x\in S} \chi(f(x))\psi(bg(x))\right| \geq \sqrt{|S|\left(1 - \frac{|S|-1}{(q-1)^2}\right)}.$$

*Proof.* Let $S = \{x_1, x_2, \ldots, x_{|S|}\}$. As in the proof of Lemma 2.5 we calculate

$$\sum_{a\in D_n(q)} \sum_{b\in\mathbf{F}_q^*} \left|\sum_{\nu=1}^{|S|} \chi(f(x_\nu))\psi(bN_n(x_\nu - a))\right|^2$$
$$= \sum_{\nu_1,\nu_2=1}^{|S|} \chi(f(x_{\nu_1}))\overline{\chi}(f(x_{\nu_2})) \sum_{a\in D_n(q)} \sum_{b\in\mathbf{F}_q^*} \psi(b(N_n(x_{\nu_1} - a) - N_n(x_{\nu_2} - a)))$$
$$= \sum_{\nu_1=1}^{|S|} \left(|D_n(q)|(q-1) + \frac{|D_n(q)|}{q-1} \sum_{\nu_2=1;\nu_2\neq\nu_1}^{|S|} \chi(f(x_{\nu_1}))\overline{\chi}(f(x_{\nu_2}))\right)$$
$$= |S||D_n(q)|(q-1) + \frac{|D_n(q)|}{q-1} \sum_{1\leq\nu_1<\nu_2\leq|S|} 2\mathrm{Re}\left(\chi(f(x_{\nu_1}))\overline{\chi}(f(x_{\nu_2}))\right)$$
$$\geq |S||D_n(q)|\left(q - 1 - \frac{|S|-1}{q-1}\right).$$

Now the theorem follows analogously to the previous theorems. □

*Remark.* For additional lower bounds see [8] and [9, p. 85/86] for finite prime fields and [7] for arbitrary finite fields.

## 3 Upper Bounds

In this section we prove upper bounds for some special subsets $S$.

**Lemma 3.1.** *Let $S \subset \mathbf{F}_q$ and $g \in \mathbf{F}_q[x]$. Then*

$$\left|\sum_{z\in S} \psi(g(z))\right| \leq \frac{1}{q} \sum_{y\in\mathbf{F}_q} \left|\sum_{x\in\mathbf{F}_q} \psi(g(x) - xy)\right| \left|\sum_{z\in S} \psi(yz)\right|.$$

*Proof.* We have

$$\psi(g(z)) = \frac{1}{q} \sum_{x \in \mathbf{F}_q} \psi(g(x)) \sum_{y \in \mathbf{F}_q} \psi(y(z-x))$$

and obtain by summing over $z$

$$\sum_{z \in S} \psi(g(z)) = \frac{1}{q} \sum_{x,y \in \mathbf{F}_q} \psi(g(x))\overline{\psi}(xy) \sum_{z \in S} \psi(yz).$$

The lemma follows by the triangular inequality. □

Now we can apply A. Weil's results on complete additive character sums (see e.g. [5, Theorem 5.38]) to estimate $\left|\sum_{x \in \mathbf{F}_q} \psi(g(x) - xy)\right|$.

Our main problem is to find subsets $S$ for which we can non-trivially estimate the sum $\sum_{y \in \mathbf{F}_q} \left|\sum_{z \in S} \psi(yz)\right|$.

There are at least three suitable kinds of subsets.

(i) Let $q = p^t$ where $p$ is prime and $\{b_0, \ldots, b_{t-1}\}$ be a basis of $\mathbf{F}_q$ over $\mathbf{F}_p$. For $1 \le k_i \le p$; $i = 0, \ldots, t-1$, we consider the boxes

$$B = B(k_0, \ldots, k_{t-1})$$

$$= \{\nu_0 b_0 + \ldots + \nu_{t-1} b_{t-1} | 0 \le \nu_i < k_i,\ i = 0, \ldots, t-1\}. \quad (2)$$

(ii) For $\nu = \nu_0 + \nu_1 p + \ldots + \nu_{t-1} p^{t-1}$; $0 \le \nu_i < p$, $i = 0, \ldots, t-1$, let

$$x_\nu = \nu_0 b_0 + \nu_1 b_1 + \ldots + \nu_{t-1} b_{t-1}.$$

For $1 \le k \le q$ we choose

$$S = S(k) = \{x_0, \ldots, x_{k-1}\}. \quad (3)$$

(iii) We consider $S = V$, where $V$ is an additive subgroup of $\mathbf{F}_q$.

**Lemma 3.2.** *For $1 \le k_i \le p$; $i = 0, \ldots, t-1$, let $B = B(k_0, \ldots, k_{t-1})$ be the subset of $\mathbf{F}_q$ defined by* (2). *Then*

$$\sum_{y \in \mathbf{F}_q} \left| \sum_{z \in B} \psi(yz) \right| \le \prod_{i=0}^{t-1} (k_i + p \log p).$$

*Proof.* [11, Lemma 7]. □

**Lemma 3.3.** *For $1 \le k \le q$ let $S = S(k)$ be the subset of $\mathbf{F}_q$ defined by* (3). *Then*

$$\sum_{y \in \mathbf{F}_q} \left| \sum_{z \in S} \psi(yz) \right| \le k p^{t-l} + q \log p^l, \text{ where } l = \left\lceil \frac{\log k}{\log p} \right\rceil.$$

*Proof.* [11, Lemma 8]. □

**Lemma 3.4.** *Let $V$ be an additive subgroup of $\mathbf{F}_q$. Then*

$$\sum_{y\in\mathbf{F}_q}\left|\sum_{z\in V}\psi(yz)\right|\le q.$$

*Proof.* We consider $V$ as an $l$-dimensional subspace with basis $\{b_0,\dots,b_{l-1}\}$. Hence we have

$$\sum_{z\in V}\psi(yz)=\prod_{i=0}^{l-1}\left(\sum_{\nu=0}^{p-1}\psi(\nu y b_i)\right).$$

The right hand side is 0 if $\psi(yb_i)\ne 1$ for an $0\le i<l$ and $p^l$ if $\psi(yb_i)=1$ for all $0\le i<l$. Since the number of $y$ with $\psi(yb_i)=1$ for $0\le i<l$ is $p^{t-l}$ we get the assertion. □

Lemma 3.1 with $f(z)\equiv 1$, Weil's theorem and Lemma 3.2, 3.3 or 3.4, respectively, yield the following theorems.

Let $g\in\mathbf{F}_q[x]$ of degree $\ge 2$ with $(\deg(g),q)=1$.

**Theorem 3.5.** *For any $1\le k_i\le p$; $i=0,\dots,t-1$, we have*

$$\left|\sum_{z\in B(k_0,\dots,k_{t-1})}\psi(g(z))\right|\le(\deg(g)-1)q^{1/2}\prod_{i=0}^{t-1}\left(\frac{k_i}{p}+\log p\right)$$
$$\le(\deg(g)-1)q^{1/2}(1+\log p)^t.$$

**Theorem 3.6.** *For any $1\le k\le q$ we have*

$$\left|\sum_{\nu=0}^{k-1}\psi(g(x_\nu))\right|\le(\deg(g)-1)q^{1/2}\left(\frac{k}{p^l}+\log p^l\right)\le(\deg(g)-1)q^{1/2}(1+\log q),$$

*where* $l=\left\lceil\frac{\log k}{\log p}\right\rceil$.

**Theorem 3.7.** *For any additive subgroup $V$ of $\mathbf{F}_q$ we have*

$$\left|\sum_{v\in V}\psi(g(v))\right|\le(\deg(g)-1)q^{1/2}.$$

*Remarks.*

(i) For $\deg(g)=1$ the method yields only the trivial bound.
(ii) For prime fields Korobov [4] proved that

$$\left|\sum_{\nu=0}^{k-1}\psi(g(\nu))\right|\le c_0k^{1-\gamma/m^2},$$

where $\deg(g)=m$, $k=p^{1/r}$, $1\le r\le m$, $\min(r,m-r+1)>\varepsilon m$ for any $0<\varepsilon<\frac{1}{2}$, and $c_0$ and $\gamma$ depend only of $\varepsilon$.

For multiplicative character sums we can prove the following results (cf. [11, Theorem 2]):

Let $\chi$ be a non-trivial multiplicative character and $f \in \mathbf{F}_q[x]$ a polynomial which is not an ord$\chi$th power and has $m$ distinct zeroes in its splitting field over $\mathbf{F}_q$. Then

(i) for any $1 \le k_i \le p$; $i = 0, \dots, t-1$, we have

$$\left| \sum_{z \in B(k_0, \dots, k_{t-1})} \chi(f(z)) \right| \le mq^{1/2} \left( \prod_{i=0}^{t-1} \left( \frac{k_i}{p} + \log p \right) - \frac{1}{m} \prod_{i=0}^{t-1} \frac{k_i}{p} \right) < mq^{1/2}(1 + \log p)^t,$$

(ii) for any $1 \le k \le q$ we have

$$\left| \sum_{\nu=0}^{k-1} \chi(f(x_\nu)) \right| \le mq^{1/2} \left( \frac{k}{p^l} + \log p^l - \frac{k}{mq} \right) < mq^{1/2}(1 + \log q), \tag{4}$$

where $l = \left\lceil \frac{\log k}{\log p} \right\rceil$, and

(iii) for any additive subgroup $V$ of $\mathbf{F}_q$ we have

$$\left| \sum_{v \in V} \chi(f(v)) \right| \le mq^{1/2} \left( 1 - \frac{|V|}{mq} \right) \le mq^{1/2}.$$

These results can be extended to incomplete Gaussian sums.

Let $\chi$ be a non-trivial multiplicative and $\psi$ a non-trivial additive character of $\mathbf{F}_q$. Let $f \in \mathbf{F}_q[x]$ a polynomial which is not an ord$\chi$th power and $g \in \mathbf{F}_q[x]$ a polynomial with $(\deg(g), q) = 1$. Let $m$ be the number of distinct roots of $f$ in its splitting field over $\mathbf{F}_q$. Then

(i) for any $1 \le k_i \le p$; $i = 0, \dots, t-1$ we have

$$\left| \sum_{z \in B(k_0, \dots, k_{t-1})} \chi(f(z))\psi(g(z)) \right| \le (m + \deg(g) - 1)q^{1/2} \prod_{i=0}^{t-1} \left( \frac{k_i}{p} + \log p \right) \le (m + \deg(g) - 1)q^{1/2}(1 + \log p)^t,$$

(ii) for any $1 \le k \le q$ we have

$$\left| \sum_{\nu=0}^{k-1} \chi(f(x_\nu))\psi(g(x_\nu)) \right| \le (m + \deg(g) - 1)q^{1/2} \left( \frac{k}{p^l} + \log p^l \right) \le (m + \deg(g) - 1)q^{1/2}(1 + \log q),$$

where $l = \left\lceil \frac{\log k}{\log p} \right\rceil$, and

(iii) for any additive subgroup $V$ of $\mathbf{F}_q$ we have

$$\left|\sum_{v\in V}\chi(f(v))\psi(g(v))\right| \le (m+\deg(g)-1)q^{1/2}.$$

*Remark.* For prime fields and linear polynomials see also [2].

# 4 Applications

## 4.1 On the Distribution of Irreducible Polynomials

Let $\mathrm{Tr}(x) = x + x^p + \ldots + x^{p^t-1}$ denote the trace of $\mathbf{F}_{p^t}$ on $\mathbf{F}_p$.

**Theorem 4.1.** *Let $n \ge 2$ and $q = p^t$ a prime power with $(n, q-1) = 1$. For $x, y \in \mathbf{F}_q$, $x \ne y$, denote by $a_i$ the number of irreducible polynomials $f$ over $\mathbf{F}_q$ of degree $n$ with*

$$Tr(f(x) - f(y)) = i;\ i = 0,\ldots,p-1.$$

*Then*

$$a_0 - \frac{|I_n(q)|}{q-1} = a_1 = \ldots = a_{p-1} = \frac{|I_n(q)|}{p}\left(q-1-\frac{1}{q-1}\right),$$

*where $I_n(q)$ denotes the set of monic irreducible polynomials of degree $n$ over $\mathbf{F}_q$.*

*Proof.* Let $\zeta = e^{2\pi i/p}$. Since

$$\mathrm{Tr}(f(x)-f(y)) = i \text{ if and only if } \psi(f(x))\overline{\psi}(f(y)) = \zeta^i$$

we have for $j = 1,\ldots,p-1$

$$\begin{aligned}
&a_0 - a_{p-1} + (a_1 - a_{p-1})\zeta^j + (a_{p-2} - a_{p-1})\zeta^{j(p-2)}\\
&= a_0 + a_1\zeta^j + \ldots + a_{p-1}\zeta^{j(p-1)}\\
&= \sum_{f\in I_n(q)}\sum_{b\in\mathbf{F}_q^*}\psi(bf(x))^j\overline{\psi}(bf(y))^j\\
&= \frac{1}{n}\sum_{a\in D_n(q)}\sum_{b\in\mathbf{F}_q^*}\psi(jbN_n(x-a))\overline{\psi}(jbN_n(y-a))\\
&= \frac{1}{n}\sum_{d|n}\mu(n/d)\sum_{a\in\mathbf{F}_{q^d}}\sum_{b\in\mathbf{F}_q^*}\psi(bN_n(x-a))\overline{\psi}(bN_n(y-a))\\
&= \frac{1}{n}\sum_{d|n}\mu(n/d)\frac{q^d-q^2}{q-1} = \frac{|I_n(q)|}{q-1}
\end{aligned}$$

by Lemma 2.3.

Since the coefficient matrix of this system is an invertible Vandermonde matrix we get

$$a_0 - a_{p-1} - \frac{|I_n(q)|}{q-1} = a_1 - a_{p-1} = \ldots = a_{p-2} - a_{p-1} = 0$$

which implies the result. □

An analogous result with multiplicative characters can similarly be shown (see [11, Theorem 3]). This result can be used to construct some difference matrices.

## 4.2 DC-Constrained Codes

A. Barg [1] studied codes $C$ defined as follows:

Let $\mathcal{P} \subset \{f \in \mathbf{F}_q[x] | 1 \leq \deg(f) \leq n < q-1\}$, $\chi$ a multiplicative character of order $s|q-1$, $\{u_0, \ldots, u_{q-1}\}$ an ordering of the elements of $\mathbf{F}_q$, and

$$C = \{\mathbf{a}_f | f \in \mathcal{P}\} \text{ where } \mathbf{a}_f = (\chi(f(u_0)), \ldots, \chi(f(u_{q-1}))$$

and $\chi(0)$ is substituted by a certain $s$th root of unity.

A central problem for such codes $C$ is to estimate the *maximum running digital sum* (MRDS)

$$S(C) = \max_{\mathbf{a} \in C} \max_{1 \leq k \leq q} \left| \sum_{\nu=0}^{k-1} a_\nu \right|, \text{ where } \mathbf{a} = (a_1, \ldots, a_q).$$

A code $C$ is called *dc-constrained* if $S(C) \leq c(q)$, where $c(q)$ is a slowly growing function in $q$.

*Example 4.2.* Let $\mathcal{P}$ be the set of monic non-constant square-free polynomials $f$ of degree $\leq n$ and $s = 2$. Let $u_1 = x_1, \ldots, u_{q-1} = x_{q-1}$ as defined in (3) and substitute $\chi(0)$ in the $k$th coordinate of a codeword $\mathbf{a}$ by $-1$ if $\sum_{\nu=0}^{k-1} a_i \geq 0$ and by 1 if $\sum_{\nu=0}^{k-1} a_i < 0$. Then we get

$$S(C) < nq^{1/2}(1 + \log q)$$

by (4).

Now we investigate dc-constrained codes with additive characters.

**Theorem 4.3.** *Let* $2 \leq n < q^{1/2} + 1$, $(n, p) = 1$ *and*

$$C = \{\mathbf{a}_g | g \in \mathcal{P}\} \textit{ where } \mathbf{a}_g = (\psi(g(x_0)), \ldots, \psi(g(x_{q-1}))$$

*and*

$$\mathcal{P} = \{a_0 + a_1 x + \ldots + a_l x^l \in \mathbf{F}_q[x] \mid 2 \leq l \leq n,\ (l, p) = 1,\ a_l \neq 0,$$

$$a_0 = a_p = \ldots = a_{\lfloor l/p \rfloor p} = 0\}.$$

*Then $C$ is a $p$-ary code of length $q$ with*

$$|C| = (q-1) \sum_{l=2,(l,p)=1}^{n} q^{l-\lfloor l/p \rfloor -1},$$

$$\textit{minimum distance } d(C) \geq \frac{p-1}{p} q^{1/2}(q^{1/2} - n + 1)$$

*and MRDS $S(C) \leq (n-1)q^{1/2}(1 + \log q)$.*

*Proof.* We have to show that $|C| = |\mathcal{P}|$ which means that two vectors $\mathbf{a}_f$ and $\mathbf{a}_g$ are distinct if $f$ and $g$ are distinct. Let $d(\mathbf{a}_f, \mathbf{a}_g)$ be the distance of $\mathbf{a}_f$ and $\mathbf{a}_g$. Then we have

$$(p-1)q - d(\mathbf{a}_f, \mathbf{a}_g)p \leq \sum_{i=1}^{p-1} \left| \sum_{x \in \mathbf{F}_q} \psi^i(f(x) - g(x)) \right| \leq (p-1)(n-1)q^{1/2}$$

by Weil's theorem. Hence for $n < q^{1/2} + 1$ we have

$$d(\mathbf{a}_f, \mathbf{a}_g) \geq \frac{p-1}{p} q^{1/2}(q^{1/2} - n + 1) > 0$$

and thus

$$|C| = |\mathcal{P}| = (q-1) \sum_{l=2,(l,p)=1}^{n} q^{l-\lfloor l/p \rfloor -1}.$$

Theorem 3.6 estimates $S(C)$. □

*Remark.* For $p = 2$ a similar result was obtained by using the ordering $\{1, g, g^2, \ldots, g^{q-2}\}$, where $g$ is a primitve element of $\mathbf{F}_q$, by [6].

## 4.3 A Variant of Waring's Problem in Finite Fields

For $k \geq 1$ and $a \in \mathbf{F}_q$ we denote by $g(n,q,k,a)$ the smallest $s$ such that for any $b \in \mathbf{F}_q^*$ the diagonal equation

$$(a + x_{\nu_1})^n + \ldots + (a + x_{\nu_s})^n = b, \; 0 \leq \nu_i \leq k-1, \; i = 1, \ldots, s \qquad (5)$$

has a solution. For $k = q$ the problem of finding $g(n,q) = g(n,q,q,a)$ is called Waring's problem in finite fields. For a survey see [10].

*Remarks.*

(i) $g(n,q,k,a)$ exists if and only if $\{(a + x_i^n) | 1 \leq i < k\}$ contains a basis of $\mathbf{F}_q$ over $\mathbf{F}_p$. Hence $g(n,p,k,a)$ exists.
(ii) For $k_1 \leq k_2$ we have $g(n,q,k_2,a) \leq g(n,q,k_1,a)$.

The following relation between character sums and Waring's problem is well-known (cf. [3]):
For the number $N$ of solutions of (5) we have

$$N = \frac{1}{q} \sum_{y \in \mathbf{F}_q} \sum_{\nu_1, \dots, \nu_s = 0}^{k-1} \psi(y((a + x_{\nu_1})^n + \dots + (a + x_{\nu_s})^n - b))$$

$$= \frac{k^s}{q} + \frac{1}{q} \sum_{y \in \mathbf{F}_q^*} \psi(-yb) \left( \sum_{\nu=0}^{k-1} \psi(y(a + x_\nu)^n) \right)^s .$$

By Theorem 3.6 we have for $n \geq 2$

$$N \geq \frac{k^s}{q} - \frac{q-1}{q}(n-1)^s q^{s/2}(1 + \log q)^s.$$

It follows:

**Theorem 4.4.** *$g(n,q,k,a)$ exists for $k > (n-1)q^{1/2}(1 + \log q)$, $n \geq 2$, $(n,q) = 1$, and we have*

$$g(n,q,k,a) \leq \left\lfloor \frac{\log q}{\log k/((n-1)q^{1/2}(1 + \log q))} \right\rfloor + 1.$$

We prove two additional results.

**Theorem 4.5.** *Let $G \subset \{(a + x_i)^n | 0 \leq i < k-1\}$ be a multiplicative group. If $g(n,q,k,a)$ exists then we have*

$$g(n,q,k,a) \leq \frac{q-1}{|G|}.$$

*Proof.* Let $A_j = \{a_1 + \dots + a_j | a_i \in G\} \cup \{0\}$. If $A_j \subsetneqq A_{j+1}$ then $y \in A_{j+1} \backslash A_j$ implies $xy \in A_{j+1} \backslash A_j$ for each $x \in G$, so that

$$|A_{j+1}| \geq |A_j| + |G|.$$

Hence in the chain $A_1 \subset A_2 \subset \dots \subset A_s = \mathbf{F}_q$ there are at most $\frac{q-1}{|G|} - 1$ strict inclusions and therefore

$$g(n,q,k,a) \leq \frac{q-1}{|G|}. \ \square$$

**Theorem 4.6.** *Let $t \geq 2$. If $g(n,p^t,k,a)$ exists then*

$$g(n,p^t,k,a) \leq t(p-1).$$

*Proof.* If $g(n,p^t,k,a)$ exists then there exists a basis $\{\beta_1, \dots, \beta_t\} \subset A_1$ of $\mathbf{F}_{p^t}$ over $\mathbf{F}_p$. All elements of the form $c\beta_i$, $c \in \mathbf{F}_p^*$, $i = 1, \dots, t$, are in $A_{p-1}$. Thus an arbitrary element $x = c_1\beta_1 + \dots + c_t\beta_t \in \mathbf{F}_{p^t}^*$; $c_i \in \mathbf{F}_p$, $i = 1, \dots, t$, is a sum of $t(p-1)$ powers. $\square$

*Example 4.7.* Consider $\mathbf{F}_9 = \mathbf{F}_3(\alpha)$ with $\alpha^2 = -1$. Then we have

$$g(2,9,2,\alpha) = 4.$$

Hence Theorem 4.5 and Theorem 4.6 are in general best possible.

## References


1. A. Barg, Exponential sums and constrained error-correcting codes, Cohen, G. et al. (ed.), Algebraic coding, *Lect. Notes Comput. Sci.* **573** (1992), 16-22.
2. D.A. Burgess, Partial Gaussian sums, *Bull. Lond. Math. Soc.* **20** (1988), 589-592.
3. S.V. Konyagin, On estimates of Gaussian sums and Waring's problem for a prime modulus, *Trudy Mat. Inst. Steklov* **198** (1992), 111-124; translation in *Proc. Steklov Inst. Math.* **1994**, no.1, 105-117.
4. N.M. Korobov, Double trigonometric sums and their applications to the estimation of rational sums, *Math. Notes* **6** (1969), 472-478.
5. R. Lidl and H. Niederreiter, "Finite fields", Encyclopaedia of Mathematics Vol. 20, Cambridge University Press, 1983.
6. S. Litsyn and A. Tietäväinen, Character sum constructions of constrained error-correcting codes, Appl. Algebra Engrg. Comm. Comput., **5** (1994), 45-51.
7. F. Özbudak, On lower bounds for incomplete character sums over finite fields, *Finite Fields Appl.* **2** (1996), 173-191.
8. S.A. Stepanov, On lower bounds for incomplete character sums with polynomials, *Trudy Mat. Inst. Akad. Nauk SSSR* **143** (1977), 175-177 (in Russian).
9. S.A. Stepanov, *Arithmetic of algebraic curves*, Plenum, New York, 1994.
10. A. Winterhof, On Waring's problem in finite fields, *Acta Arith.* **87** (1998), 171-177.
11. A. Winterhof, Some estimates for character sums and applications, *Des. Codes Cryptogr.*, to appear.

# Applications of Algebraic Curves to Constructions of Codes and Almost Perfect Sequences

Chaoping Xing[1]* and Harald Niederreiter[2]

[1] Department of Mathematics
National University of Singapore
2 Science Drive 2
117543, Singapore
e-mail: matxcp@nus.edu.sg

[2] Institute of Discrete Mathematics
Austrian Academy of Sciences
Sonnenfelsgasse 19
A-1010 Vienna
Austria
e-mail: niederreiter@oeaw.ac.at

**Abstract** We survey some recent constructions of linear codes and almost perfect sequences based on algebraic curves over finite fields.

## 1 Introduction

### 1.1 Motivation and Outline

Since Goppa's discovery of algebraic-geometry codes in 1981 (see [2]), Koblitz' construction of the elliptic-curve cryptosystem in 1987 (see [3]) and Lenstra's factorization algorithm for large integers based on elliptic curves over finite fields in 1987 (see [5]), algebraic curves over finite fields have been considered as one of the most fascinating tools in information theory and computational number theory. In 1996 Niederreiter and Xing [13,14] found another interesting application of algebraic curves over finite fields to quasirandom points that are of great significance for quasi-Monte Carlo methods and uniform distribution. Thus, since 1981, algebraic curves over finite fields have found more and more applications in important areas.

Recently, several new constructions of linear codes and almost perfect sequences were proposed based on algebraic curves over finite fields [4,16,17,23] [24,25,26,27,28]. In this paper we give a survey of these constructions. In the remaining subsections of Section 1 we review the necessary background on algebraic curves and their function fields. Section 2 presents the new constructions of linear codes and Section 3 those of almost perfect sequences.

* Research supported by the NUS grant RP3991621

## 1.2 Background on Algebraic Curves

For the finite field $\mathbb{F}_q$, let $\mathcal{X}$ be a smooth, absolutely irreducible, projective algebraic curve defined over $\mathbb{F}_q$. We express this fact by simply saying that $\mathcal{X}/\mathbb{F}_q$ is an algebraic curve. The genus of $\mathcal{X}/\mathbb{F}_q$ is denoted by $g$. A point on $\mathcal{X}$ is called *rational* if all its homogeneous coordinates belong to $\mathbb{F}_q$. For our purposes, we are interested in algebraic curves with many rational points or closed points of small degree.

A divisor $G$ of $\mathcal{X}$ is called *rational* if $G^\sigma = G$ for any automorphism $\sigma \in \mathrm{Gal}(\overline{\mathbb{F}}_q/\mathbb{F}_q)$. In this paper we always mean a rational divisor whenever a divisor is mentioned.

We denote by $\mathbb{F}_q(\mathcal{X})$ the function field of $\mathcal{X}$. An element of $\mathbb{F}_q(\mathcal{X})$ is called a function. We write $\nu_P$ for the normalized discrete valuation corresponding to the point $P$ of $\mathcal{X}/\mathbb{F}_q$. Let $x \in \mathbb{F}_q(\mathcal{X})\backslash\{0\}$ and denote by $Z(x)$ the set of zeros and by $N(x)$ the set of poles of $x$. We define the *zero divisor* of $x$ by

$$(x)_0 = \sum_{P\in Z(x)} \nu_P(x)P$$

and the *pole divisor* of $x$ by

$$(x)_\infty = \sum_{P\in N(x)} (-\nu_P(x))P.$$

Then $(x)_0$ and $(x)_\infty$ are both rational divisors. Furthermore, the *principal divisor* of $x$ is given by

$$\mathrm{div}(x) = (x)_0 - (x)_\infty.$$

The degree of $\mathrm{div}(x)$ is equal to zero, i.e.,

$$\deg((x)_0) = \sum_{P\in Z(x)} \nu_P(x) = \sum_{P\in N(x)} (-\nu_P(x)) = \deg((x)_\infty).$$

For an arbitrary divisor $G = \sum m_P P$ of $\mathcal{X}$, we denote by $\nu_P(G)$ the coefficient $m_P$ of $P$. Then

$$G = \sum \nu_P(G)P.$$

For such a divisor $G$ we form the vector space

$$\mathcal{L}(G) = \{x \in \mathbb{F}_q(\mathcal{X})\backslash\{0\} : \mathrm{div}(x) + G \geq 0\} \cup \{0\}.$$

Then $\mathcal{L}(G)$ is a finite-dimensional vector space over $\mathbb{F}_q$, and we denote its dimension by $\ell(G)$. By the Riemann-Roch theorem (see [20, Chapter 1]), we have

$$\ell(G) \geq \deg(G) + 1 - g,$$

and equality holds if $\deg(G) \geq 2g - 1$.

### 1.3 Local Expansions

For a rational point $P$ of $\mathcal{X}$, an element $t$ of $\mathbb{F}_q(\mathcal{X})$ is called a *local parameter* at $P$ if $\nu_P(t) = 1$. Such a local parameter always exists for any rational point.

Now we choose a sequence $\{t_r\}_{r=-\infty}^{\infty}$ of elements in $\mathbb{F}_q(\mathcal{X})$ such that

$$\nu_P(t_r) = r$$

for all integers $r$. For a given function $f \in \mathbb{F}_q(\mathcal{X})$, we can find an integer $v$ such that $\nu_P(f) \geq v$. Hence

$$\nu_P\left(\frac{f}{t_v}\right) \geq 0.$$

Put

$$a_v = \left(\frac{f}{t_v}\right)(P),$$

that is, $a_v$ is the value of the function $f/t_v$ at $P$. Then $a_v$ is an element of $\mathbb{F}_q$. Note that the function $f/t_v - a_v$ satisfies

$$\nu_P\left(\frac{f}{t_v} - a_v\right) \geq 1;$$

hence we know that

$$\nu_P\left(\frac{f - a_v t_v}{t_{v+1}}\right) \geq 0.$$

Put

$$a_{v+1} = \left(\frac{f - a_v t_v}{t_{v+1}}\right)(P).$$

Then $a_{v+1}$ belongs to $\mathbb{F}_q$ and $\nu_P(f - a_v t_v - a_{v+1}t_{v+1}) \geq v + 2$.

Assume that we have obtained a sequence $\{a_r\}_{r=v}^{m}$ $(m > v)$ of elements of $\mathbb{F}_q$ such that

$$\nu_P(f - \sum_{r=v}^{k} a_r t_r) \geq k + 1$$

for all $v \leq k \leq m$. Put

$$a_{m+1} = \left(\frac{f - \sum_{r=v}^{m} a_r t_r}{t_{m+1}}\right)(P).$$

Then $a_{m+1} \in \mathbb{F}_q$ and $\nu_P(f - \sum_{r=v}^{m+1} a_r t_r) \geq m+2$. In this way we continue our construction of $a_r$; then we obtain an infinite sequence $\{a_r\}_{r=v}^{\infty}$ of elements of $\mathbb{F}_q$ such that

$$\nu_P(f - \sum_{r=v}^{m} a_r t_r) \geq m + 1$$

for all $m \geq v$. We summarize the above well-known construction in the formal expansion

$$f = \sum_{r=v}^{\infty} a_r t_r .$$

This is called the *local expansion* of $f$ at $P$.

## 2 Constructions of Linear Codes

### 2.1 Codes and Curves

Goppa's construction of algebraic-geometry codes makes use of algebraic curves over finite fields with many rational points. This construction was a breakthrough in coding theory because Goppa's construction yields sequences of codes beating the Gilbert-Varshamov bound [15,22]. The length of algebraic-geometry codes from Goppa's construction cannot exceed the number of rational points of the curve. However, for some curves the number of rational points is not big enough to allow us to construct good codes. Especially, curves over small fields $\mathbb{F}_q$ such as $q = 2, 3, 4, 5$ have fewer rational points compared with their genera. This forces us to look for other constructions that make use of closed points of higher degree. A previous improved Goppa's construction is given in [1].

In this section, we review some recent constructions of linear codes based on algebraic curves over finite fields. Constructions 1 and 2 below provide powerful methods of obtaining new types of linear codes, whereas Constructions 3 and 4 offer new approaches to Goppa's algebraic-geometry codes.

### 2.2 Review of Goppa's Construction

In this subsection we briefly recall some results about Goppa's construction of algebraic-geometry codes. We fix some notations for this subsection.

$\mathcal{X}/\mathbb{F}_q$: an algebraic curve over $\mathbb{F}_q$;
$g$: the genus of $\mathcal{X}$;
$P_1, P_2, \ldots, P_n$: $n$ $(n > g)$ distinct rational points of $\mathcal{X}$;
$G$: a divisor of $\mathcal{X}$ with $\operatorname{supp}(G) \cap \{P_1, P_2, \ldots, P_n\} = \emptyset$.

Consider the $\mathbb{F}_q$-linear map

$$\phi : \mathcal{L}(G) \longrightarrow \mathbb{F}_q^n, \quad f \mapsto (f(P_1), f(P_2), \ldots, f(P_n)).$$

Then the image of $\phi$ forms a subspace of $\mathbb{F}_q^n$ that was defined as an algebraic-geometry code by Goppa. The image of $\phi$ is denoted by $C(P_1, P_2, \ldots, P_n; G)$. If $n$ is bigger than the degree of $G$, then $\phi$ is an embedding and the dimension $k$ of $C(P_1, P_2, \ldots, P_n; G)$ is equal to $\ell(G)$. The Riemann-Roch theorem makes it possible to estimate parameters of the code $C(P_1, P_2, \ldots, P_n; G)$. This leads to the following well-known result (see e.g. [21, Theorem 3.1.1]).

**Proposition 2.1.** *Let $\mathcal{X}/\mathbb{F}_q$ be an algebraic curve of genus $g$ and let $P_1, P_2, \ldots, P_n$ be $n$ distinct rational points of $\mathcal{X}$. Choose a divisor $G$ with $g \leq \deg(G) =: s < n$ and $\text{supp}(G) \cap \{P_1, P_2, \ldots, P_n\} = \emptyset$. Then $C(P_1, P_2, \ldots, P_n; G)$ is a linear $[n, k, d]$ code over $\mathbb{F}_q$ with*

$$k \geq s - g + 1, \qquad d \geq n - s.$$

*Moreover, the dimension $k$ is equal to $s - g + 1$ if $s \geq 2g - 1$.*

## 2.3 Construction 1

We fix some notations for this subsection.

$\mathcal{X}/\mathbb{F}_q$: an algebraic curve over $\mathbb{F}_q$;
$g$: the genus of $\mathcal{X}$;
$P_1, P_2, ..., P_n$: $n$ distinct closed points of degree 2 of $\mathcal{X}$;
$P$: a rational point of $\mathcal{X}$.

For a nonnegative integer $m$ consider the $\mathbb{F}_q$-linear space

$$C_m(P_1, \ldots, P_n) := \{(f(P_1), ..., f(P_n)) | f \in \mathcal{L}(m(q+1)P)\}$$

and the intersection

$$C_m(P_1, \ldots, P_n)|_{\mathbb{F}_q} := C_m(P_1, \ldots, P_n) \cap \mathbb{F}_q^n.$$

Then $C_m(P_1, \ldots, P_n)$ is an $\mathbb{F}_q$-linear space but not a linear code over $\mathbb{F}_q$, while $C_m(P_1, \ldots, P_n)|_{\mathbb{F}_q}$ is a $q$-ary linear code with parameters stated in the following theorem which is analogous to results of Xing and Ling [24,25].

**Theorem 2.2.** *If $g \leq m < \min\{q, 2n/(q+1)\}$, then $C_m(P_1, \ldots, P_n)|_{\mathbb{F}_q}$ is a linear $[n, k, d]$ code over $\mathbb{F}_q$ with*

$$k \geq (m+1-g)(m+2-g)/2 \quad \text{and} \quad d \geq n - m(q+1)/2.$$

*Example 2.3.* Many improvements on Brouwer's table for linear codes on the web page at the URL

`http://www.win.tue.nl/~aeb/voorlincod.html`

can be obtained from Theorem 2. In this example we just list several improvements based on projective lines. In the following table, $n, k, d$ are parameters of linear codes obtained from Theorem 2, whereas $d_B$ stands for the lower bound in Brouwer's table on the optimal value of $d$ for given length $n$ and dimension $k$.

Improvements on Brouwer's Table

| $q$ | genus | $m$ | $n$ | $k$ | $d$ | $d_B$ |
|---|---|---|---|---|---|---|
| 7 | 0 | 2 | 21 | 6 | 13 | 12 |
| 8 | 0 | 2 | 28 | 6 | 19 | 18 |
| 8 | 0 | 3 | 28 | 10 | 15 | 14 |
| 8 | 0 | 4 | 28 | 15 | 10 | 9 |
| 8 | 0 | 5 | 28 | 21 | 6 | 5 |
| 9 | 0 | 2 | 36 | 6 | 26 | 24 |
| 9 | 0 | 3 | 36 | 10 | 21 | 20 |
| 9 | 0 | 4 | 36 | 15 | 16 | 15 |
| 9 | 0 | 5 | 36 | 21 | 11 | 10 |

### 2.4 Construction 2

We fix again some notations for this subsection.

$\mathcal{X}/\mathbb{F}_q$: an algebraic curve over $\mathbb{F}_q$;
$g$: the genus of $\mathcal{X}$;
$P_1, ..., P_m$: $m$ distinct closed points of $\mathcal{X}$;
$G$: a divisor of $\mathcal{X}$ with

$$\mathrm{supp}(G) \cap \{P_1, ..., P_m\} = \emptyset;$$

$C_i$: linear $[n_i, k_i = \deg(P_i), d_i]$ codes over $\mathbb{F}_q$ for $1 \leq i \leq m$;
$\phi_i$: fixed $\mathbb{F}_q$-isomorphisms between the residue class field of $P_i$ and $C_i$ for $1 \leq i \leq m$.

By concatenation we define the $q$-ary linear code

$$C(P_1, ..., P_m; G; C_1, ..., C_m) := \{(\phi_1(f(P_1)), ..., \phi_m(f(P_m)))|f \in \mathcal{L}(G)\}.$$

This construction was introduced and analyzed by Xing, Niederreiter, and Lam [27]. The following result gives parameter bounds, where the bound for the minimum distance is stated in the improved form pointed out later in [17].

**Theorem 2.4.** *If $\sum_{i=1}^m k_i > \deg(G) \geq g$, then the code $C(P_1, ..., P_m; G; C_1, ..., C_m)$ has parameters $[n, k, d]$ with*

$$n = \sum_{i=1}^{m} n_i, \quad k \geq \deg(G) - g + 1, \quad d \geq \delta,$$

*where $\delta$ is the minimum of $\sum_{i \notin M} d_i$ taken over all subsets $M$ of $\{1, 2, \ldots, m\}$ with $\sum_{i \in M} k_i \leq \deg(G)$.*

*Remark 2.5.* (a) If $k_i \geq d_i$ for $1 \leq i \leq m$, then from Theorem 4 we get

$$d \geq \sum_{i=1}^{m} d_i - \deg(G).$$

In particular, if
(i) all $P_i$ are rational points and
(ii) $n_i = k_i = d_i = 1$ for $1 \leq i \leq m$,
then the code $C(P_1, ..., P_m; G; C_1, ..., C_m)$ is a Goppa's algebraic-geometry code.

(b) This construction also yields many improvements on Brouwer's table. The detailed improvements will be given in a separate paper. Note that the curves chosen for this construction do not necessarily have many rational points. We also allow some closed points of small degree. Many optimal linear codes can be obtained from this construction. In the following example we just show one optimal binary linear code.

*Example 2.6.* Consider an elliptic curve $\mathcal{X}/\mathbb{F}_2$ with 4 rational points. Then this curve has 2 closed points of degree 2 and 2 closed points of degree 4. Taking
$[n_i, k_i, d_i] = [1, 1, 1]$ for $1 \leq i \leq 4$,
$[n_i, k_i, d_i] = [3, 2, 2]$ for $5 \leq i \leq 6$,
$[n_7, k_7, d_7] = [8, 4, 4]$,
we obtain binary linear $[18, m, 12-m]$ codes for any $1 \leq m \leq 11$. In particular, the binary linear $[18, 4, 8]$ code is optimal.

Özbudak and Stichtenoth [17] have shown that this construction includes as a special case the construction introduced earlier by Niederreiter, Xing, and Lam [16]. The latter construction corresponds to the special case $n_i = k_i = \deg(P_i)$ and $d_i = 1$ for $1 \leq i \leq m$.

## 2.5 Construction 3

We need the following notations for this subsection.

$\mathcal{X}/\mathbb{F}_q$: an algebraic curve over $\mathbb{F}_q$;
$g$: the genus of $\mathcal{X}$;
$P_0, P_1, ..., P_n$: $n + 1$ distinct rational points of $\mathcal{X}$;
$D$: a positive divisor of $\mathcal{X}$ of degree $g$ with $\ell(D) = 1$ (such a divisor exists if $n \geq 3$, see [14]).

For each $1 \leq j \leq n$, choose a function

$$f_j \in \mathcal{L}(D + P_j) \backslash \mathcal{L}(D).$$

This is possible since

$$\ell(D + P_j) \geq \deg(D + P_j) + 1 - g = 2 > \ell(D).$$

Choose a local parameter $t$ at $P_0$, i.e., $\nu_{P_0}(t) = 1$, and consider the local expansion of $f_j$ at $P_0$ which has the form

$$f_j = t^{-v} \sum_{i=0}^{\infty} b_{i,j} t^i,$$

where $v = \nu_{P_0}(D) \geq 0$ and all $b_{i,j} \in \mathbb{F}_q$. For each $1 \leq j \leq n$ put

$$a_{i,j} = \begin{cases} b_{i-1,j} & \text{for } 1 \leq i \leq v, \\ b_{i,j} & \text{for } i \geq v+1. \end{cases}$$

For a fixed positive integer $m$ with $m < n$, we define the $m \times n$ matrix

$$H := (a_{i,j})_{m \times n}$$

and denote by $C_m(P_0, P_1, ..., P_n; D)$ the linear code with parity-check matrix $H$. The parameters of the above code are given by the following result in [26].

**Theorem 2.7.** *If $g \leq m < n$, then $C_m(P_0, P_1, ..., P_n; D)$ is a linear $[n, k, d]$ code over $\mathbb{F}_q$ with*

$$k \geq n - m, \quad d \geq m - g + 1.$$

This construction looks different from Goppa's construction of algebraic-geometry codes. However, the codes have similar parameter bounds. This suggests to look for possible relations between these two kinds of codes. To our surprise, they are equivalent.

Two codes $C_1, C_2 \subseteq \mathbb{F}_q^n$ are said to be *equivalent* if there exist $\lambda_1, \lambda_2, \ldots, \lambda_n \in \mathbb{F}_q^*$ such that

$$C_2 = \{(\lambda_1 c_1, \lambda_2 c_2, \ldots, \lambda_n c_n) | (c_1, c_2, ..., c_n) \in C_1\}.$$

Obviously, equivalent codes have the same parameters. The following result was shown in [26].

**Theorem 2.8.** *The code $C_m(P_0, P_1, ..., P_n; D)$ is equivalent to Goppa's algebraic-geometry code $C(P_1, P_2, .., P_n; G)$, where the divisor $G$ is equivalent to the divisor*

$$D + \sum_{i=1}^{n} P_i - (m+1)P_0.$$

## 2.6 Construction 4

This construction is due to Özbudak and Stichtenoth [17] and generalizes Construction 3 as well as another construction in [26]. We fix some notations for this subsection.

$\mathcal{X}/\mathbb{F}_q$: an algebraic curve over $\mathbb{F}_q$;
$g$: the genus of $\mathcal{X}$;
$P_1, P_2, ..., P_n$: $n$ distinct rational points of $\mathcal{X}$;
$B$: a nonspecial divisor of $\mathcal{X}$, i.e., $\ell(B) = \deg(B) + 1 - g$;
$A$: a positive divisor of $\mathcal{X}$ satisfying $\text{supp}(A) \cap \{P_1, ..., P_n\} = \emptyset$.

Since $B$ is nonspecial, we have $\ell(B + P_i) = \ell(B) + 1$, and so we may choose elements

$$f_i \in \mathcal{L}(B + P_i)\backslash\mathcal{L}(B) \text{ for } i = 1, ..., n.$$

Any $f \in \mathcal{L}(B + \sum_{i=1}^n P_i)$ has a unique representation

$$f = \sum_{i=1}^{n} c_i f_i + w \text{ with } c_i \in \mathbb{F}_q \text{ and } w \in \mathcal{L}(B)$$

and the $\mathbb{F}_q$-linear map

$$\alpha: \ \mathcal{L}(B + \sum_{i=1}^{n} P_i) \to \mathbb{F}_q^n, \quad f \mapsto (c_1, c_2, ..., c_n),$$

is surjective with kernel $\mathcal{L}(B)$. Then the code $C(B; P_1, ..., P_n; A)$ is defined to be the image of $\mathcal{L}(B + \sum_{i=1}^n P_i - A)$ under the map $\alpha$, i.e.,

$$C(B; P_1, ..., P_n; A) := \alpha(\mathcal{L}(B + \sum_{i=1}^{n} P_i - A)).$$

The following parameter bounds are shown in [17].

**Theorem 2.9.** *If* $\deg(A) > \deg(B)$, *then* $C(B; P_1, ..., P_n; A)$ *is a linear* $[n, k, d]$ *code over* $\mathbb{F}_q$ *with*

$$k \geq \deg(B) - \deg(A) + n + 1 - g, \quad d \geq \deg(A) - \deg(B).$$

In fact, it is pointed out in [17] that $C(B; P_1, ..., P_n; A)$ is also equivalent to a Goppa's algebraic-geometry code.

# 3 Constructions of Almost Perfect Sequences

## 3.1 Background on Almost Perfect Sequences

In this section we apply algebraic curves over finite fields to construct sequences that are relevant in the theory of keystreams. Keystreams are pseudorandom sequences of elements of $\mathbb{F}_q$ that are used for encryption and decryption in well-known symmetric cryptosystems called stream ciphers. The security of a stream cipher hinges on the quality of the keystream. A good keystream must possess satisfactory statistical randomness and complexity properties. The assessment of the quality of keystreams is a crucial task in the practice of stream ciphers. We refer to [19] for background on stream ciphers.

An important method of measuring the complexity of keystreams is based on the extent to which the keystream can be simulated by linear recurring sequences in $\mathbb{F}_q$. In the following, we view the zero sequence as a linear recurring sequence of order 0. If $n$ is a positive integer and $S$ a sequence $s_1, s_2, \ldots$ of elements of $\mathbb{F}_q$, then the *nth linear complexity* $L_n(S)$ is defined to be the least $k$ such that $s_1, s_2, \ldots, s_n$ form the first $n$ terms of a $k$th-order linear recurring sequence in $\mathbb{F}_q$. We always have $0 \leq L_n(S) \leq n$. The extreme values of $L_n(S)$ correspond to highly nonrandom behavior, since it is clear that $L_n(S) = 0$ if and only if $s_i = 0$ for $1 \leq i \leq n$, whereas $L_n(S) = n$ if and only if $s_i = 0$ for $1 \leq i \leq n-1$ and $s_n \neq 0$.

The sequence $L_1(S), L_2(S), \ldots$ is called the *linear complexity profile* of $S$. Since $L_n(S) \leq L_{n+1}(S)$, the linear complexity profile is a nondecreasing sequence of nonnegative integers. It may be thought of as describing the evolution of $L_n(S)$ as $n$ varies. The linear complexity profile allows refined statistical testing and a complexity analysis of keystreams. To this end, one requires as a benchmark a description of the behavior of the linear complexity profile of random sequences of elements of $\mathbb{F}_q$.

A detailed investigation of the linear complexity profile of random sequences was carried out by Niederreiter [8]. There is a natural probability measure on the set of sequences of elements of $\mathbb{F}_q$, namely the product probability measure induced by the uniform probability measure on $\mathbb{F}_q$ which assigns the measure $1/q$ to each element of $\mathbb{F}_q$. The main result of [8] says that with probability 1, i.e., for random sequences of elements of $\mathbb{F}_q$, we have

$$L_n(S) = \frac{n}{2} + O(\log n) \quad \text{for all } n \geq 2,$$

where the implied constant may depend on the sequence $S$. A survey of the probabilistic theory of the linear complexity profile can be found in [10]. A briefer, but more recent account of results on the linear complexity profile is available in [11].

There has been a strong interest in sequences $S$ for which the deviation of $L_n(S)$ from $n/2$ is bounded. Such sequences were already considered by Rueppel [18, Chapter 4] and Niederreiter [6]. We follow the terminology of Niederreiter and Vielhaber [12] and call such a sequence *almost perfect.* More precisely, if $d$ is a positive integer, then a sequence $S$ of elements of $\mathbb{F}_q$ is called *d-perfect* if

$$|2L_n(S) - n| \leq d \quad \text{for all } n \geq 1.$$

An alternative characterization of $d$-perfect sequences is based on continued fractions. To a sequence $S$ of elements $s_1, s_2, \ldots$ of $\mathbb{F}_q$ we assign its generating function

$$G = \sum_{i=1}^{\infty} s_i x^{-i} \in \mathbb{F}_q((x^{-1})),$$

where $\mathbb{F}_q((x^{-1}))$ is the field of formal Laurent series over $\mathbb{F}_q$ in the variable $x^{-1}$. Now $G$ has a unique continued fraction expansion

$$G = 1/(A_1 + 1/(A_2 + \ldots)) =: [A_1, A_2, \ldots],$$

where the partial quotients $A_j$ are polynomials over $\mathbb{F}_q$ of positive degree. This expansion is finite if $G$ is a rational function and infinite if $G$ is irrational. The following proposition is implied by results in Niederreiter [7].

**Proposition 3.1.** *A sequence $S$ of elements of $\mathbb{F}_q$ is $d$-perfect if and only if its generating function $G$ is irrational and the partial quotients $A_j$ in the continued fraction expansion of $G$ satisfy* $\deg(A_j) \leq d$ *for all* $j \geq 1$.

*Example 3.2.* Let $q = 2$ and let $S$ be the sequence $s_1, s_2, \ldots$ of elements of $\mathbb{F}_2$ defined by $s_i = 1$ if $i = 2^h - 1$ for some integer $h \geq 1$ and $s_i = 0$ otherwise. Then the generating function $G$ of $S$ satisfies the formal Laurent series identity $G^2 = xG + 1$, hence $G = x + G^{-1}$, which leads to the infinite continued fraction expansion

$$G = [x, x, x, \ldots].$$

Therefore, the sequence $S$ is 1-perfect by Proposition 10. This example is a special instance of a family of 1-perfect binary sequences which were introduced in [6] and called *generalized Rueppel sequences.* A generalized Rueppel sequence is determined by a sequence $n_1, n_2, \ldots$ of integers given by

$$n_1 = 1 \quad \text{and} \; n_{h+1} = 2n_h + c_h \quad \text{for } h = 1, 2, \ldots,$$

where $c_h \in \{0, 1\}$ for all $h \geq 1$. Then a generalized Rueppel sequence $s_1, s_2, \ldots$ is defined by $s_i = 1$ if $i = n_h$ for some $h \geq 1$ and $s_i = 0$ otherwise.

### 3.2 Construction 1

Here and in the following subsections we present methods of constructing almost perfect sequences from algebraic curves over finite fields. All constructions work with the following ingredients:

$\mathcal{X}/\mathbb{F}_q$: an algebraic curve over $\mathbb{F}_q$;
$P$: a rational point on $\mathcal{X}$;
$t$: a local parameter at $P$ with $\deg((t)_\infty) = 2$;
$f$: a function in $\mathbb{F}_q(\mathcal{X}) \backslash \mathbb{F}_q(t)$.

We first treat the case $\nu_P(f) \geq 0$ which was considered by Xing and Lam [23]. The local expansion of $f$ at $P$ has the form

$$f = \sum_{i=0}^{\infty} s_i t^i$$

with all $s_i \in \mathbb{F}_q$. From this expansion we read off the sequence $S_1$ of coefficients $s_1, s_2, \ldots$. The following result was shown in [23].

**Theorem 3.3.** *If $\nu_P(f) \geq 0$ and the integer $d$ is such that $d \geq \deg((f)_\infty)$, then the sequence $S_1$ constructed above is $d$-perfect.*

The case $\nu_P(f) < 0$ can be reduced to the situation considered above, as was pointed out in Xing, Niederreiter, Lam, and Ding [28]. Indeed, if we put $v = -\nu_P(f) > 0$, then $\nu_P(t^v f) = 0$ and we have a local expansion at $P$ of the form

$$t^v f = \sum_{i=0}^{\infty} s_i t^i$$

with all $s_i \in \mathbb{F}_q$. From this expansion we read off the sequence $S_1'$ of coefficients $s_1, s_2, \ldots$. If $d \geq \deg((f)_\infty)$ as in Theorem 12, then

$$\deg((t^v f)_\infty) \leq \deg((f)_\infty) + \deg((t^v)_\infty) - \deg(vP) \leq d + v,$$

and so Theorem 12 implies the following result (see also [28]).

**Theorem 3.4.** *If $\nu_P(f) = -v < 0$ and the integer $d$ is such that $d \geq \deg((f)_\infty)$, then the sequence $S_1'$ constructed above is $(d+v)$-perfect.*

A systematic investigation of examples of almost perfect sequences that can be obtained from Theorem 12 was carried out by Kohel, Ling, and Xing [4]. This paper also discusses the efficient computation of local expansions by means of Hensel's lemma.

*Example 3.5.* Let $\mathcal{X}$ be the projective line over $\mathbb{F}_2$. Its function field is the rational function field $\mathbb{F}_2(x)$. We choose $P = 0, t = x^2 + x$, and $f = x/(x+1)$. Then we have the expansion

$$f = \frac{x^2}{t} = \sum_{h=1}^{\infty} t^{2^h - 1}.$$

The sequence $S_1$ of coefficients $1, 0, 1, 0, 0, 0, 1, \ldots$ is 1-perfect by Theorem 12. This is the same sequence as the generalized Rueppel sequence in Example 11.

*Example 3.6.* Let $\mathcal{X}, P$, and $t$ be as in Example 14, but now we choose $f = x$. Then we have the expansion

$$f = \sum_{h=0}^{\infty} t^{2^h}.$$

The sequence $S_1$ of coefficients $1, 1, 0, 1, 0, 0, 0, 1, \ldots$ is 1-perfect by Theorem 12. This sequence is also a generalized Rueppel sequence.

*Example 3.7.* Let $\mathcal{X}$ and $P$ be as in Example 14, but now we choose $t = x/(x^2+x+1)$ and $f = x$. Then we have the expansion

$$f = t + t^2 + t^5 + t^6 + \dots .$$

The sequence $S_1$ of coefficients $1, 1, 0, 0, 1, 1, \dots$ is 1-perfect by Theorem 12. This sequence is *not* a generalized Rueppel sequence.

The following result from [28] discusses the behavior of the sequences $S_1$ in Theorem 12 under the change of some initial terms.

**Theorem 3.8.** *If the sequence $S_1$ is as in Theorem 12 and the conditions of that theorem are satisfied, then any sequence $S$ obtained by changing at most the first $m$ terms of $S_1$ is $(d+2m)$-perfect. Moreover, if the divisor inequality $m(t)_\infty \le (f)_\infty$ holds, then $S$ is still $d$-perfect.*

### 3.3 Construction 2

Let $\mathcal{X}/\mathbb{F}_q, P, t$, and $f$ be as in Section 3.2 and put $v = \nu_P(f)$. Then the local expansion of $f$ at $P$ has the form

$$f = t^v \sum_{i=0}^{\infty} s_i t^i$$

with all $s_i \in \mathbb{F}_q$. From this expansion we read off the sequence $S_2$ of coefficients $s_0, s_1, \dots$. This construction and the following theorem are due to Xing, Niederreiter, Lam, and Ding [28].

**Theorem 3.9.** *If $v = \nu_P(f)$ and the integer $d$ is such that $d \ge \deg((f)_\infty)$, then the sequence $S_2$ constructed above is $(d+v-1)$-perfect if $v > 0$ and $(d-v+1)$-perfect if $v \le 0$.*

*Example 3.10.* Let $\mathcal{X}$ be the projective line over $\mathbb{F}_3$. Its function field is the rational function field $\mathbb{F}_3(x)$. We choose $P = 0, t = x^2 - x$, and $f = x$. Then we have the expansion

$$f = -t + t^2 + t^3 - t^4 + t^5 + 0 \cdot t^6 + \dots .$$

The sequence $S_2$ of coefficients $-1, 1, 1, -1, 1, 0, \dots$ is 1-perfect by Theorem 18.

### 3.4 Construction 3

Let $\mathcal{X}/\mathbb{F}_q, P, t$, and $f$ be as in Section 3.2 and suppose that $\nu_P(f) \le 0$. Put $v = -\nu_P(f) \ge 0$. Then the local expansion of $f$ at $P$ can be written in the form

$$f = \sum_{j=1}^{v} r_j t^{j-v-1} + \sum_{i=0}^{\infty} s_i t^i$$

with all $r_j \in \mathbb{F}_q$ and $s_i \in \mathbb{F}_q$. From this expansion we read off the sequence $S_3$ of coefficients $s_1, s_2, \ldots$. This construction and the following theorem are due to Xing, Niederreiter, Lam, and Ding [28].

**Theorem 3.11.** *If $\nu_P(f) \leq 0$ and the integer $d$ is such that $d \geq \deg((f)_\infty)$, then the sequence $S_3$ constructed above is $d$-perfect.*

*Example 3.12.* Let $\mathcal{X}$ be the projective line over $\mathbb{F}_3$. Its function field is the rational function field $\mathbb{F}_3(x)$. We choose $P = 0, t = x^2 - x$, and $f = 1/x$. Then we have the expansion

$$f = -t^{-1} - 1 + t + t^2 - t^3 + t^4 + 0 \cdot t^5 + 0 \cdot t^6 + \ldots .$$

The sequence $S_3$ of coefficients $1, 1, -1, 1, 0, 0, \ldots$ is 1-perfect by Theorem 20.

Finally, we mention an open problem. Let $S$ be a $d$-perfect sequence consisting of the elements $s_1, s_2, \ldots$ of $\mathbb{F}_q$. For an integer $k \geq 0$ let $S^{(k)}$ be the shifted sequence $s_{k+1}, s_{k+2}, \ldots$. Then it follows from [9, Corollary 2] and Proposition 10 that the sequence $S^{(k)}$ is $(d+k)$-perfect. This suggests the following task.

**Problem.** *Construct a sequence $S$ of elements of $\mathbb{F}_q$ such that, for some integer $d \geq 1$, all shifted sequences $S^{(k)}, k = 0, 1, \ldots$, are $d$-perfect.*

## References

1. G.-L. Feng and T.R.N. Rao, Improved geometric Goppa codes. Part I: Basic theory, *IEEE Trans. Inform. Theory* **41**, 1678–1693 (1995).
2. V.D. Goppa, Codes on algebraic curves (Russian), *Dokl. Akad. Nauk SSSR* **259**, 1289–1290 (1981).
3. N. Koblitz, Elliptic curve cryptosystems, *Math. Comp.* **48**, 203–209 (1987).
4. D. Kohel, S. Ling, and C.P. Xing, Explicit sequence expansions, *Sequences and Their Applications* (C. Ding, T. Helleseth, and H. Niederreiter, eds.), pp. 308–317, Springer, London, 1999.
5. H.W. Lenstra Jr., Factoring integers with elliptic curves, *Ann. Math.* **126**, 649-673 (1987).
6. H. Niederreiter, Continued fractions for formal power series, pseudorandom numbers, and linear complexity of sequences, *Contributions to General Algebra 5* (Proc. Salzburg Conf., 1986), pp. 221–233, Teubner, Stuttgart, 1987.
7. H. Niederreiter, Sequences with almost perfect linear complexity profile, *Advances in Cryptology – EUROCRYPT '87* (D. Chaum and W.L. Price, eds.), Lecture Notes in Computer Science, Vol. **304**, pp. 37–51, Springer, Berlin, 1988.
8. H. Niederreiter, The probabilistic theory of linear complexity, *Advances in Cryptology – EUROCRYPT '88* (C.G. Günther, ed.), Lecture Notes in Computer Science, Vol. **330**, pp. 191–209, Springer, Berlin, 1988.

9. H. Niederreiter, Keystream sequences with a good linear complexity profile for every starting point, *Advances in Cryptology - EUROCRYPT '89* (J.-J. Quisquater and J. Vandewalle, eds.), Lecture Notes in Computer Science, Vol. **434**, pp. 523–532, Springer, Berlin, 1990.
10. H. Niederreiter, Finite fields and cryptology, *Finite Fields, Coding Theory, and Advances in Communications and Computing* (G.L. Mullen and P.J.-S. Shiue, eds.), pp. 359–373, Dekker, New York, 1993.
11. H. Niederreiter, Some computable complexity measures for binary sequences, *Sequences and Their Applications* (C. Ding, T. Helleseth, and H. Niederreiter, eds.), pp. 67–78, Springer, London, 1999.
12. H. Niederreiter and M. Vielhaber, Linear complexity profiles: Hausdorff dimensions for almost perfect profiles and measures for general profiles, *J. Complexity* **13**, 353–383 (1997).
13. H. Niederreiter and C.P. Xing, Quasirandom points and global function fields, *Finite Fields and Applications* (S. D. Cohen and H. Niederreiter, eds.), London Math. Soc. Lecture Note Series, Vol. **233**, pp. 269–296, Cambridge University Press, Cambridge, 1996.
14. H. Niederreiter and C.P. Xing, Low-discrepancy sequences and global function fields with many rational places, *Finite Fields Appl.* **2**, 241–273 (1996).
15. H. Niederreiter and C.P. Xing, Towers of global function fields with asymptotically many rational places and an improvement on the Gilbert-Varshamov bound, *Math. Nachr.* **195**, 171–186 (1998).
16. H. Niederreiter, C.P. Xing, and K.Y. Lam, A new construction of algebraic-geometry codes, *Appl. Algebra Engrg. Comm. Comput.* **9**, 373–381 (1999).
17. F. Özbudak and H. Stichtenoth, Constructing codes from algebraic curves, *IEEE Trans. Inform. Theory* **45**, 2502–2505 (1999).
18. R.A. Rueppel, *Analysis and Design of Stream Ciphers*, Springer, Berlin, 1986.
19. R.A. Rueppel, Stream ciphers, *Contemporary Cryptology: The Science of Information Integrity* (G.J. Simmons, ed.), pp. 65–134, IEEE Press, New York, 1992.
20. H. Stichtenoth, *Algebraic Function Fields and Codes,* Springer, Berlin, 1993.
21. M.A. Tsfasman and S.G. Vlădut, *Algebraic-Geometric Codes,* Kluwer, Dordrecht, 1991.
22. M.A. Tsfasman, S.G. Vlădut, and T. Zink, Modular curves, Shimura curves, and Goppa codes, better than Varshamov - Gilbert bound, *Math. Nachr.* **109**, 21–28 (1982).
23. C.P. Xing and K.Y. Lam, Sequences with almost perfect linear complexity profiles and curves over finite fields, *IEEE Trans. Inform. Theory* **45**, 1267–1270 (1999).
24. C.P. Xing and S. Ling, A class of linear codes with good parameters, preprint, 1999.
25. C.P. Xing and S. Ling, A class of linear codes with good parameters from algebraic curves, *IEEE Trans. Inform. Theory,* to appear.
26. C.P. Xing, H. Niederreiter, and K.Y. Lam, Constructions of algebraic-geometry codes, *IEEE Trans. Inform. Theory* **45**, 1186–1193 (1999).
27. C.P. Xing, H. Niederreiter, and K.Y. Lam, A generalization of algebraic-geometry codes, *IEEE Trans. Inform. Theory* **45**, 2498–2501 (1999).
28. C.P. Xing, H. Niederreiter, K.Y. Lam, and C.S. Ding, Constructions of sequences with almost perfect linear complexity profile from curves over finite fields, *Finite Fields Appl.* **5**, 301–313 (1999).

# Author Index

MIX
Papier aus verantwortungsvollen Quellen
Paper from responsible sources
FSC® C105338

If you have any concerns about our products,
you can contact us on
**ProductSafety@springernature.com**

*In case Publisher is established outside the EU,
the EU authorized representative is:
**Springer Nature Customer Service Center GmbH**
**Europaplatz 3, 69115 Heidelberg, Germany**

Printed by Libri Plureos GmbH
in Hamburg, Germany